The Ecology of Plants

The Ecology of Plants

Jessica Gurevitch
State University of New York at Stony Brook

Samuel M. Scheiner
Arizona State University

Gordon A. Fox
University of South Florida

Sinauer Associates, Inc., Publishers
Sunderland, Massachusetts U.S.A.

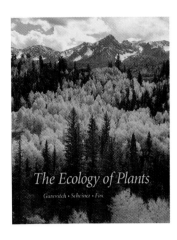

THE COVER

Morning light on aspens, San Juan Range, Uncompahgre National Park, Colorado. Photograph copyright © Willard Clay.

BACK COVER

Photograph of *Rosa blanda* (smooth rose, Rosaceae) by Samuel M. Scheiner.

ABOUT THE BOOK

Editor: Andrew D. Sinauer
Project Editor: Carol J. Wigg
Copy Editor: Norma Roche
Index: Grant Hackett
Production Manager: Christopher Small
Illustrations: Imagineering Scientific and Technical Artworks, Inc.
Book and Cover Design: Jefferson Johnson
Page Layout and Production: Janice Holabird
Book and Cover Manufacture: Courier Companies, Inc.

THE ECOLOGY OF PLANTS

Sinauer Associates, Inc.
23 Plumtree Road, Sunderland, MA 01375 USA
fax: 413-549-1118
email: orders@sinauer.com; publish@sinauer.com
www.sinauer.com

Library of Congress Cataloging-in-Publication Data
Gurevitch, Jessica, 1952-
    The ecology of plants / Jessica Gurevitch, Samuel M. Scheiner, Gordon A. Fox.
        p. cm.
    Includes bibliographical references (p.   ).
    ISBN 0-87893-291-7 (hardcover)
    1. Plant ecology. I. Scheiner, Samuel M., 1956- II. Fox, Gordon A., 1952-
III. Title.
QK901.G96 2002
581.7—dc21

Contents in Brief

Contents

PART *I* The Individual and its Environment

PART II Evolution and Population Biology

P A R T *III* *From Populations to Communities*

P A R T *IV* *From Ecosystems to Landscapes*

PART V Global Patterns and Processes

Preface

This book grew out of expressions of interest by ourselves, friends, and colleagues, for a comprehensive textbook for an upper-level course in plant ecology. It has been our goal to write such a text from a fresh, contemporary perspective, emphasizing a conceptual approach to the subject.

In writing this book, we have assumed that students using it will have had an introductory course in biology, but may or may not have had advanced biology courses, including general ecology. Recognizing that plant ecology may be the only ecology course a student takes, we strive to be inclusive, albeit from a uniquely plant perspective. The book covers the entire span of ecology, from individual plants to populations and communities to large-scale patterns and global issues. Although topics are introduced at a basic level, there is sufficient depth and coverage for more advanced students as well.

Plant ecology touches and builds on many other subject areas that may not be covered in a typical introductory biology course. Therefore, we include background information that might be considered beyond the subject of plant ecology in its strictest sense. In the chapters on the ecology of individuals, for example, we cover aspects of plant anatomy and physiology, returning to these subjects when we address herbivory and ecosystem ecology. We cover soils and belowground interactions, paleoecology, and nutrient cycling in greater depth than might be expected in an ecology text, and address global climate change from the perspective of the roles and responses of both plants and people.

We do not assume that any given course will necessarily follow the order in which topics are presented in this book; we recognize that each faculty member covers the topics in a course in plant ecology in a somewhat different order, or skips some topics entirely. To accommodate that reality, the book is organized into five parts, each of which can stand alone. Because ecology is a discipline about interrelationships, we provide abundant cross-references for topics introduced or covered in other chapters. This book should therefore be usable in courses that begin with biomes or communities, for instance, as well as those that start with the ecology of individual plants.

Each part of this book considers a different aspect of the relationship between plants and their environments. In Part I, we consider the interaction of individuals with the physical environment, including light (Chapter 2), water and temperature (Chapter 3), and soils, mineral nutrients, and belowground interactions (Chapter 4). Plants have developed an impressive variety of ways of coping with the challenges of their physical environment, many of which are explored in these chapters.

In Part II, we turn to plant populations. While the subject of this book is ecology, one cannot begin to understand ecology in any depth without a basic understanding of evolution. Thus, we first consider some of the ways in which the environment acts to shape plants through the process of evolution (Chapter 5), and then look at particular examples of the outcomes of this process (Chapter 6). We continue our exploration of the ecology of populations by considering the processes of population growth (Chapter 7), reproduction and dispersal (Chapter 8), and life cycles (Chapter 9) as they function within populations of single species.

In Part III, we consider interactions among populations of different species. We first explore competitive interactions within and among plant species (Chapter 10), then look at herbivory and plant-pathogen interactions (Chapter 11). Our focus then shifts to local aggregates of species: communities. The interactions between individuals and their physical environment, and both intra- and interspecific interactions among individuals, combine to create larger-scale patterns and processes. We explore the measurement of community properties (Chapter 12), the processes of disturbance and succession (Chapter 13), and the processes that determine the abundance and distribution of species within and among communities (Chapter 14).

Part IV addresses still larger-scale patterns and processes. These chapters consider ways in which populations and communities are linked at the level of the landscape, beginning with energy and nutrient flows within ecosystems (Chapter 15), then continuing with communities as parts of landscapes (Chapter 16), and finally landscape-level processes (Chapter 17). We discuss how the processes detailed in Parts I, II, and III ramify to produce patterns at these scales.

Finally, in Part V, we turn to the largest scales of all: patterns of climate across the globe (Chapter 18), an overview of the world's biomes (Chapter 19), and global patterns of plant diversity (Chapter 20). We end by looking at patterns in time, first looking backward at changes that have occurred over hundreds of millions of years (Chapter 21), then peering into the future to consider the interactions between plants, the atmosphere, and human activities as they affect global climate change and other global changes affecting vegetation and plant diversity (Chapter 22). In various chapters throughout the book, and particularly in the final chapter, we look at the role of humans as an integral part of the environment as well as our own contribution to ecological patterns.

An appreciation for the organism is critical for understanding its ecology. That appreciation includes both a sense of what the organism looks like and its taxonomic placement. To this end, we have included numerous photographs and drawings of the plants discussed in the book. We refer to each species primarily by its scientific name, but also provide its common name and family. For angiosperms, the family designation is based on the most recent comprehensive classification by The Angiosperm Phylogeny Group (1998).

Science has a language of its own, and acquiring that language can sometimes be daunting. We have placed words that may be unfamiliar in **bold**, and defined them in the text and in the Glossary. Ecologists in particular have been accused of needlessly proliferating jargon. (There is an old joke that an ecologist is "someone who is not afraid to call a spade a geotome.") While there is some truth to this, in many cases scientific terminology performs a necessary function by providing a concise and precise vocabulary that facilitates clarity and communication. In some cases, though, these definitions are presented not because we approve of the proliferation of jargon in ecology, but because the terms are commonly used, and students need to be familiar with them to understand the scientific literature.

Throughout the book we have provided an entry to the scientific literature through the use of examples and key references. In addition, at the end of each chapter we highlight particular papers and books. We especially encourage students to search out the classic references.

Many of these studies are still very influential as the origin of many key ideas, concepts, and questions. An appreciation of both classic and contemporary work helps convey the sense of plant ecology as a vibrant, dynamic, and exciting field of study.

Acknowledgments

As we have discovered, writing a textbook is a complex and sometimes very stressful undertaking that would not have been possibility without the help and support of many people. First we thank Carol Wigg, our tireless editor, who shepherded this project to completion. Her good humor and patience were necessary ingredients in its successful outcome. She also made many valuable suggestions for improving both the intellectual content and visual appeal of the book. We thank Andy Sinauer for his willingness to wait for this book. Our copyeditor, Norma Roche, did an outstanding job improving our prose, making sure that we were consistent throughout the book, and otherwise finding and fixing various errors. The entire staff at Sinauer Associates has worked to make the book look as good and read as well as it does.

Many people provided feedback at various stages of this project. Early on we shared several chapters with our colleagues to make sure that our coverage and style were appropriate for our intended audience. We thank all of our colleagues who used draft chapters in their classes for their willingness to read chapters and answer our questions, and their students—and our own—who read and commented on earlier versions of the book.

We thank our many colleagues who provided formal reviews of one or more of the chapters, some under very tight deadlines: Robyn Burnham (University of Michigan), Margaret Carreiro (University of Louisville), Walter Carson (University of Pittsburgh), Scott Collins (National Science Foundation), Peter Curtis (Ohio State University), John Damuth (University of California Santa Barbara), Lynda Delph (Indiana University), Henry Gholz (University of Florida), Deborah Goldberg (University of Michigan), James Grace (USGS Wetland Research Laboratory, Lafayette, LA), Sultan Hameed (State University of New York at Stony Brook), David Hooper (Western Washington University), Manuel Lerdau (State University of New York at Stony Brook), Svata Louda (University of Nebraska, Lincoln), James McGraw (West Virginia University), Tom Miller (Florida State University), Juliana Mulroy (Denison University), Robert Peet (University of North Carolina), Lindsey Rustad (USDA Forest Service, Northeastern Research Station, ME), Mark Schwartz (University of California Davis), Ruth Shaw (University of Minnesota), Jonathan Shurin (National Center for Ecological Analysis and Synthesis), Judith Skog (George Mason University), Thomas Wentworth (North Carolina State University), and Mark Westoby (Macquarie University). They all made valuable comments and offered thoughtful suggestions for improving the text, and caught many of our errors. Any remaining errors and omissions are our own.

Many of our colleagues provided valuable discussions and help with various topics including: Veronique Delesalle, Lynda Delph, Philip Dixon, Tom Ebert, Tim Howard, Colleen Kelly, Jeff Klopatek, Manuel Lerdau, Earl McCoy, Bill Platt, José Rey Benayas, Jim Rodman, Judy Skog, Art Stiles, Peter Stiling, Sharon Strauss, Daniel Taub, Mark Westoby, Ian Wright, and Richard Wunderlin. Wei Fang, Laura Hyatt, and Eliza Woo provided invaluable technical assistance as the book was coming to completion. S.M.S. thanks his co-worker Scott Collins whose library of books and journals was invaluable.

A picture is worth a thousand words. We thank our many colleagues (as acknowledged in the photo credits and listed on page 469) who freely handed over their very valuable slides and photographs and allowed us to use them in our book. We promise that they will all be returned unharmed.

Finally, we thank our long-suffering families for putting up with us during the many hours devoted to this project; J.G.: Todd Postol, Nathaniel Postol, and Julia Postol; S.M.S.: Judy Scheiner and Kayla Scheiner; G.A.F.: Kathy Whitley – and Bart and Maggie. We promise: no more textbooks for a long time!

JESSICA GUREVITCH
SAMUEL M. SCHEINER
GORDON A. FOX
May, 2002

CHAPTER 1

The Science of Plant Ecology

The biological science of **ecology** is the study of the relationships between living organisms and their environments, the interactions of organisms with one another, and the patterns and causes of the abundance and distribution of organisms in nature. In this book we consider ecology from the perspective of terrestrial plants. Plant ecology is both a subset of the discipline of ecology and a mirror for the entire field. In *The Ecology of Plants,* we cover some of the same topics that you might find in a general ecology textbook, while concentrating on the interactions between plants and their environments over a range of scales. We also include some subjects that are unique to plants, such as photosynthesis and the ecology of plant-soil interactions, and others that have unique aspects in the case of plants, such as the acquisition of resources and mates.

Ecology as a Science

Ecologists study the function of organisms in nature and the systems they inhabit. Some ecologists are concerned in particular with the application of ecological principles to practical environmental problems. Sometimes the distinction between basic and applied ecology becomes blurred, as when the solution to a particular applied problem reveals fundamental knowledge about ecological systems. In both basic and applied ecology, the rules and protocols of the sciences must be rigorously followed.

What ecology is *not* is environmental or political activism, although ecologists are sometimes environmental activists in their personal lives, and environmental activists may rely on ecological research. Ecology is not about one's feelings about nature, although ecologists may have strong feelings about what they study. Ecological systems are complex things, with a great many parts, each of which contributes to the whole in different ways. Nevertheless, ecology is indeed a science, and it works like other scientific disciplines.

How do we know whether something is true? Science is one way of knowing about the world—not the only way, but a spectacularly successful one. In contrast to some of the other ways of knowing that are part of our lives, the legitimacy of science is based not on authority, or opinion, or democratic principles, but on the weight of credible, repeatable evidence.

The Genesis of Scientific Knowledge

Throughout this book we examine how ecologists have come to their current knowledge and understanding of organisms and systems in nature. Ecology has both a strong and a rich theoretical basis and has developed from a foundation based on an enormous collective storehouse of information about natural history.

Ecology, like all of science, is built on a tripod of pattern, process, and theory. **Patterns** consist of the relationships between pieces or entities of the natural world. **Processes** are the causes of those patterns. **Theories** are the explanations of those causes. When ecologists carry out original scientific research, they seek to document patterns, understand processes, and ultimately to put together theories that explain what they have found out.

There is a distinction between the kind of research a scientist does and the kind of research done for a term paper, or by any member of the public trying to gather information about a topic using library books or material posted on Web sites. Although there are exceptions, research carried out by students or the general public is usually **secondary research**: gathering data or confirming facts that are already known. This sort of research is not only useful, it is essential: every scientific study must begin by assessing what is already known. But the heart of what research scientists do is **primary research**: gathering information or finding out facts that no one has ever known before. That experience of discovery is what makes doing science so incredibly exciting and fun.

Scientists gain knowledge by using the **scientific method**. They carry out a series of steps, although not always in a fixed order (Figure 1.1). In ecology, these steps can be summarized as follows: observation, description, quantification, posing hypotheses, testing those hypotheses using experiments (in a broad sense of the word, as discussed below), and verification, rejection, or revision of the hypotheses, followed by retesting of the new or modified hypotheses. Throughout this process, ecologists gather various kinds of information, look for patterns or regularities in their data and propose processes that might be responsible for those patterns. They often put together some sort of model to help in advancing their understanding. Eventually, they construct theories, using assumptions, data, models, and the results of many tests of hypotheses, among other things. The building of comprehensive scientific theories proceeds simultaneously from multiple directions by numerous people, sometimes working in synchrony and sometimes at cross-purposes. Science in operation can be a messy and chaotic process, but out of this chaos comes our understanding of nature.

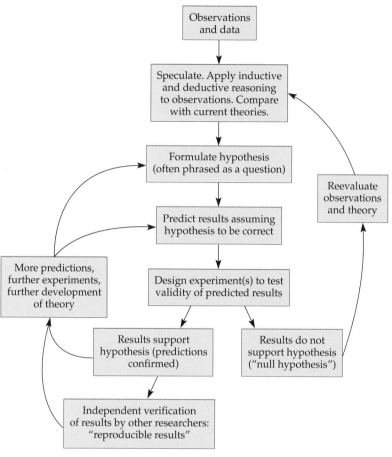

Figure 1.1
The scientific method. The cycle of speculation, hypothesis, and experimentation is circular, as new questions constantly emerge from the answers scientists obtain.

The word "theory" has a very different meaning in science than it does in common usage. A scientific theory is a broad, comprehensive explanation of a large body of information that, over time, must be supported and ultimately confirmed (or rejected) by the accumulation of a wide range of different kinds of evidence (Table 1.1). In popular usage, the word "theory" usually refers to a limited, specific conjecture or supposition, or even a guess or "hunch." Equating the meaning of a scientific theory with "a guess" has caused no end of mischief in the popular press and in public debates on politically charged issues.

When a theory is buttressed over many years with strong evidence, with new findings consistently supporting and amplifying the theory while producing no serious contradictory evidence, it becomes an accepted framework or pattern of scientific thought from which new speculation can spring. This is what has occurred with Newton's theory of gravity, Darwin's theory of evolution, and Einstein's theory of relativity. Scientists use such overarching theories to organize

Table 1.1 The components of a theory

Component	Description
Domain	The scope in space, time, and phenomena addressed by the theory
Assumptions	Conditions or structures needed to build the theory
Concepts	Labeled regularities in phenomena
Definitions	Conventions and prescriptions necessary for the theory to work with clarity
Facts	Confirmable records of phenomena
Confirmed generalizations	Condensations and abstractions from a body of facts that have been tested
Laws	Conditional statements of relationship or causation, or statements of process that hold within a universe of discourse
Models	Conceptual constructs that represent or simplify the natural world
Translation modes	Procedures and concepts needed to move from the abstractions of a theory to the specifics of application or test
Hypotheses	Testable statements derived from or representing various components of the theory
Framework	Nested causal or logical structure of the theory

Source: Pickett et al. 1994.

their thinking and derive additional predictions about nature.

A scientific **hypothesis** is a possible explanation for a particular observation or set of observations. A hypothesis is smaller in scope than a fully developed theory. Hypotheses must be testable: they must contain a prediction or statement that can be verified or rejected using scientific evidence. Experiments are the heart of science, and we discuss their design and use in more detail later in this chapter. A crucial characteristic of science is the need to revise or reject a hypothesis if the evidence does not support it. Science does not accept hypotheses on faith.

Some of the most important tools in the scientist's toolkit are models. A **model** is an abstraction or simplification that expresses structures or relationships. Models are one of the ways in which the human mind attempts to understand complex structures and relationships, whether in science or in everyday life. Building a model airplane from a kit can tell you a lot about the basic form of an airplane; civil engineers often build small models of structures such as bridges or buildings (either physical models or three-dimensional images on a computer) before construction is begun. You have no doubt seen models of DNA and of chemical reactions, and you may have heard about global climate models, which we will discuss at length in Chapter 22.

Models can be abstract or tangible, made of plastic or words. They can be diagrams on paper, sets of equations, or a complex computer program. In science, models are used to define patterns, summarize processes, and generate hypotheses. One of the most valuable uses of models is to make predictions. Ecologists deal almost exclusively with abstract models that can range from a simple verbal argument to a set of mathematical equations.

One reason the models often rely on mathematics is that ecologists are so often concerned with the numbers of things (for instance, is a species' population size so small that it is becoming endangered? How rapidly is an invasive species spreading? How many species can coexist in a community, and how does this change as conditions change?). Mathematical models offer well-defined methods for addressing questions in both qualitative and quantitative terms.

All models are necessarily based on simplifications and rest on a set of assumptions. Those (implicit and explicit) simplifications and assumptions are critical to recognize because they can alert you to the limitations of the model, and because faulty assumptions and unjustified simplifications can sink even the most widely accepted or elegant model.

Objectivity, Subjectivity, Choice, and Chance in Scientific Research

When you read a typical scientific paper, it may at first seem arcane and dull. The format follows a rigid protocol, designed for efficiently conveying essential information to other scientists. Ideas are tightly packaged, with a clear logical line running from start to finish. It may seem as if the researchers knew exactly what they would find even before they began. We will let you in on an open secret: That is not usually how real science works. The justifications for the research presented in a paper's introduction may have been thought up or discovered long after the research project began, or even after the work was finished. Because of serendipitous discoveries, laboratory or field disasters, or unusual natural occurrences, the original purpose of a research project is sometimes modified or, occasionally, entirely discarded and replaced with something else.

Ideas in science, especially in ecology, come from a variety of sources. While everyone knows that science is objective and rational, that is only half the story. In order to reach a genuinely new understanding, subjectivity and creativity must also come into play. While one must be objective in, for example, examining the weight of evidence in support of a hypothesis, subjectivity plays a subtle but important role throughout all of scientific research. What one chooses to study, where one chooses to look for answers, and what is or is not a valid topic for scientific research are all subjective decisions scientists must routinely make. To a large extent, these choices depend on the questions one asks. While determining the answers must be objective, choosing what questions to ask, and how to ask them, is largely subjective.

Many scientific endeavors are highly creative as well. Coming up with a good experiment, looking at a seemingly intractable problem from a new perspective, switching gears after a disastrous laboratory failure to extract a successful outcome from the jaws of catastrophe, and pulling a large number of disparate facts together to build a comprehensive theory are all highly creative activities.

Many scientific discoveries start with casual observations, such as Newton's proverbial apple. Or an idea may arise as a "what if" thought: What if the world works in a particular way? Or a previous experiment may have raised new questions. What makes a scientist successful is the ability to recognize the worth of these casual observations, what-if thoughts, and new questions. From these sources, an ecologist constructs hypotheses and designs experiments to test them.

Experiments: The Heart of Research

A cornerstone of the scientific process is the **experiment**. Ecologists in particular use a wide variety of types of experiments. We use the term "experiment" here in its broadest sense: a test of an idea. Ecological experiments can be classified into three broad types: manipulative, natural, and observational. **Manipulative**, or **controlled**, **experiments** are what most of us think of as experiments: A person manipulates the world in some way and looks for a pattern in the response. For example, an ecologist might be interested in the effects of different amounts of nutrients on the growth of a particular plant species. She can grow several groups of plants, giving each of them a different nutrient treatment, and measure such things as their time to maturity and their final size. This experiment could be done in a controlled environment such as a growth chamber, in a greenhouse, in an experimental garden, or in a natural community in a field setting.

This range of potential settings for the experiment comes with a set of trade-offs. If the experiment is conducted in a laboratory or growth chamber, the ecologist is able to control most of the possible sources of variation so that the differences among treatments can be clearly attributed to the factors being studied in the experiment. These sorts of controlled experiments exemplify the scientific method as it was first laid out by Frances Bacon in the seventeenth century. Baconian experiments are the mainstay of most of molecular and cellular biology as well as the physical sciences. By working in a controlled environment, however, the ecologist sacrifices something. The controlled environment is highly artificial so that it compromises realism, and it is also narrow in scope (the results apply only to a limited range of conditions), sacrificing generality.

If an experiment is conducted in a field setting, it is more realistic or more natural, but now many factors may vary in an uncontrolled fashion. In a field experiment, the only factors that are controlled are the ones being studied. Instead of attempting to control all variation, variation due to factors other than the experimental ones is randomized among replicates, and conclusions are based on the use of statistical inference (see the Appendix). Such experiments can be carried out in many settings and are not restricted to the field. This sort of experiment was first developed by R. A. Fisher in the early twentieth century. Fisherian experiments are a mainstay of ecology and evolutionary biology as well as the social sciences. They are typically less narrowly defined than Baconian experiments, and thus their results may be more readily generalized. Where along this continuum of control versus realism the ecologist carries out her experiment depends on both her scientific goals and practical considerations.

Experiments are usually designed as tests of hypotheses. If the hypothesis is partially or wholly falsified, the scientist goes back, revises his ideas, and tries again. If the hypothesis is not falsified by the outcome of the experiment, the scientist gains confidence that his hypothesis might be correct. Sometimes, however, experiments are done without an explicit hypothesis in mind. These are "poke-at-it-and-see-what-happens" experiments designed to find out more about the world. Such experiments are common throughout the biological sciences, including ecology. Scientists have studied in detail only a few hundred of the quarter-million terrestrial plant species; of these, only a few [*Zea mays* (maize), *Arabidopsis thaliana*, and possibly *Oryza* spp. (rice)] approach being "thoroughly studied." An ecologist beginning the study of a new species, community, or ecosystem must do many of these general types of experiments. Of course, he is guided by his knowledge of other similar species and ecosystems. Each species or ecosystem is unique, however, which is why each study expands our ecological knowledge.

Manipulative experiments are powerful tools for two major reasons: first, because the scientist can control which parts of the natural world will be altered to study their effects, and second, because she can separate factors that typically occur together to test them individually. Such experiments have limitations, however. Sometimes manipulative experiments are plagued by artifacts—outcomes caused by a side effect of the experimental manipulation itself rather than being a response to the experimental treatment being tested. Good experiments avoid artifacts or take them into account in evaluating the results.

Another limitation is that of scale. Ecology is concerned with patterns and processes that occur across large scales of space and time—for example, the causes of differences in the numbers of species on different continents, or the responses of populations to climate change over the next two centuries. We cannot do

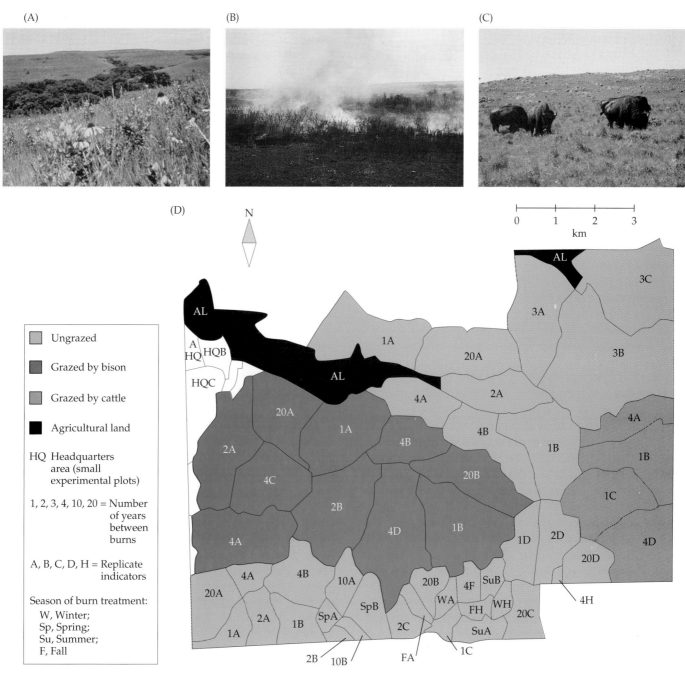

(A)

(B)

(C)

(D)

N

▢	Ungrazed
▢	Grazed by bison
▢	Grazed by cattle
■	Agricultural land

HQ Headquarters area (small experimental plots)

1, 2, 3, 4, 10, 20 = Number of years between burns

A, B, C, D, H = Replicate indicators

Season of burn treatment:
W, Winter;
Sp, Spring;
Su, Summer;
F, Fall

Figure 1.2
Large-scale manipulative experiments are being carried out at the Konza Prairie Research Natural Area (A). Prescribed burns (B) are done at various intervals to investigate the effects of fire and fire frequency on prairie communities. In addition, areas grazed by bison (C) are studied and com- pared with ungrazed areas and with plots subjected to cattle grazing. The experimental patches (D), which are watershed units, vary in size from approximately 3 to 200 hectares. In this map, each patch is designated by a code indicating the fire treatment. (After Knapp et al. 1998. Photographs courtesy of A. Knapp, Konza Prairie Biological Station, and S. Collins.)

manipulative experiments at these great scales of time and space. Ecologists are, however, increasingly making use of longer-term and larger-scale manipulative experiments (see Box 12C). One example of such a study is the long-term study of prairie ecology at the Konza Prairie Research Natural Area in Kansas, begun by Lloyd Hulbert in 1981 (Knapp et al. 1998). The reserve is divided into a series of large patches, which are subjected to different combinations of controlled burning at various time intervals and grazing by bison or cattle (Figure 1.2).

Large-scale manipulative experiments are often limited by the range of possible treatment, however; for example, at Konza Prairie, almost all of the controlled burning is carried out in the spring. Because we do not have firm data on prairie fire regimes prior to European settlement, we do not know how this spring-burning treatment compares to "natural" fire regimes.

Some types of manipulative experiments would be unethical to carry out. For example, we would not cause the extinction of a species just to study the effects of such an event. In such cases, ecologists must rely on two other types of studies. These are natural and observational studies, which may be thought of as different kinds of experiments.

Natural experiments are "manipulations" caused by some natural occurrence. For example, a species may go extinct in a region, a volcanic eruption may denude an area, or a flash flood may scour a streambed. Natural and manipulative experiments represent a trade-off between realism and precision, similar to the trade-off between laboratory and field experiments. Just as with a manipulative experiment, the ecologist compares the altered system either with the same system before the change or with a similar, unchanged system.

The major limitation of natural experiments is that there is never just a single difference before and after a change or between systems being compared. There are no guarantees, for instance, that the altered and unaltered systems were identical prior to the event. For example, if we are comparing areas burned in a major fire with others that remained unburned, the unburned areas might have been wetter, might have had a different site history or different vegetation before the fire, and so on. In other words, there are many other potential sources of difference besides the fire. Therefore, it can be difficult to determine the cause of any change.

The best natural experiments are ones that repeat themselves in space or time. If an ecologist finds similar changes each time, then she gains confidence about the causes of those changes. Another approach is to combine natural experiments with manipulative experiments. For example, the patches subjected to grazing and fire treatments at Konza Prairie are being compared with patches elsewhere that are not subjected to experimental manipulation.

Observational experiments consist of the systematic study of natural variation. Such observations or measurements are experiments if an ecologist starts with one or more hypotheses (predictions) to test. For example, one could measure patterns of species diversity across a continent to test hypotheses about the relationship between the number of plant species number and productivity (see Chapter 20). Again, the limitation of this type of experiment is the potential for multiple factors to vary together. If several factors are tightly correlated, it becomes difficult to determine which factor is the cause of the observed pattern. For example, if the number of herbivores is also observed to increase as the number and productivity of plant species increases, the

ecologist cannot tell for sure whether the increase in herbivores is a result of increased plant numbers and productivity, or whether the increased productivity is a result of increased herbivory.

As with natural experiments, observational experiments repeated in space or time add confidence to our conclusions (Figure 1.3). Other sciences, notably geology and astronomy, also rely largely or exclusively on observational experiments because of the spatial or temporal scales of their studies or because direct manipulation is impossible.

Ecological knowledge comes from combining information gained from many different sources and many different kinds of experiments. The ecologist's use of this complex variety of information makes ecology a challenging and exciting science.

Testing Theories

The testing of scientific theories, especially ecological ones, is a more subtle, nuanced, and complicated endeavor than nonscientists often realize. The popular image of the scientific method portrays it as a process of falsifying hypotheses. This approach was codified by the German philosopher of science Karl Popper (1959). In this framework, we are taught that we can never prove a scientific hypothesis or theory. Rather, we propose a hypothesis and test it; the outcome of the test either falsifies or fails to falsify the hypothesis. While hypothesis testing and falsification is an important part of theory testing, it is not the whole story, for two reasons.

First, this approach fails to recognize knowledge accumulation. In a strict Popperian framework, all theories are held to be potentially false. We never *prove* anything to be true; we merely *disprove* ideas that are false. This assumption goes against our own experience and the history of the accumulation of scientific understanding. Today we know that the Earth revolves around the sun, even though this was once just a hypothesis. We know that the universe is approximately 15 billion years old (give or take a few billion) and began with the Big Bang, even if we still do not know the details of that event. We know that life began and assumed its present shape through the process of evolution. We know that many diseases are caused by microbial infections, not by "humours," and that hereditary traits are conveyed by DNA, not by blood. While we may acknowledge that all of this knowledge has not, in a strictly philosophical sense, been proved to be true, but has only failed thus far to be falsified, we also recognize that some knowledge is so firmly established and bolstered by so many facts that the chance that we are wrong is very much less than the chance of winning the lottery several times in a row. The school of philosophy of science called realism recognizes this progressive accumulation of knowledge (Mayo 1996).

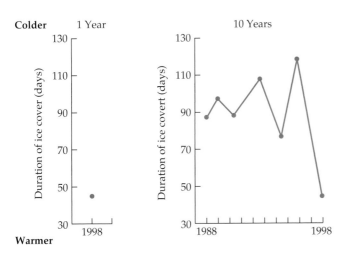

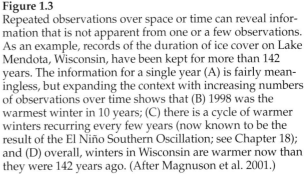

Figure 1.3
Repeated observations over space or time can reveal information that is not apparent from one or a few observations. As an example, records of the duration of ice cover on Lake Mendota, Wisconsin, have been kept for more than 142 years. The information for a single year (A) is fairly meaningless, but expanding the context with increasing numbers of observations over time shows that (B) 1998 was the warmest winter in 10 years; (C) there is a cycle of warmer winters recurring every few years (now known to be the result of the El Niño Southern Oscillation; see Chapter 18); and (D) overall, winters in Wisconsin are warmer now than they were 142 years ago. (After Magnuson et al. 2001.)

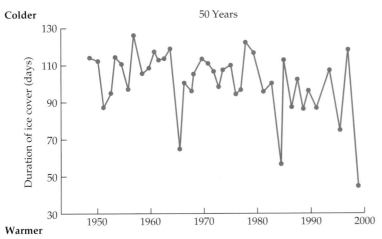

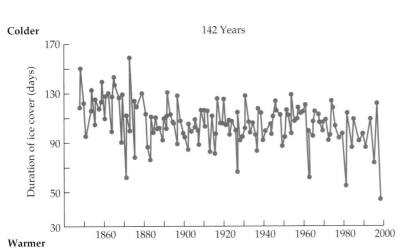

processes of competition and herbivory each contribute to shaping this community?" So, when we are building our theories about plant community structure, our activities are more akin to assembling a complex model than to falsifying a set of propositions.

Falsification does play a role in science, but a more limited one than Popper envisaged. Theory construction is like assembling a jigsaw puzzle from a pile of pieces from more than one box. We can ask whether a particular piece belongs in this spot, yes or no, by erecting a hypothesis and falsifying it. We may even conclude that this particular piece does not belong in this puzzle. Less often are we attempting to completely throw away the piece, saying that it does not belong to any puzzle.

Controversy also plays an important part in ecology, as it does in all scientific fields. During the process of amassing evidence regarding the validity of a theory, different interpretations of experimental data, and different weight given to different pieces of evidence, will lead different scientists to differing opinions. These opinions may be passionately held and forcefully argued, and discussion can sometimes become heated. As the evidence supporting a theory accumulates, some scientists will be willing to accept it sooner, while others will wait until the bulk of the evidence is greater.

If the issue under debate has political or economic implications, nonscientists will also contribute to the debate and may be able to offer valuable insight, judgment, and perspective to the discussion. But when the evidence in favor of a scientific theory becomes overwhelming and the vast majority of scientists knowledgeable in that field are convinced of its validity, then the matter becomes settled (unless startling new evidence or a new, broader theory forces a reevaluation). Ultimately, it is the judgment of scien-

Second, and more importantly, the Popperian framework fails to account for a second type of question that we very commonly ask in ecology. Often the issue is not one of falsifying a hypothesis. Rather, we ask about the relative importance of different processes. When we examine the structure of a plant community, we do not ask, "Is it true or false that competition is occurring?" Instead, we ask, "How much, and in what ways, do the

tists that must decide the answers to scientific questions. When a scientific consensus has been reached on a scientific theory, it is unreasonable to consider that theory to be just another guess or opinion and to hold that everyone's opinion is equally valid. That may work for the democratic process, but not in science. Opinions not supported by evidence are not the same as those supported by the weight of a great deal of evidence; giving them equal weight would be contrary to the way science works.

This does not mean that scientists should decide issues of public policy. For example, if scientists are in strong agreement about something—say, that if more than x% of its remaining habitat is lost, then plant species Y has a 90% chance of extinction within the next 20 years—that does not necessarily dictate any particular public policy. Policy decisions depend on how important people think it is to save species Y, and on what costs they are willing to pay to do so. While we personally hope this would never happen, we recognize that someone who wanted (for whatever reason) to exterminate species Y could use the scientific conclusions for their own ends, just as we could use those same scientific conclusions to promote conservation.

Specific Results versus General Understanding

Because ecologists work at such a variety of scales and on such a diversity of organisms and systems, the question arises as to how far one can extend the conclusions of a particular study to other organisms or places. In the fields of chemistry or physics, the results of an experiment are considered to be absolutely true for all times and places: an atom of helium is made up of two protons and two neutrons, which in turn are made up of quarks, with no qualifications needed. This is the popular image of scientific theories.

Ecology is different. Do the results of a field experiment on competition between two plant species extend to other seasons or locations, or to other pairs of species within the same families or functional groups? Experiments involving helium deal with a universal entity, the helium atom. In contrast, in experiments on plant competition, the exact composition of the entities changes (e.g., the individual plants used each time are not genetically identical), and the surroundings change as well (e.g., the weather is different this year than last year). For this reason, extremely cautious scientists take the position that no conclusion can be extended beyond the particular conditions that existed when the experiment was conducted. If that were so, however, there would be no value and no point in doing any experiments, because they would tell us nothing beyond something of such limited scope as to be worthless.

The truth is somewhere between these two extremes, creating a constant and dynamic tension in ecol-

ogy. One approach to resolving this tension is to see how the outcome of a particular experiment fits into the workings of existing models, and whether it supports or rejects the predictions of those models. Another approach is to use methods for the quantitative synthesis of the results of independent experiments. These methods, known collectively as **meta-analysis**, can be used to evaluate where the outcome of a particular experiment fits in with—or differs from—the results of other similar experiments conducted on different organisms at different places and times. This approach has been used to evaluate the broad body of experimental evidence for a number of important ecological questions in recent years (Gurevitch et al. 2001).

Scale and Heterogeneity

A great deal of recent interest in ecology has been generated by consideration of how ecological patterns and processes vary as a function of the scale at which they operate and are studied (Figure 1.4). The same phenomenon can be seen very differently when studied within a small local area and at the scale of a landscape or region—that is, at different spatial scales. Likewise, one's perspective can change dramatically when studying an ecological process over a single growing season of a few months or over a period of decades or centuries (see Figure 1.3). Different kinds of things may be going on at different scales, and expanding one's focus to more than one scale can be richly rewarding. In a study of a local community, for example, we might see that competitive interactions keep individual plants of a particular species at a distance from one another. At a larger scale, we might notice that the plants are grouped together across the landscape because individuals that are too far apart from any others never become pollinated, and fail to leave descendents; or because the seeds have limited ability to disperse. At a regional or continental scale, the plants may exist in several large but separated enclaves, determined by patterns of glaciation and species migration thousands of years in the past.

We often refer to these scale changes in terms of a hierarchy, and one can move up and down many different kinds of hierarchies in ecology. For instance, one can move from the level of molecules to tissues to organs to entire organisms. A different kind of hierarchy could expand from individual organisms to populations to communities to ecosystems and up to entire biomes; an alternative hierarchy might move from things that occur at the level of organisms to those that function at a scale of habitats, landscapes, watersheds, regions, and so on up to global-scale phenomena. These levels are not necessarily congruent: one might study the individual adaptations of plants over a range of different environments across an entire landscape or even a region, for instance,

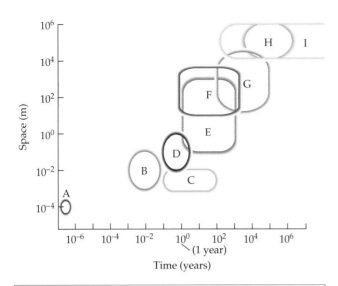

Figure 1.4
Ecologists study patterns and processes across a wide range of scales in space and time. The processes of plant physiology, such as the diffusion of CO_2 molecules in a leaf, occur over the shortest distances (10^{-4} m) and times (10^{-7} years). Moving up the hierarchy, the domain of the whole plant and its birth, growth, reproduction, and death encompasses slightly longer distances (10^0 m) and times (10^0 years). At yet larger scales (10^1–10^2 m and 10^1–10^2 years), we enter the realm of populations and communities and their changes over the course of years and decades. Finally, ecologists study patterns that stretch across the entire globe (10^5 m) and thousands or even millions of years (10^3–10^6 or more years).

or how population interactions at very local scales contribute to the global range limitations of a species. Likewise, one's interpretation of data collected over a short time period may be completely upended when the same data are examined for trends over longer periods of time.

One of the reasons that scale is now recognized as being so important is that the world is a very heterogeneous place. Even over very small distances, conditions can change in ways that may be very important to living organisms. Environmental conditions are a particular concern in plant ecology because plants cannot move. Or, at least, mature terrestrial plants generally are firmly rooted in place, although their offspring may be dispersed some distance away. So, the environment immediately surrounding an individual plant is overwhelmingly important to its survival, growth and reproduction.

The **habitat** of a population or species is the kind of environment it generally inhabits and includes the set of **biotic** (living) and **abiotic** (nonliving) factors that influence it in the places one usually finds it. But the conditions in the immediate surroundings of an individual plant—its **microhabitat**—may differ considerably from the average conditions in the general habitat (Figure 1.5). Factors operating to distinguish a microhabitat from others around it include the composition of the soil, the microclimate of the immediate area, the presence, size, and identity of neighboring plants, and other organisms in the immediate surroundings (grazers, pollinators,

seed eaters or dispersers, and mutualistic or pathogenic fungi or bacteria).

Similarly, the environment varies from moment to moment. There are no specific ecological terms for the components of temporal heterogeneity, but it also exists at many scales and has major effects on plants. Variations in conditions from day to night, summer to winter, across periods of wet years, cold years, or snowy years, and at a longer scale as climate changes over thousands of years all have important influences on plants. Depending on the ecological process being studied and the organisms involved, it may be the small-scale, moment-to-moment variation that matters most (such as fluctuations in light levels in a small forest gap on a partially cloudy day), or it may be large-scale, long-term average conditions (such as CO_2 concentration in the atmosphere), or it may be the interplay between processes occurring at different scales (such as CO_2 flux in a forest canopy over the course of a day or a season).

Groups of organisms, such as populations and species, sometimes average these sorts of microenvironmental influences over larger areas and over generations of organisms' lives. This averaging acts to counter the effects of heterogeneity, particularly over evolutionary time. At even larger scales, heterogeneity again becomes critical. As continents are carried apart on tectonic plates and climates are altered, organisms must either respond to changing conditions by evolving or by changing their distributions, or else become extinct.

The Structure and History of Plant Ecology

Ecology is a very synthetic subject. By that we do not mean that it is unnatural or artificial, but that it brings together a very wide range of other fields of science, perhaps to a greater extent than does any other subject. Some of the fields that ecology encompasses or overlaps with include geology, geography, climatology, soil science, evolutionary biology, genetics, statistics and other branches of mathematics, systematics, behavior, physiology, developmental biology, molecular biology, and

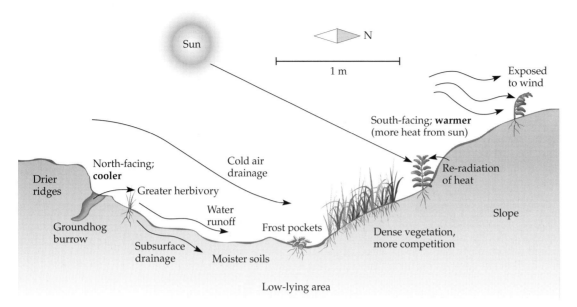

Figure 1.5
The environment in a particular microhabitat can differ in a number of ways from conditions in the surrounding area. Individual plants experience the conditions in their immediate microhabitat, not the average conditions in the general area. The grass on the north-facing slope at the left of the diagram experiences cooler temperatures, and perhaps greater herbivory due to its proximity to the groundhog burrow. The plants in the low-lying area may experience frost sooner in the fall and later in the spring than the surrounding areas due to cold air drainage; here, the soils are moister, and competition for light and soil moisture may be more intense, than on the ridges. Other potential effects of a microclimate are also illustrated.

biochemistry. We touch on many of these fields throughout this book, showing you how they fit into the toolkit of an ecologist and how familiarity with them affects the ways in which ecologists think about and study organisms in nature.

This is not the place to present a detailed and definitive history of plant ecology. Instead, we sketch some of the major milestones, with an admitted bias toward the English-speaking scientific community. Other historical details are scattered throughout the book as we discuss particular topics and subfields. While no single definitive history of plant ecology exists, several books and papers describe parts of its history (McIntosh 1985; Westman and Peet 1985; Nicholson 1990; Allen et al. 1993).

The roots of plant ecology go back to prehistoric times, when people's health and survival depended on their abilities to understand many aspects of the ecology of plants with great accuracy. Ecology as a science began with the Greeks, most notably Aristotle, in the fourth and fifth centuries B.C.E. The modern science of plant ecology began as the study of natural history in the eighteenth and nineteenth centuries, carried out by professional and amateur naturalists in Europe and North America and in their travels throughout the world. Ships on journeys of discovery and colonization often carried a ship's naturalist who cataloged the remarkable range of organisms and environments they encountered. Charles Darwin was one such ship's naturalist, and the story of his five-year voyage was published as his *Voyage of the Beagle* (Figure 1.6). Ecology as a recognized discipline coalesced in the latter half of the nineteenth century. The German biologist Ernst Haeckel coined the term "oecology" in 1866, and by early in the twentieth century the Ecological Society of America had formed.

Plant ecology as a discipline is made up of a number of different subdisciplines, some of which have quite distinct traditions and histories. Some early plant ecologists and botanists focused on whole communities, while others focused on single species and the properties of individuals. The older (now largely archaic) terms for these two subfields are **synecology** and **autecology**. Plant community ecologists, in particular, were active in the origins of ecology as a discipline in the last part of the nineteenth century and dominated plant ecology during the first half of the twentieth century. A more detailed discussion of the history of plant community ecology and some of the key figures in that history is given in Chapters 12 and 13.

Early studies in plant autecology were especially concerned with understanding unique plant adaptations

Figure 1.6
H.M.S. Beagle sailed from England December 27, 1831 on a 5-year mission to chart the oceans and collect biological information from around the world. Charles Darwin sailed with the *Beagle* as ship's naturalist; he is pictured here at the age of 24, shortly after completing the voyage. Darwin collected vast numbers of plant and animal specimens and recorded copious scientific observations that were instrumental in the creation of his most famous work, *The Origin of Species.* (Images from Science Photo Library/Photo Researchers, Inc.)

to extreme environments, such as deserts, and a number of famous studies were concerned with plant performance in the field. Although some major insights were gained, technological limitations severely hampered the development of the field. As instrumentation and methodology became more sophisticated, plant physiologists began to carry out most of their research in controlled laboratory environments.

Beginning in the middle of the twentieth century, further advances in technology made it possible for physiological studies to come out of the greenhouse and into nature, eventually leading to the creation of the fields of plant physiological and functional ecology. At about the same time, autecology began to fission into subfields that focused on single individuals and on populations. Plant population ecology as a recognizable subdiscipline had its origins in Great Britain in the 1960s, particularly with John Harper and his students. It then spread to North America in the 1970s. This is necessarily a very simplified and limited description of events; for instance, a number of individuals in many countries around the world were carrying out studies that today we would label as plant physiological ecology or plant population ecology as far back as the nineteenth century.

For the most part, plant ecology for the first three-quarters of the twentieth century developed independently of animal ecology. Animal community ecology has a long history parallel to that of plant community ecology (Mitman 1992). Substantial work in animal population ecology extends back to at least the 1920s (for example, with the work of Gause, Pearl, Lotka, and others). Plant population ecology drew on these ideas and theories as it was developing, as well as other ideas that originated among plant ecologists. Eventually new theories were needed as discoveries about the unique nature of plants made it obvious they could no longer be shoehorned into many of the theories constructed for animals.

Conversely, physiological ecology advanced earlier and more rapidly among plant ecologists than it did among animal ecologists. Undoubtedly this was because the characteristics of plants are much easier to measure, and their environments easier to characterize, than those of animals (for most purposes, one does not have to catch plants!). On the other hand, in the 1980s, animal physiological ecology joined with evolutionary biology to create the field of evolutionary physiology (Feder et al. 1987), a move that plant biologists have not yet clearly made.

The gap between the fields of plant and animal ecology was bridged in the 1970s, although distinct subfields continue to this day. Two related developments were responsible. The first was the rise of studies of plant-animal interactions, especially pollination (see Chapter 8) and herbivory (see Chapter 11). The second development was the burgeoning interest in the evolutionary aspects of ecology in the 1970s and 1980s, which transcended the traditional separation of the studies of plants and animals.

The most recent changes in the field of plant ecology have been the rise of landscape ecology and conservation ecology as recognized disciplines in the late 1980s. Landscape ecologists began their careers from various different directions, including fields as diverse as plant community ecology and remote sensing. Conservation ecologists likewise created their field from backgrounds in mathematical modeling and population, community, and ecosystem ecology. Other fields within plant ecology have seen major shifts in emphasis. Plant community ecology has seen such a shift in the past quarter century. Previously it was dominated by questions about whole-community patterns and processes; now, its main focus has shifted to questions closer to population ecology, about interactions within and among species.

A major trend in contemporary ecology, including plant ecology, is toward larger, more integrated research projects that involve many collaborators and examine phenomena across large scales of space and time or across levels of organization. Except in the subdiscipline of ecosystem ecology, which was undertaking projects with large teams of scientists in the 1970s, such multi-investigator studies were very rare in ecology until recent years. These studies may cover a range from molecular genetics up through ecosystems and social systems, and are erasing many of the traditional boundaries among subdisciplines. Plant ecology is experiencing exciting times, and we hope you will sense that excitement in this book.

Additional Readings

Classic References

Platt, J. R. 1964. Strong inference. *Science* 146 : 347–353.

Popper, K. R. 1959. *The Logic of Scientific Discovery*. Hutchinson & Co., London.

Salt, G. W. 1983. Roles: Their limits and responsibilities in ecological and evolutionary research. *Am. Nat.* 122: 697–705.

Contemporary Research

Hull, D. L. 1988. *Science as a Process*. University of Chicago Press, Chicago, IL.

Mayo, D. G. 1996. *Error and the Growth of Experimental Knowledge*. University of Chicago Press, Chicago, IL.

Nicholson, M. 1990. Henry Allan Gleason and the individualistic hypothesis: The structure of a botanist's career. *Bot. Rev.* 56:91–161.

Additional Resources

McIntosh, R. P. 1985. *The Background of Ecology*. Cambridge University Press, Cambridge.

Pickett, S. T. A., J. Kolasa and C. G. Jones. 1994. *Ecological Understanding*. Academic Press, San Diego, CA.

CHAPTER 2 *Photosynthesis and the Light Environment*

H ow does a plant "perceive" and respond to the environment that surrounds it? Why is it that some plants can survive extremes of temperature and drought while others cannot? What enables certain plants to thrive in the deep shade of the understory of a tropical rainforest, and others to succeed only in the sunniest habitats?

The **functional ecology** of plants is concerned with how the biochemistry and physiology of individual plants determine their responses to their environments within the structural context of their anatomy and morphology. Functional ecology is similar to **physiological ecology**, a subdiscipline of ecology that focuses on physiological mechanisms underlying whole-plant responses to the environment. It underlies much about the ecology of plants. Part I of this book is particularly concerned with functional ecology.

Plants must acquire energy and materials for growth, maintenance, and reproduction. They must also limit their losses; for example, if a plant loses too much water, it will wilt and eventually die. Plants must also allocate resources in ways that maximize their chances for contributing offspring to the next generation while simultaneously maximizing their chance of surviving to reproduce. In this chapter and the next two, we examine how plants capture the energy of sunlight and incorporate carbon from the atmosphere in photosynthesis, their adaptations to the light environment, their water relations, and their uptake and use of mineral nutrients. We also examine the structures in which some of these processes take place and consider some of the biochemistry involved.

While we focus here on processes occurring at the small scale of a cell, a leaf, or an individual plant, it is important not to lose sight of the forest: plants have evolved and live in an ecological context. Photosynthesis, the biochemical process by which plants capture energy from sunlight and fix carbon from the air, is usually carried out not in a laboratory, but in natural environments. The photosynthetic machinery, and the leaf in which it is housed, are both evolutionarily adapted and acclimated to the environment in which the individual plant is growing. The temperature and the amount of available light, water, and nutrients in the environment determine when and how rapidly the leaf can photosynthesize and the extent to which the plant grows and is likely to survive.

The physical conditions that the plant experiences are determined not only by the physical features of the environment, but also by other living organisms in that habitat. The amount of light available for photosynthesis may be limited by other plants competing for that light. A plant's ability to capture carbon and energy may also be reduced by herbivores eating its leaf tissue. Pathogens and pollutants may limit the plant's ability to photosynthesize. The plant responds to all of these aspects of its environment as an integrated unit, although in textbooks such as this one, we arbitrarily separate its responses into categories for convenience, treating them in different chapters. With that perspective, we begin our examination of plants' interactions with their environment with the process by which they acquire energy and carbon: photosynthesis.

The Process of Photosynthesis

Photosynthesis is the set of processes by which plants acquire energy from sunlight and fix carbon from the atmosphere. It consists of two major parts: the initial capture of light energy and the incorporation of that energy, together with carbon dioxide, into organic compounds. The organic molecules formed in photosynthesis are used by the plant to create new tissues, regulate the plant's metabolic processes, and to supply energy to those metabolic processes. Both the capture of energy (the **light reactions**) and the initial formation of carbohydrates (**carbon fixation**) take place in the chloroplasts.

The light reactions of photosynthesis occur on the **thylakoid membranes** (also called lamellae) in the interior of the chloroplasts. These double membranes exist in the form of **grana stacks** alternating with sheets of interconnecting **stroma thylakoids** (Figure 2.1). The successful capture of light energy depends on the precise spatial arrangement of these photochemical reactions within the membranes in which they occur. The architecture of the thylakoid membranes is complex and is not yet fully understood, but they appear to consist of a lipid matrix (which may be partially fluid) into which enzyme proteins and pigment-protein complexes are integrated. These highly organized components exist within a dynamic and flexible membrane system.

The pigment molecules responsible for the capture of light energy form two distinct molecular complexes in plants, **photosystem I** and **photosystem II** (Figure 2.2). Unicellular eukaryotic algae, such as the Chlorophyta, and prokaryotic cyanobacteria also have photo-

(A)

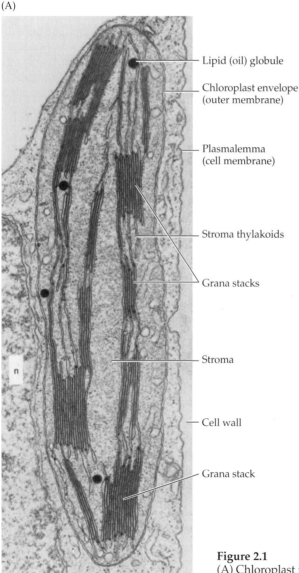

- Lipid (oil) globule
- Chloroplast envelope (outer membrane)
- Plasmalemma (cell membrane)
- Stroma thylakoids
- Grana stacks
- Stroma
- Cell wall
- Grana stack

1 μm

(B)

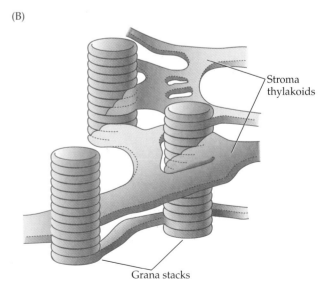

Stroma thylakoids

Grana stacks

Figure 2.1
(A) Chloroplast from a leaf of *Nicotiana tabacum* (tobacco, Solanaceae), showing grana stacks and stroma thylakoids. (B) A model of grana stacks and stroma thylakoid structure within a chloroplast. (After Esau 1977.)

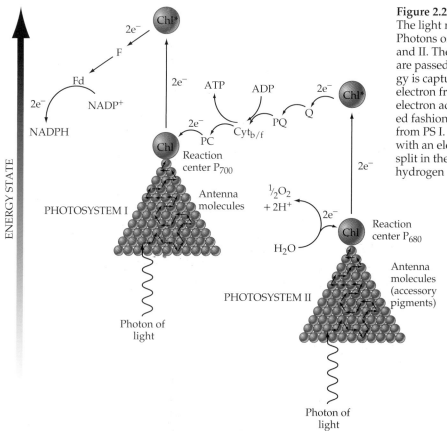

Figure 2.2
The light reactions of photosynthesis in plants. Photons of light excite electrons in photosystems I and II. The excited electrons from photosystem I are passed eventually to NADPH, where the energy is captured in a high-energy bond. The excited electron from PS II is passed through a series of electron acceptors, transferring energy in a regulated fashion to ATP, finally replacing the electron lost from PS I. The electron lost from PS II is replaced with an electron from a water molecule, which is split in the process to form oxygen gas (O_2) and hydrogen ions.

systems I and II, but certain other photosynthetic bacteria have only photosystem I. Each photosystem consists of hundreds of pigment molecules, including several forms of chlorophyll plus accessory pigments. In terrestrial plants, the accessory pigments are primarily carotenoids and xanthophylls, but eukaryotic algae and photosynthetic bacteria also use other pigments.

When a photon of light is captured by the tightly packed "antenna molecules," it is passed from one molecule to another by the process of resonance transfer until it reaches the chlorophyll molecule in the reaction center of the photosystem, where it is trapped. The excited chlorophyll molecule then passes a light-excited, high-energy electron to an electron acceptor. The energy in this high-energy electron is ultimately captured in high-energy bonds in ATP and NADPH.

Water molecules are the ultimate donors of these electrons. The oxygen that we breathe was released into the atmosphere from water molecules that were split to replace the electrons in photosystem II. Oxygen from photosynthesis was first released into the atmosphere beginning about 2 billion years ago, but was consumed by the weathering of iron and other minerals in rocks for over a billion years after that. Atmospheric oxygen reached its present levels about 400 million years ago, although there have been some fluctuations in that level since that time.

In the biochemical reactions of the **Calvin cycle** (Figure 2.3), CO_2 is taken up from the atmosphere and the carbon is ultimately incorporated into organic compounds ("fixed"). These reactions take place in the **stroma**, the watery matrix that fills the chloroplast. In C_3 plants (plants with the most common type of photosynthetic pathway; see below), large amounts of the enzymes that catalyze the reactions of the Calvin cycle are dissolved in the stroma. Energy captured by the light reactions is transferred, in the form of high energy bonds in ATP and NADPH, to the stroma, where it is incorporated, along with CO_2, into carbohydrate molecules. The fixation of carbon is thus powered by the light reactions, and the light energy captured in photosynthesis is ultimately stored in the chemical bonds of carbohydrates and other organic molecules.

Photosynthetic Rates

The rate at which a leaf can capture light energy and fix carbon is determined by several factors. Plants, like other aerobic organisms, use oxygen and release CO_2 in the process of **cellular respiration**, by which organic compounds are broken down to release energy. **Gross photosynthesis**, or the total amount of carbon captured, is reduced by the plant's respiratory release of CO_2. Photosynthetic uptake of CO_2 by plants is far greater on

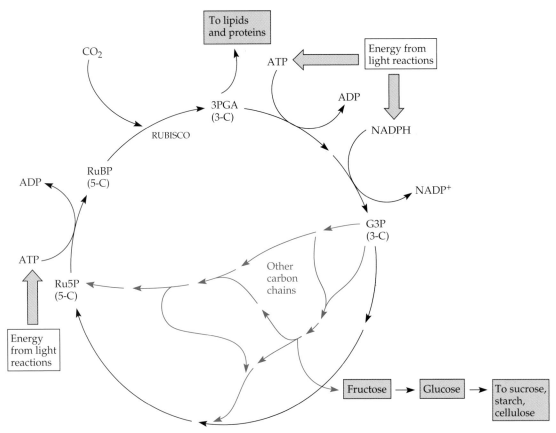

Figure 2.3
The Calvin cycle of carbon fixation in plants. CO_2 enters the stomata from the air surrounding the leaf. In a reaction catalyzed by the enzyme rubisco, CO_2 is joined with the 5-carbon molecule RuBP to form two 3-carbon molecules (3PGA). Eventually simple sugars such as fructose and glucose are formed, to be ultimately transformed into many other organic (carbon-based) molecules.

average than respiratory losses, however, resulting in a net gain of carbon by plants.

Limitations Caused by Light Levels

The most basic factor limiting photosynthesis is the total amount of light energy that reaches the thylakoid membranes. In darkness, cellular respiration results in a net loss of carbon and energy from the plant, as there is no photosynthetic capture of either light or carbon (for a partial exception, see the discussion of CAM photosynthesis below). As the light level increases, plants begin to take up CO_2. At the **light compensation point**, photosynthetic gains exactly match respiratory losses (in other words, net CO_2 exchange is zero) (Figure 2.4). Beyond that point, the more light that is available to be captured, the greater the photosynthetic rate, up to a maximum at which the rate plateaus in most plants.

The light compensation point can differ among plant species living in different parts of the environment or within a given habitat. It can even differ within individual plants depending on the structure and biochemical

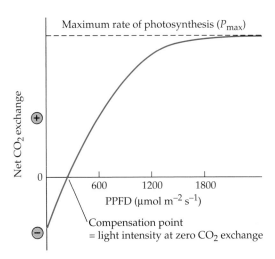

Figure 2.4
Net CO_2 exchange (per unit leaf area) for a typical C_3 leaf as a function of increasing light levels, showing the light compensation point and a plateau at a maximum rate of photosynthesis. (After Fitter and Hay 1981.)

Table 2.1 Maximum photosynthetic rates (A_{max}), light compensation points (LCP), and rubisco levels for three forest understory herbs

Parameter	Spring			Summer		Autumn
	Allium	*Viola*	*Tiarella*	*Viola*	*Tiarella*	*Tiarella*
A_{max}	15.4 ± 0.9	12.1 ± 0.7	6.8 ± 0.7	5.6 ± 0.5	3.9 ± 0.5	5.4 ± 0.3
LCP	21.6 ± 1.4	8.4 ± 1.3	9.0 ± 1.0	4.1 ± 0.9	3.2 ± 0.5	6.5 ± 0.8
Rubisco	2.83 ± 0.21	1.84 ± 0.25	1.47 ± 0.12	0.93 ± 0.07	0.50 ± 0.17	0.78 ± 0.11

Source: Rothstein and Zak (2001)

Note 1: Values are expressed on a per unit leaf area basis. A_{max} is given in μmol $CO_2/m^2/s$; LCP is given as the PPFD at which net CO_2 assimilation is zero, in μmol/m^2/s; and rubisco levels are in g/m^2. Values are means $\pm$ 1 standard error, with $n = 5$ plants per measurement.

Note 2: The duration during which each species had green leaves above ground was: *Allium tricoccum*, about 75 days; *Viola pubescens*, about 150 days; and *Tiarella cordifolia*, about 185 days.

constituents of the leaves. Rothstein and Zak (2001) contrasted the photosynthetic characteristics of three forest-floor herb species within a northern hardwood forest. In this forest, understory light levels are high in early spring before the canopy leafs out, low in midsummer, and higher again in autumn as leaves start to fall. A spring ephemeral, *Allium tricoccum* (wild leek, Liliaceae), had a constant light compensation point (Table 2.1), but was photosynthetically active only during a short period in spring. In contrast, a summer-green plant, *Viola pubescens* (downy yellow violet, Violaceae), shifted its light compensation point downward from spring to midsummer, while a semi-evergreen species, *Tiarella cordifolia* (foam-flower, Saxifragaceae), also shifted its light compensation point downward over that period, but shifted it upward again in autumn. The spring ephemeral appears to be adapted to optimize its photosynthetic uptake in the high-light environment it experiences in spring, while the other two species are both better adapted to photosynthesize under shady conditions, at least in part due to their ability to shift the light compensation point.

The quantity of light reaching the thylakoid membranes of a chloroplast can be limited by a number of factors. The location of the chloroplast within the leaf can affect the light reaching the thylakoid membranes, as can the angle at which sunlight hits the leaf. In a typical C_3 leaf, photosynthesis takes place in the spongy and palisade parenchyma cells that make up the **mesophyll** (the photosynthetic tissue between the upper and lower epidermis of a leaf). There are many chloroplasts in each photosynthetic cell (Figure 2.5). On a larger scale, self-shading by other leaves on the same plant, or shading by competitors, can also limit the amount of light available to be captured. We will examine some of these factors in more detail in other chapters.

Robin Chazdon (1985) studied the efficiency of light capture in two understory dwarf palms in the rainforests of Costa Rica, *Asterogyne martiana* and *Geonoma cuneata* (both in the Arecaceae). Both species have narrow, spi-

rally arranged leaves that minimize self-shading. *A. martiana* was found in locations with somewhat higher light levels, and had a greater number of leaves and a greater total leaf area, compared with *G. cuneata*. As a result, *G. cuneata* had greater efficiency of light interception (the proportion of incident light intercepted by the plant

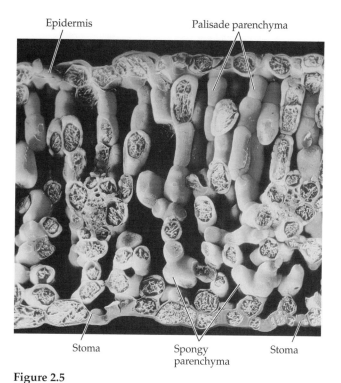

Figure 2.5
Scanning electron micrograph of a cross section of a leaf of *Brassica septiceps* (turnip, Brassicaceae), showing the palisade parenchyma and spongy parenchyma cells inside which most chloroplasts are found and in which most of the plant's photosynthesis takes place. Many of the cells in this micrograph have been broken open to expose their internal structure. The upper epidermis is visible, as are several stomata on the underside of the leaf, along with the substomatal cavities they open into on the inside of the leaf. (Photograph © J. Burgess/Photo Researchers, Inc.)

(A)

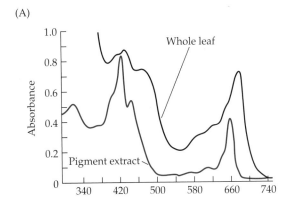

(B)

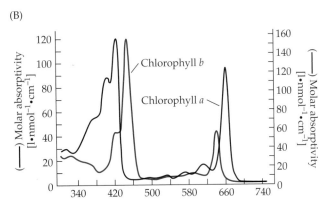

(C)

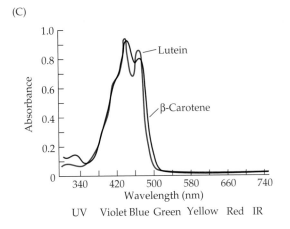

Wavelength (nm)

UV Violet Blue Green Yellow Red IR

Figure 2.6
(A) Absorption spectra of (wavelengths of light absorbed by) a whole leaf and an extract of all photosynthetic pigments from that leaf. (B) Absorption spectra of purified extracts of chlorophylls *a* and *b*. (C) Absorption spectra for two of the most important accessory pigments, lutein and β-carotene. (After Mohr and Schopfer 1995.)

leaves to mature plants with many leaves, this shift reduced self-shading to a minimal level and optimized light capture for individuals of very different forms and total leaf areas.

The quality of light, or the wavelengths of light available to be captured, can also limit photosynthetic rates (Figure 2.6). Blue and red wavelengths are preferentially captured by the light reactions. Perhaps paradoxically, given our image of the beautiful green world, green wavelengths are particularly ineffective for photosynthesis. We see nature as green because green light is reflected or transmitted—"discarded" rather than used—by plants. The wavelengths of light that can be used in photosynthesis are termed **photosynthetically active radiation**, or **PAR**. The amount of usable light energy impinging upon a leaf per unit time is called the **photosynthetic photon flux density** (**PPFD**).

Limitations on Carbon Uptake

Plants take up CO_2 from the atmosphere as air moves through the stomata and into the intercellular spaces surrounding the photosynthetic cells within a leaf. Carbon uptake is driven by a concentration gradient of CO_2, set up by the biochemical reactions in the chloroplasts that remove CO_2 from the intercellular spaces. The uptake of CO_2 is regulated by the conductance to CO_2 diffusion on the pathway from the air into the leaf and into the chloroplast. The concentration of CO_2 in the intercellular spaces depends on how rapidly CO_2 is removed by being fixed into organic compounds, and on how readily CO_2 comes into the leaf to replace that CO_2.

The **leaf conductance** to CO_2 is the rate at which CO_2 flows into the leaf for a given concentration difference between ambient and intercellular CO_2. The inverse of conductance is resistance. Low conductance or high resistance at a particular point in the pathway of CO_2 movement will limit its movement along that pathway. If the overall leaf conductance to CO_2 is high and CO_2 concentrations in the intercellular spaces are being continually drawn down by the rapid fixation of carbon, then CO_2 influx from the air surrounding the leaf will be high.

The rate of CO_2 uptake can be modeled with a flux equation. **Flux equations** are used to model flow rates, and are of the general form

$$\text{flux} = (\text{conductance}) \times (\text{driving force})$$

canopy, which depends upon leaf arrangement and display angle), but *A. martiana* had a greater total capacity to capture light (where the light interception capacity, or effective leaf area, is the product of total leaf area and light interception efficiency). Takenaka and associates (Takenaka et al. 2001) followed up this work by analyzing the effects of leaf display on light capture efficiency in another understory palm, *Licuala arbuscula* (Arecaceae), which grows in lowland rainforests in southeast Asia. This species has compound, fan-shaped leaves with long petioles. The authors found that the angle at which the petioles are held changed as the number of leaves increased. As plants grew from juveniles with few

For CO_2 uptake, the driving force is a difference in CO_2 concentration, and the flux equation can be stated as

CO₂ uptake rate = (leaf conductance to CO_2 diffusion) × (difference in CO_2 concentration between air and chloroplast)

or, using conventional symbols,

$$A = g_{leaf} \times (C_a - C_i)$$

The term A is the **assimilation rate** (in µmol/m²/s); this is the rate at which CO_2 is taken up by the leaf. The terms C_a and C_i are the ambient and intercellular concentrations of CO_2, respectively; that is, the concentrations in the surrounding air and at the surface of the photosynthetic cell. The term g_{leaf} is the total conductance of the leaf to CO_2.

We can separate leaf conductance into its two major components, g_s and g_a—the conductances to CO_2 through the stomata and through the boundary layer of air surrounding the leaf, respectively—so that

$$1/g_{leaf} = 1/g_a + 1/g_s$$

Generally, g_a is large, since CO_2 readily passes through the boundary layer, and so does not contribute greatly to the regulation of CO_2 flux. The conductance to CO_2 through the stomata (g_s), however, is highly variable and is under the control of the plant. Stomatal conductance regulates leaf CO_2 flux under most conditions. Thus, plants are not merely passive recipients of CO_2, but regulate its uptake closely. This regulation occurs over short time scales (seconds to minutes), as stomata are opened or closed, and over longer time scales (days to months), as leaf morphology and chemistry are altered. Over much longer time scales (centuries to millennia or longer), natural selection acts to alter the capacity of plant populations in different environments to take up carbon under different conditions as morphology, physiology, and other plant characters evolve.

Why would plants ever restrict their uptake of CO_2? We examine this question in more detail in Chapter 3, but briefly, it is largely because CO_2 gain is linked inextricably with the loss of water through the same stomatal openings in the leaf through which CO_2 is taken up.

A different formulation for photosynthetic rate is sometimes employed to describe net photosynthesis at **light saturation**, A_{sat}, the light level at which the maximum photosynthetic rate is reached, when CO_2 uptake is not limited by stomatal conductance:

$$A_{sat} = g_m \times (C_i - C_c)$$

where C_c is the compensation point for CO_2 and g_m is the **mesophyll conductance** or **intracellular conductance**, the conductance to CO_2 through the leaf mesophyll cells and cell walls. An enormous amount of air must be processed by the leaf in the course of photosynthesis. To make a single gram of the carbohydrate glucose, a plant needs 1.47 grams of CO_2, which is the amount in about 2500 liters of air. Looked at another way, the air needed to fill a structure the size of the Houston Astrodome could supply enough CO_2 to fix about 590 kilograms (1300 lbs) of glucose (Figure 2.7).

When the stomata of a leaf are fully open, its conductance to CO_2 is generally high. The exact value depends on the number and size of the stomata, and it varies among species, individual plants, and even leaves on the same plant. (We will return to the issue of stomatal number and size in the next chapter.) When the stomata are closed, leaf conductance to CO_2 approaches zero, although sometimes small amounts of CO_2 may "leak" through the cuticle.

The stomata of many plants are very dynamic. The guard cells that determine the degree of stomatal open-

Figure 2.7
The Houston Astrodome in its glory days. Given the amount of air that filled the Astrodome at any one moment, had its builders used real Bermuda grass instead of inventing Astroturf and not painted the dome to make it opaque, the turf could have turned the CO_2 in that air into roughly 590 kilograms of glucose—and that's without a single player or spectator breathing into it. The advent of retractable domes has returned natural, carbon-fixing turf to its former prominence at ball fields. (Photograph © Charles E. Rotkin/ CORBIS.)

ing are continually in motion, widening and narrowing the stomatal pores to regulate CO_2 entering the leaf and water leaving it. Some of the stomata may begin to close while others remain open (Figure 2.8). Such patchy stomatal closure may be more common when plants are experiencing stress, but this phenomenon is still under active study (Terashima 1992). The guard cells are under a complex set of controls that respond to both internal and external factors.

Variation in Photosynthetic Rates within and between Habitats

Photosynthetic rates sometimes vary among plants within a habitat, and across habitats, in ways that seem to make sense because they are correlated with species composition, habitat preferences, or growth rates. In other cases, photosynthetic rates may have little role in determining population processes or species' distributions. Even growth rates may be minimally related to photosynthetic rates. The total carbon accumulated by a plant depends not only on the rate of photosynthesis on a leaf area basis, but also on the total leaf area of the plant, as well as on other factors, such as the length of time the leaves are maintained and are photosynthetically active.

An example of a case in which differences in photosynthetic rates correlate with species and habitat differences comes from a study of dwarf shrubs in a subalpine heath community in the Apennine Mountains of northern Italy (Gerdol et al. 2002). In sheltered locations, two deciduous shrubs, *Vaccinium myrtillus* (blueberry, Ericaceae) and *V. uliginosum* (bog-bilberry, Ericaceae) are dominant, or at least abundant, and the vegetation is very dense. In exposed habitats, the vegetation is more open, and a variety of shrubs are dominant, including *V. uliginosum* and an evergreen species, *Empetrum hermaphroditum* (crowberry, Ericaceae). Renato Gerdol and colleagues found that the evergreen *E. hermaphroditum* had the lowest photosynthetic rates of the three species. The deciduous *V. uliginosum* had intermediate rates, which did not differ among habitats. The deciduous species restricted to the more favorable sites, *V. myrtillus*, had the highest photosynthetic rates. The differing photosynthetic rates of these plants did not, however, correspond to their relative growth rates, nor did they explain the relative responses of the plants to removal of competing neighbors or to fertilizer additions.

In the study of northeastern forest understory species discussed above (Rothstein and Zak 2001), maximum photosynthetic rates were correlated with the growth environment of the three species studied (see Table 2.1). The spring ephemeral, *Allium*, which grew only in the highest light period, had the highest maximum photosynthetic rates overall. During spring, the summer-green *Viola pubescens* had intermediate photo-

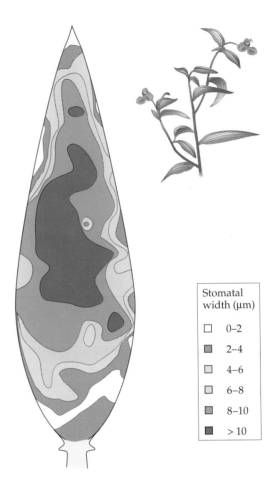

Figure 2.8
Stomatal widths in different parts of a leaf of *Commelina communis* (day-flower, Commelinaceae) at midday. Some stomata are wide open, while others are partially open or fully closed. (After Larcher 1995.)

Stomatal width (μm)	
☐	0–2
◼	2–4
▧	4–6
▨	6–8
◼	8–10
◼	> 10

synthetic rates, and the evergreen *Tiarella cordifolia* had the lowest rates. In midsummer, when light levels were lowest, photosynthetic rates declined substantially for both *Viola* and *Tiarella*, but *Tiarella* still had the lower rates of photosynthesis. In autumn, only *Tiarella* was photosynthetically active, and its maximum photosynthetic rate increased again in the higher light environment. These differences among species and seasons were positively correlated with the levels of rubisco (the enzyme that catalyzes the initial capture of CO_2; see below) in the leaves and patterns of plant growth as well as with the duration of time that each plant is photosynthetic. *Allium* gained all of its biomass during the spring high light period and lost biomass after that time. *Viola* and *Tiarella* also gained biomass rapidly during spring, and continued to increase in biomass during the summer. *Viola* sharply declined in biomass from late summer through winter, but *Tiarella* continued to accumulate biomass through early winter. Only 25% of the

Box 2A

Photorespiration

The mitochondria of plants carry out cellular respiration much as those of animals do, consuming O_2 and releasing energy to be used by the cells. Plants also carry out another kind of respiration, called photorespiration. Like ordinary cellular respiration, photorespiration consumes O_2 and releases CO_2, but unlike cellular respiration, it depends on light. It takes place in cells that contain chloroplasts, but involves two additional organelles: mitochondria and peroxisomes.

In photorespiration, rubisco catalyzes the binding of O_2 to ribulose bisphosphate (RuBP). This process thus competes with photosynthesis both for RuBP, the substrate of both reactions, and for rubisco, the enzyme that catalyzes both reactions. High partial pressures of oxygen, low CO_2 concentrations, and warm temperatures favor photorespiration over photosynthesis.

Although photorespiration is often considered to be disadvantageous because it competes with photosynthesis, it may have a protective function. Photorespiration may "soak up" excessive electron flow in bright light, thereby protecting photosystem II from damage when the leaf's carboxylation capacity is not capable of keeping up with the energy captured in the light reactions (e.g., when drought forces stomata to close partially or fully, limiting or cutting off the supply of CO_2).

The biochemical reactions of photorespiration take place in chloroplasts, mitochondria, and peroxisomes as discussed on page 24.

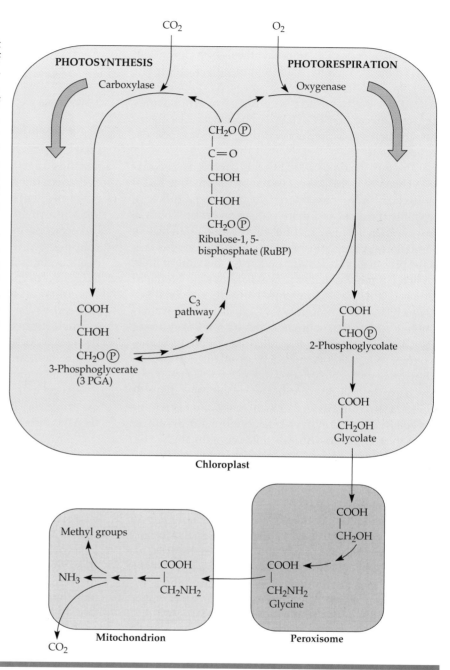

biomass gain for *Tiarella* came during the low light period of summer.

The Three Photosynthetic Pathways

Plants fix carbon using one of three different photosynthetic pathways: C_3, C_4, or CAM (crassulacean acid metabolism). C_3 photosynthesis and C_4 photosynthesis are named for the three-carbon and four-carbon molecules that are the first stable products of photosynthesis in these pathways, while CAM is named after the plant family Crassulaceae (the stonecrops), in which it was first discovered. The vast majority of plants use C_3 photosynthesis, and C_3 plants are found everywhere that plants exist. C_3 photosynthesis was the first pathway to evolve and the first to be understood by scientists. C_4 and CAM photosynthesis are modifications of C_3 photosynthesis, and evolved from it.

C_3 Photosynthesis

In the Calvin cycle of **C_3 photosynthesis** (see Figure 2.3), CO_2 is joined with a five-carbon molecule, RuBP (ribulose bisphosphate), to form a six-carbon compound that instantly separates into two three-carbon molecules (phosphoglycerate; 3PGA). In C_3 photosynthesis, therefore, the first stable product of carbon fixation is a three-carbon chain.

The initial step in which CO_2 is captured—the **carboxylation** of RuBP—is catalyzed by the enzyme RuBP carboxylase/oxygenase (mercifully nicknamed **rubisco**), probably the most abundant protein on Earth. Rubisco, however, is curiously inept at capturing CO_2, which is particularly strange considering how important this task is for primary productivity on Earth—one might have expected a more efficient process to have evolved and replaced it in plants long ago. Not only does rubisco have a relatively low affinity for CO_2, it also has an alternative function that competes with its role in capturing CO_2. Besides catalyzing the initial step of photosynthesis, rubisco can also catalyze a process called **photorespiration**, in which oxygen is taken up instead of carbon dioxide (see Box 2A on the preceding page). At higher temperatures, rubisco increasingly favors the oxygenation reaction over carboxylation, or photorespiration over photosynthesis. Likewise, the higher the concentration of O_2 and the lower the concentration of CO_2 reaching the chloroplast, the more O_2 is taken up in preference to CO_2. These properties of rubisco limit photosynthetic CO_2 uptake.

The limitations of rubisco are not especially important for plants whose leaves are shaded, because in their case photosynthesis is limited mainly by light levels rather than by the efficiency of CO_2 uptake. However, for plants growing in warm, bright environments, the limitations posed by the properties of rubisco can have major ramifications for photosynthetic rates, and ultimately for growth. Even under the best conditions, C_3 plants must maintain large quantities of rubisco to maintain adequate rates of photosynthesis. Rubisco, like all enzymes, contains a substantial amount of nitrogen. Between 10% and 30% of the total nitrogen in the leaves of C_3 plants is in rubisco.

Because of the limitations of rubisco, photosynthetic rates are also limited by the concentration of CO_2 in the atmosphere. Consequently, at elevated CO_2 concentrations, C_3 plants can achieve much higher photosynthetic rates, all else being equal. Plant growers sometimes make use of this response by growing plants in greenhouses with artificially high concentrations of CO_2

in the air. Plants evolved under atmospheric CO_2 levels very different from those of today, as we will shortly. The current rapid increases in atmospheric concentrations of CO_2 due to human activities may have long-term consequences for CO_2 uptake by plants. We will return to this issue in Chapter 22.

There is an evolutionary "solution" to the dilemma posed by photorespiration and the limitations of rubisco as a catalyst for CO_2 uptake. That solution is C_4 photosynthesis.

C_4 Photosynthesis

Like C_3 photosynthesis, **C_4 photosynthesis** ultimately depends on the Calvin cycle to convert CO_2 into carbohydrates. However, C_4 photosynthesis contains an additional step that is used for the initial capture of CO_2 from the atmosphere (Figure 2.9). In this additional step, a three-carbon molecule called PEP (phosphoenolpyruvate) is joined with CO_2 to form a four-carbon acid, OAA (oxaloacetate). Thus, the first product of carbon fixation in C_4 photosynthesis is a molecule with four carbons. The initial capture of CO_2 is catalyzed by the enzyme **PEP carboxylase**, which functions only to fix CO_2. It has a much higher affinity for CO_2 than does rubisco. Because it does not also catalyze photorespiration, PEP carboxylase can maintain high rates of CO_2 uptake even at warm temperatures.

After its formation, the four-carbon molecule is decarboxylated (the CO_2 is removed), and the CO_2 is then incorporated into organic molecules via the Calvin cycle. Rubisco functions to fix this internally liberated CO_2 molecule in C_4 plants, just as it acts to fix CO_2 coming in from the external atmosphere in C_3 plants. There are three different subtypes of C_4 photosynthesis, each with its own enzyme for decarboxylation: NADP-me, which uses NADP-malic enzyme, NAD-me, which uses NAD-malic enzyme, and PEP-ck, which depends on PEP carboxykinase. While it has been suggested that there are ecological differences among these C_4 subtypes, the evidence for this is still ambiguous because these physiological differences are confounded with other differences among the species possessing them.

C_4 photosynthesis depends on specialized leaf anatomy (Figure 2.10). In the typical Kranz (meaning "wreath") anatomy found in C_4 plants, there is a spatial separation of the C_4 and C_3 reactions. The initial capture of CO_2 from the atmosphere takes place in the mesophyll cells just under the epidermis and adjacent to the substomatal cavities, while its incorporation into carbohydrates via the Calvin cycle takes place deep inside the leaf in the bundle sheath cells. In C_3 plants, the concentration of oxygen in the chloroplasts is typically about a thousand times greater than the concentration of carbon dioxide, resulting in substantial rates of photorespiration. In C_4 plants, rubisco is located (along with the

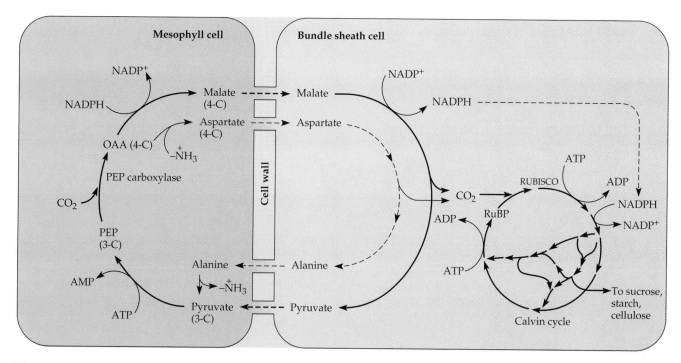

Figure 2.9
The C_4 photosynthetic pathway, with the biochemical steps that take place in the mesophyll cells shown on the left, and those that take place in the bundle sheath cells shown on the right.

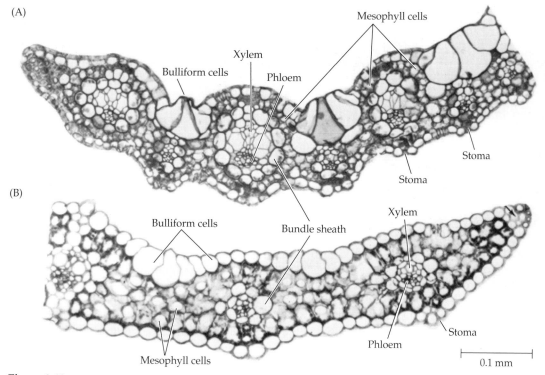

Figure 2.10
Anatomy of (A) a leaf of *Saccharum officinarum* (sugarcane, Poaceae), a C_4 grass, and (B) a leaf of *Avena* sp. (oats, Poaceae), a C_3 grass, both in cross section, showing the differences in their architectures. Note the tight packing of the mesophyll cells in the outer ring surrounding the bundle sheath cells of the C_4 leaf, which themselves tightly encircle the vascular bundle (xylem and phloem), in contrast to the more loosely packed photosynthetic cells in the C_3 leaf. There are large numbers of chloroplasts in the bundle sheath cells (clustered at the outer edge within each cell) as well as in the mesophyll cells of the C_4 leaf, in contrast with the absence of chloroplasts in the bundle sheath cells surrounding the vascular bundle in the C_3 leaf. The bulliform cells act as hinges to allow the leaf to roll up during drought (see Chapter 3). (From Esau 1977.)

Stable Isotopes and Photosynthesis

Chemically identical carbon compounds fixed via C_4 photosynthesis differ from those fixed via the C_3 pathway in a subtle way: their ratios of the two stable isotopes of carbon, ^{12}C and ^{13}C. Most molecules of CO_2 in the air contain ^{12}C, the stable carbon isotope with an atomic weight of 12. Approximately eleven out of every thousand CO_2 molecules instead have the slightly heavier stable isotope ^{13}C, with an atomic weight of 13. (^{14}C, which is unstable and thus radioactive, is much rarer still.)

When the initial capture of CO_2 is made by rubisco, $^{13}CO_2$ molecules, being slightly heavier, are disproportionately left behind, unfixed, in the atmosphere (i.e., the process discrimi-

nates against ^{13}C). PEP carboxylase is more effective at "mopping up" all of the CO_2 molecules it encounters ($^{13}CO_2$ as well as $^{12}CO_2$), and thus discriminates less against ^{13}C.

The relative proportion of the two stable carbon isotopes in carbon compounds is expressed in reference to a standard (dolomite from the PeeDee formation):

$$\delta^{13}C = \frac{\left[(^{13}C/^{12}C)_{sample} - (^{13}C/^{12}C)_{standard} \right] \times 1000}{(^{13}C/^{12}C)_{standard}}$$

For terrestrial C_3 plants, the mean ^{13}C value is $-27\permil$ (range $-36\permil$ to $-23\permil$), while C_4 plants have a mean of $-13\permil$ (range $-18\permil$ to $-10\permil$). (CAM plants have a range of ^{13}C values,

depending on the relative proportion of carbon fixed initially by rubisco and by PEP carboxylase.) Because the ranges of the ^{13}C values for C_3 and C_4 plants are nonoverlapping, the carbon isotope ratio provides a convenient way to distinguish between C_3 and C_4 plants and plant tissues. These carbon isotope ratios also persist in the food chain. The fossilized carbon of herbivores that ate largely C_4 plants, for example, can be distinguished from that of those that ate C_3 plants. This property has been used by paleontologists and archeologists to distinguish food sources; for example, it has been used to determine when grazers first began to depend heavily on C_4 grasses.

---

other enzymes of the Calvin cycle) in the bundle sheath cells, which are not exposed directly to the external atmosphere. The four-carbon acid OAA travels directly through thin strands of living tissue, **plasmodesmata**, from mesophyll cells to bundle sheath cells, where it is decarboxylated. The concentration of CO_2 in the bundle sheath cells of C_4 plants is an order of magnitude higher than its concentration in the photosynthetic cells of C_3 plants, and the ratio of oxygen to carbon dioxide is greatly reduced. Thus, rubisco is 'fed' a concentrated stream of CO_2 molecules, resulting in the effective elimination of photorespiration in C_4 plants.

The consequences of overcoming the limitations of rubisco are enormous (see Box 2B for a discussion of one less obvious consequence). C_4 plants generally have higher maximum rates of photosynthesis than do C_3 plants. The temperature optimum for photosynthesis is usually much higher for C_4 than for C_3 species (Figure 2.11). C_3 species typically become light-saturated at levels well below full sunlight, whereas for C_4 species photosynthesis often does not become light-saturated even in full sunlight (Figure 2.12) because CO_2 uptake is not limited by the high oxygenase activity of rubisco. C_4 plants contain only one-third to one-sixth the amount of rubisco that C_3 plants have and yet are able to maintain the same or greater photosynthetic rates, resulting in higher **nitrogen use efficiency** (maximum photosynthetic rate per gram of nitrogen in the leaf). Because the C_4 pathway makes CO_2 available to the Calvin cycle

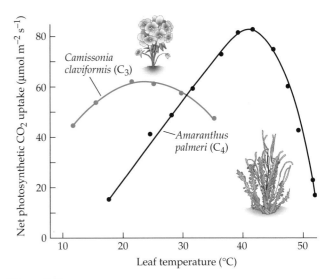

Figure 2.11
Photosynthetic rates at different temperatures in a desert C_3 plant, *Camissonia claviformis* (Onagraceae), that grows mainly during the winter and early spring, and in *Amaranthus palmeri* (Amaranthaceae), a C_4 plant from the same habitat that grows mainly in the summer. The maximum rate of photosynthesis occurs at about 23°C in the C_3 plant and at about 42°C in the C_4 plant. (After Ehleringer 1985.)

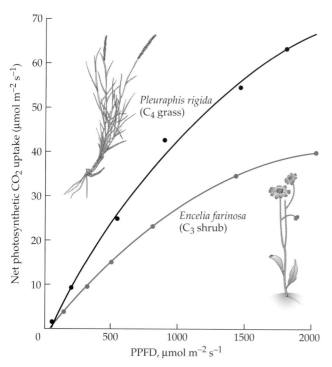

Figure 2.12
Photosynthetic response to light intensity (PPFD, photosynthetic photon flux density) in a C_4 desert grass, *Pleuraphis rigida* (Poaceae), and in a C_3 desert shrub, *Encelia farinosa* (brittlebush, Asteraceae), from the western United States. At full sunlight (approximately 2000 μmol m^{-2}s^{-1}), the leaves of the C_3 shrub are light-saturated, while those of the C_4 grass are not. (After Nobel 1983.)

with great efficiency, the ratio of CO_2 fixed to water lost (water use efficiency) is greater for a C_4 than a C_3 leaf at equal stomatal conductance.

The price for these advantages is also substantial: it takes additional energy in the form of ATP to run the C_4 pathway. When light levels are high and conditions are optimal, the energy invested is more than compensated by the additional photosynthetic gains. At lower light levels, however, this expensive machinery must continue to be paid for, resulting in potential disadvantages for plants that possess it.

Crassulacean Acid Metabolism (CAM Photosynthesis)

Crassulacean acid metabolism (CAM) uses essentially the same biochemistry as C_4 photosynthesis to overcome the limitations of rubisco and eliminate photorespiration. However, CAM plants accomplish this in a very different way than C_4 plants (Figure 2.13). In C_4 plants, rubisco is found only in the bundle sheath cells, which are spatially segregated from external air. In CAM plants, rubisco is found in all photosynthetic cells. Instead of using spatial separation, CAM plants temporally separate the capture of light energy and the uptake of CO_2.

Figure 2.13
Biochemical reactions involved in CAM photosynthesis. The reactions all take place in the same photosynthetic cell, but the reactions shown on the left take place at night, and those on the right take place in the daylight.

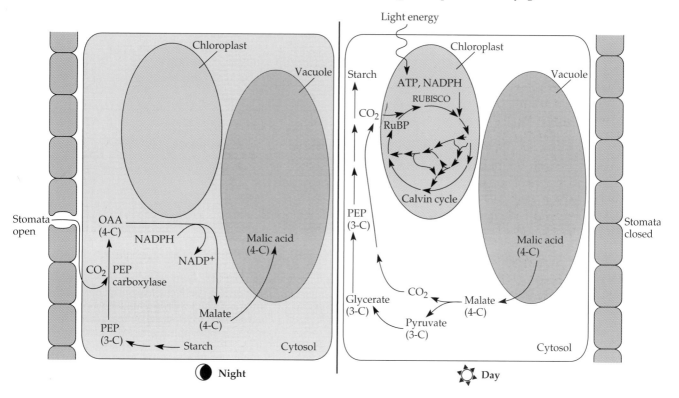

In C_3 and C_4 plants, stomata are open in the day, when light capture and carbon uptake and fixation occur. CAM plants open their stomata at night. During this time, CO_2 is captured by PEP carboxylase, and large pools of organic acids are accumulated in the vacuoles of the photosynthetic cells. In daylight, the light reactions occur just as they do in other plants, but with the stomata closed. The organic acids that accumulated overnight are decarboxylated, providing CO_2 to the Calvin cycle. Like C_4 photosynthesis, CAM photosynthesis allows rubisco to function in a high CO_2 and low O_2 environment, in which photosynthesis is favored over photorespiration.

CAM plants must have thick, succulent photosynthetic tissues with sufficient physical capacity to accumulate large amounts of organic acids overnight. The amount of photosynthate that can be accumulated by CAM photosynthesis over a 24-hour period is limited by the amount of space available in the vacuoles of the photosynthetic cells. The temporal separation of the C_4 and C_3 reactions also slows maximum photosynthetic rates. CAM plants generally cannot accumulate carbon as rapidly as can either C_4 or C_3 plants.

This photosynthetic pathway excels in one area, and that is **water use efficiency**, the grams of carbon fixed in photosynthesis per gram of water lost in transpiration. Because stomata are open only at night, when temperatures are cool and the air is more humid, water use efficiency is far higher in CAM plants than in C_3 or C_4

plants because water loss is much lower in CAM plants. Unlike C_3 and C_4 plants, in which the photosynthetic pathway is obligate (cannot be turned on or off), some CAM plants utilize both nighttime CO_2 uptake, with carbon fixation via CAM, and daytime CO_2 uptake via the C_3 pathway (Figure 2.14). Some species behave in a facultative manner, using CAM photosynthesis when water is most limiting, and incorporating CO_2 during the day using C_3 photosynthesis when conditions are more favorable so as to achieve higher photosynthetic rates.

Evolution of the Three Photosynthetic Pathways

Phylogeny of the Photosynthetic Pathways

C_4 photosynthesis and CAM have arisen independently from C_3 photosynthesis many times over the course of evolution. Approximately 2000 angiosperm species in about 18 families use C_4 photosynthesis, including both monocot and dicot species. Approximately half the species in the grass family, the Poaceae, are C_4 plants. Other families in which C_4 is well represented are the Amaranthaceae, Euphorbiaceae, Cyperaceae, and Portulaceae. Typically, all of the species in a genus are either C_3 or C_4, but there are several genera with both C_3 and C_4 members (for example, *Atriplex* in the Amaranthaceae and *Panicum* in the Poaceae). A few species are intermediate between C_3 and C_4 in their structure and func-

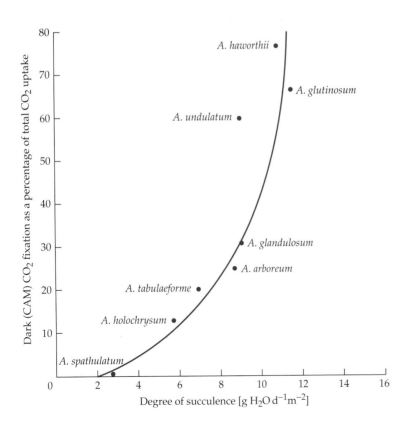

Figure 2.14
Relationship between leaf succulence (grams of water in fresh leaves per unit leaf area) and the percentage of CO_2 uptake that occurs via CAM among species in the genus *Aeonium* (Crassulaceae). Plants with thicker, more succulent leaves exhibit a greater proportion of their total CO_2 uptake at night, using CAM. (After Larcher 1995.)

Figure 2.15
Flaveria linearis (Asteraceae), a species with a photosynthetic pathway that is intermediate between the C$_3$ and C$_4$ pathways. It is shown here growing along a beachfront on the coast of Florida. (Photographs courtesy of P. Teese.)

Flaveria linearis

tion. *Flaveria* (Asteraceae), for example, contains several intermediate species (Figure 2.15), as well as species that are completely C$_3$ and completely C$_4$.

The steps that have occurred in the evolution of C$_4$ photosynthesis are not definitively known for most species that have evolved this pathway. In the genus *Flaveria*, however, it appears that the first step in the evolution of C$_4$ photosynthesis may have been the initiation of biochemical differentiation between the bundle sheath and mesophyll cells, with recycling of photorespired CO$_2$ (using rubisco to catalyze the recapture of carbon) in the bundle sheath cells. The next step may have been an increase in the activity of PEP carboxylase, followed by increasing activity of the other enzymes of the C$_4$ pathway. Finally, full anatomical separation of the C$_4$ and Calvin cycle pathways was achieved.

The evolutionary transition to CAM from C$_3$ ancestors is apparently much easier than the transition to C$_4$, and more than 20,000 angiosperm species in approximately 25 families have evolved this pathway, including both monocots and dicots. There are also CAM species among the ferns, as well as in the primitive vascular plant genus *Isoetes* (Isoetaceae). It is not unusual for a plant family to have both C$_3$ and CAM members; the genus *Euphorbia* (Euphorbiaceae) includes C$_3$, C$_4$, and CAM species. There are also species that appear to represent intermediates between fully expressed CAM and more primitive CAM cycling (that is, C$_3$ photosynthesis with the ability to recycle CO$_2$ from ordinary cellular respiration at night).

Photosynthesis through Evolutionary Time

Photosynthesis first evolved in bacteria in the Precambrian era, perhaps 2500 million years ago. Atmospheric CO$_2$ levels at that time, and for eons during the early evolution of land plants, were much greater than they are now (see Figure 21.7). The evolution of stomata about 400 million years ago was a milestone in the early history of terrestrial plants because they permitted the regulation of water loss relative to carbon uptake. Stomata appear to have evolved from simple pores in the epidermis of plants that possessed a cuticle and intercellular gas spaces. Even before the evolution of functional stomata, these pores would have facilitated higher photosynthetic rates for the plants occupying a given area (Raven 2002).

During the Mesozoic era (248 to 65 million years ago), gymnosperms dominated, dinosaurs arose, became dominant, and went extinct, and flowering plants (the angiosperms) first appeared. At this time, atmospheric CO$_2$ levels were about 3.5 times greater than they are today (1260 parts per million partial pressure of CO$_2$). These levels would have resulted in relatively high CO$_2$:O$_2$ ratios and limited photorespiration for C$_3$ plants, even at warm temperatures (Ehleringer and Monson 1993).

In the Eocene epoch (54 to 38 million years ago), when angiosperms began to dominate the landscape, CO$_2$ levels in the atmosphere became much lower (perhaps 700 parts per million). By the Recent epoch, atmospheric CO$_2$ had dropped to very low levels, measuring between 180 and 280 parts per million over the past 160,000 years. As CO$_2$ levels dropped, C$_4$ photosynthesis would have been increasingly favored, particularly in warm climates. The first evidence for C$_4$ photosynthesis comes from plants that existed 57 million years ago, at the end of the Miocene epoch. C$_4$ plants became dominant in some environments at about this time

(Ehleringer and Monson 1993), although climate change may have played a more significant role in the rise of these plants than the decline in atmospheric CO_2 (see Chapter 21). Today, as industrialization, technology, and human population growth push CO_2 levels up rapidly and drastically (see Chapter 22), some ecologists predict that plants with C_4 photosynthesis may lose their advantages over C_3 plants, perhaps leading disproportionately to the extinction of existing C_4 species.

Even among C_3 species, there may be marked differences in the responses to elevated CO_2. In a field experiment, Travis Huxman and Stanley Smith (2001) compared photosynthetic responses to elevated CO_2 in two Mojave Desert species, both with C_3 photosynthesis. One of the species, *Bromus madritensis* ssp. *rubens* (red brome, Poaceae) is an invasive annual grass; the other, *Eriogonum inflatum* (desert trumpet, Polygonaceae), is an herbaceous native perennial. *Bromus* responded to elevated CO_2 by increasing its net photosynthetic rates throughout the growing season, as well as by extending the period of high photosynthetic activity. In contrast, *Eriogonum* experienced increased photosynthetic rates for only a brief period at the peak of photosynthetic activity, and may have even curtailed photosynthetic activity more steeply over the growing season in the enhanced CO_2 treatment. Maximum stomatal conductances over the course of the year were sharply reduced in the high CO_2 treatment for *Eriogonum*, but were relatively unaffected for *Bromus*. The ultimate effects of rising CO_2 levels on the persistence or spread of these species is as yet unclear, but differences in responses to elevated CO_2 could potentially alter species' distributions and community composition over the next century.

Growth Form, Phenology, and Distribution of C_3, C_4, and CAM Plants

Growth Forms and Habitats

The most striking evidence for the adaptive significance of the three photosynthetic pathways comes from patterns of species distribution and abundance. C_3 plants are by far the most abundant in terms of both number of species and total biomass. While there are more CAM species overall than C_4 species, CAM plants are much less abundant than C_4 plants in terms of both biomass and worldwide distribution. CAM plants are generally found in particular habitat types, and their importance, whether it is measured as the percentage of species in a regional flora or as the proportion of biomass in the vegetation, is small in most ecosystems. Each photosynthetic pathway, as we have just seen, offers advantages under particular conditions, but has costs associated with it that put it at a disadvantage under other conditions.

C_4 grasses dominate, or are major components of, grassland ecosystems covering vast portions of the globe. Several important crop species are C_4 grasses (e.g., maize, millet, and sugarcane). It is no accident that C_4 photosynthesis is so important in the grasses. Grassland environments are often characterized by warm, bright conditions, and even outside of grasslands, grasses are commonly found in high light environments. These are precisely the conditions under which C_4 photosynthesis is favored. While there are many C_4 annuals, perennials, and even shrubs, very few tree species use C_4 photosynthesis; likewise, while there are C_4 species that are native to shady, cool, and aquatic habitats, they are the exceptions.

CAM plants are found in two distinct habitat types with two distinctive growth forms: as succulent terrestrial plants growing in deserts or other arid habitats, and as epiphytes growing in the canopies of trees in tropical and subtropical habitats. CAM photosynthesis has evolved in many different families in arid environments (Figure 2.16), including cacti (Cactaceae), agaves (Agavaceae), species in the Euphorbiaceae from Africa, and species in the Crassulaceae. Succulents in this group often serve as examples of convergent evolution due to their similar external morphologies (see Figure 6.11).

Epiphytes are plants that rely for support on other plants, generally living on trees, but depend on minerals deposited from the atmosphere for nitrogen and other mineral nutrients (Figure 2.17A). Epiphytes have at least one thing in common with desert plants, being subjected frequently to severe shortages of water. Even in a tropical rainforest, life high in the canopy can be surprisingly dry (see Figure 19.3B). A number of families have epiphytic CAM species. Some familiar examples are Spanish moss (*Tillandsia usneoides*; Figure 2.17B) in the Bromeliaceae and many orchids (the family Orchidaceae) and ferns.

Perhaps surprisingly, CAM plants are rare in the very hottest, driest deserts, where rainfall is not only limited, but cannot be counted on in any given year. For example, there are few CAM plants in Death Valley, California, or in the Sahara of North Africa. In the Atacama Desert of Peru, where it may be years between rains in some places, cacti are almost entirely restricted to the zone that receives moisture as coastal fog. Why might CAM plants be kept out of the most arid deserts? Under these extreme conditions, rapid growth during the brief, unpredictable periods when water is available would be strongly favored (see Figure 18.25), and CAM plants do not generally grow rapidly. Another factor may be survival during very prolonged drought. Both cellular respiration and water loss would be less in a completely dormant plant with no aboveground photosynthetic parts than in a succulent plant. Even with closed stomata, tiny amounts of water are lost from any

(A)

(B)

(C)

Figure 2.16
Some CAM species. All of these succulents display different forms of fleshy leaves or other tissues. (A) *Sempervivum tectorum* ("hen-and-chicks," Crassulaceae) (Photograph © D. Matherly/Visuals Unlimited.) (B) *Sedum* sp. (Crassulaceae) (Photograph by S. Scheiner). (C) *Opuntia humifusa* (smooth mountain prickly pear, Cactaceae) (Photograph courtesy of W. S. Judd.)

(B)

(A)

Figure 2.17
CAM epiphytes. (A) A bromeliad, *Guzmania monostachya* (Bromeliaceae), growing as an epiphyte on the trunk of a tree. (Photograph courtesy of W. S. Judd.) (B) *Tillandsia usneoides* (Spanish moss, Bromeliaceae) is an epiphyte that grows hanging from the branches of large trees in the southeastern United States. (Photograph by S. Scheiner.)

green tissue, which may add up to a substantial amount over an extended time without any water. Succulents, particularly those with CAM photosynthesis, are thus most abundant and diverse in deserts with both relatively higher levels and greater predictability of rainfall. In somewhat drier deserts with less predictable rainfall, perennials with either C_3 or C_4 photosynthesis are more likely to be dominant, depending on the season in which rain is most abundant and likely (see below). In the driest deserts, the perennials decline in importance and annual species tend to become more prevalent.

Phenology

The three photosynthetic types also tend to differ in their **phenology** (the seasonal timing of life history events, such as initiation of growth, flowering, and dormancy). C_4 photosynthesis is well represented among annual species that complete most of their growth in the summer; such summer annuals are common in semiarid environments with seasonal rainfall, as well as among weedy species that colonize disturbed environments, including roadsides and agricultural fields. In areas with mediterranean climates (such as those of California, southern France, or western Australia, for instance) where most of the rainfall occurs in winter, the native species are overwhelmingly C_3 plants, but the weeds that invade fields irrigated in the summer may be C_4 species.

In the southwestern deserts of the United States, distinct patterns exist for C_3 and C_4 annual plants. In the deserts of eastern California, where the rainy season is in winter and early spring, the annuals are C_3 species and grow when temperatures are cool. In New Mexico and Texas, where there is a summer rainy season, the desert annuals are largely C_4 species. In the Sonoran Desert of Arizona and northern Mexico, there are two separate rainy seasons, and annual C_3 species, including many beautiful wildflowers, dominate in the early spring, while C_4 species are found in the summer. The periods of greatest growth for C_3 and C_4 perennials in these environments follow a similar pattern.

Similarly, the "cool season" perennial grasses and **forbs** (broad-leaved herbaceous plants) that predominate in spring and again in autumn in the mixed-grass prairie of South Dakota are largely C_3 species, while the "warm season" species that grow most actively in midsummer have a much greater representation of C_4 plants. However, in the shortgrass prairie farther south, while there are differences in phenology between the C_3 and C_4 species, the period of maximum growth for the two photosynthetic types does not differ (Hazlett 1992).

Geographic Distributions

Various studies have considered the broad geographic distributions of C_3, C_4, and CAM species and the climatic factors associated with their relative predominance. Not surprisingly, the physiological differences between C_3 and C_4 photosynthesis are associated with differences in distribution between plants of the two types. Across North America, the percentage of C_4 grass species is most strongly determined by summer temperatures: C_4 grasses are most common where summers are warm (Figure 2.18).

Paruelo and Lauenroth (1996) examined the relative abundances (in terms of biomass rather than number of species) of C_3 grasses, C_4 grasses, and shrubs in the western United States (Figure 2.19). C_4 grasses are increasingly dominant where temperature, total rainfall, and the proportion of rain falling during summer are higher. C_3 grasses are favored by cool temperatures and by

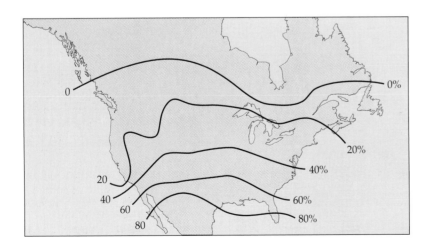

Figure 2.18
Distribution of C_4 grass species as a percentage of the total grass flora across North America. (Data from Teeri and Stowe 1976.)

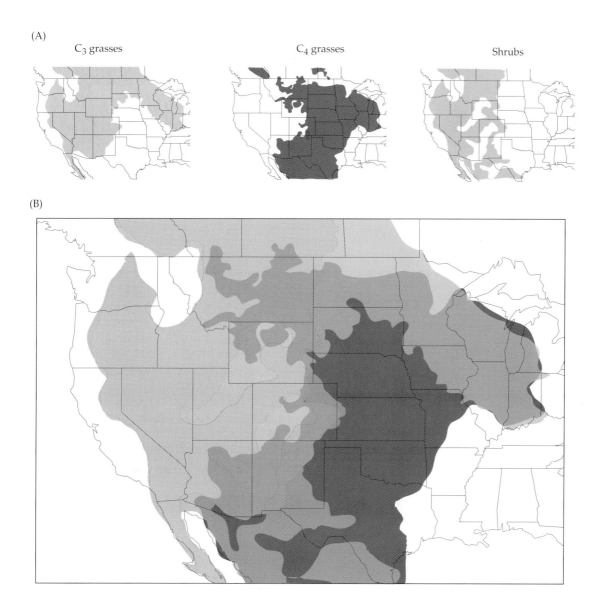

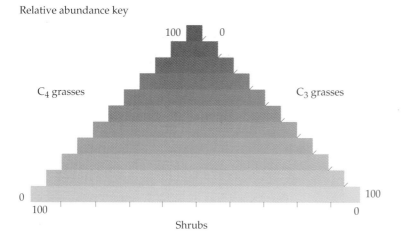

Figure 2.19

C_3 grasses, C_4 grasses, and shrubs are the dominant functional groups in the central part of North America, which is dominated by grasslands and shrublands. The maps indicate the relative abundances of each as an approximate percentage of the total biomass. (A) C_4 grasses are the most abundant species in the southern plains, while C_3 grasses predominate farther to the west and north. Shrubs are most abundant in the far west. (B) The combined abundances of C_4 grasses, C_3 grasses, and shrubs, as indicated by the key. (After Paruelo and Lauenroth 1996.)

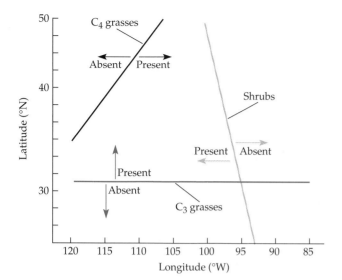

Figure 2.20
Results of a model with lines showing the latitude and longitude where C_3 grasses, C_4 grasses, and shrubs are predicted to be absent or present within the central region of North America. The distributions of C_4 grasses and shrubs are largely influenced by longitude, while C_3 grasses are influenced by latitude. The predictions of this model were largely in agreement with the distribution patterns shown in Figure 2.19. (After Paruelo and Lauenroth 1996.)

a greater proportion of rainfall during the winter months. Shrubs, with deep roots that increase their access to water stored deep in the soil, are most abundant as total rainfall decreases and as the proportion of winter rainfall increases. Consequently, the southeastern portions of the great North American grasslands (west of 90° longitude) are dominated by C_4 grasses. C_3 grasses increase to the north and to the west, and the westernmost, driest part of this region is dominated by (largely C_3) shrubs (Figure 2.20; see also Figure 18.25).

Other research has documented changes with altitude, from a vegetation dominated by C_4 species in the warm, often dry habitats at low altitudes to a C_3-dominated flora in the cooler, moister conditions at high altitudes, in Kenya, Costa Rica, Hawaii, Arizona, and Argentina. CAM species are, as one might expect from their superior water use efficiencies, most closely associated with aridity in temperate and semiarid regions.

These studies of correlations between climate and photosynthetic type all share a common limitation, however. It is possible that their results are due, to some unknown extent, to historical phylogenetic factors that link the distributions of related species to climate and biogeography, regardless of photosynthetic type. C_4 photosynthesis, for example, is most likely to have evolved in families that are found in warm environments, and therefore C_4 species in those families, like their C_3 relatives, should have numerous adaptations to living in those environments, irrespective of the pathway used to fix CO_2. Tropical C_4 grasses do not perform well in cold temperatures; neither do tropical C_3 grasses. These factors make it somewhat difficult to disentangle the adaptive significance of photosynthetic type from all of the other adaptations of species of contrasting types. This is a particularly big problem in disentangling the possible ecological significance of the C_4 subtypes, which are almost totally confounded with phylogenetic groups. We discuss this problem in a more general context in Chapter 6.

Nevertheless, there is abundant evidence, derived from a variety of approaches, that C_4 and CAM photosynthesis are adaptations to environmental factors, and we have a fairly clear idea of what those factors are, as well as the mechanistic bases for the adaptations. Keep in mind that the above discussion concerns general patterns of association between physiology and climate. There are many C_3 species that live in hot, dry places, and a few C_4 species that live where it is cold most of the

Table 2.2 Contrasts between sun leaves and shade leaves of *Fagus sylvatica* (beech, Fagaceae)

Character	Sun leaves	Shade leaves
Stomatal density (number/mm^2)	214 ± 26	144 ± 11
Leaf thickness (μm)	185 ± 12	93 ± 5
Leaf area (cm^2)	29 ± 4	49 ± 7
Fresh weight (g)	0.5 ± 0.1	0.4 ± 0.1
Dry weight (g)	0.24 ± 0.03	0.12 ± 0.02
Water content (% fresh weight)	53 ± 4	70 ± 5
Total chlorophyll (mg/g dry weight)	6.6 ± 2	16.1 ± 2
Total chlorophyll on a leaf area basis (mg/100 cm^2)	5.5 ± 1.8	3.9 ± 0.4

Source: Lichtenthaler et al. 1981.
Note: Values are means of 9 leaves ± 1 standard deviation.

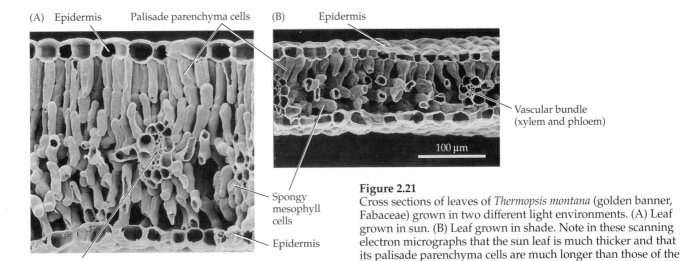

(A) Epidermis   Palisade parenchyma cells   (B) Epidermis

100 µm

Vascular bundle (xylem and phloem)

Spongy mesophyll cells

Epidermis

Vascular bundle

Figure 2.21
Cross sections of leaves of *Thermopsis montana* (golden banner, Fabaceae) grown in two different light environments. (A) Leaf grown in sun. (B) Leaf grown in shade. Note in these scanning electron micrographs that the sun leaf is much thicker and that its palisade parenchyma cells are much longer than those of the shade leaf. (Photographs courtesy of T. Vogelmann.)

time. While general patterns may be very revealing, they cannot explain everything, and photosynthetic type is only one of a host of factors that determine how a plant interacts with its environment.

Adaptations to the Light Environment

Sun and Shade Leaves

A leaf that functions well in bright sunlight usually will not perform well in deep shade, and the reverse is also true. Many species of plants produce different kinds of leaves in the sun and in the shade (Figure 2.21). Leaves at the top of a tree, for example, may differ in a variety of ways from those deep within the canopy. "Sun leaves" are typically smaller in size (leaf area) and thicker than "shade leaves," with greater leaf mass per leaf area (**LMA**, also called **specific leaf mass** or **weight**, dry mass per unit leaf area; the inverse, **specific leaf area**, or leaf area per gram of dry mass is often reported). They also tend to have higher concentrations of rubisco, chlorophyll, and other key components of both the light reactions and carbon fixation, per unit leaf area (Table 2.2).

In some plants, sun and shade leaves can also have different shapes, with deeper lobing, for example, in leaves produced in bright sunlight. As a consequence of the greater investments made in the machinery of photosynthesis, sun leaves usually have much higher light saturation levels and greater maximum photosynthetic rates than shade leaves (Figure 2.22). Shade leaves often have lower rates of cellular respiration, perhaps because there is less "machinery" to maintain, and they may have higher net rates of photosynthesis at low light levels, with lower light compensation points.

The ability of individual plants to produce different kinds of leaves in different light environments is an example of **phenotypic plasticity**, the ability of an individual of a given genotype to produce different structures (such as leaf tissues) or to function differently under different environmental conditions (see Chapter 5). This flexibility in response can also be called an **acclimation** (a type of plasticity that involves potentially reversible adjustments to environmental conditions).

Not all acclimation to changes in light levels is through the production of different types of leaves. In the study of northeastern forest understory species discussed above (Rothstein and Zak 2001), the shifts in light

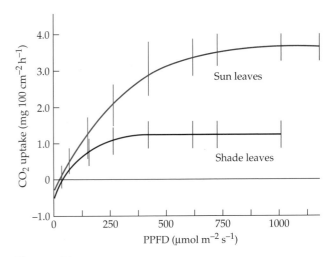

Figure 2.22
Photosynthetic CO_2 uptake in response to light intensity for sun leaves and shade leaves of *Fagus sylvatica* (beech, Fagaceae) on a leaf area basis (mg CO_2/100 cm^2/h), showing the midpoint and range for nine different sun leaves and nine different shade leaves. The sun leaves become light-saturated at much higher light intensities, and the maximum photosynthetic rates are much higher for sun leaves. (After Lichtenthaler et al. 1981.)

compensation point in *Viola pubescens* and *Tiarella cordifolia* were due to physical and chemical changes within individual leaves, rather than the plant producing different kinds of leaves in different seasons.

Species' Adaptations to High and Low Light Habitats

Contrasts between the leaves of plants that are native to high light environments and the leaves of plants from shaded habitats are similar in many ways to contrasts between sun leaves and shade leaves produced by the same individual. Species adapted to bright habitats frequently have costs or specializations that would put them at a disadvantage in deep shade, and the reverse is also true; in many cases plants could not survive in environments with drastically different light regimes. For example, the high investments in rubisco and chlorophyll and higher respiration rates that generally characterize the leaves of plants from high light environments would not be sustainable in deeply shaded habitats.

Unlike the more general differences between sun and shade leaves on the same plant, species native to high light or low light habitats also have a number of unique features. One such character is **solar tracking**. Irwin Forseth and James Ehleringer (1982) found that many desert annuals and some perennials from warm deserts—a high light environment—have leaves that track the sun across the sky during the day. This strategy increases the amount of light available to the chloroplasts over the course of the day (Figure 2.23), thereby increasing photosynthetic rates and presumably subsequent growth rates.

Why should solar tracking be advantageous in the bright, sunny environment of a desert? In desert environments, water is available for growth for only a small part of the year, and desert annuals, in particular, often germinate, grow, and reproduce in a very short time before water becomes unavailable and they die. Solar trackers are adapted to maximize their light capture in the short growth period available to them. In North American deserts, the proportion of species that are solar trackers becomes greater as the length of the growing season gets shorter. We discuss adaptations to desert environments in more detail in Chapter 3.

Some plants that live in deeply shaded habitats also have unique features. The small understory species that live in the extreme shade of humid tropical forests possess a number of unique adaptations to the quality and quantity of light available to them. Light intensities at the rainforest floor may be less than 1% of the PPFD at the top of the canopy. Characters such as velvety or satiny leaf surfaces and blue iridescence are found in widely divergent species in understory habitats (Lee and Graham 1986). Each of these characters appears to enhance the light available for photosynthesis slightly. That slight advantage in such an extreme environment

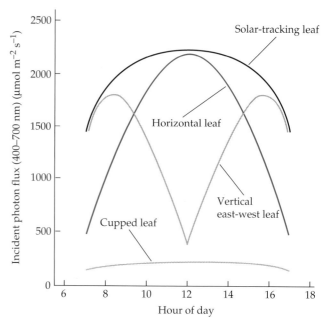

Figure 2.23
The amount of light received over the course of a midsummer day in the desert by a solar tracking leaf, a leaf held horizontally (in which case the light received in the early morning and late afternoon is sharply reduced), a leaf held vertically (in which case the light received during the middle of the day is sharply reduced), and a cupped leaf. The solar tracking leaf receives at least 38% more light than the fixed horizontal or vertical leaves. (After Ehleringer 1985.)

may, however, make the difference between life and death.

In some species, a velvety sheen is produced by epidermal cells, which act as lenses to focus light onto chloroplasts. Lee (1997) studied the intense, metallic blue iridescence found in the leaves of many plants in equatorial rainforests and found that it may serve to help them to absorb light more efficiently. In this environment, light in the photosynthetically most usable range, 400–700 nm, is particularly depleted because it has been absorbed by the plants in the canopy above. Working somewhat like the coating on a camera lens, the color of these leaves may enable them to absorb more light at the longest photosynthetically useful wavelengths, which are less depleted than the shorter wavelengths. The blue iridescence in such leaves is not produced by blue pigments, but is a structural effect (Box 2C).

All understory environments are not uniformly dark, however. The light environment is changing constantly, dim one moment and illuminated by the brilliant flash of a sunfleck the next (Figure 2.24). Shade-adapted species appear to be capable of using brief sunflecks with unusual efficiency. Studies by Robin Chazdon and Robert Pearcy (1991) and their colleagues suggest that,

Box 2C

Leaf Iridescence and Structural Coloration

Structural coloration is caused by the way light bounces off physical surfaces with particular optical properties. It can be created by simple or very complex microscopic structures in both plants and animals. In the case of iridescent blue leaves, the optical effect is caused by thin-film interference within the leaf, which results from multiple layers (usually in the cell walls) of materials that alternatively transmit or reflect particular wavelengths of light.

Similar optical interference or, alternatively, scattering of blue light by small particles is responsible for the structural blues found in many animals as well. The blues of butterfly wings, the brilliant blue colors found in some beetles, the amazing blue feathers of peacocks and the more modest blues of blue jays, the blue rump of the baboon, and the blue eyes of some humans and cats are all due to these optical phenomena rather than to blue pigment. Blue colors in plants may also be produced by blue pigments such as anthocyanins. No blue pigment has ever been found in a vertebrate.

---

while understory plants in both temperate and tropical forests are capable of surviving well in very low light, they depend on the energy captured from occasional sunflecks to be able to grow and reproduce. A large proportion (up to half) of the carbon gained by understory plants may come from energy captured from sunflecks.

The ability of these forest floor species to capture the energy available in brief, unpredictable episodes of direct bright light depends on several adaptations that make them able to respond quickly and with great efficiency to that light. These adaptations include stomata that open even in dim light, and a high electron transport capacity (a function of the light reactions) relative to carboxylation capacity (the capacity to carry out the initial fixation of CO_2). These plants also exhibit a slow loss of **photosynthetic induction** (the necessary start-up time for plants to reach maximum photosynthesis after exposure to bright light), so that once "primed" by a sunfleck, they remain capable of more fully responding to other sunflecks.

The ability of some shade-adapted species to capture and use sunflecks and light coming into the canopy at steep angles early in the morning and in the late afternoon may be enhanced by their having irregularly oriented grana stacks, in contrast to the neatly aligned grana of sun leaves, which are stacked perpendicular to the leaf's surface. That irregular orientation would mean that at least some grana would be directly aligned with

Figure 2.24
Sunflecks on leaves of *Geonoma cuneata* (Arecaceae) in the understory of a tropical forest. (Photograph courtesy of R. Chazdon.)

incoming light, regardless of the direction from which the light came. These plants may also have various adaptations that prevent them from being damaged by the intense sunlight in patches and flecks that may briefly illuminate the deep shade.

Do Sun and Shade Adaptations Exist within Species?

More controversial than the existence of species adapted to high and low light conditions is the question of whether, within a single species, there exist ecotypes (genetically distinct populations within a single species, adapted to local conditions; see Chapter 6) adapted to sunny spots and other ecotypes adapted to shaded habitats. This idea was first proposed in the 1960s based on comparisons of populations of several species collected from sunny and shaded habitats, including *Solidago virgaurea* (Asteraceae), *Rumex acetosa* (Polygonaceae), *Geum rivale* (Rosaceae), *Dactylis glomerata* (Poaceae), and *Solanum dulcamara* (Solanaceae). These studies reported that genetically distinct populations of these plants existed, with some adapted to high light environments and others adapted to shade, and went on to describe the physiological adaptations of these ecotypes in detail (for example, see Björkman and Holmgren 1966; Björkman 1968). The results of this research became widely accepted.

However, there were several basic problems with the research upon which these results were based. We discuss these problems here not only because this question is inherently interesting, but also because this case provides some insight into the scientific process and the role controversy sometimes plays in that process. While the original physiological work was innovative and of high caliber, not enough attention was paid to the population-level aspects of the question. First, only one to three individuals were sampled from each population. These sample sizes were so small that it was impossible to know whether the differences found were real biological differences between sun and shade populations or were due merely to chance. How reliably did these few individuals represent the populations from which they were drawn? A second limitation of the original studies was that the sun and shade populations sampled were often located far apart (sometimes by hundreds of kilometers) from one another. Were the hypothesized differences due to sunny versus shaded habitats or to other differences between the habitats?

To overcome these methodological difficulties, John Clough and his colleagues (Clough et al. 1979) sampled 15 individual plants of *S. dulcamara* (Figure 2.25) that were growing in sunny sites (100% full sun) and 15 individuals from shady locations (< 10% of full sun) separated by only 25 meters in northern Illinois. Cuttings were made from each plant and grown in an array of different light, water, and temperature treatments. While

Figure 2.25
Solanum dulcamara (nightshade, Solanaceae), a European species that has become a common weedy vine in the eastern and central United States. Its common name hints at its poisonous leaves and berries. In England it is also called snakeberry and witchflower, and was reputed to be helpful against witchcraft. (Photograph © J. Wexler/Visuals Unlimited.)

these researchers did find genetic differences in photosynthetic traits among individual plants, the sun and shade populations did not differ from each other on average, and did not differ at all with regard to traits associated with hypothesized sun/shade differences. Rather, all individuals adjusted their physiology and morphology to their growth conditions. Research has continued in the effort to find evidence for the existence of within-species sun and shade ecotypes in these and other species (for example, Sims and Kelley 1998).

Day Length: Responses and Adaptations

Another aspect of the light environment that is important to plants is day length. Plants can detect, quite precisely, not only how long the day (or actually, the night) is, but they can also determine whether day length is increasing (as it does in spring) or decreasing (as it does in autumn). Day length is often a more reliable indicator of season than is temperature or other more proximate cues, and plants can make use of this information to time important phenological events, including flowering, the onset of dormancy, or leaf fall. Even many seeds are capable of detecting day length (usually via the seed coat) and using this information to determine when to remain dormant and when to begin the process of germination. Day length is often used together with other environmental cues, including temperature and moisture.

In some species, genetic variation in both the sensitivity and the nature of the response to day length exists.

In the widespread arctic/alpine species *Oxyria digyna* (alpine sorrel, Polygonaceae), populations from California, Wyoming, and Colorado flower when the daylength is about 15 hours. In populations from near the Canada–United States border, some plants will flower in 15 hour daylength, but many will not. In populations from north of the Arctic Circle, flowering requires a daylength of over 20 hours (Mooney and Billings 1961; Billings and Mooney 1968). Thus, the onset of flowering occurs in mid-May both at the Canada–United States border and north of the Arctic Circle, although at the border daylengths are 15 hours at this time, while at the Arctic Circle they are 20 hours. The more southern, high elevation populations do not reach 15 hour daylengths until mid- to late-June. These alpine populations may still be under snow in mid-May, with a high probability of frost, and do not flower until about a month later than the more northern populations.

Dwight Billings

Although plant sensitivity to day length was first discovered in the 1920s, its mechanistic basis is still not fully understood, and depends in part on some unknown biological clock. It is known, however, that phytochromes—bluish pigments that exist in several different forms that respond to light—play a major role in the photoperiodic timing mechanism. Phytochromes play a role in a number of the ways in which plants detect and respond to various aspects of light in their environments (see Chapter 5).

Summary

The process of photosynthesis consists of two parts, the light reactions, in which light energy is captured, and the Calvin cycle, in which carbon is fixed. The rate at which photosynthesis proceeds depends on the amount and quality of the light reaching the chloroplasts, the CO_2 concentration gradient between the air and the chloroplasts, and conductance to CO_2.

There are three different photosynthetic pathways: C_3, C_4, and CAM photosynthesis. C_3 photosynthesis is ancestral to the other types and is the most common. C_3 plants rely on the enzyme rubisco to capture CO_2 from the atmosphere. Because rubisco also catalyzes photorespiration, photosynthetic rates can be severely limited for C_3 plants under warm, bright conditions. C_4 and CAM photosynthesis are evolutionary specializations that overcome the limitations of rubisco in such environments. Each offers advantages under particular environmental conditions and is at a disadvantage under other conditions.

Numerous adaptations exist among species adapted to high or low light environments. These adaptations confer advantages in the environment to which the species are adapted, but cause them to function more poorly in other environments. Some individual plants can produce leaves with different properties in the sun and in the shade. The ability to use the light available in brief sunflecks is an important feature of many understory forest species. Day length is another feature of the light environment that plants are often capable of detecting and which they may use as a reliable indicator of changing seasons.

Additional Readings

Classic References

Kramer, P. J. 1981. Carbon dioxide concentration, photosynthesis, and dry matter production. *BioScience* 31: 29–33.

Mooney, H. A. and W. D. Billings. 1961. Comparative physiological ecology of arctic and alpine populations of *Oxyria digyna*. *Ecol. Monogr.* 31: 1–29.

Von Caemmerer, S. and G. D. Farquhar. 1981. Some relationships between the biochemistry of photosynthesis and the gas exchange of leaves. *Planta* 153: 376–387.

Contemporary Research

Cushman, J. C. and A. M. Borland. 2002. Induction of Crassulacean Acid Metabolism by water limitation. *Plant Cell Environ.* 25: 295–310.

Thomas, S. C. and W. E. Winner. 2002. Photosynthetic differences between saplings and adult trees: An integration of field results by meta-analysis. *Tree Physiol.* 22: 117–122.

Wookey, P. A., C. H. Robinson, A. N. Parsons, J. M. Welker, M. C. Press, T. V. Callaghan and J. A. Lee. 1995. Environmental constraints on the growth, photosynthesis and reproductive development of *Dryas octopetala* at a high Arctic polar semi-desert, Svalbard. *Oecologia* 102: 478–489.

Additional Resources

Chazdon, R. L. and R. W. Pearcy. 1991. The importance of sunflecks for forest understory plants. *BioScience* 41: 760–766.

Jones, H. G. 1992. *Plants and Microclimate: A Quantitative Approach to Environmental Plant Physiology.* Cambridge University Press, Cambridge.

Lambers, H., T. L. Pons and F. S. Chapin III. 1998. *Plant Physiological Ecology.* Springer, New York.

Larcher, W. 1995. *Physiological Plant Ecology.* 3rd ed. Springer, Berlin.

Nobel, P. S. 1999. *Physiochemical and Environmental Plant Physiology.* Academic Press, New York.

3

Water Relations and Energy Balance

*P*lants, like everything else alive, need water to live, grow, and reproduce. Like animals, plants use water as the medium in which all of their biochemical reactions occur; unlike terrestrial animals, most plants also rely on water to a greater or lesser extent for support of their nonwoody parts and for maintaining their physical structure. Much of the evolutionary history of plants on land is the story of increasing success in obtaining water, moving it to parts of the plant far from the source of the water, and being able to reproduce in a very dry atmosphere. We briefly review this history, pointing out some of the essential structures that evolved during the course of the history of land plants.

Terrestrial plants not only must obtain water, but are also faced with the problem of restricting their water losses. Intimately related to this problem is the challenge of maintaining their leaf temperatures within an acceptable range. In this chapter, we examine the evolutionary "solutions" to these interrelated problems and take a look at the physics of how plants exchange water and energy with their environments.

Adapting to Life on Land

Photosynthetic organisms originated in marine environments, where they were continuously bathed by water in which mineral nutrients were dissolved. The ancestors of terrestrial land plants were single-celled aquatic organisms that depended on having each cell in direct contact with water, obtaining nutrients from and being kept moist by the water in their immediate surroundings. Like modern photosynthetic aquatic organisms, these ancestors of terrestrial plants also depended on water for sexual reproduction, releasing their gametes into the water.

Early land plants did not differ dramatically from their aquatic ancestors in many of these characteristics. One of the earliest adaptations of the first land plants was the evolution of a cuticle, a waxy, nonliving covering over the exposed epidermal cells that prevents desiccation by the air. In bryophytes such as mosses, water moves by diffusion from cell to cell. There are no specialized organs to take up water, nor to transport it to distant parts of the plant. The process is slow and not very efficient; as a consequence, none of the cells in a moss plant can get very far away from the substrate (soil or rock, for instance) that provides the water, so mosses are constrained to be very short.

The dominant generation in bryophytes, including mosses, is the haploid gametophyte (see Chapter 8). Male gametophytes produce male gametes, which must swim to the female gametes. Because these plants are terrestrial, this is possible only when the moss shoots are covered with a film of water from rainfall or dew. The sperm swim along this film of water from a male plant to a female plant, eventually making their way to an archegonium containing an egg. This will work successfully only under certain restricted weather conditions, and males and females must grow close to each other.

Early vascular plants, which branched off from the ancestors of bryophytes in the Paleozoic era (see Chapter 21), had an enormous advantage over bryophytes for living in a terrestrial environment: their vascular systems allowed the transport of water much more rapidly and efficiently throughout the plant. Because they had evolved vascular tissues, they were able to develop specialized structures to take up water from the soil (roots) and woody tissue capable of supporting a trunk and crown. Without vascular tissue to take up water and conduct it over long distances, such structures could not exist. The early vascular plants also evolved a dominant sporophyte (diploid) generation. As the sporophyte became larger and longer-lived over the course of evolution, the gametophyte became smaller and shorter-lived. However, the plants in these groups still depended on motile sperm to swim from the male to the female gametophyte, and had not evolved seeds.

The seed plants—seed ferns (now extinct), gymnosperms, and angiosperms—represent the greatest advances in the adaptation of plant life to a dry environment. The tissue that conducts water in this group of plants, the xylem, is capable of moving large volumes of water efficiently over long distances (Figure 3.1). Particularly in the angiosperms—the flowering plants—there is a great diversity of structure and physiological function related to the ability to withstand dry conditions.

The evolution of pollination removes many restrictions on the ability to reproduce in the dry conditions that dominate terrestrial life, including the requirement that male and female gametophytes be located adjacent to each other and that reproduction occur only under a very limited set of weather conditions.

The evolution of seeds was another key innovation. In non-seed vascular plants, such as ferns, dispersal occurs as haploid spores, produced by the diploid sporophyte plant. Both seeds and spores can disperse from the parent plant and can survive various environmental hazards, including dessication. However, seeds include a number of tissues, such as the seed coat, that allow for the detection of subtle variation in the environment (such as changes in daylength and temperature) and for the hormonal regulation of germination in response to environmental conditions, as well as for maternal provisioning of food for the embryo. Seed plants were, therefore, able to evolve far greater variation in dormancy and in seed provisioning than their non-seed ancestors.

Water Potential

The ability of a plant to acquire water, the ability to move it to all cells throughout the plant, and the propensity of the plant to lose that water all depend on the **water potential** of the various parts of the plant and its immediate environment (Figure 3.2). One way of thinking about water potential is that it represents the difference in potential energy between pure water (which is defined as having a water potential of zero) and the water in some system, such as that in a plant cell or in the soil.

When water is bound up in some way, or has materials dissolved in it, the potential energy it possesses is less than that of pure water, and its water potential is negative. Analyzing the movement of water in plants in terms of water potential puts the study of plant water relations into the unifying context of thermodynamics. It makes it possible to study the movement of water in soils, in plants, and between plants and the atmosphere using the same concepts, terms, and units.

Water potential (Ψ) in plants and soils can be broken down into four major components,

$$\Psi = \Psi_\pi + \Psi_P + \Psi_m + \Psi_g$$

where Ψ_π is the osmotic potential, Ψ_P is the pressure potential, Ψ_m is the matric potential, and Ψ_g the gravitational potential of the water in the system.

The **osmotic potential** is the component of water potential that is due to solutes dissolved in the water. It results from the difference in potential energy between pure water and water containing dissolved substances. Osmotic potential is either zero (in pure water) or negative, because the solutes reduce the capacity of the water to do work. The osmotic potential is capable of acting as a driving force for water movement such as when the movement of the solutes is restricted by a semipermeable membrane that allows water, but not solute molecules, to pass through it. The osmotic potential is the major component of water potential in living cells (such as those in roots and leaves), and thus is the major driving force by which water moves into those cells.

The **pressure potential** is a second key component of plant water potential, and is important both in living cells and in functioning xylem tissue (which is non-living at maturity). It can be negative, zero, or positive in value, and is a function of the hydrostatic or pneumatic pressure in the system. When water is enclosed by

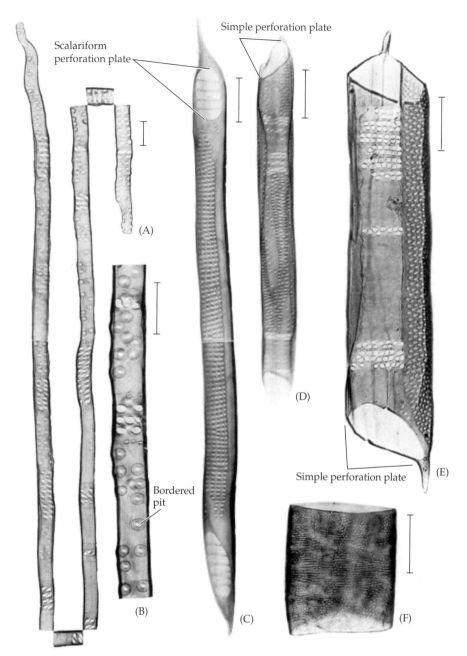

Figure 3.1
Water-conducting xylem cells from the wood of a gymnosperm and several angiosperm trees. (A) A tracheid of *Pinus lambertiana* (sugar pine, Pinaceae), bent in several places to show its full length. Notice that the ends of the tracheid are not open. Water must diffuse through these end walls as it moves up to the tracheid above it. (B) An enlarged view of the central part of the tracheid in (A), showing pits in the side walls. The pits are thinner parts of the cell wall through which some water may diffuse, but are not open. (C) Xylem vessel element of *Liriodendron tulipifera* (tulip tree, Magnoliaceae), showing openings in the end wall (the scalariform perforation plate). Water can move directly through these openings as it travels up the xylem to the next vessel element. (D) Xylem vessel element of *Fagus grandifolia* (beech, Fagaceae), with a completely open end wall (simple perforation plate). (E) Xylem vessel element of *Populus trichocarpa* (black cottonwood, Salicaceae). (F) Xylem vessel element of *Ailanthus altissima* (tree-of-heaven, Simarubaceae). Moving from A and B to E, the elements become progressively shorter, wider, and have larger openings in the end walls, all contributing to facilitating the passage of greater volumes of water more rapidly. Scale bars are all 100 µm. (From Esau 1977.)

something that restricts its volume, such as a healthy plant cell enclosed by cell walls, the water presses on the walls, generating a positive pressure potential. This is the ordinary situation in a fully hydrated, or turgid, plant cell under normal conditions, in which the action of the water is similar to that of air inflating a balloon. The opposite situation occurs when water is pulled through an open system, such as a garden hose or a

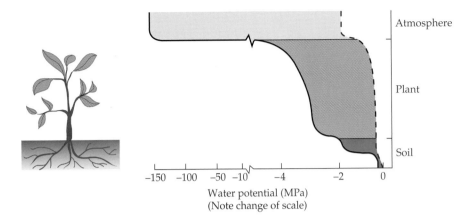

Figure 3.2
Typical values (in MPa) for water potentials in the soil, in a plant, and in the atmosphere in a mesic environment. The solid curve at the left side of the shaded region indicates the values for atmosphere, plant, and soil water potentials when the atmospheric relative humidity is low and the soil is dry. Under these conditions, soil water potentials range from close to 0 at deeper layers to about –1.8 MPa near the soil surface, plant water potentials range from about –2.0 to about –4.0 MPa, and atmospheric water potentials range from about –5.0 to –150.0 MPa. The dashed curve at the right side of the shaded region indicates the values for moist soils and high atmospheric relative humidity; under these conditions, values for soil, plant, and atmosphere would be slightly less than 0 MPa, –0.01 to –0.75 MPa, and –1.0 to < –2.0 MPa respectively. Intermediate conditions would result in values falling within the shaded regions. (After Etherington 1982.)

plant xylem vessel. In that case, the pressure potential is negative, and actually pulls the walls of the vessel inward. Both positive and negative pressure potentials exist in plants, and both are important driving forces for the movement of water within the plant and between the soil, the plant, and the atmosphere.

Gravitational potential results from the pull of gravity on water, and it is largely responsible for water draining from the largest pores of the soil in the first few days after a saturating rain (see Chapter 4). It is negative when the water is moving downward, since the water is losing potential energy. **Matric potential** is a result of the cohesive force that binds water to physical objects, such as cell walls and soil particles; it is always negative because water bound to an object has less potential energy than free water. The matric and gravitational potentials are generally much more important in soils than in plants.

The Soil-Plant-Atmosphere Continuum

Water always moves from larger to smaller values of Ψ. In plants and in the soil, this usually means moving from a region of less negative water potential to one of more negative water potential. The gradient in water

potential acts as the driving force for water movement from the soil, into and through the plant, and out to the atmosphere through transpiration.

The energy for moving water is provided without the expenditure of any energy by the plant itself. Energy is needed to lift the water up from the soil to the top of the plant, and to change the less energetic liquid water molecules inside of the plant into the more energetic molecules of water vapor in the atmosphere. The movement of water is driven by the gradient in water potential along this pathway, and the energy comes directly from the sun. In this way, plants are able to move vast amounts of water, sometimes over tremendous distances (in large trees), at no direct energy cost to themselves.

In a typical plant growing in moist soil and transpiring on a warm day, the values for water potential and its components might look like those in Figure 3.3A. Water moves from the soil into the plant roots, which have more negative Ψ, due primarily to the osmotic potential of the root cells. Then the water is drawn into the xylem vessels, which have still more negative Ψ values, due in this case to the negative pressure potential. Water moves into the leaves from the smallest veins, and out into the substomatal cavity (Figure 3.4). From there, when the stomata are open, water moves out into the atmosphere. Even in humid air (Figure 3.3B), the atmosphere surrounding the leaf has a highly negative water potential. The liquid water molecules ultimately gain the energy needed to move into a vapor state from the energy in sunlight, and the water throughout the rest of the pathway is pulled upward by the transpirational stream escaping from the leaves of the plant through the stomata.

Transpiration and the Control of Water Loss

Plants transpire vastly more water than they use directly for plant functions. A large rainforest tree with a plentiful water supply might transpire as much as a thousand liters of water in a day (weighing 1000 kg—a ton of water lifted from the soil, transported up the trunk,

(A)   Air (50% relative humidity): –68.5 MPa

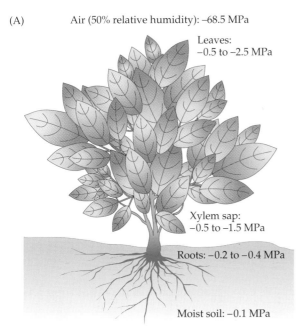

Leaves:
–0.5 to –2.5 MPa

Xylem sap:
–0.5 –1.5 MPa

Roots: –0.2 to –0.4 MPa

Moist soil: –0.1 MPa

(B)   Air (50% relative humidity): –13.7 MPa

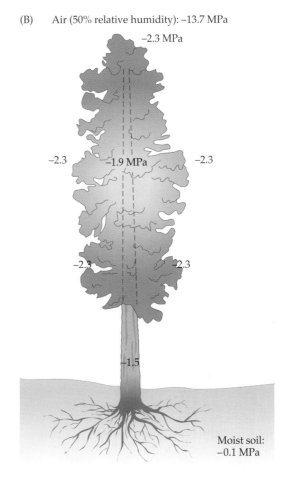

–2.3 MPa

–2.3   –1.9 MPa   –2.3

–2.3   –2.3

–1.5

Moist soil:
–0.1 MPa

Figure 3.3
(A) Typical values for water potential from roots to stem
to leaves and atmosphere in an herbaceous mesophytic
plant, with moist soil and air at 50% relative humidity.
(B) Water potential values in different parts of a *Sequoia-
dendron giganteum* (giant sequoia, Cupressaceae) tree
that is transpiring rapidly. Here, the soil is moist and the
air humid. The needles on the periphery of the canopy
have the lowest water potentials, and the trunk has a
higher water potential at the base than high in the
canopy. (After Mohr and Schopfer 1995.)

Bundle sheath   Mesophyll   Xylem

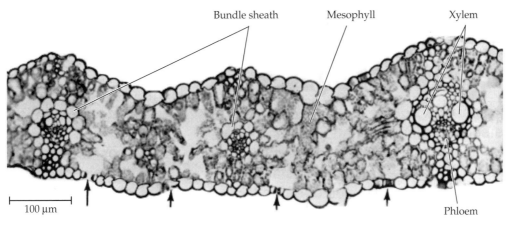

100 µm

Phloem

Figure 3.4
Cross section of a leaf of a C_3 grass *Hordeum vulgare* (barley, Poaceae), showing the
upper and lower epidermis, which are each one cell thick; photosynthetic mesophyll
cells, and a vascular bundle with xylem and phloem surrounded by a single-layered,
nonphotosynthetic bundle sheath. Stomata are shown by arrows, and the substomatal
cavities are visible inside the leaf, above the guard cells of the stomata. (From Esau
1977.)

and lost through the leaves each day!). Temperate deciduous trees may lose up to 140 liters of water, but a typical conifer loses only 30 liters in a day. While most plants transpire fairly modest amounts of water, in one study, *Salix fragilis* (brittle willow, Salicaceae) growing in moist soil lost a remarkable 463 liters per day (Cermak et al. 1984).

The loss of water through transpiration can be described by a flux equation, much like the diffusion of CO_2 into a leaf during the process of photosynthesis:

$$E = g_{\text{leaf(wv)}} \times (C_{i(\text{wv})} - C_{a(\text{wv})})$$

where E is the transpiration rate in mmol $H_2O/m^2/s$. The term m^2 refers to the surface area of the leaf, usually the area of one side of the blade, but including both upper and lower sides for leaves with stomata on both surfaces. The transpirational flux is driven by the difference between the concentration of water vapor inside the leaf, $C_{i(\text{wv})}$, and that in the external atmosphere surrounding the leaf, $C_{a(\text{wv})}$, expressed in mol/m^3. Sometimes the vapor pressure of water is used instead of vapor concentration in this equation, and the difference between water vapor in the leaf and the air is called the **vapor pressure deficit** (measured in kilopascals, KPa). The vapor pressure deficit of water is related directly to the water vapor concentration difference. The **leaf conductance** to water vapor, $g_{\text{leaf(wv)}}$, is a measure of how readily water vapor moves into or out of the stomata, and is defined as the rate at which water vapor flows into the leaf for a given difference between ambient and interc0ellular concentration; it is expressed in mmol $H_2O/m^2/s$ and is the inverse of leaf resistance. The leaf conductance for water vapor flux consists of the boundary layer conductance (g_a), cuticular conductance (due to the waxy leaf cuticle; g_c), and stomatal conductance (g_s) to water vapor. Boundary layer conductance to water vapor is generally high. The thickness of the boundary layer of air surrounding a leaf depends on leaf size and shape as well as on ambient conditions, so it is not under the immediate control of the leaf. It has little effect on impeding water vapor flux. Cuticular conductance is very small and varies little: only limited water vapor leaves the leaf through the cuticle. In contrast, stomatal conductance is highly variable and is under the control of the plant: it is high when the stomata are open, and very low when the stomata close. Stomata are therefore very important in controlling the amount of water lost by the plant through transpiration.

The loss of large volumes of water through transpiration does not have negative effects on the plant when soil water is freely available. Transpiration is what moves water, along with dissolved mineral nutrients, from the roots to the leaves without the need for the plant to expend energy. As the soil becomes drier, however, the plant faces a dilemma. As CO_2 moves into the leaves through the stomata, water moves out through transpiration. A plant that keeps its stomata closed to restrict water loss eventually faces what is essentially starvation, as it can no longer take in CO_2. Angiosperms have evolved a greater diversity of mechanisms for coping with this dilemma, and for living with drought, than any other group of plants.

Strategies for Coping with Water Availability Conditions

Plants experience drought in a variety of ways, depending on both environmental conditions where they live and on their adaptations to those conditions. Drought may be brief or prolonged, mild or extreme, occasional and unpredictable or regular in its onset and duration. It may occur when conditions are otherwise favorable for growth, or it may occur when it is too cold for plants to grow.

Plants have evolved various means of coping with limited water availability. These evolutionary solutions, or sets of coordinated adaptive traits, are known as **strategies**. Ecologists have come up with different terms to categorize some of these sets of traits, but it is worth remembering that these categories are designated for convenience rather than describing fixed, rigidly defined entities. In reality, each species has a unique set of characters for coping with the conditions in its environment.

Mesophytes are plants that live in moderately moist soils and generally experience only an occasional, mild shortage of available water. Mesophytes typically transpire as long as soil Ψ is greater than about –1.5 MPa, although some need soils wetter than –1 MPa, while others may be able to continue withdrawing water from the soil at water potentials as low as –4 MPa. When the soil becomes drier, mesophytic plants close their stomata and wait until conditions improve. In a typical mesic (moist) habitat, that may be a few hours or a few days. If the drought extends for many days (a few weeks is all that most of them can tolerate), many mesophytes will begin to die. Most crops, ornamental plants, forest trees, common meadow and forest wildflowers, and other familiar plants are mesophytes. Agricultural crops are generally the least drought-tolerant, and need soils that do not dry out too much in order to grow well.

Aquatic plants grow in standing water, and **hygrophytes** are found in permanently moist soils. Plant ecologists and amateur naturalists alike have always been fascinated with the adaptations of **xerophytes**, plants that live in regions with frequent or extended drought. **Halophytes** are plants that live in saline soils and possess unique adaptations for tolerating salt, as well as some characteristics in common with xerophytes. Halophytes experience a dilemma similar to that of xerophytes in that plants in both dry and saline soils must be able to extract water from soils with very negative water potentials. Some xerophytes continue to transpire in soils that are phenomenally dry, as low as –6 MPa. The halophyte *Atriplex polycarpa* (desert saltbush, Amaranthaceae) can continue to grow with leaf water potentials as low as –5 MPa. Even under well-watered conditions, its leaf water potentials are still at –2 MPa—a condition that would be fatal to most mesophytes.

Adaptation to environmental conditions can occur by means of either a fixed phenotype (in physiology, morphology, or anatomy) or a plastic phenotype (one that allows an individual plant to change in response to changing conditions). One type of phenotypic plasticity is the ability to **acclimate** (acclimatize) to changes in environmental conditions, such as drought, with either permanent or reversible changes in anatomy, physiology, or morphology. Permanent changes might include the production of leaves with different structural characteristics; once formed, these leaves would retain their structure. Reversible changes might include changes in the form of enzymes that function optimally at different temperatures, or changes in the osmotic potential of leaf cells. The ability to acclimate is important for plants because they are not able to move when conditions change: they can neither step into the shade when it is hot nor migrate to a warmer climate when it is cold.

Whole-Plant Adaptations to Low Water Availability

A plant experiences its environment very differently than one might expect from a cursory look at average conditions. Many desert annuals, particularly the short-lived desert ephemerals, complete their entire aboveground lives during the brief rainy season characteristic of many deserts. These plants spend most of their lives as seeds, which are able to survive extended dry periods. The seeds of these plants germinate only after prolonged rains, and the plants photosynthesize rapidly, have high rates of transpirational water loss, grow rapidly, then set seed and die as the rainy season ends. By "living fast and dying young," they experience the desert as an environment with plentiful water for most of their lives as green plants. This strategy has been called **drought avoidance**.

Many desert shrubs lose all or some of their leaves during the long dry season and grow new leaves after the rains begin again (Figure 3.5). These **drought-deciduous** plants are also practicing a form of drought avoidance, minimizing their water loss through transpiration by reducing their leaf area when water is not available. Drought-deciduous trees, such as *Ceiba pentandra* (kapok tree, Bombacaceae), are also common in seasonal tropical forests, which also experience predictable dry periods alternating with rainy seasons.

Many herbaceous perennials in deserts, arid grasslands, and other xeric (dry) environments are likewise largely dormant during the dry season, when most of the living material is found below ground or at the ground surface. The aboveground parts of many grasses, for example, die back to the ground during the dry

Figure 3.5
Fouquieria splendens (ocotillo, Fouquieriaceae), a drought-deciduous desert shrub, in Joshua Tree National Park, in the Mojave Desert of eastern California. This species may produce several sets of leaves during a single year. The photo on the left shows fully expanded leaves and an inflorescence at the end of the stem, with buds and open flowers. The right image shows newly emerging leaves. (Photographs courtesy of P. Curtis.)

season. The advantage of dormancy during the dry season is, of course, reduction of water loss, enhancing survival during the unfavorable dry season. Its major disadvantage is that there is a long period of time during which the plant cannot photosynthesize, grow, or reproduce. Drought-deciduous plants generally have fairly high maximum photosynthetic and growth rates during the time when they are active, which help to compensate for the time when they are dormant. Many of these plants also have thick roots and underground stems that are capable of storing food for long periods of time. These structures enable them to survive extended drought periods and allow rapid production of leaves and stems when conditions become favorable again.

Another way in which some species avoid drought is to grow only in the wettest places within a dry environment, such as in low spots where water temporarily collects after rains, or along temporary streams. Such plants typically lack adaptations for extreme aridity and the seedlings will die if their seeds germinate in drier microhabitats.

There are, however, plants that are able to live and grow under remarkably dry conditions. These **drought-tolerant** species, or true xerophytes, use a variety of means to "tough it out" in the desert. From the whole plant to the cellular and molecular levels, the morphology, physiology, and anatomy of these plants are adapted for life in xeric habitats.

Root morphology can be adapted to dry environments in a variety of ways. Roots can be extensive and shallow, as they are in many cactus species, enabling the plant to take up water after a brief rain that saturates only the top layers of soil. Other species have roots that extend deep into the soil profile, where the soil is slow to dry out. **Phreatophytes** have roots that extend so deeply into the soil that they reach the water table, gaining relatively permanent access to water—the ultimate drought avoidance mechanism.

The ratio of dry mass invested in roots to the mass of aboveground tissue (leaves, stems, etc.), called the **root-to-shoot ratio**, varies widely among plants. Xerophytes typically have much higher root-to-shoot ratios than do mesophytes. Consequently, they are able to take up more water, and they lose less water through transpiration. The cost of a high root-to-shoot ratio may be a reduction in maximum growth rates, both because of a reduction in the total leaf area, and therefore the amount of photosynthetic tissue, and because of the metabolic cost of maintaining the large root mass.

The form of the whole plant may also be an adaptation to limited water availability. **Succulents**, such as cacti, have the capacity to store large amounts of water in their tissues. The succulent tissue may be located in leaves, stems, or other plant parts. Cacti use stems instead of leaves as their photosynthetic organs. As a result, they have reduced surface-to-volume ratios, decreasing the total surface area capable of water loss and increasing the volume that retains water. Other desert plants also have photosynthetic stems. For example, trees in the genus *Parkinsonia* (palo verde, Fabaceae; Figure 3.6) in the Sono-

Figure 3.6
Parkinsonia microphyllum (foothill palo verde, Fabaceae) (left) has extremely small leaflets (right), which are lost during the dry season, but photosynthesis continues at low rates in the green bark. (Photographs by S. Scheiner.)

ran Desert retain their leaves only during the rainy season, but remain capable of photosynthesis at low rates all year through their green bark.

In parts of the world where the soil freezes in winter, it can be difficult for plants to obtain liquid water from the soil for many months. Plants in these environments can experience severe drought stress in winter. This is one reason why deciduous trees and shrubs in the northern temperate zone and further north lose their leaves in winter, and it is one of the factors that limits the distribution of broad-leaved evergreen species northward.

Physiological Adaptations

When a plant begins to experience a water deficit, a series of physiological events is triggered. Hormones are produced and travel throughout the plant, signaling the onset of various changes in plant function. Cell growth and most protein synthesis slows, then ceases. As the plant experiences more prolonged water deficits, the allocation of materials to roots and shoots is adjusted, the stomata begin to close, and photosynthesis is inhibited. Leaves may begin to wilt. In some species, older leaves are allowed to dry out and die, freeing up available water to preserve younger leaves for a longer period.

In some species, specific measures for **osmoregulation** are initiated under dry conditions. These plants synthesize certain soluble compounds (proline and other low-molecular-weight nitrogen compounds as well as soluble carbohydrates) during drought. The resulting increase in solutes lowers the osmotic potential of the cells, leading to an influx of water by osmosis, which prevents turgor loss and wilting. Osmoregulation is particularly important in allowing many halophytes growing in saline soils (where soil water has a negative water potential largely due to a negative osmotic potential caused by dissolved salts) to maintain a favorable water potential gradient. Other halophytes have the ability to excrete salt.

Some plants, known as "resurrection plants," have highly unusual adaptations that allow them to survive extended complete desiccation (Figure 3.7). These plants include many lichens and some mosses, ferns, and other non-seed plants, as well as a number of angiosperms (in the Scrophulariaceae, Lamiaceae, Poaceae, and Liliaceae). They are found in various parts of the world, but are most diverse and abundant in southern Africa. All of these different plants survive cellular dehydration by a highly coordinated set of processes. Drought-stable proteins are synthesized, phospholipid-stabilizing carbohydrates are incorporated into the cell membranes, and the cytoplasm may become gelated. In the dehydrated state, metabolism is brought almost to a halt. Rehydration when water is again available also occurs in a highly coordinated manner in which the cell components are reconstructed step by step.

Figure 3.7
Selaginella lepidophylla (spike moss, Selaginellaceae) is a "resurrection plant." During an extended drought, the plant has biochemical mechanisms that allow it to slow its metabolism to a virtual halt. Although the brown and desiccated plant (above) appears to be dead, the advent of rainfall rehydrates the tissues, which quickly become green and photosynthetic (below). (Photographs © W. P. Armstrong.)

At the other end of the water availability spectrum, adaptation to flooding is critical for survival in some habitats (Figure 3.8). Plants may be subjected to variation in the depth to which they are submerged (from waterlogged soil to total submergence of the shoot) as well as in the frequency, season, and duration of exposure to flood conditions. Coastal species, such as those in salt marshes, and species found in floodplains and in riparian (river and stream) margins are predictably exposed to flooding, but the frequency and severity of flooding in many such locations have been increasing over time (see Chapter 22), and may surpass the toleration of these plants. Other habitats, including low-lying areas in forests, meadows, and fields, also expose plants to occasional submersion, particularly in sites with clay soils that do not drain readily after heavy rains.

Figure 3.8
An intermittently flooded swamp in southern Illinois dominated by *Taxodium distichum* (bald cypress, Taxodiaceae). This species produces "knees" (visible in the foreground) that project from the root system upward above the water, to facilitate root gas exchange. (Photograph by S. Scheiner.)

Anatomical and Morphological Adaptations

A wide variety of anatomical and morphological adaptations enable plants to survive and grow under very dry or very wet conditions. Among the most important are variations in stomatal numbers, arrangement, size and behavior, which vary widely among plants. Stomata are the gateway for most of the water leaving a plant. Stomatal conductance varies directly with the pore width (size of the stomatal opening). Pore width is controlled by the **guard cells** on either side of the opening, which continually change shape, widening or narrowing the opening (Figure 3.9). The movement of the guard cells is controlled hormonally by the plant; it is not a simple physical consequence of the guard cells passively wilting as water is lost, as once was thought.

When the soil is flooded, plants are faced with a number of conditions that make it difficult to function normally, or even to survive. The greatest problem is lack of soil oxygen. In ordinary, well-drained soils, oxygen diffuses through the soil to the roots, which use it for respiration. In waterlogged soils, the diffusion of oxygen is severely diminished because the soil pores are filled with water rather than with air. Longer-term exposure to flooding results in compact soils, in which larger soil particles collapse, leading to a reduction in soil pore space and soils that are harder for growing roots to penetrate. The oxygen that remains in the soil is quickly consumed by plant roots and soil organisms. Oxygen-dependent microbial processes are severely inhibited, and toxic substances from bacterial anaerobic metabolism begin to accumulate.

Plants that have evolved to tolerate flooding exhibit a variety of physiological, morphological, anatomical, and life history characteristics that allow them to function in flooded environments. One physiological mechanism for surviving for some time with little or no oxygen available to the roots is an increased reliance on glycolysis. The NADH that is generated by glycolysis can be oxidized to NAD by ethanolic fermentation, allowing some ATP production to proceed. The oxygen deficiency caused by flooding also stimulates a series of metabolic responses, including the production of polypeptides called anaerobic stress proteins, whose function is not yet fully understood.

The guard cells open and close the stomata primarily in response to three factors: light, CO_2 concentration, and water availability. The action of the guard cells differs among the three photosynthetic pathways: C_3, C_4, and CAM (see Chapter 2). Light directly causes the stomata of C_3 and C_4 plants to open. The partial pressure of CO_2 in the intercellular leaf space has a major influence on the guard cells, signaling the stomata to close when it goes up and to open when it goes down. At night, when photosynthesis ceases in C_3 and C_4 plants, the CO_2 concentration in the intercellular leaf space increases, and the stomata close. During the day, photosynthetic capture of CO_2 reduces the partial pressure of CO_2 in the leaf, and the stomata are induced to open. In CAM plants (see Chapter 2), the partial pressure of CO_2 drops at night as it is captured by PEP carboxylase, and the stomata open. During the day, malate is decarboxylated (CO_2 is removed) and accumulates in the leaf, signaling the stomata to close (see Figure 2.13).

Plant water relations are critical in determining the behavior of the stomata. Stomata respond to both leaf water potential and ambient humidity (Figure 3.10). A declining water potential in the leaf will override other factors, such as CO_2 partial pressure, to close the stomata; clearly, preventing desiccation is of more immediate concern than is maintaining photosynthetic rate.

The stomata of different plants vary in their sensitivity to these three factors. Mesophytes, particularly

(A)
Guard cell   Stomatal pore

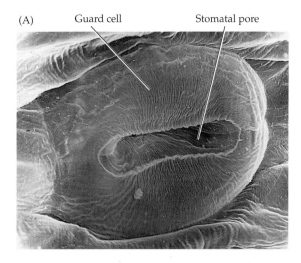

(B)

Palisade
parenchyma cells   Phloem   Xylem

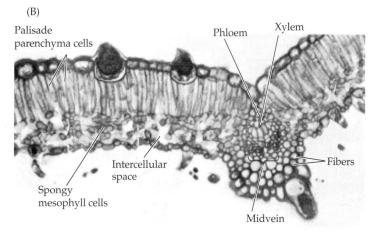

Intercellular
space   Fibers

Spongy
mesophyll cells

Midvein

Figure 3.9
(A) Scanning electron micrograph of a stoma in a leaf of *Allium* sp. (onion, Alliaceae). The view is of the inside epidermis, and shows a pair of guard cells facing the stomatal cavity. Guard cells control the width of the stomatal pore, and thus the amount of water leaving the plant. (Photograph courtesy of E. Zeiger and N. Burnstein.) (B, C) Cross sections showing leaf structure in two mesophytes. (B) *Cannabis sativa* (marijuana, Moraceae). The photosynthetic cells are the elongated palisade parenchyma cells and the smaller, irregularly spaced spongy mesophyll cells. The midvein consists of xylem and phloem cells and has a protective cap of fiber cells. (C) A leaf of *Dianthus* sp. (pink, Caryophyllaceae). The micrograph shows a larger vein (at right) and several smaller veins, and palisade and spongy mesophyll tissue. There are stomata on both the upper and lower surfaces of the leaf (arrowheads)—a condition more typical of xerophytes than of mesophytes such as this species. (From Esau 1977.)

(C)

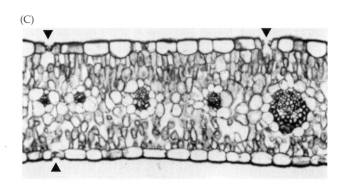

agricultural species, often close their stomata during the middle of the day when water losses are greatest, even in moist soil. The stomata of xerophytes behave differently from those of mesophytes when the soil begins to dry. Most mesophytes quickly close their stomata as the soil dries and leaf water potential drops. In contrast, many xerophytes are able to keep taking up water from the soil and photosynthesizing deep into a period of drought (see Figure 3.10). Mesophytes typically have much greater maximum pore widths, higher maximum stomatal conductances, and higher maximum rates of photosynthesis than xerophytes (Table 3.1).

In environments where dry conditions alternate with a predictable wet season, some plants are able to photosynthesize and grow rapidly during periods when water is available. By maximizing carbon accumulation when water is most available and stomata can be opened without excessive water loss, the plant gains the most carbon and energy over the course of the year without adverse consequences. Specializations in leaf anatomy can help plants to achieve high rates of photosynthetic carbon uptake during favorable periods.

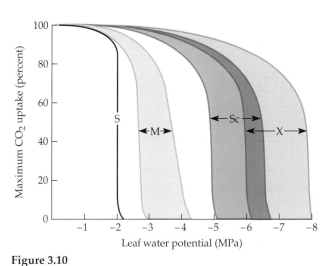

Figure 3.10
Stomata respond to a decline in leaf water potential by closing. The graph shows the decline in photosynthetic rate, expressed as the percentage of maximum CO_2 uptake, with decreasing leaf water potential in plants with different strategies. S, succulents; M, mesophytes; Sc, sclerophyllous plants (trees and shrubs of mediterranean and other semi-arid climates), X, xerophytes (which include herbaceous dicots, grasses, and shrubs of deserts and other very arid environments). Sclerophylls and xerophytes have partially overlapping values. (After Larcher 1995.)

Table 3.1 Properties of stomata on the lower leaf surface in plants from mesic and xeric environments

Type of plant	Stomatal density (number/mm² leaf area)	Pore length (μm)	Maximum (μm) pore width	Pore area (% leaf area)	Maximum conductance for CO_2 (mmol/m²/s)
Mesic environments					
Herbaceous plants (sunny habitats)	150 (200)	15 (20)	5	0.9 (1.0)	425 (700)
Herbaceous plants (shaded habitats)	75 (100)	23 (30)	6	1.0 (1.2)	130 (200)
Grasses	75 (100)	25 (30)	3	0.6 (0.7)	270 (460)
Tropical forest trees	400 (600)	18 (25)	8	2.3 (3.0)	100 (300)
Deciduous trees	200 (300)	10 (15)	6	0.9 (1.2)	150 (250)
Xeric environments					
Xerophytic desert shrubs	225 (300)	13 (15)	—	0.4 (0.5)	155 (260)
Sclerophylls	300 (500)	13 (15)	2	0.4 (0.5)	175 (250)
Succulents	33 (50)	ca. 10	10	0.3 (0.4)	95 (130)
Evergreen conifers	80 (120)	18 (20)	—	0.7 (1.0)	200 (—)

Source: Larcher 1995.

Note: Medians and average maxima (in parentheses) are shown for each stomatal characteristic. While there are broad overlaps between the values for plants from different environments, stomatal characteristics reflect environmental adaptations in a number of ways. Grasses are found in a wide range of habitats, from mesic to xeric. Succulents, many of which use CAM photosynthesis, have very different stomatal properties from other plants of xeric environments. Evergreen conifers tend to live in habitats that are dry due to reduced rainfall or low soil water holding capacities, but have unique leaves (needles) with unusual characteristics.

Many plants native to arid habitats are **amphistomatous** (have stomata on both sides of the leaf; mesophytes typically have stomata only on their undersides). Over 90% of species in North American deserts are amphistomatous. Many desert plants are also **isobilateral** (have a distinctive, symmetrical internal leaf architecture with palisade mesophyll tissue on both upper and lower sides of the leaf) (Figure 3.11). These two features together are likely to be an adaptation to the high light levels available to plants growing in arid habitats. Because intense sunlight can penetrate more deeply into the interior of a leaf than can dimmer light, maintaining these packed layers of photosynthetic cells deep inside the leaf allows the plant to achieve high maximum photosynthetic rates. Having stomata on both sides of the leaf reduces the average distance that CO_2 must diffuse before reaching chloroplasts, which also enhances photosynthesis.

Although xerophytes usually have a greater number of stomata per mm² of leaf surface area than do mesophytes, the average total pore area as a percentage of the total leaf area is about half as great for xerophytes as for mesophytes (Table 3.1). This may be an adaptation for maintaining tight control of water loss when the soil begins to dry, while still minimizing the distance CO_2 must travel to reach any photosynthetic cell.

Another feature of some plants from arid environments is stomata that are sunken below the leaf surface, or even located within deep crypts (Figure 3.12). This feature has the effect of lengthening the distance that water must travel to diffuse away from the leaf surface, increasing the leaf's resistance to water loss. Transpiration is also directly reduced because the water vapor concentration difference between the leaf and the air immediately outside the stoma (within the crypt) is greatly reduced due to the water vapor retained in the crypt.

The waxy cuticle covering the epidermis of the leaves may be especially thick in xerophytes, increasing cuticular resistance. Cuticular resistance becomes important when the stomata are closed. A plant with low cuticular resistance could leak considerable amounts of water over a prolonged dry period, even with closed stomata; preventing such loss is critical for survival in an arid habitat. The cuticle may also have a specialized construction that reflects much of the light falling on the leaf.

The leaves of many species adapted to dry conditions are **sclerophyllous**, or tough and leathery. Various structures make the leaves tough and hard, including thick cell walls (especially in the epidermis) and increased amounts of fibers and other structural tissues. Stronger, tougher leaves serve several functions, from providing mechanical strength to reduce wilting to protecting against herbivory, pathogen attack, and even damage from wind (which is likely to be stronger in open, arid habitats). Many of the features of xerophytes discussed here are also, to some degree, characteristic of the plastic responses of nonxerophytic plants to sunny, dry conditions.

There is a vast array of morphological differences among the root systems of different plants. Roots, of

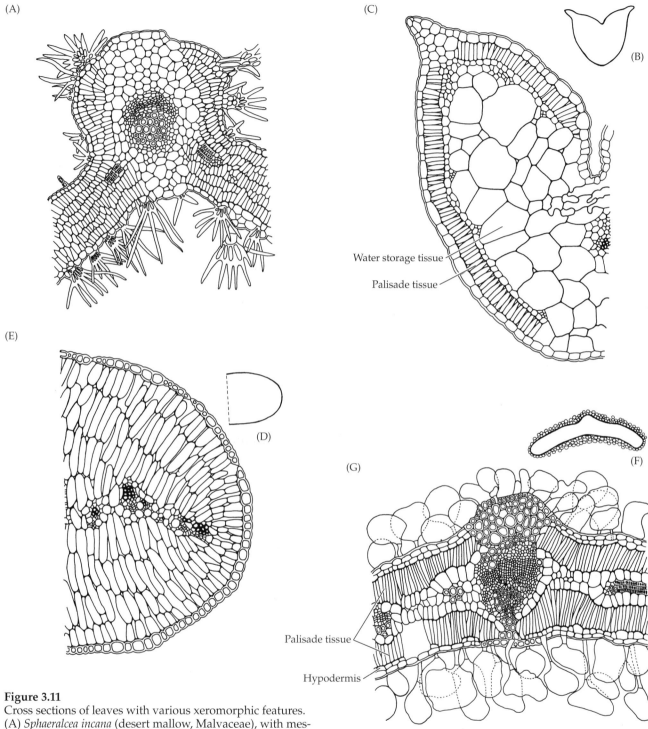

Figure 3.11
Cross sections of leaves with various xeromorphic features. (A) *Sphaeralcea incana* (desert mallow, Malvaceae), with mesophyll consisting only of elongated palisade cells; also note the numerous trichomes (needlelike hairs; see Figure 11.9). A large midvein is visible in the center of the illustration. (B, C) *Salsola kali* (tumbleweed or Russian thistle, Amaranthaceae). (B) Overall leaf outline. (C) Anatomy of the succulent leaf, showing large-celled water-storing tissue surrounded by a single layer of palisade parenchyma. (D, E) *Nerisyrenia camporum* (fanmustard, Brassicaceae). (D) Overall leaf outline. (E) Leaf anatomy, showing low surface-to-volume ratio and mesophyll consisting only of elongated palisade cells; several small vascular bundles are visible in the middle of the leaf. (F, G) *Atriplex canescens* (four-wing saltbush, Amaranthaceae). (F) Overall leaf outline. (G) Leaf anatomy, showing isobilateral mesophyll and large, abundant hairs on both the lower and upper epidermis. The large midvein is in the center of the leaf, with a cap of thick-walled fibers between it and the upper epidermis. Note that the leaves of all these plants have thickened cuticles. (From Esau 1977.)

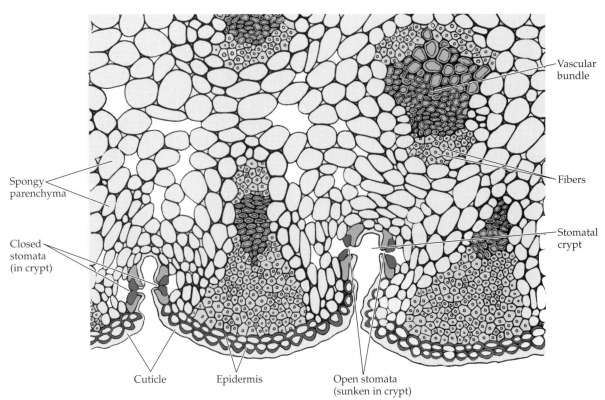

Figure 3.12
Stomatal crypts in the leaf of *Yucca* sp. (yucca, Agavaceae). There are two stomata within each crypt in this illustration; the ones in the crypt at the left are closed (the substomatal chambers are to the left and right of the crypt) and the ones in the crypt at the right are open. Other xerophytic features of this plant include thick leaves, large bundles of fibers, and a thickened, waxy cuticle. (After Weier et al. 1974.)

course, are what most vascular plants use to obtain water, and root systems differ according to taxonomic group as well as the environment to which the species is adapted and, to some extent, the environment in which the individual plant finds itself growing. There are a number of different basic kinds of root systems. **Fibrous root systems** form a dense network, capable of exploring a large volume of soil for water and nutrients. Most grasses, and many other monocots, have fibrous root systems (Figure 3.13). Other species, generally dicots, have a single **taproot t**hat extends deep into the soil, and may also be thickened and capable of storing food.

Roots also possess various anatomical adaptations to the environments in which they grow. Roots have received less study than leaves, perhaps because they are less apparent. One of the most dramatic adaptations of root anatomy to habitat conditions is the **aerenchyma** (aerated tissues) of aquatic plants and other plants adapted to flooded conditions (Figure 3.14). The formation of aerenchyma is a way of avoiding the low oxygen conditions of flooded soil by providing an interconnected system of air channels for gas movement

inside the plant. Oxygen from the external air, as well as oxygen released by photosynthesis, diffuses to the roots through the aerenchyma, while other gases (such as ethylene and methane) escape from the roots and out through the aboveground parts of the plant. Other plants growing in inundated soils, such as some mangroves, produce specialized aboveground roots, or roots that bend upward and emerge from the soil into the air, presumably to obtain access to oxygen (Figures 3.15, 3.8).

Xylem, the tissue that carries water throughout the plant in vascular plants, also differs greatly among plants depending on their phylogeny and their ecological adaptations. In all plants, mature xylem cells are dead, and the cell contents have been removed to produce hollow vessels through which water is transported. The size, shape, and connections between xylem vessels differ among taxa, and these differences strongly affect their function. Xylem evolution illustrates the adaptation of plants to moisture availability and loss, to the predictability of water stress, and to the need for mechanical strength

The water-conducting tissue in earlier-evolved vascular plants (ferns, horsetails, lycopods, and the like)

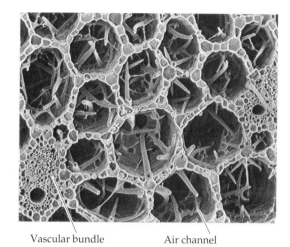

Vascular bundle   Air channel

Figure 3.14
This scanning electron micrograph shows the aerenchyma tissue of a species of *Nuphar* (yellow water lily, Nymphaeaceae). The air channels allow gases, including essential oxygen, to move through plants growing in inundated conditions. (Photograph by J. N. A. Lott/Biological Photo Service.)

Gymnosperms also depend on tracheids, although their tracheids display various advances over those of the earlier vascular plants. Angiosperms sometimes have tracheids as well, but they have also evolved another kind of cell called **xylem vessel elements** (see Figure 3.1). These cells make up what are essentially long water-conducting tubes called **xylem vessels** (Figure 3.17). In comparison with tracheids, vessel elements are shorter, and they may be fully open at the ends because of the lack of a primary cell wall, allowing more water to move

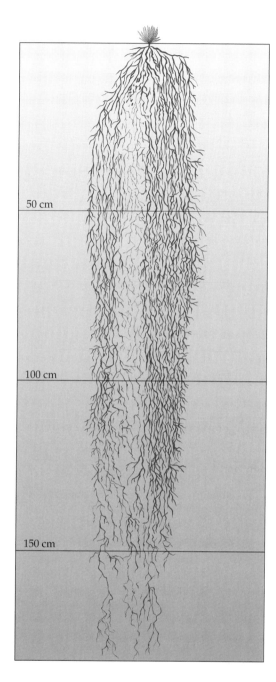

Figure 3.13
The fibrous root system of *Triticum aestivum* (wheat , Poaceae), a cultivated monocot species. The shoot of this plant has been cut off just above the soil surface. (After Mohr and Schopfer 1995.)

consists of elongated cells called **tracheids** (Figure 3.16). These cells are connected to one another by the thinner parts of their end walls, through which water diffuses. Water transport in such a system is limited; much less water is transported, and it moves more slowly, than in more recently evolved taxa.

Figure 3.15
Aerial roots of *Rhizophora mangle* (red mangrove, Rhizophoraceae), which act as props and as a means of channeling air to the submerged portions of the roots. (Photograph by S. Scheiner.)

Tracheid

Vessel elements

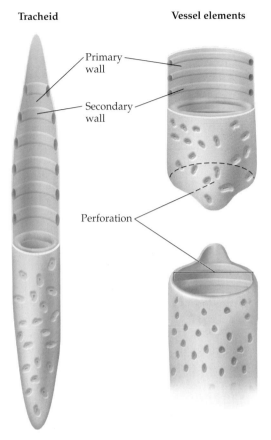

Figure 3.16
Structure of a tracheid (left) compared with that of a vessel element (right). The tracheid is not drawn to scale, and is approximately 50 times longer than shown in relation to the vessel element. Note that the outer (primary) wall of the tracheid is whole, but the end wall of the vessel element is perforated and open to water. The secondary wall is laid down inside of the primary wall, and strengthens it. The tracheid and vessel element both have pits—thinner parts of the cell wall through which some diffusion occurs—on their outer surfaces. (After Mauseth 1998.)

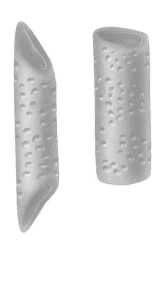

Figure 3.17
Xylem vessel elements from four different angiosperms, arranged from the earliest to evolve (left; similar to *Magnolia*) to the most recently evolved (right; similar to *Quercus* [oak]). The earliest vessel elements are long and thin, and closely resemble tracheids (see Figure 3.16). The evolutionary trend has been toward shorter, wider elements with thicker cell walls, which can conduct much larger volumes of water. Because of the greater volumes of water, these wider vessel elements are under greater stress; additional structural support is provided by surrounding fibers.

rapidly from roots to shoots. The xylem vessel elements of dicots from dry environments tend to be particularly short and narrow, with exceptionally thick walls. This structure offers an adaptive advantage for conducting water under the very negative pressure potentials necessary for maintaining a favorable water potential gradient when water is being extracted from very dry soils. The extreme negative pressure that the xylem is exposed to in these xerophytic plants would be enough to collapse the vessels of species not possessing this specialized kind of xylem.

The Energy Balance of Leaves

The survival of any living thing depends on maintaining a temperature that is neither too cold nor too warm for

its tissues and organs to function. Clearly, plants must avoid freezing or cooking. They must, in addition, maintain their leaves at temperatures at which they can carry out metabolic processes optimally, or at least adequately, most of the time. How can a plant manage to do this? One does not ordinarily think of plants as being capable of regulating their own temperatures. Animals can shiver or bask in the sun; they can retreat to the shade or to an underground burrow when it is too hot, too sunny, or too windy—but none of these options is available to plants. What is possible for a plant? To understand this better, we need to look at the way plants, particularly leaves, interact with their energy environment.

All physical objects in the universe are continually exchanging energy with their surroundings, and plants are no exception. The factors that determine the temperature of any object, such as a leaf, are best understood in terms of the **energy balance** (also called the energy budget) of the object. **Temperature** is a measure of the average random kinetic energy of the molecules in a substance. Calculating the energy balance of an object means accounting for the amount of energy absorbed by the object, the amount of energy leaving it, and the amount of energy stored within it. To do so, we must first take a more specific look at the components of the energy balance of an object such as an organism. We take an intuitive rather than a mathematical approach to this topic (see Campbell 1977 and Nobel 1983 for a detailed look at the physics of the energy balances of organisms).

In order to get an intuitive feel for the components of the energy balance of an organism—a groundhog, a leaf, or yourself—we begin by describing an ordinary experience; precise definitions will follow. Imagine that you are sitting on a rock at the beach on a warm summer day, soaking up the rays. Your skin is warmed by the sun's **radiant energy**. You can also feel heat reaching your skin by **conduction**—direct transfer—from the rock you are sitting on, which has also been warmed by the radiant energy of the sun. A breeze cools you by **convection**—heat transfer via a fluid medium.

Now suppose you jump into the water. The water is cooler than you are, so you feel chilly. You do not get cold as quickly as a smaller animal would, however, because your greater mass stores more heat energy. You swim vigorously, and the **metabolic energy** produced by your muscles burning calories stored from food warms you up. When you get out of the water, the evaporation of water from your skin makes you feel chilled: you are experiencing **latent heat loss**.

This scenario summarizes the major terms of the energy balance equation. An understanding of the values for these terms can explain many aspects of the way organisms exchange energy with their environments and their adaptations for maintaining appropriate temperatures.

Most leaves are too small to store appreciable amounts of energy, and unlike animals, most plants do not generate much metabolic heat. Because it is usually unnecessary to account for metabolic energy and energy storage, it is relatively simple to calculate the energy budget of a typical leaf. The energy balance equation for a leaf at constant temperature can be written as

$$R_n - H - \lambda E = 0$$

In other words, R_n, the net amount of radiant heat energy coming into the leaf, minus H, the **sensible heat** (conduction + convection) loss, and λE, the latent heat loss, is equal to zero: the energy gained by the leaf is exactly balanced by the energy lost. If there is more energy coming in than going out, the temperature of the leaf will rise; if more energy is leaving the leaf than coming in, its temperature will decline.

While these terms are generally adequate for leaves, they are not sufficient for determining the energy balance of certain massive plant organs, such as the thick stems of some cacti, which can store considerable heat energy. There are also a few unusual plants and plant organs, such as *Amorphophallus titanum* (voodoo lily, Araceae; see Box 8A) and *Symplocarpus foetidus* (skunk cabbage, Araceae), that are capable of generating substantial metabolic energy. In such cases, additional terms for heat storage and metabolic heat must be included in the energy balance equation.

We now explore what these terms mean, and what determines how large or how small they are under different conditions. We also discuss what a plant can do to modify these terms, on an immediate basis or over evolutionary time, to control its exchange of energy with the surrounding environment.

Radiant Energy

Radiant energy is that energy transferred from one object to another by **photons**, discrete packets of energy that travel at the speed of light. Physical objects are continually absorbing and emitting photons. The emission of a photon (by an electron undergoing a transition in energy state) produces electromagnetic radiation at a single wavelength. The photon flux from a source—the amount of energy emitted—depends on the wavelengths of the photons emitted, with shorter wavelengths having higher energy levels.

The radiant energy emitted by the surface of an object, per unit surface area, is given by the **Stephan-Boltzmann equation**,

$$\Phi = \varepsilon \sigma T^4$$

where Φ is the energy emitted (the emitted flux density, in W/m^2), ε is the emissivity of the surface, σ is the Stephan-Boltzmann constant (5.67×10^{-8} W/m^2/K^4), and T is the absolute temperature in degrees kelvin (K; equal to the temperature in °C + 273). **Emissivity** is a measure of how efficient a body is at emitting energy. If an object absorbs all the radiation that falls on it, and emits, or radiates, all of the energy possible for an object at its temperature, it is called a **perfect blackbody radiator**. The emissivity of a perfect blackbody radiator is therefore 1.0. No real objects are perfect blackbody radiators over the entire electromagnetic spectrum, but most natural objects have emissivities of between 0.90 and 0.98 (although this can vary quite a bit according to the wavelength of the emissions).

Notice that one of the most important quantities in the Stephan-Boltzmann equation is temperature, because it is raised to the fourth power. Thus, the surface of the sun, which has a temperature of approximately 6000 K, emits 73,000 W/m^2; the sun is close to being a perfect blackbody radiator. The energy reaching the surface of the Earth from the sun is reduced, among other things, by the scattering of the photons by particles in the atmosphere (Box 3A; see also Figure 18.1). The sun emits the greatest number of photons in the visible

Box 3A

Why the Sky Is Blue and the Setting Sun Is Red

Some of the visible radiant energy from the sun is scattered before it reaches the Earth's surface. The gas molecules and small particles of the atmosphere scatter light primarily at short wavelengths (Rayleigh scattering). This explains why the sky is blue: blue wavelengths are the shortest visible ones common in the solar spectrum. When we see this scattered blue light from Earth's surface, the sky appears blue to us.

At day's end, the angle of incoming sunlight is much steeper relative to the Earth's surface, and blue light is refracted (bent) high above the surface by the atmosphere. With the blue wavelengths removed, what remains are the longer wavelengths (reddish colors), which produce the reds and oranges of a sunset. In addition, large-particle scattering of light is most effective at longer wavelengths. That is why volcanic dust, particulate pollution, and other sources of suspended material in the atmosphere can result in particularly intense red sunsets.

range of the spectrum, with the peak at yellow light (explaining why the sun looks yellow), although more energy comes from higher-energy, shorter wavelengths in the ultraviolet (UV) range. The Earth, in contrast, has an average temperature of about 290 K and emits about 400 W/m^2 in the infrared part of the spectrum.

You may be surprised to learn that the Earth radiates energy. It does, as do all objects in the universe, including you and the leaves of plants. These wavelengths of these emissions are not in the visible spectrum, however, because Earth's temperature (as well as your temperature, and that of a leaf) is too cool to emit visible light.

The net radiation of a leaf (R_n) accounts for all of its incoming and outgoing energy (Figure 3.18). In sunlight, R_n is positive, but at night it will often be negative as the leaf radiates more energy than it absorbs from its surroundings. By convention, we divide incoming radiation into **shortwave radiation** (wavelengths < 700 nm, including visible and ultraviolet wavelengths) and **longwave radiation** (infrared and longer wavelengths > 700 nm). Shortwave radiation includes direct sunlight shining on the leaf, light reflected from the surroundings, and diffuse shortwave light from various sources, including the background light from the sky. A leaf held directly perpendicular to the sunlight will receive more shortwave energy than one held at a steep angle (Figure 3.18). Longwave inputs include the radiant energy emitted by the ground, the sky, and other objects sur-

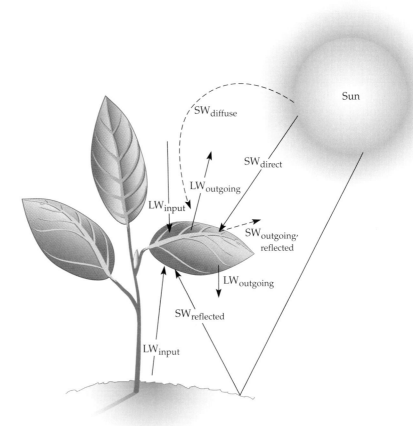

Figure 3.18
The radiant energy flux of a leaf held horizontally on a warm, sunny day. Longwave radiant energy is indicated by LW, shortwave radiant energy by SW. The leaf absorbs direct and diffuse shortwave radiant energy from the sun, as well as shortwave energy reflected from nearby objects, such as the ground (or other leaves). It also absorbs longwave radiant energy from the ground and the sky. It emits longwave radiant energy upward and downward. The leaf also reflects some of the shortwave radiant energy that reaches it. (After Campbell 1977.)

Water Relations and Energy Balance

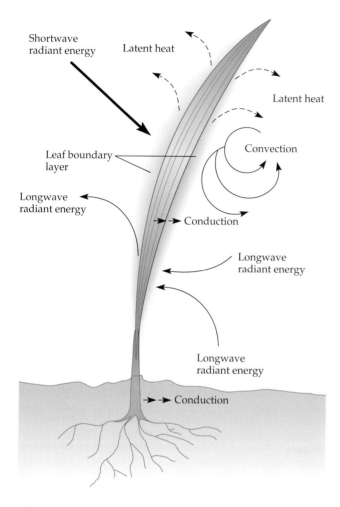

Figure 3.19
Major components of the energy balance of a grass leaf, showing the major sources of energy entering and leaving the leaf. The leaf has a net gain of shortwave radiant energy from the sun, and it also gains longwave radiant energy from the atmosphere and its surroundings (especially the ground). It loses energy by convection, by conduction (primarily through the boundary layer and underground to the soil), and by latent heat loss from transpired water, and it radiates energy out to its surroundings in the longwave part of the spectrum. The boundary layer of the leaf—a blanket of still air that surrounds it and reduces convective heat loss—is thickest at the central part of the leaf.

rounding the leaf. The leaf absorbs only some of the radiant energy impinging on it, depending especially on its own surface characteristics. The remainder is reflected back to the sky or toward other objects in the environment.

Conduction and Convection

Sensible heat exchange—energy exchange that results in a change in temperature—includes two major components, conductive and convective heat transfer (Fig-

ure 3.19). The transfer of sensible heat between an organism and its surroundings is directly proportional to the temperature difference between them, and inversely proportional to the organism's resistance to heat transfer. This relationship can be modeled using flux equations. Heat is transferred from warmer objects to cooler ones. The molecules in a warmer solid object vibrate faster than those in a cooler one (molecules or atoms in warmer liquids slide around over one another more rapidly, and those in gases bounce around more rapidly in three-dimensional space). When the molecules of two objects are in contact, the molecules with more kinetic energy speed up those with less. In the process, the faster molecules slow down a bit, until eventually both objects have the same temperature.

Conduction is the direct transfer of heat energy from the molecules of a warmer object to the molecules of a cooler one. It occurs when objects of different temperatures are in direct contact with each other. Conduction is a slow and fairly ineffective way of transferring energy, and it is not usually very important in either water or the atmosphere. It may be important in the soil, however, as well as inside leaves, in the boundary layer of air surrounding a leaf, and sometimes in other parts of the plant under certain conditions.

Convection is heat transport by a packet of fluid (air or water) moving as a unit. It is much more rapid and effective than conduction in transferring heat energy between organisms and their environment. If your body is warmer than the surrounding air, you experience convective heat loss as a cooling breeze. That is why wind chill is important in how you experience the cold: the greater the wind speed, the more rapidly your body is cooled by convective heat loss. In unusual environments in which the surrounding air is warmer than your body, the breeze will feel like a blast of heat from a furnace. In this case, you are experiencing convective heat gain.

The fur of mammals and the feathers of birds can offer considerable resistance to convective heat exchange, but in general, leaves and other plant parts do not. However, leaves are surrounded by a blanket of relatively still air, called the **boundary layer**, that surrounds objects in air or water. Heat (and mass) transfer across a boundary layer is much less efficient than heat transfer elsewhere in the fluid. The thicker the boundary layer, the greater the resistance to heat transfer posed by a leaf.

The boundary layer of a leaf is determined mainly by wind speed and by the size and shape of the leaf. The greater the wind speed, the more eroded the boundary layer becomes. Larger, rounder, more entire leaves carry a thicker boundary layer than leaves that are smaller, narrower, or have more dissected (broken-up) shapes. The **characteristic dimension** of a leaf is a measure that takes into account both its size and shape and represents the effective width of the leaf with respect to fluxes of energy and mass entering and leaving the leaf. It is directly proportional to the boundary layer resistance of the leaf.

Latent Heat Exchange

Latent heat exchange is the transfer of energy that occurs during evaporation—the process of converting water from a liquid to a gaseous state. (It is called latent—concealed—because there is no change in temperature during this energy transfer process, in contrast with sensible heat exchange, which can be "sensed" due to a change in temperature.) It is equal to the product of the evaporation rate, E, and λ, the latent heat of vaporization of water (2.45 MJ/kg at 20°C). Like convection, latent heat exchange can be a highly efficient cooling mechanism for organisms. The molecules in a substance in the gaseous state have much more energy than those in the liquid state at the same temperature. At 20°C, water vapor has about 2450 joules per gram more energy than liquid water. In contrast, warming liquid water by one degree takes only about 4 joules per gram. Where does all of the extra energy come from as water is transformed from liquid to gas in the process of evaporating? It is taken from—and thus cools—the surface that is losing the water: the tongue of a panting animal, the skin of a sweating athlete, or the tissue of a transpiring leaf.

Putting It All Together: Leaf Temperature

The ability of a leaf to survive and function depends on its ability to maintain its temperature within an acceptable range. The leaf temperature, in turn, depends on the energy balance of the leaf. The large amount of radiant energy input to a leaf in direct sunlight, for example, must be balanced by adequate convective and/or latent heat loss. Leaves that have a large characteristic dimension and a high boundary layer resistance can potentially become many degrees warmer or cooler than leaves with a small characteristic dimension (Figure 3.20).

A plant with large leaves in bright sunlight (large radiant energy input), still air (minimal convective heat loss), and dry soil is faced with a critical dilemma. Opening the stomata to allow cooling through latent heat exchange risks severe wilting and even death. Stomatal closure to restrict water loss will also reduce latent heat loss, potentially causing the leaf to heat to lethal temperatures. With adequate water available and even a light breeze, however, latent and convective heat losses can cool a large leaf to well below air temperature.

Small leaves are likely to remain close to air temperature, even with closed stomata, as the slightest breeze will minimize their boundary layer resistance. Under windy conditions, leaves of any size will have substantial rates of convective heat exchange. Conversely, at equally high stomatal conductances, a warm leaf will lose a lot more water than a cool leaf because

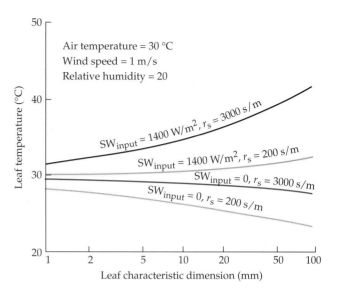

Figure 3.20
Leaf temperatures at a constant air temperature of 30°C, with a constant wind speed of 1 m/s (a fairly stiff breeze) and relative humidity of 20%. Leaf temperatures were calculated for leaves of increasing characteristic dimensions (approximately equal to leaf widths) at different shortwave radiant energy input levels (SW$_{input}$, expressed in W/m², watts per square meter of leaf surface) and stomatal diffusive resistances (r_s) expressed as s/m, seconds per meter). Small leaves (those with low values for leaf characteristic dimension) remain close to air temperature, regardless of sunlight intensity and stomatal resistance (which affects water loss, and thus latent heat loss), while large leaves can become much hotter or much cooler than the air, depending on shortwave input and stomatal resistance. (After Campbell 1977.)

water evaporates more quickly from a warm surface (the warm surface supplies more energy to the evaporating water).

Thus, on a short time scale, plants (at least those with larger leaves) can directly control latent heat loss from their leaves, and thus leaf temperature, by controlling transpiration rate. Over long periods of time, adaptation to warm, dry environments can involve the evolution of leaves of different sizes and shapes, changing boundary layer properties and altering convective heat exchange. Many desert plants, for example, have small or narrow leaves with reduced boundary layer resistance to heat transfer, resulting in leaves that remain close to air temperature even when the stomata are closed.

Radiant heat input is also, to some extent, under the control of the plant. The leaves of many desert shrubs are held at a steep angle (Figure 3.21), reducing radiant heat gain. There are also plants that change leaf angle in response to leaf temperature. Leaf coverings such as reflective white **pubescence** (hairs) and shiny, waxy

Figure 3.21
Leaves of *Simmondsia chinensis* (jojoba, Simmondsiaceae), a desert shrub, in Joshua Tree National Park, in the Mojave Desert. The thick, leathery leaves are held at an upright angle to reduce radiant heat gain. (Photograph courtesy of P. Curtis.)

coatings serve a similar function, reducing radiant heat input to the interior of the leaf (at the cost of reducing the light available for photosynthesis).

Some cacti have dense coverings of highly reflective spines; while these structures may also act to deter herbivores, their primary benefit is reducing radiant heat input. *Encelia farinosa* (brittlebush, Asteraceae) is a desert shrub that produces two different kinds of leaves, depending on the season. In summer, the plant puts out densely pubescent and highly reflective leaves. In spring, when temperatures are cooler and water is more available, it produces green, nonpubescent leaves, which absorb much more shortwave radiation and consequently have higher photosynthetic rates than the reflective summer leaves (Figure 3.22). Other plants also produce different types of leaves in different seasons, each of which has morphological or physiological adaptations for functioning optimally in its own season.

A common adaptation of many grasses (and some other plants) is to roll their leaves into narrow cylinders when facing water stress. This change serves to modify the leaf energy balance in several ways, minimizing both boundary layer resistance and radiant heat input. Also, rolling up the leaf directly reduces total water loss by reducing the amount of leaf surface

exposed to the dry surrounding air. There are also many whole-plant adaptations for managing water and energy losses, including partial or total shedding of leaves in dry or cold seasons.

Summary

Plants are descended from organisms that lived in the oceans. Their ability to live on land has required the evolution of anatomical, physiological, and other adaptations. The ability of a plant to obtain water, and the movement of water throughout the plant, depends on the water potential of plant tissues and of the surrounding soil and atmosphere. Water potential consists of various components, of which osmotic potential and pressure potential are generally most important within the plant. In any system, water tends to move from components with less negative water potential to components with more negative water potential, so we can predict the direction of water movement if the water potential of various components is known.

Plants lose much more water in transpiration than they use metabolically. Plant strategies for taking up and

Figure 3.22
Variation in the appearance of leaves of *Encelia farinosa* (brittlebush, Asteraceae). The leaves in the center of the plant are being produced as the environment is becoming warmer and drier; those surrounding it were produced earlier, when conditions were cooler and moister. The central leaves are thicker, smaller, and covered with a dense white pubescence (hairs) that reflects sunlight; those toward the outside are greener because they lack the thick pubescence. (Photograph courtesy of S. Schwinning.)

coping with the loss of water differ greatly depending on the species and the environment to which it is adapted. Mesophytes, aquatic plants, hygrophytes, xerophytes, halophytes, phreatophytes, and desert ephemerals all use different strategies to obtain water and restrict its loss. There are a rich variety of physiological, morphological, and anatomical adaptations to the typical range of water conditions experienced in an environment.

Plants are continually exchanging energy with their surroundings. The temperature of leaves and other plant organs depends on their energy balance: the difference between the energy taken up and the energy lost. Radiant energy, sensible heat exchange (conduction and convection), and latent heat loss are the major components of the leaf energy balance and together determine leaf temperature.

Plants are often perceived as passive receptors of environmental conditions. Yet, the long-term adaptations of plant species and the shorter-term functions of individuals modify the actual values they experience for many critical environmental factors affecting them, including temperature, light, and moisture.

Additional Readings

Classic References

Ehleringer, J., O. Björkman and H. A. Mooney. 1976. Leaf pubescence: Effects on absorptance and photosynthesis in a desert shrub. *Science* 192: 376–377.

Maximov, N. A. 1929. *The Plant in Relation to Water.* Allen and Unwin, London.

Odening, W. R., B. R. Strain and W. C. Oechel. 1974. The effect of decreasing water potential on net CO_2 exchange of intact desert shrubs. *Ecology* 55: 1086–1094.

Contemporary Research

Borchert, R. 1994. Soil and stem water storage determine phenology and distribution of tropical dry forest trees. *Ecology* 75: 1437–1449.

Holbrook, N. M. and F. E. Putz. 1996. From epiphyte to tree: Differences in leaf structure and leaf water relations associated with the transition in growth form in eight species of hemiepiphytes. *Plant, Cell, Environ.* 19: 631–642.

Additional Resources

Campbell, G. S. 1977. *An Introduction to Environmental Biophysics.* Springer Verlag, New York.

Lambers, H., F. S. Chapin III and T. L. Pons. 1998. *Plant Physiological Ecology.* Springer-Verlag, New York.

Kramer, P. J. and J. S. Boyer. 1995. *Water Relations of Plants and Soils.* Academic Press, New York.

Nobel, P. S. 1983. *Biophysical Plant Physiology and Ecology.* W. H. Freeman, New York.

CHAPTER 4

Soils, Mineral Nutrition, and Belowground Interactions

Most terrestrial plants are rooted in the soil and depend on it for support, water, and mineral nutrients. What is soil, and how does it affect plants? What are the characteristics and properties of soils? How is soil created, and how long does it take for soil to be made? What are some of the differences among soils in different places?

In this chapter, we look down and examine some of the surprisingly complex things going on right under our feet. We look first at the characteristics and properties of soil, and will see how those properties affect the availability of water to plants. Since plants depend on soil for mineral nutrients, we examine what these minerals are and how the plant uses them. We also introduce the ecologically important underground interactions between plants and two other group of organisms: nitrogen-fixing bacteria and mycorrhizal fungi.

Soil Composition and Structure

You might think of soil as "dirt," a sort of ground-up mixture of sand, dust, and crumbled bits of rock. If so, your image of soil would be very far from reality. Soil is a complex, often highly structured system that is a unique product of the interaction between living organisms and a physical matrix. Because soils are formed by the interaction among living things, rocks, air, water, and other materials, they occur only on Earth and nowhere else in our solar system (at least as far as we know!).

What materials are found in soil, and what is it made of? To begin with, there are mineral particles, derived from rock. These particles can range from very large ones—stones and boulders—to, in progressively smaller sizes, pebbles, gravel, sand particles (> 2.0–0.02 mm diameter), silt particles (0.02–0.002 mm), and clay particles (< 0.002 mm). Conventionally, only sand, silt, and clay are considered to be part of the soil itself. Besides these rock-derived particles, there is organic matter in varying states of decomposition. Air and water containing dissolved minerals are found in the pores between the mineral particles.

Finally, soils are, to one extent or another, the major environment of various kinds of living things: fungi, bacteria, photosynthetic prokaryotes (such as cyanobacteria), single-celled eukaryotes such as diatoms, protists, visible and microscopic insects and other arthropods (see Chapter 15), and many other kinds of small (slugs, earthworms, nematodes) and larger (groundhogs, gophers) animals. And, of course, soils are filled with the roots and rhizomes of plants. The living things in the soil continually alter its physical and chemical properties. Plants, for example, take up large volumes of water from the soil, changing the amount and distribution of water within the soil. The various soil organisms act and interact in a wide variety of ways. Some benefit one another; others consume one another or defend themselves to avoid being consumed. In the process of metabolism, soil organisms respire, altering the chemistry of their environment, and produce waste products and other substances, which become part of the soil. These activities affect plants in many ways, from making nutrients available to causing disease. We examine some of these effects in this chapter.

Soil Texture

The properties of soil, and its effects on plants, depend, first, on **soil texture**: the relative proportions of the different particles making up the soil. Soils are composed of sand, silt, and clay particles. Depending on which particle size dominates the character of a soil, the soil texture is categorized as sandy, silty, clayey, or loamy (Figure 4.1). The properties of **loamy soils** are a balance between sand, silt, and clay particles, and are generally considered the most desirable soils for agriculture. **Sandy soils**, with more than about 50% sand particles, have a coarse texture. These soils drain rapidly after a rain, and they hold water and minerals poorly. Water and air penetrate sandy soils easily. Because of these characteristics, they warm readily in spring and cool quickly in autumn.

Clay particles have very distinctive properties, and even a soil with as little as 35–40% clay will exhibit those properties (a little clay goes a long way). **Clayey soils** can hold a large volume of water, and they retain water and minerals exceptionally well. Similarly, clay-dominated soils retain pesticides, pollutants, and other substances. They are much less permeable to air and water than are sandy soils, which can result in puddling, greater runoff, poor drainage, and poor aeration (due to water filling up the pore spaces and excluding air). As a consequence, clayey soils are very slow to warm in spring and cool more slowly in

(A)

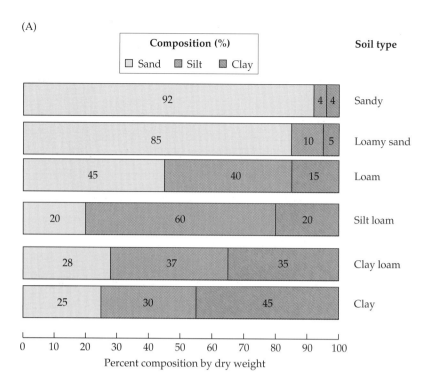

(B)

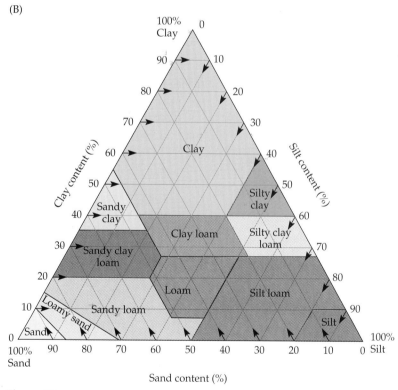

Figure 4.1
(A) Examples of soil particle size distributions of several different soil types. (B) The soil texture triangle, which shows how soil types are classified according to the percentage of sand, silt, and clay they contain (by dry weight). The values for clay are drawn parallel to the sand side of the triangle, those for silt are parallel to the clay side, and those for sand are parallel to the silt side. The points on the grid at which the three lines intersect defines the type of soil; for example, the grid lines for a soil with 20% clay, 40% sand, and 40% silt intersect in the region that shows a soil with this particular composition is loam.

(A)

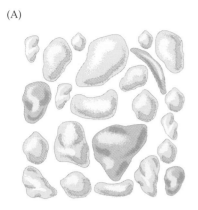

Figure 4.2
(A) Sand grains are irregular in size and shape, and are composed largely of quartz with some other secondary minerals. Silt particles are similar to sand grains in mineral composition and shape, but are smaller in size. (After Buckman and Brady 1969.) (B) Clay particles have a distinctive crystalline structure. Three different kinds of clay particles are shown here. Of the three, montmorillonite has the greatest cation exchange capacity (ability to hold cations) because in addition to binding cations on its surface, it has a large exchange surface available between its plates. Contrast the highly structured form of these clay particles with the rather formless sand and silt particles in (A). The clay particles are also far smaller. (After Etherington 1982.)

(B)

Kaolinite

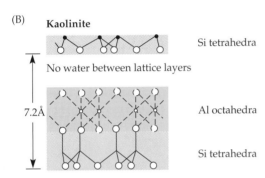

Montmorillonite

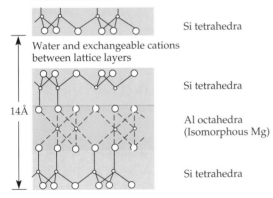

Illite

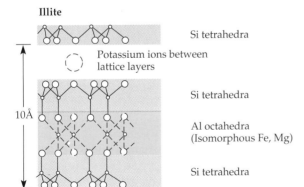

autumn; plant growth in these soils may be slow to get started early in the growing season. **Silty soils** tend to be intermediate in their characteristics and properties between sandy and clayey soils. Because soils often possess characteristics of more than one of these texture classes, they may be described by a combination of these terms, such as sandy loam, silty clay loam, or sandy clay.

To understand why different soil textures have the properties they do, we need to look more closely at the soil particles themselves. Sand and silt particles are generally irregular in shape, ranging from somewhat rectangular or blocky to chunky, spherical shapes (Figure 4.2A). In contrast, clay particles have a much more specific structure. These particles are made up of plates, whose size, shape, and arrangement depends on the minerals they contain and the conditions under which they were formed. A clay particle is made up of two or three flat crystalline plates, layered or laminated together (Figure 4.2B). These particles can be hexagonal in shape with distinct edges, or they may form irregularly shaped flakes or even rods. Soils may also contain particles similar in size to clays, but without the distinctive crystalline structure of clay particles. For example, an amorphous material called allophane, with particles about the same tiny size as clay particles, is prevalent in soils developed from volcanic ash.

Most of the clay particles in soil exist in a colloidal state (sand and silt particles are much too large to form colloids). In a **colloid**, one or more materials in a finely divided state are suspended or dispersed throughout a second material. (Other examples of colloids are gelatin, fog, cytoplasm, blood, milk, and rubber.) Clay particles have an enormous amount of external surface area because they have a large surface-to-volume ratio and because there are so many of them in a given volume of soil. In some kinds of clays, there is additional internal surface area between the plates (see Figure 4.2B). Thus, the tremendous surface area that characterizes clays is due both to the fineness of the particles and to their platelike structure.

The external surface area of a gram of fine clay is at least 1000 times that of a gram of coarse sand; the total surface area in the clay present in the top 10 cm of soil in less than a half hectare of clayey soil, if spread out, would cover the continental United States.

Because sand particles have a low surface-to-volume ratio, sandy soils have large, open pores between the mineral particles. Water drains easily from these pores because there is nothing to hold the water against the pull of gravity, and air penetrates them readily. Clay-dominated soils have a much larger number of pores than do sand-dominated soils. The total amount of pore space—the total proportion of the soil occupied by air and water—is greater in clayey soils (50–60%) than in sandy soils (35–50%), not only because of the smaller size of clay particles, but because of their arrangement. Clay particles (as well as particles of organic matter) tend to cluster together, forming porous aggregates, resulting in much more pore space than in sands, in which the particles lie close together. Soils with native vegetation often have greater pore space than those that have been cultivated because tilling damages some of the soil structure.

The greater pore space is one factor contributing to the greater amount of water that can be held in clay-dominated soils, but there is another factor that is also very important. Unlike sand and silt particles, clay particles typically bear a strong negative electrochemical charge. Thus, they act as anions in the soil, attracting **cations** (positively charged ions, including the major plant nutrients), as well as water molecules, to their surfaces. The role of clay particles in **adsorbing** nutrient cations—attracting and holding them to their surface—is one of their most important effects for plants.

Although many kinds of cations are attracted to clay particles, certain ones are most prominent and most important for plant growth. In humid regions, hydrogen and calcium ions (H^+ and Ca^{2+}) are usually most abundant, followed by magnesium (Mg^{2+}), potassium (K^+), and sodium (Na^+) ions. In soils of arid regions, hydrogen ions move to last place in this list, and sodium ions become more important. These positively charged ions attract numerous molecules of water to their surfaces, adding to the overall capacity of the clay particles to retain moisture in the soil.

The cations adsorbed on clay particles are partially available to be taken up by plants. There is a continuous, dynamic interaction between the ions adsorbed on clay and other colloidal particles and those in the **soil solution**— the water in the soil and its associated dissolved minerals. Ions that are displaced from their positions on the surface of a particle enter the soil solution, from which they can be taken up by plants, **leached** (lost from the surface soil as water drains), or adsorbed on another particle. These reactions differ greatly among soils of different origins, textures, and chemical compositions, and among regions differing in temperature and especially in rainfall.

Soil pH

The **pH** of soil—the negative logarithm of the concentration of H^+ ions in the soil solution—varies widely among different soils. The pH scale ranges from 1 to 14, where 7.0 is neutral (the pH of pure water); acid soils have lower pH numbers and higher H^+ ion concentrations. Soils in the United States can range from less than 3.5 (for example, in the pine barrens of New Jersey) to as high as 10 (for example, in arid grasslands of the southwestern United States). For a soil to have a pH above 7 (neutral soil), it must be calcareous (containing $CaCO_3$), sodic (containing Na_2CO_3), or dolomitic (containing $CaCO_3 \cdot MgCO_3$). Most crops grow best in slightly acidic soils, but native vegetation can be adapted to anything from very acid to neutral to alkaline soils.

Soil pH has enormous effects on plant growth, and indeed, in determining what species can survive and grow in the soil. It acts on plants indirectly, however, through its strong effects on the availability of mineral nutrients and on the activity of some soil organisms (such as bacteria and fungi), changing the conditions for plant growth in a complex manner. Forest trees can grow over a range of soil pH, but are especially tolerant of acid soils. Conifers and some other tree species tend to increase the acidity of the soil in which they are growing, primarily through the properties of the litter they produce (shed needles, etc.). Grasslands tend to be found on relatively alkaline (high pH) soils, but this may be an indirect effect because low rainfall results in soils with high pH (as we will see below) and also favors grass-dominated vegetation. There are other characteristic associations between soil pH and plants. Species in the Ericaceae, for example, such as heaths and blueberries, tend to grow only in very acid soils, while other taxa, such as *Larrea tridentata* (creosote bush, Zygophyllaceae), are typically found in alkaline soils.

Soil pH can strongly affect the availability of cations to plants. Cations bind loosely to clay particles, which are generally negatively charged. Under acid conditions, the excess H^+ ions tend to bind more strongly than do the nutrient ions to the clay particles, displacing these nutrients into the soil solution. Thus mild acidity, which is characteristic of many soils, promotes nutrient availability. Under extreme acidity (whether natural or due to acid precipitation), however, the nutrient cations are so mobile that they are easily leached, and are carried off in **groundwater**—water found underground in aquifers, rock crevices, and so on.

What determines the pH of a soil? Two cations, hydrogen and aluminum, tend to increase soil acidity, while the other cations have the opposite effect, increasing soil alkalinity. Hydrogen ions are continually added to the soil by decaying organic matter, roots, and various soil organisms. They are actively exchanged between the surfaces of colloidal particles in the soil (such as clay and organic matter) and the soil solution, contributing directly to the acidity of the soil. Aluminum ions (Al^{3+}) indirectly cause hydrogen ions to be released from colloidal particles by reacting with water to form $Al(OH)^{2+}$ and $Al(OH)_2^+$ plus H^+. These hydrogen ions are then added to the H^+ ions in the soil solution. High rainfall levels favor the predominance of aluminum and hydrogen ions because these ions are held very strongly by colloidal particles, while the other cations are more readily leached and thus lost from the soil.

Most of the other cations, called exchangeable bases, contribute to making the soil more alkaline. The **cation exchange capacity (CEC)** of a soil is a measure of the total ability of the soil colloids to adsorb cations (in units of centimoles of positive charge per kilogram of dry soil [$cmol_c/kg$]), and the **percentage base saturation** is the proportion of the CEC that is occupied by exchangeable bases (Figure 4.3). In arid regions, the bases are not leached out of the soil by rainfall, so the percentage base saturation is very high (90–100%), H^+ concentration is low, and the soils tend to be alkaline. In areas with higher rainfall, the bases are leached more easily while H^+ and Al^{3+} ions are retained, so the percentage base saturation is much lower (50–70%), and the soils tend to be acid.

Plants and soil organisms are an important source of soil acidity. When roots or soil organisms respire, they generate CO_2. In wet soils, this CO_2 goes into solution immediately, creating a weak acid, carbonic acid. This is also the reason why rainwater is naturally a bit acid: CO_2 from the atmosphere dissolves into the raindrops. In tropical rainforests, where conditions favor massive amounts of respiration and there are often few clay particles to bind cations, this added CO_2 can help make the soils quite acid and promotes the mobility of cations. The consequence is rapid uptake of nutrients by plants in undisturbed forests, and rapid loss of those nutrients when forests are cleared.

Horizons and Profiles

Soils are not homogenous; they contain characteristic layers, or **horizons**, that differ from one soil type to another. If you look at roadcuts or excavations dug for construction, you can often see the horizons quite clearly. The sequence of horizons that characterizes a soil is called the **soil profile** (Figure 4.4). The horizons are grouped under four categories: O, A, B, and C. Subcategories of these four categories are numbered according to their particular characteristics (Figure 4.5).

The O horizons consist of organic material formed above the mineral soil, derived from decaying plant materials, microbial matter, and the remains and waste products of animals. The A horizons—the surface layer of mineral soil— represent the region of maximum leaching, or **eluviation**. The uppermost A horizon, the A_1, is often darker than the rest of the soil and may contain

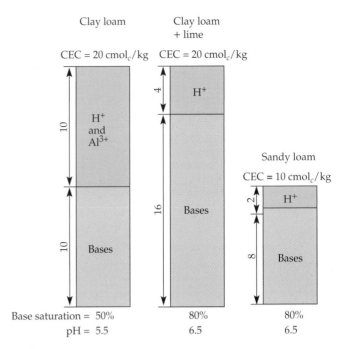

Figure 4.3
Percentage base saturation in three soils: a clay loam (left); the same clay loam with agricultural lime (calcium carbonate) added to raise the pH (center); and a sandy loam with low cation exchange capacity (right). The CEC for each soil is indicated. CEC is measured in the SI units, centimoles of positive charge per kilogram of dry soil ($cmol_c/kg$), indicating the numer of centimoles (1/100th of a mole) of positive charge adsorbed per unit mass of the soil; 1 mole of negative charge attracts 1 mole of positive charges whether the charges come from H^+, K^+, Ca^{2+} or Al^{3+}. See a current soils text for further explanation. (After Buckman and Brady 1969.)

highly decayed organic matter. The B horizons, deeper in the soil, represent the region of maximum **illuviation**, or deposition of minerals and colloidal particles leached from elsewhere. Clays, iron, and aluminum oxides often accumulate in the B_2 horizon. The C horizon is the undeveloped mineral material deep in the soil; it may or may not be the same as the material from which the soil develops. Below that may be bedrock, or just deep accumulations of mineral material deposited by wind, water, or glaciers.

No one soil has all of the horizons pictured in Figure 4.5. Some horizons may be much more distinct and better developed than others. Plowing and the activity of earthworms may obscure the distinction between the upper horizons. The soil profile may also not be fully developed, and some horizons may be absent or indistinct, if the soil is relatively young. Upper horizons may have been lost through erosion, leaving the deeper horizons exposed at the surface.

Erosion is a particular problem where forests have been clear-cut, on soils that have been long cultivated

Figure 4.4
Profile of a forest soil at the edge of the Adirondack Mountains in northern New York State (in the spodosol soil order). This soil is a stony loam, forested with birch, hemlock, and spruce, and is quite acidic throughout the profile. There is a thick layer of organic material at the top of the profile, and a number of distinct horizons, each with particular properties and a characteristic appearance. Roots are shown reaching down to the top of the B3 horizon at about 45 cm deep, although the deeper roots of large trees would certainly penetrate more deeply into this soil. (After Buckman and Brady 1969.)

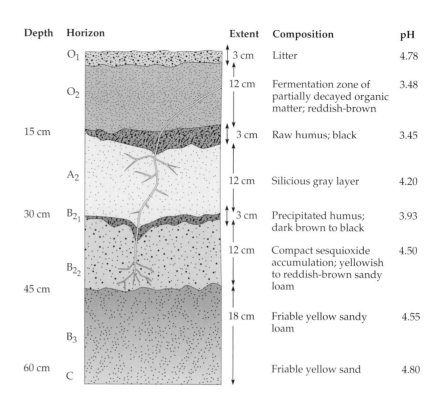

Depth	Horizon		Extent	Composition	pH
	O_1		3 cm	Litter	4.78
	O_2		12 cm	Fermentation zone of partially decayed organic matter; reddish-brown	3.48
15 cm			3 cm	Raw humus; black	3.45
	A_2		12 cm	Silicious gray layer	4.20
30 cm	B_{2_1}		3 cm	Precipitated humus; dark brown to black	3.93
	B_{2_2}		12 cm	Compact sesquioxide accumulation; yellowish to reddish-brown sandy loam	4.50
45 cm					
	B_3		18 cm	Friable yellow sandy loam	4.55
60 cm	C			Friable yellow sand	4.80

using poor farming practices, on slopes, and on lands that have been severely overgrazed. Soil erosion has been enormously important in human history with both economic and social consequences and has even precipitated the collapse of several Old and New World civilizations. It remains one of the most critical (and most widely ignored) environmental problems today (Figure 4.6). It occurs in many different parts of the world and affects many natural and semi-natural ecosystems. Rebuilding soils to replace losses to erosion can take thousands of years. The consequences of soil erosion for our ability to raise food crops and for the sustainability of natural ecosystems can be devastating.

Soils vary in depth, from thin layers of soil barely covering a rock substrate (for example, in many alpine areas) to very deep soils of close to 2 meters (for example, in some well-developed prairie soils). Soil depth has a great deal of influence on vegetation and plant growth: The greater the soil depth, the more favorable the soil for plant growth, all other things being equal. Deeper soils can hold more water and nutrients and can retain water for a longer period of time without rainfall, and they allow greater development of plant root systems.

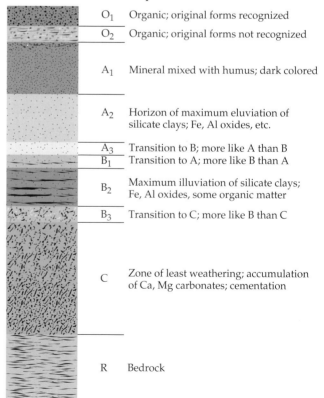

Horizon	Properties
O_1	Organic; original forms recognized
O_2	Organic; original forms not recognized
A_1	Mineral mixed with humus; dark colored
A_2	Horizon of maximum eluviation of silicate clays; Fe, Al oxides, etc.
A_3	Transition to B; more like A than B
B_1	Transition to A; more like B than A
B_2	Maximum illuviation of silicate clays; Fe, Al oxides, some organic matter
B_3	Transition to C; more like B than C
C	Zone of least weathering; accumulation of Ca, Mg carbonates; cementation
R	Bedrock

Figure 4.5
An abstract, general soil profile showing the major horizons that might be present in particular soils. No one soil is likely to have all of the horizons shown, and particular soils may have greater development of subhorizons than what is shown. Only the upper part of the C horizon is considered to be part of the soil proper. The depth of soils varies tremendously with the location and nature of the soil, but to gain some perspective, readers might picture this illustrated profile as being about a meter deep to the top of the bedrock. (After Buckman and Brady 1969.)

Figure 4.6
Soil erosion is a critical and current problem in many areas of the world. This photograph was taken near Fort Benton, Montana, U.S.A. in early May of 2002. The tractor has been partially covered by drifting topsoil on the edge of a farm field that in previous years would have been ready for spring planting. Unusually dry conditions combined with high, persistent winds have turned the cultivated topsoil of this agricultural region into a present-day "dust bowl." (Photograph by Liz Hahn © 2002.)

Origins and Classification

Soils are formed by the action of weathering on **parent material**, the upper layers of the heterogeneous mass that is left over after the action of weathering and other forces on rocks. The physical and chemical actions of temperature, rainfall, and other climatic factors further grind up, move, and leach the parent material until it begins to develop into true soil. Parent material is originally derived from rocks, which are classified as **igneous** (of volcanic origin), **sedimentary** (from the deposition and recementation of material derived from other rocks), or **metamorphic** (changed by the action of great pressures and temperatures on igneous or sedimentary rocks deep underground).

The parent material can be broken up by the mechanical action of temperature changes, including repeated freezing and thawing, by the direct action of water, wind, and ice in abrading and eroding rock fragments, and by the actions of plants, fungi, and animals. Chemical reactions occurring in the disintegrating rock material accelerate the physical breakdown of the fragments and begin the process of decomposition and chemical alteration (Figure 4.7). Large and small fragments of rock and rock-derived debris are moved by ice (particularly by glaciers), wind, and water. They are then redeposited in the form of glacial till and outwash, **loess** (pronounced "luss") and other aeolian (wind) deposits, alluvial (river and stream) deposits, and lake and marine sediments. These deposits may eventually form soil parent material.

The process of soil formation from parent material takes thousands of years. A young soil may be 10,000 years old; an old soil may be 100,000 years old or more. As a soil ages, the primary minerals that came from the parent material undergo chemical changes, and secondary minerals are formed. The structure of the soil develops, and then changes, as materials are leached, redeposited, and lost from the soil, as well as being physically and chemically altered. The mineral constituents of the soil shift, as minerals that are most stable at Earth's surface gradually come to dominate. Young soils are found where the parent material is still present, as in areas that were glaciated relatively recently (such as much of the northern part of North America) or in areas of recent alluvial deposits (such as the bottomlands of the world's great river systems). Old soils are most commonly found in the Tropics and Subtropics. Some of the world's oldest soils are found, for instance, in parts of Africa.

Five major factors are responsible for determining the kinds of soils that develop in an area: climate, the nature of the parent material, the age of the soil, topography (which acts to alter the effects of climate), and living organisms. These factors do not operate in isolation, but rather interact in complex ways with one another. Vegetation, with its associated soil organisms, has an especially strong influence on soil development, but the nature of the vegetation that is present is also dependent on both soil and climate.

Soils are classified according to a comprehensive taxonomic system developed in the United States, although some soil scientists and ecologists still rely on earlier classification systems. The broadest category in the modern system is the **soil order**. There are ten soil orders worldwide (Table 4.1).

(A)

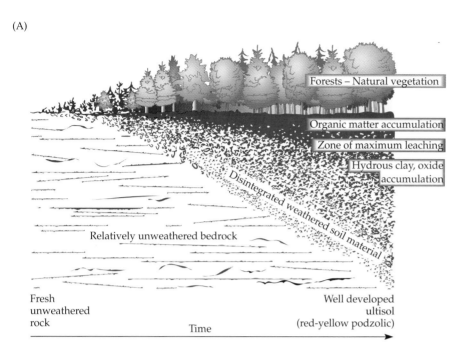

Forests – Natural vegetation

Organic matter accumulation

Zone of maximum leaching

Hydrous clay, oxide accumulation

Disintegrated weathered soil material

Relatively unweathered bedrock

Fresh unweathered rock

Time

Well developed ultisol (red-yellow podzolic)

(B)

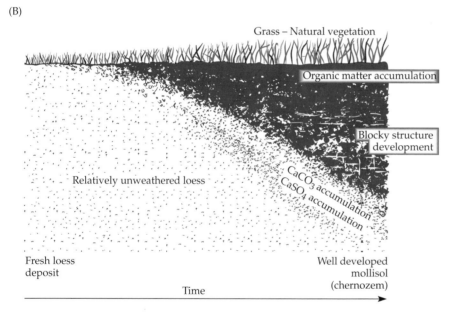

Grass – Natural vegetation

Organic matter accumulation

Blocky structure development

CaCO₃ accumulation
CaSO₄ accumulation

Relatively unweathered loess

Fresh loess deposit

Time

Well developed mollisol (chernozem)

Figure 4.7
Development of soil structure from two different parent materials: (A) bedrock and (B) loess deposited by wind. These soils are also developing under two different climatic and vegetation types. Over time, organic matter accumulates in the upper horizon, depending on the type of vegetation present. In the lower horizons, clay, iron oxides, or $CaCO_3$ are deposited and accumulate, and characteristic structures develop. The soil orders (ultisol and mollisol) are described in Table 4.1. (After Buckman and Brady 1969.)

The soil classification categories most commonly used by plant ecologists are **soil series** and **soil type**. These are used to classify local soils. A soil series is usually named after some local geographic area or feature, and generally consists of a dominant type with several associated other types. Soil types are based on topography, parent material, and the vegetation under which the soils were formed. Soil series occur at regional scales, and soil types characterize landscape-scale features. In the United States, soil series are mapped for almost every county; detailed maps are available for most locations showing the soil series and type, and often providing extensive information on the characteristics of

local soils. Such data can be highly valuable for many kinds of ecological studies, particularly at the community and ecosystem levels. The availability of information on local soils in other countries varies widely.

Organic Matter and the Role of Organisms

So far we have emphasized the role of physical processes in soil development and character. However, we began this chapter by emphasizing that soils are the unique product of living things acting on the physical environment (and in turn being affected by that environment). What are some of the ways in which organisms affect soils?

Table 4.1 Soil orders and some of their characteristics

Soil order	Development and characteristics	Found in
Mollisols	Develop under prairie vegetation; many of these soils are under cultivation, and they constitute some of the most productive agricultural soils	Great Plains of the United States and Canada; large areas in the heartland of Russia, Mongolia, and northern China; northern Argentina, Paraguay, and Uruguay
Spodosols	Generally develop in cold-temperate humid regions under northern forests; usually strongly acid and of low fertility	Northeastern and northern midwestern United States and adjacent areas in Canada; northern Europe, Siberia; also some soils in southern South America
Alfisols	Develop in humid regions under deciduous forests; weathered somewhat less than spodosols but more than inceptisols; contain silicate clay-dominated horizon; often cultivated and highly productive agriculturally	Large area from Baltic states to western Russia; southern half of Africa; eastern Brazil; England, France, central and parts of southern Europe; Michigan and Wisconsin to Pennsylvania and New York in the United States
Oxisols	Usually tropical soils, often supporting tropical rainforest vegetation. The oldest and most highly weathered soil order, with a deep subsurface horizon high in clays and other colloidal particles; intense leaching has removed most of the silica, leaving a high proportion of oxides of iron and aluminum, often with a red color	One of the most widespread soil orders globally, and among the most important in terms of the size of the human population supported; also supports the greatest biodiversity of many groups of plants and animals. Covers much of Africa and South America
Entisols	Recent soils with lack of significant profile development; highly variable in fertility	Sand hills of Nebraska; areas of northern and southern Africa; northern Quebec; parts of Siberia and Tibet; many areas that were covered by the most recent glaciation
Inceptisols	Young soils exhibiting limited weathering, but with greater profile development than entisols; many of these soils are under agricultural production	All continents—north Africa, eastern China, western Siberia, Spain, central France, central Germany, northern South America, northwestern United States
Aridisols	Develop in arid regions where soils are dry through most of the year and there is little leaching	Southwestern United States and northern Mexico; southern and central Australia; Bobi and Taklamakan Deserts in China; Sahara Desert and also southwestern Africa; Pakistan; other arid regions
Vertisols	High content of certain clays; sticky and plastic when wet and hard when dry, with extensive cracking; unstable for building and difficult to farm	Large regions of India, Sudan, and eastern Australia; small areas in southeastern Texas and eastern Mississippi in the United States
Ultisols	Develop under warm to tropical climates under forest or savanna vegetation; generally highly weathered but still retaining some minerals; somewhat acid, with clay horizons; these are fairly old soils	Southeastern United States; northeastern Australia; Hawaii; southeastern Asia; southern Brazil
Histisols	Wetland and bog soils; develop in a water-saturated environment; very high in organic matter content; sometimes also high in clay. Important soils not only in present-day wetlands, but in areas of ancient swamps and bogs that are currently forested, cultivated, or used for mining peat	Not widespread, but may be locally important, as in the Everglades in southern Florida and in many areas of northern Europe. The largest areas are in Canada southwest of Hudson and James Bay and in northwestern Canada into eastern Alaska

Note: These soil orders reflect the modern classification system currently used in the United States. Some of the major Great Soil Groups of the older classification system, with the soil orders with which they roughly correspond, are: podzols (spodosols and alfisols), chernozems and brunizems (mollisols), latosols and lateritic soils (oxisols), lithosols, solonchak soils, and desert soils (aridisols), azonal soils (entisols), and wetland/bog soils (histisols).

Organic matter is the decaying and decomposed material in soil that comes from living things. It includes substances secreted by plants, microorganisms, and animals, products of excretion by animals, parts shed by animals and plants, and dead organisms and parts of organisms. Microscopic organisms, while tiny individually, collectively contribute a great deal of organic matter to the soil. The activities of various different kinds of microorganisms affect soil properties in many additional ways, including the recycling of mineral nutrients used by plants (see Chapter 15).

A property of soils that is of vital importance to plant life (and to animals, fungi, and the other organisms living in soils) is soil structure. **Soil structure** describes the physical arrangement of soil particles into larger clusters, called aggregates (or peds). One of the most important aspects of soil structure is the **porosity** of the soil, which measures the total volume of pores and their size, shape, and arrangement between and within the aggregates (the aggregates are not solid, and may be "fluffy" or packed to varying degrees). Soil porosity determines how much, and how easily, water and air can be held by and move around in the soil, and thus is an important factor in the water-holding capacity of the soil.

Porosity and organic matter content also determine how easily roots or fungal hyphae can penetrate soil, and how easy it is for animals to burrow and tunnel through it. The actions of organisms (from animals to roots and fungal hyphae) in moving through the soil create pores and soil structure. Organic matter is amorphous (it does not have a well-defined structure, unlike clay particles, for example), and particles of organic matter typically have very large surface areas. Organic materials bind mineral particles together to stabilize the aggregates so that the soil porosity is maintained throughout the physical action of wetting, drying, freezing, thawing, and other processes.

Organic matter also supplies H^+ ions, determining soil pH. Furthermore, soil organic matter is critically important in plant nutrition, both by supplying essential nutrients and by providing physical particles that act, like clay particles, to attract and retain ions because they are negatively charged and have very large surface areas. Fresh and decayed organic matter in the soil also contains compounds, including various organic acids, that chemically alter essential plant mineral nutrients such as calcium (Ca), iron (Fe), manganese (Mn), copper (Cu), and zinc (Zn), making them more readily available to plants.

Water Movement within Soils

Imagine a meadow in the summertime that has been without rain for some time. A thunderstorm sails in, offering a solid drenching. (A similar set of events would take place in a suburban lawn, a forest, an agricultural field, and even, to some extent, an urban vacant lot.) What happens within the soil? Pores that had been filled with air fill with water, first in the upper layers of the soil profile, then deeper as more rain falls. Eventually almost all of the pores are filled with water. Some water runs off the soil surface, the amount depending on the vegetation, slope, and other factors (Figure 4.8). The soil is now **saturated**. Depending on the soil texture, soil structure, and soil depth, soils may hold very different amounts of water when they are saturated.

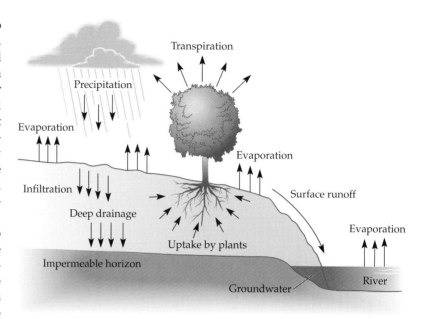

Figure 4.8
The fate of water falling on the ground from a rainstorm. (After Kramer 1983.)

Rainwater immediately begins to drain downward due to the pull of gravity, entering the groundwater and draining into the streams, rivers, and lakes of the watershed. After about a day, this rapid downward movement of water slows, and many of the large macropores in the soil refill with air (Figure 4.9A) The tiny micropores, however, remain filled with water at this point, and the soil is at **field capacity**. The water potential (see Chapter 3) of the soil is now at –0.01 to –0.05 MPa (Figure 4.9B).

Water is held in the soil largely by attraction to the surfaces of soil particles, particularly clays and organic matter. It moves through small pores by **capillary action**, a process that occurs within narrow tubes or in a surface film of water (such as within soil pores, the xylem, and the substomatal chambers of leaves). Water is pulled upward (or horizontally) by the attraction of the water molecules to the charged particle surfaces and to one another. As the plants in the hypothetical meadow of our example transpire, a water potential gradient (see Figures 3.2 and 3.3) is established, and soil water moves toward the roots and is absorbed by them.

As the soil continues to become drier, the small pores begin to empty of water, filling with air. As the small pores empty, the once continuous film of water is broken in many places, and water movement is greatly slowed. Water can continue to move as water vapor through these emptied pores, but only small amounts of water can be transported in this way. Each day, as the plants transpire and the soil progressively dries, the water potential of the plants declines (becomes more negative), and each night, as stomata close, the water

(A)

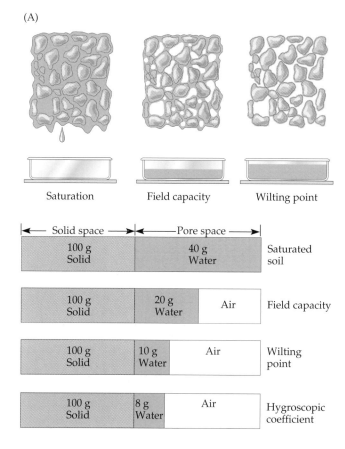

| Saturation | Field capacity | Wilting point |

← Solid space →	← Pore space →	
100 g Solid	40 g Water	Saturated soil
100 g Solid	20 g Water	Air	Field capacity
100 g Solid	10 g Water	Air	Wilting point
100 g Solid	8 g Water	Air	Hygroscopic coefficient

Figure 4.9
(A) A silt loam soil at saturation, at field capacity, and at the wilting point, showing the soil particles and pores filled with water and/or air (above). At saturation, the soil is holding all the water it can accommodate, and water will drain from the soil due to gravity. At field capacity, a considerable amount of water has been removed from the soil (shown in the beaker below the soil), and at the wilting point, still more water has been removed. The bar graph below shows the relative amounts of solid particles, water, and air in the soil for each of these states. A further reduction in soil moisture is reached at the hygroscopic coefficient, when water is held mostly by the soil colloids, and is completely unavailable to plants. (Buckman and Brady 1969.) (B) The relative amounts of available and unavailable water in soils of various textures, expressed both as percentage of soil volume and centimeters of water per centimeters of soil. (After Kramer 1983.)

potential of the plants rises somewhat as the plants equilibrate with the soil (Figure 4.10).

When the soil reaches a water potential of about –1.5 MPa, most mesic plants no longer can extract water from the soil, and exist in a permanently wilted condition, about to die. This state is called the **permanent wilting point** of the soil, and the soil moisture content at this point is called the **wilting coefficient** of the soil. However, plants adapted to dry environments can continue to remove water from soils that are much drier. The

(B)

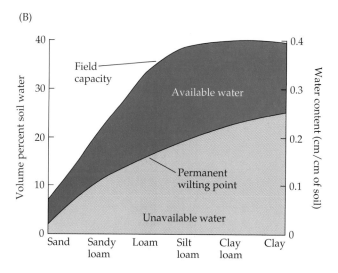

amount (by weight or volume) of water held in the soil at any point in this drying curve will vary greatly with the soil texture and also with the proportion of organic matter in the soil (Figure 4.11).

Water moves upward in the soil as well as downward. As water evaporates from the soil surface, capillary action pulls water from deeper in the soil upward against gravity. The roots of transpiring plants pull water upward from deeper in the soil as well. Sometimes large amounts of water can be transferred from deep horizons in the soil to dry surface layers by deeply rooted plants. During the day, roots generally have more negative water potential than soils. At night, the dry upper layers of the soil may have more negative water potential than roots, and may be capable of removing water from the roots. Deeply rooted plants may take up water from the moist deeper horizons, from which it travels up the root system and moves out of the upper part of the root

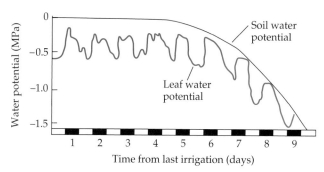

Figure 4.10
Daily changes in soil and leaf water potential for a cultivated plant growing in a clay loam soil over the course of nine days and nights after receiving irrigation. Dark portions of the *x*-axis indicate nighttime conditions; open portions indicate daytime. (After Etherington 1982.)

system into the dry soil in the upper horizons. There, the water can become available to other, more shallowly rooted plants. This phenomenon, called **hydraulic lift**, may in some cases allow the survival of plants with shallow roots during periods drought or in arid environments, because deeply-rooted neighbors may increase the availability of soil moisture closer to the surface.

Plant Mineral Nutrition

What is a plant made of? A large tree is clearly many orders of magnitude larger than the seed from which it grew. Where did all of that "stuff" in the tree come from? A man or woman is much larger than a child largely as a result of his or her cumulative food intake (of course,

Table 4.2 Elements essential for plant growth and survival

Element	Symbol	Principal form in which absorbed	Concentration in plant		Important functions
			Average mass (g/kg plant dry weight)	Percent of total mass (range)	
Found in all organic compounds					
Carbon	C	CO_2	450.0	~44%	Major component of all organic compounds, including cellulose in cell walls, which makes up much of the plant's dry weight; other carbohydrates (sugars, starches) and lipids store and transport the energy captured in photosynthesis
Oxygen	O	H_2O or O_2	450.0	~44%	Component of all major organic compounds; cellular respiration
Hydrogen	H	H_2O	60.0	~6%	Component of all major organic compounds; essential to all major biochemical reactions and to acid/base balance
Macronutrients					
Nitrogen	N	NO_3^- or NH_4^-	15.0	1–4%	Essential component of nucleotides, nucleic acids, amino acids, proteins (including structural proteins and enzymes), and chlorophylls
Potassium	K	K^+	10.0	0.5–6%	Involved in osmosis, ion balance, pH regulation, and opening and closing of stomata; activator of many enzymes; protein synthesis
Calcium	Ca	Ca^{2+}	5.0	0.2–3.5%	Strengthens cell walls and some plant tissues; involved in cell division and cell elongation, membrane permeability, cation-anion balance; structural component of many molecules; second messenger in signal conduction between environment and plant growth and developmental responses
Magnesium	Mg	Mg^{2+}	2.0	0.1–0.8%	Essential component of the chlorophyll molecule; activator of many enzymes; involved in cation-anion balance and regulation of cytoplasm pH
Phosphorus	P	$H_2PO_4^-$ or HPO_4^{2-}	2.0	0.1–0.8%	Essential structural component of nucleic acids, proteins, ATP, and $NADP^+$; critical for energy transfer and storage
Sulfur	S	SO_4^{2-}	1.0	0.05–1%	Component of some amino acids and proteins, coenzymes (including coenzyme A), and secondary metabolic products (including defensive compounds)

much of the food one eats is used to provide energy, which is used up along the way). But plants clearly do not eat. In order to understand where all of the material that makes up a plant comes from, we first have to see what a plant is made of.

Most of the material in a plant is made up of compounds of carbon, hydrogen, and oxygen (Table 4.2). The carbon is obtained by fixing atmospheric CO_2 during photosynthesis. The oxygen also comes from the atmosphere, while the hydrogen comes from water taken up by the roots. With one noteworthy exception—the nitrogen fixed by symbionts of some plants—all the rest of the elements in a plant are taken up from the soil solution by the plant's roots.

Element	Symbol	Principal form in which absorbed	Concentration in plant		Important functions
			Average mass (g/kg plant dry weight)	Parts per million (range)	
Micronutrients					
Chlorine	Cl	Cl–	0.1	100–10,000 ppm	Used in stomatal regulation, proton pumps, and osmoregulation; critical for splitting water molecules in photosystem
Iron	Fe	Fe^{3+}, Fe^{2+}	0.1	25–300 ppm	Component of heme proteins (such as cytochromes, essential molecules in light reactions, cellular respiration, and nitrate reduction; leghemoglobin, used in N fixation) and iron-sulfur proteins; necessary for chlorophyll synthesis and in the electron transport chain in light reactions
Manganese	Mn	Mn^{2+}	0.05	15–800 ppm	Present in several enzymes; needed for activation of many enzymes
Boron	B	$B(OH)_3$, $B(OH)_4^-$	0.02	5–75 ppm	Poorly understood; cell wall synthesis, nucleic acid synthesis, and plasma membrane integrity
Zinc	Zn	Zn^{2+}	0.02	15–100 ppm	Present in several enzymes; needed for activation of many enzymes; protein synthesis and carbohydrate metabolism; structural component of ribosomes; important but not well understood role in plant hormone metabolism
Copper	Cu	Cu^{2+}	0.006	4–30 ppm	Present in some proteins and in plastocyanin (necessary for light reactions); enzyme activation; pollen formation and ovule fertilization; lignification of secondary cell walls (in wood formation)
Nickel	Ni	Ni^{2+}	~0.0001	~0.1 ppm	Necessary for function of some enzymes; nitrogen metabolism
Molybdenum	Mo	MoO_4^{2-}	0.0001	0.1–5 ppm	Enzyme cofactor for N fixation and some other reactions; pollen formation and seed dormancy
Essential to some plants					
Sodium	Na	Na^-	Trace		Osmotic and ionic balance, especially in some desert and salt-marsh species
Cobalt	Co	Co^{2+}	Trace		Required by nitrogen-fixing microorganisms
Silicon	Si	SiO_3^-	Variable		Not well understood; appears to have a role in disease resistance; important structural component of cells in grasses, *Equisetum* spp. (horsetails), and other plants; may reduce leaf water loss

Source: Compiled in part from data in Barbour et al. 1987, Marschner 1995, and Raven et al. 1999.

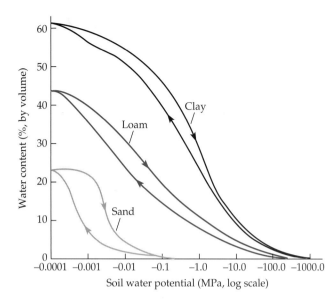

Figure 4.11
Illustration of general relationships between water content (by volume) and soil water potential (MPa) for clay, loam, and sand. These curves differ depending on whether the soil was measured as water content was increasing or decreasing (the direction of change in moisture shown by arrows). (After Rendig and Taylor 1989.)

How do plants use these elements? There are broadly two roles that an element can play in an organism: it can be part of the material that makes up the organism's structure, or it can be essential for the operation of the organism's metabolism. All the elements except C, O, and H are usually called the **essential mineral nutrients**—those chemical elements that plants need to live, grow, and reproduce. Table 4.2 lists some of the main functions of each of these elements. The essential nutrients needed in larger quantities are called **macronutrients**; those needed in very small quantities are known as **micronutrients**. Another group of minerals is the **beneficial mineral elements**; these minerals either are essential only for certain plant species, or they are not essential but do stimulate growth. Beneficial mineral elements are sodium, silicon, and cobalt.

While plants cannot grow and function without the essential mineral nutrients, more is not necessarily better. Most of these minerals are necessary in small to moderate amounts, but toxic in large quantities. Some species have evolved tolerance of high concentrations of minerals that are toxic to most other species (see Figure 6.1).

Nitrogen in Plants and Soils

The two most important essential mineral nutrients for plants are nitrogen and phosphorus. Plants need large amounts of nitrogen to photosynthesize (see Chapter 2), grow, and reproduce, and the amount of nitrogen avail-

able in the soil frequently limits the ability of the plant to carry out these functions. Plant form and development are also influenced by nitrogen availability. Increased nitrogen supply may delay the senescence of leaves or of the entire plant. It may increase the proportion of biomass allocated to shoots relative to roots and the proportion allocated to vegetative growth versus sexual reproduction. Nitrogen availability may also affect the plant's biochemical composition in a complex manner, and may alter the concentrations and types of defensive compounds present. The ratio of carbon to nitrogen in plant tissues is affected by nitrogen supply rates, and in turn affects many other aspects of plant life, including susceptibility to herbivores and litter decomposition rates.

Nitrogen exists in the soil in inorganic and organic forms. Plants take up both inorganic nitrate (NO_3^-) and ammonium (NH_4^+) from the soil, but nitrate is generally more abundant and more readily taken up. Some plants are able to take up ammonium, particularly when they are young, and plants adapted to acid soils or soils with limited oxygen potential (such as waterlogged soils) preferentially take up ammonium. The costs and benefits of nitrate versus ammonium uptake are complex, and depend on their concentrations in the soil as well as other factors. Nitrate must be reduced to ammonia before it can be incorporated into organic compounds in the plant, but as nitrate it is highly mobile, and can be transported readily in dissolved form in the xylem and stored in cell vacuoles. Ammonium is largely incorporated into organic compounds in the roots directly after it is taken up. Ammonium can be toxic to plants, particularly when converted to ammonia (NH_3).

Biological Nitrogen Fixation

Nitrogen is needed in large quantities by plants, and its availability in the soil is almost always low. Ironically, plants are surrounded by an ocean of nitrogen in the atmosphere in the form of elemental N_2, which they cannot tap directly. You might wonder why plants cannot use this form of nitrogen. It turns out that some plants are able to do so—not directly, but by forming symbioses (Box 4A) with nitrogen-fixing prokaryotes (bacteria).

Only prokaryotes can directly use N_2 and "fix" it in biologically usable forms. These nitrogen-fixing organisms are remarkably diverse, and encompass nineteen families of bacteria, including eight families of cyanobacteria (the blue-green bacteria, which are photosynthetic). Some nitrogen fixers are free-living, some are symbiotic with terrestrial plants, and some are associated with the root surfaces and intercellular spaces of plants. The biological reduction of N_2 to NH_3 is catalyzed by an enzyme complex, nitrogenase, which is found in all N_2-fixing prokaryotic microorganisms, and nowhere else in the living world. The reaction requires

Box 4A

Symbioses and Mutualisms

The word "symbiosis" is rooted in two Greek words, *bios* (life) and *sym* (together). **Symbioses** are associations between members of two different species that live in intimate contact with each other. There are a variety of symbiotic associations among organisms.

A symbiotic association may benefit members of both species involved, or it may benefit only one member, with no effect or a negative effect on the other. If both members benefit, the symbiosis is a **mutualism**. If one member benefits

while the other is unaffected, the relationship is called a **commensalism**. And if one member benefits while the other is harmed (sometimes even destroyed), it is **parasitism**. Any of these three types of symbioses may be **obligate**—necessary for the survival of one or both members—or **facultative**—existing only under some conditions. The nature of a particular symbiosis may even change over time, or as environmental conditions change.

Two important symbioses between plants and other organisms are symbiotic nitrogen fixation by certain species of bacteria living in or on plant roots, and the formation of mycorrhizae by plants and certain species of fungi. Both of these symbioses are often, but not always, mutualistic. Other common symbioses include interactions between some plants and animals that act as pollinators or seed dispersers.

---

a great deal of energy, and the source of that energy is the key to understanding the different ecological roles played by different nitrogen fixers.

The free-living nitrogen fixers are either photosynthetic cyanobacteria, such as *Anabaena* (Figure 4.12), or are heterotrophic and obtain their energy from the decomposition of plant residues in the soil. The volume of nitrogen fixation by the heterotrophs is generally low because they are limited by the amount of carbon and energy they can extract from dead material. In contrast, the photosynthetic free-living nitrogen fixers can provide substantial amounts of nitrogen to the systems in which they live because they use the far more abundant energy of sunlight. For example, free-living (as well

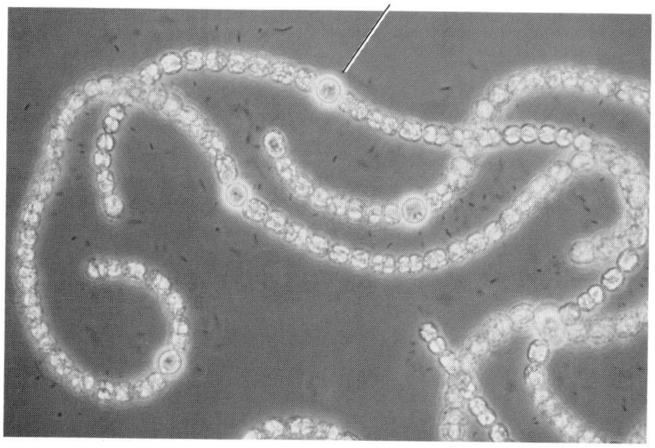

as symbiotic) nitrogen-fixing cyanobacteria are important sources of nitrogen in flooded rice paddies (see Figure 15.14) and in some surface soils.

Symbiotic nitrogen fixation is more important in most ecosystems than fixation by free-living organisms. There are a great variety of symbiotic associations between nitrogen-fixing prokaryotes and terrestrial plants. There is generally a strong specificity in these symbioses, in which particular bacterial species preferentially infect particular plant species. Three of the most common general types of nitrogen-fixing symbioses between prokaryotes and plants are nodulated legumes, nodulated non-legume symbioses, and looser associations between plants and cyanobacteria. The first is the most familiar, particularly because of its agricultural importance, but it is not the only one with ecological importance.

Symbioses with legumes (plants in the family Fabaceae, such as peas, beans, clover, and acacia trees) are formed by bacteria belonging to either the fast-growing species in the genus *Rhizobium* (Figure 4.13A) or slow-growing *Bradyrhizobium* species. *Rhizobium* and *Bradyrhizobium* bacteria live in nodules on the roots of legumes (Figure 4.13B), providing the plant with a source of NH_3 and receiving carbohydrates (providing carbon and energy) from the host plant. The nodules provide the anaerobic conditions that the bacteria require to fix nitrogen.

The Fabaceae is one of the largest and most ecologically diverse plant families. Many important crop species are members of this family, and its crops are major protein sources for civilizations worldwide (e.g., soybeans in eastern Asia, lentils in southern Asia, pigeon peas and peanuts in Africa, pintos and other beans in the Americas, and fava beans and garbanzos in the Middle

Heterocyst

Figure 4.12
Free-living nitrogen-fixing cyanobacteria such as *Anabaena azollae* fix nitrogen in specialized cells called heterocysts. (Photograph © J. R. Waalund/Biological Photo Service.)

(A)

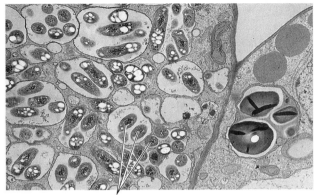

Rhizobium

(B)

Figure 4.13
Symbiotic nitrogen-fixing bacteria of the genus *Rhizobium* cannot fix nitrogen until they form an association with the plant's root cells. (A) The cell on the right has not been "infected" with *Rhizobium*. Bacteria have entered the cell on the left, where they take a nitrogen-fixing form called bacteroids. Root cells with bacterioids form nodules. (Photograph © E. Newcomb and S.Tandon/Biological Photo Service.) (B) Root nodules on *Glycine max* (soybean, Fabaceae). The anaerobic conditions inside the nodules provide the necessary environment for *Rhizobium* to fix nitrogen. (Photograph © K. Wagner/Visuals Unlimited.)

fixed by this symbiotic association can be very important in these successional ecosystems (see Chapter 13).

In the third type of plant-prokaryote association, that of nitrogen-fixing cyanobacteria with plant species, there may be high specificity between the host plant and the bacterial species, but the bacteria remain external to the root cells. In these loose associations, the bacteria obtain energy from root exudates from the plant, but provide little nitrogen to the plant until they die. These associations are not always mutualistic, and sometimes one of the partners in the relationship obtains all or most of the benefits of the association without providing much to the other partner.

External energy input is required for N fixation. The symbiotic bacteria obtain this energy from photosynthate supplied by the plant host, which must be diverted from other plant functions such as growth or reproduction. Factors that increase photosynthetic rate tend to increase nitrogen fixation rates, and factors (such as

East; Figure 4.14). Many tropical trees and shrubs (such as *Acacia* spp.) belong to the Fabaceae as well; these plants can play a critical role in ecosystem function by providing nitrogen to soils and by providing high-protein food for wildlife (particularly seeds and fruits). There are also members of the Fabaceae that do not form symbioses with nitrogen-fixers.

The second major type of nitrogen-fixing symbiosis is that of non-legumes with actinomycete bacteria in the genus *Frankia*. As in the legumes, the bacteria live in nodules on the plants' roots and obtain carbon and energy from the plants while providing them with NH_3. There are about 200 plant species in the Tropics and Subtropics that form such associations, but they also occur in some temperate systems. *Alnus* spp. (alder, Betulaceae), for example, are symbiotic with *Frankia*, and are common early colonizing trees in many riparian communities; the nitrogen that is

Figure 4.14
A wide variety of dried beans are for sale in this market stall in Ecuador. Legumes such as beans, members of the family Fabaceae, are crucial sources of protein for people throughout the world. (Photograph from Corbis archive.)

light and water limitation) that reduce photosynthetic rates tend to decrease nitrogen fixation.

The presence of large amounts of mineral nitrogen in the soil inhibits nitrogen fixation by suppressing nitrogenase activity and severely reducing the number of nodules formed. The mechanisms by which this suppression occurs are complex, and beyond the scope of this book. As nitrogen becomes more readily available in the soil, however, the uptake of soil nitrogen becomes energetically cheaper than providing carbohydrates to bacterial symbionts. From the perspective of plant fitness, the switch from using symbiotically fixed nitrogen to direct uptake from the soil should occur at this point.

Phosphorus

Like nitrogen, phosphorus is needed by plants in larger quantities than are generally available in soils. Next to nitrogen, phosphorus is likely to be the most limiting mineral nutrient for plant growth. Phosphorus is also needed to support symbiotic nitrogen fixation.

Soils vary greatly in the amount of phosphorus they contain. In the United States, total soil phosphorus tends to be greatest in the Pacific Northwest, extremely low in the Southeast, and low to moderate in the Northeast, Midwest, and Southwest. Soils derived from high-phosphate limestones are generally highest in phosphorus, and those derived from sandstones and acid igneous rocks lowest. Arid regions with calcareous soils usually are high in phosphorus because substantial leaching has not occurred. Very old soils, particularly in the Tropics and Subtropics, such as those in Australia and parts of Africa, tend to be exceptionally low in available phosphorus (organic forms readily accessible to plants) due to steady, slow losses by leaching over an extended period of time (as well as to the slow transformation to less-available organic forms).

However, the total amount of phosphorus in the soil is only a small part of phosphorus limitation. The availability of phosphorus to plants is not necessarily related to the total amount present in the soil. Unlike nitrogen, most phosphorus present in the soil is bound in ways that make it unavailable to plants. Soil phosphorus chemistry is exceedingly complex and depends on pH and many other factors. Phosphorus in the soil exists in various organic and inorganic forms, and both organic and inorganic phosphates can be tightly bound in ways that make them unavailable to plants (see Figure 15.19). Phosphorus can be tightly bound to clay particles, or complexed with Ca, Fe, Al, and silicates, or bound up in organic matter. Because much of the phosphorus in the soil is tightly bound, it leaches much less readily and more slowly than does nitrogen. Microbial activity both removes available forms of phosphorus from the soil solution and transforms them to forms unavailable to plants (that is, it immobilizes them), as well as trans-forming them back to available inorganic forms, and releases them back into the soil solution, where they can be taken up by plants once again (see Figure 15.19).

Phosphorus in soils is derived from apatite minerals in the parent material. In natural systems, most of the phosphorus in plants is recycled by microbial decomposition. While some systems have considerable reserves of phosphorus in the soil, in others, such as many grasslands and some tropical rainforests, essentially all of the available phosphorus exists in living plant tissue, litter, and decomposing organic matter, from which it is rapidly recycled or lost. Most of the phosphorus taken up by plants comes from the **rhizosphere**, the microbe-rich environment within about a millimeter of and surrounding the root.

It is sometimes possible for plants to directly influence the availability of nutrients in the rhizosphere. Some plants, for example, can create their own locally acidic environment by secreting organic acids that promote the uptake of cations. Plants of the family Proteaceae—a prominent group in the old Gondwanan continents (Australia, South Africa, and South America)—which often grow in very phosphorus-poor soils, may form specialized **proteoid roots**. These dense clusters of short, fine lateral rootlets produce organic acids or other chelating agents, which act to release phosphorus from insoluble complexes with Ca, Fe, and Al. This ability allows these plants to acquire phosphorus (and sometimes other mineral nutrients) in soils where it is unavailable to other plants. Some unrelated plants—such as a *Lupinus* sp. (lupine, Fabaceae)—have also been observed to form proteoid roots.

Mycorrhizae

While proteoid roots are an unusual and highly specialized adaptation to plant growth in phosphorus-limited conditions, plants have another approach to phosphorus limitation and other low-nutrient conditions that is extremely common, widespread, and of overwhelming ecological importance. Symbioses between various fungi and the roots of terrestrial plants, called **mycorrhizae** (singular mycorrhiza; the word refers to the plant-fungal symbiosis, not strictly to the fungus), confer a number of benefits on the plants. These interactions depend on a fungal species successfully infecting and living in or on the roots of a plant. Mycorrhizae are often (but not always) mutualistic symbioses (see Box 4.1), in which the fungus aids the plant in obtaining certain materials from the soil, and is in turn provided with carbon and energy in the form of carbohydrates by the plant.

Mycorrhizae are not only common, they are ubiquitous—they are found on almost all terrestrial plants and in almost all ecosystems. The ability of plant species to survive in many environments depends on mycorrhizae. They are critical components of ecosystem function in many environments as well, and play a major role in nutrient uptake by plants and in nutrient recycling in soils. About 80% of angiosperms and all gymnosperms are involved in some form of mycorrhizal symbiosis. Mycorrhizae occur in all plant divisions, including some mosses and ferns. There are also some plants that never form symbioses with mycorrhizal fungi; mycorrhizae are almost never found in plants in the families Brassicaceae and Amaranthaceae, and are rarely found in the Proteaceae (or in other species with proteoid roots).

The overwhelming importance of mycorrhizae in natural ecosystems and throughout the plant kingdom is often overlooked. Mycorrhizae are sometimes considered only within the general context of symbioses formed between plants and soil organisms. These plant-fungal symbioses differ in important ways from the symbioses between plants and nitrogen-fixing bacteria. The latter, of course, involve bacteria (prokaryotes), while the former involve fungi (complex, multicellular eukaryotes). Mycorrhizae are found on most plants in most places, while nitrogen-fixing symbioses are found only on particular plant species, and only in particular places. Nitrogen fixation is the process of transforming elemental N_2 from the atmosphere into ammonia, and thus involves a single nutrient, nitrogen. Mycorrhizal symbioses, on the other hand, function in a wide variety of ways. They greatly facilitate the availability and absorption of multiple nutrients and soil water as well as protecting plants from pathogens. It is possible for a species to have both symbiotic nitrogen-fixing bacteria and mycorrhizae. For legumes growing in phosphorus-deficient soils, mycorrhizae can increase nodulation and nitrogen fixation by the bacteria as well as host plant growth.

Major Groups of Mycorrhizae

There are two major, common groups of mycorrhizae: **ectomycorrhizae (ECM)** and **endomycorrhizae**. The endomycorrhizae are subdivided into three groups. The most important of the endomycorrhizae are the **vesicular-arbuscular mycorrhizae** (**VAM** or **AM**; not all arbuscular mycorrhizae make vesicles). **Ericoid** (specialized for symbioses with plant species such as heathers, blueberries, and rhododendrons) and **orchidaceous mycorrhizae** (specialized for symbioses with orchids) are sometimes classified on their own and sometimes as members of the endomycorrhizae. The **ectendomycorrhizae** have characteristics of both the ECM and ericoid groups. The kinds of plants with which each of these groups is associated, and some of the fungal taxa involved, are listed in Table 4.3. Plants can be hosts to one or many fungal species; some plant species have both VAM and ECM.

The most common type of mycorrhizae is the VAM, which occur in both herbaceous and woody species. VAM are most abundant in systems where phosphorus is limited and in warmer or drier climates. They predominate in tropical ecosystems and are especially important to many crops. The body of the VAM fungus grows branched inside the cortex cells of the roots, forming a structure called an arbuscule, and in the intercellular spaces between root cells, with hyphae extending out several millimeters (or more) into the soil (Figure 4.15).

Ectomycorrhizae are most commonly found on woody plant species, including those in the Pinaceae, Betulaceae, Fagaceae, and Salicaceae, and in some tropical and subtropical trees. ECM are particularly common in northern coniferous and temperate deciduous forests and in soils where nitrogen is particularly limiting. In fact, coniferous trees cannot grow in nature without these fungal symbionts. ECM are distinguished by two structures, one inside the root and one outside it: the Hartig net, a complex of mycelia that grows in between the root cortical cells, enmeshing them, and the mantle, a dense network of hyphae that partially or fully ensheaths the

Table 4.3 Major groups of mycorrhizal associations

Mycorrhizal group	Plant taxa involved	Fungal taxa involved
Ectomycorrhizae (ECM)	Dipterocarpaceae (98%), Pinaceae (95%), Fagaceae (94%), Myrtaceae (90), Salicaceae (83%), Betulaceae (70%), Leguminosae (16%), and some others	Basidiomycetes, some ascomycetes, some zygomycetes
Endomycorrhizae	Most plant families	
VAM	By far most common mycorrhizae in gymnosperms except Pinaceae, and in angiosperms, except for non-mycorrhizal families—Brassicaceae, Portulaceae, Caryophyllaceae, Proteaceae, etc.)	Glomales (common ones are zygomycetes belonging to the genera *Glomus*, *Acaulospora*, *Gigaspora*, and *Sclerocytis*)
Ericoid	Many species in the Ericales	Basidiomycetes and some ascomycetes
Orchidaceous	Members of the Orchidaceae (orchids)	Basidiomycetes and some ascomycetes
Ectendomycorrhizae	Some species in the Ericales; some gymnosperms	Basidiomycetes and some ascomycetes

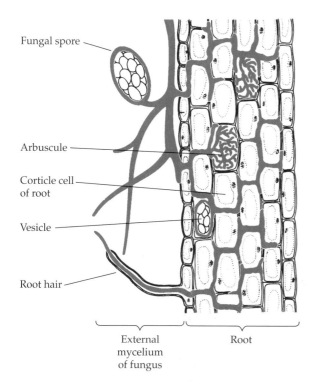

Fungal spore

Arbuscule

Corticle cell
of root

Vesicle

Root hair

External   Root
mycelium
of fungus

Figure 4.15
Association of a vesicular-arbuscular (VA) endomycorrhizal fungus with a plant root. The fungal hyphae, shown in green, grow in the intercellular spaces and may penetrate cells of the root's cortex. During this process, the fungus can form structures that are either oval (vesicles) or branched (arbuscules). Fungal mycelium growing on the outside of the root can form reproductive spores. (After Mauseth 1988.)

The Role of Mycorrhizae in Plant Phosphorus Nutrition

There are several mechanisms by which both VAM and ECM improve phosphorus uptake. VAM in particular are highly efficient at scavenging available phosphorus from the soil, due in part to the phenomenally large surface area of the many small hyphae that extend into the soil. In addition, a number of active metabolic processes enhance nutrient uptake. For example, both VAM and ECM produce enzymes (acid phosphatases), which are secreted from the external hyphae into the surrounding soil. These enzymes release phosphorus that is tightly bound to organic matter and otherwise unavailable to the plant. The phosphorus, once it is made available, is taken up by the fungus, and what is not used by the fungus itself is transferred to the plant.

Plants with mycorrhizae can, on average, take up several times the amount of phosphorus as nonmycorrhizal plants, and as a consequence, can grow much more successfully on soils in which phosphorus is low. However, the phosphorus uptake role of mycorrhizae is highly variable. Some species of mycorrhizal fungi are highly effective at taking up phosphorus and transferring it to their hosts, while others provide very little

outside of the root (Figure 4.16A). Plant species with ECM grow unusual-looking, short, stubby roots (Figure 4.16B); this growth form may be controlled hormonally by the fungus.

The different branched and finely divided structures of VAM and ECM, including hyphal strands growing into the soil, function to increase nutrient uptake from the soil. Nutrients are taken up from the soil by the fungal mycelium and transferred from the fungal cells to the root cells of the host plant.

(A)

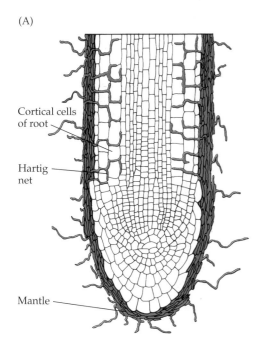

Cortical cells
of root

Hartig
net

Mantle

(B)

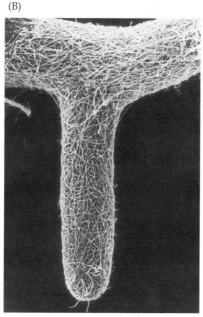

Figure 4.16
(A) The hyphae of an ectomycorrhizal fungus, shown here in green, form an interior Hartig net between the cortical cells of a plant root. In addition, a mantle of hyphae covers the exterior of the root. (After Rovira et al. 1983.) (B) A eucalyptus root mantled in fungal hyphae. Such a covering benefits the plant by greatly increasing the amount of area available for the absorption of water and mineral nutrients. (Photograph © R. L. Peterson/ Biological Photo Service.)

phosphorus to their hosts. Environmental factors are also important in determining whether or not phosphorus uptake is enhanced by the fungus. At extremely low levels of phosphorus, the fungi are apparently not able to function well, and the degree to which roots become infected with the mycorrhizal symbionts is low. At slightly higher phosphorus levels, the fungi colonize roots more successfully and become highly effective at enhancing phosphorus uptake for the plant. At high levels of available soil phosphorus, fungal growth inside the root is apparently actively suppressed by the plant, and the roots take up phosphorus directly on their own.

Other Functions of Mycorrhizae

High levels of nitrogen in the soil may also act to suppress infection by mycorrhizal fungi. Other factors, such as aridity, toxic minerals, and inadequate or excess concentrations of other minerals (such as, for example, too little boron), may also inhibit the formation and function of mycorrhizae because these conditions either harm the fungal partner or function in some more complex way to inhibit the symbiosis.

In addition to enhancing phosphorus availability, VAM increase plant access to some other minerals, such as copper and zinc. ECM tend to have a much more diverse role. If the fungal mantle of ECM totally encases the roots, all water and mineral uptake must take place through the fungus. Because the surface area of the mass of hyphae is far greater than that of the root system, the uptake of water, nitrogen, and other dissolved minerals is greatly enhanced, Some ECM produce and excrete organic acids, increasing the solubility and uptake of various minerals, somewhat like proteoid roots. Some ECM and ericoid mycorrhizae secrete enzymes that degrade proteins, making the nitrogen locked in these complex organic molecules accessible to their hosts. ECM fungi can enable their hosts to tolerate toxic heavy metals in the soil, protecting the roots by sequestering and binding heavy metals and other toxic ions in their cell walls.

Plants in the heath family (Ericaceae) and in some other families in the Ericales generally live in acidic, extremely nutrient-poor habitats. The formation of ericoid mycorrhizae is apparently necessary for them to survive, grow, and reproduce in their natural habitats, and they clearly depend on these associations to obtain mineral nutrients from these infertile soils. Few, if any, nonmycorrhizal plants live in these habitats.

Mycorrhizae can also exert a substantial effect on plant water relations. VAM often decrease the negative effects of drought on the host plant. It is possible that this is an indirect effect of the fungus on plant morphology and nutrient status. In addition, the fine hyphae extending into the soil from VAM may change the soil structure itself, increasing the movement of water toward the root and helping to bind the soil particles together, decreasing erosion (which is especially important in sands and sandy soils, as in beach dunes). ECM fungi form long, thick mycelial structures called rhizomorphs that can extend several meters into and across the soil. These rhizomorphs are capable of carrying water rapidly from the soil to the root.

One of the most surprising effects of mycorrhizae is their protection of the host plant from bacterial and fungal plant diseases, as well as nematode attack. Both ECM and VAM can offer this protection. The mechanisms for this protective function are diverse, particularly in ECM. The fungus may either secrete pathogen-killing chemicals directly into the rhizosphere or stimulate the root to produce such compounds. The fungal sheath of ECM can directly shield the root from pathogens and plant root-consuming invertebrates as well.

Orchids and Their Mycorrhizal Associations

Orchids may have a special obligate relationship with mycorrhizal fungi, depending on them for ordinary growth and development. Orchid seeds are exceptionally tiny, with no stored food reserves. The presence and activity of mycorrhizal fungi are necessary for seed germination and early seedling development and growth. Particularly when the plant is young, the fungus may supply it with carbon and energy in the form of carbohydrates, rather than receiving them from the plant. Some orchids never become photosynthetic and depend on mycorrhizae throughout their life span for all of their nutrients, including carbohydrates. It is not known where these fungi obtain their carbon, but they may get it by parasitizing other plants or by breaking down organic matter in the soil (that is, by functioning as saprophytes, like many nonmycorrhizal fungi).

Not enough work has been done on orchid-mycorrhizal interrelationships (particularly in the Tropics, home to the majority of orchids) to understand the extent and nature of this extreme form of plant dependence. If only because the orchids are the most highly diverse group of plants on Earth, orchidaceous mycorrhizae present an interesting and important topic for future ecological and evolutionary research.

Mutualism or Parasitism?

You may wonder, why would any self-respecting fungus do so many nice things for a plant? Soil fungi are heterotrophic, like animals, and typically degrade organic matter or parasitize living things to make a living. (Others are predators, particularly on soil nematodes). Mycorrhizal associations are a way in which fungi can obtain carbon and energy from a living host plant. Generally, they appear to be mutualistic associations (although how generally is not yet known), with the plants supplying carbohydrates obtained by photosynthesis to the fungi and receiving mineral nutrients and other benefits from the fungi.

However, like any relationship, this one is not inevitably benign. One partner (either the plant or the fungus) can become parasitic on the other, receiving benefits without supplying them. Sometimes mycorrhizal fungi do not provide any minerals or other detectable benefits to their hosts, yet continue to receive nutrients. Sometimes only minor amounts of nutrients are provided by the fungus, the nutrients made available are not needed or do not enhance plant growth, or the benefits provided are vastly outweighed by the carbon and energy drain on the plant. Sometimes the plant provides little benefit to the fungus, yet receives benefits from it, as in the case of the orchid mycorrhizal association (at least some of the time).

In some cases, the plant may be capable of discarding the fungus when phosphorus is available, accepting the fungus only when it is of benefit to the plant. The fungus may, on the other hand, be the controlling member of the pair, not only determining the association but controlling the plant's growth, development, and allocation patterns by producing hormones that override the plant's own hormonal signals. Nonmycorrhizal plants have been observed to use a variety of mechanisms to actively reject fungal infection.

Plant Interconnections via Mycorrhizal Fungi

Among the host of strange and interesting phenomena associated with mycorrhizae, one that has fascinated many plant ecologists is the potential interconnection between different individual plants via mycorrhizal hyphae. Two plants that are infected with the same mycorrhizal fungus can be connected by a hyphal strand running through the soil. Even plants of completely unrelated species or genera can be connected in this way. Phosphorus, nitrogen, water, and even carbohydrates have been found to move between plants, apparently through these hyphal connections. However, it is difficult in practice to distinguish such a "pipeline" effect from that of substances leaking out of the roots or hyphae of one plant and being immediately picked up by the roots or hyphae of another.

The major unresolved question is, to what extent is this kind of transfer important in natural ecosystems? Some ecologists have argued that nutrients may be transferred from dominant individuals to seedlings, to understory species in forests, or to competitively inferior plants, allowing them to survive and persist in the plant community. These plants would then be, at least to some extent, parasitizing the plants from which they were drawing carbon or mineral nutrients. We do not yet know whether this phenomenon is a rare curiosity or a common occurrence, or how important it is in determining plant community composition.

A closely related issue is the potential for such interconnections to alter competitive relationships among individuals. Mycorrhizal nutrient transfer could potentially intensify such competitive interactions or ameliorate them. It could alter the competitive hierarchy, making inferior competitors more successful, or it could help the dominant competitor. We do not yet know how to predict the effects of such interconnections on competition, nor do we have any idea how common or how important such effects are in nature. Future research on mycorrhizae is certain to come up with many surprises.

Summary

Soil is a complex product of the interaction between living organisms and their terrestrial substrate. It includes physically and chemically altered products derived from rock and from organic materials, along with air and water. Soils vary from one region to another, and within local environments, in many different aspects. There is a tremendous diversity of soil properties and characteristics, and these differences affect the vegetation found growing in the soil as well as affecting the growth of individual plants.

Soil texture describes the relative proportions of clay, silt, and sand particles in the soil. These particles differ from one another in size, shape, and mineral composition, and they impart different characteristics to soils. The pH of soil also has an important influence on plant growth and on what plants are found growing in the soil. One of the important factors in characterizing local soils is the soil profile, which consists of often distinct soil horizons, or layers at different depths. In addition to the physical factors responsible for soil properties, soil organic matter is a critical component in determining soil structure.

Plants depend on the soil to obtain water. Different soils hold different amounts of water in the macropores and micropores of the soil structure. As a soil dries after a rain, water first drains out of the macropores, and then moves by capillary action through the micropores to plant roots. Eventually, as the soil water potential declines, water becomes less and less available to plants, causing wilting and ultimately death unless more water enters the soil.

Plants also depend on the soil for the mineral nutrients that are essential to their survival and growth. Nitrogen and phosphorus are generally the most limiting mineral nutrients for plants. Both are needed by plants in fairly large quantities. Nitrogen availability, in particular, depends on bacterial activity, including symbiotic nitrogen fixation. The availability of these two nutrients varies greatly in soils because nitrogen is easily lost from soils, while phosphorus is often present in unavailable forms.

Plants may depend on a symbiotic relationships with fungi living on or in their roots, called mycorrhizae, to obtain phosphorus and other minerals from the soil. Plants may obtain other benefits from these associations as well. Most terrestrial plants have mycorrhizae and depend on them to survive, grow, and reproduce in their natural habitats. Two of the most important types of mycorrhizae are ectomycorrhizae (ECM) and endomycorrhizae, particularly vesicular-arbuscular mycorrhizae (VAM). The relationships between plants and mycorrhizal fungi are complex and variable, and are important in the function of many ecosystems.

Additional Readings

Classic References

Lyon, T. L. and H. O. Buckman. 1922. *The Nature and Properties of Soils.* Macmillan, New York. [Subsequent editions, H. O. Buckman and N. C. Brady 1952, 1960, 1969.]

Current edition:
Brady, N. C. and R. R. Weil. 2001. *The Nature and Properties of Soils.* 13th ed. Prentice-Hall, Upper Saddle River, NJ.

Contemporary Research

Batty, A. L., K. W. Dixon, M. Brundrett and K. Sivasithamparam. 2001. Constraints to symbiotic germination of terrestrial orchid seed in a mediterranean bushland. *New Phytologist* 152: 511–520.

Brundrett, M. 1991. Mycorrhizas in natural ecosystems. *Adv. Ecol. Res.* 21: 171–313.

Newsham, K. K., A. H. Fitter and A. R. Watkinson. 1995. Multi-functionality and biodiversity in arbuscular mycorrhizas. *Trends Ecol. Evol.* 10: 407–411.

Fitzhugh, R. D., C. T. Driscoll, P. M. Groffman, G. L. Tierney, T. J. Fahey and J. P. Hardy. 2001. Effects of soil freezing disturbance on soil solution nitrogen, phosphorus, and carbon chemistry in a northern hardwood ecosystem. *Biogeochemistry* 56: 215–238.

Additional Resources

Jeffrey, D. W. 1987. *Soil-Plant Relationships: An Ecological Approach.* Croom Helm, London, and Timber Press, Portland, OR.

Pimentel, D. (ed.). 1993. *World Soil Erosion and Conservation.* Cambridge University Press, Cambridge.

Rendig, V. V. and H. M. Taylor. 1989. *Principles of Soil-Plant Interrelationships.* McGraw Hill, New York.

Schulte, A. and D. Ruhiyat (eds.). 1998. *Soils of Tropical Forest Ecosystems: Characteristics, Ecology, and Management.* Springer Verlag, Berlin.

PART II Evolution and Population Biology

CHAPTER 5 *Processes of Evolution*

*I*n Part I, as we explored the various ways in which plants interact with their environments, we looked at the features of plants that permit them to cope with different environments. All of those features arose through the process of evolution by natural selection. The most amazing aspect of this process is how a few simple consequences of biology and natural laws have resulted in the vast diversity of life. All of the species that compose ecological communities arose through evolution, and ecological processes provide the context for evolution. The ecologist G. Evelyn Hutchinson emphasized this intimate relationship in his seminal book, *The Ecological Theater and the Evolutionary Play* (1965). The intertwining of evolution and ecology is greatest in the process of natural selection.

This chapter and the next summarize the basic principles and processes of evolution and explore its outcomes. In this chapter, we cover the process of natural selection as well as other processes that affect evolution. In Chapter 6, we examine a series of specific examples of evolution by natural selection.

Natural Selection

Natural selection occurs when individuals with differing traits leave different numbers of descendants because of those differences. Evolution by natural selection occurs when those differences are **heritable** (have a genetic basis). **Adaptive traits** are ones that have come about through the process of evolution by natural selection. The suite of traits associated with CAM photosynthesis (see Chapter 2), for example, is adaptive in hot, dry climates because individuals that have those traits are able to leave more descendants than individuals that do not. Arctic plants are short-statured because by remaining close to the ground they remain warmer, grow more, and ultimately leave more offspring.

The principles of natural selection were first proposed by Charles Darwin in his book *On the Origin of Species by Means of Natural Selection*, published in 1859. Natural selection is one of the four central processes of evolution. The others are mutation, migration, and genetic drift.

Variation and Natural Selection

An important starting point for any discussion of evolution and natural selection is variation. Variation is ubiquitous in nature. Nearly all natural phenomena vary at some level. This principle is expressed in the adage that "no two snowflakes are exactly alike." Living beings are much more complex than snowflakes, with an even greater potential for different kinds and amounts of variation. Evolution requires two kinds of variation: phenotypic variation and genetic variation.

Plants, like most living organisms, vary phenotypically. The term **phenotype** refers to all of the physical attributes of an organism. These properties include aspects of outward appearance (such as height, leaf size and shape, flower color, or fruit number), behaviors (such as growth of roots toward water), life history characteristics (such as being an annual), internal anatomy, and the content of cells (such as protein composition).

Phenotypic variation can be extensive for traits such as seed size, which can vary 20-fold within a single population of a wildflower species (Figure 5.1), or biomass, which can vary over several orders of magnitude in trees. Other kinds of traits, such as flower size and solute concentrations in cells, tend to vary much less among individuals in a population. Still other traits, such as petal number or photosynthetic pathway, tend to be invariant within a species. However, even these traits vary within some species (see Chapter 2). Furthermore, the particular pattern of growth and development exhibited by an individual plant depends on a complex interaction between its genes and its environment. All of these differences among individuals result in phenotypic variation.

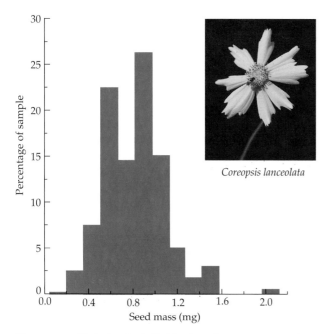

Figure 5.1 Variation in individual seed mass in a population of *Coreopsis lanceolata* (Asteraceae) growing on an inland dune south of Lake Michigan. (Data from Banovetz and Scheiner 1994.)

The **genotype** of an individual is the information contained in its **genome**—the sequence of its DNA. This information is expressed through the processes of transcription of DNA to RNA and translation of the RNA into protein. Gene expression, and the development of the plant as a whole, is controlled by the information contained within the genome itself through various feedback mechanisms. A seed contains all of the information necessary for the growth and development of the adult plant. Development can be thought of as the unfolding of the information contained in the genome.

The Components of Natural Selection

Natural selection requires three components: *phenotypic variation* among individuals in some trait; *fitness differences* (some individuals must leave more descendants than others as a result of the phenotypic differences), and *heritability* (the phenotypic differences must have a genetic basis). If these three components are present for a trait within a population, then the frequency of that trait will change in that population from one generation to the next: evolution by natural selection will occur.

Fitness differences include differences in mating ability, **fecundity** (number of gametes produced), fertilizing ability, **fertility** (number of offspring produced), and **survivorship** (chance of surviving). What causes these fitness differences? It could be chance. But often these differences are associated with differences in some trait of the plants. **Phenotypic selection** occurs when individuals with different trait values have consistent differences in fitness.

A useful and convenient way to study the process of natural selection is to subdivide it into two parts: that which occurs within a single generation—phenotypic selection—and that which occurs from one generation to the next—the genetic response (Endler 1986). Phenotypic selection consists of the combination of phenotypic variation and fitness differences. This part is what we often think of as natural selection. But for evolution to occur, the other part—the genetic response—is equally important. The **genetic response** is the change in the genetic makeup of the population that occurs from one generation to the next. The genetic response depends on the heritability of a trait. If a trait is heritable, then if phenotypic selection favors that trait in one generation, the next generation will have a greater proportion of individuals with that trait.

In a population of the grass *Danthonia spicata*, for example, individuals with longer leaves had higher survivorship and fecundity than individuals with shorter leaves (Figure 5.2A). Variation in leaf length is known to be at least partially under genetic control. So the next generation, on average, will be likely to have longer leaves. From one generation to the next, the change in the population may be small. But if the process of natu-

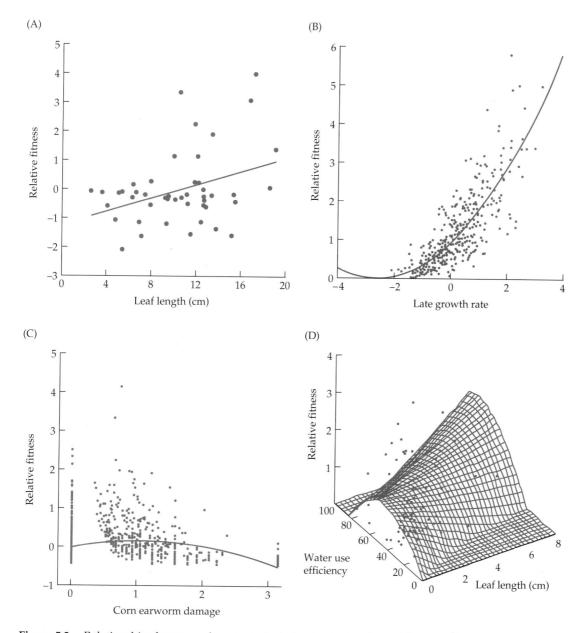

Figure 5.2 Relationships between phenotypic variation in plant traits and fitness differences result in phenotypic selection. If genetic variation for these traits also exists, then evolutionary change may occur. (A) Directional selection for longer leaf length in the perennial *Danthonia spicata* (poverty grass, Poaceae) in a natural population growing in a white pine–red oak forest in northern lower Michigan. Fitness was measured as the total number of spikelets produced over five years and was corrected for correlations with other traits. (Data from Scheiner 1989.) (B) Directional selection for greater growth rate late in the season in the annual *Impatiens capensis* (jewelweed, Balsaminaceae) in a natural population growing in a deciduous forest in Wisconsin. Fitness was measured as the final dry weight of the plant, which is correlated with the number of seeds produced. The fitness function is curved, but still monotonically increasing. (After Mitchell-Olds and Bergelson 1990.) (C) Stabilizing selection for the amount of damage caused by the corn earworm to the annual *Ipomoea purpurea* (morning glory, Convolvulaceae) in an experimental population growing in an old field in North Carolina. Fitness was measured as the number of seeds produced by the end of the growing season. (Data from Simms 1990.) (D) Stabilizing selection for water use efficiency and directional selection for smaller leaf size in the annual *Cakile edentula* (sea-rocket, Brassicaceae) in an experimental population growing on a beach along southern Lake Michigan. Fitness was measured as the number of fruits produced by the end of the growing season. The optimal water use efficiency depends on leaf size, which is an example of correlational selection. (After Dudley 1996.)

ral selection goes on for many generations, the population may become very different from its ancestor. Much of the striking variation among species is due to such long-term evolutionary responses to natural selection.

Directional selection occurs when individuals with the most extreme value for a trait have the highest fitness (Figure 5.2A,B). In this case, the population will continue to evolve in a single direction over time, if no other processes interfere. The primary change will be in the mean value of the trait. **Stabilizing selection** occurs when individuals with intermediate trait values have the highest fitness (Figure 5.2C). Under stabilizing selection, the mean value of the trait will not change, but the variability in that trait will decline. Under both directional selection and stabilizing selection, all additive genetic variation for the selected trait will eventually disappear, unless other processes introduce new variation. Phenotypic variation may remain, however, due to two other factors: plasticity and errors of development. **Correlational selection** occurs when the pattern of selection on one trait depends on the value of another trait (Figure 5.2D). Correlational selection can result in highly coordinated suites of traits and complex adaptations.

Phenotypic Variation and Phenotypic Plasticity

Phenotypic differences among individuals can result from three types of variation: genetic, environmental, and developmental. The genotypes of individuals usually differ, and this genetic variation often results in phenotypic variation among individuals. Individuals also experience different environments as they develop and grow. The environment can vary in many ways, even over very small distances. Each grass shoot in a seemingly uniform meadow, for example, experiences a somewhat different environment. Individuals with different genotypes may respond differently to their environment, and their development may proceed differently. Even two individuals with identical genotypes growing in identical environments, however, do not necessarily look alike or function identically. Small, random differences in when and how genes are expressed, called **errors of development**, can lead to measurable differences in the adult plant.

Phenotypic plasticity is the capacity of a genotype to give rise to different phenotypes in response to different environmental conditions. In the species *Impatiens capensis* (jewelweed, Balsaminaceae), for example, plants grown in the shade tend to be tall and spindly, while those grown in the sun tend to be short and stout (see Figure 6.6; Schmitt et al. 1995). Shady conditions can be due to competition from neighboring plants. Under those conditions, it is better for a plant to grow taller quickly so that it can shade out its neighbors rather than be shaded itself. On the other hand, in sunny conditions, it is better for a plant to use its resources to produce more and larger leaves and more flowers.

Plants need some way to sense how shady their environment is. A key difference between sunny areas and shady areas is the ratio of red light (wavelength = 665 nm) to far-red light (wavelength = 730 nm). In open areas, this ratio is about 1.15, while under a leafy canopy it ranges from 0.15 to 0.97. Plants sense differences in light quality through a group of pigments called phytochromes (Smith and Whitelam 1990). Phytochrome A is labile, changing its form in the presence of light of different wavelengths. The Pr form, when exposed to red light, changes to the Pfr form. This form, in turn, responds to far-red light, which changes it back to the Pr form. The other four phytochromes are stable in form, but interact with phytochrome A. Together, these phytochromes turn different suites of genes on or off, resulting in the different growth forms.

Phytochromes are also used by plants to determine the length of the night. The Pfr form of phytochrome A slowly, but spontaneously, changes back to the Pr form after some time in the dark. So the Pfr:Pr ratio is a cue to day length, and thus season, in temperate parts of the globe. Flowering, seed dormancy, and seed germination are among the physiological responses that are mediated by phytochromes in many species.

Besides affecting gene expression, the environment can influence plant growth directly. The amount of water available in the soil, for example, has a direct effect on plant size. Opening the stomata to take up CO_2 during photosynthesis results in the loss of water (see Chapter 3). If a plant does not have an adequate supply of water, it will close its stomata. This closure causes a reduction in the amount of carbon fixed by the plant and can result in reduced growth. Likewise, in shady environments, the number of photons of light energy available to be captured for photosynthesis is limited, and this limitation affects both photosynthesis and plant growth.

Heritability

Although at times it may seem as if the world consists of an infinite variety of species, that is not the case. It is easy to imagine organisms that could exist but do not, such as unicorns or dragons. Why not? One part of the answer is that the genetic basis of traits constrains evolution. For evolution to occur, there must be appropriate genetic variation. What limits this variation? To answer that question, we must first understand what we mean by genetic variation in an evolutionary context. Then we can explore the ways in which the environment interacts with genes to determine that variation.

Resemblance among Relatives

Heritability (h^2) is the amount of resemblance among relatives that is due to shared genes. Offspring tend to resemble their parents and their siblings because the phenotype of an individual is determined, in part, by its genotype, and an individual receives its genes from its parents and shares those genes with its siblings. Consider a trait such as height in an annual plant at the end of the growing season. We might do the following experiment with *Brassica campestris* (Brassicaceae): We choose pairs of plants, pollinate one individual with the other in each pair, and cover the flowers to prevent any other plant from pollinating that individual. When we collect the seeds at the end of the growing season, we thus know exactly who the parents were. Before the parental plants die, we measure their heights. Then we plant the seeds, let them grow, and measure their heights at the end of the next growing season.

We then plot the height of the offspring against the height of their parents (Figure 5.3). In this case, we find that taller parents tend to produce taller offspring. We measure this tendency using a statistical technique called correlation (see Appendix). If offspring always exactly matched their parents, the correlation between parental height and offspring height would be 1.0. If there were no relationship, the correlation would be 0.0. In our example, the correlation is 0.41 and the slope of the line is 0.21;

there is a resemblance, but some offspring are taller than their parents, while others are shorter. Negative correlations are also possible for some traits in some species, although they are very unusual.

One measure of the heritability of a trait is the slope of the line of a regression of offspring trait values on parental values. In the above example, because we used information from both parents, the slope is exactly equal to the heritability. If we had measured only one parent, we would have information about only half of the genes being contributed to the offspring, and the slope would be one-half the heritability.

The other common way of measuring heritability is to measure the correlation among siblings. Imagine if we took two seeds from each of many plants. We could germinate the seeds and grow the pairs of siblings, measure their heights, and construct a graph much like Figure 5.3, except that now the axes would be the heights of the two siblings, and each point would represent a sibling pair. Again, the slope would measure heritability, with the exact relationship depending on whether the plants shared both parents or just one parent. We can do such an analysis with cousins or with any individuals that were related as long as we knew their relationships. Nor are we restricted to using pairs of individuals. Various statistical techniques can be used to measure heritability in groups of related plants with different degrees of relatedness.

There is a critical distinction between the heritability of a trait and whether that trait has a genetic basis. Heritability requires that phenotypic differences among individuals be due, at least in part, to genetic differences among those individuals. In Box 5A we describe a case in which height is genetically determined. In that example, some individuals have a genotype of *AA*, some *Aa*, and some *aa*. Instead, imagine that all individuals in the population have the same genotype, *AA*. Assume, however, that height also depends on the amount of nitrogen in the soil. If the population is growing in a field that varies in soil nitrogen from spot to spot, then individuals will differ in height. However, none of those phenotypic differences will be due to differences in genotypes. If we were to measure these plants, collect seeds, and raise the offspring in that same field, the correlation between parental height and offspring height would be 0, and the heritability of height in that population would be 0. Yet, there is still a gene that determines height.

This example also demonstrates that the heritability of a trait depends on the frequencies of its alleles in the population. When the frequency of *A* is 1.0—all individuals have the *AA* genotype—the heritability of the trait is 0. Thus, heritability estimates for the same trait can differ among populations, or for the same population measured at different times. Heritability estimates

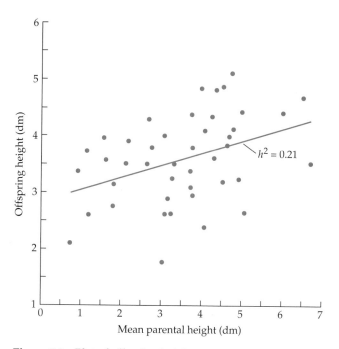

Figure 5.3 Plot of offspring height against mean height of the parents in the annual *Brassica campestris* (Brassicaceae) for plants grown in a greenhouse. The heritability of this trait (the slope of the line) is 0.21. (Unpublished data courtesy of Ann Evans.)

A Simple Genetic System and the Resemblance of Relatives

This example is based on the work that Gregor Mendel did in the nineteenth century with garden peas (*Pisum sativa*, Fabaceae), although the details have been modified for illustrative purposes. Although it is based on a simple, one-locus system, in fact nearly all traits of ecological interest are based on several to many loci. However, the same principles hold no matter how many loci affect a trait.

Consider a simple genetic system in which plant height is determined by a single diploid locus. We assume that individuals with genotype *AA* are tall (100 cm) and those with genotype *aa* are short (20 cm). We also assume that there are no environmental effects ($V_E = 0$ and $V_{G \times E} = 0$) or errors of development ($V_e = 0$).

Case 1: Strict Additivity

If individuals with genotype *Aa* have a phenotype that is exactly intermediate between *AA* and *aa* individuals, then genetic variation is strictly additive. In this case, *Aa* individuals would be intermediate in size (60 cm tall). Suppose we want to predict the phenotypes of the offspring of a cross. Because the effects of the alleles are strictly additive, we can do so. If both parents are tall, the cross will be *AA* × *AA*, and all offspring will be tall. If both parents are short, the cross will be *aa* × *aa*, and all offspring will be short. If one parent is tall and the other short (*AA* × *aa*), all offspring will be 60 cm tall (*Aa*). That is also the height that we get by averaging the parental phenotypes; the mean offspring phenotype equals the mean value of the parents' phenotypes. If one parent is 100 cm tall and the other is 60 cm tall (*AA* × *Aa*), half the offspring will be 100 cm tall and half will be 60 cm tall. Again, the mean

value of the parents' phenotypes, 80 cm, exactly equals the mean value of the offspring phenotype. Note that for this cross, no parent or offspring is actually at the mean height; the mean is a descriptor of the group, not a property of any particular individual. A graph of mean parental phenotype against mean offspring phenotype (part A of the accompanying figure) has a slope of 1.0. That is, the heritability of this trait is 1.0, because we can perfectly predict the average offspring phenotype from our knowledge of the parental phenotypes.

Case 2: Dominance

Now assume that *A* is dominant to *a*, such that *Aa* individuals are 100 cm tall. In this case, predicting offspring phenotypes becomes more difficult. If both parents are short, all offspring will be short. But if both parents are tall, their genotypes could both be *Aa*, or both be *AA*, or one *AA* and the other *Aa*. In the latter two instances, all offspring will be 100 cm tall. But if both parents are *Aa*, then 1/4 of the offspring will be *aa* and will be short. The mean offspring phenotype will be 80 cm (3/4 × 100 + 1/4 × 20), even though the mean phenotype of the parents was 100 cm. If we assume that both alleles exist with equal frequency in our population, then a graph of mean parental phenotype against mean offspring phenotype will have a slope of 0.67 (see the accompanying figure). The heritability of the trait is less than 1.0 because some of the genetic variation is nonadditive due to the dominance relationship. In other words, some offspring differ phenotypically from their parents because of the effects of dominance; if they do not inherit the dominant allele from either of their parents, they do not resemble

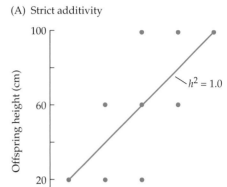

(A) Strict additivity

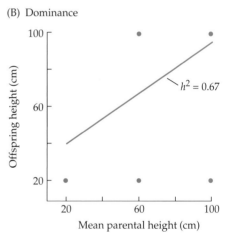

(B) Dominance

Plot of offspring phenotype against mean parental phenotype for two genetic systems. The slope of the regression line is the heritability of the trait. (A) Strictly additive; slope = 1. (B) Complete dominance; slope = 0.67. These heritabilities assume equal frequencies of the two alleles in the population.

their parents. The exact heritability of a trait in a population depends on both the degree of dominance and allele frequencies in the population.

are always specific to the population and environment in which they are measured.

Partitioning Phenotypic Variation

Another way of thinking about heritability is to con-

sider the various sources of phenotypic variation described above. If we measure height in a population of annual plants, it will vary due to differences in genotype, environment, and errors of development. The percentage of that variation that is caused by genetic differences is heritability. Mathematically, we can express this idea as follows. First, consider the total phenotypic variation, which we symbolize as V_P. By a combination of experimental and statistical techniques, we can determine how much of that variation (statistical variance; see the Appendix) is due to different causes. Breaking up variation into its components is called variance partitioning. In the simplest case, we might be concerned with partitioning the phenotypic variation only into that part due to genetic variation (V_G) and that part due to all other causes (V_e): $V_P = V_G + V_e$. Then, heritability would be $h^2 = V_G/V_P$, the percentage of phenotypic variation that is due to genetic differences among individuals. This concept of heritability is identical to that of heritability defined as the resemblance among relatives, just a different way of measuring it.

Genetic variation can be further partitioned, however. In a diploid organism, **dominance** occurs when the expression of an allele depends on the properties of the other allele at that locus. In both diploid and haploid organisms, **epistasis** occurs when gene expression depends on the properties of alleles at other loci. The amount of genetic variation that is manifested in a population can be a result of differences in the direct expression of an allele (additive variation, V_A), or to differences in combinations of alleles at each locus (dominance variation, V_D), or to differences in combinations of alleles at different loci (epistatic variation, V_I) (see Box 5A). Again, by various experimental and statistical procedures, we can partition this variation into its causes: $V_G = V_A + V_D + V_I$. If heritability is calculated as just the percentage of additive genetic variation ($h^2 = V_A/V_P$), then we are speaking of **narrow-sense heritability**. If it is calculated as the total genetic variation (V_G), then we are speaking of **broad-sense heritability**. This distinction is important because the response of a trait to natural selection depends on its narrow-sense heritability.

Heritability values tell us whether there is genetic variation for a trait in a population, and if so, whether there is just a little variation or a lot of variation. In terms of evolution, the amount of genetic variation may impose a constraint on evolution. If there is no genetic variation ($h^2 = 0$), the constraint is strong. No matter how much natural selection there is on a trait, there will be no genetic response. If there is a little bit of genetic variation, the constraint is weak; there will be a genetic response, but it will be small, and evolution will proceed slowly. If there is a lot of genetic variation, there is almost no constraint.

Genotype-Environment Interactions

How does the environment influence heritability measures? Our original definition of heritability assumes that differences among individuals with different genotypes do not depend on their environment. That is, we assume that if an individual is 10% taller than another when growing in one environment, it will still be 10% taller in a different environment. But what happens when this assumption does not hold? Suppose, for example, that when a certain plant species is grown under shady conditions, all individuals are small and about the same size, but when it is grown in a sunny spot, some of those individuals are much taller than the others due to genetic differences. In other words, the genetic differences are apparent in some environments, but not in others.

These types of differences in genetic expression as a function of the environment are referred to as **genotype-environment interactions**. Genotype-environment interactions are the genetic component of phenotypic plasticity. Adding environmental effects, a full partition of the phenotypic variation of a population is

$$V_P = V_E + V_{G \times E} + V_G + V_e$$

where V_E is variation due to the environment, $V_{G \times E}$ is variation due to genotype-environment interactions, and V_e now refers only to variation due to errors of development (Figure 5.4).

The presence of variation resulting from genotype-environment interactions can have large effects on heritability. Heritability is not simply a result of the genet-

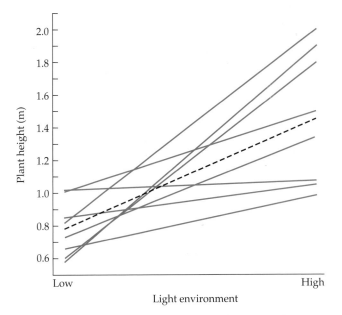

Figure 5.4 An example of phenotypic plasticity and genotype-environment interaction for plants growing in high light and low light environments. In this example, cuttings were taken from individual genets and grown in each environment. Each shaded line connects the mean height for each genet in the two environments. The dashed line connects the overall means in each environment and represents V_E. Variation in average heights among genets represent V_G. The extent to which the shaded lines are not parallel represents $V_{G \times E}$.

ic differences among individuals: those genetic differences must result in phenotypic differences. Some kinds of genetic differences among individuals never result in phenotypic differences; for example, some types of variation in noncoding regions of the DNA do not do so. In other cases, whether genetic differences result in phenotypic differences depends on the environment. When genotype-environment interaction variation is present, then the amount of expressed genetic variation may differ among environments. In the example given above, plants grown in the shade were all of similar size; in other words, phenotypic differences were minimized in that environment. If the heritability of size were measured only in the shade, we would conclude that it was very low because the amount of phenotypic variation would be low. On the other hand, if heritability were measured only in a sunny environment, it would be larger. Thus, evolution would be constrained in the shady environment because of a lack of heritable variation.

Gene-Environment Covariation

The final term that must be included in accounting for the total phenotypic variation of a trait in a population is **gene-environment covariation**, which is abbreviated Cov(G,E). A nonrandom relationship between any two factors is called covariation; such a relationship can be either positive or negative. Covariation is closely related to correlation; one is a mathematical transformation of the other. In this case, we are interested in covariation between genetic and environmental effects on a trait. Such covariation is often found in nature because individuals are usually not randomly distributed across environments. This is especially true of plants, as seeds often end up close to the maternal parent plant.

Positive covariation between genetic and environmental effects occurs when genetic traits are positively associated with responses to the environment. For instance, more vigorous competitors might dominate small patches of rich soil. In this case, plants that are already genetically capable of growing more rapidly will also be growing under better conditions, while slow growers will be relegated to poorer conditions. Thus, genetic differences in growth rate are exaggerated by environmental effects.

Nonrandom distribution of genotypes can act instead to minimize genetic differences. Negative covariation between genetic and environmental influences occurs when genetic and environmental influences have opposite effects. Consider a population of shrubs in which larger individuals produce more seeds. If you took the seeds of several individual shrubs and grew them all under optimal conditions in a garden, you would find a positive correlation between parental size and offspring size. In a natural population, however, most of the seeds will germinate in the shade of the parent plant. Because large plants produce many seeds, those seeds will be growing under crowded conditions. Plants that are genetically capable of growing larger (the offspring of the large shrubs) will be growing under poorer conditions, so genetic and environmental influences will be acting on size in opposite directions. Therefore, a negative covariation might exist.

Adding the term for covariation to the equation for phenotypic variation given above, we end up with our final equation describing phenotypic variation in a natural population:

$$V_P = V_E + V_{G\times E} + V_G + \text{Cov(G,E)} + V_e$$

Since V_P is the denominator in calculating heritability, changes in any of its components will affect the heritability of the trait.

Patterns of Adaptation

Natural selection can result in three different patterns of adaptation. First, individuals may become specialized to perform best in different environments. Within a population, for example, individuals of one phenotype might have the highest survival during periods of drought or in the driest spots, while individuals of another phenotype might survive best during wetter periods or in microhabitats with moister soil. Second, individuals might become phenotypically plastic, changing their form in different environments so as to match the most fit trait value for each set of environmental conditions. In response to changes in water availability, for example, individuals might produce leaves of different sizes and shapes. This pattern would allow them to have high survivorship during both wet and dry periods. Finally, all individuals in a population might converge on a single, intermediate phenotype that does at least passably well in all environments. This pattern is sometimes referred to as a jack-of-all-trades strategy, as in the saying "a jack-of-all-trades is a master of none."

Which of these patterns of adaptation will prevail depends on a complex combination of factors, including how variable the environment is, how that variation is distributed in time and space, and how much genetic variation exists in the population. If the environment varies over short time periods, then phenotypic plasticity is often favored. Usually we think of phenotypic plasticity with regard to traits that become fixed during development, such as leaf shape. In aquatic plants from many unrelated families, submerged leaves are characteristically feathery and highly dissected, while emergent leaves on the same plant may have very different shapes (Figure 5.5). In water, CO_2 diffusion is much slower than in air, and the highly dissected underwater leaves have increased surface areas and a smaller boundary layer (see Chapter 3), allowing

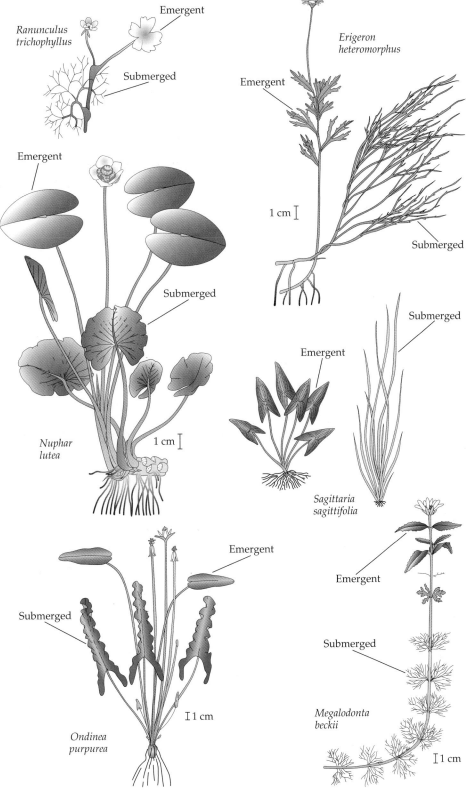

Figure 5.5
Phenotypic plasticity results in adaptive differences in form between emergent and submerged leaves of several aquatic plant species. Clockwise from top left: *Ranunculus trichophyllus* (threadleaf crowfoot, Ranunculaceae), *Erigeron heteromorphus* (fleabane, Asteraceae), *Sagittaria sagittifolia* (Hawaii arrowhead, Alismataceae), *Megalodonta beckii* (Beck's water marigold, Asteraceae), *Ondinea purpurea* (purple water lily, Nymphaeaceae), and *Nuphar lutea* (yellow pond lily, Nymphaeaceae).

better CO_2 uptake rates. When environmental changes occur over much longer time spans, however, phenotypic plasticity is much less likely to be favored. Instead, specialization is often favored, giving rise to patterns such as the contrasting leaf shapes among ecotypes of *Geranium* and *Achillea* discussed in Chapter 6.

Levels of Selection

Natural selection usually occurs at the level of the individual. Individuals have different fitnesses, and genetic responses to natural selection are measured by looking at differences among the individuals in a population from generation to generation. However, none of the three components of natural selection requires that individuals be the only focus of the process. Plants, in particular, have growth forms that can result in other units being the focus of natural selection.

The first question we must address is, what is an individual? In many plant species, an individual is obvious. There is a central trunk or stem that grows, reproduces, and dies. But many other plants are **clonal**, existing as a set of genetically identical, possibly interconnected, but semi-autonomous to fully autonomous units. Many grasses are good examples. From a single grass seed grows a plant that first sends up leaves and puts down roots. Later, it produces specialized roots called **rhizomes**, which cause the plant either to grow larger or to spread over a wider area. New leaves and roots are produced at nodes along the rhizome called **tillers**. Water and nutrients pass along the rhizomes to different parts of the plant. However, once the new units are large enough, the connections between them can be cut and each unit can survive as an independent plant. Even if the connec-

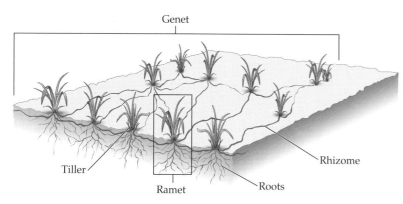

Genet

Tiller

Ramet

Roots

Rhizome

Figure 5.6 Diagram of a grass plant that grows by vegetative spread. The entire plant is a single genet; each tiller and its roots comprise a single ramet.

tions remain, the transfer of water and nutrients among different units may be small.

Deciding exactly what constitutes an individual, then, is not simple. The British ecologist John Harper proposed the following distinction (Figure 5.6): A **genet** is a genetic individual, the product of a single seed. A **ramet** is a potentially physiologically independent unit

John Harper

of a genet, such as the tillers of a grass plant. A single genet may consist of many separately functioning ramets. While individual ramets may come and go, a genet can exist for a long time. Quaking aspen (*Populus tremuloides*, Salicaceae) is a tree that can spread by sending up new trunks from its roots. While individual trunks usually live only 50 or 60 years, one quaking aspen genet was estimated to be more than 10,000 years old (Mitton and Grant 1996; Figure 5.7). Other common examples of plants that grow clonally, or vegetatively, are kudzu (*Pueraria*, Oxalidaceae), strawberries (*Fragaria*, Rosaceae), and cattails (*Typha*, Typhaceae).

Natural selection can occur whenever its three components—phenotypic variation, fitness differences, and heritability—are present. While most selection occurs at the level of the individual, selection at other levels has been found (Lewontin 1970). Selection can occur below the level of the individual. A **somatic mutation**—a mutation in a non-gamete-producing cell—could occur in the meristems of a grass tiller (see Figure 8.1), resulting in genetic variation among the ramets within a genet. If those genetic differences led to phenotypic differences that affected fitness, then selection could act on the ramets. Selection can also occur above the level of the individual, among populations or among species. We describe an example of this type of selection in Chapter

6. When trying to determine whether a particular feature of a plant is a result of natural selection, we should consider the possibility that the feature is due to selection at a level other than the individual. Because evolution occurs over very long stretches of time, even very slow processes, or ones that occur very rarely, can be important.

Other Evolutionary Processes

One way of classifying processes that contribute to evolution is to group them into those that add genetic variation to a population and those that eliminate that variation. Natural selection acts to eliminate genetic variation as alleles that lead to increased fitness become fixed in a population and other alleles are lost. An important constraint on evolution is the lack of sufficient genetic variation for natural selection to act upon. For evolution by natural selection to continue, genetic variation must reenter a population.

Processes that Increase Variation

The ultimate source of all genetic variation is **mutation**: change in DNA sequences. Mutation includes not only

Figure 5.7 This quaking aspen genet, growing in the Rocky Mountains of Colorado, is estimated to be more than 10,000 years old. (Photograph courtesy of J. Mitton.)

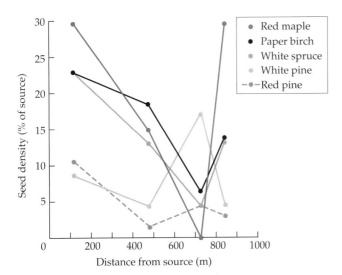

Figure 5.8 Density of seeds of five tree species found at varying distances from their source. (After Greene and Johnson 1995.)

changes in single base pairs (e.g., AT → GC), but also deletions, insertions, rearrangements of parts of chromosomes, and duplications of part or all of a chromosome. Some mutations, such as third-position codon substitutions, may have no effect on the phenotype. Other mutations, such as deletions, may destroy the function of a gene. Duplications of whole genes may have no immediate effect on the phenotype, but may allow the evolution of a new function in one of the two copies. In order to produce a permanent change in a population, a mutation must occur in gametes or gamete-producing cells.

Mutations are relatively rare events. Mutations of single base pairs occur at a frequency of 10^{-8} to 10^{-10} per base pair per generation. For the average-sized gene, mutations occur at a frequency of 10^{-5} to 10^{-7} per gene per generation. On average, in a given population for most species, one individual in ten has a new mutation somewhere in its genome each generation.

Another source of genetic variation is **migration**, which can introduce new alleles to a population. Migration in plants occurs primarily by the movement of spores, pollen, and seeds. For seed plants, the movement of pollen is particularly important. Pollen can be carried by the wind for tens of kilometers. Seed movement tends to be more restricted, although studies of wind-dispersed seeds show that fairly long-distance dispersal is common. A study of five Canadian tree species, for example, found substantial densities of seeds nearly a kilometer from their source (Figure 5.8). Pollen and seed movement is discussed in more detail in Chapter 8.

Because gene expression can depend on the presence of alleles at other loci, new combinations of alleles at different loci can create new variation. For this reason, the process of sexual reproduction, which includes cross-

ing-over, meiosis, and recombination, is an important source of new genetic variation.

Processes that Decrease Variation

Several processes can act to eliminate genetic variation from a population. Probably the most important of these, after natural selection, is genetic drift. **Genetic drift** refers to changes in gene frequencies due to random sampling effects (Figure 5.9). We can illustrate genetic drift with a simple example: Imagine a population that consists of yellow-flowered and white-flowered plants. In this population, flower color has a genetic basis. In one generation, by chance alone, the more yellow-flowered plants happen to be growing in richer soil. Because

(A)

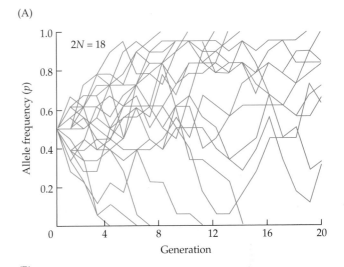

(B)

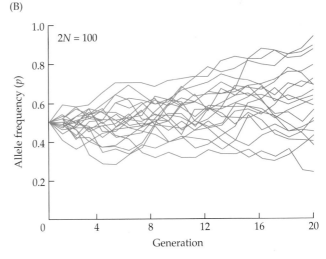

Figure 5.9 Changes over 20 generations in gene frequencies due to random sampling. (A) An allele frequency of $p = 0.5$ implies that there are 9 copies of the A allele and 9 copies of the a allele. (B) An allele frequency of 0.5 implies 50 copies of each allele. The much larger population size in (B) results in smaller oscillations in allele frequencies. As population size decreases, the effects of genetic drift increase.

of this, the yellow-flowered plants, on average, produce slightly more seeds than the white-flowered plants. In the next generation, the frequencies of the two flower colors in the population will have changed, but not because of natural selection. Rather, the change in trait frequency is due to random seed establishment.

Genetic drift can occur at many life history stages, including meiosis, pollination, seed germination, and maturity. In most populations, changes due to genetic drift will be small. Generally, genetic drift is important only in populations smaller than 100 individuals; in populations greater than 1000 individuals, its effects are usually much smaller than those of other evolutionary processes. Thus, population size is a critical parameter in determining how evolution will proceed.

Ipomoea purpurea (morning glory or bindweed, Convolvulaceae) provides an example of the effects of population size on genetic drift. This species is an annual weed of agricultural fields in the eastern United States. Its flowers vary in color; some individuals have blue flowers and some have pink flowers. Flower color is a simple one-locus trait, and blue is dominant to pink. The chief pollinators of this plant are bees, and gene flow among populations is primarily by pollen. Seeds are passively dispersed and stay close to the parent plant, although long-distance seed dispersal probably occurs in conjunction with agricultural or horticultural activities. Population densities vary among fields, and distances between plants range from 1 to 6 meters.

A study of flower color was initiated by Michael Clegg and his associates (Ennos and Clegg 1982; Brown and Clegg 1984; Epperson and Clegg 1986) to determine the evolutionary processes responsible for variation within and among populations. These studies looked at patterns both within fields and among fields across the states of Georgia, South Carolina, and North Carolina. The percentage of individuals with pink flowers varied greatly from field to field, from 5% in some fields to 55% in others. This variation occurred over a number of different distances from as small as 50 meters (the shortest distance between fields) to as great as 560 kilometers (the farthest distance between samples). Fields that were located close to one another were no more similar in their color morph frequencies than distant fields. This pattern most likely means that long-distance seed dispersal occurs through human activity.

In contrast, distance did matter within fields (Figure 5.10). Within a field, there were clusters of plants with different flower colors. Some clusters had only plants with pink flowers, some clusters had only plants with blue flowers, and yet other clusters had some plants with pink and some with blue flowers. On average, each

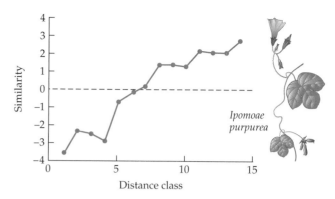

Figure 5.10 Spatial autocorrelation in clusters of *Ipomoea purpurea* (morning glory, Convolvulaceae) across a field near Athens, Georgia. Similarity is measured as the likelihood that a pink-flowered plant will be found next to a blue-flowered plant. At close distances, the similarity is negative, indicating that pink-flowered plants tend to be found next to other pink-flowered plants, and the same for blue-flowered plants. In contrast, at longer distances, clumps dominated by blue-flowered plants alternate with those dominated by pink-flowered plants. The point at which the line crosses zero indicates the size of the average cluster, approximately 50 meters in this field. (After Epperson and Clegg 1986.)

cluster contained 10 to 15 plants. The pollinators, bees, tend to travel from one plant to the next nearest plant. So alleles in these populations tend to move very short distances from one generation to the next.

Within fields, variation in flower color was most clearly due to the combination of genetic drift and migration by pollen movement. Random sampling caused gene frequencies to become different between clusters. Migration caused nearby clusters to be similar, but was not sufficient to keep more distant clusters from becoming different. **Spatial autocorrelation** (resemblance as a function of distance) is one hallmark of genetic drift, although spatial autocorrelation can also be caused by other processes.

Population size is of particular concern for species conservation. In a widespread species, even if each population is small, genetic variation can still be maintained across the species as a whole as long as there is some migration between populations. If the entire species is reduced to a single population or a small number of populations, then maintaining genetic variation can become difficult. Genetic variation can be lost quickly through natural selection and genetic drift. New variation through migration will not be available because the few populations will all contain the same alleles. New variation through mutation will be rare because the rate of new mutations depends on the number of individuals. In such populations, if the environment changes such that new alleles would increase fitness, those alleles may not be present, and the population may be unable to adapt to the new conditions. Such concerns are even greater today because of human-caused changes to the environment (see Chapter 22). Global warming may cause large changes in local environments in just a few decades. Small populations, unable to adapt, will be faced with extinction.

Variation among Populations

So far, we have discussed evolutionary processes that affect variation within a single population. We can also consider how evolutionary processes affect variation among populations. Natural selection can both increase and decrease variation among populations. If natural selection favors different trait values in different locations, different populations of a species may become adapted to different conditions. Eventually, the species may split into two species (see Chapter 6). Conversely, if natural selection favors the same trait values in different locations, the populations will tend to remain similar.

Mutation and genetic drift both tend to increase variation among populations. It is very unlikely that exactly the same mutations will occur in different populations. Because genetic drift is random, it is unlikely that different populations will experience exactly the same random changes. These tendencies for populations to diverge will be greatest in small populations.

Migration tends to decrease genetic variation among populations. Migration brings alleles into a population from nearby populations. These alleles could represent genetic variation previously lost from the population, or they could represent a mutation that appeared in another population. If migration rates between two populations are high enough, the populations will have the same alleles at almost the same frequencies.

All of these processes—natural selection, mutation, genetic drift, and migration—can act together to determine the evolution of a set of populations. Imagine a set of populations that initially are genetically the same. If the populations are relatively small, they may start to diverge from one another by genetic drift and mutation. One result is that new alleles, and new combinations of alleles, will appear in each population. These changes can result in several possible evolutionary outcomes. First, natural selection may favor different combinations of alleles in each population, resulting in the populations becoming different from one another. Sufficiently high rates of migration among populations, however, can sometimes prevent a population from evolving adaptations to the local environment. Therefore, another possible evolutionary outcome is that all of the populations will evolve to the average condition. A third outcome may result if the populations differ in how well they each adapt to local conditions. The most successful population may send out many more migrants than the other populations. Eventually, all of the populations will resemble that successful one.

Summary

Natural selection is the central evolutionary process determining the forms of living organisms. It comes about through the interaction of three components: phenotypic variation in traits, fitness differences linked to those trait differences, and heritable variation in those traits. All three components are necessary for evolution by natural selection to occur. The ecology of a plant plays a role in all three components, influencing the phenotype by affecting the course of growth and development, determining fitness differences, and influencing the expression of genes.

Natural selection acts on variation among phenotypes. Almost all plant traits vary among individuals within populations and species. Heritability of variation means that at least some of the difference among individuals in a population is due to genetic differences. Heritability is measured by studying resemblances among relatives. Heritability values depend on gene frequencies within populations and genotype-environment interactions, and they can be affected by the nonrandom distribution of genotypes. Thus, heritability estimates are specific to the population and environment in which they are measured.

Other processes—mutation, migration, and genetic drift—are also important in evolution. The first two processes are important for supplying genetic variation. Genetic drift can act in two ways, decreasing genetic variation in a population and causing evolution in the absence of natural selection. Not all evolutionary change is due to natural selection, although natural selection is the most important evolutionary process and is key to creating adaptations.

Natural selection is ubiquitous. Substantial amounts of phenotypic selection have been found in nearly every instance that it was looked for. Additive genetic variation has been found for most traits. Thus, the necessary components for natural selection often exist. These observations are the basis for our conclusion that natural selection has shaped the characteristics and nature of the plants that we see around us, even though that process occurred long in the past, or is occurring now at rates too slow to measure directly. In the next chapter we examine how natural selection has operated.

Additional Readings

Classic References

Darwin, C. 1859. *On the Origin of Species by Means of Natural Selection.* John Murray, London.

Dobzhansky, Th. 1937. *Genetics and the Origin of Species.* Columbia University Press, New York.

Stebbins, G. L. 1950. *Variation and Evolution in Plants.* Columbia University Press, New York.

Contemporary Research

Lande, R. and S. J. Arnold. 1983. The measurement of selection on correlated characters. *Evolution* 37: 1210–1226.

Mitchell, R. J. 1994. Effects of floral traits, pollinator visitation, and plant size on *Ipomopsis aggregata* fruit production. *Am. Nat.* 143: 870–889.

Winn, A. A. and T. E. Miller. 1995. Effect of density on magnitude of directional selection on seed mass and emergence time in *Plantago wrightiana* Dcne. (Plantaginaceae). *Oecologia* 103: 365–370.

Additional Resources

Futuyma, D. J. 1997. *Evolutionary Biology.* 3rd ed. Sinauer Associates, Sunderland, MA.

Hartl, D. L. and A. G. Clark. 1989. *Principles of Population Genetics.* 2nd ed. Sinauer Associates, Sunderland, MA.

Provine, W. B. 1986. *Sewall Wright and Evolutionary Biology.* University of Chicago Press, Chicago, IL.

Wilson, E. O. and W. H. Bossert. 1971. *A Primer of Population Biology.* Sinauer Associates, Sunderland, MA.

CHAPTER 6 *Outcomes of Evolution*

How do we know that a particular plant trait was shaped by natural selection? Answering this question is not simple. There is no single method for doing so, and all methods require assembling multiple kinds of information. Historically, ecologists have sometimes been guilty of not documenting natural selection. Rather, they have simply *assumed* that organisms were optimally adapted to the environments in which they were found. This attitude was termed "Panglossian" by Stephen Jay Gould and Richard Lewontin (1979), after the character Dr. Pangloss in the novel *Candide* by Voltaire. In that story, the naïve Dr. Pangloss goes through life confidently declaring his assumption that "All is for the best in this, the best of all possible worlds."

Sometimes ecologists have constructed scenarios to explain what processes may have led to particular adaptations, and then simply accepted those scenarios without substantiation. For example, some plant species produce seeds with a sticky, mucilaginous coating. Some ecologists originally claimed that this coating was an adaptation for dispersal; seeds getting stuck on the feet of ducks was one particular scenario they proposed. Such scenarios are termed "just-so stories," after the title of the book in which Rudyard Kipling recounts fanciful tales about the origins of various animals. Today we understand that in most species the coating has more to do with water retention by the seed.

The problem with "just-so" scenarios constructed by scientists is not that they are necessarily wrong; it is that they are based on scant evidence or on unfounded extrapolation from what is known. For example, while we cannot justifiably make the leap to sticky seeds having evolved by the process of natural selection acting to increase the dispersal success of all seeds, it has been documented with substantial evidence based on careful study that the sticky seeds of some species, such as *Phoradendron californicum* (desert mistletoe, Santalaceae), are in fact spread by birds.

Thus we must be cautious in either accepting or dismissing such adaptive speculations. Speculations are an indispensable first step in posing hypotheses to test whether a character has come about by natural selection. As ecologists are usually knowledgeable about the organisms and systems they study, such speculations often prove eventually to be correct. But until evidence has been assembled that firmly supports these suppositions, they remain just speculations.

In the last chapter we described how natural selection works. In this chapter we explore examples of the results of natural selection and the process of speciation. The examples in this chapter were chosen because they are based on more than just speculation. Most of them are based on multiple lines of evidence. As we examine each, we discuss the reasoning behind the conclusions concerning the shaping of the trait by natural selection.

This brings us back to the question we posed in the first sentence of this chapter: Why is it so difficult to decide whether a trait was shaped by natural selection? While natural selection is a primary process in shaping plant form and function, other processes may also be responsible for shaping particular traits. In some instances these processes act in concert with natural selection. Mammals have backbones not only because a backbone is a handy thing to have, but because they are vertebrates, and they have inherited this trait from vertebrate ancestors: all vertebrates share this trait for that reason. Likewise, angiosperm cells exhibit cellular respiration not because it is especially advantageous to angiosperms per se, but because the ancestors of angiosperms possessed this trait, which they have inherited, and thus share with a much broader group of living organisms.

In addition, natural selection is only one of the factors that cause evolutionary change (see Chapter 5). Other processes may act in concert with natural selection, or they may act against it, or they may act independently. Mutation, for example, is a necessary source of the variation that natural selection requires. Migration of individuals from other environments can prevent adaptation to local conditions in response to natural selection. Genetic drift can also cause changes in gene frequencies. Often each of these three processes can produce results that look like the outcome of natural selection. Because we are reconstructing a historical event, our conclusions must be based on indirect inference, rather than direct observation. All of these factors make the determination of adaptation by natural selection a challenging task.

Heavy-Metal Tolerance

We begin with one of the best-documented cases of natural selection and local adaptation in plants. This example is also important historically: it was the first demonstration of fine-scale genetic differentiation in response to selection by a known factor in the environment. The example involves a **cline**, a gradient in genetic composition, resulting from genetic differentiation and adaptation. Clines can occur over very short distances as well as over larger geographic scales.

In the 1960s, A. D. Bradshaw and his students (e.g., Jain and Bradshaw 1966; McNeilly and Antonovics 1968;

Antonovics and Bradshaw 1970) began to study local adaptation of grasses to differences in soil conditions due to mine waste contamination. In Great Britain, zinc and copper have been mined for centuries. The soil left over from the diggings and ore extraction, known as tailings, was simply dumped outside the mines. Although most of the ore had been removed, these tailings still contained high concentrations of the metals, concentrations too low to be worth extracting but still high enough to be toxic to most plants growing on the soil. Some plants, however, managed to grow on these tailings—in particular, the grasses *Anthoxanthum odoratum* and *Agrostis tenuis*. Although some of the mines had been abandoned for only a century—probably fewer than 40 generations for these perennial species—the researchers found that populations of these grasses on the mine tailings were tolerant of the heavy metals, while plants growing in ordinary adjacent pastures were not. The tolerance of heavy metals was achieved by the evolution of biochemical mechanisms that prevented the uptake of the toxic ions. These differences in genetic composition occurred over very short distances—within a meter of the mine boundary (Figure 6.1).

A. D. Bradshaw

The abrupt nature of the genetic boundary illustrates several aspects of the evolutionary process. First, the alleles for heavy-metal tolerance did not spread outward from the population on the mine tailings. Why? Experiments were performed in a greenhouse in which plants from both the mine tailings and the adjacent pasture were grown in soil with and without the heavy-metal contaminants. As expected, plants from the pasture population died when grown with the heavy metals, usually as seedlings. However, when grown in the absence of the heavy metals, the pasture plants grew much more quickly than those from the mine tailings population. Something about the mechanism of tolerance reduced growth under uncontaminated conditions. This interaction is an example of a **trade-off**, a reduction in fitness in one feature of an organism due to an increase in fitness in another feature. In this case, heavy-metal tolerance and growth in uncontaminated soils are traded off against each other. Thus, different alleles are favored on and off the mine tailings, and the populations have diverged genetically due to natural selection.

Other features of the populations have also evolved as a consequence of this divergence. All grasses are wind pollinated, so pollen from the adjacent pastures can easily land on the stigmas of plants on the mine tailings, and vice versa. The abruptness of the genetic boundary was

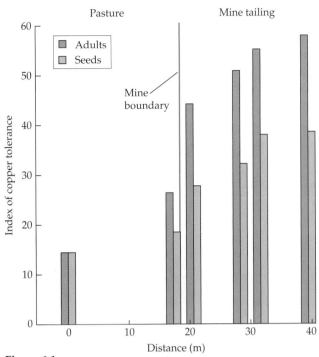

Figure 6.1
Ability of adults and plants grown from seeds of *Agrostis tenuis* collected from mine tailings and from adjacent uncontaminated pasture to grow in the presence of copper. Some 100 years after the mine was abandoned, the genetic composition of the grass population differs dramatically at the mine boundary, where alleles for enzymes that prevent the uptake of toxic ions are suddenly favored. (After McNeilly 1968.)

found to be related to the prevailing wind direction, with more of the "wrong" genotype being found on the downwind side. If an individual is pollinated by a plant from the other population, then its offspring will be less well adapted to the local conditions, as shown by the lower tolerance of the plants grown from seed in Figure 6.1. The problem of "wrong" pollinations results in selection for mechanisms that reduce cross-pollination. In these populations, two mechanism were involved in reducing outcrossing (Antonovics 1968; McNeilly and Antonovics 1968). First, individuals on the mine tailings tended to flower earlier than those in the pasture, with the difference being greatest for plants just at the boundary of the mine tailings. Second, plants near the boundary tended to self-pollinate more than plants farther away from the boundary. Both changes reduced the chance of an individual receiving the "wrong" pollen from plants adapted to the alternate conditions (toxic or uncontaminated soils). These changes in mating patterns were a secondary consequence of the primary selection for heavy-metal tolerance. They acted to reinforce the genetic differentiation of the populations. Over a long enough period of time, such reinforcement can lead to total genetic isolation of the populations, and ultimately to speciation.

These studies clearly demonstrate, for several reasons, that the differences among populations were due to natural selection. First, the investigators demonstrated the presence of all three necessary components of natural selection: phenotypic variation, fitness differences, and heritability (see Chapter 5). Second, they found that secondary predictions were also met. The change in flowering time, for example, was a consequence of natural selection for heavy-metal tolerance. Third, they found the pattern of adaptation repeated both among populations of the same species at different mines and among different species. Such repeated patterns of differentiation would be very unlikely to have occurred by chance alone. For all of these reasons, we conclude that the differences in heavy-metal tolerance between populations on and off the mine tailings were due to evolution by natural selection.

Ecotypes

Plant ecologists and taxonomists have recognized since before the time of Linnaeus, well over 250 years ago, that the same kind of plant may have a different appearance depending on where it is growing. Early in the twentieth century, researchers in Europe and in the United States began careful experimental work to better understand the causes of some of these differences. In Sweden, Turesson (1922) collected plants belonging to several dozen species from habitats all over Europe and grew them in a **common garden experiment** at a single site in Sweden. He found that many of the differences that he noticed when the plants were growing in their natural habitats were retained when they were all grown in a single environment. Turesson coined the term **ecotypes** to describe populations of a species from different habitats or locations that possess genetically based differences in appearance and function. Ecologists usually use the term "ecotype" to refer to such differences that appear to be adaptive.

At about the same time as Turesson was doing his experiments in Sweden, Clausen, Keck, and Hiesey (1940) were doing similar work in California. These researchers established three common garden sites at low, intermediate, and high altitudes, from the Carnegie Institute south of San Francisco, where they worked, up into the high Sierra (the spectacular Sierra Nevada that John Muir, the famous conservationist, called "the range of light"). Plants from a large number of species were collected along a **transect**—a long line along which samples are taken—running from the coast of California inland, up to the highest part of the Sierra Nevada and down the arid eastern side of the mountains (Figure 6.2). The plants were vegetatively propagated, and genetically identical copies of many different individual plants were grown in each of the three gardens. Clausen, Keck, and Hiesey

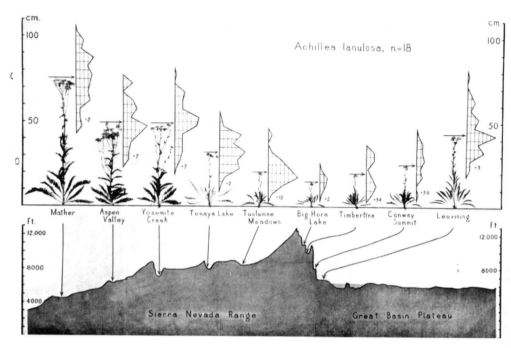

Figure 6.2
Part of Clausen, Keck, and Hiesey's transect across California from the foothills of the
Sierra Nevada to the Great Basin desert, showing (below) the height in altitude of the
sites from which plants were collected and (above) the appearance of *Achillea* plants
when grown in a common garden at Stanford near sea level. The small graphs show the
variation in height among individual plants from each site, with the specimens pictured
representing plants of average height and the arrows indicating mean plant height.
Note that horizontal map distances are compressed. (From Clausen et al. 1948.)

studied two groups of species most intensively: several
closely related species of *Achillea* (yarrow, Asteraceae),
and *Potentilla glandulosa* (cinquefoil, Rosaceae).

Like Turesson, the researchers found that many of
the differences in **morphology** (growth form) and **phenology** (the timing of seasonal events, such as flowering
and dormancy) among plants from different sites were
still present when those plants were compared in the
common gardens (Figure 6.3). *Achillea* plants from lower
sites were larger than plants from higher altitudes, with
longer leaves and taller flower stalks. Plants from higher altitudes generally entered dormancy sooner and
began growing later than plants from lower, warmer
sites. These alpine (high-mountain) plants also tended to
flower later than those from lower-altitude sites. Plants
generally had the highest survival rate and best performance in the garden with the conditions that most
closely resembled those in which they grew naturally
(Clausen et al. 1948). The researchers also noted striking
differences in the appearance of the leaves from different
sites along the transect. Leaves from the high-altitude
populations were not only smaller, but tended to have a
dense gray pubescence (a mat of short hairs on the leaf
surface), and were much more compact in shape. Leaves

from lower-altitude plants were smooth and green, and
were highly **dissected** (had a blade divided into small
but connected parts, making the leaf look very feathery).

Many years after these original studies, Gurevitch
and colleagues (Gurevitch 1988, 1992a,b; Gurevitch and
Schuepp 1990) returned to the original sites to carry out
further studies on the same populations that Clausen,
Keck, and Hiesey had worked on. As anyone who has visited or lived in California might expect, many of the original sites where plants had been collected are now under
asphalt. However, the Carnegie Institute still maintains
the sites at Mather and Timberline, where two of the
common gardens once were located, and plants from
these intermediate- and high-altitude populations were
collected for study. The populations differed genetically
in photosynthetic characteristics. There were also genetic differences between the populations in both leaf size
and leaf shape, particularly in the degree to which these
complex leaves were dissected. As the early researchers
had noticed, the Mather population had much more feathery, highly dissected leaves, while those from the Timberline plants were more compact (Figure 6.4).

Could there be an adaptive reason for these differences in leaf shape? Energy budget studies demonstrat-

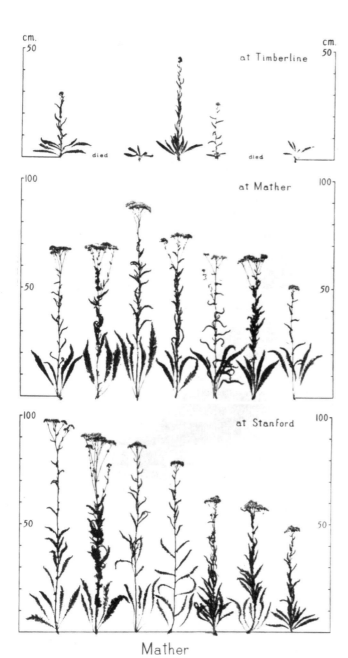

Figure 6.3
Appearances of seven clones from the Mather population of
Achillea lanulosa (collected at 1400 m elevation) when grown
in the three transplant gardens. Timberline (top; 2100 m),
Mather (center), and Stanford (bottom; 30 m elevation).
(From Clausen et al. 1948.)

dry environment at the lower altitude, leaves would
remain relatively cool, while high-altitude plants might
be able to warm up above the chilly air temperatures
common in their environment to maximize photosynthesis and growth.

Another study taking a similar approach was carried out on the widespread European species *Geranium
sanguineum* (Geraniaceae; Lewis 1969, 1970). This species
grows in a variety of habitats. As Turesson himself and
other researchers had noticed for other species, plants of
this species from the most **xeric** (dry) sites had the most
highly dissected leaves, with the narrowest lobes, while
those from more **mesic** sites (with more water available)
had larger, less dissected leaves (Figure 6.5). Lewis found
genetic differences in leaf shape and size among plants
collected from a large number of different sites in
Europe. There was a cline in the degree of leaf dissection, with the most highly dissected leaves from the driest environments and the more compact leaves from
moister, more oceanic sites. A different geographic cline
in leaf size correlated with the openness of the habitat,
with the largest-leaved plants from the least open, most
tree-shaded habitats. Based on physiological and energy budget studies, he hypothesized that differences in
leaf energy budgets due to the differences in leaf size and
shape would lead to differences in leaf temperatures.
The more dissected leaves from the drier habitats would
remain closer to air temperature, thus potentially avoiding overheating when drought forces stomatal closure.
Other physiological advantages were also predicted for
leaves of each ecotype in its "home" environment.

Originally, it was believed that there was a strict
dichotomy between clines and ecotypes. Clines represented gradual differences in genetic characters, responsible for gradual morphological and physiological differences over geographic space. Ecotypes, in contrast,
represented sharp differences among populations that are
distinct genetically and in various phenological traits.
Most plant ecologists no longer see this distinction as
being very useful. Genetic variation responsible for morphological, physiological, and phenological differences
may occur over many different scales, sometimes with
sharp distinctions among populations and sometimes
with gradual change. The abruptness of the change
depends on the nature of the evolutionary processes acting on these groups of plants: selection, genetic drift, and
migration all shape the rate of genetic change over space.
We may choose to call distinctive populations "ecotypes"
for convenience, or we may wish to emphasize the gradual nature of change and call the gradient a "cline." As
natural habitats grow ever smaller and more fragmented,
more and more cases of gradual genetic and phenotypic
change in plant and animal populations will become isolated, thus resembling ecotypes even though their origins
were as members of a continuously varying cline.

ed that the highly dissected leaves of the Mather plants
had a thinner boundary layer with greater heat conductance (see Chapter 3), making it more likely that they
would remain close to air temperature. The compact
leaves of the high-altitude plants had a thicker boundary layer and less heat conductance, and energy budget calculations suggested that these leaves could warm
considerably above air temperatures. Thus, in the warm,

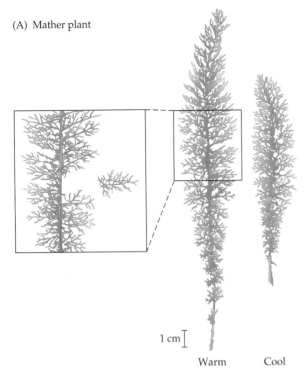

Figure 6.4
Leaves from the Mather (A) and Timberline (B) populations of *Achillea lanulosa* when grown together in common greenhouse environments under warm and cool conditions. An expanded part of the midsection of a Mather leaf is shown to show the complex shape. The longer Mather leaf is approximately 11.5 cm. long. (After Gurevitch 1992.)

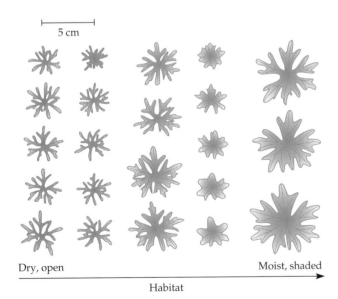

Figure 6.5
Differences in leaf shape in *Geranium sanguineum* collected from the field from a variety of habitats from northern to central and eastern Europe, then grown in a common garden. The sites ranged from dry, open sites (alvar, xeric limestone, left) to intermediate sites (steppe, woodsteppe–open woodlands, and coastal clifftops) to moist, shaded sites (woodland, right). (After Lewis 1972.)

Adaptive Plasticity

For the most part, plants are not mobile. This observation, while obvious, has profound implications for the evolution and adaptation of plants. If the environment becomes unfavorable in one spot, a mobile animal can move to a more favorable location. But what can a plant do?

Because plants cannot move, they must be able to tolerate environmental variation. As discussed in Chapter 5, one solution to this problem is for individuals to be phenotypically plastic, able to change their form so as to match the most fit trait value for the environment they find themselves in (Bradshaw 1965). For example, plants produce leaves with different characteristics when grown in the sun than when grown in the shade (see Chapter 2). Plants in general are far more plastic than animals; for example, while well-fed animals are taller and heavier than those with poorer diets, plants grown under optimal conditions can be orders of magnitude larger than genetically identical plants grown in poor conditions.

But how do we demonstrate that plasticity itself is adaptive, or investigate the conditions under which it is most advantageous? Many characteristics of plants vary with the environment. While some of this variation is adaptive, some changes in plant appearance and function are merely unavoidable consequences of plant physiology, such as having yellow leaves when

Figure 6.6
Impatiens capensis (Balsaminaceae) grown in the sun (high red:far-red ratio) and in the shade (low red:far-red ratio). Short, stout plants are favored in low-density conditions, while tall, thin plants are favored in crowded conditions. (Photograph courtesy of J. Schmitt.)

finding itself in this situation will benefit by elongating its stem as it grows. If it can detect the fact that it is growing in an uncrowded environment, it may be better off not elongating its stem very much. The chlorophyll in the leaves of neighboring plants absorbs heavily in the red part of the spectrum. Therefore, a plant that receives light with a low red:far-red ratio can use that information as a signal that it is surrounded by other plants.

Is the elongation response truly adaptive? Schmitt addressed this question by manipulating plant form and growing conditions. Seedlings were initially grown in a greenhouse under two conditions: (1) with filters that blocked the red part of the light spectrum and (2) with full-spectrum controls that reduced light by the same amount without changing the red:far-red ratio. These treatments created two sets of plants, elongated and suppressed. The experimental plants were then transplanted into a forest at either high or low density. Fitness was measured as the total number of seed capsules present at the end of the summer. As predicted by the adaptive plasticity hypothesis, the elongated form had higher fitness under crowded conditions, while the short, stout form had higher fitness at low densities (Figure 6.7).

The pattern of plasticity also differed in an adaptive way among populations growing in different environments. The researchers compared three populations

deprived of sufficient nitrogen. Demonstrating that phenotypic plasticity is adaptive requires, first, that there be more than one possible response, and second, that there is a consistent relationship between plasticity and fitness. The following study demonstrates how one can amass evidence on these points.

Johanna Schmitt and her colleagues studied populations of *Impatiens capensis* (jewelweed, Balsaminaceae) growing in different light conditions (Schmitt 1993; Dudley and Schmitt 1995, 1996). This species is an annual that grows in deciduous forests of the northeastern and north-central United States. Jewelweeds grow in conditions from full sun to forest understories, often along streams, and can reach very high densities.

A typical response to crowding in plants is stem elongation. This strategy allows a plant to overtop its close neighbors, thus gaining the most light (Figure 6.6). In uncrowded situations, however, plants remain short, because elongation has costs as well as advantages. More resources must be put into support structures rather than flowers and seeds, and the elongated stem is thinner and in greater danger of falling over. The elongation response is controlled by the ratio of red to far-red light that a plant receives (see Chapter 5).

Is it possible that a plant can tell whether or not it is growing near other plants, and thus at risk of eventually becoming shaded by its neighbors? If so, a plant

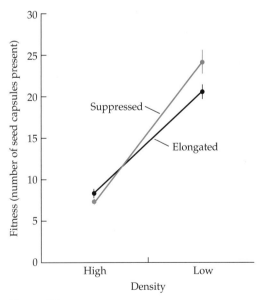

Figure 6.7
A test of adaptive plasticity of stem elongation in *Impatiens capensis*. Elongated plants were initially grown in a greenhouse with filters that blocked the red part of the light spectrum; suppressed plants were grown with full-spectrum controls that reduced light by the same amount without changing the red:far-red ratio. The two sets of plants were then grown at high or low density. The bars indicate 1 standard error. (After Dudley and Schmitt 1996.)

growing within 1 kilometer of one another, one in a woodland clearing and the other two in the woods. They grew plants from all three populations together in a greenhouse. The clearing population had a stronger elongation response than the other two populations. This result is predicted on adaptive grounds: In the woods, a plant that grows taller than its neighbors will still be in the shade. On the other hand, in a sunny site, a plant that overtops its neighbors will be in bright light. Thus, being more plastic is adaptive for the population in the clearing. As in our first example, here again the researchers showed the existence of the three necessary components of natural selection: plants varied in how plastic they were, differences in plasticity led to fitness differences, and plasticity was shown to be heritable. Finally, they showed that differences in plasticity among populations were in the direction expected if plasticity were evolving by natural selection.

The term "plasticity" is used in two different contexts by ecologists. We have already discussed one context, the plasticity of an individual. The other context is the plasticity of a species, meaning the range of ecological conditions that a species will grow in. This range is also called the species' **niche** (Grinnell 1917). The niche of a species is determined by the sum of the niches of its members (Figure 6.8). Those individuals may be plastic and may have wide or narrow niches. They may vary genetically in their optima. If they vary sufficiently, they may be referred to as ecotypes. So, a species may consist of a combination of generalist and specialist genotypes, while at the same time we speak of generalist and specialist species. In plant ecology, the idea that the total variation in a species can result from some combination of plastic and nonplastic genotypes traces to an influential paper by Bradshaw (1965). However, only in the past 20 years has this distinction begun to be more widely recognized (see reviews by Schlichting 1986; Sultan 1987; West-Eberhard 1989; Scheiner 1993; Via et al. 1995; Pigliucci 2001).

We also speak of two versions of the niche, the **fundamental niche** and the **realized niche** (Hutchinson 1957). The fundamental niche is the range of conditions that a species is physiologically capable of growing in. The realized niche is the range it is actually found in. Factors such as competition, herbivory, and lack of pollinators or seed dispersers all act to reduce a species' realized niche from the potential of its fundamental niche; these factors will be covered in more detail in Chapters 8 through 11, and Leibold (1995) offers a comprehensive review of niche concepts.

Selection in Structured Populations

The fitness of an individual depends on its environment. One component of that environment is population structure: the phenotypes of the other individuals in the population and their frequencies. In Chapter 7 we describe two types of population structure, age structure and stage structure, but there are also other sources of phenotypic variation among the individuals in a population, such as the colors or shapes of their flowers. When the fitness of an individual depends on the number and kinds of other phenotypes in the population, selection can take various forms. Two of these forms are frequency-dependent selection and group selection; here we explore examples of both types.

Frequency-Dependent Selection

Frequency-dependent selection occurs when the fitness of a genotype depends on whether it is common or rare in a population. An example is seen in the evolution of mating systems in *Lythrum salicaria* (purple loosestrife, Lythraceae; see Figure 14.3). This species is an invasive weed in wetlands of northern North America. It is still in the process of invading the continent, which makes it an ideal species for studying evolution in action. In the study described here, scientists were able to directly observe populations evolving over a five-year period.

Lythrum salicaria has an unusual mating system, called tristyly. The style is the female reproductive structure that supports the stigma, upon which pollen lands to fertilize the ovules (see Figure 8.6). This mating system gets its name because individual plants in a population have one of three different forms for the style: short, medium, or long (Figure 6.9A). The form of the style is genetically determined. The style lengths are matched by three stamen lengths (the stamen supports

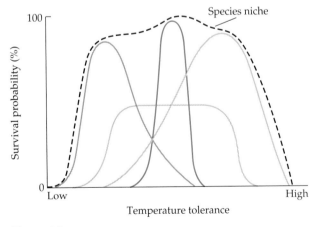

Figure 6.8
The niche of a species is the sum of the niches of its members. Here we consider one aspect of a niche, temperature tolerance. In this graph, each solid curve is the survival probability of a single individual as a function of temperature. The temperature range of the curve is one measure of plasticity of that individual. The dashed curve is, then, the niche of the species.

(A)

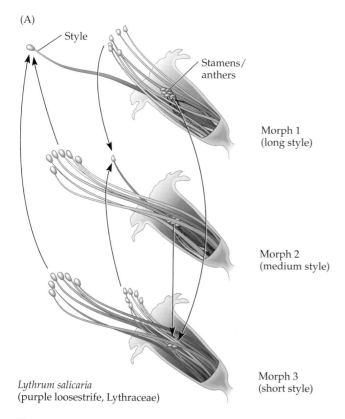

Lythrum salicaria
(purple loosestrife, Lythraceae)

(B)

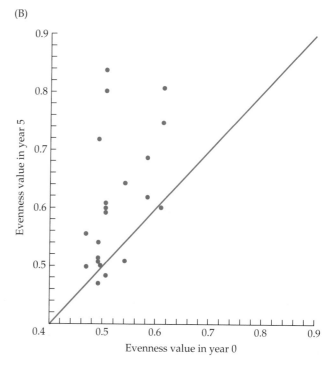

Figure 6.9
(A) Tristyly in *Lythrum salicaria* (purple loosestrife, Lythraceae). The plant has three different forms, or morphs, of stamen and style, a system that promotes outcrossing. An individual with a given style length can mate only with the other two stamen types. (B) Change in style length evenness over a 5-year period in 24 populations of *L. salicaria*. An evenness value of 1 would indicate that the three lengths were equally frequent. In all but 4 of the populations, the frequencies were more even in year 5 than in year 1, indicating that the rare morph had increased in frequency. (B after Eckert et al. 1996.)

the anthers, which hold the pollen). A single individual has a style of one length and stamens of the other two lengths. This system is a way of enforcing outcrossing. An individual with one style length cannot mate with another individual with the same style length; it can mate only with the other two types. (See Chapter 8 for more details on mating systems and tristyly.)

Frequency-dependent selection comes about because of this inability to mate with other individuals of the same type. Imagine a population of 99 long-styled individuals and 1 individual with short styles. That short-styled individual would be able to mate with all the other plants, as it could fertilize all the other plants in the population with its pollen. The long-styled plants would be limited to pollinating just the one short-styled individual. So in the next generation, the genes of the short-styled individual would end up in the offspring of 100 plants (itself through its female gametes and the

other 99 plants through its pollen). In contrast, each of the long-styled individuals would have offspring through at most 2 plants (themselves and the short-styled plant). As a result, the genes of the short-styled plant would increase in frequency. In general, the rarest mating type (the one with the least common style length) will always enjoy this sort of mating advantage. The population will therefore continue to evolve until all three types are at equal frequency, with one-third of the population having each of the style types.

Just such a pattern of evolutionary change was observed by Christopher Eckert and colleagues (1996), who surveyed 24 populations of *L. salicaria* across the Canadian province of Ontario. These populations were newly established as part of the invasion of *L. salicaria* into North America. During the creation of each population, some style length types had a high frequency while others had a low frequency due to chance alone. Sometimes it was the short type that was rare, sometimes it was the medium type, and sometimes it was the long type. The researchers predicted that over time the populations should evolve to an equal frequency of the three types. They tested this prediction by returning to the populations 5 years later. The prediction was met: In all cases, the rarest style length had increased in frequency (Figure 6.9B). Most of the populations were still not yet at equilibrium, with one-third of each type; that will require several more decades. However, the populations had consistently evolved in the predicted direction, something that would be highly unlikely to occur by chance.

This example is an instance in which natural selection is strong enough that we can observe its effects in action. It is unusual to be able to measure a consistent change in a natural population in just 5 years. Part of the strength of this study is that this change was observed across 24 different populations. Thus, it is highly unlikely that the change is due to anything other than natural selection. So, we can be confident that what we are observing is evolution by natural selection.

Group Selection

Group selection occurs when variation in group characteristics causes differences in fitness among groups. We briefly described the basis of group selection in Chapter 5 in the discussion of levels of selection. Group selection is one term used for selection above the level of the individual. A common form of group selection is **kin selection**, which occurs when the groups in question consist of closely related individuals.

The concept of group selection is somewhat controversial among evolutionary biologists. Kin selection is less controversial because the individuals in a group are closely related, and thus share large numbers of genes. No one disputes that group characteristics can affect individuals' fitness. However, group selection also implies that group-level characteristics can be subject to natural selection. Whether selection on group-level traits has occurred outside of kin groups is a matter of dispute.

An example of group-dependent fitness effects was found in populations of *Impatiens capensis* (Stevens et al. 1995). These plants often grow in dense patches, or groups of plants, as we saw in Schmitt's work on adaptive plasticity. *Impatiens* can produce both closed, self-pollinated flowers (**cleistogamous** flowers) and open, potentially cross-pollinated flowers (**chasmogamous** flowers) on the same plant. In this study, plant size and fitness were measured; fitness was measured as the number of either cleistogamous or chasmogamous flowers produced on each plant by the end of the growing season. The researchers then compared the flower production among groups (patches). Some patches had mostly small individuals, and thus a small mean plant size. Other patches had a few large individuals surrounded by small neighbors; in these patches the average plant size was large, due to the very large size of the dominant individuals. Both individual-level and group-level selection were found (Figure 6.10). Individual-level selection occurred within each patch, because larger individuals within a patch had greater numbers of both kinds of flowers. Group-level selection occurred among patches, because individuals in patches with a small mean plant size produced, on average, greater numbers of closed flowers than did individuals in patches with a large mean plant size. (There was no group-level selection on the number of open flowers.) That is, the char-

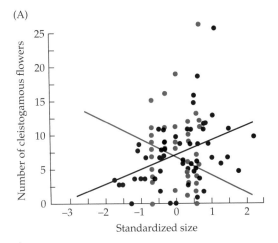

(A)

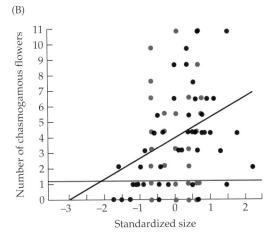

(B)

Figure 6.10
Individual- and group-level selection on size in *Impatiens capensis*. Fitness was measured as the number of cleistogamous flowers (A) and chasmogamous flowers (B) produced. The black lines show the relationship between an individual's flower production and the individual's size—a result of individual-level selection. The green lines show the relationship between an individual's flower production and the average size of the other individuals in its patch—a result of group-level selection. (After Stevens et al. 1995.)

acteristics of the group determined the number of flowers, and thus the fitness of individuals.

How could this occur? Individuals with small neighbors had greater numbers of cleistogamous flowers than those with large neighbors. In patches where the mean size of individuals was large, only the largest individuals were able to produce cleistogamous flowers. These very large individuals depressed the flower production of their smaller neighbors by shading them and reducing their access to resources. Yet the large individuals did not completely use up the resources in the patch; some resources went unused. In contrast, in patches where the mean size of individuals was small, more individuals had access to the resources, and because more of the total resources were used, all individuals had greater flower production. This effect of asymmetries in size leading to suppression of growth and flowering is discussed in detail in Chapter 10.

The researchers' conclusions about the existence of group selection and the mechanisms responsible were based on a combination of studies of natural populations and manipulative experiments involving clipping of plants (Stevens et al. 1995; Kelly 1996). The clipping experiments were especially important for providing information on the causative agent of natural selection. Manipulative experiments are an important complement to studies of selection in natural populations (see Chapter 1).

Convergent Evolution

One way to determine whether the form of a trait is due to adaptation by natural selection is to ask the question, "Do we find the same form repeated in similar environments?" Heavy-metal tolerance, for example, has been found repeatedly in populations from different mines. But how do we decide about adaptations if an entire species, or group of species, shows one form? In the Cactaceae, all species have CAM photosynthesis (see Chapter 2). These species also all grow in dry habitats or microhabitats. Is CAM photosynthesis an adaptation to low water availability? One alternative explanation is that plants in the Cactaceae all have CAM photosynthesis not because it is adaptive, but only because the ancestors of the Cactaceae had CAM photosynthesis. The adaptation of species in the Cactaceae to conditions

in dry habitats may have nothing to do with their possession of CAM photosynthesis; this association could just be a coincidence.

How do we separate such chance associations from those due to adaptation by natural selection? As before, we look for repeated associations. In this instance we look for repetition across different, unrelated groups of species. One type of such repetition is **convergent evolution**, the independent evolution of similar features in unrelated taxa. Because plants in many unrelated families found in arid environments or microhabitats have CAM photosynthesis, and because CAM photosynthesis evolved independently in many of them, we begin to become convinced that CAM photosynthesis offers adaptive advantages in these environments. Because we know the physiological mechanisms involved, we understand that CAM photosynthesis results in restricted water use and greater water use efficiency in plants that have this photosynthetic pathway. These multiple lines of evidence make it more believable that CAM photosynthesis is an evolutionary adaptation.

Another example of convergent evolution is found in the morphology of some of the cacti of the Sonoran Desert of North America and species in the family Euphorbiaceae growing in the deserts of southern Africa. Consider *Lophocereus schottii* and *Euphorbia stapfii*; these two plants look remarkably alike (Figure 6.11). Both are leafless, have green photosynthetic stems, are columnar in shape, have spines along the ribs of the stem, and have CAM photosynthesis. These species are very distantly related, so their similar features must be due to independent evolutionary events. Other mem-

(A)

(B)

Figure 6.11
An example of convergent evolution: two similar-looking desert-dwelling species. (A) *Lophocereus schottii* (Cactaceae), found in the Sonoran Desert of North America. (Photograph by D. Sokell/Visuals Unlimited.) (B) *Euphorbia spiralis* (Euphorbiaceae), found in Africa. (Photograph by J. N. Trager/Visuals Unlimited.)

bers of the genus *Euphorbia*, far more closely related to *Euphorbia stapfii*, look very different; the genus includes plants with a wide variety of growth forms, ranging from small herbaceous species to shrubs to large plants (the widely cultivated poinsettia is a member of this genus; *E. pulcherrima*). Another way that we know that the striking similarities between *L. schottii* and *E. stapfii* are due to convergent evolution is that some of the features have different developmental origins. The spines on cacti evolved from leaves, which can still be seen on primitive cacti. In contrast, the spines on the *Euphorbia* evolved from stipules (appendages to petioles). The independent evolution of such a large number of different features by chance alone is very unlikely. Therefore, we can be confident in our conclusion that these features are adaptations to conditions in desert environments.

Another striking example of convergent evolution is found among alpine plants in the families Campanulaceae and Asteraceae. *Lobelia telekii* grows on Mount Kenya in Africa and *Espeletia pycnophylla* grows in the Andes of South America. They are both unusual-looking plants, standing up to 2 meters tall, with a thick stem topped by a rosette of leaves (Figure 6.12). Their tall size is an adaptation to large day/night temperature swings. At night the temperature near the ground can get below freezing because the ground is losing large amounts of radiant energy to the very cold night sky. But the upper

parts of the plants, at 2 meters above the ground, stay above freezing due to longwave radiant energy input from the ground (see Chapters 3 and 18). Thus, the sensitive flowers and meristems never experience freezing temperatures. Both of these species evolved from herbaceous, ground-dwelling species growing at lower elevations. Again, the similarity of traits in species that are both distantly related and geographically separated, but grow in similar environments, is strong evidence that these features are adaptations shaped by natural selection.

Speciation

One outcome of evolution and natural selection is **speciation**, the production of new species. Speciation may occur when two or more populations of the same species become isolated from each other and adapt to different environmental conditions over a long time. Eventually, the populations' responses to differences in natural selection in their disparate environments can result in populations that have become so different that they are **reproductively isolated**, meaning they can no longer interbreed. Most biologists consider reproductive isolation to signify that two populations have become different species—which can also occur through processes other than adaptation. Usually, such differentiation happens in populations that are geographically distant from each other (**allopatric speciation**) (Figure 6.13), but

(A)   (B)

Figure 6.12
An example of convergent evolution: two similar-looking alpine tundra species.
(A) *Lobelia telekii* (Campanulaceae), found on Mount Kenya in Africa. (Photograph by
E. Orians.) (B) *Espeletia pycnophylla* (Asteraceae), found in the Andes of South America.
(Photograph by P. Jorgensen.)

Figure 6.13
Geographic ranges of three species in the genus *Gilia* (Polemoniaceae). The most likely explanation for the pattern of adjacent ranges is allopatric speciation of an originally widespread species.

it can occur in adjacent populations (**parapatric speciation**), or even within a single population (**sympatric speciation**). These modes of speciation are common to both animals and plants; however, plants undergo some unique speciation processes.

The **biological species concept** defines a species as a group of actually or potentially interbreeding organisms that are reproductively isolated from other such groups (Mayr 1942). A species thus forms an evolutionary unit within which populations are linked in their genetic changes through time. In most cases, species defined in this fashion are identical to species defined by taxonomists. **Taxonomic species** are generally defined based on shared morphological characters. One problem with using morphological criteria to define species, however, is the existence of **cryptic species**, individuals that appear to belong to the same species and yet are reproductively isolated from each other. In plants, cryptic species occur among polyploids (see below). More common are individuals that appear to be different, especially if they come from distant locations, yet are still able to interbreed. If those individuals are connected by others that are morphologically and geographically intermediate, they would probably be classified as members of the same species.

The reproductive barriers between species are not always impermeable, however. **Hybridization** occurs when members of different species of the same genus mate and produce viable offspring (Figure 6.14). ("Hybridization" as used by evolutionary biologists has a different meaning to agronomists, who use the same term to refer to crosses between different strains of a single species.) The extent of hybridization varies greatly among genera. In most genera, hybridization is a very rare event, if it occurs at all. On the other hand, some genera are notorious for hybridizing frequently, including *Quercus* (oaks, Fagaceae), *Crataegus* (hawthorns, Rosaceae), and *Atriplex* (saltbush, Amaranthaceae). In some cases species identities become very difficult to discern as the boundaries between species become indistinct, and we speak of the group of hybridizing species as a **hybrid swarm**. Hybridization appears to be much more common in plants than in animals. This difference may be due in part to the more passive nature of mating in plants, which must rely on wind, water, or animals to transport pollen from one flower to another. Hybrid swarms tend to be more common in wind-pollinated genera.

Another feature of evolution common to plants is **polyploidy**, the duplication of the entire set of chromosomes, resulting in two or more copies of the genome in each cell. Polyploidy can come about in two ways. First, consider a diploid ($2n$) individual that by some error produces pollen and ovules without meiosis (the reduction division), resulting in diploid gametes with $2n$ chromosomes instead of haploid gametes with $1n$ chromosomes. If that individual self-pollinated, its offspring would have four copies of each chromosome instead of the usual two; these offspring would be **tetraploid**, with $4n$ chromosomes. Such doubling does not require self-pollination; if two separate individuals both produced unreduced, diploid gametes and they mated with each other, again a tetraploid offspring would result. This process is referred to as **allopolyploidy** if the gametes come from individuals of different species in conjunction with hybridization. It is referred to as **autopolyploidy** if both gametes come from individuals of the same species.

Polyploidy does not have to stop at $4n$; further doublings can take place. Hybridization with $2n$ members of the same species can result in $6n$ offspring. In one species of saltbush, *Atriplex canescens*, individuals ranging from $2n$ to $20n$ are known. Ferns, which are a very ancient lineage, often have very large numbers of chromosomes. The *Polypodium vulgare* (Polypodiaceae) group has individuals with $2n = 74$, 148, and 222. The high base number is undoubtedly due to much older polyploidy events. The highest chromosome number known in any plant occurs in the fern *Ophioglossum reticulatum* (Ophioglossaceae), with $2n = 1260$. The modular developmental system of plants, at least in some taxa, is apparently able to handle these ploidy changes without difficulty. A typical consequence of polyploidy is an individual that is larger or more vigorous than its parents, but otherwise similar in appearance. The larger size comes about in part because cell size is proportional to

Genus *Clarkia*

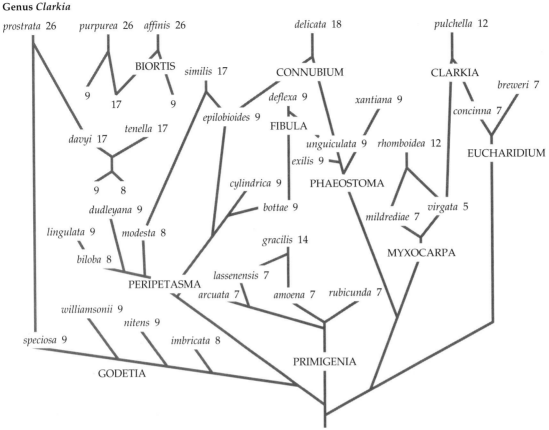

Figure 6.14
Hybridization among species in the genus *Clarkia* (Onagraceae). Names in uppercase letters are sections of the genus. Names in lowercase letters are species, given with their haploid chromosome number. Polyploid forms have arisen from hybridization between species. (After Lewis and Lewis 1995.)

nucleus size; more chromosomes result in a larger nucleus and, thus, a larger cell.

Individuals with odd ploidy numbers (e.g., $3n$ or $5n$) are also possible. Such individuals cannot produce gametes through typical meiotic processes because the chromosomes cannot pair properly. Through a process known as **agamospermy**, some plants can produce seeds without fertilization or meiosis (see Chapter 8); dandelions (*Taraxacum officinale*, Asteraceae) and grapefruit (*Citrus paradisi*, Rutaceae) are two such plants. (Sometimes pollination is required to initiate seed development, even though fertilization does not occur.) Seeds produced by agamospermy contain an exact duplicate of the parent's genome. Agamospermy is not restricted to species with odd ploidy numbers. Species with odd ploidy also can evolve to have even ploidy through chromosome doubling or hybridization.

Aneuploidy, the gain and loss of individual chromosomes, is another common way in which plant lineages change their chromosome number. Aneuploidy is especially common in lineages that are already polyploid.

In the genus *Hesperis* (Brassicaceae), for example, species are known in which $n = 7, 14, 13$, and 12. In this instance, the most likely evolutionary pathway was first a doubling of the number of chromosomes from 7 to 14, then a subsequent loss of first one, then a second chromosome.

One result of polyploidy is individuals that are instantly reproductively isolated. A $4n$ offspring cannot mate with its $2n$ progenitors, except by a new, rare hybridization event. Thus, under the biological species concept, such a polyploid individual is a member of a new species. In order for the species to become established, however, the polyploid individual must reproduce, and the population must grow. It is only because most plants produce both male and female gametes and are self-compatible that such individuals are likely to reproduce. *Atriplex* provides an interesting exception; as mentioned above, this genus contains many hybrids and polyploids, yet it is **dioecious** (male gametes and female gametes are produced on different individuals). Despite the reproductive isolation caused by polyploidy, taxonomists tend to classify individuals with different ploidy levels as members of the same species if their overall appearance is similar—an instance of taxonomic species differing from biological species.

Summary

Explaining the origin and maintenance of current plant traits is not a simple matter. We can look at a plant and its environment and make an educated guess about the evolutionary processes that shaped it. But elevating that guess to the level of scientific confidence can be difficult. To some extent, all traits are shaped by natural selection. However, other processes, such as mutation and migration, are also important in shaping traits. Sorting out which processes are responsible for current traits is difficult because different processes can create similar-looking outcomes. Evolutionary processes often work very slowly over long periods of time, making them difficult to observe in the short time span of most scientific studies.

Nevertheless, evolutionary biologists have been able to observe natural selection in action and infer its past action. Phenotypic selection has been measured in dozens of populations and species. By examining instances in which selection is strong (e.g., heavy-metal tolerance), we are able to find evidence for adaptation by natural selection occurring over just the past few hundred years. Much of our evidence comes from observing current patterns of variation among populations and species. These observations range from comparisons among nearby populations of the same species growing in different environments, to comparisons among populations separated by large geographic distances, to comparisons among species that are distantly related and geographically separated.

The sum of these evolutionary processes creates the amazing diversity of the living world that we see around us. This diversity can be seen on scales ranging from individuals to populations to species. Ecologists study this diversity to see how it fits together in today's world. Evolution provides the context within which this diversity arose, while ecology provides the context for the operation of evolution in the past, present, and future.

Additional Readings

Classic References

Clausen, J. D., D. Keck and W. M. Hiesey. 1940. *Experimental Studies on the Nature of Species*. I. *Effects of Varied Environments on Western North American Plants*. Carnegie Institute of Washington, Washington, DC.

Turesson, G. 1922. The species and the variety as ecological units. *Hereditas* 3: 100–113.

Contemporary Research

Kalisz, S. 1986. Variable selection on the timing of germination in *Collinsia verna* (Schrophulariaceae). *Evolution* 40: 479–491.

Lechowicz, M. J. and P. A. Blais. 1988. Assessing the contributions of multiple interacting traits to plant reproductive success: Environmental dependence. *J. Evol. Biol.* 1: 255–273.

Van Tienderen, P. H. and J. Van der Toorn. 1991. Genetic differentiation between populations of *Plantago lanceolata*. II. Phenotypic selection in a transplant experiment in three contrasting habitats. *J. Ecol.* 79: 43–59.

Additional Resources

Endler, J. A. 1986. *Natural Selection in the Wild*. Princeton University Press, Princeton, NJ.

Primack, R. B. and H. Kang. 1989. Measuring fitness and natural selection in wild plant populations. *Annu. Rev. Ecol. Syst.* 20: 367–396.

Travis, J. 1989. The role of optimizing selection in natural populations. *Annu. Rev. Ecol. Syst.* 20: 279–296.

CHAPTER 7

Population Structure, Growth, and Decline

*I*s a particular plant population increasing, decreasing, or staying constant in number? What is its genetic composition? How are the plants in the population arranged spatially? Questions of this kind are central to both basic and applied ecology. Foresters may be interested in maximizing the survival of trees that have been attacked by bark beetles, while conservation ecologists may seek to prevent the decline of a rare plant species. Evolutionists want to understand the action of natural selection on plant traits. Land managers want to control the spread of invasive weeds. All of them are asking questions that fall into the domain of **population ecology**—the study of population growth, composition, and spatial dispersion. Changes in a population, such as changes in the number of individuals alive, changes in the genetic composition or age structure of a population, or movement among a population's individuals are all part of a population's **dynamics**.

In studying population dynamics, we consider the properties of populations rather than individuals. For example, the wind may blow an individual seed a certain distance from its parent plant; but an entire plant population has the property *mean seed dispersal distance*. Individuals germinate and die on particular days, but populations have germination and death rates. These examples may be simple, but the distinction between individual and population properties can sometimes seem complicated. For example, the net transpiration and the photosynthetic rates of trees in a closed canopy depend both on the properties of the individual trees and on population properties such as the number, sizes, and spacing of individuals. Not only do populations have properties that individuals do not, but population-level properties can be more than just the sum of the properties of individuals. Thus, we have to study the populations themselves.

Plant population ecology dates only from the 1960s. Prior to that time, ecologists studied plant communities and the physiological ecology of plants, but did not generally think of plant populations as being composed of individuals that varied. In this way, they lagged behind animal ecologists, who had been acquiring demographic data and modeling population growth since the 1940s. John L. Harper, working at the University of Bangor in Wales, first introduced a strong population-level perspective into plant ecology. Harper and his students published a large number of studies in the 1960s and 1970s examining aspects of the growth of plant populations. Indeed, the first study to use a

José Sarukhán

population growth model tailored for plant populations was published in 1974 by Harper's student Jose Sarukhán, working with Madhav Gadgil (Sarukhán and Gadgil 1974). In the decades since this early work, there has been an explosion of work in plant population ecology; hundreds of papers are now published annually. Harper's students have in turn trained many plant ecologists all over the world. Sarukhán, for example, returned to Mexico and helped develop the Universidad Nacional Autónoma de México into an important center for ecological research, especially on plants.

Ecologists often distinguish between density-dependent and density-independent population growth. Some factors affecting population growth depend on the density of a population. For example, competition might be important in a crowded stand of plants, but much less important when there are greater distances between individuals. Thus, competition is usually thought of as a **density-dependent** factor. By contrast, if 10% of all juveniles were killed by a severe storm, regardless of population density, ecologists would say that the weather is acting as a **density-independent** factor.

Although this distinction between density dependence and density independence seems simple enough, nature is often more complicated. If a severe storm kills 10% of all the individuals *not* growing in sites protected from high winds, and protected sites are in limited supply, is the mortality density-dependent or density-independent? Clearly, the distinction between the two types of mortality is not very meaningful in such cases. Nevertheless, this distinction is often quite useful,* and in any case density independence is a good starting point for thinking about population growth.

This distinction must be applied cautiously to plants. In discussing the assessment of competition, Harper (1977) pointed out that the average density over a whole plant population is not the important quantity to measure. Because plants are sessile, they usually compete only with their neighbors, not with more distant plants in the same population (see Chapter 10). Thus, when assessing competition, one needs to think of density dependence on a very localized scale. Other spatial scales may be important for other phenomena—for example, many pathogens can disperse over considerable distances, so average density over a larger area may be a useful quantity to measure to understand the progress of disease in a population. A more general way to categorize these dif-

ferent kinds of phenomena would be to distinguish between changes in the population that depend on interactions among individuals (such as competition) and changes that do not require such interactions. Because ecologists use the terminology of density to describe population change, we will use it here as well.

This chapter concerns density-independent changes in population size. We discuss density-dependent changes in populations in Chapters 10 and 11.

Some Issues in the Study of Plant Population Growth

What controls the growth and decline of plant populations? Change in population size is determined by the numbers of births and deaths, plus the number of individuals immigrating into the population, minus the number emigrating from the population. Ecologists usually get data on changes in population size by conducting regular **censuses**—say, by counting individuals every year or every week. If we think of the first census as occurring at time t, and the next census as occurring at time $t + 1$, then we can describe the change in population size as

$$n_{t+1} = n_t + B_t - D_t + I_t - E_t \qquad (7.1)$$

Here the n's are the population sizes at sequential censuses. For example, if we were tracking a population of *Pinus ponderosa* (Ponderosa pine, Pinaceae), n_t would be the number of individual trees in the first year of our census, and n_{t+1} would be the number of trees in the following year. B_t is the number of births between the two censuses, D_t the number of deaths, I_t the number of individuals arriving (immigrating) from other populations, and E_t the number dispersing away (emigrating) from the population.

The relationship in Equation 7.1 may seem to be a truism. Though simple, it is a useful starting point because it helps us to focus on some major factors affecting plant population growth. It also points to some major differences between plant and animal populations and to some major conceptual problems in plant ecology.

Equation 7.1 implies several equivalent ways in which plant population size can be affected. It implies, for example, that the same population growth can be assured by increasing the number of births or by decreasing the number of deaths. As we will see in much of the rest of this chapter, life is not generally so simple unless all of the individuals in a population are demographically equivalent, and that is rarely the case. Large plants usually have different effects on the population than small ones, and old plants usually have different effects than young ones. Therefore, adding x births is not usually equivalent to reducing the number of deaths by x.

* Many important scientific ideas are like this: valuable in one context but not in another. That is because such ideas are simplified descriptions of nature— useful but limited models of patterns or processes (see Chapter 1).

But Equation 7.1 raises still more issues. What exactly do we mean by "birth" and "death" in plant populations? That depends on whether we mean births and deaths of genetically distinct individuals (**genets**) or of individuals that may be physiologically independent but are not necessarily genetically distinct (**ramets**; see Figure 5.6). One useful way to keep this distinction straight is to remember that every new genet is also a new ramet, but the reverse is not necessarily true: a new (potentially) physically independent individual is not necessarily a new genetic individual.

New genets are formed by the fertilization of an ovule and the maturation of the resulting seed (see Chapter 8). In plant species that never spread clonally (such as most annual herbs), each genet is a single ramet. Plant ecologists usually think of birth as occurring either when the seed is mature (i.e., the embryo is fully formed) or when the maternal plant sheds the seed. Seed germination is not considered birth because seeds are already new individuals—physiologically independent and (usually) genetically distinct from their parents long before they germinate. Moreover, they often spend a substantial amount of time, sometimes years, living in the soil before germination. Plant populations whose seeds have such a dormant phase are said to have a **seed bank**.

Many plants can spread clonally (also called vegetatively) by producing structures that contain tissues capable of producing an entire new plant, which is genetically identical to the parent. In these species, each genet is usually made up of multiple ramets, some of which may be physiologically independent from one another. Thus, discussing births and deaths in populations of these species means that first we must decide whether we are studying change in the number of genets or in the number of ramets. Studies of genet number would use a definition like the one above: newborns would consist only of new seeds. Studies of ramet number would count all new individuals, including both new seeds and new shoots, because all of them are new ramets. Which definition is best? That depends on the questions one asks.

This distinction between ramets and genets affects the way we think about dispersal between populations (the I_t and E_t terms) as well. Most mature plants are sessile—rooted in place—but (by the above definition) seeds are certainly plants, and they are usually mobile. Similarly, pollen grains are plants that are usually mobile (a pollen grain is the male gametophyte of gymnosperms and angiosperms, and is multicellular and genetically different from the sporophyte plant that produced it; see Chapter 8). It should not be surprising, therefore, that seeds and pollen, and their dispersal, have been and continue to be the subject of much ecological research. But some plants can disperse by clonal fragmentation as well. Many cholla cacti (*Opuntia* spp., Cac-

taceae Figure 7.1), for example, lose joints that are then moved by animals or runoff from storms; these may take root and grow into mature plants. At the other end of the moisture availability gradient, while dispersal of clonal fragments is rare in marine angiosperms such as *Zostera* spp. (eelgrasses, Zosteraceae; Barrett et al. 1993), it may be quite important in the establishment of new populations. Since most living plant tissues can, at least potentially, generate entirely functional new plants, dispersal by clonal fragmentation may be widespread.

Dispersal by clonal fragmentation has another implication. Seed dispersal means that an individual immigrating to one population must be emigrating from another. Clearly this is not so with dispersal by fragmentation: while the fragment itself may leave one population and move to another, the rest of the genet from which it was derived may stay put.

This does not mean that dispersal by fragmentation is a "free lunch." Indeed, it has an obvious consequence:

Figure 7.1
Many *Opuntia* cacti—especially the chollas, which have cylindrical joints—disperse partly by clonal fragmentation; some of the fragments root and establish new ramets. *Opuntia fulgida*, shown here, is known as "jumping cholla" because its joints very readily detach from the parent plant; unwary people walking by (including the authors) have been unpleasantly surprised by this plant! (Photograph by S. Scheiner.)

the fragmenting genet is packaged into smaller ramets. Since the survival and fecundity of ramets usually depend at least in part on their size, it is possible that clonal fragmentation might eventually affect the survival of the genet as a whole. It is an open question how the population dynamics of ramets and of genets affect one another.

Population Structure

The preceding discussion points to an important conclusion: When we study plant populations, we usually need to pay attention not only to the total number of individuals in a population, but to the classes of individuals and the relative frequencies of each class in the population as well. These classes are usually defined by age, life history stage (e.g., seed, seedling, sapling, canopy tree), size (e.g., 0–2 cm, 2–4 cm, >4 cm tall), or some combination of these categories. The **population structure** is a description of the relative frequency of each class (Figure 7.2).

Information about population structure is important for two reasons. First, different kinds of individuals have different effects on population growth. Adding 100 seeds to a population of the giant *Sequoia sempervirens* (California redwood, Taxodiaceae), for example, would obviously have a very different effect on the growth of the population than adding 100 mature redwood trees. If you were asked to census a tree population, you would almost certainly count seeds, seedlings, saplings, and mature trees as different classes, and you might make even finer distinctions. Ecologists need to know what kinds of individuals are in a population as well as their numbers. This is generally true for both animal and plant populations.

The second reason we need to know about population structure is particular to plants. Because of their extreme phenotypic plasticity, individual plants can vary over orders of magnitude in size, shape, physiological status, and consequently, in their importance for population growth (see Chapter 5). Animals certainly vary in their size and contribution to population growth as well. But we can often predict the **vital rates** of an animal—its reproductive rate and its chance of surviving—if we know only its age. Many animal populations (especially mammals and birds) can thus be thought of as **age-structured**. To describe such populations, we need to know how many individuals are in each age group. The methods used for the study of these populations are **age-based**, requiring only information on the population's age structure. Age-structured populations have another convenient property: they progress in only one direction. A bear that is x years old now will either be dead or age $x + 1$ next year.

In most plant populations, stage—a category based on size, physiology, or developmental status, such as

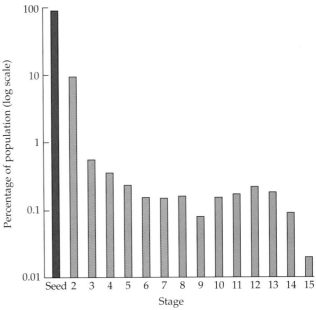

Figure 7.2
Stage structure of a population of *Pentachlethra macroloba* (Fabaceae), a tropical tree species, at Barro Colorado Island, Panama. Here, the stage classes above seed are based on size. Classes 2–7 (seedlings; light green) are in increments of 50 cm in height. Subsequent classes are based on diameter at breast height (DBH): class 8 = 2–5 cm, class 9 = 5–10 cm, class 10 = 10–20 cm, classes 11–14 are in 20 cm increments, and class 15 > 100 cm. Thus 88% of the population consists of seeds and 9% of seedlings 0–50 cm in height. The remaining stages (gray bars) comprise only 3% of the population. (After Hartshorn 1975.)

seed, seedling, juvenile, small individual, large individual—plays a stronger role in determining demographic performance than age; therefore, plant populations are generally thought of as **stage-structured** (see Figure 7.2). This should be obvious in the case of reproductive output because the number of flowers and fruits usually depends on plant size. Survival and subsequent growth usually also depend on plant size more than on age. Older plants are usually larger, but in most plants there is so much variation in the sizes of individuals of a given age (Figure 7.3) that it is often more useful to use **stage-based** methods than age-based methods to study plant demography.

That said, we hasten to add that chronological age can be important to know in plants for several reasons. Age is quite important in an evolutionary context; for example, if we want to understand how the genetic composition of a population changes over time, we need information about the ages of individuals. Similarly, if we want to study the evolution of life history traits (see Chapter 9), we often need information about the ages of individuals. As we will see in several examples in this chapter, age is sometimes an important determinant of

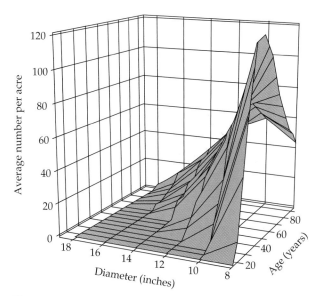

Figure 7.3
The average relationship between age and size in *Pinus palustris* (longleaf pine, Pinaceae) in the southeastern United States. The data show the average number of plants per acre of a particular size, given the age of the stand. Longleaf pine frequently grows in such even-aged stands. (After Forbes 1930.)

survivorship (the chance of surviving) and **fecundity** (the number of offspring contributed to the next generation). But age alone is not a very good predictor of these vital rates in plants; stage information is usually needed as well.

As recently as the 1980s, there was some doubt about the relative importance of age and stage in plant demography. Older textbooks, for example, often discussed the problem of "age- versus stage-based demography." It is now clear that when chronological age is important, it is usually important only within the context of a particular stage (see Caswell 2001). For example, as we can see in a life cycle graph for *Collinsia verna* (blue-eyed Mary, Campanulaceae; Figure 7.4), age plays an important role in the seed bank: one-year-

old seeds have a different chance of germinating than two-year-old seeds, and a different chance of surviving as viable seeds in the seed bank. Similarly, in *Cypripedium acaule* (pink ladyslipper orchid, Orchidaceae; Figure 7.5), the survival of corms from year to year depends on their age, and the survival of dormant plants from year to year depends on the number of years they have been dormant.

While age can be important, it is frequently impossible to determine a plant's age. Although many temperate-zone trees do produce reliable annual growth rings, most other plants—both herbs and woody plants—do not. Unless we are working with marked individuals of known age, then, we usually do not know the ages of plants (but see the discussion of long-lived plants on page 133).

Some Population Structure Issues Special to Plants

Why are plants so much more variable in their population structure than animals? A major reason is their **modular** growth form: an individual plant is a system of repeated units (nodes, lateral organs, and internodes: see Chapters 5 and 8). This modularity means that plants have very flexible growth patterns. It also means that they can lose large areas of their bodies and still survive. In other words, plants can actually shrink from year to year. In the ladyslipper orchid, for example, adults may not appear aboveground in some years (see Figure 7.5).

Stage structuring makes life complicated for plant ecologists. As mentioned above, when dealing with animals, one can be certain that in a year, all individuals now

Collinsia verna

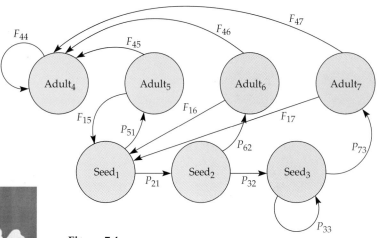

Figure 7.4
Life cycle graph of *Collinsia verna* (blue-eyed Mary, Campanulaceae), an annual plant of eastern North America with an age-structured seed bank. There are three age classes of seeds; when these germinate, the resulting vegetative plants are in different classes. The notation for life cycle graphs is explained on pages 123–124. (After Kalisz and McPeek 1992; photograph courtesy of S. Kalisz.)

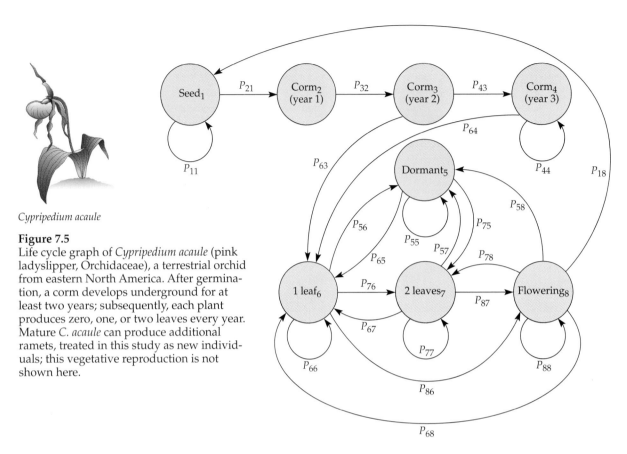

Cypripedium acaule

Figure 7.5
Life cycle graph of *Cypripedium acaule* (pink ladyslipper, Orchidaceae), a terrestrial orchid from eastern North America. After germination, a corm develops underground for at least two years; subsequently, each plant produces zero, one, or two leaves every year. Mature *C. acaule* can produce additional ramets, treated in this study as new individuals; this vegetative reproduction is not shown here.

x years old will be either dead or a year older. Even in many stage-structured animal populations, individuals can never move to an earlier stage—frogs do not become tadpoles. There are few such prohibitions in plants. Once germinated, a plant can never become a seed, but in most plant species established plants can grow, stay the same size, or shrink (Figure 7.6). Even trees, which we often think of as always growing larger, lose branches and sections of their trunks. Consequently, plant ecologists need

to keep track of many stages and possible transitions between them.

Sources of Population Structure

Individuals within plant populations vary in many characteristics, including size, morphology or developmental status, genotype, and physiological status. Size is the most commonly studied factor structuring plant populations. Size can be measured in several ways: as a quan-

Argyroxiphium sandwicense

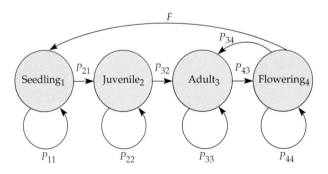

Figure 7.6
Argyroxiphium sandwicense (Asteraceae) are mainly semelparous, reproducing only once in their lives; rare individuals, however, revert to vegetative growth after flowering, or flower repeatedly. Endemic to the Hawaiian Islands, *Argyroxiphium* is one of three genera collectively known as the Hawaiian silverswords. The life cycle graph is for *A. sandwicense* ssp. *sandwicense* (Mauna Kea silversword). The photograph shows the closely related *A. sandwicense* ssp. *macrocephalum* (Haleakala silversword). (After Powell 1992; photograph courtesy of R. Robichaux.)

tity such as weight or height or as the number of modules (e.g., branches, tillers in a grass, or leaves) a plant has. Tree size is often measured as **diameter at breast height (DBH)**; curiously, "breast height" is defined as 1.3 meters in some countries and 1.4 in others. Morphological or developmental status is usually important in structuring populations: how many individuals are seeds, how many seedlings have become established, how many plants have become reproductively active? Genetic variation is often important in structuring populations, and physiological status is probably often important as well.

This list of categories may seem a bit overwhelming, but categories such as size, morphology, age, genotype, and physiological status are important for population studies only if there are demographic differences among the classes. For example, it might be easy to measure the difference in size between 30-, 60-, and 90-cm DBH trees of the same species, but it may not be useful to use these size differences as size classes if the trees of each size did not differ from the other sizes in their chances of survival, growth, or reproductive output.

Currently, there are real limits to our abilities to measure this variation, to tell how important it is in structuring a particular population, and to perform appropriate computations. Consequently, most ecological studies today concentrate on only one or a few of these categories of variation. Put differently, we know that reality is much more complex than our demographic analyses.

Studying Population Growth and Decline

Is a population growing or declining, and at what rate? This might seem like a simple question to answer: compare the number of plants last year (call it n_t) with the number this year (call it n_{t+1}). The rate of change in population size is then n_{t+1}/n_t. If the population is growing, this ratio is greater than 1; if the population is declining, the ratio is less than 1.

Unfortunately, life is this simple only when there is no population structure. Certainly it is true that if $n_{t+1} > n_t$, there are more plants now than there were last year. But in a structured population—that is, in most plant populations—individuals of different classes make different contributions to future population growth. This means that their short-term effects can differ from their long-term effects on the population, and we need to be able to ask about both kinds of effects.

Consider a population of a long-lived tree species. The oldest trees have a period of senescence that lasts up to several decades, during which they do not flower but slowly lose limbs and experience heartwood rot. Young trees cannot flower until they are several decades

old. Thus neither very old nor very young trees are reproducing, so a population of 100 very old or very young trees will simply decline in number as some die and none reproduce.

Now imagine that year-to-year survival rates are the same among young and old trees, and compare a population composed entirely of old trees with one composed entirely of young trees. In the short run, both populations will decline at the same rate. But over time, this will not be true, because the surviving trees in the young population will eventually reach maturity and begin reproducing, so the population will grow. By contrast, the old population will disappear. Therefore, studies of population growth and decline in plants must take into account the structure of the population and look at both long-term and short-term growth.

Life Cycle Graphs

To learn how population structure affects population growth, we need a picture of how individual plants can shift from stage to stage between censuses. **Life cycle graphs** such as those in Figures 7.4 through 7.7 provide useful summaries of information about transitions between stages. Each node in a life cycle graph represents a stage. Note that some stages are defined by developmental status (e.g., seeds and dormant individuals), while others are defined by their size or age within a developmental stage. Arrows between stages describe "transitions" between those stages—survival and reproduction—during the interval between censuses.

Let's look at the life cycle graph for *Coryphantha robbinsorum* (pincushion cactus, Cactaceae) (Figure 7.7). (This endangered cactus species will be used as an example throughout the next few sections of the chapter.) The arrow from large juveniles to adults represents the probability (P) of a large juvenile at one census becoming an adult by the next census. The self-loops—arrows from a node to itself—represent the probability of remaining in the same stage. Thus, the arrow from small juveniles to small juveniles in Figure 7.7 represents the probability that a small juvenile at one census will still be a small juvenile at the next census.

The arrow from adults to small juveniles in Figure 7.7 needs a more careful interpretation, which points to an important lesson about life cycle graphs and the matrix models that can be derived from them. This arrow refers to the number of small juveniles produced by the adults. Biologically, we know that every individual starts as a seed, and that there is generally some time between seed maturation and germination (there are a few viviparous plant species that germinate right on the maternal plant, but *Coryphantha* is not one of them). A complete biological diagram of the life cycle would include these steps. But we are trying to describe

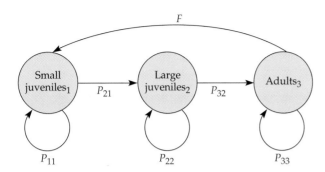

Coryphantha robbinsorum

Figure 7.7
Life cycle graph of *Coryphantha robbinsorum* (pincushion cactus, Cactaceae), an endangered cactus from Arizona and Sonora, Mexico. As this species has no seed bank and plants were censused annually, the graph does not include a seed stage. The transition between adults and small juveniles is thus the product of (rate of seed production) times (chance of a seed surviving) times (chance of germination) times (chance of a seedling surviving to the first census). (Data from Schmalzel et al. 1995; photograph courtesy of U.S. Fish and Wildlife Service.)

the demography of the life cycle as examined in a real field study. Since this species has no seed bank and the censuses are annual, no individuals in the seed class could have been counted unless the census were timed to include them. To represent *Coryphantha* demography as measured, then, we need to estimate fecundity:

F = (rate of seed production) × (chance of seed survival) × (chance of germination) × (chance of seedling surviving to the first census) =
(the rate at which adults one year produce small juveniles the next)

If we wanted to study the seed stage per se, we would need censuses that were timed differently, and possibly more frequent censuses. This may seem like a quibble, but it turns out that the number of classes we include

in a life cycle graph or matrix model can greatly affect most calculations for which the graph or matrix model is subsequently used.

Once we have a life cycle graph, we can then use field data to estimate values for the various transitions (Figures 7.4–7.7; Box 7A). Often it is impossible to observe all of the transitions directly using the same indi-

---

Box 7A

Demography of an Endangered Cactus, Coryphantha robbinsorum

Coryphantha robbinsorum is a small, cluster-forming cactus found on limestone outcroppings in southern Arizona and adjacent Sonora, Mexico. Schmalzel et al. (1995) marked plants at three sites on a hill and followed their growth, reproduction, and survival for a five-year period.

Because adults (plants that had flowered at least once), small juveniles (plants smaller than 11 mm in diameter), and large juveniles had significantly different chances of survival and reproduction, the authors used these and seeds as their stage classes. Since *C. robbinsorum* has no seed bank (see Figure 7.7) and the censuses were annual, we re-analyzed the authors' data with

a model that does not include a seed stage. The data can be used to create matrix models for each of the three sites:

$$\mathbf{A}_{\text{site A}} = \begin{bmatrix} 0.67 & 0 & 0.56 \\ 0.02 & 0.85 & 0 \\ 0 & 0.14 & 0.87 \end{bmatrix}$$

$$\mathbf{A}_{\text{site B}} = \begin{bmatrix} 0.49 & 0 & 0.56 \\ 0.01 & 0.73 & 0 \\ 0 & 0.23 & 0.99 \end{bmatrix}$$

$$\mathbf{A}_{\text{site C}} = \begin{bmatrix} 0.43 & 0 & 0.56 \\ 0.33 & 0.61 & 0 \\ 0 & 0.30 & 0.96 \end{bmatrix}$$

Analysis of these matrices reveals that the population at site C is growing, because $\lambda_1 = 1.12$. At the other two sites, the population is stable or declining slowly: $\lambda_1 = 0.998$ at site A and $\lambda_1 = 0.997$ at site B. The stable stage distributions differ among these plots, as Figure 7.9 shows. Most notable is the predicted scarcity of large juveniles at sites A and B. The plots also differ in the rates at which they approach the stable distribution. Sensitivities are shown in Table 7.2, elasticities in Figure 7.10, and reproductive values in Table 7.1.

viduals. For example, it is difficult to use the same individuals to estimate the survival and germination rates of seeds as well as subsequent transitions, because seeds usually cannot be marked, and even when they can be, accurate censusing usually requires digging them up—which can certainly bias their future chances of surviving or germinating!

In a large study of *Collinsia verna*, Susan Kalisz and Mark McPeek (1992) addressed this problem by estimating seed survival rates destructively but independently from their estimates of the vital rates of aboveground plants (see Figure 7.4). They were able to do so because *C. verna* grows in very large populations, making it possible to randomly select separate areas for studying aboveground and belowground demography.

On the other hand, in the ladyslipper orchid study (see Figure 7.5), Margaret Cochran and Stephen Ellner (1992) were forced to make several assumptions about mortality rates in dormant plants and seeds. It would have been more satisfactory to observe the belowground events directly, but this would have destroyed study areas, and might have added to the effects of poaching on these orchids. There are no simple or general solutions to this problem.

Matrix Models

How do the rates of survivorship and fecundity affect the growth of a population as a whole? Which stages have the strongest effects on population growth? Questions like these are important in many contexts, including conservation biology (on which stages should we concentrate protection efforts?), population ecology (which stage is most likely to limit population growth?), and evolution (on which stage can natural selection have the greatest effect?).

One approach to answering such questions is to use estimated survivorship and fecundity rates to model the population's growth rate, using matrix demographic methods developed over the past two decades. These models can be used to ask how the population growth rate changes as the specific survivorship and fecundity rates in each class change. All of the information needed for this method is contained in the life cycle graph.

To see this, consider the *Coryphantha* example (see Figure 7.7). The life cycle graph tells us that small juveniles are produced by adults at a rate F (as noted above, F is a composite of several factors). For convenience, we number the stages from 1 to 3, so that every symbol we use can be interpreted. Thus n_1 means the number of small juveniles, n_2 the number of large juveniles, and n_3 the number of adults. Each transition has two subscripts—the first refers to the stage this year, and the second to the stage last year. Thus P_{11} is the probability of remaining a small juvenile, P_{32} is the probability of a

large juvenile becoming an adult, and so on. Using these symbols, we can write an equation for the number of small juveniles in year $t + 1$:

$$n_1(t+1) = P_{11}\, n_1(t) + F\, n_3(t) \qquad (7.2)$$

In other words, the number of small juveniles this year is the sum of the individuals that remain small juveniles plus new juveniles; that is, the number of small juveniles last year times the chance that they survived as small juveniles, plus the number of adults last year times their rate of production of small juveniles.

Similarly, the number of large juveniles is given by

$$n_2(t + 1) = P_{21}\, n_1(t) + P_{22}\, n_2(t) \qquad (7.3)$$

That is, the number of large juveniles this year is the number of small juveniles last year times the chance that they became large juveniles, plus the number of large juveniles last year times the chance that they remained large juveniles.

Finally, the number of adults is

$$n_3(t + 1) = P_{32}\, n_2(t) + P_{33}\, n_3(t) \qquad (7.4)$$

The number of adults this year is the number of large juveniles last year times the chance that they became adults, plus the number of adults last year times the chance that they survived as adults.

A **transition matrix model** is a compact way of writing the same thing. The matrix arranges the coefficients from the life cycle graph (the F's and P's, shown as a matrix in the middle of Equation 7.5). This matrix thus represents the survivorship and fecundity rates. Next a vector is assembled consisting of the number of individuals of each stage; in Equation 7.5, this is the column on the right-hand side. To obtain a vector of the number of individuals in each stage next year—the column on the left-hand side of Equation 7.5—we multiply the vector by the matrix. In Box 7B, we explain the rules for multiplying the vector by the matrix. A matrix model of the life cycle graph in Figure 7.7 can thus be written as

$$\begin{bmatrix} n_1(t+1) \\ n_2(t+1) \\ n_3(t+1) \end{bmatrix} = \begin{bmatrix} P_{11} & 0 & F \\ P_{21} & P_{22} & 0 \\ 0 & P_{32} & P_{33} \end{bmatrix} \times \begin{bmatrix} n_1(t) \\ n_2(t) \\ n_3(t) \end{bmatrix} \qquad (7.5)$$

More general rules for manipulating matrices are given in Caswell (2001).

The principal reason for using matrices is that there are standard rules for manipulating them, and these rules make it much easier to draw some important ecological conclusions about the population. If we assume (for the moment) that the birth and survival probabilities (the coefficients) stay constant (so that each year we multiply the population vector by the same matrix), we can find

Multiplying a Population Vector by a Matrix

To understand how Equation 7.5 can say the same thing as Equations 7.2 through 7.4, you need to know how to multiply matrices and vectors. When you multiply a vector by a matrix, the result is a vector. In this case, the initial vector is the numbers of individuals in each stage this year, the resulting vector is the numbers of individuals in each stage next year, and the matrix is the survivorship and fecundity rates. Matrix multiplication is "row-by-column." To get the first element in the population vector for next year (the number of small juveniles), multiply the coefficient in the first row, first column (P_{11}), by the first element in the vector for this year [$n_1(t)$] to get $P_{11} n_1(t)$. Then multiply the coefficient in the first row, second column (0), by the second element in the vector [$n_2(t)$] to get 0. Finally, multiply the coefficient in the first row, third column (F), by the third element in the vector [$n_3(t)$] to get $Fn_3(t)$. Then sum these three products to get $n_1(t + 1) = P_{11} n_1(t) + 0 + Fn_3(t)$, which is exactly what Equation 7.2 says.

To find the second element in next year's vector (the number of large juveniles), repeat this process, but multiply each of the elements of this year's vector by the coefficients in the second row of the matrix.

Another useful way of thinking of the matrix is that it describes the transitions in "from-to" form. The matrix element in the ith row and jth column always refers to the transition from the jth stage to the ith stage—from the "column-number" stage to the "row-number" stage.

- The short- and long-term population growth rates
- The population structure at any time in the future
- The reproductive value of each age or stage class. Roughly speaking, the reproductive value of an individual in class x is its expected contribution to future population growth. In other words, reproductive value is a way of evaluating the relative demographic importance of the different classes
- The sensitivity and elasticity of population growth to changes in the specific probabilities of survival and reproduction. Sensitivity tells us how absolute changes (e.g., how adding 0.01 to a survivorship coefficient) in the survival and fecundity of each class would affect population growth rates; elasticity tells us how proportional changes (e.g., how increasing a survival coefficient by 1%) would change the population growth rate.
- The relationship between the ages of individuals and their stages

Matrix models and the estimates of growth rates, reproductive value, sensitivity and elasticity that they provide are important tools in evolutionary ecology, conservation biology, and applied ecology.

Analyzing Matrix Models

How do we get all of this information from matrix models? In this section we introduce some major ideas used in analyzing these models. The basic method used to analyze matrices depends on an important observation: if we repeatedly multiply the **population vector**—the vector of the number of individuals of each stage—by the transition matrix, after a while the population vec-tor attains a stable structure or **stable stage distribution**, at which point the proportion of individuals in each class stays constant each generation, although the population keeps growing. This observation implies that, at least once the population has reached its stable structure, we can multiply the population vector by a single number (a scalar) rather than by the entire matrix and get the same result as if we were multiplying by the matrix.

How do we find this number? Mathematically, we are looking for a number, which can be denoted by λ (the Greek letter lambda), based on the equation $\mathbf{A} x = \lambda x$, where $\mathbf{A}$ is the transition matrix and x is a vector. (In this chapter we will use standard mathematical notation: capitalized boldfaced symbols, such as $\mathbf{A}$, are matrices, and lowercase boldfaced italic symbols, such as x, are vectors.) It is always possible to find such numbers (the values for λ) for transition matrices; this is almost always done numerically on a computer. Numbers that can play this role are called **eigenvalues** or **characteristic values**. Every such number λ has a corresponding population vector x, called an **eigenvector** or **characteristic vector**. For a population with N stages, the transition matrix has N rows and columns. There are also N eigenvalues. Each eigenvalue has a corresponding eigenvector. In most cases, all of the eigenvalues are distinct, although under some circumstances there are pairs of duplicated eigenvalues.

The number λ tells us some important things about a population. If $\lambda > 1$, the population is growing, while if $\lambda < 1$, it is declining in number. The population remains at a constant size only in the special case that $\lambda = 1$. But what are these eigenvalues and eigenvectors? The eigenvalues are components of the population's

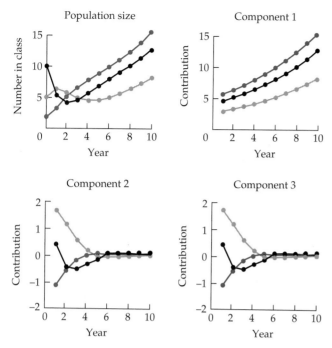

Figure 7.8

Growth of the three stage classes in the *Coryphantha robbinsorum* population at site C, and convergence to the stable stage distribution. The panel on the upper left gives the total population size for each of the stage classes. The other three panels show the contributions of each pair of eigenvalues/eigenvectors to the change in population size. By the eighth year, the size change for each class is within 0.01 of λ = 1.12; only the dominant eigenvalue is making a substantial contribution at this point. (After Fox and Gurevitch 2000.)

where the c's depend on the initial conditions (Caswell 2001), x_i is the ith eigenvector, and t is the number of time steps. Raising the eigenvalues to a power (equivalent to repeated multiplication by the matrix) makes the smaller eigenvalues decline in relative importance over time. Recent studies have taken a look at methods for studying populations that are far from their stable distribution (Fox and Gurevitch 2000).

But Real Plants Live in Variable Environments

By now you might reasonably be wondering whether this approach to plant populations makes much sense. In our discussion of matrix models, we have assumed that a population experiences fixed survival and birth rates. Then we have concentrated on what happens eventually in a constant environment. Since one of the more obvious facts about ecology is that things change, and since factors influencing populations (such as the weather) can be quite variable (see Chapter 18), the unrealistic assumptions of these matrix models might seem to render any conclusions based on them invalid (see Chapter 1).

Matrices and life cycle graphs are nevertheless quite useful, but their utility depends on how one interprets the results. We can use these models for two very different purposes: to try to predict actual population growth and composition at some point in the future, or to ask what would happen to the population if present conditions persisted. Only the latter usually makes much sense.

growth rate, and the eigenvectors are components of the population's structure. A very important result is that one eigenvalue is always larger than the others for this type of matrix. As a result, over time, the population growth rate approaches this value (which is therefore called the **stable growth rate**; Figure 7.8), and the population structure approaches the associated eigenvector (the stable stage distribution; Figure 7.9). The eigenvalues are often ranked by their size, so that λ_1 usually refers to the largest (**dominant** or **leading**) **eigenvalue** and λ_N to the smallest. When authors discuss λ without any subscripts, they usually mean the dominant eigenvalue.

Even when the population is not yet near its stable distribution, its growth can be predicted using the eigenvalues and eigenvectors (Caswell 2001). The population's size at any time in the future can be written as a weighted sum of the products of the eigenvalues and eigenvectors:

$$n(t) = c_1\lambda_1^t\boldsymbol{x}_1 + c_2\lambda_2^t\boldsymbol{x}_2 + \ldots = \sum_{i=1}^{n} c_i\lambda_i^t\boldsymbol{x}_i$$

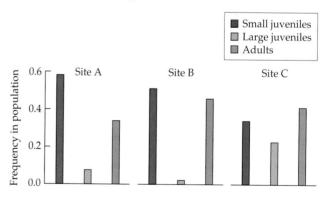

Figure 7.9

Stable stage distribution for the cactus *Coryphantha robbinsorum* at three sites. Sites A and B (on hillsides) have similar stable population structures. The stable structure at site C (on the hilltop) has many more large juveniles. At this site, 33% of small juveniles become large juveniles, as compared with only 1–2% at the other two sites. This is also the principal factor causing λ to be much larger at site C than at the other two sites (1.12 at site C; just below 1).

It follows from this that matrix approaches are helpful in studying the year-to-year variation experienced by plant populations. For example, in their study of the annual *Collinsia verna*, Kalisz and McPeek (1992) found that in one year the population grew substantially, with λ estimated as 1.80, but in the next year it declined, with λ estimated as 0.41. In another example, hurricanes were known to cause large-scale variation in the population growth rates of *Pinus palustris* (longleaf pine, Pinaceae), but matrix models showed that there is substantial variation in growth rates among even "normal," non-hurricane years (Platt and Rathbun 1993; Platt et al. 1988).

Finally, concentrating on the dominant eigenvalue and eigenvector does not mean that one is studying things that would be important only in the long run, and only in a constant environment. As noted above, short-term population growth can be analyzed as a weighted sum of the eigenvalues and eigenvectors of a matrix. The dominant eigenvalues and eigenvectors are part of this sum, so in studying them, one is always studying a major component of short-term population growth (Caswell 2001). Many plant population matrices seem to approach the stable stage distribution fairly rapidly—usually within five or ten years (see Figure 7.8). This can occur only if the dominant eigenvalue is much larger than the others, and therefore it is a very important component of short-term population growth.

Thus, one need not know the entire history of a population to understand the reasons for its stage structure. The stage structure will be the result of the average survivorship and fecundity rates even when the matrices vary randomly from year to year, as they might due to environmental variation (Tuljapurkar 1990).

Lifetime Reproduction: The Net Reproductive Rate

How many offspring does an average individual produce in its lifetime? For age-structured populations, this quantity is easily estimated from the data in a transition matrix, using some simple mathematics. Call l_x the probability of surviving from the first census to the xth census. The calculation of l_x amounts to multiplying all the individual survival probabilities up to that point. For example, the probability of a newborn reaching age 3, l_3, is the probability of its reaching age 2, P_{21}, times the probability of an individual age 2 reaching age 3, P_{32}. It follows from this definition that $l_1 = 1$. In an age-structured population, l_x cannot increase over time: if 1000 individuals are censused starting after their birth, at each subsequent census the number surviving must be the same as at the prior census, or smaller.

We can use F_x to represent the fecundity of an average individual in age class x.* This means that, averaging over the whole population, a newborn individual can be expected to have $l_x F_x$ offspring at age x, and $l_{x+1} F_{x+1}$ at age $x + 1$, and so on. Over its entire life, an individual plant can be expected to have R_0 offspring:

$$R_0 = l_1 F_1 + l_2 F_2 + \ldots + l_x F_x + \ldots = \sum_{i=1}^{\infty} l_i F_i$$

(7.6)

R_0 is called the **net reproductive rate**. In general, it is *not* equal to λ, unless the population is at equilibrium (has reached a stable stage or age distribution, with λ = 1).

R_0 is obviously a useful quantity to know. Because it is based on age, it was not possible to estimate R_0 for stage-structured populations until recent advances in the analysis of stage-structured matrices (see "Age and Stage Revisited," p. 132).

Reproductive Value: The Contribution of Each Stage to Population Growth

Individuals of different stages do not make equivalent contributions to future population growth. In the *Coryphantha* populations described in Box 7A, for example, most small juveniles died, while most adults survived for a long time. If the manager of a preserve wanted to establish a new population, it might be better to introduce *n* adult plants than *n* seeds or small juveniles (if enough adults were available, and if they could survive transplanting well enough). But since seeds are generally much easier to work with, one might be tempted to introduce them instead.

Instead of guessing how many individual seeds or adults to introduce, it would be helpful to have a way of measuring the effects of different kinds of individuals on future population growth. The **reproductive value** of the different stages or ages gives us precisely that information. The reproductive value of stage *x* is the contribution an average individual now in stage *x* will make to the next generation before it dies. For simplicity, we first discuss reproductive value for age-structured populations, then extend the concept to stage-structured populations.

It might seem at first that reproductive value could be estimated by summing the quantities in R_0 over a shorter interval—for example, for an individual of age *x*, one might start the sum at *x* instead of 1. However, this does not provide the information we need. Summing the items in Equation 7.6 from age *x* on will always give a smaller number than summing over an entire lifetime. Summing from age *x* on provides only the number of offspring a newborn can be expected to have from age *x* on, instead of the number expected over its whole life. Moreover, this sum over the shorter period does not account for the fact that the population will have changed in size by the time a newborn reaches age *x*. Its future reproduction will have a different effect on the population than its present reproduction, because a new-

*There is some confusion in ecology texts over the notation to use for fecundity; you may have seen fecundity represented as m_x. That symbol is best reserved for continuous-time models. In this book we use the notation suggested in Caswell 2001.

born that begins its life sometime in the future will be part of a larger (or smaller) population. Calculating the effect of individuals of age x on future population growth therefore requires accounting for the chance of surviving to age x, as well as for the change in population size in that interval.

This leads us to a second, and more precise, verbal definition of reproductive value (R.V.):

$$\text{R. V.} = \frac{\begin{array}{c}\text{proportion of future births}\\ \text{in the population}\\ \text{to individuals now age } x\end{array}}{\text{proportion of the population now age } x} \quad (7.7)$$

Both the numerator and denominator of this expression are related to the sum in Equation 7.6, but they also take into account the rate at which the population is growing. To see how this works, consider a population with n individuals, growing at the rate λ. A single offspring born to a parent of age x now constitutes $1/n$ of the population. A single offspring born in the next interval will constitute $1/(\lambda n)$ of the population, one born two time-steps later will constitute $1/(\lambda^2 n)$ of the population, and so on. In a growing population, offspring born later in life will contribute less to the next generation than offspring born earlier. The reverse is true in a declining population.

Thus, to calculate the numerator in Equation 7.7 (the proportion of future births to individuals now age x), we need to sum the expected future reproduction at each age (from x to death) as in Equation 7.6, but we also need to *divide* each element l_iF_i by λ^i:

$$\text{numerator} = \sum_{i=x}^{\infty} \frac{l_iF_i}{\lambda^i} \quad (7.8)$$

The denominator in Equation 7.7—the proportion of the present population that is age x—can be found using similar reasoning. Individuals now age x were first recorded $x-1$ censuses ago, and l_x of the original cohort is still alive now. Because the population has grown by a factor of λ^{x-1} since then, we have

$$\text{denominator} = \frac{l_x}{\lambda^{x-1}} \quad (7.9)$$

Putting the numerator and denominator together, the final expression for the reproductive value (the expected contribution to future population growth) of an individual of age x (Goodman 1982) is

$$v_x = \frac{\lambda^{x-1}}{l_x} \sum_{i=x}^{\infty} \frac{l_iF_i}{\lambda_i} \quad (7.10)$$

Thus, all of the information we need to calculate reproductive value is contained in a transition matrix or a life cycle graph.

Table 7.1 Reproductive values (v_x) of stage classes of *Coryphantha robbinsorum* at three sites

Site	Small juveniles	Large juveniles	Adults
A	1	17.82	19.26
B	1	39.72	45.22
C	1	2.06	3.44

Note: These reproductive values are calculated as the dominant left eigenvectors of the transition matrices in Box 7A, standardized so that small juveniles have $v_1 = 1$.

These ideas apply to stage-structured populations as well. We can think of the reproductive value of a tree that is 5 m tall, or of a dormant seed in the seed bank. But a moment's thought should suggest a problem: it is difficult to calculate the numerator in Equation 7.7 for a stage-structured model. For example, our 5m tree may grow continually, stay 5 m tall for years, or shrink for a time.

There is an alternative way of calculating reproductive value that gets around this problem. The reproductive values for each age or stage can be calculated as the dominant **left eigenvector** of the matrix model. A left eigenvector y of the matrix $\mathbf{A}$ is defined so that $y \mathbf{A} = \lambda y$. This definition is parallel to that for the (right) eigenvector that we used above in discussing the stable stage distribution. (When biologists refer to an eigenvector of a matrix without any further qualification, they usually mean a right eigenvector.)

There are some important differences between age-structured and stage-structured populations. In most age-structured populations, reproductive value at birth is very low (because reproduction is delayed for some time, and many newborns never reach maturity). Reproductive value increases to a maximum near the age of maturity, and then decreases. In stage-structured populations, in which individuals can remain in the last stage for a long time—as in many perennial plant populations—a decrease may not occur.

The *Coryphantha* data make this clear. The reproductive values of each stage at each site are given in Table 7.1. At site A, a plant that makes it to the adult stage contributes roughly 19 times as much, on average, to the next generation as a small juvenile does on average, but only slightly more than an average large juvenile. At all three sites, the reproductive value of an adult is much larger than that of a small juvenile, but only a bit larger than that of a large juvenile. This is because relatively few small juveniles survive to reproduce. Reproductive value does not decline in the adult stage because adults can survive for indefinite periods, as expected in stage-structured plant populations in which there is no senescence. This information is precisely what would be needed if a manager were planning to introduce plants to augment this population. For example, if trans-

Partial Derivatives

If you have forgotten (or never knew) what a partial derivative is, think of it as being like an ordinary derivative (that is, a rate of change over a very small change in something else, such as time), but one in which we temporarily hold everything else (all other variables) constant.

For example, if $f = ax + by^2 + cxy$, then

$$\frac{\partial f}{\partial x} = a + cy$$

In other words, f changes with respect to x by a (because a multiplies x in the first term in the sum) plus cy (because cy multiplies x in the third term; we are

momentarily acting as though y is constant). Using this logic, you should see that the partial derivative of f with respect to y is

$$\frac{\partial f}{\partial y} = 2by + cx$$

---

planting nursery-bred plants is practical, the data suggest that it is nearly as effective to introduce large juveniles as adults, allowing one to save time, money, and effort in growing the plants.

Sensitivity

How does λ change as the individual matrix elements change? We already have all the information we need to answer this question, in the form of the eigenvectors of **A** and the reproductive values of each stage. Call a_{ij} the

matrix element in the ith row and jth column. The **sensitivity** of λ to changes in a_{ij} is

$$\frac{\partial \lambda}{\partial a_{ij}} = \frac{v_i x_j}{\sum_{k=1}^{N} x_k v_k} \tag{7.11}$$

where the x's are the elements of the dominant eigenvector (so that x_j is the jth element of the dominant eigenvector), the v's are the reproductive values (so that v_i is the reproductive value of stage i), and ∂ denotes a partial derivative (see Box 7C). In other words, a small change in a_{ij} (the rate at which stage j individuals generate stage i individuals) causes a change in the long-term growth rate. The magnitude of this change is proportional to the relative importance of stage j in the stable distribution (x_j) times the reproductive value (v_i) of stage i individuals.

The sensitivities of λ to changes in the matrix elements in the three *Coryphantha* populations are shown in Table 7.2. The element in the ith row and jth column represents the rate at which λ changes as that element in the transition matrix **A** changes, with all other matrix elements held constant. The sensitivities at sites A and B were quite similar. In both cases, the largest change in λ is expected if there are increases in small juveniles' chances of surviving to either of the larger classes, and in the survivorship of adult plants. In contrast, increasing the fecundity term—the rate at which adult plants generate new

Table 7.2 Sensitivities of *Coryphantha robbinsorum* long-term population growth rates (λ) to changes in the matrix elements

Site	Sensitivities		

A

	Small juveniles	Large juveniles	Adults
Small juveniles	0.069	0.009	0.040
Large juveniles	1.239	0.152	0.720
Adults	1.339	0.164	0.778

B

	Small juveniles	Large juveniles	Adults
Small juveniles	0.023	0.001	0.021
Large juveniles	0.912	0.043	0.820
Adults	1.038	0.049	0.934

C

	Small juveniles	Large juveniles	Adults
Small juveniles	0.153	0.100	0.186
Large juveniles	0.314	0.205	0.383
Adults	0.526	0.344	0.642

Note: Sensitivities and λ are calculated from the transition matrices in Box 7A. Each entry shows the change in λ resulting from a unit change in the corresponding element of the transition matrix, while all other matrix elements are held constant. For example, a unit increase in P_{11} increases λ at a rate of 0.069 at site A.

small juveniles—has little effect on λ in either population. Site C (where the population is clearly growing) is a bit different: increasing any of the terms has a marked effect on λ, although the largest effects are still be achieved by increasing survival rather than fecundity.

Elasticity

There is a potential problem with the use of sensitivities: they compare things that are often measured on very different scales. For example, the survival terms in a matrix model must be between 0 and 1, but the fecundity terms can sometimes be orders of magnitude larger. Similarly, the survival terms for small plants are often much smaller than those for larger plants. When this is true, a small increase of, say, 0.01 actually represents a much larger *proportional change* in the smaller terms than in the larg-

er terms. A related problem with sensitivities is that matrix elements may be zero for some basic biological reason, or by definition. In the *Coryphantha* example, it is impossible for juveniles to reproduce (by definition)—but the sensitivity analysis still tells us that λ would be increased by a certain amount if that transition were increased from zero.

The constraints of sensitivity analyses can be addressed by using the **elasticity**, the proportional change in λ caused by a proportional change in a matrix element. The elasticity of λ with respect to changes in the *ij*th matrix element is (Caswell 2001)

$$\partial_{ij} = \frac{\partial \ln(\lambda)}{\partial \ln(a_{ij})} = \frac{a_{ij}}{\lambda} \frac{\partial \lambda}{\partial a_{ij}} \tag{7.12}$$

In other words, the elasticities are the sensitivities times a_{ij}/λ. Thus, when a matrix element is zero, its elasticity will also be zero.

The elasticities for the *Coryphantha* data are given in Figure 7.10. At sites A and B, the largest proportional effects on λ would be achieved by increasing adult survival. All other changes would have very small effects

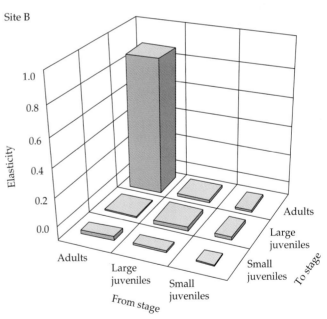

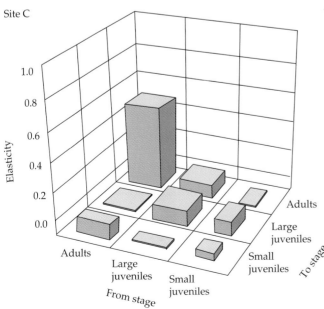

Figure 7.10
Elasticities for the three populations of the endangered cactus, *Coryphantha robbinsorum.* Each bar gives the elasticity for a particular transition corresponding to the matrices in Box 7A. The elasticities describe the proportional change in the long-term growth rate (λ) resulting from a proportional change in each matrix element, holding all other matrix elements constant. They sum to 1 for any given analysis, and can therefore be thought of as describing the proportional importance of each matrix element for λ. For example, at site A, adult-adult survival (P_{33}) accounts for nearly 80% of λ. At site C, where the population is expected to grow rapidly (λ = 1.12), adult-adult survival makes only about a 60% contribution to λ.

on λ. At site C, λ would still be most strongly affected by increasing adult survival. Increasing the other terms would have smaller, though perhaps not negligible, effects. In any case, it seems clear that at all three sites an effective plan to protect the population would involve protecting the established plants—especially adults—rather than enhancing their fertility. This pattern is common for organisms with long-lived adults.

Elasticities have an additional property that makes them useful: all the elasticities of a matrix sum to 1. Consequently, they can be interpreted as the relative contribution to λ of the corresponding matrix element (given that all other elements stay constant). This means that one can directly compare the elasticities of different matrices based on the same life cycle graph. For example, at *Coryphantha* site C, adult survival was responsible for about 55% of λ. By contrast, at the sites where the population is shrinking or just holding its own, adult survival accounted for a much greater proportion of the population growth rates: 76% at site A, and 92% at site B.

Age and Stage Revisited

Saying, as we have, that age alone is usually a poor predictor of demography in plant populations does not imply that age is unimportant. However, apart from cases in which we know that age plays a direct and important role in plant demography (e.g., see Figure 7.4), it has been difficult to relate stage to age. Nevertheless, it is sometimes important to do so. For example, in managing an endangered population, there are several important questions we might need to ask: How long does it take, on average, for a seed to reach maturity? How old is the average individual in a particular stage class? What is an individual's chance of surviving x years? What is the net reproductive rate of a population? How long will it take the population to replace itself?

Until recently, questions such as these could not generally be answered. In some groundbreaking work, Cochran and Ellner (1992) realized that these questions have answers that are, at least in principle, simple. A stage-based matrix model implicitly includes information about age because it predicts stage transitions one time-step into the future. Consider the *Coryphantha* example: A small juvenile has a chance P_{11} of surviving until next year as a small juvenile, and a chance of P_{21} of growing into the large juvenile class and surviving until next year. Its chance of being alive next year is therefore the sum $P_{11} + P_{21}$.

By generalizing this approach—summing over all possible paths between two stages—it turns out that one can calculate *any* age-based quantity from stage-based data, including R_0. Even in the relatively simple *Coryphantha* life cycle graph (see Figure 7.7), there are many possible paths from birth to maturity. One path

is for a newborn plant to be a small juvenile for one year, a large juvenile for one year, and then mature to become an adult. The matrix model tells us that the proportion $P_{21}P_{32}$ of newborns following this path become adults after two years. Another path is for individuals to spend two years at each stage before making the transition to the next stage; the model tells us that the proportion $P_{11}P_{21}P_{22}P_{32}$ of newborns follow this path to maturity. Cochran and Ellner (1992) developed methods for summing over all the possible paths and thereby estimating age-based information from a useful-based model.

Table 7.3 shows some useful age-based data for *Coryphantha*; estimated survivorship curves for the species are shown in Figure 7.11. It is straightforward to calculate these and other quantities. Because the calculations involve many summations over all possible pathways, the mathematical expressions are complex in appearance, and we refer interested readers to Cochran and Ellner (1992).

This approach also makes it possible to calculate the **life table** for a stage-structured population. A life table is a list of the estimated mortality, cumulative survival, and related rates for a **cohort** (a group that germinated, reached a particular size, or entered a study at about the same time). In age-based populations, the survival

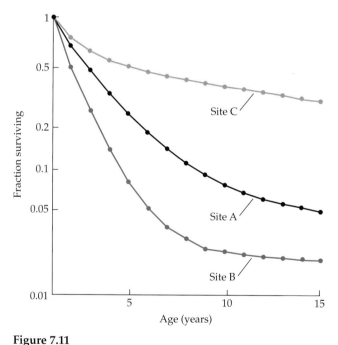

Figure 7.11
Estimated survivorship curves for *Coryphantha robbinsorum* at three sites, calculated using the methods of Cochran and Ellner (1992) to analyze the matrix models in Box 7A. Survival is much greater at site C than the other two sites, because 76% of small juveniles ($P_{11} + P_{12} = 0.76$) survive there; the comparable figures are 69% and 50% at sites A and B, respectively.

Table 7.3 Some age-based quantities calculated from the stage-based *Coryphantha* matrices[a]

Site	Site position	Probability of reaching maturity	Average age maturity (years)	Net reproductive rate (R_0)	Generation time[b]
A	Northeast exposure	0.05	10.7	0.92	42
B	Southwest exposure	0.02	6.7	0.81	72
C	Hilltop	0.46	5.3	5.84	27

[a] Quantities calculated using the methods of Cochran and Ellner (1992).
[b] Mean age at which a cohort of newborns produces offspring.

information in a transition matrix is the same as the information in a life table. But it is not obvious how one could use the survival data in a stage-based matrix to predict a quantity such as the fraction of newly germinating plants that will still be alive in 10 years. Fortunately, Cochran and Ellner (1992) have provided formulas that allow one to estimate the life table for stage-structured populations.

Other Approaches to Modeling Plant Demography

The matrix approach can be much more sophisticated than the basics we have outlined here. For example, matrix models can be arranged so that they include both environmental and demographic stochasticity (Caswell 2001). They can also be made to be density-dependent (Caswell 2001). Doing so requires more advanced mathematics than we will use in this book, and the analysis no longer depends simply on the eigenvalues and eigenvectors, but it is important to know that this method has much broader applications than those we have so far discussed.

There are also situations in which matrix models are not the best choice. For example, if the spatial arrangement of individuals proves to be important in population dynamics, a matrix model can be misleading. In such cases, models that take spatial location into account to simulate the birth, growth, and death of individuals and their particular interactions with their neighbors can be useful. There is, of course, a catch: to estimate the parameters of such spatially explicit individual-based models accurately, one needs substantially more data than for a matrix model. A series of models developed by Stephen Pacala and associates (Pacala and Silander 1985; Pacala 1986a, 1986b, 1987) illustrates both the strengths of this approach (the authors were able to successfully predict many features of the dynamics of their study population) and some of its difficulties and limitations (large quantities of data were needed to estimate model parameters, sophisticated computer programming was necessary to study the models, and analyses such as sensitivity and elasticity would be difficult to perform).

Finally, some size-structured plant populations may be better modeled with a method that treats size as a continuous variable, rather than with a matrix approach. Michael Easterling et al. (2000) have suggested a general approach that is analogous to the matrix approach, but treats size as a continuous variable. The mathematics involved are a bit more complicated than with matrix approaches because the method uses integrals, but these can be estimated numerically on a computer. Easterling et al. (2000) argue that their approach is more natural than the matrix approach; an added bonus, they suggest, is that one can often use their method with fewer parameters than are typically used in a matrix model. There have been few studies to date using this method, so all of its strengths and drawbacks are not yet clear, but it is certainly worthy of further attention.

Demographic Studies of Long-Lived Plants

A small number of plant species reach very great ages. Some *Larrea tridentata* (creosote bush, Zygophyllaceae) genets in the Mojave Desert are estimated to be about 14,000 years old. Obviously it is difficult to study the demography of populations when some individuals have been alive for nearly as long as humans have occupied North America! Fortunately, most plants do not present quite such difficulties for demographic study. More typically, trees and shrubs live for a few decades to a few centuries. But even these life spans are difficult to study when the average researcher is active for only a few decades.

Considerable insight into the demography of long-lived plants can be gained by following marked individuals. William Platt and his associates (1988, 1993) have been studying populations of *Pinus palustris* (longleaf pine, Pinaceae) since the 1970s. Year-to-year variation in environmental conditions—sometimes due to large disturbances such as hurricanes—has been a hallmark of these studies, and it appears to play a major role in the dynamics of these populations. Similarly, Robert Peet and Norman Christensen (1987) have studied large plots of trees established in the early twentieth century in the Duke Forest at Duke University. The results of these studies suggest that it is often reasonable to study long-lived plants using such methods. Clearly, such studies can have some weaknesses. Measurements of survival are often taken at longer intervals than one might like—such as 10 years, or even multiples of 10 years. Seeds, seedlings, and small saplings are usually not measured. Because of the long intervals between censuses, the variation between years is generally undersampled. But the information obtained even with such poor resolution is still useful.

One approach to studying long-lived plants, used widely in forestry and forest ecology, has been to develop **static** or **vertical life tables**. This method assumes that the population has a stable age distribution, that the population's transition matrix has been roughly constant for a long period, and that λ = 1. It then asks what pattern of survival and reproduction would have generated a population with the (present) observed structure.

Joan Hett and Orie Loucks (1976) used this approach to infer survival curves for several populations of *Tsuga canadensis* (eastern hemlock, Pinaceae) and *Abies balsamea* (balsam fir, Pinaceae) in Michigan. First, they measured tree sizes and determined the ages of a sample of trees by counting their rings (in adults) or terminal bud scars (in seedlings and saplings). Then they used the observed relationship between age and size to estimate the population age structure. Hett and Loucks first considered two models for these data: one that assumed the same rate of mortality for all ages, and one that assumed decreasing mortality rates with age. The fit of these models to the data is shown in Figure 7.12. Noting that the data systematically depart from both regression lines, Hett and Loucks then fit a more complex model to the data, in which the pattern of decreasing mortality with age was overlain by a periodic function. The latter model appeared to provide the best fit, and Hett and Loucks proposed several hypotheses that might explain the apparent oscillation, including population cycling and periodic outbreaks of pests.

As this example illustrates, the use of static life tables requires strong assumptions that cannot usually be justified. Static life tables require the assumption of constant vital rates, which is unlikely over long periods. Many trees in the hemlock study lived through part of the Little Ice Age—a period of low solar activity, cold temperatures, and glacial advances from 1500 to 1850—as well as relatively warm interludes in 1540–1590 and 1770–1800. But is this not a criticism that would apply equally well to matrix models? A principal use of matrix models is to study present conditions by asking what their consequences would be if they remained constant. By contrast, the static life table approach asks what set of constant conditions would have generated the current population structure. As Jonathan Silvertown (1987) points out, many quite different models, assuming constant or varying conditions, can be fitted to data of this kind equally well. The consequences of constant conditions can be calculated, but there is no unique answer to the question asked by static life table approaches.

This does not mean that to study a long-lived plant one must begin by marking vast numbers of individuals and hoping for a long career, along with plentiful descendants to continue the study. There are reasonable inferences that can be drawn from a population's structure without invok-

(A)

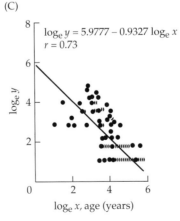

(B)

$\log_e y = 2.9927 - 0.0099\,x$
$r = 0.61$

(C)

$\log_e y = 5.9777 - 0.9327 \log_e x$
$r = 0.73$

(D)

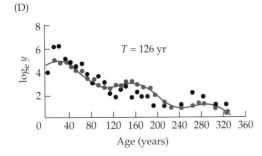

$T = 126\ \text{yr}$

Figure 7.12
(A) Estimated age structure of a *Tsuga canadensis* (eastern hemlock, Pinaceae) population in Michigan. (B–C) Three models attempting to explain the data with static life table approaches. (B) Model assuming equal mortality rates for all ages. (C) Model of declining mortality with age. (D) Model of declining mortality with age, overlain by a periodic function with period *T*. (After Hett and Loucks 1976.)

ing the strong assumptions of vertical life tables. If one finds that a population consists of widely separated age classes (as occurs in many forest trees and in saguaro cacti), for example, it is reasonable to suggest that opportunities for successful recruitment to the population occur only rarely. If there is also a tendency for plants of a certain age class to be located near one another, it is reasonable to infer that the process creating these recruit-

ment opportunities is spatially patchy. Many forest trees, for example, occur in nearly even-aged stands because wildfires create patches for recruitment. By constructing fire histories—inferred from examining the ages of fire scars on trees and from dating charcoal in the soil—investigators have gained insight into the frequency and severity of these disturbances. Although methods like these do not provide estimates for matrix elements, they produce strong inferences if well done.

Several researchers have gained insight into the demography of long-lived plant populations by finding creative ways to obtain data on survivorship retrospectively. Deborah Goldberg and Raymond Turner (1986) first used maps of plots established by Forrest Shreve in the 1930s to document survival in several species of long-lived cacti and desert shrubs. Perhaps the most unusual studies have been those in which James Hastings and Raymond Turner (1965; Turner 1990) used landmarks to match old photographs (some dating to the nineteenth century) with new ones and identify surviving desert trees and large cacti (Figure 7.13).

1903

1961

1996

Figure 7.13
Comparison of repeated photographs has provided important insight into the demography of large plants and vegetation change in arid country. Pioneered by James Hastings and Raymond Turner (1965), the technique requires landmarks for identifying the sites of old photographs. These repeated photographs show changes in the population of *Pachycereus pringlei* (cardón, Cactaceae) on Isla Melisas in Guaymas Bay, Sonora, Mexico. (A) This photograph, taken in 1903, shows a population of old, many-branched cardón. (B) In this photograph, taken in 1961, most of these old individuals have been replaced by much younger plants, although several large individuals can still be seen along the ridgeline. (C) In this photograph, taken in 1996, a dense stand of cardón can be seen; most of the plants are much larger than in the 1961 photo. Three large, old individuals are still apparent along the ridgeline. (Photographs courtesy of R. Turner.)

Random Variation in Population Growth and Decline

In the real world, population growth rates vary because of **stochastic** (randomly varying) factors. Hurricanes and fires, for example, play important roles in the growth of many populations of forest trees, and occasional wet years may be crucial to the persistence of some populations of desert plants. Furthermore, small populations of endangered plants may go extinct largely due to chance: If individuals have a 50% chance of surviving a winter, for example, it is likely that the actual fraction surviving will be different from half—just as repeated coin tosses do not usually result in 50% heads. There-

fore, it should not surprising, in a small enough population, if all the individuals happen to die.

Fortunately, many of the tools described above can be modified for use in a variable world. First, however, it is important to be clear about how random variation affects population growth.

Causes of Random Variation

There are two ways in which population growth can be affected by stochastic factors. First, in the case of **environmental stochasticity**, vital rates vary due to environmental factors that affect all the individuals in a stage class (or in a population) in roughly the same way. Desert envi-

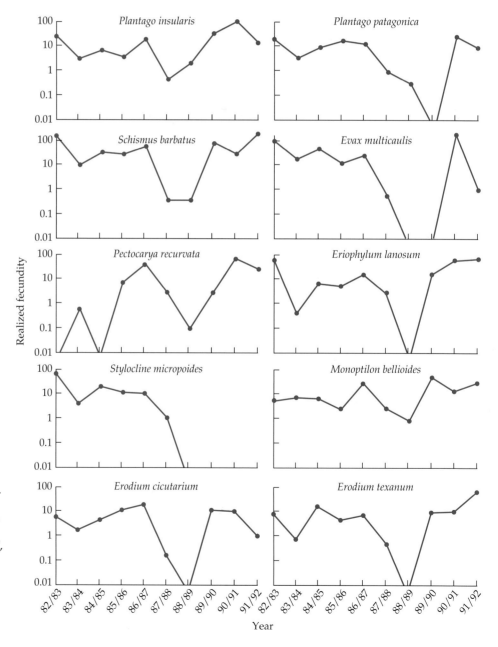

Figure 7.14
The effect of environmental stochasticity on the realized fecundity of Sonoran Desert winter annuals. Increased rainfall due to the El Niño phenomenon occurred in 1982–1983, 1986–1987, 1990–1991, and 1991–1992—all years of high realized fecundity for most species. All species except *Erodium cicutarium* are native to the study area. (After Venable and Pake 1999.)

ronments, for example, have highly variable rainfall. In a 10-year study of winter annuals in the Sonoran Desert, Lawrence Venable and Catherine Pake (Venable and Pake 1999) documented major variation from year to year in the **realized fecundity** (the chance of surviving to maturity times the fecundity of survivors) of ten species of these plants (Figure 7.14). This variation underlies fluctuations in the population sizes of these species (Figure 7.15). Interestingly, years in which realized fecundity is high appear to be associated with El Niño events (Venable and Pake 1999; see Chapter 18).

Second, in the case of **demographic stochasticity**, the chance variation in the fates of individuals reduces the average long-term growth rate of the population. Con-

sider a population of eight plants with a probability $P = 0.25$ of an individual surviving until next year. There is a fair chance that next year's population will not be two, the product of P and n (this year's population). (For further insight, see the discussion of genetic drift in Chapter 5, with which demographic stochasticity shares many properties.)

Some important features of demographic stochasticity are shown in Figure 7.16. Figure 7.16A shows the observed number of survivors in a hypothetical population in which individuals have a probability of 0.25 of surviving. You might notice that there are more extremely low values than extremely high values. There is a reason for this: Even in a population of 1000 plants,

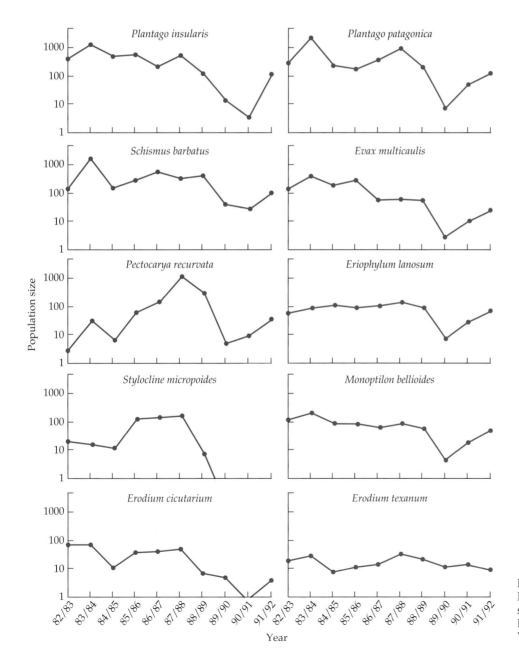

Figure 7.15
Fluctuations in population size in ten species of Sonoran Desert winter annuals. (After Venable and Pake 1999.)

(A)

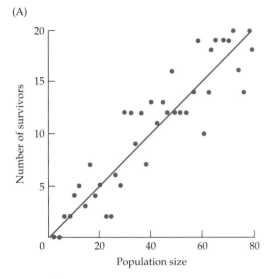

(B)

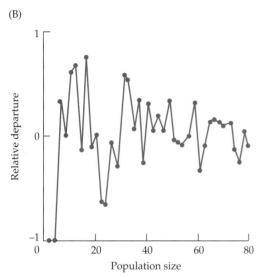

Figure 7.16
The effect of demographic stochasticity on survival in a hypothetical population with a survival probability of 0.25. (A) The observed number of survivors given a population size of N where survivorship rates were determined by the average survival probability plus a deviation due to random chance. The line shows the expected number of survivors: 0.25 times the population size. (B) The relative size of the departure from the expected value: (number of survivors – expected survivors)/expected survivors.

it is possible that, by chance, no plants will survive—but it is never possible for more than 1000 to survive. Figure 7.16B shows the relative departure from the expected number of survivors, which tends to be greater in small populations.

Both forms of stochasticity occur in all populations. Environmental stochasticity can have substantial effects in populations of any size. As the examples above suggest, demographic stochasticity is important mainly in small populations.

Long-Term Growth Rates

Imagine a population of annual plants (without a seed bank) that grows, in good years, at the rate $\lambda_g = 1.01$, and in bad years, at the rate $\lambda_b = 0.99$. To keep things simple, imagine that good and bad years alternate. It may surprise you to realize that this population is slowly going extinct! How can this be, when the average rate of growth is obviously 1? It is because the long-term growth rate is not an ordinary arithmetic average, but is equal to the square root of 1.01×0.99, which is less than 1. Why? Population growth is a multiplicative process: seeds produced in good years then germinate in bad years, and vice versa. The square root thus gives us the appropriate average over time.

Given an average rate of population growth, variation in that rate among years reduces the long-term rate of population growth. Consider a group of unstructured populations—such as annual plants without seed banks—having the possible values of λ that are given in the second column of Table 7.4, and assume that these values occur with equal probability. The average value of λ is the same for populations 1–3, but λ varies more in population 3 than in population 2, and it does not vary at all in population 1. The average value of λ is called the **arithmetic mean** and is symbolized $\bar{\lambda}$. The arithmetic mean of n numbers is their sum divided by n (see the Appendix).

Although populations 1–3 all have the same average growth rate, their long-term growth rates differ, because population growth is a multiplicative process. Next year population 1, whose growth rate does not vary, will be 1.03 times its current size, and in 8 years it will be $1.03^8 = 1.267$ times its current size. If the two different values of λ occur for population 2 with equal probability, that population will grow at some different rate, such as $1.05 \times 1.01 \times 1.01 \times 1.01 \times 1.05 \times 1.01 \times 1.05 \times 1.05$. In this example, population 2 will be about 1.264 times its current size in 8 years—it will have grown less than population 1.

The long-term growth rate of an unstructured population is given by the **geometric mean** of the annual λ's, symbolized as a in the fifth column of Table 7.4. (The geometric mean of n numbers is the nth root of their product.) The geometric mean has an important property: it is guaranteed never to exceed the arithmetic mean, and is equal to the arithmetic mean only if the n numbers are all the same—in other words, if the variance of λ is 0 (see the Appendix for a discussion of variance). For a given average growth rate $\bar{\lambda}$, then, increasing the variance of λ always reduces the long-term growth of the population.

This discussion might be interpreted to mean that increasing the variance of λ always reduces a, but this is true only if the average growth rate $\bar{\lambda}$ stays constant. Population 4 has a larger arithmetic mean and a larger

Table 7.4 Relationship between annual and long-term population growth rates in a variable environment

Population	Values of λ	Arithmetic mean (λ)	Variance (σ_λ^2)	Geometric mean a
1	1.03	1.03	0	1.03
2	1.01, 1.05	1.03	0.0008	1.0298
3	1, 1.01, 1.05, 1.06	1.03	0.0009	1.0297
4	1, 1.01, 1.05, 1.07	1.0325	0.0011	1.0321
5	0.9, 1.01, 1.05, 1.17	1.0325	0.0124	1.0280

Note: The long-term growth rate depends on the mean and variance of yearly rates. In unstructured populations, the long-term rate is the geometric mean of the yearly rates.

variance in λ, and its long-term growth rate is larger than any of populations 1–3. On the other hand, population 5 has still greater variance in λ, and its long-term growth rate is the smallest of the five populations.

These examples should help to convince you that both the average conditions and their variation are important in determining long-term rates of population growth. There are few studies on this kind of long-term variation. There are many anecdotal accounts of plants in highly variable environments (such as desert annuals) that usually have high mortality and low fertility, but in an occasional "good year" produce vast quantities of seed. Thus it is possible for a plant population to have positive growth in the long run, but be declining in numbers during most years (Venable and Pake 1999).

These general ideas hold for age- and stage-structured populations as well. To study population growth in a stochastic environment, we use the stochastic growth rate a rather than λ_1. In structured populations, a is the long-term average growth rate of the population. Generally, a must be estimated by simulation. Because matrix multiplication is not commutative, a is not a simple geometric mean of the estimate annual growth rates (see Caswell 2001 for a discussion of methods for estimating a). Just as the geometric mean of a sequence of numbers is smaller than the arithmetic mean, a is less than λ_1 unless there is no stochasticity, in which case the two quantities are equal. Calculations of the long-term growth rate are actually done with logarithms analogous to $r = \ln(\lambda)$.

Studying Variable Population Growth

To study the effects of environmental stochasticity, one needs the same kinds of data used to estimate average population growth rates, recorded over enough years to estimate the variances and covariances of the matrix elements—the survivorship and fecundity rates. Given such data, one can perform computer simulations of population growth that is subject to random variation. There are many studies reporting variation in matrix elements (e.g., Horvitz and Schemske 1990; Bierzychudek 1982).

Few of these studies, however, are of sufficient length to provide good estimates of variance, and fewer still provide estimates of the covariance among matrix elements.

Even a population with $a > 1$ can sometimes go extinct due to random variation. A population's **extinction probability** is estimated as the fraction of replicate populations that can be expected to go extinct. This value is estimated by simulation. Extinction probabilities for several California species of *Calochortus* (mariposa lily, Liliaceae) are shown in Figure 7.17. Note that extinction probabilities increase as the amount of environmental stochasticity increases, and that this increase is fastest for populations with the smallest value of λ.

An idea that is very closely related to extinction probability is the **minimum viable population (MVP)**. The MVP is the minimum size necessary to give a population a probability x of surviving t years. Typically the probability is taken as 0.95, and the time as 50 or 100 years. Plans for the management of endangered species increasingly require estimates of MVP. Eric Menges (1992) conducted simulations to estimate the MVP for the Mexican palm *Astrocaryum mexicanum* (chocho, Arecaceae), using published matrices from the extensive study of Piñero et al. (1984); the results are shown in Figure 7.18. Populations subject to only demographic stochasticity could begin with only 50 plants and still have a 95% chance of persisting 100 years. Populations subject to environmental stochasticity required larger initial sizes to have this chance of persisting. As environmental stochasticity became large (i.e., there was more year-to-year variation), the chance of a population going extinct increased rapidly in smaller populations.

Most demographic studies of plants have lasted only a few years, and consequently have focused on average values of growth rates $\bar\lambda$. These are useful quatities, but it is important to realize that $\bar\lambda$ overestimates a (just as the arithmetic mean is always larger than the geometric mean, unless the variance is zero). There have been few attempts to estimate a in plant populations. Hal Caswell (2001) estimated a for two populations of *Arisaema triphyllum* (jack-in-the-pulpit, Araceae) in New

(A)

(B)

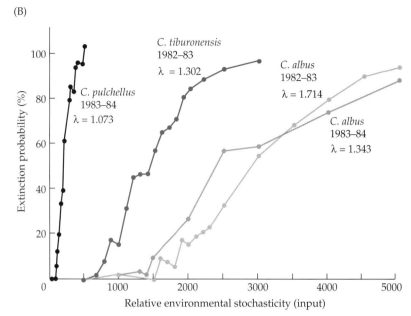

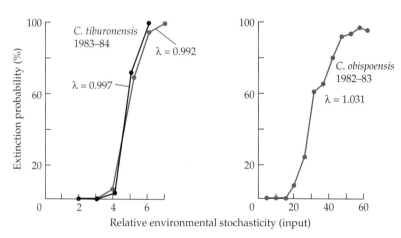

Figure 7.17
(A) *Calochortus howellii* (Mariposa lily, Liliaceae), which is endemic to southwestern Oregon, especially serpentine areas. (Photograph courtesy of J. Sainz.) (B) Extinction probabilities of several California populations of *Calochortus* spp. Environmental stochasticity was modeled by taking 1% of the variance/mean for seed production, and 0.01% of the variance/mean for other vital rates, and then multiplying these values by the numbers shown on the horizontal axis. (After Menges 1992.)

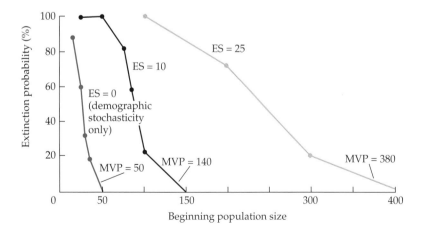

Figure 7.18
Results of a simulation conducted to estimate the minimum viable population (MVP) size for a population of the Mexican palm *Astrocaryum mexicanum* (chocho, Arecaceae) under different levels of environmental stochasticity (ES). Environmental stochasticity was modeled using the methods described in Figure 7.16. MVP is the smallest number giving an extinction probability of 5% or less in 100 years. (After Menges 1992.)

York state studied by Paulette Bierzychudek (1982). In this study, *a* and its 95% confidence intervals were 1.2926 ± 0.0025 in the Fall Creek population, and 0.8979 ± 0.0028 in the Brooktondale population.

Summary

Studies of change in plant populations require defining an "individual." In many plants, unlike most animals, genetic and physiological individuals are different because many genetic individuals can reproduce vegetatively as well as sexually. The "right" kind of individual to study depends on the questions being asked.

Survival and reproductive rates in plants usually depend much more on stages (size, life history stage, physiological status) than on ages. An important goal of plant population studies is identifying the factors causing variation in survival and reproductive rates. Most population models use stages to study changes in population size.

Matrix models are useful tools for studying change in population size and composition. They make it possible to estimate the population's eventual growth rate and composition—an important way of examining the consequences of current demographic conditions. Reproductive value, which can be calculated from a transition matrix, is useful in comparing the importance of different stage classes for future population growth. It is now possible to calculate age-based quantities (such as the average age of first reproduction) from stage-based data.

Plant populations vary in their growth rates. One cause of this variation is random variation in the environment, which affects all plant populations. Variation in population growth rates is also caused by random events involving individuals. This kind of random variation is most important in small populations. Generalizations of matrix modeling approaches make it possible to estimate long-term population growth rates, extinction probabilities, and minimum viable population size.

Additional Readings

Classic References

Harper, J. L. 1977. *Population Ecology of Plants*. Academic Press, London.

Hartshorn, G. S. 1975. A matrix model of tree population dynamics. In *Tropical Ecological Systems*, F. B. Golley and E. Medin (eds.), pp. 41–51. Springer-Verlag, New York.

Contemporary Research

Horvitz, C. C. and D. W. Schemske. 1990. Spatiotemporal variation in demographic transitions for a tropical understory herb: Projection matrix analysis. *Ecol. Monogr.* 65: 155–192.

Platt, W. J. and S. L. Rathbun. 1993. Dynamics of an old-growth longleaf pine population. In *The Longleaf Pine Ecosystem: Ecology, Restoration and Management*, S. M. Hermann (ed.), pp. 275–297. Proceedings of the Tall Timbers Fire Ecology Conference, No. 18.

Kalisz, S. and M. A. McPeek. 1992. Demography of an age-structured annual: Resampled projection matrices, elasticity analyses, and seed bank effects. *Ecology* 73: 1082–1093.

Additional Resources

Caswell, H. 2001. *Matrix Population Models*. 2nd ed. Sinauer Associates, Sunderland, MA.

Cochran, M. E. and S. Ellner. 1992. Simple methods for calculating age-based life history parameters for stage-structured populations. *Ecol. Monogr.* 62: 345–364.

Easterling, M. R., S. P. Ellner and P. M. Dixon. 2000. Size-specific sensitivity: Applying a new structured population model. *Ecology* 81: 694–708.

CHAPTER *8* *Growth and Reproduction of Individuals*

The growth and reproduction of individuals are critical components of the growth (or decline) of populations. Growth and reproduction are also key ways in which plants interact with other plants, animals, and the physical environment around them through uptake of nutrients, use of space, pollination relationships, and the dispersal of fruits and seeds. The movements of pollen, seeds, and fruits are the main factors responsible for initiating spatial patterns in plant populations. The evolution of specific reproductive traits marks key episodes in the history of terrestrial plants. We begin this chapter at the within-plant level of anatomy and physiology, but we then make the transition to considering growth and reproduction at the level of the whole plant and of plant populations.

Plant Growth

The growth of plants is **modular**; that is, plants grow by adding repeated units, or modules, to their bodies. We can compare growing plants to children's construction toys in which complex structures are built up by adding simple repeated units to the original structure. As with such toys, there are only a few types of repeated units in modular plant growth; all plant structures are iterations of these units.

The basic rules of plant growth have important consequences for the ecology of plants. In particular, the ability to add (or lose) individual modules means that plants have a great deal of plasticity in their size and shape. This plasticity affects their ability to respond to damage, to capture resources, and to interact with one another. Many plants can survive the loss of large portions of their bodies to herbivores. Commonplace as this may seem, it is remarkable: few animals could survive a similar loss of body parts! Similarly, many plants regularly lose parts of the **shoot**—the aboveground part of a plant—to abiotic factors such as freezing temperatures, drought, and fire—and nevertheless survive.

The addition of new modules depends on the activity of **meristems** (meristematic tissues). Meristems are collections of undifferentiated cells (cells that do not yet have specialized functions). At the tip of a growing stem or branch is an **apical meristem** (Figure 8.1). The growing stem is created by the production of new cells behind the apical meristem. Eventually, these new cells become differentiated into the many different kinds of cells making up the shoot, such as epidermal, xylem, phloem, and parenchyma cells. Certain cells

Figure 8.1
The basic structures involved in plant growth. The shoot apical meristem differentiates, generating lateral organs such as leaves, as well as axillary meristems. The points at which these lateral organs or buds are attached are called nodes. The leaf primordia expand to become leaves. (After Purves et al. 2001; photographs by J. R. Waaland/Biological Photo Service.)

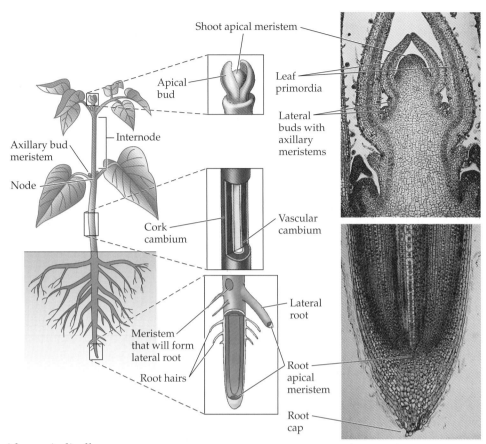

within the apical meristem divide, periodically generating a **node**—a point of proliferation for the development of leaves or flowers. Nodes, therefore, are the places where leaves and flowers are attached to the stem. In many plants, the stem is visibly swollen at each node, and a primordium (bud) forms there for each potential new leaf or flower.

Internodes—the stem segments between nodes—are created in a separate step. When the apical meristem generates a node, it may create a group of cells called **intercalary meristems**. The internodes grow when these intercalary meristems are activated. Some plants have short or even unnoticeable internodes. These plants, called **rosette plants**, include cabbage, dandelions, many alpine plants, and many low-growing weedy herbaceous plants. In many of these species, the onset of flowering activates the intercalary meristems, and the plant "bolts," or lengthens due to the expansion of the internodes. They have little in the way of intercalary meristems, or never activate them.

At each node there is also an **axillary meristem** in the **leaf axil**, the place where the leaf and stem join. When the axillary meristems grow, they may become branches or flowers (which are actually special types of branches). The axillary meristems are generally inactive while the apical meristem is intact because the apical meristem produces hormones that inhibit the proliferation of cells in the axillary meristems; this phenome-

non is called **apical dominance**. If the apical meristem is removed—for instance, when an herbivore chomps off the tip of the plant—the axillary meristems are freed from apical dominance, and many of them begin to create new branches below the damage. This basic modular growth response to damage is a major way in which plants recover from herbivory and continue to grow, although now in a lower, bushier form.

A young plant or new shoot begins to increase in size by the process described above, called **primary growth**. All perennial gymnosperms and angiosperm dicots (and many annuals in these groups) continue their growth by another process, called **secondary growth**. Secondary growth is largely a process of increasing girth (thickness) by producing woody tissues (secondary xylem). In a new stem, the vascular bundles are separated from one another, but as the stem matures, the vascular bundles grow together and coalesce to form a sheath surrounding the inner part of the stem. In the center of this vascular tissue, between the xylem and the phloem, a new meristem, the **vascular cambium**, is formed just inside the stem, under the phloem and epidermis. The vascular cambium forms a thin sheath that extends for much of the length of the stem. A similar process occurs in roots.

Secondary xylem, or **wood**, is typically much stronger and better protected than primary xylem and is a major source of the structural strength that enables perennial plants to survive for more than a short time. In monocots, increase in girth is more complicated. Palms, for example, have no vascular cambium, but early in life they add numerous **adventitious roots** (roots that originate from the stem), leading to an increase in girth. Once this process stops, the width of the palm is set. That is why palms stay about the same girth over their entire height.

Root apical meristems control root growth. The apical meristem of a growing root is located behind, and protected by, a cap of loosely held cells that sloughs off as the growing tip of the root pushes through the abrasive soil. Roots do not have systems of nodes, internodes, or most of the different meristems characteristic of shoots. Roots branch when special differentiated cells resume active division, and each branch forms its own root apical meristem. One important feature of roots is their very fine root hairs. Most uptake of water and minerals occurs through these hairs (or through mycorrhizal structures; see Chapter 4), but the hairs usually function for only a few days. Thus, continued growth of roots is often important for uptake of water and nutrients.

Ecology of Growth

Plant Architecture and Light Interception

Plants have a great variety of **architectures** (arrangements of regenerating parts) and shapes. The growth forms of plants can reflect their environments in at least three ways. First, plant growth form is in part a result of evolutionary adaptations to the timing of unfavorable seasons. A widely used classification of perennial growth forms is shown in Figure 8.2. Each growth form has different consequences for survival and resource retention. An excellent example of the evolution of a diversity of growth forms is the morning glory genus, *Ipomoea* (Convolvulaceae). Some Mexican and Central American species are drought-deciduous trees (i.e., phanerophytes), while others are vines with succulent root crowns in which the leaves and stems die back during the dry season (i.e., both cryptophytes and hemicryptophytes). Farther north, in the United States, *Ipomoea* species are annual vines and other herbaceous forms.

Variation in plant shape can also be a result of phenotypic plasticity (see Chapter 5) in response to environmental conditions. Plants growing under poor resource conditions often have a different shape than plants of the same species growing in richer microhabitats. In some species, this variation reflects evolutionary adaptation to variation in the environment; in other species, however, the fact that plants have different shapes under poor conditions is a simple reflection of the stress induced by those conditions. Not all plasticity is adaptive.

Some forms of plasticity are adaptations that enable individuals to acquire more resources, disperse, or improve their competitive ability. One of the best-studied examples of such plasticity is tree architecture. In many forest trees, shaded individuals have a characteristic shape that is different from that of the same species when it reaches the forest canopy, or when it grows in a gap or along an edge. Many species have patterns of branching and growth that vary in response to light availability, and light interception is often a key factor affecting plant shape. Forest trees typically shed lower branches that are shaded and fall below the light compensation point (see Chapter 2). In some cases, the same species retain their lower branches when grown in the open because those branches are above the light compensation point under those circumstances. Most species have a characteristic pattern of **self-pruning**—shedding of branches—that tends to keep branches above the light compensation point.

Self-pruning, however, is not necessarily just a passive response to shading; it can also be an important part

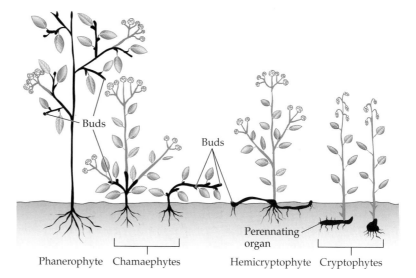

Figure 8.2
The Raunkaier system of perennial growth form classification based on the position of buds or regenerating parts of a plant. Perennating tissues are shown in black; deciduous tissues are unshaded. The four categories are phanerophytes, trees or tall shrubs with buds more than 25 cm above the ground; chamaephytes, shrubs with buds less than 25 cm above the ground; hemicryptophytes, perennial herbs with buds at the ground surface; and cryptophytes, perennial herbs with perennating organs below the ground surface.

of plant strategies for competition. When plants grow in dense stands (as in forests), selection often favors allocation of resources to upper branches rather than to lower branches because the upper branches not only gain more light, but also can also overtop neighbors and deprive them of light. Such competition for light appears to have driven the evolution of trees. Trees are plants that have invested most of their biomass in wood, which allows vertical growth, rather than in leaves or reproduction. The principal selective advantage of producing a long trunk seems to be competitive superiority, although woody tissue also contributes to the ability to live for a long time.

Growth of Clonal Plants

In **clonal** plants—those with numerous ramets—growth form can vary in more complex ways. The spatial distribution of ramets can vary, as can the size and shape of each ramet. The two most important ecological factors affecting the spatial distribution of ramets are competition among genets (including individuals of different species) and spatial variation in the distribution of resources. These factors are not necessarily independent of each other. If resources (say, available soil nitrogen) are concentrated in distinct patches, individual genets are subject to selection for their abilities both to acquire these resources and to prevent their neighbors from getting them.

One attempt to classify clonal growth forms was Lesley Lovett Doust's (1981) characterization of plants as having either "phalanx" or "guerilla" growth. The idea was that some plants spread like a classic advancing army, with ramets grouped tightly together in a distinct front, while others spread like guerilla forces, usually through isolated stolons or rhizomes penetrating their competitors' turf (Figure 8.3; see also Figure 5.6). Although Lovett Doust pointed out that these categories were only the end points of a continuum, some ecologists devoted considerable effort to typing clonal plants as either phalanx or guerilla species. One problem with such typology is scale: the categories are subjectively determined based on what individual ecologists see as being tightly grouped. If this kind of metaphor is to be truly useful, it must be applied on the scale of the plant and the dispersion of resources.

Why do clonal plants have variable patterns of spatial dispersion? Several kinds of explanations, none of which are mutually exclusive, have been investigated. These explanations involve the consequences of four different factors: the mechanics of clonal spread, the degree to which plants grow by simply following the distribution of resources, the degree to which plants actively explore the environment seeking resource patches ("foraging"), and the degree to which the clone is integrated as a strategy for minimizing environmental variability.

(A) "Phalanx"

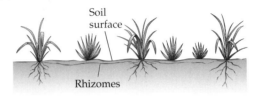

(B) "Guerilla"

Soil surface

Rhizomes

Figure 8.3
Spatial patterns of the spread of ramets within a clone are sometimes described as being like the phalanx of a classic army (A), because the ramets are spatially clumped, or like a guerilla army (B), because the ramets are spread widely in space. Although this characterization is dichotomous, the spacing of ramets actually varies continuously. How one characterizes their spacing is dependent on scale: the image in part B might be viewed as an example of either extreme, depending on whether one views it from near or afar.

Some of the variation may be caused simply by differences in the biological mechanisms of clonal spreading. Each method of spread can potentially result in a wide range of patterns of ramet dispersion. Spread by vegetative (nonsexual) connections, for example, can lead to the slow growth of a compact patch of one genet, or it can lead to a more diffuse structure, with a number of smaller patches of ramets. The spread of genets by the dispersal of **plantlets**—small plants created vegetatively—or asexual seeds (described below)—units that obviously are not physiologically integrated—may also create a variety of spatial patterns.

It is also clear that some variation in clonal form may be a more or less passive response to spatial variation in resource availability. Once a resource-rich (or toxin-rich) patch is encountered, roots and new ramets tend to follow the concentration gradient of increasing resources (or avoid the gradient of increasing toxins). Amy Salzman (1985) showed that a clonal ragweed, *Ambrosia psilostachya* (Asteraceae), was able to avoid concentrations of saline soil in this manner.

This kind of clonal growth has been called "foraging" as an analogy to the way in which animals search for food. Colleen Kelly (1992) made a useful distinction, however, between plants passively following a resource gradient and those actively "seeking" new patches of resources. She suggested that one can distinguish the two responses in cases in which the plant growth occurs

(A)

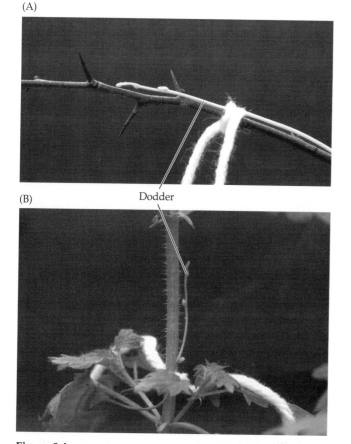

Dodder

Does not coil   Coils

Figure 8.4
Cuscuta europaea (dodder, Convolvulaceae) is an obligate parasite. Like the other 400 or so *Cuscuta* species, *C. europaea* can use a number of host species, which vary in quality as resources. To attack a host plant, it coils around the host and then sinks absorptive tissues into the host's phloem. The "decision" to attack a particular host is made before any resource uptake, and it is made within minutes of contact with the host, as shown in this choice experiment. (A) A strand of dodder tied to a branch (left) of *Crateagus monogyne* (hawthorn, Rosaceae) fails to coil on this host (right); *Crateagus* is sometimes accepted as a host, but dodder grows poorly on it. (B) A strand of dodder tied to *Urtica dioica* (stinging nettle, Urticaceae) coils readily; dodder grows well on *Urtica*. Host choices are predicted well by models of optimal foraging from animal ecology. (Photographs courtesy of C. Kelly.)

before any resource is actually taken up. Her studies of *Cuscuta europaea* (dodder, Convolvulaceae)—a parasitic plant that coils around and sinks absorptive tissues into its host before any resource uptake can occur—showed that this plant does forage in her strict sense. Dodders attacked different host species at different rates. Remarkably, the rate at which each host was attacked was predicted well by optimal foraging models derived from animal behavior studies (Figure 8.4). Kelly suggested that this kind of strict-sense foraging might occur in many other plants.

Finally, clonal plants vary considerably in the extent to which they remain physiologically integrated. Ramets connected by a root or rhizome can share water, carbohydrates, and other nutrients. Such sharing may benefit the plant in several ways. It may allow a genet to spread into microhabitats where a seedling could not survive. The genet may grow faster by increasing the number of ramets. If there are very small-scale differences in the environment (for example, greater availability of nitrogen in one spot and phosphorus in another), the differences can be averaged by the plant, sometimes leading to improved overall vigor for the genet.

Plant Reproduction

What is reproduction in a plant? It may seem surprising to begin a discussion of reproduction by asking what it is—after all, we have a pretty good idea of what our own reproduction involves. But the biology of plants means that their reproduction is not so clearly defined. Reproduction must, by definition, involve the formation of a new individual. But as we saw in Chapter 5, the concept of an individual is not as well defined in plants as it is in animals (especially vertebrates). A genet may be made up of many ramets. While everyone would include sexual reproduction in the category of reproduction, production of new ramets is sometimes best thought of as asexual reproduction and sometimes as growth, depending on the ecological context. For example, if we are concerned with natural selection on *Populus tremuloides* (quaking aspen, Salicaceae), which forms gigantic clones (see Figure 5.7), we look at the entire clone as the individual, and reproduction occurs only when new genets are formed. But if we are interested in the spread of aspen on a mountain, it may be useful to think of aspen as reproducing in two ways, through new ramets (asexual or vegetative reproduction) and through new genets (sexual reproduction).

Vegetative Reproduction

Vegetative reproduction (reproduction by vegetative growth of a new ramet) is extremely widespread among plants. It occurs in most herbaceous perennials as well as in many shrubs and a few trees, and it can take a number of forms. Often the result is a collection of ramets that are physiologically integrated with other ramets, at least initially. For example, some plants generate new ramets by sending out modified shoots. *Fragaria* (strawberries, Rosaceae) spread by **stolons** (runners)—branches that spread at or just above the surface of the soil and generate ramets at nodes touching the ground. *Eichhornia crassipes* (water hyacinth, Pontederiaceae), a South American native, has become a serious pest in many tropical and subtropical regions after human introduction because of its ability to spread rapidly through stolons. It is currently a severe problem in many waterways in Africa, the southern United States, and subtropical Australia. Many grasses (bamboos, for instance), gingers, and irises spread by **rhizomes**, which are underground horizontal stems growing near the soil surface. The dense rhizomes of *Hedychium gardnerianum* (Kahili ginger, Zingiberaceae), a Himalayan plant introduced into Hawaii, make it a serious pest that crowds out many native species. Other species, including familiar plants in the lily family (Liliaceae) such as tulips (*Tulipa*), onions (*Allium*), and daffodils (*Narcissus*), spread by division of **bulbs**, underground rosette stems that store nutrients. Many woody plants spread by the formation of buds on some of their near-surface roots, often called **suckers**. *Populus tremuloides* forms huge clones in this way, spreading over hectares in western and northern North America.

Some plants (such as some seagrasses and cholla cacti; see Figure 7.1) spread by **clonal fragmentation** (pieces of the plant break off and are capable of rooting to form new independent plants), and the new ramets are thus never physiologically integrated with other ramets. *Elodea canadensis* (Canadian pondweed, Hydrocharitaceae), a common aquarium plant, was introduced into Great Britain in the 1800s. In almost every part of Britain, only male plants were introduced, so reproduc-

tion is necessarily asexual. *Elodea* spread so extensively by clonal fragmentation that British canals became severely clogged, and it is thought that this plant may have contributed to the predominance of railways rather than barges in British commerce (Simpson 1984). Some plants, such as *Kalanchoe daigremontianum* (mother of thousands, Crassulaceae), produce little plantlets on the edges of leaves. Physiologically independent ramets are also created in plants that reproduce with **bulbils** (tiny, bulblike organs vegetatively produced in inflorescences or leaf axils), which are formed above ground in monocots such as *Agave* (century plants, Agavaceae*), *Allium* (onions and garlics, Liliaceae), and *Lilium* (lilies, Liliaceae), dicots such as *Polygonum viviparum* (Polygonaceae) and *Saxifraga cernua* (Saxifragaceae*), and ferns such as *Cystopteris bulbifera* (Athyriaceae).

Seeds Produced Asexually

Apomixis (a general term for asexual reproduction) occurs in many plant taxa. In many cases this asexual reproduction occurs as vegetative reproduction. However, **agamospermy**, the production of seeds asexually, also occurs in a number of plant taxa (many people use the term "apomixis" to refer specifically to this process). There are a number of mechanisms by which asexual seed production can occur, described by a complex terminology (Gustafsson 1946, 1947a,b). Most of these mechanisms involve partial meiosis without a reduction division, but in some cases there is no meiosis at all. In most cases, however, recombination is prevented, so that the new embryos are **clones**—genetic duplicates—of the plants on which they are formed.

Agamospermy is fairly widespread. It occurs in some familiar plants, including members of the Asteraceae such as dandelions (*Taraxacum*) and hawkweeds (*Hieracium*); citrus plants (Rutaceae); a number of plants in the Rosaceae such as raspberries and their relatives (*Rubus*) and cinquefoils (*Potentilla*); some buttercups (*Ranunculus*, Ranunculaceae); many grasses (Poaceae); and nettles (*Urtica*, Urticaceae).

Sexual Life Cycles of Plants

Plant life cycles are fundamentally different from those of animals because all plants have **alternation of generations**: they have a haploid generation that produces haploid gametes, alternating with a diploid generation that produces diploid spores (Figure 8.5). Alternation of generations set the stage for the diversification and success of plants on land (see Chapter 3). Most of the plants that we see around us are the diploid generation, because that is what predominates in gymnosperms and angiosperms, the most abundant plants in the modern world. The individuals that actually mate are haploid. In some plant phyla, such as the Bryophyta (mosses), however, the haploid generation is actually the larger, longer-lived, dominant generation; the familiar moss plants you see growing in forests are haploid.

The haploid generation is called the **gametophyte** because it produces gametes. The diploid generation is called the **sporophyte** because it produces spores (haploid somatic cells). Only the gametophytes have a gender (are male or female), which is reflected in the type of gametes they produce. However, in flowering plants, we sometimes speak loosely of a male or a female plant (sporophyte) if the individual produces spores that result in only one type of gamete, either male (pollen) or female (embryo sacs). Figure 8.5 gives a general picture of plant life cycles; consult an introductory biology or botany text for a more complete discussion.

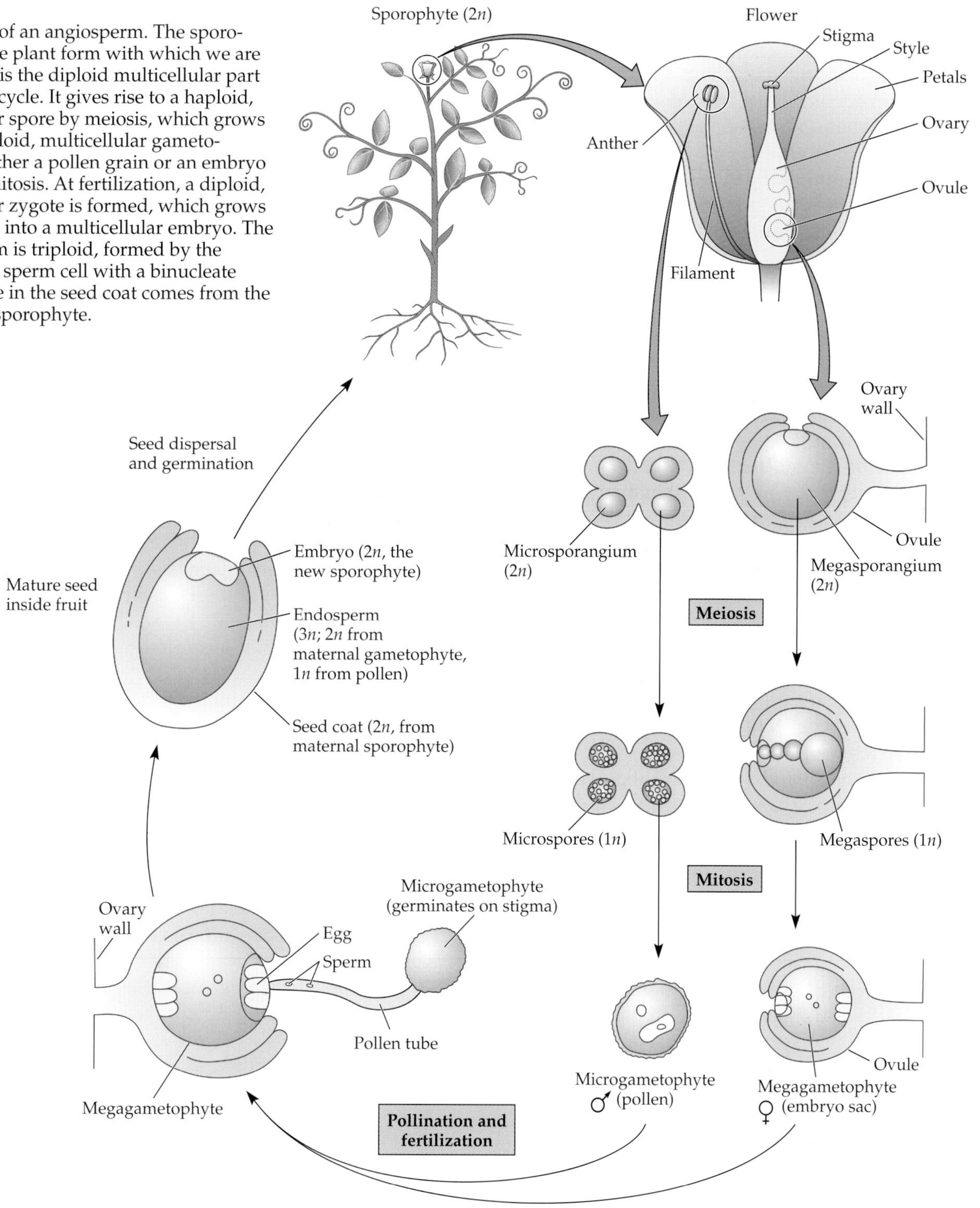

Figure 8.5
Life cycle of an angiosperm. The sporophyte—the plant form with which we are familiar—is the diploid multicellular part of the life cycle. It gives rise to a haploid, unicellular spore by meiosis, which grows into a haploid, multicellular gametophyte—either a pollen grain or an embryo sac—by mitosis. At fertilization, a diploid, unicellular zygote is formed, which grows by mitosis into a multicellular embryo. The endosperm is triploid, formed by the union of a sperm cell with a binucleate cell. Tissue in the seed coat comes from the maternal sporophyte.

There are trade-offs between the two plant generations (Niklas 1997). If gametophytes live independently (as they do in mosses and ferns), they can live only in moist places. Free-living gametophytes must be large enough to nourish the new sporophyte, at least initially, but must be low enough to the ground to permit the transfer of swimming sperm. Moreover, they must be near one another if the sperm are to get from plant to plant. If sporophytes grow to a large size, they cause the demise of the parental gametophyte, limiting its size (as in ferns). By contrast, gametophytes maturing inside the sporophyte (as in the seed plants) can get their nutrition from the sporophyte. This arrangement allows the gametophytes to be much smaller. It also allows reproduction to occur in dry environments, and frees them from the constraint of having to be near one another for fertilization to occur.

Most familiar plants are the sporophytes of gymnosperms (such as the conifers) or angiosperms (flowering plants). The reproductive structures of gymnosperms are cones or conelike structures. The reproductive structures of angiosperms are flowers. Within either the cone or flower the gametophyte is produced and fertilization occurs. Both gymnosperms and angiosperms produce seeds. In angiosperms, fertilization results in the ovary maturing into a fruit with seeds inside. Thus, although not all angiosperms produce showy flowers or fruit that is edible by humans, they all produce flowers, seeds, and fruit as their basic reproductive structures. Flowers can be large and conspicuous or small and inconspicuous, depending largely on the mechanisms by which pollination occurs in a particular plant group. Fruits can be large or small, colorful, sweet, and fleshy or dry and hard, toxic or nutritious, depending on the evolution and ecology (especially the seed dispersal mechanisms) of the plant producing them. The tremendous diversity of seed plants and their dominance over terrestrial habitats are due largely to these reproductive structures, which permit them a great deal of independence from water for mating and dispersal.

In seed plants, the new embryo (sporophyte) begins life in a protected structure, the seed, which can be dispersed and can sometimes remain dormant for a long time. Flowers sometimes act to attract certain animals (and discourage others) as pollinators. The fruits of angiosperms further enhance the protection and dispersal of the seeds enclosed within them. Finally, the interaction of several different kinds of tissues—from both gametophyte and sporophyte—has allowed the evolution of mechanisms for limiting matings within genotypes. In sum, these characteristics of plant life cycles are basic to plant reproductive ecology and to the evolutionary success of the seed plants, especially the angiosperms.

When cells in flowers undergo meiosis, the haploid daughter cells that result are called spores. These spores grow mitotically into either **pollen** (also called microgametophytes) or **embryo sacs** (megagametophytes). An embryo sac usually consists of seven cells, one of which ultimately becomes the egg cell. It develops within a sporophytic structure called an **ovule**, which is located inside the **ovary**. The spores that become pollen grains undergo two further mitotic divisions. One division creates a vegetative cell and a generative cell. The latter divides again, forming two sperm cells. These three cells are haploid, and make up the entire male gametophyte in angiosperms. The pollen grains develop in a sporophytic structure called an **anther**, located at the end of a stalk called the **stamen**.

Pollen germinates after landing on a stigma, the receptive surface of the flower (Figure 8.6A). Upon germination, the vegetative cell grows into a pollen tube (see Figure 8.10), which penetrates the stigma and style, and, ultimately, the ovary and an ovule inside the ovary. The sperm cells are transported down the pollen tube, where one of them fertilizes the egg cell, forming a new diploid zygote. This zygote grows and divides by mitosis, forming the embryonic tissues of a new sporophyte. In almost all angiosperms, the second sperm cell fuses with a specialized cell in the embryo sac that has two nuclei. This triploid cell grows to become **endosperm**, which serves as the primary nutritive source for the developing embryo. Once fertilized, an ovule is called a **seed**. In angiosperms, this seed grows inside a **fruit**, which is the mature ovary of a flower produced by the maternal sporophyte. The tissue in the seed coat comes from the ovule. Thus the tissues in a fruit, and some of the seed tissues, come from the maternal sporophyte; only the embryo inside the seed is a new genetic individual. In the other seed plants, the gymnosperms, the basic pattern of gametophyte development and seed development are similar, with the key differences being that the female gametophyte consists of many more cells, endosperm is absent, and development usually takes substantially longer.

Pollination Ecology

Wind Pollination

When most people think of flowers and pollination, they think of showy flowers that are visited by birds or insects. But for a large number of plants, pollen is carried from flower to flower mainly by wind. There are two quick ways to convince yourself that wind pollination is important. First, almost all pollen transfer in such important plant groups as grasses, rushes, and most temperate-zone trees (including all conifers, oaks, beeches, maples, elms, willows, and many others) is by wind. Second, one of the most common human medical complaints (at least in developed countries) is allergy to

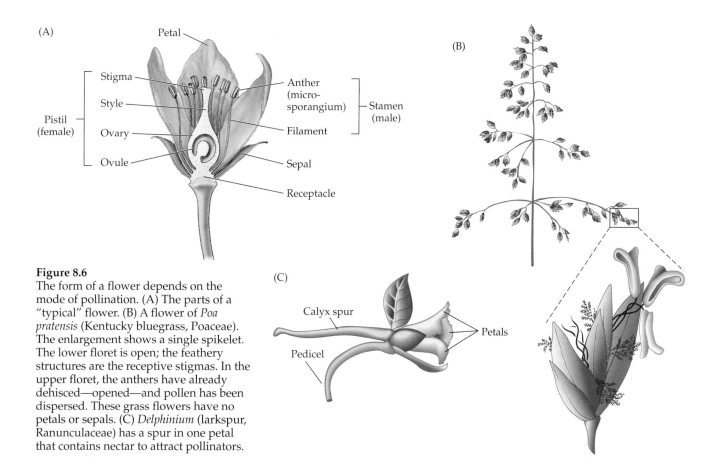

Figure 8.6
The form of a flower depends on the mode of pollination. (A) The parts of a "typical" flower. (B) A flower of *Poa pratensis* (Kentucky bluegrass, Poaceae). The enlargement shows a single spikelet. The lower floret is open; the feathery structures are the receptive stigmas. In the upper floret, the anthers have already dehisced—opened—and pollen has been dispersed. These grass flowers have no petals or sepals. (C) *Delphinium* (larkspur, Ranunculaceae) has a spur in one petal that contains nectar to attract pollinators.

pollen ("hay fever" is one type)—the overreaction of our immune systems to the foreign proteins in the huge amounts of pollen in the air.

The majority of plant species in the modern world are angiosperms, and all angiosperms produce flowers. However, while many animal-pollinated species produce attractive flowers, wind-pollinated flowers are rarely showy. There is generally no selective advantage to producing metabolically costly large petals, scents, nectar, and floral pigments if they are not needed to attract pollinators. In fact, there is often a disadvantage to making large petals and other organs that interfere with pollen transfer by wind. The grass family, for example, is almost entirely wind-pollinated, and grass flowers lack petals and sepals (Figure 8.6B).

Wind-pollinated plants produce massive quantities of pollen because they cannot exert much influence over where their pollen blows. The quantity of pollen produced can be expressed as the ratio of pollen to ovules produced by an individual or population. Wind-pollinated plants typically have much higher pollen:ovule ratios than insect-pollinated species. Many wind-pollinated species also have very large, feathery stigmas to enhance pollen capture. Pollen in these plants is light and rarely sticky, whereas insect-pollinated plants often have heavy, sticky pollen. Although these patterns are

striking, a note of caution is in order: In many plants, pollen is carried both by wind and by animals. The presence of petals does not prohibit wind pollination, nor does their absence mean that animal pollination is necessarily unimportant.

Pollen transfer by wind is much reduced in sheltered habitats where air movement is reduced. Thus wind pollination is most common in plants of open habitats. Wind-pollinated temperate forest trees usually flower well before their leaves emerge in spring. While this timing has sometimes been explained as a mechanism to reduce the amount of foreign pollen received on stigmas, it is worth noting that pollen movement is generally greatest when trees are leafless, which also often happens to be the windiest time of the year. Wind can carry pollen very long distances. On June 5, 1998, some local hunters at Repulse Bay in extreme northern Canada along the Arctic Ocean noticed large amounts of pollen at the edges of ponds (Campbell et al. 1999). It proved to be pollen from *Pinus banksiana* (jack pine, Pinaceae) and *Picea glauca* (white spruce, Pinaceae) that had been blown on June 1 from Quebec, Canada, over 3000 kilometers away!

It is tempting to think that wind pollination is evolutionarily primitive. It seems so simple and inelegant when compared with some of the complex adaptations some plants have for animal pollination. And it is clear that adaptation to animal pollinators has been a driving force in plant evolution. However, both ancient lineages (such as gymnosperms) and more recently evolved taxa (such as grasses) have wind pollination. Some wind-pollinated taxa have given rise to animal-pollinated taxa, while others have been derived from animal-pollinated ancestors.

Attracting Animal Visitors: Visual Displays

Flowers exist for one reason only: sex. Everything about them has to do with sex. This is obvious with the floral parts that are directly involved in producing or receiving pollen, or in the growth of seeds and fruits. But it is also true of the other floral parts, such as petals, that function mainly to influence the transfer of pollen.

Animals visit flowers because they generally contain rewards—mainly nectar, which is rich in sugars and sometimes amino acids, and pollen, which is rich in protein. Most people think of floral traits such as shape, color, and scent as being important in attracting animal pollinators—as advertisements of the rewards the animal can expect to acquire at the flower. But the interaction between plants and their floral visitors often involves a great deal of mutual cheating, false advertising, and theft. As we will see below, some plants appear to advertise rewards, but do not deliver them. Some animals visit flowers and remove rewards, but do not pollinate the flowers. Moreover, one species' attractant may be another species' repellant.

Displays of bright red flowers catch our eye. Although most people also enjoy looking at white flowers, we often do not notice them as easily as we notice red ones. Animals vary in their ability to see and discriminate among colors (Kevan 1978). For example, humans see colors between the wavelengths of about 380 nm (violet) and 780 nm (deep red), and our eyes are most

sensitive to blue (436 nm), green (546 nm), and red (700 nm). Insects, on the other hand, can usually see colors at shorter wavelengths than humans. Bees see wavelengths mainly from 300 nm to 700 nm. Thus, bees are less sensitive than we are to long-wave light and have difficulty discriminating among shades that we perceive as red, but there are colors that they can see—those between 300 and 380 nm—that are invisible to us. Birds' eyes are most sensitive to colors in the middle and red parts of the spectrum. Thus, flowers that are visually attractive to bees may not be as noticeable, attractive, or easily distinguishable to birds, and vice versa. Many bee-pollinated flowers have yellow hues, while bird-pollinated flowers are often red or orange. Flowers pollinated by moths are usually white to pale yellow, while bat-pollinated flowers are usually white to brown. Butterflies are often attracted to yellow or blue flowers. A note of caution is again in order, however: Many flowers reflect light in several different wavelengths. Flowers that we perceive as, say, red may be easily distinguished by bees because they also reflect in the blue wavelengths.

Flower color plays a more complicated role in pollination than these observations might suggest. Many flowers have contrasting colors in the **perianth** (the petals and sepals together). For example, the central portion of the flower may contrast sharply with the rest of the flower, as in many irises (Iridaceae). In most species that have been studied, these contrasts seem to have one of two functions. In some cases, the contrast itself is the attractant for pollinators. In other cases, one of the colors functions as a "nectar guide"—pollinators are first attracted to the flower as a whole, and are then attracted by the color contrast to the areas containing nectaries, anthers, and stigma (Figure 8.7).

Flower colors can vary over time, and numerous studies have shown that these variations make a difference in attracting pollinators. In most cases, older flowers fade,

(A)

(B)

Figure 8.7
Caltha palustris (marsh marigold, Ranunculaceae) under white (left) and ultraviolet (right) light. The color contrast that we see only under ultraviolet light is visible to bees in ordinary sunlight, and serves as a nectar guide. (Photographs courtesy of T. Eisner.)

(A) Determinate inflorescences

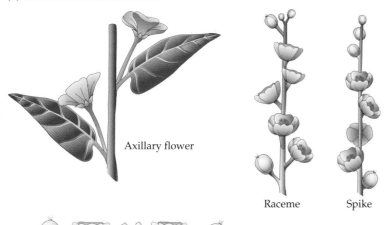

Cyme, opposite branching

Umbel (in bud)

Ray florets

Disk florets

Cyme, alternate branching

Head (capitulum)

(B) Indeterminate inflorescences

Axillary flower

Raceme

Spike

Corymb

although there are some, such as those of the prickly pear and cholla cacti, that grow darker over time. These changes in flower color seem to function much like nectar guides. The old flowers add to the attractiveness of the floral display as a whole, but once an animal has been drawn to the plant, it tends to be attracted to young, fresh flowers; these are the ones containing nectar or pollen.

Other parts of plant bodies have also been modified to act as attractants. **Floral bracts**—specialized leaves found below many flowers—are often colored and appear to act as part of the plants' display. A well-known example is *Euphorbia pulcherrima* (poinsettia, Euphorbiaceae), whose red bracts and upper leaves are far more colorful than its tiny yellow flowers. Many vase-forming bromeliads also have leaves that are more brightly colored than their flowers (see Figure 2.17).

Flowers are often arranged in aggregations called **inflorescences**. There are a number of types of inflorescences, ranging from the capitulum (head) of composites (Asteraceae) to the catkins of beeches and oaks (Fagaceae) to some rather loosely arranged racemes in plants such as buttercups (Ranunculaceae) (Figure 8.8). The terminology describing inflorescences is extensive and somewhat obscure; an introductory botany or plant morphology text can be consulted for more information. Inflorescence structure is important in pollination ecology. Having flowers gathered into an inflorescence may act to increase the size of the attractive display without altering the flowers themselves. It is likely that the form of an inflorescence also affects the behavior of pollinators.

Attracting Animal Visitors: Floral Odors and Acoustic Guides

Many flowers have noticeable odors, which clearly act as attractants. Why have scent as well as visual attractors? It is generally thought that scent acts over much

Figure 8.8
Some forms of inflorescences. A principal division is between (A) determinate inflorescences (which have a flower at the end of each flowering branch) and (B) indeterminate inflorescences (which do not). In a determinate inflorescence, the terminal flower normally blooms first, and the maximum number of flowers in the inflorescence is strictly limited.

longer distances than visual attraction, since many animals are able to detect extremely low concentrations of molecules in the air. On the other hand, scents are hard to locate precisely. Once an animal is in the general area of a patch of flowers, it is likely to find them much more quickly by vision than by scent.

Much less is known about floral odors than about colors, partly because the science of measuring odors is not nearly so well developed as the science of optics. Nevertheless, a few general points can be made. Perhaps the most important is that odors can vary independently of colors. Genes that make for yellow or red flowers usually have no effect on flower odor. All of us have been surprised to find no odor, or an unpleasant one, coming from a visually attractive flower. Pollinators similarly experience floral odors quite separately from the way they experience colors.

The second point should, upon reflection, be obvious: odors that attract one group of animals may repel others. Casual observation of the responses of humans, dogs, and flies to various odors should convince you of this! Bees are often attracted to what we perceive as sweet odors, and bats to musty odors. A number of plants, especially the milkweeds in the family Apocynaceae and the aroids (Araceae), have odors similar to animal dung or rotting tissue, which are strongly attractive to the flies and gnats that pollinate these plants. Some plants have odors that mimic insect mating pheromones, which cause male insects to visit the flowers and sometimes attempt to mate with them.

Possibly the most unusual flower attraction system known is that of the tropical vine *Mucuna holtonii* (Fabaceae), a bat-pollinated species (von Helversen and von Helversen 1999). The upper petal on the flower is raised when the flower opens. This petal is concave and acts as a sound mirror—an acoustic guide—reflecting back bats' sonar signals and letting them know that the flower is open. To reach the nectar, a bat has to press its snout deep into the flower. When it does so, another petal, which is holding the stamens, bursts open, throwing the pollen onto the bat's rump. This plant species is the first known to have an acoustic guide, but other bat-pollinated species may also have them.

Limiting Unwanted Visits

Many plants have adaptations not just for attracting certain visitors, but for repelling others. Why would a plant want to limit visitors? Once an animal visitor has removed pollen from a flower, that pollen has to reach another flower of the same species for it to contribute to the reproductive success of the plant producing it. If the next flower the animal visits is of another species, then the pollen is likely to be deposited in the wrong place. From the perspective of the plant, the best pollinator visits only one species, ensuring that the pollen gets to the right place. Limiting visits to only a small group of pollinators would increase the chance that the next plant visited will be of the same species. Numerous animals that are not effective pollinators are nevertheless attracted to flowers. From the animal's viewpoint, pollination is almost always an accidental consequence of this visitation. Thus many plants have undergone selection to deter certain floral visitors—but, as we all know from personal experience, it can be hard to get rid of unwanted guests!

Changes in flower shape are the main way in which access to flowers is restricted. The earliest flowers are thought to have been pollinated by insects unspecialized for this role, such as beetles. Not surprisingly, flowers that are typically beetle-pollinated today are often open and bowl-shaped. On the other hand, many flowers are tubular, and some have long spurs or smaller cones behind the main flower tube (see Figure 8.6C) in which nectar collects. Only animals with long tongues (or other long mouthparts), such as birds, butterflies, or moths, can reach the nectar and thus pollinate these flowers.

The type of reward a plant offers also restricts visitations. Beetles, for example, are generally unable to use nectar, whereas birds, bats, and bees use it easily. Some orchid bees (family Euglossidae) collect floral scents (by scraping the floral cells that secrete odor-producing oils), which they use in mating displays. Others use the oils for nest building. The timing of reward availability is important as well; bat- and moth-pollinated flowers usually produce nectar at night.

Mechanisms such as these do not, of course, provide any guarantees to the plant against wasted nectar and pollen. Recent studies have shown that many flowers are visited by large numbers of animals that transfer little pollen. In a study of *Calathea ovadensis* (Marantaceae), a tropical Mexican herb, Douglas Schemske and Carol Horvitz (1984, 1988) showed that moths and butterflies are the most common visitors, but bees account for the most pollination. Among the bees, the most efficient pollinators turned out to be relatively unimportant because they were uncommon. An extreme case of wasted resources is that of "nectar robbers," animals that insert their mouthparts between petals, or bore or chew through the side of the flower, and drink the nectar without ever coming into contact with the pollen. Very small insects such as ants also sometimes act as nectar robbers because they can visit the nectaries without contacting stamens or stigma.

Pollination Syndromes

Color, odor, reward type, timing of rewards, and shape all affect the numbers and types of floral visitors. It is not surprising, therefore, that certain combinations of these attributes often seem to be associated with particular types of pollinators. From what we have said, one might expect that bee-pollinated flowers would be yellowish, sweet-smelling, broad enough for the bee to contact the anthers and stigma, and produce nectar during the daytime. Bird-pollinated flowers might be expected to be red or orange, produce copious nectar during the daytime (but little scent), and have long tubes or spurs—and so forth. Such associations of floral attributes with particular types of pollinators (usually taxonomic classes or orders) are called **pollination syndromes** (Faegri and van der Pijl 1979). These syndromes are usually thought of as being matched by adaptive suites of pollinator traits—plant–pollinator interactions are thought of as tightly coevolved mutualisms (Howe and Westley 1997). Some ecologists have stood the idea of pollination syndromes on its head, treating floral traits as diagnostic of particular pollinators: if flowers have traits X, Y, and Z, the reasoning goes, then the plant's important pollinators are in animal taxon A.

There is little doubt that many plants and pollinators are coevolved—that the plants have strongly influenced the evolution of their pollinators, and vice versa. There are some well-known cases that offer evidence for tight coevolution of particular pairs of plants and pollinators (Box 8A). But there also are reasons to think that the strength and generality of pollination syndromes have been overstated. In a recent critique, Nickolas Waser and his collaborators (Waser et al. 1996) pointed to evidence suggesting that animals outside these syndromes accomplish much pollination—birds at "bee flowers," bees at "butterfly flowers," and the like. A principal criticism of the use of the pollination syndrome idea is that the syndromes are tendencies—and often rather weak ones—rather than laws. The critics have suggested that by concentrating on these tendencies, ecologists may be missing variation that is important. They have also argued that, under some fairly broad conditions, both plants and pollinators can be expected to evolve to be pollination generalists, rather than specialists. Thus, focusing on syndromes has caused us to neglect these generalist plant-pollinator interactions.

Aquatic Plants and Pollination

Some flowering plants, such as waterlilies (*Nymphaea* or *Nuphar*, Nymphaeaceae), live in the water. In most of these, the flowers are borne above the water's surface, and insects or wind pollinate the plants. But in plants of some 31 genera in 11 families, the flowers are borne underwater. As with wind pollination, it is inherently unlikely that pollen grains in water will find receptive stigmas. Furthermore, tides and currents are likely to make pollen dispersal quite unpredictable.

Aquatic plants have generally reduced this unpredictably by making their pollen dispersal units larger (and thus more likely to contact a stigma) and, in many cases, by dispersing them mainly at the water's surface (so that they travel in a two-dimensional plane, rather than a three-dimensional volume). A remarkable example is the freshwater plant genus *Vallisneria* (Hydrocharitaceae; Figure 8.9). In this genus, staminate (male) flowers containing pollen are released underwater and float to the surface. The petals curve downward to keep the flower floating on the water, and two sterile anthers act as tiny sails. The flowers are thus blown around until they encounter pistillate (female) flowers containing ovules, which are held at the surface by long pedicels (stalks). The pistillate flowers create small depressions in the water surface, and the staminate flowers fall in, pollinating the pistillate flowers.

In most aquatic species, pollen, rather than a whole flower, is the dispersal unit. In many cases, the pollen

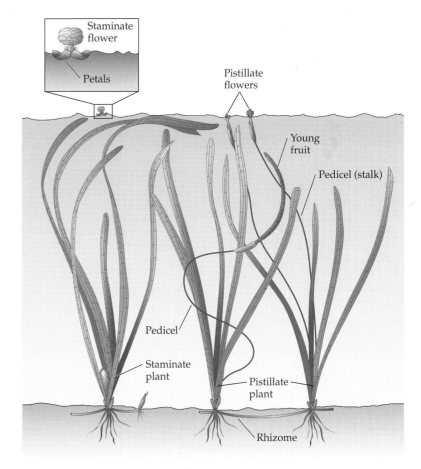

Figure 8.9
The aquatic plant *Vallisneria* (Hydrocharitaceae) has special adaptations for pollination at the water's surface.

Box 8A

Specialized Plants and Pollinators

Plant relations with their pollinators range from being highly specialized—with the plant depending on visits from only one or a few species of animals—to quite generalized. There are many interesting cases of specialization and adaptations by plants or their pollinators. In most cases, the animals are not as specialized as the plants, as they are pollinating plants for longer periods than any single plant species is in flower.

Flowers of *Yucca* (Agavaceae) are pollinated exclusively by *Tegeticula* moths. The moths pollinate the yucca flowers while laying their eggs in an ovary, and the larvae eat a fraction of the seeds in that ovary. If too many eggs are laid in a particular ovary, the plant will abort it. Some yuccas abort a sizeable number of pollinated flowers.

In the orchid family (Orchidaceae), pollen grains are clumped together in structures called **pollinia**. The size of the pollinia makes it relatively difficult to move them from flower to flower. In some tropical epiphytic orchids, the flowers are shaped so that an insect—usually a bee—must crawl through a particular opening to get to the nectar or oil rewards, and in doing so, cannot avoid hooking the pollinia in a way that facilitates their transport to the stigma of the next flower visited. Perhaps the most famous case is that of *Ophrys* (first studied by Charles Darwin [1877]), which has an odor similar to the mating pheromones of female bees. Males visit these flowers and "pseudocopulate" with them, thus transferring pollen.

Amorphophallus titanum (voodoo lily, Araceae), a plant native to Sumatra, has the largest single inflorescence of any plant. The man in the photograph is Hugo de Vries, an important early geneticist and one of those responsible for bringing Gregor Mendel's work to the attention of biologists. (Photograph by C. G. G. J. van Steenis, Hugo de Vries archive, Faculty of Science, University of Amsterdam.)

Many plants in the family Araceae have foul-smelling flowers that attract gnats and other small flies. Two common North American examples are *Symplocarpus foetidus* (skunk cabbage) and *Arisaema triphyllum* (jack-in-the-pulpit). The most extreme case is that of *Amorphophallus titanum* (voodoo lily), a plant native to Sumatra. Not only does this plant have the largest inflorescence in the plant kingdom, but it may also have the most powerful odor, which is reputed to resemble a combination of rotting fish and burnt sugar (none of the authors has yet had the pleasure). There are reports of people fainting from the odor.

Many African succulents in the dogbane family (Apocynaceae, another family that often has pollen grains clumped together in pollinia) are fly-pollinated. These plants attract flies by having flowers that smell much like decaying meat or, in some cases, animal dung.

Wild figs (*Ficus*, Moraceae) are mainly pollinated by specialized wasps. The flowers in this family are borne in a unique way: they are on the inside of the inflorescence, so that a wasp can reach them only by boring a hole in it. The wasp then lays eggs in the inflorescence. When the brood hatches, they mate (usually, this is a brother-sister mating), and the wingless males then die (so every wild fig contains dead wasps). As the young females leave the inflorescence, they are dusted with pollen, and when they visit another inflorescence to lay their eggs, they pollinate it.

Fascinating as these examples are, it is easy to get the wrong impression: as noted in the text, most plants have much less specialized mechanisms of pollinator attraction.

---

is sticky and tends to form rafts that float on the surface until they contact pistillate flowers. Pollen release in some subtidal marine species occurs only near very low (spring) tides, which also increases the chances of contact between pollen and stigmas. In species with entirely underwater pollination, such as the turtle grass *Thalassia* (Hydrocharitaceae), pollen tends to be elongated or embedded in strands of mucilage—features that also increase the chances of contacting stigmas.

A remarkable case of elongation is found in the seagrass genus *Zostera* (Zosteraceae). The pollen germinates before release, forming a pollen tube in the process. The pollen tube greatly increases the length of the pollen and enhances its chance of becoming caught in a pistillate flower. Pistillate flowers of aquatic plants also show a number of features, such as changes in stigma shape, that act to increase the probability of contact with pollen.

Who Mates with Whom?

Plants have remarkably complex sex lives. In this section we look at plant **mating systems**—biological factors that govern who can mate with whom. Some species are obligately **self-fertilizing** (individuals can only pollinate themselves), other species are obligate outcrossers (individuals must be pollinated by others), while some species may both self and outcross. This multiplicity of mating patterns is accomplished by a variety of mechanisms. The conditions under which one mating system or another should be favored by selection constitute an area of active research, both theoretical (for example, Charlesworth and Charlesworth 1978; Barrett and Harder 1996) and experimental (for example, Carr and Dudash 1996; Husband and Schemske 1996; Vogler and Kalisz 2001).

Plant Gender

Gender expression in plants is complex and varied. Some individuals are **cosexual**—they function as both males and females simultaneously. The most common form of cosexuality is **hermaphroditism**. Hermaphroditic individuals have **perfect flowers** containing both functional stamens and functional stigmas. Under **monoecy**, plants have some flowers with functional stamens only (**staminate flowers**) and some with functional stigmas only (**pistillate flowers**); sometimes perfect flowers also occur on the same individual. By contrast, under **dioecy**, at least some plants in a population have only pistillate or only staminate flowers.

These three systems of gender expression can exist in combination with one another. Under **gynomonoecy**, both pistillate and perfect flowers occur on the same individuals. By contrast, under **andromonoecy**, both staminate and perfect flowers occur on the same individuals. Andromonoecy is much more common than gynomonoecy. Under **gynodioecy**, some plants in a population have only pistillate flowers, while others have either perfect flowers or a mixture of staminate and pistillate flowers. By contrast, under **androdioecy**, some plants in a population have only staminate flowers, while others have either perfect flowers or a mixture of both staminate and pistillate flowers. Gynodioecy is much more common than androdioecy.

Still another form of gender expression is **sequential hermaphroditism**: beginning life as one gender (almost always male) and then switching to the other. In some sequential hermaphrodites the switch is not instantaneous, and individuals may function as hermaphrodites for a time.

Some cosexuals are capable of self-fertilization (sometimes within the same flower and sometimes only between flowers), while others are not. What is even more remarkable is that some of these different patterns of variation among individuals can occur simultaneously within a single population.

Having a particular set of floral organs, however, does not mean that a plant population of hermaphrodites necessarily uses them. Within a population of hermaphrodites, for example, some individuals may mainly contribute pollen and set few seeds, acting mainly as males. The pollen on what appear to be hermaphrodites may be inviable or cannot germinate, and those individuals are functionally strictly female. When individuals that appear to us to be hermaphrodites actually are functionally unisexual, we speak of **cryptic dioecy**. The functional gender of a plant may be very different from its apparent physical gender; in other words, the ratio of an individual's male reproductive output to its female reproductive output may be quite different from 1.

Since this kind of variation can greatly affect the genetic composition of populations, it should be no surprise that plant gender have been the subject of much study among evolutionary ecologists, conservation biologists, and population geneticists. Eric Charnov (1982) developed the basic theory that has motivated many studies of the evolution of gender. The central idea is that of fitness maximization: in different settings, fitness may be maximized by increasing the allocation of resources to both male and female function, or by increasing one and decreasing the other. The genetic basis of variation in gender is understood in some plant species; there are well-known cases in which the controls are quite simple.

Discussion of plant gender obviously involves a lot of terminology. To make matters worse, different biological disciplines use these terms to mean somewhat different things. Animal biologists usually use "hermaphroditic" to mean that both male and female sex organs are produced by the same individual (since there is no animal analogue to a flower). Geneticists use "monoecious" to mean the same thing. This terminology is not imprecise. Scientists all mean something quite specific when using these terms, but each subdiscipline means something a bit different. In practice, this is not usually a problem: one just needs to be aware of these differences and use the context to determine the meaning an author intends.

Competition for Pollinators and among Pollen Grains

As Oscar Wilde remarked, "Nothing succeeds like excess." Many plants appear to have followed this advice. Almost all plants start out with many ovules that never turn into viable seeds. These excess ovules may exist for three reasons: because pollen is a limiting resource, because the flowers have another function, or because the plant is hedging its bets (producing more ovules than can usually be matured, because under unusually favorable conditions these ovules can be matured). We consider the first two reasons here; bet

Box 8B

Pollination Experiments

Many questions about plant reproductive ecology can be answered with very simple experiments. Often we have two questions about a particular species: Are the plants self-incompatible? Are the plants pollen-limited? To answer these questions, we can do various combinations of three types of manipulations: cover flowers or inflorescences with mesh bags to exclude pollinators, remove stamens from (emasculate) plants, and hand-pollinate plants. The results of these treatments are then compared with results from unmanipulated plants, allowing us to estimate the typical rate of seed set. Bagged flowers indicate how many seeds in the unmanipulated plants came from outcrossed pollination. Emasculated plants show how many seeds in the unmanipulated plants came from self-pollination. If we get no seeds from bagged flowers, hand pollination with pollen from the same or different individuals tells us whether the species is self-incompatible or simply requires animals for pollination. Doing both bagging and emasculating tells us whether the species is capable of producing seeds by agamospermy. If hand-pollinated plants produce more seeds than naturally pollinated plants, the plants are pollen-limited.

---

hedging is discussed more fully in Chapter 9.

For animal-pollinated plants, pollen may be limiting because of a lack of pollinators (Box 8B). If there are more flowers than pollinators to visit them, or if the pollinators are inefficient at transferring pollen, the plants may be competing for pollinators. This competition can occur both among plants of the same species and among species in a community. These two types of competition can have different consequences for flowering patterns, as we will see in Chapter 9.

Competition for pollinators among individuals of the same species is analogous to competition among males (or females) for mates in animals. When this concept was first developed in the 1980's—notably by Mary Willson (Willson 1983; Burley and Willson 1983)—it was met with some skepticism because the analogy between pollen grains on a stigma and male birds competing for the attention of females seemed strained to some botanists. However, there is now a large body of evidence that pollen donors and pollen grains do indeed compete. Some individuals in a population may end up with many matings (lots of pollen), while others may get very few. This variation results in **sexual selection**—selection by means of differential success at mating. Most evolutionary ecologists consider sexual selection as a subset of natural selection, although common usage sometimes treats this area of selection theory separately.

There are two broad categories of sexual selection: male-male competition, and female choice. In plants, male-male competition is often thought to result in extremes in floral displays, such as large inflorescences in which most of the flowers func-

Mary Willson

tion only to attract pollinators. These flowers are, thus, functionally male; they reproduce only by having their pollen fertilize the ovules of flowers on other plants, and not by being pollinated themselves. Such cryptic dioecy is particularly common in legumes (Fabaceae) and in the Bignoniaceae. Competition among pollen grains after they land on the stigma can also result in sexual selection. There is a race for the ovules as pollen tubes grow down through the style (Figure 8.10). Experimental studies have shown that in some species, the maternal sporophyte "chooses" among its pollinators through biochemical interactions that affect pollen germination and the rate of pollen tube growth, resulting in more fit offspring (Marshall and Folsom 1991). Maternal sporophytes are also known to selectively abort certain seeds before they mature (such as *Cucurbita lepo* [zucchini, Cucurbitaceae]; Stephenson and Winsor 1986).

Pollen Dispersal and Its Consequences

Some degree of **inbreeding** (mating among relatives; the opposite is **outcrossing**) is common in plant populations. The reason is simple: The largest number of matings occur among neighboring plants, which tend to be relatives because most seeds do not travel far.

Most pollen ends up near the sporophyte on which it originated. In animal-pollinated species there are several reasons for this pattern. First, most plant species have clumped spatial distributions. Moreover, some patches of plants are typically more rewarding to pollinators than others, and the pollinators usually concentrate on the more rewarding patches, reducing the average distance traveled between plants. Finally, most pollen is transferred to the next flower visited (Richards 1986). In *Lupinus texensis* (bluebonnet, Fabaceae), for example, most pollinator movements between plants are less than 0.6 m, and over 90% of movements are less than 1.2 m (Figure 8.11). Gene flow does not precisely match

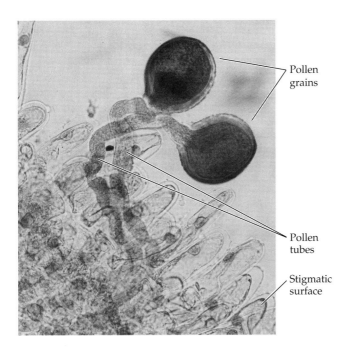

Figure 8.10
When pollen grains germinate, the pollen tubes grow, penetrate the stigmatic surface, and grow down the style towards the ovules. Shown here are two germinated pollen grains of *Erythronium grandiflorum* (glacier lily, Liliaceae). (Photograph courtesy of A. Snow.)

pollinator movement (again suggesting that pollinator visitation is not equivalent to pollination), but gene movement also declines rapidly with distance from the source.

Pollen movement by wind also usually results in most pollen being distributed near its source, but for different reasons. Wind usually blows in short gusts, most of which have low velocity, so most pollen movements will be over short distances. In trees, pollen is released at varying heights, and the pollen released at the greatest height usually travels the farthest (Richards 1986). However, most trees are shaped so that little pollen is released from the treetops; thus, most pollen travels short distances. Since most wind-pollinated plants (such as grasses) occur in large stands, most pollen tends to land on neighboring plants. In some populations, plants have variable characteristics that can affect the distance that pollen travels. In *Plantago lanceolata* (Plantaginaceae), Stephen Tonsor (1985) found that individuals differed in the extent to which they produced pollen as single grains or as clumps of two, three, or up to fifty or more grains. A wind-tunnel experiment showed that single grains travel much farther than larger clumps (Figure 8.12). Thus, some plants are more likely to mate

(A)

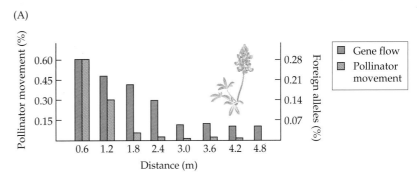

(B)

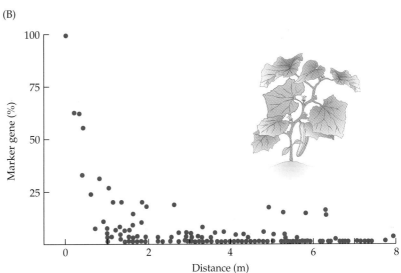

Figure 8.11
(A) The distance traveled by pollinators (open bars) and the appearance of marker genes (solid bars) in seedlings of *Lupinus texensis* (bluebonnet, Fabaceae). While the general pattern of distribution is the same, using pollinator movement alone to estimate gene flow would underestimate it, because pollen from one flower may be deposited on the second, or third (and so forth) flower that is visited. (After Schaal 1980.) (B) The appearance of marker genes in seedlings of *Cucumis sativus* (Cucurbitaceae). (After Handel 1983.)

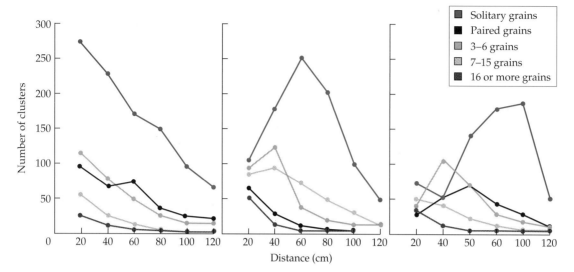

Figure 8.12
Variation in the distance traveled by *Plantago lanceolata* (Plantaginaceae) pollen clumps in a wind tunnel as a function of size in three trials. In each trial, a plant was placed at one end of the wind tunnel, and rows of Vaseline-coated microscope slides were placed at 20 cm intervals along the floor of the tunnel. Each line shows the number of solitary grains or clumps that traveled each distance. (After Tonsor 1985.)

with distant individuals, and some are more likely to mate with close neighbors.

Assortative Mating

Plants with similar phenotypes—for example, flower color or flowering time—are sometimes more likely to mate with one another than they would by chance alone. This phenomenon is called **assortative mating**. In the case of flowering time, this is especially obvious: early flowerers are more likely to mate with other early flowerers than with late flowerers. Assortative mating for flower color often occurs because many pollinators have preferences (sometimes on an individual basis) for colors. A bee, for example, may travel mainly from yellow flower to yellow flower or red flower to red flower.

When individuals tend to mate with others with dissimilar phenotypes, we refer to **negative assortative mating** or **disassortative mating**. Two common forms of this phenomenon in plants are **self-incompatibility** (meaning that matings can occur only between individuals with certain genetic differences) and **heterostyly** (flowers have styles and stigmas of different lengths, and matings within morphs occur only rarely; see Figure 6.9). These types of negative assortative mating cause obligate outcrossing.

Understanding Plant Mating Systems

Under what environmental circumstances would these different mating systems be favored? Complex as plant mating systems are, it seems that only a few major factors have been important in their evolution. These factors include the fitness consequences of inbreeding versus outcrossing, the relative success of male and female floral functions, the ability to adjust resource allocation to male and female functions, and the availability of pollinators.

There is almost always some inbreeding among plants because the individuals in a population are likely to share common ancestors. As in a small human village, everyone is a cousin of some sort. But strong inbreeding occurs when plants pollinate themselves or mate mainly with closely related neighbors. Reproductive traits can promote or limit such inbreeding.

If inbreeding decreases fitness (**inbreeding depression**), it is advantageous to limit matings with relatives. Self-incompatibility (SI) systems can accomplish this. There are two general types of SI systems. Typically, the self-incompatibility loci are referred to as S loci and individual alleles are numbered S_1, S_2, and so forth. In **gametophytic SI**, the pollen grain's haploid genotype at the S locus prohibits it from growing down the stigma and style of a sporophyte sharing the same allele By contrast, in **sporophytic SI**, the diploid genotype of the pollen parent prohibits it from mating with any sporophyte sharing either of its S alleles. The two systems act differently because of differences in the time the S alleles are expressed. In gametophytic SI, many matings are "semi-compatible": if pollen from an S_1S_2 plant is deposited on the stigma of an S_1S_3 plant, the S_2 pollen will grow and can fertilize both S_1 and S_3 ovules. In sporophytic SI, pollen from an S_1S_2 plant cannot fertilize any ovules on an S_1S_3 plant or an S_2S_3 plant. Thus, while self-fertilization cannot occur under either system, sporophytic SI limits mating with more types of relatives than does gametophytic SI.

Table 8.1 Ratio of the number of pollen grains produced per flower to the number of ovules produced per flower in species with different levels of outcrossing

Mating system	Number of species examined	Mean pollen: ovule ratio (standard error)
Cleistogamous	6	4.7 (0.7)
Obligate selfer	7	27.7 (3.1)
Facultative selfer	20	165.5 (22.1)
Facultative outcrosser	38	796.6 (87.7)
Obligate outcrosser	25	5859.2 (936.5)

Source: Cruden 1977.

Self-pollination and nondispersing seeds often co-occur in plants. If a particular genotype is successful in a particular spot, it may be advantageous to produce some nondispersing offspring of the same genotype. This may be part of the explanation for why many plants have some **cleistogamous** flowers (which never open and always self-pollinate), but evidence for this is equivocal. In some species, cleistogamous flowers are located near ground level; in annual grasses of the genus *Amphicarpum* (Poaceae), they are actually underground. More familiar open flowers are **chasmogamous**. Cleistogamous and chasmogamous flowers may be produced simultaneously, as in many species of the temperate-zone herbs *Viola* (Violaceae), *Oxalis* (Oxalidaceae), and *Impatiens* (Balsaminaceae). In some species, cleistogamy appears to be environmentally determined. Flowers of the cactus genus *Frailea* (Cactaceae) are cleistogamous except in extreme heat, and so are considered to be facultative selfers. In several British plants, including *Lamium amplexicaule* (Lamiaceae) and *Hesperolinon* (Linaceae), flowers produced in the dry season are cleistogamous (Richards 1986).

Plant mating systems often strongly affect the number of pollen grains that are produced by an individual or a population relative to the number of ovules (the pollen:ovule ratio). As Table 8.1 shows, self-fertilizing species tend to have much lower pollen:ovule ratios than outcrossing species, and the greater the assurance of selfing—as in cleistogamy—the lower the ratio.

The Ecology of Fruits and Seeds

Since most plants are rooted in place, they disperse as fruits, seeds, or related structures. These dispersal units are sometimes referred to collectively as **diaspores**. All flowering plants have fruits, and fruits have a tremendous range in size, shape, and number of seeds, characteristics of other tissues, color, odor, and chemical properties. A fruit is just a mature ovary, so even plants such as grasses have fruits, called **grains**, although these fruits contain little maternal tissue beyond the seed itself. Fruits have two major functions: they protect the developing seeds, and they can affect the dispersal of the seeds. These functions often conflict with each other, and variation in their relative importance leads to much of the variation in fruit characteristics.

Many fruit characteristics are best understood as traits that affect the behavior of animals feeding on fruits and seeds. On the one hand, most seeds are quite nutritious, and seeds are therefore important animal foods. Most human caloric intake, for example, comes from the grains of grasses (for example, rice, wheat, and corn). There is, therefore, a selective advantage to plants in protecting seeds from being eaten before they are mature—otherwise those offspring are simply lost. On the other hand, after seed maturity, there is often a selective advantage in attracting animal consumers to disperse the seeds. Plants have various ways of balancing these two functions. Some seeds can pass through animals without being destroyed; for example, while humans consume vast quantities of wheat and rice seeds, tomato seeds pass through our digestive tracts without any loss of viability.

Many fruits attract **frugivores** (fruit eaters) after seed maturity. Unripe fruits typically have little in the way of attractive odors, flavors, or colors, and sometimes these may even be repellent. Moreover, in many taxa, unripe fruits contain compounds that act as defenses; animals eating them tend to become sick, especially when they are eaten in large quantities (Herrera 1982a,b). You may have experienced this phenomenon yourself if you have ever eaten unripe fruit. As they ripen, however, most fruits change color, begin to give off attractive odors, and become palatable through the conversion of starches to sugars and other chemical and physical transformations.

If seed dispersal is to occur, animals must not spend too much time feeding at any one plant. If the plant and its fruits are too attractive to the animals, the animals will not move away, and the seeds will not be dispersed. How do plants encourage frugivores to move along? First, the fruits on an individual plant may mature asynchronously. For example, flocks of birds or troops of monkeys often travel from tree to tree, eating only the ripe fruits. Second, if ripe fruit remains very mildly toxic, an animal will eat only a few fruits of a given species at once.

Seed Dispersal Patterns

Seeds can be dispersed by animals, by wind, or, less commonly, by water. Animals can disperse seeds internally or externally. Seeds can be dispersed singly or in groups. Furthermore, seeds can be dispersed in time as well as space. All of these factors have different effects on the pattern of seed dispersal. The pattern of seed dispersal is important because it determines the demographic and genetic structure of a plant population, the amount of competition within and among species, and species migration patterns.

There are two modes of dispersal by wind: Some diaspores float on the breeze, and some flutter to the ground. Dandelion seeds (*Taraxacum*, Asteraceae), which children enjoy blowing around, are the archetype of the former sort (Figure 8.13A). This form of dispersal is very

common among plants that colonize disturbed habitats. Many of them are annuals that survive by being very good dispersers, but poor competitors. We discuss such trade-offs in more detail in the next chapter.

Maples (*Acer*, Sapindaceae) are a good example of plants that produce fruits that flutter to the ground (Figure 8.13B). Clearly most maple fruits are unlikely to travel nearly as far as most dandelion seeds. Most maple seeds end up falling very close to the parent tree, although a few may be blown a substantial distance. Carol Augspurger and Susan Franson (1987) showed that the shapes of diaspores can have a large effect on how far they travel (Figure 8.14). Even after initial dispersal, significant diaspore movement can occur. For example, seeds of *Daucus carota* (wild carrot or Queen Anne's lace, Apiaceae) disperse very late in autumn and can then be blown across the snow in winter.

Carol Augspurger

Dispersal by animals involves more varied mechanisms and leads to more varied patterns. In species that rely on animal dispersal through frugivory, the seeds have thick, hard seed coats and are either regurgitated by the animal or passed through the digestive tract in their feces. For the seeds, the latter may be an ideal location because they get dispersed to a location that is provided with fertilizer. Some fruits contain a mild laxative to speed their passage through the gut, thereby preventing their digestion and speeding their dispersal. Some seeds require passage through the gut of an animal to facilitate germination. In other cases, the fruit may be eaten and the seeds spit out (as in watermelon: *Citrullus lanatus*, Cucurbitaceae).

An interesting problem in frugivory was discovered in the 1970s when scientists noted that *Sideroxylon grandiflorum* (tambalacoque, Sapotaceae), a native of Mauritius, was near extinction. There were only a handful of individuals left, and these all appeared to be centuries old. The seeds of this plant have very thick coats, requiring **scarification** (abrasion) to germinate. An early hypothesis was that much of this scarification was provided by dodos (*Raphus cucullatus*, large flightless pigeons), which were driven extinct in the seventeenth century by hunting by European sailors. As a result of the dodo extinction, the hypothesis went, the seeds were unable to germinate. Subsequent research has suggested an alternative culprit: introduced monkeys readily eat the fruits before they are ripe, thus eliminating most recruitment. Whatever the cause, the species remains near extinction.

Granivores (animals that eat seeds or grains) sometimes act as important seed dispersers. Animals intend-

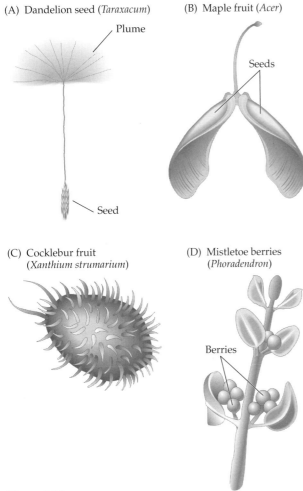

Figure 8.13
Seeds often have structures that facilitate particular dispersal mechanisms. (A) Some wind-dispersed seeds, such as those of *Taraxacum* (dandelions, Asteraceae), float on the breeze. (B) Other wind-dispersed diaspores, such as those of *Acer* (maples, Sapindaceae), flutter to the ground. (C) Some animal-dispersed seeds, such as those of *Xanthium* (cocklebur, Asteraceae), are carried externally, stuck to the animal's fur or feathers. (D) Other animal-dispersed seeds, such as those of *Phoradendron* (mistletoe, Santalaceae), pass through the gut of a vertebrate before being dispersed. *Phoradendron* seeds are coated with a sticky mucilage, readily adhering to the branches of potential host trees; frugivorous birds often rub their cloacas on the branches to remove the seeds from their bodies, thus effectively planting the seeds.

(A) Dandelion seed (*Taraxacum*)
Plume
Seed

(B) Maple fruit (*Acer*)
Seeds

(C) Cocklebur fruit (*Xanthium strumarium*)

(D) Mistletoe berries (*Phoradendron*)
Berries

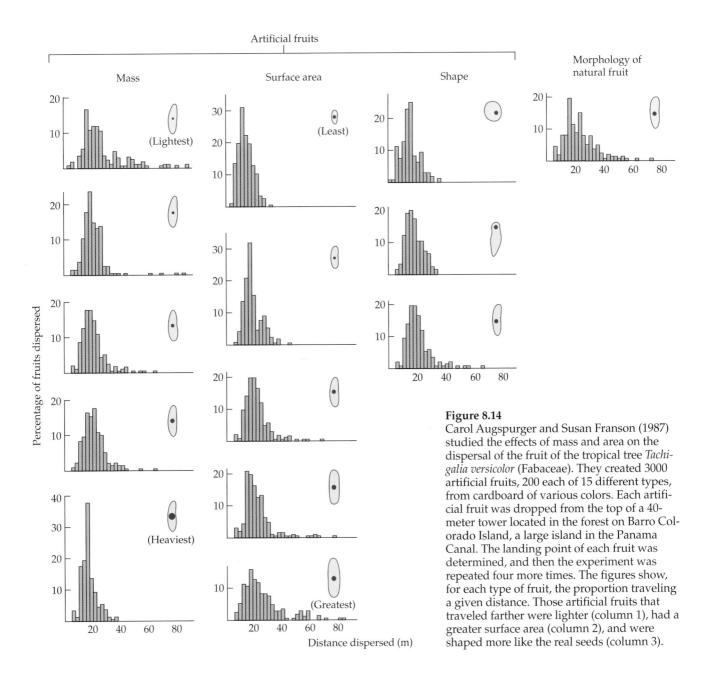

Figure 8.14
Carol Augspurger and Susan Franson (1987) studied the effects of mass and area on the dispersal of the fruit of the tropical tree *Tachigalia versicolor* (Fabaceae). They created 3000 artificial fruits, 200 each of 15 different types, from cardboard of various colors. Each artificial fruit was dropped from the top of a 40-meter tower located in the forest on Barro Colorado Island, a large island in the Panama Canal. The landing point of each fruit was determined, and then the experiment was repeated four more times. The figures show, for each type of fruit, the proportion traveling a given distance. Those artificial fruits that traveled farther were lighter (column 1), had a greater surface area (column 2), and were shaped more like the real seeds (column 3).

ing to eat seeds often drop them accidentally. Animals that bury seeds for later use also act as dispersers if they fail to return. Both birds and mammals cache seeds in this fashion. For example, *Pinus albicaulis* (whitebark pine, Pinaceae) in the mountains of the northwestern United States and southwestern Canada is mainly dispersed by granivorous birds, especially *Nucifraga columbiana*, Clark's nutcracker. The seeds of this tree are not released from the cones when they are mature, and the tree depends on birds to remove and disperse them. The nutcrackers pry the seeds loose from the cones, remove them, and cache them, storing from 1 to 15 seeds together in each of many caches just under the soil surface. They later retrieve the seeds and eat them. Although Clark's

nutcrackers have excellent memories, they occasionally forget where some of their caches are, or they may be killed or driven away, leaving their uneaten seeds to germinate and grow into trees (Lanner 1996).

Sometimes the mode of animal dispersal can be subtle. Many species of ants collect seeds for food. In some plant species, however, the seeds are attached to a lipid body called an **elaiosome**. Instead of eating these seeds, ants carry the seeds to their nest, remove and eat the elaiosomes, and toss the seeds outside the nest. Thus seed dispersal by animals may be highly clumped, with many seeds ending up in a refuse heap near an ant nest or in the dung pile below the feeding tree of a monkey troop. Other seeds disperse by getting stuck in the hair or feathers of an animal (Figure 8.13C), which can be a very effective mode of long-distance dispersal.

While most seeds end up relatively close to the parent plant, a small percentage of the seeds may be dispersed long distances, and this long-distance dispersal can have important demographic consequences. Note, however, that a long distance for one species could be a short distance for another. When we speak of "long-distance dispersal," we mean that individuals are moving far enough from their parent population that they can no longer interact with individuals in that population. It also helps to have a rough idea of the actual distances involved; for many plants, dispersing more than a few hundred meters can move them out of their parent population.

In Chapter 21, we discuss the migration of trees northward in North America following the retreat of the glaciers. Some tree species are estimated to have migrated as fast as 400 meters per year. Since it takes several years for a tree to become large enough to produce seeds, migration at this rate must have depended on very long distance dispersal events. The tail of the distribution of dispersal distributions is therefore quite important, but estimating the shape of the tail presents some important statistical problems (Portnoy and Willson 1993). Interestingly, wind-dispersed species migrated faster than animal-dispersed species (see Table 21.1); one might expect the reverse, since animal-dispersed species often appear to have more long-distance dispersal events.

Colonization of newly disturbed habitats or new islands must also rely on long-distance dispersal. Once a plant becomes established in such a habitat, short-distance dispersal is responsible for local population growth. For example, the seeds of *Cocos nucifera* (coconut, Arecaceae) disperse from island to island by floating in seawater—one of the few species whose seeds can survive this experience. On a given island, however, most seeds travel only very short distances, as you might imagine would be so for such large seeds (although they travel from tree to ground very fast!).

Seed Banks

Dispersal in time, as well as in space, can be important for some plant species. Once a seed is buried in the ground, it can remain there for years or decades. The **seed bank** is the collection of seeds in the soil; this term is sometimes used to refer to the seeds of a single species and sometimes to the seeds of an entire community. There is a tendency for short-lived plants (such as annuals) to have long-lived seeds, while long-lived plants (such as trees) and plants with large seeds tend to have short-lived seeds, although this is not a hard-and-fast rule. As a result, the abundances of species in the seed bank may have little relationship to those of the plants

growing in the same spot. Many species growing above ground have few seeds in the soil. Other species may exist only in the seed bank at a given time. Species that specialize in colonizing disturbed places by wind dispersal, for example, may exist above ground for only a few years following a disturbance.

Seed banks also occur above ground in populations that have serotinous fruits or cones. **Serotinous** fruits or cone scales are bonded shut with resins and open only when temperatures are sufficiently high (see Figure 13.4). Normally this occurs only in response to fire. The seeds are thus dispersed only under conditions suitable for germination (bare mineral soil and increased nutrient availability) and are protected from granivorous animals at other times (Groom and Lamont 1997; Hanley and Lamont 2000; Lamont and Enright 2000). Serotiny is especially important in many conifers and in a number of species of Australian Proteaceae such as *Banksia* and *Hakea*.

We actually know very little about how long the seeds of many species can remain dormant because the time scales necessary for such experiments are so long. In 1879, Edward Beal began a famous experiment at the University of Wisconsin. He took a set of 20 glass jars and in each put a mixture of soil and 50 seeds of each of 19 common weed species and one crop species. He then buried the jars uncapped, but upside down. The idea was to allow moisture, small arthropods, microbes, and so forth access to the seeds, while not allowing new seeds to get in. Beal and his successors dug up one jar every 5 years for the first 40 years, and at 10-year intervals after that time, and tested the seeds to see whether they were still viable (Priestley 1986). For 16 species, all seeds that germinated did so within 50 years, while the seeds of 4 species continued to germinate after that time. For two of these species—*Malva pusilla* (round-leaved mallow, Malvaceae) and *Verbascum* spp. (Scrophulariaceae; common mullein, *V. thapsus*, and moth mullein, *V. blattaria*, were not distinguished)—some germination occurred among 100-year-old seeds. There were also interesting patterns of germination and dormancy. For example, seeds of *M. pusilla* germinated after 5, 20, and 100 years of burial, but not at other time intervals. As the experiment was designed to look for maximum longevity rather than variation in germinability, we cannot estimate the latter. However, half the species showed gaps in germination like that seen in *M. pusilla*, albeit not so dramatic, suggesting that complex patterns of dormancy may exist in many species.

The existence of a seed bank can have important demographic consequences for a population. It can also help to buffer a population against year-to-year variation in demographic rates (bet hedging; see Chapter 9). Seed banks can also affect evolution in two ways. First, they slow the response to selection by maintaining genotypes produced in previous years. Second, mutation rates in seeds can actually be substantial, and seed banks can thus act as a source of genetic novelty (Levin 1990).

Summary

Growth and reproduction are key ways in which plants interact with the animals and the physical environment around them through their uptake of nutrients, use of space, pollination relationships, and dispersal of fruits and seeds. Plants grow by adding repeated modules such as leaves, flowers, and branches. This mode of growth leads to plasticity, allowing plants to vary their growth pattern in response to light availability or other environmental factors. There are a variety of ways of using this capacity, and clonal plants use it to explore their environments in varied ways—sometimes by following concentrations of resources, and sometimes by more active "foraging."

Both sexual and asexual reproduction are common in plants. Some asexual reproduction is vegetative, but some is from seeds. Most seeds are produced sexually, however, and the evolution of pollen and seeds laid the basis for much of the diversification of plants. Many adaptations affect pollen transfer, including changes in the sizes, shapes, colors, and odors of flowers as well as in many pollen characteristics. These changes have led to some remarkable floral adaptations and cases of coevolution with specific pollinators, but most pollination in most species is probably by generalized animal taxa.

Plants have complex mating systems. The main causes of this complexity appear to be that inbreeding sometimes has important fitness consequences, that male and female floral functions may have different relative success rates, and that it is often beneficial to be able to adjust the relative allocation of resources to male and female functions.

The dispersal of fruits and seeds is affected by a large number of traits. Again, interactions with animals can be an important mode of dispersal, although plants must accomplish seed dispersal while avoiding seed consumption. Most seed dispersal is local, but rare long-distance dispersal events can be demographically important. The seeds of some species remain dormant in the soil for many years. The movements of pollen, seeds, and fruits are key causes of spatial patterns in plant populations.

Additional Readings

Classic References

Darwin, C. 1877. *The Various Contrivances by Which Orchids are Fertilized*. John Murray, London.

Horn, H. S. 1971. *The Adaptive Geometry of Trees*. Princeton University Press, Princeton, NJ.

Contemporary Research

Barrett, S. C. H. and L. D. Harder. 1996. Ecology and evolution of plant mating. *Trends Ecol. Evol.* 11:73–79.

Schemske, D. W. and C. Horvitz. 1988. Plant-animal interactions and fruit production in a Neotropical herb: A path analysis. *Ecology* 69:1128–1137.

Waser, N. M., L. Chittka, M. V. Price, N. M. Williams and J. Ollerton. 1996. Generalization in pollination systems, and why it matters. *Ecology* 77:1043–1060.

Additional Resources

Charnov, E. 1982. *The Theory of Sex Allocation*. Princeton University Press, Princeton, NJ.

Lanner, R. M. 1996. *Made for Each Other: A Symbiosis of Birds and Pines*. Oxford University Press, Oxford and New York.

Willson, M. F. 1983. *Plant Reproductive Ecology*. Wiley, Chichester, UK.

Willson, M. F. and N. Burley. 1983. *Mate Choice in Plants: Tactics, Mechanisms, and Consequences*. Princeton University Press, Princeton, NJ.

CHAPTER **9** *Plant Life Histories*

Why do plant species have such a wide variety of life cycle patterns? **Annual** species, such as most weeds and field crops, begin reproducing within a relatively short time after germination, produce many seeds in one or a few bouts of reproduction, and then die within the year in which they germinated. Other plants—the giant *Lobelia telekii* (Campanulaceae; see Figure 6.12) of Mt. Kenya, *Argyroxiphium* (Hawaiian silverswords, Asteraceae; see Figure 7.6), and *Digitalis* (foxgloves, Scrophulariaceae; Figure 9.1), wait for years before they reach a sufficient size for reproducing, then pour their resources into reproduction, producing seeds in a single bout before they die. Still other plants, woody **perennials** such as most trees and shrubs and herbaceous perennials such as spring-flowering garden plants, must reach a certain minimum size before reproducing (which can take months to years), and produce relatively few seeds each time they flower, but survive for a long time and reproduce many times over their life span.

Organisms that reproduce in a single bout are called **semelparous** (the botanical term is **monocarpic**, because the plant produces flowers once), while those that reproduce repeatedly are **iteroparous** (**polycarpic**). A plant's schedule of birth, mortality, and growth is called its **life history**. Variation in plant life histories has evolved over time, and this variation has important consequences for population dynamics.

In this chapter we discuss schedules of birth, mortality, and growth over a plant's life span and the causes and consequences of variation in those schedules—for example, the implications of being perennial versus being annual. We also discuss the timing, or **phenology**, of growth and reproduction within a year.

Size and Number of Seeds

The idea of **trade-offs** caused by limited resources (that increasing one thing necessarily means decreasing something else) is central to most thinking about selection on life histories. To see why this is so, ask yourself what would be the most favorable plant life history if no trade-offs were necessary. Such a plant would produce unlimited quantities of seeds, each of which was large enough to maximize its chance of becoming established and growing quickly. New cohorts of seeds would be produced continually. The plant itself would grow rapidly and live indefinitely. Clearly, this combination of traits is not possi-

Figure 9.1
Digitalis purpurea (foxglove, Schrophulariaceae) is a "biennial"—a short-lived monocarpic perennial. These plants take two or more years to reach maturity, then reproduce in a single bout in which an enormous fraction of their available resources is devoted to reproduction. Relatively few plant species are strict biennials; in most so-called biennials, some plants reproduce in two years, and some in more. (Photograph courtesy of M. Rees.)

ble. Even if we substituted definite numbers for words such as "unlimited" and "indefinitely," it would not be possible for a plant to maximize all of these components of fitness simultaneously; trade-offs would be necessary. A good way to see this is to consider the size and number of seeds.

The number of seeds per reproductive bout can vary tremendously. There are some plants that mature only a few dozen or so seeds per flowering episode, such as *Cocos nucifera* (coconut palms, Arecaceae). A mature cottonwood tree (*Populus fremontii*, Salicaceae), on the other hand, may produce tens of thousands of seeds per episode; some orchids may produce hundreds of thousands to millions of seeds. If natural selection favors those genotypes that leave behind the most representatives over time, why don't all plants produce huge numbers of seeds?

Most plant ecologists studying this question have focused on the idea of a trade-off between the number of seeds a plant can make and the sizes of those seeds. Larger seeds generally contain more endosperm (or other nutritional stores), and thus generally have a greater chance of establishment success, than seeds provided with fewer resources. Large amounts of endosperm in the seeds of some arid-zone perennials, for example, allow them to sink deep roots quickly, increasing their chances of survival.

Thus, in fitness gained through maturing seeds (often called a plant's **female fitness**), there is a trade-off in the use of limited resources. Producing more (but smaller) seeds probably reduces the chances of success for each seed because smaller seeds typically grow more slowly. On the other hand, there are likely to be diminishing returns in provisioning only one seed. The joint chance of success of, say, two medium-sized seeds can be greater than the chance of success of a single large seed. Thus, natural selection on maternal plants favors those that can divide resources among seeds so as to optimize the number of descendants left behind. Is there a single best seed size, or is it best to produce seeds in a range of sizes? What is the best size or range of sizes?

John Harper (1977) suggested that variation in seed size is quite limited in most plant species, as was suggested by the data available at the time. However, studies in the 1980s and 1990s revealed much variation in seed size within populations, even when natural selection was acting to constrain seed size because a particular size was optimal for a species (Mazer and Wolfe 1992; Platenkamp and Shaw 1993; Mojonnier 1998). In fact, individual plants often produce seeds of quite varied size (Venable 1985; Winn 1991; see Figure 5.1)

Using a graphical model, Christopher Smith and Stephen Fretwell (1974) showed that there is a single optimal size for seeds (Figure 9.2) if environmental conditions are predictable. McGinley et al. (1987) showed that Smith and Fretwell's conclusion holds under fairly broad circumstances. However, their analysis found that variable seed size is sometimes favored when the environment varies over time and the geometric mean of fitness (see Chapter 7) is used as the criterion for fitness. The geometric mean gives the average rate of population growth over a series of variable years, and is therefore the appropriate measure of fitness.

Individual plants actually produce seeds of variable size, as noted above, so this theoretical result seems to differ from what we actually see in nature. McGinley et al. (1987) suggested that this difference might be a consequence of evolutionary constraints that limit the abilities of plants to reduce variation in seed size. For example, many species in the sunflower family (Asteraceae)

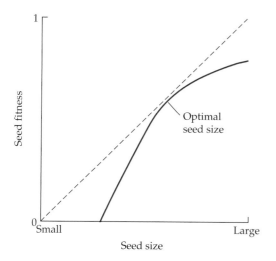

Figure 9.2
Model for finding the seed site that maximizes fitness of the parental plant. The curve gives the fitness of seeds of a particular size under particular environmental conditions. Increasing seed size increases fitness, but above a certain size, there are diminishing returns in producing large seeds. Below a minimum size, the seeds have no fitness. The optimal seed size is the value of the curve tangent to the 45-degree line. (After Smith and Fretwell 1974.)

have flowering heads with two kinds of flowers, disk florets and ray florets, which can be constrained to produce different-sized seeds by their physical shape (Figure 9.3; Venable 1985). Seeds within a legume pod can differ in size because those more proximal (closer) to the maternal plant are able to garner more resources than more distal (farther) seeds (Silvertown 1987). It is important to realize that other factors besides natural selection determine the outcome of evolution (see Chapter 5). Even when selection favors a single seed size and when natu-

ral selection is the dominant evolutionary force, it may not be possible to eliminate variation in seed size within or between plants, because this variation is also determined by environmental and developmental factors.

In determining seed size, there is an inherent conflict between the fitness of the maternal sporophyte and the fitness of the individual seeds. Natural selection on maternal plants favors those that leave the largest number of descendants, but it acts differently on seeds: the favored seeds are those that have the best individual size—often, the largest. An analogy may be helpful: if a pair of parents have a large sum of money to leave their children, they may best assure the success of the largest number of descendents by dividing the money evenly among the children—but an individual child may best assure his or her success by getting as large an individual inheritance as possible. This sort of competition among offspring—and conflict between parents and offspring—also occurs in plants.

Individual seeds can vary in their efficiency at sequestering a share of the maternal sporophyte's resources. Genetic differences, positional effects, differences in the timing of fertilization, or environmental variation can lead to differences in seed size. Selection on seeds favors increased ability to garner maternal resources whenever the fitness gained by doing so exceeds the inclusive fitness lost by depriving siblings of those resources. An individual's **inclusive fitness** is a measure of how well that individual passes on its genes as well as how well its relatives pass on theirs, weighted by how closely they are related. For example, if every seed in a fruit has the same pollen parent, then, selection favors an individual's increasing its seed size if doing so increases its fitness by more than the fitness lost by two siblings, because, on average, the seeds share half their genes with one another.

Disk florets

Ray florets

Figure 9.3
Helianthus annuus (sunflower, Asteraceae) is a widespread North American annual plant that is widely cultivated for its seeds, but also has many wild populations. Sunflower inflorescences (also called heads) are composed of two types of flowers (florets), each of which can produce a single seed. Disk florets compose the center of the head, while ray florets—which have a single very large petal and four very small petals—occur around the outside of the head. In some species in this family, seeds produced by different kinds of florets can differ in size, shape, and dispersal and dormancy properties. (Photograph by S. Scheiner.)

Classic studies of seed size were conducted by Edward Salisbury in England (1942) and Herbert Baker in California (1972). Both of them focused on the ecological correlates of seed size. Baker, for example, concluded that plants with larger seeds were characteristic of more open habitats, such as deserts and coastal dunes. These classic studies have stimulated much research, partly because this kind of study makes it difficult to distinguish the effects of selection from the effects of phylogeny. For example, seed size may have different ecological consequences in different environments, as Baker concluded. But it may also be true that open habitats have many large-seeded plants simply because the plants that colonized those habitats are from taxa with large seeds. In an extensive study of the flora of the Indiana Dunes at the southern end of Lake Michigan (where Cowles conducted his pioneering studies of succession; see Chapter 13), Susan Mazer (1989) was able to control for these different effects statistically. She found that habitat accounted for only 4% of the variance in seed size by itself. In the Indiana Dunes, the relationship between seed size and habitat is thus largely caused by the history of the taxa that occur in each habitat.

Another factor that can affect selection on seed size and number is **granivory**, or seed predation. If a substantial fraction of seeds are eaten by granivores, there can be selective advantages to reducing seed size and increasing seed number. While tiny seeds may have reduced chances of establishment, they may also be too small to be worth the attention of granivores. There is a large body of theory concerning animal behavior, called "optimal foraging theory," concerned with how animals should be expected to search for, select, and handle food items. This theory predicts that granivores should not pay attention to seeds that are too small, and this prediction is supported by much empirical evidence. In addition, small seeds, scattered widely, are far more difficult for vertebrate seed predators such as rodents or jays to find than are a few large seeds (this is not true for fungal seed pathogens, however).

Seed size may also be dictated by dispersal strategy. If seeds are dispersed by wind (see Figure 8.13A), then there may be an optimal size to maximize the distance traveled. It is not just the resources invested in the seed that matter, but also the resources invested in the dispersal structure (see Figure 8.14). Variation in seed size may ensure that seeds are dispersed different distances away from the parent plant, resulting in less competition among siblings.

In this discussion of the ecology and evolution of seed size and number, we have assumed that there is a fixed total amount of resources allocated to reproduction at any time. When this is true, the problem reduces to the trade-off between number and size of seeds. This assumption is useful in gaining an initial insight. But it

is clear that both the timing of reproduction and the amount of resources allocated to reproduction are often genetically variable and have evolved over time. This realization leads to the more general problem of the evolution of life histories.

Life History Strategies

We now consider the more general problem: how does selection act on the schedule of reproduction and survival throughout a plant's life? Martin Cody (1966) suggested that throughout their lives, organisms must apportion resources among competing demographic functions—survival, growth, and reproduction. How an organism does this determines its fitness. Many studies have been conducted using the assumption that these three demographic functions really are competing. It has become clear, however, that this is not always the case, because the functions of plant body parts do not correspond in any simple way to the three demographic categories (see "Reproductive allocation" below). However, there is no doubt that life history traits are subject to natural selection.

Arabidopsis thaliana (Brassicaceae; Figure 9.4) is a weedy European plant that is now widely naturalized throughout the temperate world. It is perhaps best known as the "fruit fly" of plant molecular biology—

Figure 9.4
Arabidopsis thaliana (mouse-ear cress, Brassicaceae) is a small annual mustard. Native to central Europe, it has become naturalized in much of the temperate world. *A. thaliana* has one of the smallest genome known in any angiosperm, and therefore is widely used in genetic studies. Because laboratory strains have been selected for extremely short life cycles—as little as a few weeks—the species is widely used in a number of other studies, including a growing number of ecological studies. (Photograph courtesy of K. Scheiner.)

that is, as an important model organism—because it has a small genome and reproduces quickly. Under some circumstances, some *Arabidopsis* genotypes can begin flowering three weeks after germinating. There is also remarkable variation in *Arabidopsis* life spans. Individuals from some wild European populations live as long as a year. *Arabidopsis* from other populations can have two generations within a year—one of the few plants that do so. In some places with mild winters, such as the mid-latitude U.S. states of Kentucky and North Carolina, this plant is a winter annual—germinating in the fall, growing slowly through the winter, and flowering in the spring. Where the winters are colder, as in the northeastern United States, it is a spring annual—germinating in the early spring, growing quickly, and flowering by late spring or early summer. There is a genetic basis for much of this variation within *Arabidopsis* (Kuittinen et al. 1997). Those genotypes with the shortest life histories have been the focus of most research because of their convenience, but evolutionary ecologists are now exploring the molecular basis of life history variation in this species.

Seed Germination

A number of different environmental factors help to trigger germination. In many plants, temperature plays a key role in regulating germination. For example, winter annuals in the Sonoran Desert do not germinate during the summer rainy season, and the reverse is true of summer annuals. At least in some plants, some cell membranes in seeds undergo temperature-dependant phase transitions that render cells either more or less permeable to water, thus allowing (or preventing) **imbibition** (uptake of water by seeds) depending on the season (Bewley and Black 1985). Light also plays a key role in many species. For example, many weedy species require light to germinate; this is why many species germinate following soil disturbance. There are also species such as *Eschscholzia californica* (California poppy, Papaveraceae) in which germination is inhibited by light.

When should a seed germinate? This might seem to be obvious: it should germinate when conditions are favorable. But can plants use environmental conditions at germination time to predict subsequent conditions—that is, can plants evolve predictive germination? The answer depends critically on the predictability of the environment. In a maritime climate (see Figure 18.14C), for example, summers are often warm and wet. Germination in late spring usually provides a fairly good chance of surviving to maturity. In contrast, rainfall in desert environments can be highly unpredictable (see Figure 18.19), and many plants that germinate do not survive to maturity (Fox 1989; Venable and Pake 1999).

In a study of the annual *Plantago insularis* (Plantaginaceae) growing in the Sonoran Desert, Maria Clauss

and Lawrence Venable (2000) found only a slight correlation between rain at germination time and subsequent rain during the growing season. This correlation was greatest at mesic sites (that is, favorable microhabitats). However, populations at xeric sites (unfavorable microhabitats) would be expected to experience the strongest selection for being able to predict when adequate rainfall would be available. Clauss and Venable pointed out that predictive germination might still be occurring in these plants, but if so, it is a complicated phenomenon. For example, heavy rain at germination time does predict subsequent rain in xeric (but not mesic) sites in El Niño years (see Chapter 18).

Predictive germination may involve factors other than climate. Annuals in fire-prone shrublands in California are stimulated to germinate by compounds in smoke (Keeley and Fotheringham 1997). Fire reduces aboveground biomass and the density of competitors, and minerals in the ash add nutrients to the soil; therefore, conditions following fire are more favorable for growth of annuals. Many species classified as weedy (ruderal) use cues of recent disturbance as signals for germination.

Life Span

Annuals, as we saw above, are plants in which the vegetative life cycle is completed in less than a year. Individual plants are "born" as seeds; since most annual plants have seed banks, this means that most annuals that you see growing are actually more than a year old. In temperate-zone environments, many annuals can be categorized as either winter annuals (like *Arabidopsis* in the southern United States) or summer annuals. These terms can be confusing. **Winter annuals** germinate in the autumn, overwinter as vegetative plants, and flower and die in the spring. **Summer annuals** complete their aboveground life cycle in the warm months. Research with annuals in the Sonoran Desert (Mulroy and Rundel 1977) has showed that winter and summer annuals differ considerably in their developmental and physiological traits, and this is probably true of winter and summer annuals in other environments as well.

Annuals are especially common in certain environments, including deserts, many dunes, and recently disturbed sites. Most common herbaceous weeds are annuals. Annuals in warm deserts have been the subject of much ecological investigation. In years favorable to plant growth, these plants produce many seeds. Subsequent years may well have poor rainfall, leading to poor germination, survival, and growth for these plants. It is only in "good" years that population growth may actually be positive (see Figure 7.14).

Semelparity includes plants with a range of life spans. **Biennials** (Figure 9.5; see also Figure 9.1) are semelparous plants that flower after two or more years.

Figure 9.5
Cirsium canescens (Platte thistle, Asteraceae) is a "biennial." (Photograph courtesy of S. Louda.)

The term itself is something of a misnomer, as it takes most biennial species longer than two years to reach flowering size, and individuals in most biennial populations vary in this respect. **Semelparous perennials** are plants that live for a number of years before flowering. There is no firm boundary between biennials and semelparous perennials (Young and Augspurger 1991), although generally biennials die back to ground level in winter (or the unfavorable season), persisting only underground, while semelparous perennials have substantial aboveground structures year-round.

Both theoretical and empirical studies point to two factors that might select for semelparous life histories: a critical size for reproduction and diminishing returns from retaining resources for subsequent bouts of reproduction. Biennials and short-lived semelparous plants often occur in early successional habitats, where canopy closure after several years causes a decrease in the quality of the environment. Long-lived semelparous plants are characteristic of relatively unproductive but persistent habitats, such as deserts, many grasslands, and alpine regions. Factors such as competition for pollinators may be more important in selecting for semelparity in these settings.

In discussing semelparity, it is important to be clear as to whether it is the genet ot the ramet that is semelparous. Monocarpic ramets may be favored by selection when production of inflorescences is very costly and it takes plants long periods to acquire enough resources to flower. This can be the case when plants are pollen-limited (see Chapter 8) and large inflorescences receive disproportionately many pollinator visits (Schaffer and Schaffer 1979). It is important not to confuse monocarpic ramets with true semelparity: a genet with monocarpic rosettes, such as an agave (Figure 9.6) or a bromeliad, can reproduce many times in its life, while a semelparous plant (such as a Hawaiian silversword) has a much riskier life history in that it can reproduce only once.

In most habitats, it is the perennial plants that give the landscape its characteristic appearance. The category "perennial" includes a great diversity of plants. Some live for only a few years, while others live for centuries

Flowering stalk

Flowering ramet

Nonflowering (vegetative) ramets

Figure 9.6
Agave (century plant, Agavaceae) has semelparous rosettes that flower only once, mobilizing all the energy stored during a lifespan of a few decades to produce a large inflorescence with hundreds of flowers. The genet itself is iteroparous, because vegetative reproduction precedes flowering. (Photograph © Frans Lanting/Photo Researchers, Inc.)

or even longer. While most woody plants live for at least a few years, the reverse is not necessarily true. Some herbaceous plants, such as many tundra plants, live for centuries. Among forest trees there are species that live for decades and others that live for centuries, even in the same type of habitat. In pine flatwoods in the southeastern United States, for example, *Pinus palustris* (longleaf pine, Pinaceae) can live for 300 or more years, while *Cersis canadensis* (eastern redbud, Fabaceae) lives for only a few decades. Typically, long-lived perennials have short-lived seeds and a small or nonexistent seed bank. Most perennials, however, have the ability to become dormant as vegetative plants during winter or the dry season. Perennials have more ability to withstand variable and harsh environmental conditions in their aboveground phase than do annuals; this ability increases with the longevity of the species (long-lived temperate trees, for example, are more likely to withstand a hard freeze or a drought than a short-lived herbaceous perennial would be).

How do we explain all of these different life histories, as well as the variation within each? What are their consequences for population growth? Early efforts at such explanations focused on the theory of *r*- and *K*-selection, while recent work has centered on demographic approaches.

r- and K-selection

Robert MacArthur and his students and colleagues developed their ideas about life history evolution, focused on the concept of *r*- and *K*-selection, during the 1960s. This work stimulated much of the early research on the evolution of life histories. The concept of *r*- and *K*-selection plays little role in current research on life histories. Nevertheless, it continues to be influential in popular writing on ecology, in textbooks, and in some aspects of community ecology, so we explain it briefly here.

Using the logistic model of population growth,

$$\frac{dN}{dt} = rN\left(\frac{K-N}{K}\right)$$

where *r* is the intrinsic rate of population growth, *K* is the carrying capacity, and *N* is the population size, MacArthur and Wilson (1967; MacArthur 1972) suggested that at low population densities, selection will be strongest on traits that increase the intrinsic rate of population growth, *r*. At high densities, they suggested, selection will be strongest on traits that increase population size, or the carrying capacity, *K*. Loosely, they argued that there is stronger selection for productivity than for efficiency at low densities, and vice versa at high densities. Selection for increasing *r* and *K* was widely interpreted as counterposed. There is, however, no reason why there must be a trade-off between *r* and *K*. It is

mathematically straightforward to show that selection under logistic growth always favors increases in both parameters (Boyce 1984; Emlen 1984).

Using the assumption that high population densities lead to reduced survival of juveniles, Eric Pianka (1970) developed detailed predictions for animal populations, which were later extended to plants: *K*-selection should favor delayed reproduction and a reduced number of offspring, while *r*-selection should favor early reproduction and an increased number of offspring. This prediction is what most people mean when they refer to *r*- and *K*-selection: weedy plants, for example, are taken to be *r*-selected, while forest trees are taken to be *K*-selected. As Boyce (1984) put it, references to "*r* strategists" generally mean small organisms that reproduce quickly and are poor competitors, and thus should be favored in disturbed habitats, while references to "*K* strategists" generally imply that organisms have the opposite characteristics, although these characterizations were not derived from the original model.

Perhaps the greatest problem with *r*- and *K*-selection theory is that its predictions are qualitative and relative, rather than quantitative, and so are difficult to test. Plants that are *r*-selected, for example, are predicted to evolve an earlier age at first reproduction and greater seed numbers than those that are *K*-selected. But what can one measure to test this? Plants within one community might be classified as *r*-strategists or *K*-strategists based on which reproduced at an earlier age, but this classification might be different in another community, or in a comparison of one of these species with another species. What is more, such a classification says nothing about whether the differences among the species are actually a result of selection on age of first reproduction, and certainly not whether the differences are part of an entire life history strategy. Finally, the emphasis on the distinction between plants that live mainly in disturbed habitats (*r*-strategists) and those that live in stable habitats (*K*-strategists) may have been misplaced. Many "stable" habitats, such as mature forests, have been shown to depend on disturbance for their maintenance (see Chapter 13). Previous notions of climax communities and stable equilibria have changed dramatically in recent years, reducing the utility of this equilibrium-based theory.

Grime's Triangular Model

Responding to some of the limitations of *r*- and *K*-selection theory, plant ecologist Philip Grime (1977, 1979) proposed an extension of that theory, classifying life histories as being influenced primarily by selection for one of three traits: colonizing ability (which he called *R* selection, for ruderal plants), competitive ability (*C* selection), or stress tolerance (*S* selection). Grime's theory was not derived from any particular model of population growth or evolution. He assumed that there are necessary trade-offs between these syndromes. In fact, Grime (1979) pointed out that plants could have any combination of traits associated with *R*, *S*, and *C* selection. Like *r*- and *K*-selection theory, then, Grime's theory made predictions that do not lend themselves to critical tests. While *C-S-R* theory made assumptions and predictions about life histories, it had a particular emphasis on the relative importance of competition in different habitats (see Chapter 10). Consequently, this theory has received a fair amount of attention and generated considerable controversy among plant community ecologists, but has not had much influence in the study of life history evolution.

Demographic Life History Theory

The demographic approach to life history theory centers on studying how variation in survival and fecundity as individuals vary in size, age, or other life history stages affects population growth. Its utility can be seen in its solution of "Cole's paradox." In studying the consequences of schedules of reproduction, Lamont Cole (1954) developed a model that compared the fitnesses of a perennial phenotype with an annual phenotype in the same species. If a fraction *p* of the plants survive to reproduce each year, and produce *F* seedlings if they do reproduce, then the perennial's fitness is $\lambda_P = p(F + 1)$, while the annual's fitness is $\lambda_A = pF$. Cole concluded that annuals need only increase their reproductive output by one new seedling per year to do as well as perennials. He pointed out that this finding was paradoxical because most organisms are perennial, not annual, suggesting that there must be other advantages to being perennial.

Using a demographic approach, Eric Charnov and William Schaffer (1973) pointed out a flaw in Cole's reasoning: his result depends on the assumption that rates of survival and fecundity do not depend on age. If his model is modified slightly to give first-year plants a different chance of surviving than older perennials, the paradox disappears. Charnov and Schaffer demonstrated this as follows: If the fraction of first-year plants surviving is *c*, and the surviving annuals produce F_A seeds while the surviving perennials produce F_P seeds, then $\lambda_P = cF_P + p$ and $\lambda_A = cF_A$. Thus the population growth rates of both phenotypes are given by the chance of individuals surviving to reproduce times their fecundity if they do survive; for the perennials, this rate is increased by the chance of surviving from year to year. Therefore, the annuals have greater fitness only when $\lambda_P < \lambda_A$, that is, when $F_A > F_P + p/c$, or when their fecundity exceeds the perennials' fecundity plus the survival chances of adults relative to those of juveniles. This means that where survival of adults is much greater than survival of juveniles, the annuals' reproductive output would need to exceed that of the perennials by a considerable

amount for the annuals to have a greater lifetime fitness—thus resolving the paradox.

Reproductive Allocation

If it matters when a plant reproduces, and how much it reproduces at each life stage or age, the obvious question becomes, how does selection act on the allocation of resources at each stage? This question, especially with the added assumption of a trade-off between reproduction and survival, has provided the focus for much research in life history evolution.

Key to addressing this problem is the concept of reproductive value (*v*; see Chapter 7). William Schaffer and Michael Rosenzweig (1977) showed that fitness is maximized at every age if the sum of present reproduction plus future reproduction weighted by relative reproductive value,

$$F_i + \frac{p_i v_{i+1}}{v_0}$$

is maximized at every age. The fraction of resources allocated to reproduction at age or stage *i* is the **reproductive effort** at that age or stage. Schaffer and Rosenzweig analyzed selection on life histories graphically by plotting current and future reproduction against reproductive effort.

Figure 9.7 shows two simple alternatives. In Figure 9.7A, the curves for both present reproduction (F_i) and future reproduction [$(p_i v_{i+1})/v_0$] are convex, so their sum is maximized when reproductive effort is either 0 or 1 at any given age or stage. These curves can differ for different ages or stages. Since a reproductive effort of 1 means putting all resources into reproduction and none into survival, this graph describes selection for a semelparous life history. In Figure 9.7B, both curves are concave, so selection favors intermediate reproductive effort at every age after the onset of reproduction—an iteroparous life history. Combinations of curves of different shapes can thus describe a wide variety of life histories.

Testing this kind of theory has proved difficult. One problem, pointed out by Schaffer and Rosenzweig (1977), is that in many cases more than one life history is optimal. But there is a deeper problem in actually measuring reproductive effort: How does one assign the fraction of resources used for reproduction? Little is known about the underlying developmental and biochemical processes of reproduction. An early study (Hickman and Pitelka 1975) found that the dry weight of plant parts, except for some unusual tissues, is highly correlated with their energy content. This result led to hundreds of studies comparing the dry weights of plant parts under varied circumstances.

But how does one assign a plant part to "current" versus "future" reproduction? For flowers and fruits, the choice is clear, but this is not so for organs such as roots.

(A) **Semelparity**

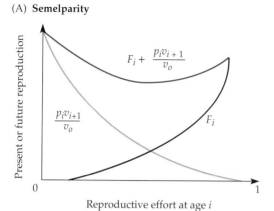

(B) **Iteroparity**

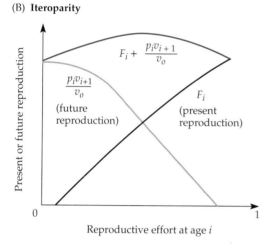

Figure 9.7
Selection for semelparity versus iteroparity. Fitness is greatest when the sum (green curves) of present reproduction plus future reproduction is maximized at every age. (A) Semelparity is favored when the curves for both present and future reproduction are convex—fitness is greatest when reproductive effort is always either 0 or 1. (B) Iteroparity is favored when both curves are concave—fitness is greatest at intermediate reproductive effort once maturity is reached. (After Schaffer and Rosenzweig 1977.)

One attempt at handling this problem was a study by Edward Reekie and Fakhri Bazzaz (1987). These investigators measured the amount of root mass present during flowering of a grass and the amount present at other times, and found no difference. They concluded that none of the root mass counted toward reproduction.

The problem with testing theory of this kind is that the parts of a plant do not relate in any simple way to their demographic consequences. Moreover, reproductive effort is a ratio. As such, it does not uniquely describe the life history (Schaffer 1983) because there are many different life histories that could yield a measurement of, say, 75% allocation to reproduction at a particular time. This observation led to models using more

sophisticated mathematics, such as optimal control theory, which can handle the technical problem of optimizing reproductive allocation throughout the life span. Additional work along these lines has been slow because meaningful predictions require that the models be much more biologically realistic. Further work in developmental biology and physiology is thus needed to make much more progress in this approach to life history theory.

Matrix models (see Chapter 7) play an important role in both theoretical and empirical studies of life histories. In the simple case of the Charnov-Schaffer model of perennial life histories, the original model stated that the numbers of juveniles and adults (J and A, respectively) next year will be $J_{t+1} = cF_pJ_t + pF_pA$ and $A_{t+1} = cJ_t + pA_t$. Rewritten in matrix form, this is

$$\begin{bmatrix} J_{t+1} \\ A_{t+1} \end{bmatrix} = \begin{bmatrix} cF_p & pB_p \\ c & p \end{bmatrix} \times \begin{bmatrix} J_i \\ A_i \end{bmatrix} \quad (9.3)$$

Analysis of this model tells us that the population will eventually grow at the rate $\lambda_p = cF_p + p$, the dominant eigenvalue of the matrix. At that time, the population will be composed of F_p times as many juveniles as adults. In this simple case, it is easy to see the consequences of varying the survival and fecundity rates. For example, increasing juvenile survival c increases λ_p at the rate F_p. With more realistic models, it is not so easy to do this sort of thing visually. However, the methods described in Chapter 7 make it possible to completely analyze the effects of varying any survival or fecundity rate on population growth or composition.

Bet Hedging in a Variable Environment

Year-to-year variation in survival and fecundity also affects fitness. How can fitness variation among years be reduced when some of that variation is inherently unpredictable? Many plants appear to achieve this by "hedging their bets" (spreading the risk) among years. To see why variation among years changes fitness, compare a hypothetical annual species that always produces an average of 1.5 seeds every year (that is, half of the plants in the population produce 1 seed, and half produce 2) with another annual that produces 0.5, 1, and 3 seeds in dry, average, and wet years. If these types of years occur with equal frequency (1/3), the average number of descendants of the variable species will be $0.5^{1/3} \times 1^{1/3} \times 3^{1/3} \approx 1.14$, or 0.36 fewer than the number produced by the invariant species. In general, increasing the variance in fitness among years decreases its long-term geometric mean. This implies that there is a trade-off between mean and variance in fitness: gaining added fitness in "good" years can come at the expense of reducing long-term average fitness. For matrix models, the problem is somewhat more complex, because the order of multiplication matters. Analyses of both structured and unstructured population models suggest that selection often acts to increase fitness by reducing among-year variation in fitness.

Spreading reproduction among more years or more evenly among years, increasing the area over which seeds are dispersed, and increasing dispersal through time by means of seed banks are all examples of bet hedging. Seed banks are perhaps the best studied of these mechanisms. Most annual plants have seed banks. Without environmental variation, among-year seed dormancy is opposed by selection to germinate immediately, since seeds that remain dormant in the soil cannot reproduce, but those that germinate can reproduce. A number of studies (e.g., Brown and Venable 1991; Kalisz and McPeek 1993) suggest that seed banks play an important role in buffering populations from environmental variation.

Difficulties in Measuring Trade-Offs

There are an enormous number of studies comparing species or higher taxa that show a negative correlation between traits such as longevity and fecundity. Nevertheless, a substantial number of studies at the population level—where selection must be acting on these trade-offs—not only fail to show a negative correlation, but often show the opposite. This kind of result led a few scientists in the 1980s to argue that there are really no trade-offs. The problem was clarified when Arie van Noordwijk and Gerdien de Jong (1986) pointed out that a positive correlation between, say, reproductive allocation and survival-related traits could actually be expected if the total amount of resources available varied among individuals. They suggested a useful analogy: for people operating under a fixed income, the amount of money available to spend on housing is negatively correlated with the amount available to spend on cars. Nevertheless, if you looked at the data, you would find that people who spend more on housing also spend more on cars—because they have greater total wealth. The trade-offs are real, but they operate in the context of variable access to resources.

A critical distinction must be made between those traits that must be traded off because the physical rules of the universe demand it, and traits that are traded off because selection has molded species that way. An example of the first category is resource allocation to leaves versus flowers: limiting resources (carbohydrates, nitrogen, or water) used to produce and maintain leaves cannot also be used to produce flowers. A possible example of the second category is an apparent trade-off between speed of wood growth and wood toughness. Fast-growing trees such as quaking aspens (*Populus tremuloides*, Salicaceae) have weak wood, and individual trees generally live only 50 or 60 years. Consequently, the quak-

ing aspen tends to be an early successional species (see Chapter 13). We cannot tell whether the growth rate versus wood density trade-off is inherent in the biochemistry of wood, or if the right mutation has simply not come along, or if selection always favors just these two combinations of traits.

Thus, we need to know more about variation among individuals in resource availability, as well as about the mechanisms determining resource allocation to different functions. Using a genetic model, David Houle (1991) showed that even when trade-offs do exist, genetic correlations between life history traits could evolve to be either negative or positive, depending on the genetic details. This finding points to the importance of increasing our understanding of the biological basis of life history traits.

Phenology: Within-Year Schedules of Growth and Reproduction

Phenology—the timing of growth and reproductive activity within a year—can vary greatly among species, populations, and even individuals. In temperate climates, there are usually some plants flowering at all times between the first and last frost. Some woody species are evergreen, while others are deciduous. Even within groups that we usually think of as all of one type—such as conifers, which we think of as evergreen, there may be exceptions—such as the deciduous conifer species *Larix laricina* (the larch or tamarack, Pinaceae; Figure 9.8). In general, plant phenologies are constrained by seasonality, mainly by temperature or moisture availability. Phenologies have a more complex relationship with animal pollinators, seed dispersers, and herbivores because the plants and animals can be agents of selection on each other.

Vegetative Phenology

In temperate deciduous forests, many herbaceous plants on the forest floor expand their leaves and flower before the canopy trees begin leaf expansion. The consequence is that these plants do most of their growing and reproducing in temperatures that are far colder than those experienced by the canopy trees during the same life stages. Many studies have shown that this timing is actually advantageous to the forest-floor herbs. Once the canopy closes, very little sunlight—sometimes as little as 1%—gets through to the forest floor. Thus, the forest-floor herbs' growth and reproduction would be even more limited by light availability than they are by cool temperatures.

In most temperate species, it appears that temperature and photoperiod (day length) are the main factors determining vegetative phenologies. It is important to realize that temperature plays this role in a particular

Figure 9.8
Larix (larch, Pinaceae) is a deciduous conifer found in taiga around the world. Left: *L. kaempferi* (ochudo) at 3500 m elevation on Mt. Fuji, Japan. Right: A branch of *L. laricina* (tamarack) in a bog at 200 m elevation in northern Michigan. (Photographs by S. Scheiner.)

way. It is generally the number of **degree-days**—the sum of temperatures experienced over some period of time—and not the temperature on a particular day that determines the timing of leaf expansion. In cold years, then, leaf expansion (sometimes called "leaf flushing") is delayed. Plants, like many other organisms, often use photoperiod (see Chapter 2) as a reliable predictor of the average temperature. If they relied exclusively on temperature, a warm spell in midwinter would typically cause many plants to expand their leaves. By using temperature as a cue, plants can respond to an early spring by flushing their leaves early, but because they also use photoperiod as a cue, this response is limited. These plants, in other words, respond to environmental fluctuations, but they do so cautiously.

As the example of temperate forest-floor herbs suggests, there are at least two main types of selective factors that act on vegetative phenologies. Abiotic factors limiting growth—such as the timing of killing frosts or seasonal droughts—often determine the beginning or end of growth episodes (or both). Biotic factors—in particular, competition for light or water—are quite important for many species.

This observation raises an interesting question: Why do the canopy trees in temperate forests wait so long before expanding their leaves? Why don't they use the early spring to add to their growth, as the forest-floor herbs do? There are a number of selective forces affecting the timing of leaf expansion. First, the canopy is elevated, and temperatures in the treetops can be considerably colder than those at ground level (see Chapter 3). Thus, leaf expansion for trees occurs later on the calendar, but may not really be much later in terms of degree-days. Second, late frosts are a common occurrence. Leaves on trees are much more vulnerable to frost damage than those on forest-floor herbs because the latter are partly sheltered by the trees and because the ground-level temperatures are higher. Third, all metabolic enzymes have defined temperature ranges over which they can operate and are most efficient at particular temperatures. Enzymes adapted for peak functioning at warm temperatures are unlikely to be efficient in the early spring; it is possible that earlier leaf expansion might reduce the total annual growth of the trees, not increase it. Finally, many temperate-zone trees are wind-pollinated, and pollination in most of these species occurs while the trees are leafless. The presence of leaves earlier in the season would be likely to limit pollen transfer.

The phenologies of temperate forest-floor herbs are actually more complex than we have implied so far. Most of these plants in northeastern North America have phenologies like those we have described, but there are others that are able to capture and use light at other times. For example, some species can use light flecks on the forest floor to grow during the warmer months (see Figure 2.24). Still others use the additional light available in autumn, when some canopy species have begun shedding their leaves (Mahall and Bormann 1978).

In warmer habitats, it is usually the availability of moisture that has the strongest effect on vegetative phenologies. In tropical dry forests and tropical rainforests, as well as temperate and tropical deserts, rainfall is seasonal. Where it is strongly so—as in deserts and tropical dry forests—many species are **drought-deciduous**, shedding their leaves during dry periods. Is the timing of leaf expansion and leaf drop a plastic response to the environment or an evolved strategy based on the average environmental conditions? To a great degree, the answer is "both." There are some species that retain their leaves if given additional water, and others that shed them at the usual time of the onset of the dry season, regardless of current water availability. There are species that expand their leaves weeks before the rains begin and others that do so only after abundant rainfall. All of these strategies are evolved traits. Perhaps the most extreme case of a drought-deciduous phenology is that of *Fouquieria splendens* (ocotillo, Fouquieriaceae; see Figure 3.5). This desert plant of the southwestern United States and Mexico is drought-deciduous. However, it produces new sets of leaves whenever water is available, as long as temperatures are high enough. Thus ocotillos can have four or five sets of leaves over a year.

Evergreen versus Deciduous Habits

Why do plants shed their leaves at all? At first glance, it might seem more advantageous to be evergreen, so that the plant could photosynthesize whenever light was available. In actuality, there are major trade-offs involved in being evergreen versus being deciduous. The leaf traits of evergreens include being thick, tough, waxy, and generally reduced in area. Although these traits reduce the rate at which the plant loses water and nutrients (Aerts 1995), they also reduce the plant's rate of net photosynthesis (see Chapter 2). Thus, they are advantageous under only some conditions.

Evergreens are typically found in nutrient-poor soils, as well as in habitats that are subject to drought over much of the year. For example, pines and other conifers in North America are most common in acidic soils, which have low nutrient availability, and at high elevations and latitudes, which are subject to temperature extremes. They are also found in sandy or shallow, rocky soils (which also tend to be acidic and low in nutrients). They frequently experience drought during winter because of the limited water-holding capacity of such soils or, in northern and alpine habitats, because most of the water is in the form of snow. They also undergo summer drought in many such habitats, either because of limited summer rains or due to soil condi-

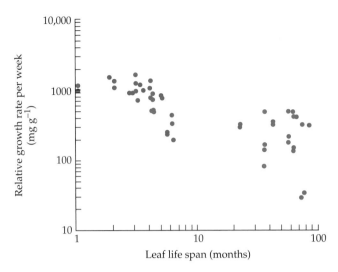

Figure 9.9
The relationship between the relative growth rate (RGR) of young plants and the life span of leaves is negative. Species with long-lived leaves tend to have low RGR. (After Reich et al. 1992.)

tions. A number of pine and juniper species are characteristic of the margins of North American deserts. Similarly, evergreen oaks (*Quercus*, Fagaceae) in North America are found where soils are very poor in nutrients, such as the coastal plain of the Gulf of Mexico, and on the edges of deserts and some grasslands. Oaks from higher elevations and more northern latitudes are typically deciduous.

The leaves or needles of an evergreen may live from one year to over forty years (Aerts 1995), although more commonly they last between one and four years. A negative relationship has been demonstrated between leaf life span and the growth rate of young plants (Figure 9.9). Leaf life span is also negatively correlated with photosynthetic rate, leaf nitrogen concentration (on a weight basis), and **specific leaf area** (leaf area per unit leaf weight). It has been hypothesized that the costs of constructing evergreen leaves are higher than those for deciduous leaves. This hypothesis has been shown not to be true; rather, different evergreen species have different leaf construction costs due to different ratios of biochemical constituents in their leaves (Figure 9.10). Evergreen leaves tend to have relatively high investments in materials that protect the leaf and contribute to its long life span, such as lignin and fibers, and lower investments in materials that contribute to high photosynthetic and growth rates. Deciduous leaves tend to have high investments in the latter materials, particularly photosynthetic enzymes and other proteins, and lower investments in lignin and fibers. Overall, the carbon costs of producing leaves are roughly comparable in evergreen and deciduous plants.

(A)

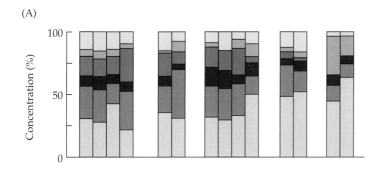

(B)

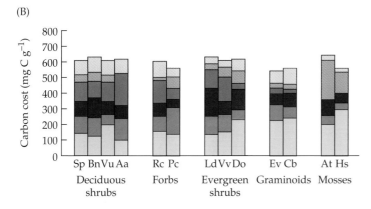

- ☐ Protein
- ◩ Lignin
- ◼ Tannin
- ◼ Lipid
- ◩ Carbohydrate
- ☐ Fiber

Figure 9.10
Carbon costs of the chemical components of leaves in 13 species of tundra plants. Costs vary substantially among species and growth forms; for example, fiber and lignins are relatively expensive for mosses, and lipids are relatively expensive for evergreen shrubs. Deciduous leaves contain somewhat more protein, most of which is RuBP-carboxylase (rubisco), which is essential to photosynthesis. There is no evidence in these data that evergreen leaves are more costly to produce. Species shown are *Salix pulchra* (Sp), *Betula nana* (Bn), *Vaccinium uliginosum* (Vu), *Arctostaphylos alpina* (Aa), *Rubus chamaemorus* (Rc), *Pedicularis capitata* (Pc), *Ledum decumbens* 1st and 2nd year leaves (L1 and L2, respectively), *Vaccinium vitis-idaea* (Vv), *Dryas octopetala* (Do), *Eriophorum vaginatum* (Ev), *Carex bigelowii* (Cb), *Aulocomnium turgidum* (At), and *Hylocomium splendens* (Hs). (After Chapin 1989.)

As a consequence of retaining their leaves for long periods of time and having low tissue nutrient concentrations, evergreen plants tend to have low nutrient loss rates—that is, they excel at retaining the nutrients they obtain from the soil. This ability appears to confer substantial advantages in low-nutrient soils, allowing these plants to achieve annual dry matter production at least equal to that of deciduous species under nutrient-limited growth conditions. In fertile soils, evergreen species cannot accumulate as much total growth as deciduous species.

Litter decomposition in both nutrient-poor and arid soils is quite slow. Because conifer needles and oak leaves decompose very slowly and tend to acidify soils under moist conditions, the presence of large numbers of these trees tends to favor plants with leaf traits that reduce nutrient loss. Thus, on nutrient-poor soils, the evergreen habit tends to be self-reinforcing (Aerts 1995). Indeed, the dominance of conifers in boreal forests plays a key role in the acidification and nutrient depletion of these soils (see Chapter 4). This phenomenon is unlikely to be important in arid soils.

Reproductive Phenology: Abiotic Factors

While we often think of flowers as something we see in the spring, plants can flower under an amazing range of conditions. *Ranunculus adoneus* (snow buttercup, Ranunculaceae) is a plant of alpine snowfields in the Rocky Mountains of Colorado (Figure 9.11). This perennial herb

begins flowering while still covered with snow (Galen and Stanton 1991). The North American desert annual *Eriogonum abertianum* (Polygonaceae, Figure 9.12) has a remarkably variable reproductive phenology: individuals may flower in either or both of two rainy seasons separated by as much as six months of drought (Fox 1989). In both of these plants, flowering time is affected by physical factors: temperature and light availability in *Ranunculus* and temperature and moisture availability in *Eriogonum*. Both plants are unable to reproduce in winter. *Ranunculus* is covered by deep snow then, while *Eriogonum* germinates only during the cool desert winter, so it is usually much too small to flower in winter. In both cases, there is apparently strong selection to flower at what appear to be unusual times.

Eriogonum may flower during either the spring or the summer rainy season, or both. Like most plants, *Eriogonum* produces more flowers and fruits the larger it gets—and it gets very large only after the summer rains. The fact that many or most individuals, depending on site and year, die before summer does not necessarily provide sufficient selection for earlier flowering. For example, at one site, well under 1% of the plants alive in the spring managed to survive and flower in the summer, but they accounted for about 25% of the seeds produced (Fox 1989).

In both species, there is substantial variation in flowering time within a population. Much of this variation

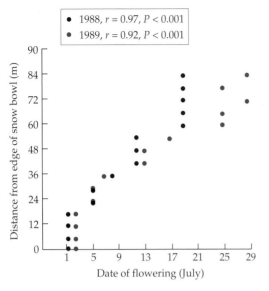

Figure 9.11
Ranunculus adoneus (snow buttercup, Ranunculaceae) is found in alpine snowfields in the Rocky Mountains; it begins flowering while still covered by snow. Flowering time depends on the plant's position in a snowfield. Plants on the edge of the snowfield tend to flower earlier. (Data after Galen and Stanton 1991; photograph courtesy of M. Stanton.)

appears to be due to phenotypic plasticity rather than genetic variation within populations (Fox 1990a; Stanton and Galen 1997). This is especially clear in *Ranunculus*. Plants on the edge of the snowfield flower much earlier, mainly because the snow melts earlier there, so that they receive more light earlier in the year (see Figure 9.11). With *Eriogonum*, small differences in soil mois-

ture lead to big differences in reproductive timing. Plants that reach sufficient size and have access to sufficient moisture begin flowering in the spring; these plants have a greater chance of surviving to the summer rainy season than plants that do not flower in the spring (Figure 9.12). Despite this plasticity, it is clear that differences between populations also reflect local adaptation. Populations that experience a greater risk of dying before summer have more early-flowering plants than those at a lesser risk, (Table 9.1), and there is much genetic variation for life history traits between these populations (Fox 1990a).

It is widely believed that moisture limitation—and perhaps other types of physiological stress—tends to induce flowering in many plants. There is no evidence for this effect in annual plants (Fox 1990b). In perennials, the problem is more complicated, because flower buds are often formed one or more years before they open. The effect of drought or other stress therefore probably depends on its timing. Drought that occurs

Table 9.1 Frequencies and sizes of spring-flowering classes of *Eriogonum abertianum* in several populations

Site and year	Plants alive in spring that flowered in spring (%)	Mean (standard error) height to highest node		n
		Spring flowerers	Non-spring flowerers	
Chihuahuan Desert				
Portal 1985	2.6	4.4 (0.58)	2.2 (0.05)	494
Portal 1986	2.6	4.8 (0.43)	1.7 (0.03)	1404
Sierra Vista 1986	0.0	—	1.3 (0.04)	200
Sonoran Desert				
Tucson Mts. 1986	56.7	2.8 (0.10)	1.7 (0.06)	201
Waterman Mts.1986	31.0	4.2 (0.43)	2.0 (0.05)	200
Organ Pipe 1985	21.3	2.5 (1.9)	1.5 (0.06)	150
Organ Pipe 1986	23.1	3.1 (0.7)	1.7 (0.02)	1382

Source: Data from Fox 1992.
Note: More plants flower in spring in the Sonoran Desert (which receives less rain, and less predictable rain, in the summer) than in the Chihuahuan Desert. Spring-flowering plants are larger on average, and have improved chances of surviving to summer (see Figure 9.11), even when compared with same-sized plants (Fox 1989). Moisture availability, as well as size, strongly influences phenological variation within populations; the differences between the Portal and Organ Pipe populations, at least, have a strong genetic basis (Fox 1990a).

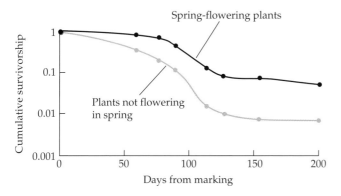

Figure 9.12
Eriogonum abertianum (Polygonaceae) is an annual plant of the Sonoran and Chihuahuan Deserts of the southwestern United States and Mexico. Its flowering time depends on plant size and moisture availability. Therefore spring-flowering plants are more likely to survive to the summer rainy season, when most seed is set. These data are for a population at Organ Pipe Cactus National Monument, Arizona, in 1985. (Data from Fox 1989; photograph courtesy of K. Whitley.)

when buds are forming may have a different effect than drought that occurs just as the buds are opening.

How tightly is reproductive phenology coupled with vegetative phenology? Evidence for temperate forest trees suggests that their coupling is generally weak (Lechowicz 1995). The main reason is that plants can store resources such as photosynthate and remobilize them at other times. This mechanism is what allows so many temperate trees to flower when they are leafless.

Reproductive Phenology: Biotic Factors

Biotic factors may also influence reproductive phenology. Because of the inherent difficulties in studying multispecies interactions, many studies of these influences have relied heavily on circumstantial evidence—but the circumstantial evidence in some cases is quite strong.

If pollinators are only briefly available, there may be strong selection for flowering while these animals are present and active. Although this seems clear in principle, it is important to realize that such selection will be strong only if the plants require animal pollination and seed set is limited by the availability of pollen. Some plants that are mostly self-fertilizing still require animals for pollination.

A number of authors in the 1970s inferred that flowering schedules of entire communities were structured by competition for pollinators. They concluded that this competition was a crucial factor in shaping phenologies, causing the evolution of separation in phenologies among overlapping species. Other authors suggested that many plant species act as mutualists by having similar phenologies and thus jointly attracting more pollinators than any would by flowering alone.

Clearly both phenomena are possible, but there is remarkably little evidence for either. Analyses by Robert

Poole and Beverly Rathcke (1979) showed that many of the claims that observed phenologies showed strong correlations among plant species were based on inappropriate data analyses. For example, many of the effects that had been claimed disappeared when Poole and Rathcke took into account the fact that the seasons limit when plants can flower. If one analyzes temperate-zone data without accounting for winter, or tropical data without accounting for dry seasons, it is easy to wrongly conclude that many species' flowering times are clustered more than one would expect by random chance.

Frugivores (fruit-eaters) and **granivores** (seed eaters) may also exert considerable selection on phenologies. In most cases, however, it appears that the animals' phenologies are more strongly dependent on plant phenologies than the reverse. Animals may also act as important causes of phenotypic plasticity in reproductive phenology. For example, herbivory (like many other stresses, including crowding and drought) may cause an individual's flowering time to be delayed (Lyons and Mully 1992).

Finally, other plants may influence reproductive phenology, either directly or indirectly. Crowding by neighbors, for example, may act to delay flowering (Lyons and Mully 1992). Such delays, in turn, may result in selection for faster growth or earlier flowering.

Summary

Plant life histories vary considerably. Life history traits have strong effects on plant fitness and have consequently evolved over time. Studies of the evolution of these traits—both singly and as syndromes of associated traits—are an important part of plant ecology.

Trade-offs–the necessity of decreasing one thing if something else is increased—are thought to be important in the evolution of life history traits. They occur because of some basic constraints. If the amount of resources allocated to reproduction were unlimited, a plant could produce an infinite number of seeds and make them as large as possible. Since resources are limiting, however, there is a trade-off between seed size and number of seeds. An important group of models attempts to understand the direction of natural selection by asking what combination of traits (such as seed size and number) confers the greatest fitness on plants. There have been several important attempts to develop broad theories to describe how trade-offs shape syndromes of life history traits. The theory of *r*- and *K*-selection, and Grime's extension of that theory, do not lend themselves to critical tests because their predictions are relative. Models of optimal reproductive effort have also encountered difficulties with testability. One important conclusion is that we need more information about the underlying biology to make much progress in understanding the constraints governing life history evolution.

In confronting environmental uncertainty, plants are often subjected to a different kind of trade-off, because variation in fitness among years acts to reduce long-term fitness. Many plants appear to reduce among-year fitness variation by bet hedging—spreading risk by evolving traits such as iteroparity and seed banks.

Phenology, the within-year timing of growth and reproductive events, is also affected by various trade-offs. Evergreen plants are able to photosynthesize under a wider variety of conditions than deciduous plants, but to be evergreen they must have a number of characteristics—such as smaller leaf surface areas—that act to reduce the rate at which they can fix carbon under favorable conditions. Plants living on forest floors frequently must either experience most of their growth when the canopy trees are bare or evolve the ability to effectively use light flecks that travel across their leaves when the canopy is in leaf. The timing of flowering involves trade-offs between the availability of abiotic resources such as water and light and biotic factors such as pollinators and frugivores.

Additional Readings

Classic References

Baker, H. G. 1972. Seed mass in relation to environmental conditions in California. *Ecology* 53: 997–1010.

Cole, L. C. 1954. The population consequences of life-history phenomena. *Q. Rev. Biol.* 29: 103–137.

Monk, C. D. 1966 An ecological significance of evergreeness. *Ecology* 47: 504–505.

Salisbury, E. J. 1942. *The Reproductive Capacity of Plants.* Bell, London.

Contemporary Research

Brown, J. S. and D. L. Venable. 1991. Life history evolution of seed-bank annuals in response to seed predation. *Evol. Ecol.* 5: 12–29.

Kalisz, S. and M. A. McPeek. 1993. Extinction dynamics, population growth and seed banks. *Oecologia* 95: 314–320.

Mazer, S. J. 1989. Ecological, taxonomic, and life history correlates of seed mass among Indiana Dune angiosperms. *Ecol. Monogr.* 59: 153–175.

Stanton, M. L. and C. Galen. 1997. Life on the edge: adaptation versus environmentally mediated gene flow in the snow buttercup, *Ranunculus adoneus. Am. Nat.* 150: 143–178.

Additional Resources

Aerts, R. 1995. The advantages of being evergreen. *Trends Ecol. Evol.* 10: 402–407.

Roff, D. A. 2002. *Life History Evolution.* Sinauer Associates, Sunderland, MA.

Stearns, S. C. 1976. Life-history tactics: A review of the ideas. *Q. Rev. Biol.* 51: 3–47.

PART III From Populations
to Communities

CHAPTER 10 *Competition*

Competition has been studied intensively by plant ecologists, and even more intensely debated. Key questions have included: What are the effects of competition on individuals, populations, and communities? What determines the outcome of competition among different individuals? What are the resources for which plants compete, and how does the type of resource competition affect the nature or outcome of competitive interactions? Does the intensity of competition vary with productivity? How important is competition relative to herbivory and other factors, and how do these factors modify competitive interactions?

Competition can be defined as a reduction in fitness due to shared use of a resource in limited supply. Other definitions emphasize different aspects of competition, such as the mechanism of resource exploitation. More importantly, there are different ways of measuring competition, each of which reveals something different about the nature of competitive interactions.

Ecologists who study animal populations have argued a great deal about the importance of competition in determining population structure and abundance. Plant ecologists, on the other hand, have generally accepted that the effects of competition are obvious and pervasive. Anyone who has ever grown a garden knows how much weeds can affect the plants one is trying to grow, and the millions of dollars spent every year on herbicides by farmers argue that competition has major effects on plant productivity. However, because its effects are so complex, plant ecologists have debated just about everything else regarding plant competition, from how to define and measure it to when and where it is important.

A fair amount is known about the effects of competition on individual plants. Competitors can reduce a plant's biomass and growth rate and decrease its ability to survive and reproduce. Among plants of the same species in the same environment, the number of seeds produced is highly correlated with maternal plant mass, so successful competitors that accumulate more mass will have more resources to put into reproduction.

We begin this chapter by examining what is known regarding the effects of competition on seedling survival and growth, primarily from greenhouse experiments, but also from studying single-species (monospecific) plantings of trees for timber. Next we look at the effects of competition on populations, population distributions, and community composition. It is these aspects of competition that have been the subject of the greatest controversy. We exam-

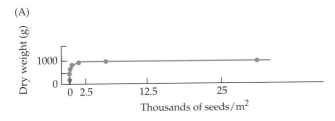

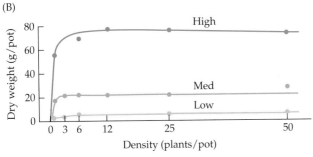

Figure 10.1
Relationships between yield of dry matter (g) and plant density for two pasture plants. (A) *Trifolium subterraneum* (Fabaceae), measured after flowering (density is expressed as thousands of seeds sown/m^2). (B) *Bromus unioloides* (Poaceae) at low, medium, and high levels of fertilization with nitrogen (density is expressed as plants/pot). (After Donald 1951.)

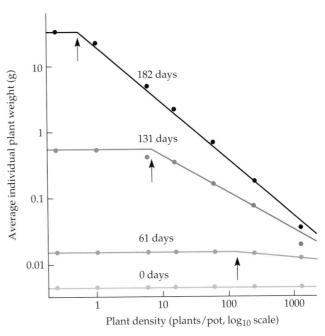

Figure 10.2
Average individual plant weight (g) for *Trifolium subterraneum* (Fabaceae) planted over a range of densities (expressed as number of plants/pot) and harvested at 0, 61, 131, and 182 days after seeds were planted. Note that both axes are on a log$_{10}$ scale. The arrows show the densities at which plants began to reduce one another's growth at different ages. At early harvests, only very dense plantings showed reductions in average plant weight, but by the last harvest, all but the lowest-density planting demonstrated weight reductions related to density. Decline in weight with density is linear on a log-log scale. (Data from Donald 1951 and Kira et al. 1953, after Harper 1977.)

ine some of these debates, and take a look at the evidence ecologists have gathered on the role and importance of competition.

Competition at the Level of Individuals

Size and Density

Plant growth is highly plastic, and the mass, height, number of leaves, and reproductive output of an individual plant can vary over orders of magnitude depending on growth conditions. When plants are grown without close neighbors, they are generally much larger than similar individuals surrounded closely by others, and often have a very different morphology, or form (see Figure 6.6).

In a classic experiment, C. M. Donald (1951) showed that British annual pasture plants sown over a wide range of densities had a remarkably consistent total final dry weight in a given area (Figure 10.1). Whether seeds were planted sparsely or very densely (above a certain minimum density), the total aboveground dry matter at final harvest was constant. The total yield was increased when more resources were supplied, but the same relationship held. While the average plant size at low densities was relatively large, the average plant size became smaller as density increased. In a re-analysis of these data, Tatuo Kira and his colleagues (Kira et al. 1953)

showed that as plant density increased, the mean weight of individual plants decreased in a linear fashion when both were expressed on a log scale (Figure 10.2).

Mean plant size can be a misleading measurement, however. Size relationships among individuals in even-aged, dense single-species plantings have been well studied in greenhouse and in a few field studies. Individual plant sizes are generally extremely uneven in such stands. Typically, a few large individuals dominate the available area, while most individuals remain very small. These highly unequal size distributions are called **size hierarchies**.

It has been hypothesized that size hierarchies are caused by **asymmetric competition** (Weiner 1990; Schwinning and Fox 1995), in which the largest individuals have disproportionate negative effects on their smaller neighbors. It has been suggested that small initial differences in access to light are responsible for progressively greater inequality in size over time, as long as density remains constant (Figure 10.3). Among a group of seedlings germinating together, a small head start may confer

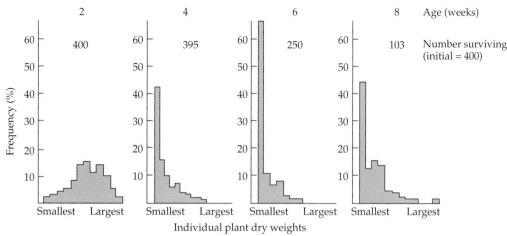

Figure 10.3
Frequencies of dry weights of individual seedlings of *Tagetes patula* (marigold, Aster-aceae), an annual plant, grown in a greenhouse experiment at 2, 4, 6, and 8 weeks. The number of surviving plants is shown above each graph. At 2 weeks, the distribution is close to a normal (bell) curve, but the distribution of dry weights becomes increasingly unequal (hierarchical) as the population becomes older, with many small plants and a small number of very large individuals. Over time, death removes the smallest individuals from the population (self-thinning), so that by 8 weeks the population is somewhat less unequal than at 6 weeks. (After Ford 1975.)

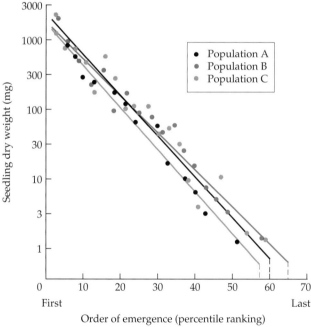

Figure 10.4
The effect of relative order of emergence (percentile ranking) on seedling dry weight (mg). Each line shows the relationship (the regression) for one of three different populations of *Dactylis glomerata* (orchard grass, Poaceae) in a greenhouse experiment. The consequences of emerging sooner or later than one's neighbors can be enormous: seedlings with the lowest percentile ranking—those that germinated and appeared aboveground first—were more than 1000 times larger than those emerging last. Seedling dry weight (mg) is graphed on a log scale. (After Ross and Harper 1972.)

a large advantage. Ross and Harper (1972) showed a strong relationship between the order of emergence and plant size for densely planted *Dactylis glomerata* (orchard grass, Poaceae; Figure 10.4). In natural populations, this advantage can be offset by variation in hazards (such as frost) and resource availability in early spring.

While size hierarchies have been found in many greenhouse studies and in some tree plantations, much less is known about them in natural populations. Wilson and Gurevitch (1995) examined the spatial relationship of plant sizes in a dense natural stand of *Myosotis micrantha* (forget-me-not, Boraginaceae), a small winter annual. These plants had extremely unequal sizes (Figure 10.5). The researchers hypothesized that if asymmetric competition was the cause, then large individuals should be surrounded by small, suppressed neighbors.

They found instead that the opposite was true. Large plants had large immediate neighbors, and small plants were associated with small neighbors. Individual plant mass was also highly correlated with the combined mass of neighbors, so that the population formed a mosaic of patches of large plants and patches of small plants. Plants without close neighbors were much larger than plants with neighbors, however, so competition was probably important in affecting plant size. The researchers concluded that asymmetric competition was unlikely to have caused the extreme size hierarchy found in this natural population. Rather, the size hierarchy was probably caused by variation in plant density or patchy resource distribution.

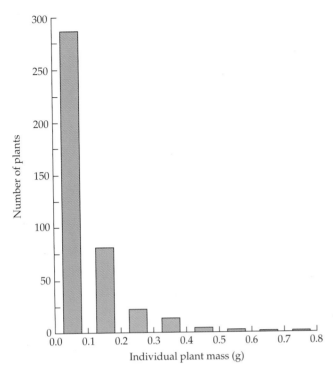

Figure 10.5
Distribution (number of plants) of individual plant dry weights (g) in a natural population of the annual *Myosotis micrantha* (forget-me-not, Boraginaceae), harvested at flowering. There were large numbers of very small plants, with a steep drop in numbers of plants in the larger categories, and very few of the largest individuals. (After Wilson and Gurevitch 1995.)

Mortality, Size, and Density

If seeds of herbaceous plants (usually annuals) are planted in dense monospecific stands and their survival and weights are followed through early life history stages, the seedlings grow until they begin to crowd one another. As crowding becomes severe, some individuals eventually begin to die; usually the smaller and weaker plants succumb most readily. More densely planted stands begin to experience mortality sooner and at smaller individual plant sizes than more sparsely planted stands. Factors such as greater soil fertility or greater initial seed size that result in larger individuals also increase mortality because larger plants cannot maintain as high a density as smaller plants.

This process of density-dependent mortality is known as **self-thinning**. In a crowded monospecific, even-aged stand of plants, a log-log plot of density against total plant mass will have a negative slope—plant growth can occur only as some individuals die. It is important to realize that the term "self-thinning" does not imply a voluntary, altruistic self-sacrifice of the weaker individuals for the general good (which is certainly not what is happening!). Since the weight of a plant is directly related to its volume (a cubic measure)

and density is determined by area (a squared measure), Yoda and colleagues (1963) proposed what they called the **–3/2 thinning law**:

$$w = cN^{-3/2}$$

where w is the mean dry weight per plant, c is a constant that differs among species, and N is density. On a log-log scale, the relationship between mean dry weight and density is predicted to have a slope of –1.5 (Figure 10.6).

Although the –3/2 thinning law was widely accepted until the 1980s, it has become a matter of considerable debate since then. There is no doubt that mortality occurs in crowded stands; what is in dispute is whether the process is so regular that it can be described by a single numerical relationship for all populations. Studies of single populations have been plagued by serious statistical problems (Weller 1987), as have review papers attempting to evaluate the generality of the relationship (Lonsdale 1990; Weller 1991). On the other hand, Silvertown and Lovett Doust (1993) make a case for the relationship holding well as an upper limit.

Mechanisms of Competition

Plants compete for light, for water and mineral nutrients from the soil, for space to grow and to acquire resources, and for access to mates. Competition for animal vectors

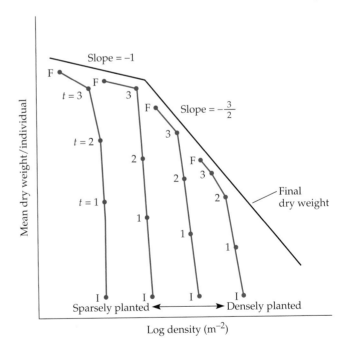

Figure 10.6
Effects of planting at different densities on the mean dry weights of individuals as seedlings age. Each of the green lines represents a different initial planting density. The lowermost point on each line is the initial dry weight at germination (I), and the uppermost point is the final dry weight (F), with weights shown at time intervals $t = 1, 2,$ and 3. (After Kays and Harper 1974.)

(A)

(B)

Figure 10.7
(A) A strangler fig (*Ficus* sp., Moraceae), a tropical plant that begins life as a vine, killing the tree that it grows on and subsequently becoming a tree itself. (Photograph courtesy of J. Thomson.) (B) *Celastrus orbiculata* (Oriental bittersweet, Celastraceae), a vine native to eastern Asia that is now invasive in the eastern United States, is shown here completely cloaking a tree. (Photograph by J. Gurevitch.)

for pollen and seed dispersal is a special kind of competitive interaction because these mobile biotic resources can respond to plants in ways abiotic resources do not.

Light is in some ways the most peculiar resource for which plants compete (see Chapter 2). In one sense, plants do not really compete for light, because no matter how much light plants take up, an essentially infinite supply is always available directly above their canopies. Yet plant canopies can very effectively mop up photons, reducing light at ground level to less than 1% of the incident sunlight. It can be very dark in many communities, not only for the sapling struggling to grow in a dense forest understory, but also for the seedling emerging in a grassy meadow. One might argue that plants are competing for access to light, or for sunlit space, rather than for light itself. After all, light is available in unlimited supply, but space is not. Perhaps it is most useful to think of plants as competing for light, but in a manner different from that in which they compete for soil resources.

Overgrowing or overtopping neighbors is an obvious means of competing for light. Overtopping is probably a common mechanism leading to successional change from old fields to forests, in which shorter herbaceous plants are replaced by shrubs and eventually by trees (see Chapter 13). Another dramatic example of overtopping is the overgrowth of trees by vines. In the

Tropics, strangler figs (*Ficus*, Moraceae) grow on trees, encircling and eventually killing their hosts (Figure 10.7A). In temperate forests in the eastern United States, trees along forest edges or gaps are sometimes overgrown by either native or invasive exotic vines (Figure 10.7B). It is not always clear whether vines cloak and weaken otherwise healthy trees or overwhelm only trees previously weakened by other factors, such as insects or disease. Once covered by vines, trees become much more likely to be toppled in windstorms and killed. This vulnerability probably results both from the weakening of the root system as a consequence of loss of photosynthate as the vines increasingly shade the leaves and from the tremendous extra weight of the vines.

Most of the competitive interactions experienced by a plant occur on a very local scale. The shade of larger neighbors reduces the plant's ability to photosynthesize, while the roots of the plants immediately sur-

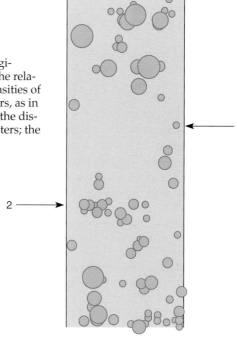

Figure 10.8
A transect from a natural population of *Myosotis micrantha* (forget-me-not, Boraginaceae). Each circle represents an individual plant; the size of the circle shows the relative plant size. Plants in the same population may experience relatively low densities of neighbors, as in the part of the transect shown at 1, or high densities of neighbors, as in the part of the transect shown at 2. The x- and y-axes are not on the same scale: the distance along the vertical dimension from region 1 to region 2 is about 35 centimeters; the width of the transect is 20 centimeters. (Data from Wilson and Gurevitch 1995.)

rounding it may absorb some of the water and nitrogen that it needs to function. Plants even a short distance away, however, may experience no effects at all from those particular plants. For this reason, the density experienced by an individual plant is almost always strictly the density of the plants in the immediate patch surrounding it. The average density of plants in the population is irrelevant to the degree of crowding it experiences (Figure 10.8). This pattern is quite different from the effects of density on mobile animals, which may be competing with many other animals in the same general area for food. One exception to the very local nature of plant competition is competition for animal vectors (pollinators and dispersers). Because these animals may be highly mobile, competition for pollinators may occur between plants some distance apart, rather than strictly between immediate neighbors.

Resources such as light, water, nitrogen, and phosphorus can be very patchy. One of the ways plants respond to this patchiness by preferential growth into the areas where these resources are available. Forest trees, for example, respond to newly created gaps in the canopy by the growth of branches into the gap. Young saplings in a newly created gap shoot up rapidly in height. Some individuals and some species are much better at responding to gaps and other patches of high resource availability than are others, and this ability may make them much better competitors under some conditions.

Dense masses of roots can develop in nutrient-rich patches of soil (Robinson et al. 1999). Root proliferation may, under certain conditions, allow a plant to monopolize the nitrogen in the soil. Little is known about how species differ in their abilities to monopolize soil resources through root proliferation, but such differences could lead to differences in competitive ability. The roots of different species are known to differ in their nutrient uptake rates, at least under laboratory conditions, and these rates may also affect their competitive ability.

Is Bigger Better?

Size generally confers an advantage in competition—bigger is usually better. Larger plants can affect smaller neighbors without being much affected themselves. Among young plants growing side-by-side in an abandoned field, for example, the larger neighbor will produce disproportionately more leaves and roots, have more flowers, and set more seed.

However, plants of extremely different sizes or ages may not be competing at all, or they may be competing in different wys than more similar-sized or similar-aged individuals. A forest tree may not compete directly with an herbaceous understory plant growing beneath it: understory plants commonly grow in the shade, and the roots of trees and herbs are often found at very different depths in the soil, drawing on different supplies of soil resources.

The assumption that plants of very different growth forms do not compete can be deceptive, however. Such competition may be occurring, although it may be one-sided. A small sapling growing beneath a canopy tree may have no measurable competitive effects on the mature tree, but the shade of the tree may completely prevent the sapling from growing larger. Cut down the tree, and the sapling is likely to rapidly increase in height and biomass. But things are not always so simple: foresters have found that suppression of understory plants with herbicides can lead to great increases in the growth rates of mature lumber trees, presumably because the understory plants intercept water or reduce soil nutrients available to the trees.

Being larger is not inevitably an advantage, however, and in some cases, larger plants become susceptible to other factors that reduce their effectiveness in competition. A plant with a greater amount of leaf surface area, for example, loses much more water by transpiration than a similar plant with less leaf area. Consequently, the larger plant may be affected more by drought than its smaller neighbors, increasing its likelihood of dying or reducing its future effectiveness in competition.

Apparent Competition

One type of animal-mediated interaction among plants is **apparent competition**—density-dependent negative interactions between species that might at first appear to be due to competition for resources, but are actually due to a shared predator or herbivore (Holt 1977). Apparent competition might occur if, for example, as the combined density of two plant species increased, they increasingly attracted the attention of herbivores. Thus, both species would suffer from the other's presence and abundance, but because of herbivory rather than competition.

Apparent competition among plants has been demonstrated only rarely. A recent study showed that a native New Zealand fern, *Botrychium australe* (adder's tongue, Ophioglossaceae) was declining in abundance due to apparent competition with an introduced grass, *Agrostis capillaris* (Sessions and Kelly 2002). The grass may serve as a refuge for an introduced slug. The native fern survives well after fire. However, fire also increases populations of the introduced grass, which leads to an increase in slug density. Increased slug populations following a severe fire led to severe defoliation and mortality of the fern due to the positive effects of the greater grass cover on the slugs.

Populations and Competition

Competitive Hierarchies

Plant ecologists differ strongly on the question of whether consistent competitive hierarchies exist among plant species. Are some species always competitively superior, and is the rank order of subordinates relatively fixed? Some ecologists argue that consistent competitive hierarchies exist. Others argue that fluctuating dominance in competitive networks is the general rule for plant interactions.

What are the implications of these two opposing views? The issue of competitive hierarchies is important because it touches on questions regarding the basic structure of plant communities. If competitive hierarchies were the general rule, ecologists could predict the competitive abilities of plant species based on their traits, such as growth form and size (Herben and Krahulec 1990; Shipley and Keddy 1994). Explanations of species diversity would require an understanding of the factors that prevent competitive exclusion, such as nonequilibrium processes.

In contrast, if communities consisted of competitive networks with frequent reversals of competitive dominance, an understanding of community structure would require studies of niche partitioning and species packing and of how life history or other trade-offs permit stable coexistence. In that case, species diversity in ecological communities would be maintained primarily by competitive interactions, implying that communities are generally deterministic and equilibrial in character.

Connolly (1997) pointed out that most of the data that support the existence of competitive hierarchies are based on two-species replacement series experiments (described below), which he argued are biased. Biased or not, extrapolation of the results of these greenhouse experiments to natural plant communities is necessarily limited. We need to have more information, particularly from field experiments, before reaching a conclusion as to the general existence of competitive hierarchies in nature. Unfortunately, while there have been hundreds of studies of plant competition in the field, few have addressed this issue, and thus the data on this question are still too limited to allow any reliable conclusions to be drawn (Goldberg and Barton 1992).

Quantifying Competition

How one measures the intensity of competition will affect the interpretation of the outcome of plant competition studies. Some of the disagreements about the nature of competition among plants can be traced to differences in the way competition has been studied and measured (Grace 1991, 1995). Goldberg and Werner (1983) distinguished between the **competitive effect** of a plant on its neighbors and the **competitive response** of a plant to its neighbors. Recognition of these two distinct components of plant competitive ability can help us to understand the ways in which plants interact. The rankings of competitive effect and competitive response among the plants in a community are not necessarily correlated (Figure 10.9; Goldberg 1990).

The interpretation of the results of a competition experiment depends on the units in which the outcome is expressed. In many experiments, the results are calculated on the basis of area (such as biomass per unit area). Alternatively, one can examine the effects of competition per gram of biomass of competitors, or on a per-individual basis.

Not only does each of these measures mean something different, but their usefulness depends on the purpose of the measurement. For example, some models of competition require results that are expressed per individual, while others make the most sense per unit area. Furthermore, while assessing competitive results per individual might be straightforward for many trees or even annuals, it is likely to be impossible (and not even very interesting) for clonal perennials. These issues are clearly a consequence of the great plasticity of plants.

One argument for measuring competition on a per-gram basis is that this practice eliminates the effects of size. But is eliminating size always desirable? If two individuals begin at more or less the same size, and one acquires resources more quickly and efficiently than the other and becomes the dominant competitor, is the greater size of that individual a bias that we want to get rid of, or is it the essence of the competitive interaction? Of course, if per-gram effects are calculated at the beginning of such an experiment, the faster-growing

(A)   (B)

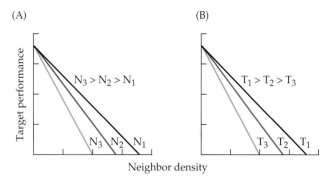

Figure 10.9
Hypothetical examples of different strengths of competitive effects and competitive responses for different target plant species (T_1, T_2, and T_3) with different neighbor species (N_1, N_2, and N_3). In this example, neighbors are grown at a range of densities. (A) As neighbor density becomes higher, target plant performance (across all species) declines; the rate of decline is steepest when species N_3 is the neighbor, less steep for N_2, and least for N_1. Plants of neighbor species N_3, therefore, have the greatest competitive effect on the target plants. (B) Increasing neighbor densities have the greatest effect on target species T_3 and the least effect on T_1— plants of species T_1 decline less in performance as each individual of the neighbor species is added and neighbor density increases (the slope is shallower). Target species T_1 has the best competitive response because it is least sensitive to competition. (After Goldberg and Landa 1991.)

plant in this scenario will have greater per-gram effects, but if measured in terms of final size (as is often done), its per-gram effects will be artificially low.

Goldberg and Landa (1991) compared the competitive effects and competitive responses of seven species during the early stages of competition among seedlings in a greenhouse experiment. They used regressions of target plant biomass on neighbor biomass to determine the per-unit-biomass effects of neighbors, and regressions of target plant biomass on neighbor density to determine the per-individual-plant responses to neighbors. They found consistent competitive hierarchies, but the ranks differed for effect and response, and for per-plant and per-gram measures of competitive ability.

There are several approaches to quantifying the intensity of competition. One of the most common is to use an index, usually a ratio that standardizes responses across species and environments so that they can be compared on the same scale. One of the most common indices is the **relative competition index** or relative competition intensity, RCI:

$$RCI = \frac{P_{monoculture} - P_{mixture}}{P_{monoculture}}$$

where $P_{monoculture}$ and $P_{mixture}$ are plant performance in monoculture (single-species stands) or in mixture (stands with two or more species), and performance is usually measured as dry mass or growth rate. Another

index of performance is the **absolute competition index** (ACI), which is simply the difference:

$$ACI = P_{monoculture} - P_{mixture}$$

Using these two different indices as measures of the outcome of an experiment may lead to very different conclusions.

Despite its popularity for measuring competition intensity, the RCI is subject to a number of limitations. Ratios expressed on an arithmetic scale have poor statistical properties and are asymmetric (changing the numerator affects the ratio differently than changing the denominator by the same amount). The ACI is also subject to both conceptual and statistical problems. One alternative is to use the **log response ratio** (LRR):

$$LRR = \ln \frac{P_{mixture}}{P_{monoculture}}$$

This index has the advantage of expressing performance relative to potential performance, as RCI does, but has better statistical properties, including symmetry (Hedges et al. 1999). LRR is very similar to the difference in relative growth rates between monoculture and mixture if the initial sizes of the plants are similar and small in comparison with the final sizes.

The debate regarding the superiority of one index of competition over the others rests on the assumption that an index can accurately represent the essential features of a competitive interaction. Any index simplifies reality, however, and thus has certain drawbacks as a tool for understanding competitive interactions (as do indices expressing any kind of relationship, from species diversity to stock market performance). Any attempt to compress complex data into a single index suffers from a loss of information. Other limitations to competitive indices include that they unrealistically assume linear responses to neighbor density; that many different outcomes can result in the same index; and that competition indices tend to be strongly influenced by initial plant size (Grace et al. 1992.)

In making comparisons of competition intensity across environments or species, one alternative is to examine interaction diagrams and test for statistical interactions, using performance directly without converting to indices (Box 10A). Better still would be information over time on the performance of the competitors, following their trajectories to determine the progress of the interaction.

Experimental Methods for Studying Competition

Greenhouse and Garden Experiments

The majority of plant competition experiments have been conducted in greenhouses. Greenhouse experiments offer the advantages of relatively controlled conditions and precise manipulation of the factors of interest. Limitations are the uncertainty in extrapolating results to natural communities, as well as various artifacts of greenhouse conditions, including very different wind and humidity conditions. An important potential artifact is that plant water relations and root structure are drastically affected by growth in pots.

Three basic experimental designs are commonly used: substitutive (replacement), additive, and additive series designs (Gibson et al. 1999). **Substitutive designs** test the relative strength of intraspecific versus inter-

specific competition by altering the frequencies of two hypothesized competitors while keeping the total density constant (Harper 1977). One of the most common forms of this type of experiment is the de Wit replacement series (de Wit 1960), in which the relative percentages of two competing plants are varied from pure monocultures to 50/50 mixtures at a constant density, often with intermediate mixtures as well (Figure 10.10). These experiments were once the most important tool for studying plant competition, but have been criticized on a range of issues. They are subject to theoretical and statistical limitations, suffer from restrictive assump-

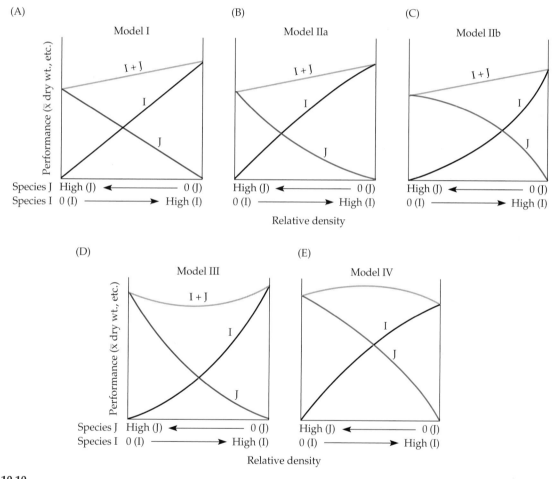

Figure 10.10
Different possible outcomes in a de Wit replacement series competition experiment. In each experiment, plants of two species (I and J) are grown in trays with different relative mixtures of the two, while total plant density is kept constant. (A) In an experiment with a model I outcome, the effects of individuals of species I and J are the same on themselves (intraspecific competition) as on each other (interspecific competition). The combined performance in mixture (top line) is the sum of the performances of the two species in the mixture; neither species is a superior competitor. (B,C) In a model II outcome, plants of one of the species (the superior competitor) have a greater interspecific than intraspecific effect, while the other species (the inferior competitor) has a lower interspecific than intraspecific effect. I is the better

competitor in (B) and J is the superior competitor in (C); in each case, the superior competitor affects members of the other species more than members of its own species, and is affected less by the other species than by its own. Combined performance is again the sum of the performances of the two species. (D, E) In model III and IV outcomes, the performance in mixture is not additive. In (D) the combined performance is less, and in (E) it is more, than the sum of the performances of the two species when grown alone. (D) Here each species affects the other species more than it affects its own, and both suffer more in mixture than in monoculture; neither is competitively superior. (E) Here both species have a greater effect on their own species than on the other species (perhaps due to having somewhat different niches).

Box 10A

The Implications of How Competition is Measured

Imagine that an ecologist is studying two co-occurring species; we will call them Species A and Species B. Species A contributes most of the biomass in the area being studied, while species B contributes only a small proportion of the biomass. Plots are cleared, and seedlings of the two species are planted together in experimental plots. One seedling of each species is planted per plot, each seedling weighing on average 1 gram. After some period of time, the plants are harvested and weighed, with the following results (the numbers shown are plot means):

Species A

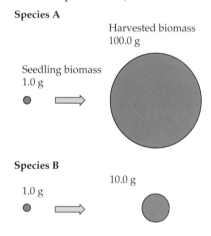

Species B

Now imagine that the ecologist wishes to test the hypothesized effects of each species on the other. Single seedlings are planted without neighbors to assess what their performance would be without competition. The results are:

Species A

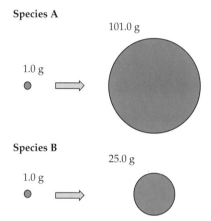

Species B

To proceed, we next calculate the natural logs of all biomass values (ln 1.0 = 0, ln 10.0 = 2.30, ln 25.0 = 3.22, ln 100.0 = 4.61, ln 101.0 = 4.62). How should these competitive effects be assessed? First, we calculate the **relative growth rate** (RGR) over some time interval for each species as

$$RGR = \ln W_2 - \ln W_1$$

where W_2 is the dry weight at the end of the time interval and W_1 is the dry weight at the start of the time interval (Hunt 1990). (We could also have chosen to use an absolute growth rate as a measure of performance.) Thus

$$RGR_{A,mono} = 4.62 - 0 = 4.62$$
$$RGR_{A,mix} = 4.61 - 0 = 4.61$$

In this case, RCI for the intensity of competition experienced by species A, based on RGR as the measure of performance, is

$$RCI_A = (4.62 - 4.61)/4.62 = 0.002$$

The intensity of competition experienced by species B (RCI_B) is 0.28, reflecting the large effect of species A on species B and the very small effect of species B on species A.

Alternatively, an interaction diagram using RGR would look like this:

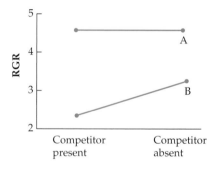

The graph likewise reflects the very small effect of competition on species A and the much larger effect on species B. A statistically significant interaction

---

tions, and offer limited ability to extrapolate their results (Connolly 1986), although they may be useful in particular circumstances (Firbank and Watkinson 1990; Shipley and Keddy 1994). Perhaps their most important limitation is that they consider only a fixed density of plants, making it impossible to determine the circumstances under which each population could be expected to grow or decline (Inouye and Schaffer 1981).

Simple **additive designs** manipulate the total density of neighbors, usually over a range of densities, while keeping the density of the target species constant (usually a single individual). These experiments have been faulted for confounding density with species propor-

tions and for too often basing their conclusions on a single measure of final yield. They offer various advantages, however, for addressing a number of questions (Gibson et al. 1999). For example, the experiment by Goldberg and Landa (1991) discussed above used an additive design to compare competitive effects and responses on a per-individual and per-unit-biomass basis.

Several authors have discussed some of the advantages of **additive series** and "complete additive" designs (also called **response surface experiments**), in which both densities and frequencies are varied (e.g., Firbank and Watkinson 1985; Silvertown and Lovett Doust 1993).

would confirm that interspecific competition has a greater effect on species B than on species A.

Connolly (1987) suggested the use of the difference between the relative growth rates of the two species in mixture to predict the dynamics of the mixture. This index, the **relative efficiency index** (REI), in this case is

$$REI = RGR_A - RGR_B = 4.61 - 2.30 = 2.31$$

Grace (1995) suggests that comparing REI in mixture and in monoculture allows one to assess competitive interactions and predict the "winner" in mixture. Here, $REI_{mono} = 4.62 - 3.22 = 1.40$; the greater value of REI in mixture indicates that species A is competitively dominant and is predicted to replace B in mixture.

If, as is commonly done, RCI were calculated on the basis of final biomass rather than RGR, the results would be

$$RCI_A = 101 - 100 / 101 = 0.0099$$
$$\text{and}$$
$$RCI_B = 25 - 10 / 25 = 0.60$$

showing less of a difference between the two species in response to the presence of competitors than when expressed on the basis of RGR. One would not want to express RCI on a per unit biomass basis; if we divided by the mass of the competitor at the end of the experiment, it would appear that the competitive effects of the two species on each other were very similar. Measured as log response ratios (LRR), the values are

$$LRR_A = \ln (101/100) = 0.01$$
$$LRR_B = \ln (25/10) = 0.92$$

A problem arises when the comparison of the absolute and the relative differences differ (Grace 1995). Grace gives the example of species X reaching 30 grams in competition and 40 grams alone, and species Y growing to 2.5 grams in competition and 10 grams alone. The absolute difference (biomass alone – biomass in competition) for species X is greater than for species Y, but the proportional effect of competi-

tion is greater for species Y. In such cases, it is particularly important to be aware that additive effects can be different from proportional effects, and that ACI will give the former and RCI the latter.

An interaction diagram graphed on the basis of the (untransformed) final biomass will emphasize additive comparisons and give similar results to ACI. An interaction diagram based on relative growth rates will emphasize proportional differences (assuming that the initial sizes were similar and small relative to the final sizes for both species) and will therefore give similar qualitative results to RCI, perhaps with somewhat more information. LRR will also emphasize proportional differences, like RCI, but again, has better statistical properties than RCI. It may be informative to carry out these calculations and draw the graphs for yourself; you may wish to use Grace's example and assume various initial sizes for the two species.

These designs offer more information than simple additive designs, but are large and complex, and they have not commonly been used, mostly for that reason. The information gained may not always be worth the additional effort and cost involved. Multispecies competition experiments, likewise, are complex and are done only occasionally.

A few other experimental approaches have been used on occasion in greenhouse competition studies. Experiments in which root and shoot competition are decoupled (Figure 10.11) can be useful in determining the ways in which competitors affect one another more precisely. Fan designs were created to efficiently study the effects of density and frequency simultaneously (Fig-

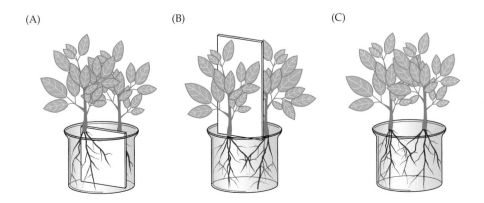

(A) (B) (C)

Figure 10.11
One type of experiment designed to separate root and shoot competition while keeping total soil volume constant. (A) Shoot competition alone (note barrier to root interaction). (B) Root competition alone (note barrier to shoot interaction). (C) Both roots and shoots are allowed to compete. One could also combine root and shoot barriers to eliminate both aboveground and belowground interaction.

(A)

(C)

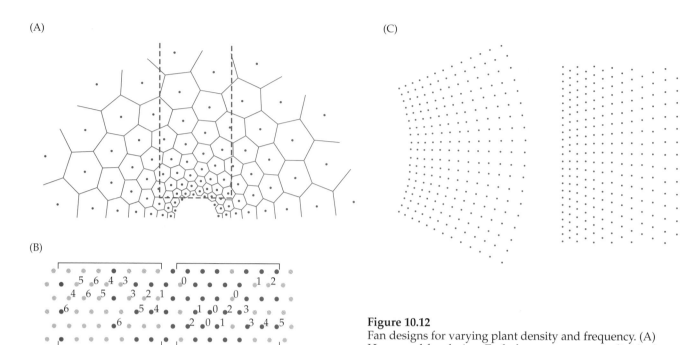

(B)

Figure 10.12
Fan designs for varying plant density and frequency. (A) Hexagonal fan design. Each dot represents an individual plant. (B) Two-species variation on fan design. Shaded and solid circles represent individual plants of two different species; numbers indicate the number of shaded-circle neighbors of plants of either the shaded-circle or solid-circle plants. The number of solid-circle neighbors is equal to 6 minus the number of shaded-circle neighbors. (After Antonovics and Fowler 1985.) (C) Two other types of fan designs. (After Mead 1988.)

ure 10.12). Hexagonal fans consist of a series of contiguous hexagons of increasingly greater sizes, in which each plant is surrounded by between zero and six neighbors at different distances. In addition to efficiency of design, these experiments provide information on density and spacing effects as well as on the effects of the proportions of inter- and intraspecific competition. However, they have some statistical drawbacks, including lack of independence, certain kinds of bias, and other limitations (Gibson et al. 1999).

Using yet a different approach, Gurevitch et al. (1990) compared the performances of individual plants grown alone (no competition), with an intraspecific neighbor, and with an interspecific neighbor (as well as at higher densities of neighbors). They emphasized the importance of comparing intraspecific and interspecific competition with no competition, rather than weighing interspecific competition against intraspecific effects, as is commonly done. They also grew all species combinations in a range of pot sizes to separate the effects of sharing space with a competitor from those of having less absolute space available.

Many of the same kinds of experiments carried out in greenhouses can also be carried out in gardens, or in pots set in the ground or above ground in gardens or

natural communities. There is no strict demarcation between greenhouse, garden, and field experiments. Recent technical advances have extended the ability to manipulate many environmental factors in field experiments, from soil temperatures to atmospheric CO_2.

Field Experiments

Competition has been extensively studied in field experiments (reviewed by Goldberg and Barton 1992). The majority of plant competition experiments in natural communities have been concerned with simple questions regarding whether neighbors affect biomass, growth, or some other fitness component in a single location at one point in time, but a few studies have addressed more complex questions (see the section below on competition along environmental gradients).

The most common experimental design in natural communities involves the removal of all or some neighbors of a target individual (Goldberg and Scheiner 2001). Removals of different species or different functional groups (e.g., woody versus nonwoody neighbors) may be contrasted. While there is no reason that neighbor densities could not be increased rather than reduced, this has rarely been done in plant competition experiments. This approach would almost certainly yield interesting results.

Other experiments manipulate target species abundance over natural gradients of productivity or neighbor density, usually by transplanting individuals into native vegetation. Generally, only the growth of mature plants is measured in such experiments, although a few studies have examined population responses. Both seed additions and adult removals were used by Norma Fowler (1995), for example, to examine the density dependence of demographic responses in two perennial grasses, *Bouteloua rigidiseta* and *Aristida longiseta*, in Texas. She found that the density dependence of demographic parameters for both *Bouteloua* and *Aristida* was weak—competition was much less important than other factors for population regulation in these plants. In other experimental approaches, target species may be transplanted into communities that differ in species composition rather than in productivity or density.

Mechanisms of Interspecific Competition

What happens when plants compete with one another? What are the superior competitors doing that makes them superior competitors, and what is happening to the losers? We have already touched upon some of the answers to these questions above. These factors begin to account for possible mechanisms of competition, although we do not yet have a full understanding of the various mechanisms by which plants compete.

Resource-Based Competition

One general approach to thinking about mechanisms of plant competition is to consider the effects plants have on resources. The extent of resource depletion can be used as a measure of the competitive effect of one species on others. Suppose we call the level of a limiting resource at equilibrium R^*. That is, R^* is the amount of the resource left when a population of a single species grown alone reaches its equilibrium density. It has been proposed that the outcome of competition should be determined by the R^* rule: Over the long term, in a constant environment, the species with the lowest R^* when grown alone will competitively displace all other species (Armstrong and McGehee 1980; Tilman 1982; Grover 1997). This is predicted to happen regardless of the relative initial numbers of the competing species. The resources typically suggested to be involved are light and soil nitrogen.

The R^* rule implies that a relatively uniform habitat should be dominated by a single species, and that different species should dominate in different environments. Since monospecific stands of superior competitors are uncommon, it must be that the R^* rule is inadequate, that every microhabitat is sufficiently different from the others around it that each can be competitively dominated by a different species at very small

scales, or that competitive equilibrium is rarely reached, with herbivory, physical disturbance, or other processes preventing the superior competitor from displacing other species. The R^* rule has been subjected to only a few experimental tests (reviewed by Grover 1997), and carrying out good long-term tests presents some practical challenges. Nevertheless, more extensive testing of this hypothesis would certainly be valuable.

Another means by which plants can compete is by "getting there first"—arriving at a newly available locale before other **propagules** (seeds, vegetative reproductive units, or other dispersal units) and being able to hold that space against later arrivals. It is clear that this is one of the means by which some species dominate some habitats, to the exclusion of other species. Peter Grubb (1977) hypothesized that differences in the conditions and circumstances required for germination and establishment among species, which he called the species' **regeneration niche**, might be an important factor in species coexistence. We discuss disturbance, colonization, and succession further in Chapter 13.

Trade-offs and Strategies

Do certain traits generally make for better competitors, or are there trade-offs that make some traits advantageous in certain kinds of competitive situations but a handicap in others? This is yet another area in which the predictions made by different ecologists sharply diverge. Seventy years ago, Weaver and Clements (1929) observed that in prairies, taller grasses had a competitive advantage, and at least some later researchers in herbaceous perennial communities also found that plant height was associated with competitive dominance (Keddy 2001). But traits that confer an advantage to individuals or species competing for light may not be advantageous in competing for soil nutrients.

David Tilman and other ecologists have assumed that particular traits have associated costs that put organisms possessing them at a competitive advantage only under certain conditions or in certain environments. In any environment, the best competitors **under the conditions present** will be dominant.

In contrast, J. Philip Grime (1977) and others have argued that certain traits, particularly those that confer rapid growth rates under favorable conditions, are always associated with competitive superiority. In favorable environments, such species will always dominate, whereas in unfavorable or disturbed environments, characteristics other than competitive superiority will determine species dominance and persistence.

To examine the relationship between plant traits and competitive effect and response, Paul Keddy and colleagues (Keddy et al. 1998) carried out a large garden experiment with 48 wetland species. They found that competitive effect was strongly related to relative growth

(A)

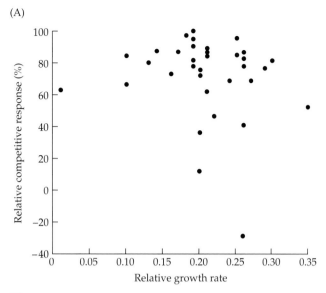

(B)

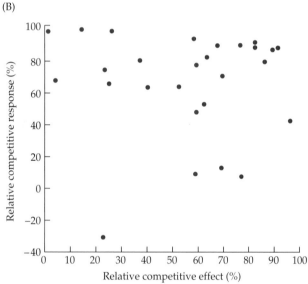

(C)

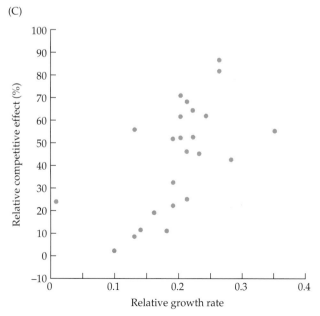

Figure 10.13
Relationships between (A) relative competitive response and relative growth rate, (B) relative competitive response and relative competitive effect, and (C) relative competitive effect and relative growth rate for 48 wetland species from Ontario, Canada. Relative growth rate and relative competitive effect have little influence on relative competitive response. Relative growth rate has a strong and fairly tight relationship to relative competitive effect. (After Keddy et al. 1998.)

rate. Competitive response, however, was not related to either relative growth rate or competitive effect (Figure 10.13). Also, species' competitive responses were similar across a range of very different communities; that is, the response of a species was consistent even when its neighbors were very different. These results suggest that fast-growing, tall species generally are better competitors, and that in natural communities, a few dominant plant species should account for most of the biomass. However, there may be few communities in which this is true (herbaceous temperate systems are one of the few examples), and there is little evidence as to whether the most abundant species in a community are generally competitively superior.

Relative allocations to roots, photosynthetic organs, stems, and reproduction are traits that can affect plant competitive abilities. Shoot architecture has a large effect on the shading of neighbors. Root structure and distribution in the soil (including lateral extent, depth, and the degree to which roots fill space) affect water and nutrient uptake relative to neighbors' abilities to access these resources. Relative allocations to nonwoody versus woody structures and to perennating versus storage organs also affect competitive interactions. It is difficult to make generalizations about how different allocation strategies determine competitive interactions, however, particularly because their consequences differ in different environments.

For many years, plant ecologists have tried to identify plant **strategies**: characteristic suites of traits that should be most successful in certain environments or under particular circumstances. Defining generally useful strategies has been difficult, but the basic idea of strategies continues to come up in various forms (see Chapter 9), and it has intuitive appeal (reviewed by Grover 1997). Mark Westoby and colleagues (Westoby 1998; Westoby et al. 2002) have attempted to overcome objections to previous strategy schemes by proposing objective, widely applicable schemes that can be used to make worldwide comparisons of plant species and communities. Westoby's strategy schemes are based on the assumption that there are trade-offs along a small number of axes that define major, easily quantified traits such as specific leaf area, leaf life span, leaf size, twig size, canopy height, seed mass, and seed number. His goal in developing these schemes is similar to that of people seeking to define plant functional groups (see Box 12A): reducing the vast diversity of terrestrial plants to conceptual categories so that plants can be grouped in ways that enable one to pose testable hypotheses and make predictions. If the ecology of every plant species must be described uniquely, only very limited progress will ever be made in understanding the ecology of any multi-species units, from plant communities and ecosystems to biomes. It is as yet unclear whether this goal is attainable.

Allelopathy

Do plants engage in "chemical warfare" with other plants, releasing toxins to poison neighbors? If this purported phenomenon, called **allelopathy**, could be demonstrated, it would be a potential mechanism by which plants could secure uncontested access to resources. However, allelopathy has been notoriously difficult to demonstrate unambiguously in the field. The major problem is in separating the effects of allelopathic chemicals from the effects of other factors affecting neighbors' performances and distributions, particularly resource competition and herbivory. Another difficulty is detecting the toxic compounds in the field and determining whether they are released naturally in sufficient quantities to harm neighboring plants. To further complicate the situation, the compounds released by the plant may not be harmful in themselves; rather, their degradation products may be the effective agents. There is no doubt that many plants contain toxic substances that can harm other plants when extracted and applied to those plants, but these compounds may be serving strictly or primarily antiherbivore, antibacterial, or antifungal functions.

A good bit of the early research on allelopathy was motivated by the observation that in some systems, bare patches of ground often surrounded plants of certain species. These plants often seemed to be ones with pungent aromatic scents, indicating the release of volatile chemicals into the environment. In the early 1960s, Muller and his colleagues and students claimed to have found evidence for allelopathy among shrubs of the California chaparral (Muller 1969). These aromatic woody plants are often striking in the lack of any herbaceous understory, whereas herbaceous plants are abundant just outside of the bare halos surrounding the shrubs. Another area of allelopathy research was based on the suggestion that dominant grassland species could inhibit succession to woody vegetation for many decades by allelopathic interactions with soil microorganisms (Rice 1974).

In a very clever experiment testing the claims of Muller, Bartholomew (1970) placed small wire exclosures in the halos under chaparral shrubs purported to be allelopathic to keep out herbivorous small mammals.

Inside the exclosures, a lush growth of herbaceous plants appeared. Apparently the animals spent most of their time foraging under the cover of the shrubs, seeking protection from predators (particularly birds). Bartholomew concluded that the bare zones were created by herbivory, not by allelopathy.

Another blow to allelopathy came from Lawrence Stowe (1979; Stowe and Wade 1979), who showed that allelopathic effects were more likely to be found in highly artificial conditions (such as in a petri dish) than in more natural conditions (such as in a pot with soil), that these effects were unrelated to previous claims of allelopathy in the species tested, and that laboratory and greenhouse results had no relationship to the likelihood that two species grew together in an old field. Other problems with allelopathy include the question of how allelopathy could evolve in the face of autotoxicity (Newman 1978; Harper 1977). Ecologists have become largely skeptical of the literature on allelopathy. This does not imply that allelopathy cannot exist, only that it has yet to be convincingly demonstrated to be operating in real communities.

What appears to be allelopathy may sometimes reflect more ordinary mechanisms of competition. *Centaurea diffusa* (diffuse knapweed, Asteraceae) is an important invasive weed of arid grasslands—especially rangelands—in western North America; it presently infests about 1.2 million hectares in the United States. Ragan Callaway and Erik Aschehoug (2000) grew *C. diffusa* in combination with a number of species from its native Eurasian range and its range in North America. They found that *C. diffusa* had stronger negative effects on the growth of the North American plants; moreover, the Eurasian plants, but not the North American plants, reduced the size of *C. diffusa* when grown in common pots. While their study pointed to competition for phosphate between *C. diffusa* and the other species, Callaway and Aschehoug argued that allelopathy was also involved. In one treatment, they added activated carbon to the soil to absorb any hypothesized allelopathic chemicals. This treatment improved the relative competitive ability of the North American plants, but not their ability to absorb phosphate. Besides absorbing chemicals, however, activated carbon has other effects on soil and soil microbes, as the researchers noted. However, Katherine Lejeune and Timothy Seastedt (2001) have suggested that *C. diffusa* and four congeners (all important invasives in western North America) may achieve competitive dominance by altering the balance of soil nutrients, rather than through allelopathic interactions.

Modeling Competition

Models of plant competition seek to explain the vast diversity of plant life by identifying the circumstances under which competitors can coexist. Models of plant competition may also try to explain the mechanisms by which competitors interact and the characteristics of winners and losers. Many basic textbooks in ecology introduce a single model of competition, the Lotka-Volterra model of two-species competition, or at least use this model as a departure point for their discussion of the way species compete. We do not do so in this book, as this model is not very useful for describing competition among plants.

Unfortunately, no comparably simple or generally accepted model exists for competition among plants. We discuss several widely cited models. The first two of these are **equilibrium-based models**, in that they depend on the populations reaching a stable end point at which the outcome of competition is determined. The other models do not make this assumption, and in general depend on chance events in one way or another in determining competitive outcomes.

Two Equilibrium Approaches

One model that describes competition in additive design experiments was developed by Watkinson and his colleagues (Firbank and Watkinson 1985; Law and Watkinson 1987) to generalize the self-thinning equation for two-species mixtures. Recall that the self-thinning law describes a linear, negative relationship between average plant weight and plant density when expressed on a log-log scale. In the two-species model, the intensity of competition is estimated by the competition coefficients α and β. For species A, mean weight per plant in mixture, w, is predicted to be

$$w_A = \frac{w_{mA}}{\left(1 + a_A(N_A + \alpha N_B)\right)^b}$$

and density-dependent mortality is described by

$$N_{sA} = \frac{N_{iA}}{1 + m_A(N_{iA} + \beta N_{iB})}$$

where the subscripts A and B specify which of the two species a given variable represents; w_m is the average weight per plant in the absence of competition (i.e., w_{mA} is the mean weight for species A in absence of competition); a_A and b are fitted parameters; α and β are competition coefficients that measure the average effect of an individual of species B on an individual of species A and vice versa; and N_i and N_s are the initial and final densities (Firbank and Watkinson 1990). Analogous equations are used for species B.

This model expresses the mean weight for species A as being equal to the mean weight in the absence of competition, divided by one plus the total competitive effects of neighbors. The $(N_A + \alpha N_B)$ term gives the standardized number of neighbors because α allows us to equate individuals of the two species in terms of their effects on species A. Raising this term to the power b allows density to have nonlinear effects, and the entire quantity is multiplied by a_A, which measures the effect of these standardized neighbors on mean weight. The equation for numbers of individuals is similar.

Because mortality changes in response to density, the number of individuals of species A at the end of the experiment will be equal to the number at the start, divided by one plus the standardized initial number of plants of both species (by using β, the per-individual effect of species A on species B), together multiplied by the effect of neighbors on the survival of species A, m_A.

Law and Watkinson (1987) extended this model by relaxing the requirement for a constant value for the competition coefficients and allowing those coefficients to vary with the frequency and density of the two species. Their model describes rather than predicts competitive outcomes because the outcomes are determined by fitting parameters empirically—the values used in the equations are determined by the results obtained in the experiment.

In an attempt to model resource competition mechanistically, Tilman (1982) introduced the resource ratio model for plant competition (Figure 10.14). This model predicts that stable coexistence among competing plants is possible for plants that are competitively superior at different ratios of essential resources. However, the resource ratio model has not been well supported by experimental tests. This may be because plant communities are likely rarely at equilibrium, and many other processes, such as herbivory, contribute to determining the outcome of competition. The few experimental tests of one component of the model, the R* rule (discussed above), offer some support for this one component, particularly when plants are competing for soil nitrogen (summarized by Grover 1997).

Patch-Based and Nonequilibrium Models

A number of conceptual and mathematical models of competition and its role in community structure have used nonequilibrium and nondeterministic approaches, in which chance events have a key role in determining competitive outcomes. Alexander Watt (1947) introduced the idea of the plant community as a mosaic of patches. These patches, he maintained, are dynamically related to one another. There is a tension between predictable order and chance events that tend to disrupt that order, resulting together in the structure of the community. Competitive interactions played a major role in Watt's view of patch dynamics (Figure 10.15). Watt specifically

Eddy van der Maarel

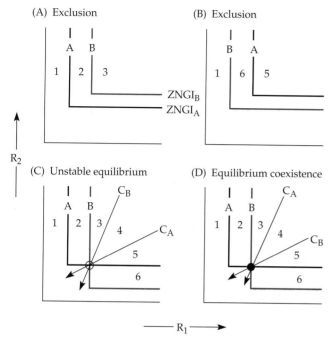

Figure 10.14
The possible outcomes in David Tilman's resource ratio model of competition in which species A and B are competing for resources R_1 and R_2 with resource uptake vectors C_A and C_B. Each point on the graph represents the supply (availability) of the two resources in a particular environment. Growth is positive for a species in the region above or to the right and negative below or to the left of the zero net growth isocline (the line where growth is 0) for that species ($ZNGI_A$ for species A and $ZNGI_B$ for species B). The numbers indicate different resource supply combinations in which neither, either, or both species are able to exist (alone and/or in competition with the other). (A) Competitive exclusion of species B by the competitively superior species A, where species A can reduce resource levels below those needed by species B. (B) Competitive exclusion of species A by the competitively superior species B. (C) Unstable equilibrium occurs when resources are drawn down to the point indicated by the open circle, at which there is unstable coexistence of the two species in region 4 of the graph and competitive displacement of one or the other species in the other regions. (D) Stable equilibrium, in which there is coexistence of the two species in region 4 of the graph and competitive displacement of one or the other species in the other regions. Stable equilibrium is reached at the point indicated by the closed circle where C_A and C_B cross. The resource uptake (or consumption) vectors C_A and C_B indicate the ratio with which resources R_1 and R_2 are used by species A and B; they define the region where coexistence is possible. (After Tilman 1982.)

excluded successional change from this conceptual model of the maintenance of community structure.

Contemporary views of plant communities are remarkably similar to Watt's perspective (Pickett and White 1985). Eddy van der Maarel and coworkers (van der Maarel and Sykes 1993) extended Watt's perspective in the "carousel model" they developed to

describe the dynamics of species on very small scales in species-rich grasslands on the island of Öland in southern Sweden. van der Maarel emphasized the small-scale mobility of plants both as individuals, as parts of plant die and shoots colonize adjacent gaps, and through propagules. The "carousel" is the cycle of species replacement, which may go around quickly or slowly.

A formal model that incorporated competitive interactions and environmental patchiness was developed by Avi Shmida and Stephen Ellner (1984). This model had a strong influence on ecologists' thinking about competition and species coexistence. Shmida and Ellner's lottery model assumed that the outcome of competition among juveniles for space is determined by chance (a "lottery"). This lottery for space can be biased, in that some juveniles may have better chances than others. The model incorporates asymmetric competition (juveniles cannot displace adults) for microsites, nonuniform seed dispersal, and larger-scale spatial and temporal patchiness (Figure 10.16). In contrast to most previous models, this one suggested mechanisms for species coexistence without requiring differences in habitat or resource use.

Recent advances have been made in modeling plant competition and species coexistence, particularly the explicit inclusion of the effects of spatial location. Using

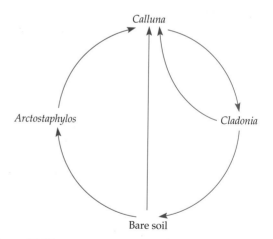

Figure 10.15
Watt's model of vegetation dynamics in the Dwarf Callunetum. This plant community, at about 800 meters altitude in the Cairngorm Mountains of Scotland, has a natural pattern of double strips of *Calluna* (heather, Ericaceae) and *Arctostaphylos* (bearberry, Ericaceae) separated by wind-swept bare soil. Watt explained the dynamics of this pattern as follows: The young shoots of *Calluna* exclude lichens, but as they age, their competitive ability diminishes, and the area is invaded by *Cladonia* (a mat-forming lichen). The constant wind prevents the *Cladonia* mat from persisting, however, and in time it disintegrates, leaving bare soil. *Arctostaphylos* then invades by vegetative spread, eventually achieving complete occupation of the patch. Young shoots of *Calluna* in time spread in from the margins, eventually competitively replacing the *Arctostaphylos*. (After Watt 1947.)

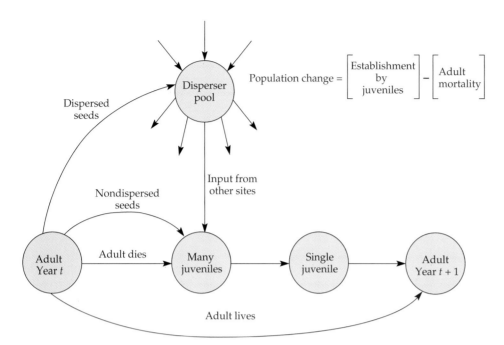

$$\text{Population change} = \begin{bmatrix} \text{Establishment} \\ \text{by} \\ \text{juveniles} \end{bmatrix} - \begin{bmatrix} \text{Adult} \\ \text{mortality} \end{bmatrix}$$

Figure 10.16
"Lottery model" of the role of microsite occupation and dispersal in a community. The diagram pictures what happens within a single microsite over the course of one year. Adults within the population and from outside the population contribute seeds to the disperser pool. Some of these seeds leave the population, but some proportion of the incoming and nondispersed seeds germinate to produce juveniles. Seeds produce successful juveniles only when an adult dies, because they cannot survive under an adult. A juvenile can grow to become an adult in the following year; an adult may die or survive as an adult in the next year. (After Shmida and Ellner 1984.)

invasion analysis models, Lavorel and Chesson (1995) simulated competition between two species with detailed attention to spatial pattern and dispersal strategies under different disturbance regimes. They found the greatest species coexistence at intermediate levels of habitat suitability and disturbance frequency. Spatial patterns interacted with species dispersal patterns and regeneration strategies in a complex fashion.

Cellular automata models represent another approach to modeling competition spatially (Balzter et al. 1998). These dynamic models have been used for a variety of processes at a variety of scales, from spatially dependent molecular interactions to the development of spiral galaxies, as well as spatially dependent ecological processes. Balzter et al. (1998) demonstrated the usefulness of this approach for plant communities with a preliminary model of the dynamics of three plant species—*Lolium perenne* (ryegrass, Poaceae), *Trifolium repens* (white clover, Fabaceae), and *Glechoma hederacea* (ground ivy, Lamiaceae)—growing in a lawn over a three-year period. Eventually this approach may prove valuable in making predictions about species' abundances and conditions for coexistence in a spatially explicit context.

Effects of Competition on Community Composition

Is competition important in most environments? What is its role in determining community composition, and how important is it relative to other processes such as herbivory? These are some of the most important ques-

tions concerning competition among plants, and we do not yet have good answers for them. Within a topic that generates a great deal of controversy, these questions seem to be the most vigorously debated, perhaps because they bring together all of the other issues discussed in this chapter about which there is disagreement. There is insufficient evidence as yet to either accept or reject any of the general theories of plant community organization or of the role of plant competition in determining community composition.

We must first distinguish between the importance, intensity, and significance of competitive interactions. Competition is *important* if it plays a major role in determining community composition. Competition can be *intense*—that is, have major effects on various aspects of individuals' performances—without having much effect at the community or even at the population level. Competition is ecologically *significant* if it plays a substantial role in determining species coexistence within a community.

In one of the earliest experimental studies of species interactions, Arthur Tansley sought to determine the role of competition in the distribution of two species of *Galium* (bedstraw, Rubiaceae), both small herbaceous perennials (Tansley 1917). *Galium saxatile*, the heath bedstraw, is found primarily on sandy soils, while *Galium sylvestre* (probably the species now called *G. sterneri*), the limestone bedstraw, is found primarily on calcareous, limestone-derived soils.

Tansley grew plants of both species alone and together in each of the two soil types in large wooden boxes outdoors. When grown alone in calcareous soil,

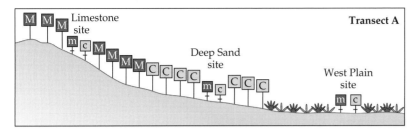

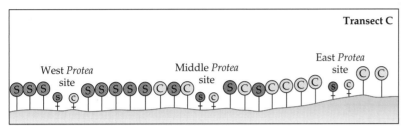

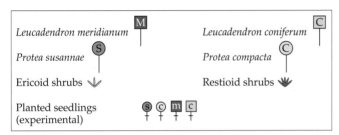

Figure 10.17
Diagram of two transects, A and C, in the fynbos of southern Africa, showing the locations of the three experimental study sites along each transect. Dominant species and plants of some other growth forms are also shown. (After Richards et al. 1997.)

limestone bedstraw grew normally, as expected. Heath bedstraw survived, but grew slowly and had yellowish leaves, indicating nutrient deficiency. When grown alone in sandy soil, heath bedstraw grew vigorously, while limestone bedstraw survived, but grew poorly. When the two species were grown together in calcareous soil, limestone bedstraw overtopped heath bedstraw and eliminated it from the mixture. In sandy soil, heath bedstraw became dominant, but did not completely eliminate limestone bedstraw during the course of the experiment. Tansley concluded that while each species appeared to be adapted to the soil in which it lived in nature, competition also played an important role in determining the restriction of the two species to different soil types.

A recent study in a different system found quite different results. M. B. Richards and his colleagues at the university of Cape Town (Richards et al. 1997) examined the relative importance of competition and adaptation to soil type among six shrubs in South Africa belonging to the Proteaceae. These species grow in the fynbos of southern Africa, one of the most species-rich places on Earth for plants (see Box 20A). The fynbos is dominated by an astonishing diversity of shrubs as well as other growth forms. The environment is characterized by nutrient-poor soils, summer drought, and rolling, dissected terrain. Large numbers of apparently ecologically similar species coexist in the fynbos, and there is great species variation among communities. What determines the typically sharp discontinuities among communities dominated by different species, and what is the role of competition in determining these boundaries?

Richards and his colleagues chose three transects, each of which crossed a sharp community boundary with a transition from one distinct soil type to another. Each of the transects contained a different species pair in which one dominant species replaced the other (Figure 10.17). In a three-year experiment to compare the influences of soil type and interspecific competition, seeds of both species were planted in monoculture and in mixture at three sites along each transect. Interspecific competition consistently decreased growth, but the magnitude of its effect was small compared with that of soil type. Adaptation to soil conditions strongly affected both seedling growth and survival at two of the three sites (Figures 10.18 and 10.19) and the researchers suggest that soil type, not competition, may be the critical factor determining species distribution.

Competitive interactions may change over time, and the ultimate outcome of competition may be different than initial observations suggest. *Lythrum salicaria* (purple loosestrife, Lythraceae) is an invasive species in North America that appears to be displacing native wetland species (see Chapter 11 and Figure 14.3). Some ecologists have argued, however, that evidence that *Lythrum* is actually displacing native species is weak. Mal and col-

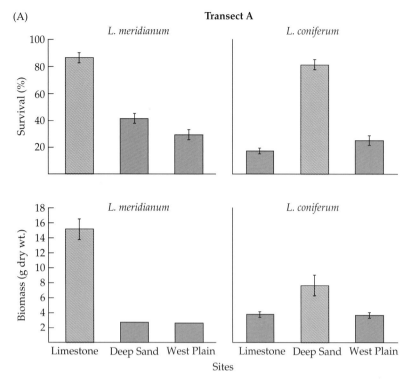

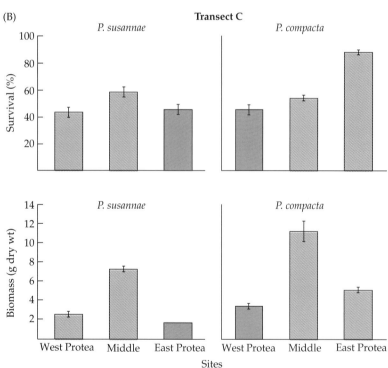

Figure 10.18
Mean survival (%) and biomass (g dry weight ± 1 standard error; some error bars too small to be visible) after three years for experimental seedlings of (A) *Leucadendron meridianum* (Proteaceae) and *L. coniferum* at transect A; and (B) *Protea susannae* (Proteaceae) and *P. compacta* at transect C when grown at sites within their natural distribution (green bars) or outside of it (gray bars). In transect A, survival and biomass were both far greater within the natural distributions of both species than outside them. In transect C, survival and biomass were lowest outside the natural distributions of both species for most sites. (After Richards et al. 1997.)

leagues (Mal et al. 1997) carried out a four-year field competition experiment between *Lythrum* and *Typha angustifolia* (cattail, Typhaceae), a dominant native wetland species. *Typha* was initially competitively superior, but in the second and third years of the experiment, the species were relatively evenly matched. By the fourth year of the study, *Lythrum*, the invader, became competitively dominant, displacing *Typha*. The researchers attribute this result to differences in life history strategy between the two species. *Typha* has large rhizomes with substantial stored resources, which might give it an initial competitive advantage, but the high costs of producing new ramets and the strong suppressive effects of *Lythrum* led to the eventual competitive replacement of *Typha*.

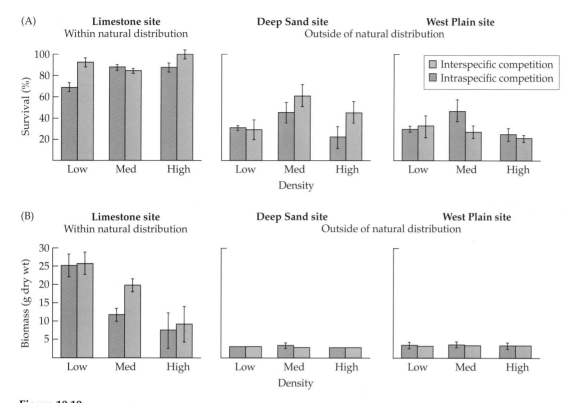

Figure 10.19
(A) Mean survival (%) and (B) biomass (g dry weight ± 1 standard error; some error bars too small to be visible) after three years for experimental seedlings of *L. meridianum* planted at three densities, in monoculture and in mixture, at three sites along transect A. Competition did not appear to have much effect on biomass or survival, or may have had inconsistent effects that were far smaller than the effects of being within or outside the natural distribution. *L. meridianum* did not decrease in survival at increased densities, and decreased in biomass at increased densities only at the Limestone site. Performance in mixture (green bars) was generally better than performance in monoculture (gray bars). Patterns were similar for other species and transects. (After Richards et al. 1997.)

Competition along Environmental Gradients

One of the most critical and contentious issues concerning the importance of competition in plant communities is whether there are particular kinds of habitats in which competition is predictably strong, determining community composition, or predictably weak and unimportant. Ecologists agree that competition is intense in productive, nutrient-rich habitats, at least when disturbance and herbivory are low. In these environments, plants are able to develop large canopies quickly, and competition is thought to be primarily for light. The relative intensity and importance of competition in productive and unproductive habitats, however, remain matters of debate.

Conceptual Models of Competition in Habitats with Differing Productivities

Grime (1977, 1979) proposed that competition is unimportant in unproductive environments, and that success in these environments is dependent largely on ability to tolerate abiotic stress (low nutrients, drought, or cold, for example), rather than on competitive ability. He further argued that in environments where disturbance frequently reduces plant mass, competitive exclusion should be prevented. The dominant plants in such environments should not be competitively superior, but rather should possess traits that allow them to withstand disturbance or recolonize rapidly following disturbance.

Most subsequent discussion has focused on unproductive environments. Newman (1973) disagreed with Grime's characterization, arguing that competition is important in low-resource as well as high-resource environments, but that the resources for which plants compete differ—light in productive environments, nutrients and water in unproductive environments. Later work by Tilman (1987) reinforced and developed Newman's ideas. Tilman argued that competition in low-productivity environments would be for belowground resources (primarily nitrogen), whereas competition in high-productivity environments would be primarily for aboveground resources (light). Aerts (1999) reiterated Grime's argument in part, maintaining that selection in nutrient-poor habitats would favor traits that reduce nutrient losses rather than those that enhance the ability to compete for nutrients, resulting in slow growth rates.

Keddy (1990) adapted to plant community organization the "centrifugal theory of community organization" that Rosenzweig and Abramsky (1986) had proposed for desert rodent communities. The centrifugal model proposes that there is a core habitat type preferred by all species in a region, presumably with ideal growing conditions. Other habitat types, called peripheral habitats, are defined by particular negative conditions (stresses or disturbance) to which only some of the species are adapted (Figure 10.20). Interspecific competition is most intense, Keddy suggests, in core habitat, and is more relaxed in the peripheral habitats because fewer species are adapted to the particular conditions in each of them. Thus the peripheral habitats serve as refuges, preventing competitive exclusion. In wetlands in Ontario, Canada, for example, all species prefer sites with high fertility and low disturbance rates (core habitat), while the peripheral habitats are defined by infertile soils and disturbances such as ice scouring. The core habitat is dominated by *Typha latifolia*, while different species dominate as one moves toward more extreme conditions in each peripheral habitat type.

One of the problems with trying to resolve the debate about the relative intensity and importance of competition in productive and unproductive habitats is that environments can be unproductive for very different reasons. Many of the hypotheses about competition in unproductive environments implicitly focus on low nutrient levels. But low productivity may also be due to cold temperatures, short growing seasons, saline soils, or toxic materials in the soil that inhibit growth, such as heavy metals. Inadequate water supply is one of the most important factors limiting productivity globally.

Plants are not only likely to have very different adaptations to cold, heavy metals, and drought than to low nutrients, but species' interactions under each of these conditions are also likely also to be very different. Goldberg and Novoplansky (1997) predicted that the effects of competition in nutrient-poor environments would differ substantially from those in water-stressed habitats. Water availability, they pointed out, is pulsed: even in arid environments, water

is freely available for short periods of time, but there are long interpulse intervals in which water is partly or completely unavailable. They hypothesized that growth and competition should be limited to the periods of high water availability (the pulses) in arid environments. Growth and competition in low-nutrient soils, in contrast, should not be limited to pulses of short duration.

Experimental Evidence

The evidence for and against variation in the intensity of competition along productivity gradients remains confused and ambiguous. Wilson and Keddy (1986) compared the competitive abilities of six species that are dominant at different points along a productivity gradient on the shore of Axe Lake, Ontario. The gradient ranged from wave-disturbed, nutrient-poor beaches with low standing biomass to sheltered, nutrient-rich sites with dense vegetation. Plants were grown in competition in plastic beakers at a protected site in the field, using substrate from the favorable (sheltered, nutrient-rich) end of the gradient. The researchers found that competitive ability (measured as both competitive response and competitive effect) was positively correlated with mean position along the productivity gradient (Figure 10.21). They interpreted these results as evidence that species with high competitive ability occupied nutrient-rich, undisturbed sites, while species

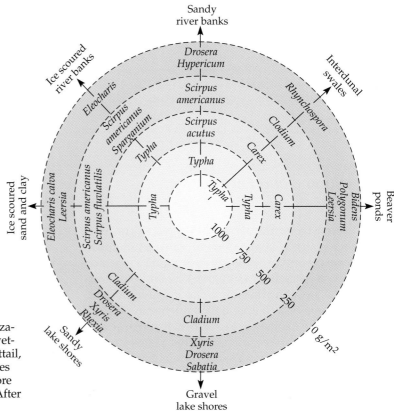

Figure 10.20
The centrifugal model of plant community organization, applied to the distributions of a number of wetland species in Ontario, Canada. *Typha latifolia* (cattail, Typaceae) occupies core habitat, while other species become more prominent as one moves toward more extreme conditions in more peripheral habitats. (After Keddy 1990.)

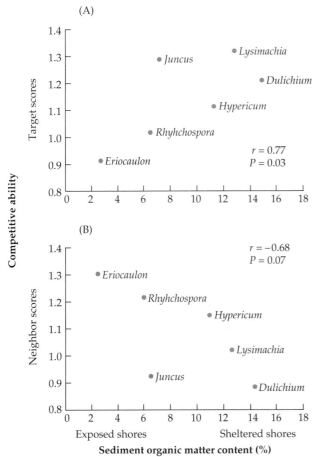

Figure 10.21
Competitive abilities of six wetland plant species and the locations where they are naturally found along a gradient from exposed, nutrient-poor to sheltered, nutrient-rich shores, corresponding to a gradient in the percentage of organic matter contained in the sediment. (A) Competitive abilities expressed as target scores, defined as the mean relative growth (increase in dry mass) of the target species when grown in the presence of all neighbor species; this score is similar to competitive response. (B) Competitive ability expressed as neighbor scores, defined as the mean relative growth (increase in dry mass of all neighbor species in the presence of the target species; this score is similar to competitive effect. (After Wilson and Keddy 1986.)

with low competitive ability were displaced to disturbed sites with poor soils, where competitive exclusion was prevented by wave action and low soil nutrients.

Gurevitch (1986) carried out a field study of competition along an environmental gradient in southeastern Arizona. She hypothesized that *Hesperostipa neomexicana*, a C_3 grass, was limited to arid ridgetops by competitively superior C_4 grasses. This is precisely the opposite of what one would expect if physiology were determining species distributions, as C_4 species should be better able to tolerate the unfavorably hot, dry conditions on the ridge crests. She removed neighbors from

around target *Hesperostipa* individuals at three sites along a gradient from a ridge crest to a moister lower slope. Competition affected growth of mature plants, flowering, seedling establishment, and survival.

When competitors were removed, growth and flowering for the mature *Heterostipa* plants were greatest on the lower slope, where its abundance was lowest. Competition had the smallest effect on estimated population growth rates on the ridge tops, where *Heterostipa* was most abundant, and increasingly greater effects downslope. The largest effects were at the lowest sites where *Heterostipa* was present (Figure 10.22). These results strongly suggested that competition was a major factor in determining the distribution of *Hesperostipa* along this gradient of productivity and environmental favorability.

Theodose and Bowman (1997) suggested the existence of the opposite pattern, in which competition prevented a species from a more productive area from growing in a resource-poor site. The perennial *Deschampsia caespitosa* (hair grass, Poaceae) is common in moist alpine meadows in the tundra of the Front Range of Colorado, but is rare in dry meadows. The dry meadows are dominated by a sedge, *Kobresia myosuroides* (Cyperaceae). The authors hypothesized that *Deschampsia* was prevented from growing in the dry environment by competition with *Kobresia*. An earlier study had demonstrated that *Kobresia* was kept out of the moist meadows by deep winter snow.

Theodose and Bowman transplanted individuals of each species, as well as two-species pairs of plants to a dry meadow, either clipping the existing vegetation (largely *Kobresia*) at ground level or leaving it intact. *Deschampsia* had a greater increase in survival in response to vegetation clipping than did *Kobresia*, and soil moisture was substantially depressed in plots with intact vegetation compared with those in which vegetation was clipped (Figure 10.23). The researchers concluded that interspecific competition with *Kobresia* excluded *Deschampsia* from the dry meadows by depressing soil moisture below the drought tolerance of *Deschampsia*. *Kobresia*, with greater tolerance of drought, was able to survive during periods of low moisture.

One problem with this conclusion, however, is that the mortality of *Kobresia* (28%) was actually higher than the mortality of *Deschampsia* (15%) in intact vegetation. Furthermore, most of the *Deschampsia* plants survived in the intact vegetation of the dry meadow. Therefore, it is difficult to argue convincingly that competitive exclusion was the result of high mortality in *Deschampsia*. The growth of *Deschampsia* in intact vegetation was also greater than the growth of *Kobresia*, and the growth of both species was about equally affected by vegetation clipping. Nothing is known about the effects of competition on reproduction or establishment for these species in these environments. So, although this study clearly

	Ridge crest		Midslope		Lower slope		Wash	
	Hes	C₄'s	*Hes*	C₄'s	*Hes*	C₄'s	*Hes*	C₄'s
Biomass (g/m²)	140	16	76	111	1	282	0	High, not measured
% Cover	20.5	9.5	8.5	24.5	3.0	37.0	0	~100%

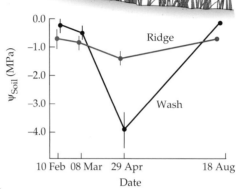

Hesperostipa:	Ridge crest		Midslope		Lower slope	
	Control	Removal	C	R	C	R
Seedlings/m²	4.55	9.36	0.58	2.04	0.10	0.35
Seedling survival	0.47	0.79	0	0.56	0	0.50
Growth $\left(\frac{\text{final size}}{\text{initial size}}\right)$	0.77	1.97	0.69	2.58	1.38	4.28
Number of flowers	1.50	13.30	4.10	28.70	35.50	58.90
Population growth (λ)	0.93	1.04	0.83	0.92	0.59	0.88

Figure 10.22
Results of a removal experiment at three topographic positions in a southeastern Arizona grassland (elevation 1400 m). The table at the top shows means for biomass and cover (*N* = 40; confidence limits available in Gurevitch 1986) at the three experimental sites (ridge crest, midslope, and lower slope) and at a lower topographic position below them (wash). The lower table gives experimental results for Control (neighbors not removed) and Removal (mature neighbors removed) treatments at the three experimental sites. Means for seedlings/m² (in 1980 plus 1981), mature plant growth, and number of flowers produced by mature plants (*N* = 20) over the 20 months of the experiment are shown. Seedling survival was based on the total number of seedlings surviving, and population growth was estimated as described in the original paper. The graph at right shows soil water potential at the top (ridge crest) and bottom (wash) of the topographic gradient at the 15–20 cm depth (where the bulk of the roots were located) over a drying and rewetting cycle of 6 months in 1980. (After Gurevitch 1986.)

demonstrated intense and statistically significant effects of competition in this resource-poor habitat, more work needs to be done to conclusively demonstrate that competition leads to the exclusion of *Deschampsia* in the dry meadows.

Evidence from Research Syntheses

While the results of individual studies are apparently contradictory, there have been several attempts to gain a better overview of competition along environmental gradients. In a cross-continental set of field experiments, the intensity of competition was compared for transplanted *Poa pratensis* (bluegrass, Poaceae) individuals at

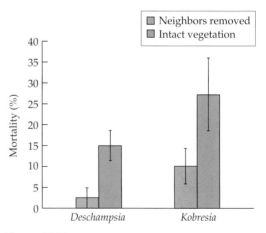

Figure 10.23
Mortality of *Deschampsia caespitosa* (Poaceae) and *Kobresia myosuroides* (Cyperaceae) in a dry meadow when transplanted into intact vegetation (gray bars) and with neighbors clipped (green bars). Values are mean mortality (± 1 standard error) based on four plots, each of which had 10 experimental plants per species. (After Theodose and Bowman 1997.)

(A)

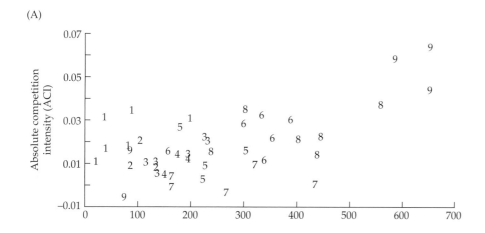

(B)

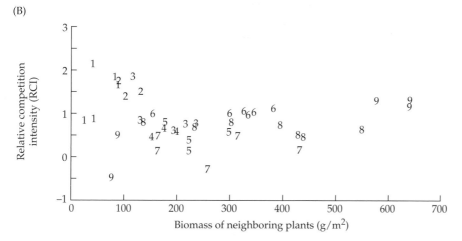

Biomass of neighboring plants (g/m²)

Figure 10.24
Results from a large set of field competition experiments in which perennial bluegrass (*Poa pratensis*) was planted in 44 plots across nine sites in locations across the world. Each point represents a plot; numbers indicate particular sites. At each site, competition was studied over a gradient of neighbor biomass, so there is more than one experimental outcome shown for each site. Competition was measured as (A) absolute competition intensity (ACI) and (B) relative competition intensity (RCI). (After Reader et al. 1994.)

12 locations in Europe, North America, and Australia (Reader et al. 1994). The intensity of competition was compared across sites over the range of standing biomass (a surrogate for productivity) within each site. The authors reported the results using two indices of competition intensity, relative competitive intensity (RCI) and absolute competitive intensity (ACI). There was some suggestion that ACI increased as neighbor biomass increased, but RCI showed no clear relationship to neighbor biomass (Figure 10.24). The authors concluded that there was no convincing evidence to support the hypothesis that competition increased along a gradient of increasing neighbor biomass, when measured across a wide range of sites and productivities. Since there are flaws in both ACI and RCI, it would be interesting to examine the results using other approaches to measuring competition intensity.

In another approach to synthesizing results over a greater range than is possible in an individual experiment, Goldberg et al. (1999) carried out a meta-analysis, or quantitative synthesis, of 14 papers reporting a large number of outcomes of competition experiments. The synthesis examined competition and facilitation among plants along productivity gradients, using vegetation bio-

mass as a surrogate for productivity. The meta-analysis had surprising results: when the measure of competition intensity used was the log response ratio (LRR), the researchers found a strong *negative* relationship—the opposite of what was expected—between competition intensity and productivity for final plant biomass and survival, but not for growth (Figure 10.25). When competition intensity was measured using RCI, they found no clear relationship between RCI and vegetation biomass except, again, a negative relationship for survival. It is difficult to know what these results mean for our understanding of the role of competition in environments of differing productivity. Clearly, however, the hypothesis of increasing competition intensity with increasing productivity is not supported as a general pattern, at least when competition is measured using these indices.

Resolution of Differing Results

What explains the lack of agreement among these studies, and what can be done to achieve a better understanding of the importance of competition over a range of environments? Performing experiments in a relatively consistent way across many sites, as Reader and colleagues did, is a big step in the right direction, as is using

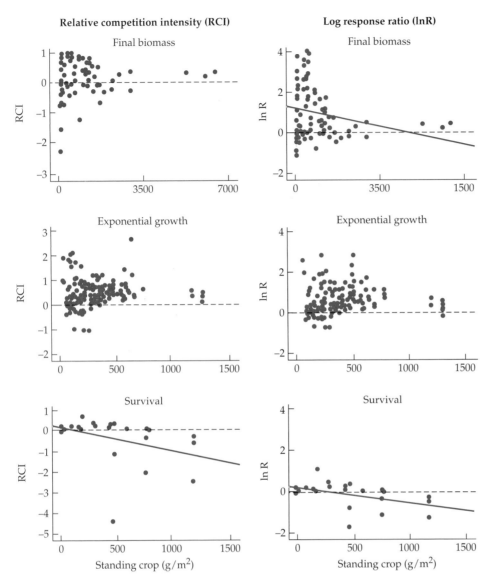

Figure 10.25
Values for two measures of response to competitors—relative competition intensity (RCI) and log response ratio (LRR)—across 14 separate published studies over a range of natural biomass (standing crop, g/m²). Regression lines are shown only where there was a statistically significant relationship between the variables. Positive values of RCI or LRR indicate that competition is occurring, and negative values indicate that neighbors have a beneficial effect on target plant performance. (After Goldberg et al. 1999.)

modern quantitative synthesis techniques to bring together large numbers of independent studies, as Goldberg and her colleagues did. It is clear from these two efforts at synthesis that the conclusions can vary greatly depending on how the experiments are designed, how long they are maintained, what is measured, and how the results are analyzed.

Some of the inconsistencies in the conclusions among different researchers are almost certainly a result of these artifacts. More profoundly, most studies on plant competition look at individual growth (or other measures of the responses of individual plants) in response to competition, rather than estimating population responses. Comparisons of RCI, for example, cannot show whether competition restricts where a species is found; it just does not provide enough information. If the question is what limits the distributions of species or determines community composition, studying population-level responses is the only way to get an accurate answer.

Summary

The importance of competition in shaping species distributions and community composition has been debated by ecologists for many years. Competition among plants is a reduction in fitness due to shared use of a resource that is in limited supply. The resources for which plants compete include light, water, mineral nutrients, space, pollinators, and seed dispersers. Most competition occurs among adjacent individuals. Thus, local density, not the total density of a population or community, determines the intensity of competition for an individual.

Competition often results in large size differences among individuals, ultimately translating into differences in survival and reproduction. Larger individuals usually have a competitive advantage over smaller individuals. However, larger size is not always advantageous.

The outcome of competition can be determined by a variety of mechanisms. One attempt to predict competitive outcome is the R* rule: over the long term, in a constant environment, the species with the ability to deplete resources to the lowest level when grown alone will competitively displace all other species. There is limited experimental evidence to support this proposition. Allelopathy—the release of toxins to poison competing neighbors—has long been hypothesized to explain some species distributions, although the experimental evidence for its existence is weak.

Whether and how competition structures communities may depend on whether dominance hierarchies are consistent among species and environments. If a species is dominant in one environment, will it continue to be dominant in other environments? Is the intensity of competition the same in all environments, or is it weaker in low-resource or stressful environments? Resolving these questions is difficult, in part because the answers depend on which measure of competition is used and whether one measures competitive effect—the effect of an individual on its neighbors—or competitive response—the effect of the neighbors on the individual. Many experimental designs are used to assess competition, including various ways of manipulating densities and frequencies. These experiments can be carried out in greenhouses, transplant gardens, and natural populations.

Models of competition are built on one of two assumptions: (1) that populations reach a stable end point at which the outcome is determined, or (2) that chance events, such as disturbance and herbivory, determine the outcome of competitive interactions. In the next chapter we explore the role of herbivory and other types of species interactions in shaping plant populations and communities.

Additional Readings

Classic References

Newman, E. I. 1973. Competition and diversity in herbaceous vegetation. *Nature* 244: 310.

Tansley, A. G. 1917. On competition between *Galium saxatile* L. (*G. hercynicum* Weig.) and *Galium sylvestre* Poll. (*G. asperum* Schreb.) on different types of soil. *J. Ecol.* 5: 173–179.

Watt, A. S. 1947. Pattern and process in the plant community. *J. Ecol.* 35: 1–22.

Yoda, K., T. Kira and K. Hozimu. 1963. Self thinning in overcrowded pure stands under cultivated and natural conditions. *J. Biol., Osaka City Univ.* 14: 107–129.

Contemporary Research

Aerts, R. 1999. Interspecific competition in natural plant communities: Mechanisms, trade-offs and plant-soil feedbacks. *J. Exp. Bot.* 50: 29–37.

Balzter, H., P. W. Braun and W. Kohler. 1998. Cellular automata models for vegetation dynamics. *Ecol. Modelling* 107: 113–125.

Goldberg, D. and A. Novoplansky. 1997. On the relative importance of competition in unproductive environments. *J. Ecol.* 85: 409–418.

Grace, J. B. 1995. On the measurement of plant competition intensity. *Ecology* 76: 305–308.

Lavorel, S. and P. Chesson. 1995. How species with different regeneration niches coexist in patchy habitats with local disturbances. *Oikos* 74: 103–114.

Additional References

Keddy, P. A. 2001. *Competition.* 2nd Ed. Kluwer Academic, Dordrecht (Netherlands) and Boston.

Grace, J. B. and D. Tilman (eds.). 1990. *Perspectives on Plant Competition.* Academic Press, New York.

Grover, J. P. 1997. *Resource Competition.* Chapman and Hall, London.

CHAPTER *11* *Herbivory and Plant-Pathogen Interactions*

Green plants are the foundation of almost all terrestrial food webs. All animals (including humans, of course) ultimately depend on plants for their existence. Yet casual observation seems to reveal a green world teeming with (uneaten) plants. Why is the world so green? Conversely, what are the consequences of herbivory from the plants' perspective?

Herbivory is the consumption of all or part of a living plant. Some ecologists use the term "predation" when an herbivore eats and kills an individual. **Seed predators,** or **granivores,** are herbivores that consume seeds or grains, killing the individual within. **Grazers** are herbivores that eat grass and other ground-growing plants, while **browsers** eat leaves from trees or shrubs. **Frugivores** are herbivores that consume fruits, sometimes without damage to the seeds.

Plants are consumed by organisms from a variety of kingdoms: animals, fungi, bacteria, and even other plants. Herbivory can have ecological effects at the level of the individual plant, the population, the community and landscape (e.g., patterns of coexistence of plant species), and the ecosystem (e.g., nutrient cycling). Herbivores can also influence the evolution of plants. Ecologically and evolutionarily, some of the most important herbivores are grazing mammals and insects. However, other types of herbivores, such as birds, mollusks, and nematodes, can be very important in particular systems.

In this chapter we look at the consequences of herbivory and of plant-herbivore interactions for plant population dynamics and for the structure of plant communities. Can herbivory affect the trajectory of a plant population? Can the presence of herbivores—or their exclusion—have a decisive role in determining what plant species are able to coexist in a community? We begin by examining the effects of herbivory on individuals, populations, and communities, and briefly touch on its landscape effects. In addition, we look at how plants defend themselves against herbivores and at how those defenses evolved. Then we turn to the evolutionary consequences of herbivory and the responses of plants. Finally, we take a look at what is known about the role of plant-pathogen interactions in plant communities.

Herbivory at the Level of Individuals

What do herbivores do to individual plants? Herbivores can consume an entire plant, causing the death of the individual. By eating seeds, granivores—such as some ants, rodents, and birds—kill individual plants. Alternatively, herbivores can eat (or manipulate) only some of the parts of a plant, damaging, removing, or destroying those parts, but not necessarily immediately killing the plant. Deer, for example, frequently restrict their consumption to only the newest leaves or parts of shoots. Herbivores can also live on or within a plant and consume some of the plant's resources. Insects called aphids, for example, extract dissolved sugars and other nutrients directly from the phloem; pathogenic microorganisms can parasitize a plant, depleting its resources over time. Some plants are parasites on other plants, tapping water, sugars, proteins, and other resources for their own use.

The effects of herbivory on the plant depend, among other things, on what parts of the plant are consumed. Removal of or damage to roots can reduce or prevent the plant's uptake of water and mineral nutrients and can make the plant more vulnerable to being toppled by wind, flooding, or soil erosion. Consumption of leaves reduces the photosynthetic surface area, while removal of phloem sap may reduce the energy and materials available for growth and reproduction. Consumption of leaves, stems, and twigs may alter the competitive relationships among neighboring plants. Removal of meristems may alter the growth form of the plant. Consumption of flowers, fruits, and seeds may reduce the potential contribution of the plant to the next generation. Of course, to the new individual in each seed, consumption means death. Alternatively, fruits are often consumed without damage to the seeds, in which case the frugivore may disperse the seeds to potentially favorable locations.

Also important is the life history stage at which the plant is attacked or damaged. Seedlings are particularly vulnerable to herbivores. One mouthful for the herbivore can kill a seedling, but hardly affect a more mature plant. Grazing on grasses that have just begun to flower can critically affect their ability to produce seeds, whereas heavier grazing after seeds have been shed may have less of an effect on population dynamics.

How much do herbivores eat, on average? It has been estimated that about 10% of the leaves of forest trees are lost every year to herbivores (Coley and Barone 1996). Herbivory is greatest in dry tropical forests, somewhat less in tropical rainforests, and least in temperate forests. Young leaves tend to be eaten more readily than mature leaves in the Tropics. As you might expect, there is tremendous variation among species, locations, and years in the degree of damage caused by herbivores.

Can herbivory ever actually help plants to grow or reproduce? In the 1980s and early 1990s, a group of scientists postulated the existence of **overcompensation**, in which plants purportedly respond to herbivory by growing more (McNaughton 1983). (If a plant's hypothesized extra growth in response to herbivory resulted in no net difference between grazed and ungrazed

individuals, it would be called **compensation**.) The researchers suggested that overcompensation was due to the coevolution of plants and herbivores, particularly grasses and mammalian herbivores. Buffalo saliva and urine, for example, were thought to contain growth stimulants for grasses (Detling et al. 1980). These ideas were highly controversial and received a great deal of attention; it seemed difficult for many ecologists to believe that being eaten could actually be a good thing for plants.

While there was some experimental evidence for these ideas, when taken as a whole, they were not well supported by the available data (Belsky 1986; Belsky et al. 1993). One of the possible explanations for reported overcompensation was that researchers had measured only aboveground plant parts, while underground reserves may have been depleted to stimulate the observed aboveground growth. Long-term herbivory might result in significant depletion of these underground reserves, ultimately harming the ability of the plant to recover from subsequent bouts of herbivory. Another explanation was that in dense plant stands, if herbivores eat only shaded, unproductive understory leaves, there may indeed be no reduction in the photosynthetic capacity of the plant, and thus no negative effects of herbivory. Joy Belsky and her associates (Belsky et al. 1993) argued that overcompensation seemed to occur mainly in the experimental treatments most favorable to plant growth, such as the combination of high nutrient availability and reduced competition.

Ecologists now generally believe that herbivory usually harms individual plants (Hawkes and Sullivan 2001). While there are cases of overcompensation, on average, herbivory reduces both growth and reproductive output, resource addition increases both, and there is no evidence of an interaction between the two effects (Figure 11.1). One interesting observation is that plant responses to herbivory depend to some degree on their growth forms and phylogenies: for example, monocots with basal meristems (such as grasses) have more regrowth after herbivory under high-resource conditions, while dicot herbs and woody plants regrow more after herbivory under low-resource conditions. Unfortunately, most studies have used only short-term measures of plant response to herbivory, not measures of lifetime fitness.

Herbivory and Plant Populations

The extent to which herbivores affect plant population dynamics is a highly controversial and unresolved question. Two different explanations have been put forward for why herbivores should not be important. The "top-down" school of thought argues that herbivores are maintained at such low densities by their own predators that they rarely exert negative effects on

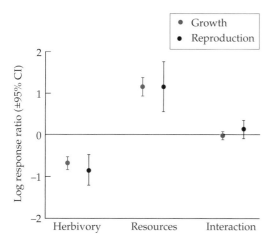

Figure 11.1
The effect of herbivory on both growth and reproduction is generally negative. The graph shows the average effects on growth in 82 studies and reproduction in 24. The response ratio is the plant performance when subjected to herbivory or treated with resources divided by plant performance without additional resources or herbivory; error bars are 95% confidence intervals. A negative ratio means that the treatment decreased plant performance, while a positive value indicates increased performance. Resource addition improved plant performance, but there is no evidence of an interaction between resource addition and herbivory. (After Hawkes and Sullivan 2001.)

plant populations (e.g., Strong et al. 1984). The "bottom-up" school argues that plant populations are limited by abiotic factors such as water, light, and soil nutrients, not by herbivores (Hairston et al. 1960; Slobodkin et al. 1967; Hairston and Hairston 1993).

In contrast, others have argued that herbivores do play an important role in controlling plant populations. In all likelihood, herbivory regulates plant population dynamics in some cases, while "top-down" or "bottom-up" processes dominate in others. More work is needed to determine not only under what circumstances each of these processes is important, but also what factors lead to the predominance of one kind of regulatory process over others, and when their interaction becomes important.

One of the most obvious situations in which plant populations are dramatically affected by herbivores is the killing of large stands of forest trees by insects. Outbreaks of lepidopteran larvae can cause massive defoliation, in some instances resulting in tree mortality. Repeated defoliation of oaks in the northeastern and midwestern United States by gypsy moths has caused die-offs of large numbers of trees (Davidson et al. 1999; see Figure 13.8). Trees in more mesic areas appear to be more likely to die from a single episode of gypsy moth defoliation than those in more xeric sites.

Bark beetles (family Scolytidae) are an important cause of conifer mortality in western North America and

the southeastern United States (Figure 11.2; Powers et al. 1999). While most bark beetles inhabit branches and trunks of trees already undergoing stress, such as those damaged by a lightning strike, some species attack and kill healthy trees. The beetles enter the tree by chewing a hole through the bark into the **cambium**, the actively growing layer under the bark, and lay their eggs there. After hatching, the larvae feed on the cambium, destroying it and the vascular tissue. Coniferous trees (particularly pines) usually respond to boring by oozing sap (also called pitch) into the wound, thus either suffocating the beetle or pushing it out of the hole. However, a massive attack by a large number of beetles appears to reduce the ability of the tree to "pitch out" the beetles. Stresses created by drought or injury may have the same effect in making trees more vulnerable.

Figure 11.2
A "ghost forest" of *Pinus albicaulis* (whitebark pine, Pinaceae) killed by *Dendroctonus ponderosae* (mountain pine beetle, Scolytidae) in an extensive outbreak during the late 1920s. The young trees growing on this site are mainly *Abies lasiocarpa* (subalpine fir, Pinaceae) and *Picea engelmannii* (Engelmann spruce, Pinaceae). Scolytid beetles cause extensive damage to conifers in western and southeastern North America. (Photograph courtesy of K. Kipfmueller.)

Bark beetles that are responsible for widespread tree mortality in North America include the Mexican bark beetle, the western pine beetle, the spruce beetle, the western balsam bark beetle, and the southern pine beetle. As their names suggest, bark beetles are fairly selective, generally specializing on one or a few species of conifers. The southern pine beetle, for example, primarily attacks *Pinus echinata* (shortleaf pine, Pinaceae), *P. taeda* (loblolly pine), *P. palustris* (longleaf pine), and *P. elliottii* (slash pine); the spruce beetle kills spruces. The beetles function with symbiotic fungal partners that also attack the trees, eliciting specific biochemical responses from their plant targets. We return to the effects of these fungal symbionts of bark beetles later in this chapter.

Chronic herbivory—herbivory that occurs over long time periods—can have large effects on plant demography. *Pinus edulis* (pinyon pine) trees subjected to chronic herbivory have reduced rates of growth, an altered shape, and produce male cones almost exclusively (Whitham and Mopper 1985). In a study of *Eucalyptus* in subalpine habitats in the Snowy Mountains of Australia, Patrice Morrow and Valmore LaMarche (1978) found that trees treated with insecticide experienced large increases in growth. Given the growth history of trees on that site, they concluded that growth was suppressed by chronic herbivory. These results imply that chronic herbivory can greatly reduce lifetime fitness. By spraying insecticides on experimental plants several times a year, Waloff and Richards (1977) showed that chronic herbivory reduced the seed yield of the British shrub *Sarothamnus scoparius* (broom, Fabaceae) by about 75% over 10 years. Similarly, in a study of the semelparous plant *Cirsium canescens* (Platte thistle, Asteraceae; see Figure 9.5), Svata Louda and Martha Potvin (1995) excluded inflorescence-feeding herbivores from experimental plants, which led to increases in the total number of seeds set, seedling density, and the number of flowering adults.

Herbivory and Spatial Distribution of Plants

The spatial distribution of plants can also be affected, or even determined, by herbivores. The role of granivorous rodents in the distribution of a range grass with large, nutritious seeds, *Achnatherum hymenoides* (Indian ricegrass, Poaceae), was studied in the desert of western Nevada. Indian ricegrass is common in sandy soils, but rare in adjacent rocky habitats (Breck and Jenkins 1997). The grass was able to survive and grow in both soil types when planted experimentally, although the plants grew taller on the sandy soils. However, rodents cached seeds only in the sandy areas, and this seed dispersal behavior appears to be an important factor in determining the distribution of this plant.

The distribution of the shrub *Haplopappus squarrosus* (yellow squirrel cover, Asteraceae) in California is strongly affected by granivory. Louda (1982) showed that *H. squarrosus* was most abundant in an inland transitional zone between coastal and interior climates, but produced more seeds closer to the ocean (Figure 11.3). This pattern appears to be mainly a result of increased insect granivory on the plants growing closest to the ocean.

Donald Strong and his colleagues have unraveled a complex set of interactions that apparently underlie the population dynamics of a shrub, *Lupinus arboreus* (lupine, Fabaceae), along the central coast of California (Strong et al. 1995, 1996). Large patches of this woody perennial periodically die off and are eventually regenerated from seed, so that the population fluctuates over time. The plants are killed primarily by subterranean caterpillars of the ghost moth, *Hepialus californicus*, which

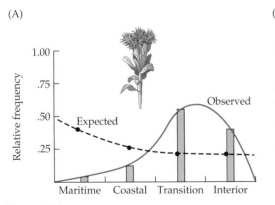
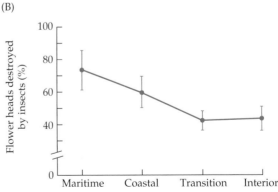

Figure 11.3
(A) Observed (bars) versus expected (dashed line) relative frequency of *Haplopappus squarrosus* (yellow squirrel cover, Asteraceae) over a climatic gradient in San Diego County, California. The observed distribution is based on the presence or absence of plants along 15,250 roadside segments, each 167 meters in length. The expected frequency is derived by assuming that relative population size should be proportional to relative seed production. (B) Percentage of flower heads destroyed by insect granivores in each climate zone; error bars are 95% confidence intervals. (After Louda 1982.)

bore into the roots. However, an insect-killing nematode, *Heterorhabditis hepialus*, with its symbiotic bacterium, *Photorhabdus luminescens*, is a highly effective predator on the ghost moth caterpillars. Consequently, lupine-dominated areas that are heavily colonized by the nematode are protected from attack by the caterpillar, while those areas without the nematode experience massive periodic die-offs. Few other studies have reported control of plant populations by underground herbivores, not to mention such trophic complexity, but then, few studies have looked for either.

Granivory

Granivory can also have important consequences for plant populations. In some populations, granivores consume a large fraction of the seeds. *Tachigali versicolor* (alazan, Fabaceae) is a monocarpic tree from Central America with unusually large (500–600 mg) seeds. In a detailed study of the seeds and seedlings produced by two large adult trees, Kaoru Kitajima and Carol Augspurger (1989) found that 51% to 83% of seeds died prior to germination, depending on the tree and the distance of the seed from the tree. The important granivores were bruchid beetles (eating 13% to 38% of seeds) and vertebrates (eating 0% to 59% of seeds). Seedlings had somewhat lower mortality rates in their first two months (24% to 47%, again depending on the tree and distance from the tree). Seedling mortality was primarily due to herbivory (6% to 17% of seedlings) and disease (3% to 25% of seedlings).

The chemical properties of seeds can deter or enhance granivory. Some seeds contain compounds that strongly deter granivores (we discuss such defensive compounds in more detail below). Seeds of *Erythrina* (coral bean trees, Fabaceae), for example, contain compounds that are strongly neurotoxic in vertebrates; as a result, they are rarely eaten.

On the other hand, a study of squirrel granivory on oaks provided some surprises. *Quercus rubra* (red oak, Fagaceae) has high tannin concentrations in its acorns, while *Q. alba* (white oak) does not. One might expect that gray squirrels (*Sciurus carolinensis*) would prefer the acorns of white oaks. However, Peter Smallwood and Michael Steele (Smallwood et al. 2001; Steele et al. 2001) showed that the interaction is more complicated. Squirrels prefer to eat white oak acorns, but prefer to cache red oak acorns because the tannins protect them from fungi and bacteria, making them less perishable. Moreover, the squirrels often remove the embryos of the white oak acorns before caching them. Thus, more red oak than white oak seedlings emerge from squirrel caches, and the squirrels are more effective as dispersers of red oaks—not because the squirrels avoid the tannin-laden red oak acorns, but because they actively prefer them!

Many plants—especially woody plants—show large and erratic variation among years in the size of the seed crop produced, and this variation is generally synchronized among most of the plants in a population. This phenomenon, called **masting**, has been widely explained as an adaptation to granivory. Various ecologists have hypothesized that during masting years, the large numbers of seeds overwhelm the capacity of granivores to eat them all, allowing at least some seeds to survive. This explanation has been questioned recently by other ecologists, who argue that other factors may be more important in selection for masting. They argue that masting can more easily be explained as an adaptation to wind pollination, which is increasingly efficient at high pollen densities (Smith et al. 1990; Kelly et al. 2001).

In a large survey of studies on variation in seed production, Carlos Herrera and associates (Herrera et al. 1998) argued that the distinction between masting species and non-masting species is false because there is a continuum of variability. Most iteroparous woody plants have variable seed production. Even when phylogenetic relationships are controlled for, there is slightly more variability in wind-pollinated species than in animal-pollinated species. However, these researchers also found more year-to-year variation in species dispersed by granivores or abiotic mechanisms (such as wind or water) than in those dispersed by frugivores. It seems clear that more studies are needed to assess the importance of pollination and granivore satiation in among-year variability in seed production.

Biological Control

Biological control is the deliberate use by humans of herbivores or pathogens to control populations of undesirable plant (or other) species. Many such biological control agents are introduced from other continents, and successful instances of biological control of plant pests offer examples of herbivores controlling plant populations. One well-known example is the introduction of the moth *Cactoblastis cactorum* to Australia to control the introduced and invasive prickly pear cacti *Opuntia inermis* and *O. stricta* (Cactaceae). The cacti had spread to cover vast areas in Australia, rendering them useless for sheep grazing (Figure 11.4). The moth was introduced in 1925 from Argentina, where its caterpillars were found to be specialist herbivores on prickly pears. By 1935, the prickly pear populations had been decimated, and they have remained at low levels since that time.

Unfortunately, the success of *Cactoblastis* in Australia is colored by recent problems with this species in North America. *C. cactorum* was introduced to the Caribbean in 1957 to control *Opuntia* species that had spread due to overgrazing. Subsequently, the moths invaded Florida and are now spreading rapidly in the southeastern United States, where they are affecting several endan-

(A)

(B)

Figure 11.4
(A) A dense stand of *Opuntia* in Queensland, Australia, prior to the release of *Cactoblastis*. (B) The same stand three years later. (Photographs by A. P. Dodd and the Commonwealth Prickly Pear Board.)

gered *Opuntia* species (Johnson and Stiling 1998). There is much concern that this invasion will spread to the southwestern United States and Mexico. *Opuntia* species are ecologically important in both countries, and are also economically important in Mexico.

A recent example of a similar phenomenon has been the introduction of several specialist insects to control purple loosestrife (*Lythrum salicaria*), an introduced, highly invasive species that now dominates vast areas of wetlands in eastern and central North America (see Figure 14.3). Bernd Blossey and his colleagues found that purple loosestrife was attacked by many different insects in central Europe, where it is native. These researchers imported populations of a root-mining weevil (*Hylobius transversovittatus*) and two leaf-feeding chrysomelid beetles (*Galerucella calmariensis* and *G. pusilla*) and released them in the United States and Canada in 1992 and 1993 (Blossey et al. 2001). These and other host-specific insects released later appear to be successfully eliminating the dense populations of purple loosestrife without attacking other species, allowing the recolonization of native wetland plants in areas where purple loosestrife had maintained monospecific stands. Purple loosestrife may be limited by herbivores in its native habitat. In Europe, it never reaches more than 5% cover and remains a minor component of wetland vegetation, in dramatic contrast to its spread in North America when it was introduced without specialist herbivores.

An important question, however, is the extent to which these and other biological control agents may affect nontarget species. The control agents released for purple

loosestrife were characterized as "specialists" because they prefer *L. salicaria* to related native North American species. However, these insects feed on native species when there is little purple loosestrife available, creating a potential for negative effects on the native species.

An agent introduced to control the severely invasive exotic plant *Carduus nutans* (musk thistle, Asteraceae) and its relatives in the midwestern United States has proved to have serious negative effects on rare native thistles. *Rhinocyllus conicus* (flowerhead weevil) was introduced to control *Carduus* species after studies indicated that this weevil prefers *Carduus* to native thistles of the genera *Cirsium, Silybum,* and *Onopordum.* However, despite this preference for the invasive musk thistle, studies by Svata Louda and her associates (Louda et al. 1997) showed that the weevil is also consuming, and is negatively affecting, three native thistle species—*Cirsium canescens, C. centaureae,* and *C. undulatum.* Ob-served levels of *R. conicus* infestation were as great or greater on the native thistles than on the exotic thistles, and were also greater than the levels of native insect infestation on the native thistles. As a result, infested plants set many fewer seeds. In the case of *C. canescens,* which is already uncommon and restricted in habitat, this new herbivore may present a danger to population persistence, as reduction in seed production is expected to reduce the population growth rate of this species (Louda et al. 1997). The best protection for the native *C. canescens* may be the maintenance of the invasive *Carduus* populations in the same habitat to attract the weevils away from the native thistle—thus the original target species may act as sort of a biological control agent for the weevils!

These studies, together with recent empirical studies (Henneman and Memmott 2001), have caused many biologists to begin reconsidering the safety of biological control in general (Simberloff and Stiling 1996). While biological control has proved strikingly successful in cases such as that of *Opuntia* in Australia, its strong point—the fact that the control agents disperse and reproduce on their own—also makes it potentially risky.

Effects of Herbivory at the Community Level

Consequences of Herbivore Behavior

Herbivores show preferences and exhibit behavior with respect to what they eat and how and when they eat it. Their behavior can have profound consequences for species richness and abundance. One important way in which herbivore behavior can affect plant community composition is the extent to which animals behave as generalists versus specialists with regard to what they eat. At one extreme, a pure generalist herbivore eats plants in the same proportions at which they are present in the community. At the other extreme, a pure specialist herbivore eats only plants (or even only specific plant parts) belonging to a single species or to a small group of closely related species. Extreme specialists are the most desirable choices for biological control agents to minimize effects on nontarget species.

Most herbivores are somewhere between these two extremes, preferring certain foods to others, but able to eat a variety of plants. Thus, it is misleading to strictly contrast specialists and generalists. However, it is often believed that generalist herbivores tend to promote or maintain species diversity because they keep faster-growing, dominant species from outcompeting

others. The effects of specialists, on the other hand, depend on the roles that their preferred food species play in the community. It is likely that a specialist on a potentially dominant species will have a very different effect on the community than a specialist on a less common species. Thus, herbivory can either increase or decrease the diversity of the plant community. The outcome depends not only on patterns of consumption, but also on interactions between herbivory, plant competition, and abiotic factors such as soil moisture, nutrients, and light levels. The effects of herbivory can vary over spatial scales and over the course of time.

Nancy Huntly (1987) studied the foraging behavior of pikas and the consequences of their foraging for the plant community. The pika, a small, territorial relative of rabbits and a strong contender for the world's cutest mammal, lives on high-altitude, rocky talus slopes in western North America. Pikas forage outward from their dens. They are generalists, but prefer certain plant foods to others. Pikas do not hibernate; instead, they collect large haypiles during the summer for winter use (Figure 11.5). Huntly experimentally excluded pikas from vegetation plots at different distances from their dens into the surrounding meadows. She found that the foraging animals had large effects on plant community composition close to their dens, where they spent the most time, with decreasing effects farther away.

Introduced and Domesticated Herbivores

One of the classic stories in the ecology of herbivory is that of rabbits, chalk grassland vegetation, and a rabbit disease, myxomatosis. Chalk grasslands are highly

Figure 11.5
Pikas (*Ochotona princeps*), relatives of rabbits and hares, are common in high-altitude habitats in western North America. Pikas feed on almost all the plant species that grow around their rockpiles. Because they forage from a central location—a den—their effect on the plant community is strong, but becomes weaker at greater distances from the den. As shown here, pikas collect haypiles during the summer for winter use. (Photograph courtesy of C. Ray.)

diverse plant communities found on limestone-derived soils in southern England. These grasslands alternate with woodlands, and they have been used for centuries to graze sheep and, beginning in the twentieth century, cattle. Rabbits were brought to England in medieval times as a source of food and for sport hunting, where they, well, bred like rabbits. They became widespread, attaining very high numbers after about 1850, undoubtedly due to direct and indirect human influences, including predator reduction. Rabbit densities varied greatly from one spot to another.

What effects have the rabbits had on the vegetation of the chalk grasslands? The father of modern experimental plant ecology, Arthur Tansley, observed that where rabbit populations were high, the turf was typically chewed down to a height of 1 to 2 centimeters, while that of sheep-grazed grasslands was typically 5 to 10 centimeters in height. He hypothesized that if sheep grazing and rabbit feeding were prevented, the grasslands would revert to forest. Tansley experimentally fenced plots of vegetation to exclude rabbits and sheep (Tansley and Adamson 1925). At first, there was a great increase in the growth of plants inside the exclosures, and many plants flowered abundantly inside the exclosures that never succeeded in reproducing outside the fences. After some time, however, perennial grasses, the preferred food of the rabbits, grew up and shaded shorter dicot species. Total biomass and average vegetation height increased substantially. The palatable grasses became more dominant, while less

Sir Arthur George Tansley

competitive plant species declined in abundance, but did not disappear altogether. In addition, plants not characteristically found in the chalk grasslands invaded some of the ungrazed areas. However, the predicted large-scale colonization by woody species did not occur, perhaps because the exclosures were too small or too far away from sources of tree seeds.

This experiment was repeated unintentionally on a larger scale some 30 years later, when a viral rabbit disease, myxomatosis, was accidentally introduced into Great Britain and nearly wiped out rabbit populations in the chalk grasslands and elsewhere. Immediately following the decimation of the rabbit population in 1954, many rare plant species were found in the chalk grasslands that had not been seen in years (Thomas 1960, 1963). These species had been selectively grazed by rabbits, never becoming large enough to flower and be noticed, or had been grazed as soon as they grew past the seedling stage. Various rare orchids and other showy species, such as *Helianthemum chamaecistus* (frostweed, Cistaceae) and *Primula veris* (primrose, Primulaceae), appeared and flowered in abundance. Some of these species had been common a hundred years before, prior to the great increase in rabbits. Other species decreased as the rabbits disappeared, either because they were outcompeted or because they had been favored by the nitrogen from the rabbits' urine. Tall grasses began to become more prominent, and woody species began to invade. When the rabbit populations recovered from the epidemic of myxomatosis in the early 1960s, the vegetation largely reverted to its previous state.

Striking pictures of the effects of grazing are offered by fence lines where one side is heavily grazed and the other ungrazed or lightly grazed (Figure 11.6). Like pikas, cattle are selective generalists, eating many

Figure 11.6
A fence line in northern Arizona. The area at the left has been grazed by cattle; the area at the right is not grazed. (Photograph by S. Scheiner.)

species, preferring some, and avoiding others. They usually avoid woody and spiny species, as well as those with toxic or noxious defensive chemicals. Cattle can poison themselves, for example, by grazing on species such as *Digitalis* (foxglove, Scrophulariaceae; see Figure 9.1) and *Astragalus* (locoweed, Fabaceae). Cattle can have dramatic effects on community composition as a result of grazing, particularly when their densities are high or when grazing occurs at particularly sensitive times of the year for plant recovery and regeneration (e.g., when grass seeds are ripening). Over time, particularly with heavy grazing, preferred plants such as palatable and nutritious grasses decline in abundance and are replaced by less edible species, drastically changing the composition and appearance of the plant community. Heavy overgrazing leads to bare patches of ground, weed invasion, and severe erosion, especially on slopes. The very landscape can be changed, with deep ravines and gullies replacing rolling, grass-covered slopes, as a result of long-term damage to the plants that once held the soil in place. These problems tend to be more severe in arid environments, but can occur even in mesic habitats. Other grazing animals, including sheep and goats, can cause similar effects. Problems caused by overgrazing are widespread in western North America, but also occur in Africa, the Mediterranean region, and Australia, among other places. Overgrazing by domestic animals has contributed to turning vast areas of grassland into shrublands or deserts.

Effects of Native Herbivores

Seed-eating and vegetation-eating small mammals have been shown to affect plant community structure in a number of arid environments. In the semiarid shrubland of northern Chile, Javier Gutiérrez and his colleagues used fences and netting to selectively exclude small mammals (principally the herbivorous degu, *Octodon degus*; Figure 11.7) and predatory birds (particularly owls) from large plots (Gutiérrez et al. 1997). The exclusion of degus resulted in an increase in the cover of shrubs and perennial grasses, an increase in the diversity of perennial species, and a decrease in annual plant cover. The researchers found some indirect effects of predator exclusion on the vegetation (presumably by allowing increases in herbivores), and they also documented strong effects of weather as well as interactions between weather and herbivore effects.

Interestingly, while the arid and semiarid vegetation of Chile and Argentina is primarily home to herbivorous and insectivorous species, with few granivores, seed-eating rodents dominate North American deserts. James Brown and his colleagues have conducted a series of very long-term experiments to exclude different small mammals (particularly heteromyid rodents) and ants from plots in the Chihuahuan Desert of eastern Arizona

Figure 11.7
The degu (*Octodon degus*) is a small rodent that is a principal herbivore in Chilean deserts. (Photograph courtesy of B. Lang.)

(Brown and Munger 1985; Brown et al. 1997). Over time, removing either rodents or ants caused substantial changes in plant species composition. Where rodents were removed, large-seeded species increased and small-seeded species decreased. Where ants were removed, the opposite results were found (Guo and Brown 1996).

Valerie Brown and her colleagues have carried out many innovative field experiments to investigate the effects of herbivores on plant communities. In one large study, they used insecticides to kill either aboveground or belowground insects in an early successional field in Great Britain (Brown and Gange 1992). The purpose of the experiment was to see how the effects of root-feeding insects on the plant community might differ from the effects of foliage feeders. They found that both aboveground and belowground herbivory by insects had major (but different) effects on the timing and direction of succession. In this field, the aboveground insects were largely sap-feeding Hemiptera, which preferred perennial grasses. Their herbivory suppressed perennial grass colonization, slowing succession. The removal of the insects led to a luxuriant growth of the grasses, which then shaded and replaced the lower-growing herbaceous dicots, leading to a steep decline in species richness. Underground, chewing insects in the Coleoptera and Diptera fed primarily on the roots of the dicots. Reducing belowground insect numbers led to the persistence of annual dicots and a great increase in colonization by perennial dicots, and consequently a great increase in species richness. The ordinary presence of these root-feeding insects thus speeds up succession by reducing the dicots and increasing the colonization of the field by grasses.

Herbivory is not the only way in which plant-eating animals affect plant communities. Other kinds of herbivore behavior can also change the environment and have strong effects on plant communities. Mammalian herbivores, in particular, create gaps and patches when making burrows and trails, and trample vegetation around available water sources. Domesticated cattle herds can cause severe damage to vegetation along streams and watercourses by repeated trampling. Elephants consume and trample enormous amounts of plant material. Tree canopy cover in Serengeti National Park in Kenya, for example, was reduced by about 50% by elephants (Pellew 1983).

Not all such effects of herbivores are negative. In tallgrass prairies in the midwestern United States, native bison create depressions where they roll in the dust. In the spring these areas contain small temporary pools of water, and in the summer they are inhabited by annual species that otherwise would be excluded by perennial grasses (Collins and Uno 1983).

Large herbivores can also affect plant distributions and species richness through strong effects on soil nutrients caused by their urination and defecation. In a study at Yellowstone National Park, David Augustine and Douglas Frank (2001) compared soil characteristics and community characteristics between ungrazed grasslands and grasslands grazed by large herbivores—elk, bison, and pronghorn. Species richness and diversity were greater in the grazed grasslands, particularly at very small scales (Figure 11.8).

Even smaller animals can have dramatic effects, particularly when they reach very high numbers. Lesser snow geese breed on the coastal marshes of the tundra of Hudson Bay and James Bay in Canada during the summer, migrating in autumn to the Gulf Coast of Louisiana and Texas. Over the past 30 years their population has grown tremendously, with midwinter counts climbing from about 0.8 million in 1969 to about 3 million in 1994. Their winter and summer habitat use has expanded greatly, both in area and in the variety of habitats they occupy (Abraham and Jefferies 1997). Much of this population growth appears to be caused by the response of the geese to changes in land use by humans. Expansion of farmland and the use of crops as food sources by the geese, both during migration and on their wintering grounds, have decreased overwinter adult mortality. Changes in predation rates may also be a factor. In summer, ever larger flocks of geese return each year to the tundra in excellent nutritional condition, capable of rearing large numbers of young. Snow geese are grazers; however, they are also capable of killing plants by pulling them up by their roots or rhizomes. Large flocks of the birds can have dramatic effects on the tundra landscape, not only changing plant species composition but also leaving entire areas bare of plants. It has been estimated that over a third of the tundra coastal marsh ecosystem has been severely overgrazed, threatening some 200 other species of water birds that summer

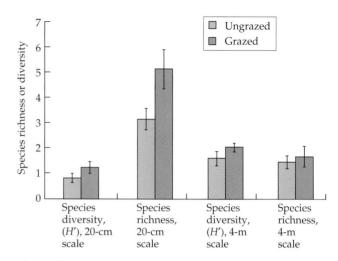

Figure 11.8
Plant species richness and plant species diversity, as measured by the Shannon-Weiner index (see Chapter 12), are greater in grazed than in ungrazed grasslands in Yellowstone National Park. The effect was greater for comparisons of 20 × 20-centimeter quadrats than for 4 × 4-meter quadrats, implying that the effects were occurring at a very small spatial scale. This scale effect was contrary to expectations because urination and defecation by large mammals is thought to lead to a very patchy distribution of soil nitrogen and nitrogen-mineralization rates. Species richness is the mean number of species per plot on the 20-centimeter scale, and the mean number of species per plot divided by 10 on the 4-meter scale. Error bars are ± 1 standard error. (After Augustine and Frank 2001.)

there (Abraham and Jefferies 1997). The greater snow geese that summer in the eastern Canadian Arctic and migrate to the southeastern United States are poised for a similar population explosion, with similar effects predicted for the vegetation in their breeding habitat.

Generality

How general and important are the effects of herbivores on plant communities? This question has been a controversial one. In a classic and much-cited conceptual paper, Hairston et al. (1960) suggested that herbivores are kept at low densities by their own predators, limiting their effects on plants. Widely referred to as the HSS hypotheses (after the authors' initials), this point of view has been controversial ever since it was proposed. William Murdoch (1966) argued that the logic of HSS is circular. Pointing to its argument that because they do not eat all the available plant material, herbivores must not be food-limited, Murdoch observed that one could apply the same logic to the next trophic level: because they do not eat all the available herbivores, carnivores must not be food-limited—exactly the opposite of what Hairston et al. argued was the case. Murdoch also argued that the HSS hypotheses were poorly defined and therefore untestable. Paul Ehrlich and L. Charles Birch (1967) argued that the HSS hypotheses hinge upon the belief that because species persist, they must be regulated at or near an equilibrium—a "balance of nature." Ehrlich and Birch argued that, to the contrary, most populations experience so much random environmental variation that their populations undergo large fluctuations.

Much research and debate has been stimulated by these ideas in the ensuing decades; to do justice to this body of work would require another book! Confining ourselves to the question with which we began, however—whether herbivory is important for plant communities—the answer seems clear. In a recent meta-analysis of a large number of experimental studies on herbivory, David Bigger and Michelle Marvier (1998) concluded that on average, herbivory causes a substantial reduction in plant biomass in natural plant communities. Contrary to the assumptions of many ecologists, invertebrates such as insects have a much greater effect than vertebrates.

Plant Defenses

Because plants cannot move, most plants have to "sit there and take it" from herbivores, resulting in natural selection for plants that are tougher, less palatable, and generally better defended. There are a great many different types of plant defenses, just as there are a great many different ways that plants can be attacked.

Plant Physical Defenses

Physical and mechanical defenses include obvious structures such as thorns and spines, which probably serve best to discourage mammalian browsers and birds, but do little to deter insects. Single-celled plant hairs—**trichomes**—serve many functions, including protection (Figure 11.9A). Insects are deterred by leaf hairiness and can be impaled by some trichomes. Other trichomes secrete noxious compounds that can deter vertebrates or sticky substances that impede insects. Species of *Urtica* (stinging nettles, Urticaceae) have brittle, elongate trichomes that break off when brushed, leaving a pointed fragment that pierces the skin and injects a painfully irritating fluid (Figure 11.9B). Trichomes of this kind have

(A)

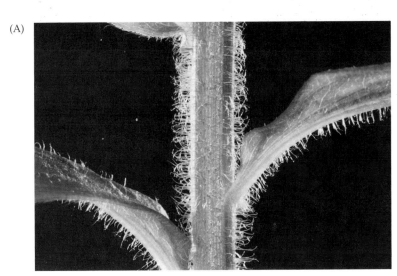

(B)

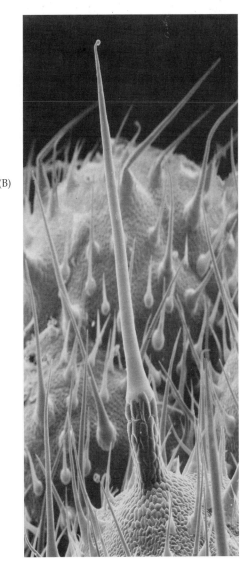

Figure 11.9
(A) A haze of trichomes covers the stem and leaves of this *Solidago* plant (goldenrod, Asteraceae). (Photograph © J. Wexler/Visuals Unlimited.) (B) Scanning electron micrograph of trichomes on the leaf of *Urtica dioica* (stinging nettle, Urticaceae). The long trichome in the center is a stinging hair; when its tip is broken off by some action (such as the touch of a human hand or foot), it easily penetrates the skin and injects the neurotransmitter acetylcholine along with histamines that produce allergic reactions. The shorter trichomes do not have these protective chemicals. (Photograph © A. Syred/Science Photo Library.)

evolved independently in four plant families: Urticaceae, Euphorbiaceae, Loasaceae, and Boraginaceae.

Other physical defenses include materials that make entry into the plant difficult, such as thick bark on trunks or roots or the tough coats that protect seeds and some fruits, such as nuts. Various cells and tissues that make up the plant body probably also serve a defensive function, such as the cap of thick **sclerenchyma** (a plant tissue with lignified cell walls) that surrounds the vascular bundles carrying food and water in young stems, and the waxy cuticle on the surface of leaves, which both reduces water loss and protects against fungal attack.

Leaf toughness presents an important mechanical barrier to chewing insects, mammalian browsers, and other herbivores, as well as interfering with tissue digestibility. Toughness is primarily a result of the content, type, and placement of sclerenchyma fibers (elongated cells), although short **sclereids** (sclerenchyma cells with thick, lignified, much-pitted walls) may be responsible for the hardness of some structures. Toughness may also depend on other thick-walled cell types, such as xylem and **collenchyma** (supporting tissue made of living elongated cells with walls thickened irregularly). Newly formed, expanding leaves are considerably less tough than older ones because the structural tissues involved in making leaves tough interfere with the leaf's ability to grow to full mature size. For this reason, new leaves are vulnerable to attack, and must depend on alternative means of protection, such as chemical defenses.

Grasses are unpalatable to most generalist herbivores because they sequester large amounts of silica, which makes them difficult to chew and difficult to digest (picture chomping down on ground-up bits of glass, or large amounts of sand). The silica is contained in specialized epidermal cells as well as in other plant parts. A few other taxa also use silica as a deterrent to herbivory, including horsetails (Equisetaceae) and palms (Arecaceae).

Another characteristic that has been suggested to act as a defense against herbivores is the nutritional quality of leaf tissue. Eating leaves with a low nitrogen and water content results in poor growth and survival for herbivores. Leaves with a higher nitrogen and water content are usually preferred by herbivores over those with a lower content (all else being equal). This fact poses an evolutionary dilemma for plants: The metabolic enzymes responsible for growth and photosynthesis contain nitrogen. Reducing the concentration of nitrogen-containing compounds in the leaves to discourage herbivory could result in reduced photosynthesis and growth.

Nitrogen is almost always limiting to plants, but it is even more limiting to herbivores. Animals need much more nitrogen to function than plants do, and they maintain much higher concentrations of nitrogen in their bodies. Plants growing in soils with low available nitrogen may have higher concentrations of carbon relative to nitrogen in their tissues, reducing their nutritive value. As global CO_2 concentrations increase because of human emissions (see Chapter 22), many herbivore populations are expected to decline because of the relative decline in carbon to nitrogen ratios, and thus the nutritional value, of many plants. Recent studies by Peter Stiling and his associates (Stiling et al. 1999) found that several species of leaf miners consumed more tissue from oak leaves grown under enriched CO_2 conditions, but nevertheless had greater rates of mortality (Figure 11.10).

Figure 11.10
An experiment comparing the performance of leaf miners of the genera *Stigmella, Cameraria,* and *Stilbosis* on *Quercus myrtifolia* (myrtle oak, Fagaceae) grown under ambient versus enriched CO_2 conditions. (A) The insects removed more tissue from the plants grown under enriched CO_2 conditions (error bars are ± 1 standard error.) (B) Nevertheless, the insects feeding on those plants had higher mortality rates from a number of causes (error bars are ± 1 standard deviation). (After Stiling et al. 1999.)

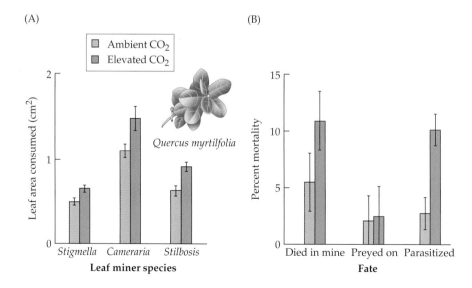

Plant Secondary Chemistry

Plants can marshal an arsenal of chemical weapons in response to herbivores. Chemical ecologists distinguish between **primary metabolites** and **secondary chemicals** (or **secondary metabolites**). Primary metabolites (such as sugars, amino acids, and DNA) are compounds necessary for the basic functioning of the plant. Secondary chemicals constitute a broad group of compounds that serve a wide variety of purposes, including defense and attraction of pollinators, rather than primary functions such as photosynthesis. Secondary compounds are generally found only in particular species or groups of species, and often only in specific organs or tissues. The distinction between primary and secondary compounds is somewhat arbitrary; some primary metabolites, for example, are also used in defense. The terms are to some extent a holdover from an earlier era when secondary chemicals were believed to be waste products by plant physiologists and biochemists unaware of their ecological functions.

The three major categories of defensive secondary chemicals are **phenolics**, **alkaloids**, and **terpenes**. These categories are not exclusive as some large molecules can contain subunits of more than one type. Additional defensive compounds include toxic proteins and amino acids, protease inhibitors, and cyanogenic compounds. Phenolics include a large variety of chemicals consisting of an aromatic ring with an attached hydroxyl group, —OH (Figure 11.11). Probably the most important defensive phenolic compounds in angiosperms and gymnosperms are the **tannins**, which reduce the digestibility of plant tissues. They are packaged in cell vacuoles and are present in high concentrations in the leaves of many woody plants, such as those in the Fabaceae, Fagaceae, Myrtaceae, and Polygonaceae. Another important group of phenolic compounds are the **lignins**, which impregnate secondary (woody) cell walls, giving them structural strength as well as providing a barrier to attack by herbivores and pathogens. Other phenolics include poisonous saponins as well as the flavonoids and anthocyanins, pigments that also give flowers and fruits their colors.

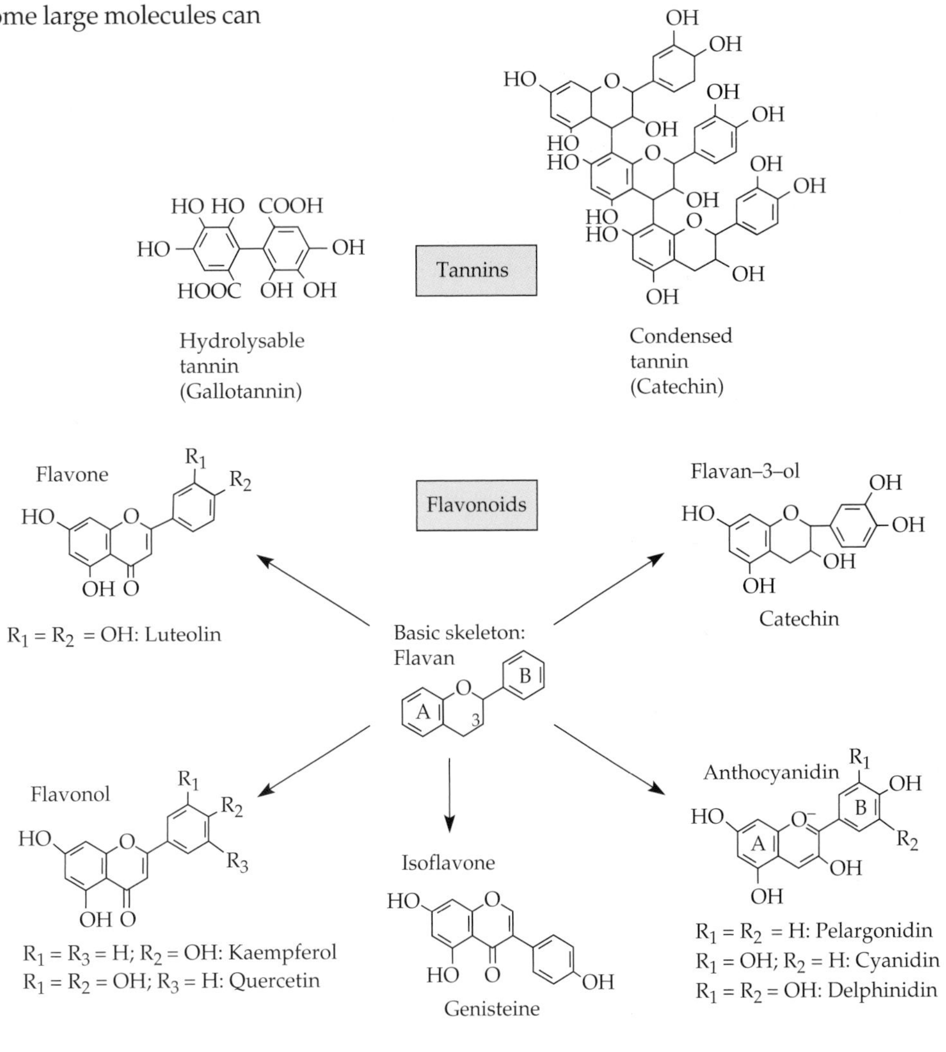

Figure 11.11
Defensive phenolic compounds include tannins and flavonoids. Examples of specific ones are given below the structures. (After Larcher 1995.)

The alkaloids are another broad group of compounds and include many that are used as pharmaceuticals (Figure 11.12). Some 10,000 alkaloids have been isolated and their structures analyzed. Alkaloids are relatively small molecules that contain nitrogen. They have a bitter taste, and many are toxic to herbivores. Alkaloids are generally highly specific to the plant species or group of species in which they are found. They are often effective in small quantities—as in the cases of cocaine, nicotine, and caffeine—although in some cases particular plants or plant parts produce high concentrations of alkaloids.

Terpenes are found in all plants, and an individual may contain many different terpenes (Figure 11.13). Terpenes play a wide variety of functional roles in plants. They are composed of multiple units of the hydrocarbon isoprene (C_5H_8), and they may be large or small, depending on how many isoprene units they contain. The compound isoprene is emitted in sunlight by the leaves of some plant species (such as *Eucalyptus*), sometimes in large quantities, and may serve to protect them from heat damage. The **latex** (milky sap) found in members of the spurge (Euphorbiaceae) and dogbane (Apocynaceae) families contains toxic terpenoid defenses. The oils that are responsible for the characteristic flavors and scents of the mints (Lamiaceae) are terpenes; they deter herbivores and reduce the growth of bacteria and fungi. Cardiac glycosides can cause heart damage in vertebrates and are poisonous to many insects; these terpenes are produced by many unrelated plant species, including species in the Scrophulariaceae (such as *Digitalis*, foxgloves) and the Apocynaceae (such as *Asclepias*, milkweeds, and *Apocynum*, dogbanes). Phytoecdysones are terpenes that mimic insect molting hormones and alter the development of insect larvae; they are produced by a number of taxa, including ferns, cycads, and some angiosperms.

Plants in the Brassicaceae (mustard family) contain characteristic secondary compounds, chiefly glucosinolates, which are effective repellents of most generalist herbivores. These mustard-oil precursors (responsible for the "bite" of the many cultivated species in this family) almost completely deter mammals and nonadapted insects. Glucosinolates are not as effective against insect herbivores that have become specialized on this family, however, and only high concentrations can offer some protection. Louda and her colleagues have found that glucosinolates not only can affect the interaction between plants and their herbivores, but also can indirectly control the distribution and abundance of the plants (Louda and Rodman 1996). *Cardamine cordifolia* (bittercress, Brassicaceae) is native to the Rocky Mountains of North America, where it grows only in shaded, moist forest edges. Bittercress may be restricted from sunny sites by greater vulnerability to chronic herbivory

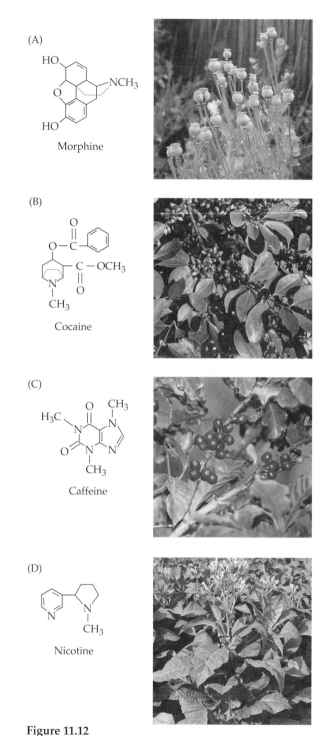

Figure 11.12
Examples of various alkaloids used by humans and their sources. (A) Opium comes from the milky sap of seed pods of *Papaver somniferum* (opium poppy, Papaveraceae); morphine is derived from opium. (Photograph © R. Shiell/Animals Animals.) (B) Cocaine is found in the leaves of *Erythroxylum coca* (coca, Erythroxylaceae). (Photograph © G. Dimijian/Photo Researchers Inc.) (C) Caffeine is found in coffee "beans," the fruits of *Coffea* (coffee, Rubiaceae), as well as many other plants. (Photograph © K. Fink/Photo Researchers, Inc.) (D) Nicotine comes from the leaves of *Nicotiana tabacum* (tobacco, Solanaceae) and other species in this genus. (Photograph © G. Grant/Photo Researchers Inc.)

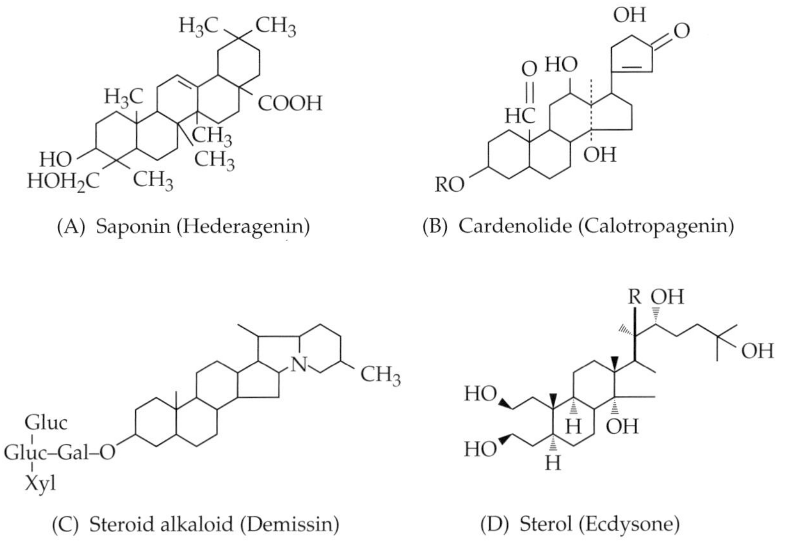

Figure 11.13
Triterpenes play important roles in plant defenses against herbivores. (A) *Hedera helix* (English ivy, Araliaceae) contains the saponin hederagenin in its leaves and fruits. (B) *Asclepias* (milkweeds, Apocynaceae) contains the cardiac glycoside calotropagenin. (C) *Solanum demissum* (potato, Solanaceae) produces the steroid alkaloid demissin in its leaves. (D) A number of taxa produce ecdysones, compounds that mimic insect hormones and interfere with insect metamorphosis. (After Larcher 1995.)

(A) Saponin (Hederagenin)

(B) Cardenolide (Calotropagenin)

(C) Steroid alkaloid (Demissin)

(D) Sterol (Ecdysone)

at those sites. In a series of experiments, plants grown in sunny sites experienced greater water stress, which reduced their glucosinolate concentrations. Consequently, those plants had more insect herbivores and suffered greater damage. Thus, herbivory may be controlling the plant's distribution by causing differential damage among microhabitats.

Constitutive versus Induced Defenses

Constitutive defenses are those that are present in a plant regardless of herbivore damage. They may be present throughout the life of a plant or may change as an individual grows and matures. **Induced responses** are elicited by an attack by herbivores. If these responses serve to protect the plant (whether or not they harm the herbivore), they are called **induced defenses**; if they have a negative effect on the attacking herbivores (whether or not they defend the plant), they are termed **induced resistances** (Karban and Baldwin 1997). Both physical and biochemical defenses can be induced. There has been a great deal of discussion regarding the relative benefits of constitutive versus induced defenses (Karban and Baldwin 1997; Lerdau and Gershenzon 1997). Historically, most research on plant responses to herbivory has focused on constitutive defenses, but beginning in the 1980s, interest in induced defenses greatly increased.

A classic example of an induced defense is the production of cyanogenic glycosides by *Trifolium repens* (white clover, Fabaceae) when the leaves are damaged; this response can also be induced by frost damage. Cyanogenic glycosides are an effective deterrent against snails, which are voracious herbivores in regions with mild, wet winters. The ability to produce these defensive compounds is controlled by two genes, and populations of white clover in Great Britain are polymorphic for the

trait (Dirzo and Harper 1982a,b). In the presence of snails, plants that had the ability to produce cyanogenic glycosides had higher survival rates than those that did not. In the absence of snails, however, those plants had reduced growth and reproduction compared with acyanogenic (non-cyanide-producing) plants. As predicted by evolutionary theory, areas with high snail densities had a higher proportion of cyanogenic individuals, while those with fewer herbivores had a preponderance of acyanogenic individuals.

In a study of *Lepidium virginianum* (pepper-grass, Brassicaceae), Anurag Agrawal (2000) showed that induced defenses were effective against a generalist herbivore, but not against an herbivore that specialized on that plant (Figure 11.14). Whether provided with a choice of food items or not, generalist caterpillars did substantially more damage to plants whose defenses had not been induced. However, specialist caterpillars produced the same amount of damage in induced and noninduced plants.

The production of sap by conifers in response to attack by bark beetles is an example of an induced response. In addition to the physical defense of the sticky, suffocating pitch, this response has a biochemical function. Coniferous sap contains substantial amounts of various terpenes (including monoterpenes and sesquiterpenes) as well as phenolics and other compounds that are particularly toxic to the symbiotic fungi that attack with their beetle partners. Raffa and colleagues (1987, 1991) found that wounding a tree mechanically caused the production of small amounts of monoterpenes, while trees attacked with living fungi responded with massive amounts of monoterpenes. Trees responded most strongly to those particular fungi associated with the bark beetle species that ordinarily attack them.

(A)

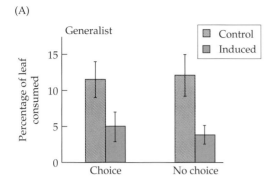

(B)

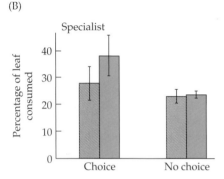

Figure 11.14
Induced defenses in *Lepidium virginianum* (pepper-grass, Brassicaceae) were effective against caterpillars of (A) generalist herbivores (noctuid moths), but not against caterpillars (B) of the specialist butterfly *Pieris rapae* (Pieridae). The defenses were induced by allowing a fixed number of larvae to feed on plants several days before the experimental trials. The amount of damage was the same whether or not the caterpillars had a choice of induced or noninduced plants. Error bars are ± 1 standard error. (After Agrawal 2000.)

Many different classes of chemicals have been found to act as induced defenses. Ian Baldwin and his associates (Baldwin 1988, 1991; Karban and Baldwin 1997) have done much to increase our understanding of the process of induction. In some cases chemical defenses are induced by highly specific cues, such as caterpillar saliva. In other cases any mechanical damage is sufficient. The toxic alkaloid nicotine, for example, is produced in response to mechanical damage to the leaves of *Nicotiana* (tobacco, Solanaceae). Nicotine mimics the neurotransmitter acetylcholine, blocking the receptor for that neurotransmitter in both insects and vertebrates. At low doses, nicotine stimulates the nervous system, but at higher doses it is a depressant, ultimately capable of causing paralysis and death. Smokers become addicted to the excitatory effects of low doses of this potent nerve toxin; caterpillars or grazing mammals may be paralyzed and killed by the same toxin. Small amounts of nicotine are present in undamaged tobacco plants. After leaf damage occurs, nicotine levels can increase

four to ten times, reaching concentrations that can kill an herbivore after a single meal.

Nicotine is synthesized in the roots of tobacco plants in response to leaf damage and transported via the xylem to the leaves. The signal indicating that the leaves have been damaged, cuing the roots to produce nicotine, is apparently a hormone transported from leaves to roots through the phloem. Greater leaf damage induces more nicotine production. Because of the long-distance nature of this response and the time needed for biosynthesis, it takes from ten hours to several days for nicotine to increase to its maximum levels. What might be the advantages of belowground synthesis of nicotine, given the vulnerability of the plant during this prolonged wait? If nicotine production occurred largely in the shoots, herbivores consuming the leaves might severely compromise the plant's ability to defend itself from further attack. The site of the synthesis of this defensive compound is thus protected belowground from leaf-eating herbivores. Even if large amounts of the aboveground parts of the plant are damaged or destroyed, the plant can continue to defend itself.

Some plants increase physical defenses in response to herbivory. Grass plants can increase their silica concentrations. *Opuntia* (Cactaceae) can secrete additional mucilage and fiber, effectively walling off insect herbivores.

Evolutionary Consequences of Plant-Herbivore Interactions

You may be wondering, after reading about the arsenal of defenses possessed by plants, why all herbivores do not simply starve to death. There are several reasons that they do not. First, fortunately for the rest of the biological world, plants are not uniformly well defended. Species, populations, individuals, life stages, and plant parts vary in their defenses, and the less well defended are often sought out and depended on for food. Second, some herbivores have evolved various means of overcoming plant defenses, leading to what has been termed a "coevolutionary arms race" between consumed plants and their would-be consumers. The behavior, biochemistry, and morphology of herbivores may contribute to their ability to overcome plant defenses. Some insects, for example, are capable of carefully avoiding the ducts and canals that carry toxic latex in certain plants. Even more surprisingly, some chrysomelid beetles neatly incise these canals "upstream" before beginning to feed, preventing the flow of the toxic compounds to the leaf they are consuming. Giraffes, with their tough mouths and phenomenally long tongues, freely browse *Acacia* (Fabaceae) in the African savanna, undeterred by the plants' abundant and sharp thorns (Figure 11.15). In the

Figure 11.15
Giraffes browse *Acacia* (Fabaceae) trees, despite the long, sharp thorns that protect these species. Browsed trees can give the impression of having been "shaped" into a topiary. (Photograph courtesy of E. Orians.)

northeastern United States, white-tailed deer avidly consume vines of *Smilax glauca* (catbrier, Liliaceae), at least when the stems are young, despite their ferocious hooked spines; it is not clear how the deer avoid being "clawed." Clearly, one cannot always depend on broad generalizations in making predictions about patterns of herbivory. One must understand the biology of the species involved.

A number of leaf-eating herbivores have symbioses with microbial species that can digest cellulose. Almost no animals can digest cellulose, a major component of plant tissue, but various bacterial species can. Ruminants (including deer, cattle, and antelopes) depend on bacterial fermentation of their food, which occurs in a specialized stomach compartment called the **rumen**; the animals receive nourishment from the fermented material and from digesting some of the bacteria. Cellulose-digesting bacteria are also symbiotic with other groups of mammals and, most notably, with termites.

Grasses and grazers provide examples of mutual adaptations between plant and herbivore. While many generalist leaf eaters are effectively discouraged from eating grasses, a large group of mammals has evolved

to depend on them as a primary food source. In most mammals, the teeth cease to grow in adults, but those that are adapted to eating grass have teeth that grow continuously. As the silica in the grass erodes their teeth, new growth replaces the worn material. Grasses are highly adapted to being grazed as well. Grasses, like all plants, grow from regions of actively dividing, undifferentiated cells called meristems (see Chapter 8). The meristems of grasses are generally located at ground level, out of the way of a passing grazer's teeth. Thus, the plants are able to rapidly regrow the tissue lost to grazers, at least under ideal conditions. Heavy grazing, however, may result in destruction of these meristems, or in plants being torn up by the roots. Low resource levels or repeated heavy grazing can also compromise the ability of the plants to produce new leaf tissues.

Many specialist herbivores are able to detoxify or sequester secondary compounds of their preferred host plants that would deter or kill other herbivores. It has long been known, for example, that the caterpillars of monarch butterflies (*Danaus*) are specialists on milkweeds (*Asclepias*, Apocynaceae) (Figure 11.16). The milky latex of milkweeds contains cardenolides, which are

Figure 11.16
A monarch caterpillar on *Asclepias syriaca* (common milkweed, Apocynaceae). (Photograph by S. Scheiner.)

highly toxic to most insects and act as cardiac poisons in vertebrates. While most herbivores avoid these plants, monarch butterfly larvae are capable of sequestering the toxins, accumulating them in their own bodies to protect themselves from the birds that are their own predators.

However, there is more to this classic evolutionary ecology story. Malcom and Zalucki (1996) found that plants of *Asclepias syriaca* (common milkweed, Apocynaceae) rapidly spike up their levels of cardenolides after they are wounded, but drop the levels immediately afterward. This specialized type of induction, reason the

investigators, allows the plants to kill small monarch larvae, which are sensitive to the cardenolides. But the subsequent decrease reduces the supply of cardenolides to the bigger, older caterpillars, which have the ability to sequester the chemical. Other specialist herbivorous insects also have evolved the ability to sequester plant secondary compounds and use them for their own protection, such as cucumber beetles that use cucurbitacin from cucumbers to deter invertebrate predators.

Plant defenses can have complex effects. Producing a defensive compound, for example, can affect not only herbivores, but also pollinators. Pointing out that herbivory and pollination can interact in complex ways, Sharon Strauss (1997) suggested a conceptual model of the effect of herbivory on plant fitness (Figure 11.17). Her ideas were tested in a study of the annual plant *Castilleja indivisa* (Indian paintbrush, Scrophulariaceae). *Castilleja* (like many species in its family) is a root **hemiparasite** (its roots facultatively tap into the roots of other plants and divert mineral resources and secondary compounds). Lynn Adler and her associates (Adler et al. 2001) grew some *Castilleja* plants on a strain of *Lupinus albus* (Fabaceae) that produces a high concentration of alkaloids and some on a strain that produces a low concentration. Half of the *Castilleja* plants from each treatment were then sprayed with an insecticide, which both deterred herbivores and limited pollinators. That is, the experiment consisted of high and low alkaloid levels combined with or without pesticides. As a result, the researchers were able to estimate the fitness effects of herbivory (seed production), both directly and indirectly through its effects on pollination. The alkaloids had a positive effect on plant fitness because they reduced herbivory on buds, leaving more flowers and increasing pollinator visitation rates.

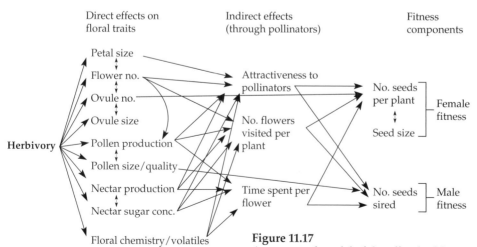

Figure 11.17
A conceptual model of the effect herbivory exerts on fitness through its effect on floral characteristics. Single-headed arrows indicate a causal relationship; double-headed arrows indicate a correlation. (After Strauss 1997.)

Pathogens

For reasons that are not fully apparent, the ecology of plant-herbivore interactions has received much greater attention than the ecology of plant-pathogen interactions. Yet plants are clearly subject to diseases, and they defend themselves against disease organisms. Pathogens can affect plants at all of the ecological levels, and in most of the ways, discussed previously for herbivores. In one of the best-known examples, chestnut blight, caused by the fungus *Cryphonectria parasitica,* virtually eliminated *Castanea dentata* (American chestnut, Fagaceae; Figure 11.18), a forest tree species that once dominated much of the eastern deciduous forests of North America. By doing so, the fungus drastically changed the composition of these communities. Today a number of oak species (*Quercus,* Fagaceae) dominate these forests.

As with herbivory, seedlings are particularly vulnerable to pathogens. A fungal disease called damping off, for example, kills many plants at the seedling stage. Damping off is an important cause of seedling mortality in tropical forests (Augspurger 1984). One explanation for the high local diversity in tropical rainforests is the effect of such fungal pathogens. If the fungal pathogens are species-specific, Connell (1978) argued, they are likely to be found at high concentrations around maternal parents, leading to strong selection for dispersal away from the parent plant.

Fungi, bacteria, and viruses are the most common infectious agents in plants. An important class of plant pathogens just beginning to be understood is the **mollicutes**, tiny wall-less bacteria that includes the somewhat more familiar mycoplasmas. The mollicutes include important pathogens not only of plants, but also of humans, other mammals, and arthropods. **Spiroplasmas** (helical, motile mollicutes) can be transmitted between insects and plants (including major crops) and can cause disease in both.

Some of the chemical defenses outlined above for herbivory also, or even primarily, serve as defenses against fungal, bacterial, and viral pathogens. Interestingly, it appears that the biochemical cascade that signals herbivore damage, mediated chiefly by a group of plant hormones called jasmonates, is quite distinct from the series of biochemical events signaling pathogen invasion, which is generally mediated by salicylic acid.

Responses of Individual Plants to Infection

Some diseases rapidly kill large numbers of plants. *Phytophthora infestans,* the fungal pest that causes potato blight, can kill a field of potatoes within a few days under the right weather conditions. Most pathogens have less severe effects, however. The bacterium *Xanthomonas axonopodis* pathovar *citri* causes citrus canker; infected *Citrus* (Rutaceae) trees drop their fruits and leaves prematurely. In other species, fungi and bacteria lead to spotting of leaves and fruits, making fruit crops less marketable.

Phytoalexins are secondary compounds that act as specific defenses against pathogens. Phytoalexins are produced at the site of an infection to kill microbes. There are a host of different classes of chemical compounds known to function as phytoalexins, including many of the chemical defenses mentioned above. Through mechanisms that are not yet fully understood, plants can also

Figure 11.18
Castanea dentata (American chestnut, Fagaceae) tree in full flower in northern lower Michigan, far outside its normal range. This individual was undoubtedly planted, and only its great distance from natural populations has allowed it to avoid infection by the chestnut blight fungus and reach full size. (Photograph by S. Scheiner.)

acquire both localized and even systemic resistance to pathogens. The systemic resistance observed in plants is analogous to the acquired immunity humans develop to some diseases, although it is very different biochemically, physiologically, and evolutionarily.

A physical defense against infection is **phloem plugging**, in which the phloem clogs up in response to damage, preventing the spread of the infectious agent through the vascular system of a plant. Similarly, in some plants, cells in and around the infected tissue die, sealing off the area so that the infection will not spread. This tissue death reduces the amount of photosynthetic surface area and can therefore reduce the plant's growth rate.

More Complex Interactions

Insects spread many plant pathogens. Insects visiting flowers of *Silene* (campion, Caryophyllaceae), for example, spread *Ustilago violacea* (anther smut fungus). The infection prevents the anthers from producing pollen; instead, they produce fungal spores (Thrall and Antonovics 1995). Similarly, infection by the rust fungus *Uromyces pisi* prevents *Euphorbia cyparissias* (Euphorbiaceae) from flowering. Instead, the plant forms pseudoflowers that shed fungal gametes, which are dispersed by insects (Pfunder and Roy 2000). Some fungal infections are thus analogous to sexually transmitted diseases in animals.

In some cases, however, there is a closer symbiotic association between an insect and a pathogen. Bark beetles are associated with fungal species that harm their target trees more than the insects do. The fungi may be present on the insects' surfaces or may be carried in specialized structures. A tree is infected with the fungi when the beetles tunnel into the cambium. The spreading fungal mycelia cause "blue-stain disease," which destroys the tree's vascular tissue, killing it. The insects and fungi appear to benefit mutually from their concerted attack on the plant. Bark beetles are responsible for the spread of Dutch elm disease in North America, caused by the introduced fungus *Ophiostoma ulmi*, which has killed millions of *Ulmus americana* (American elm, Ulmaceae). Both beetles that are native to North America and introduced beetles from Europe are involved. The fungus kills the trees by blocking water flow in the xylem and by production of a toxin.

In a study of the interaction between soil-dwelling pathogens (fungi and nematodes) and plant succession in the Netherlands, VanderPutten and Peters (1997) found that the pathogens facilitated species replacement within the plant community. On coastal sand dunes, *Ammophila arenaria* (merram grass, Poaceae) is replaced as the dunes become stabilized by another grass, *Festuca rubra* ssp. *arenaria* (sand fescue). In a series of experiments, the researchers found that the soil pathogens severely reduced the competitive ability of merram grass relative to sand fescue. In the absence of the pathogens, merram grass was not competitively inferior to sand fescue. Thus, the interaction between competition and the effects of pathogens may be responsible for the decline of the merram grass on stabilized dunes and its replacement by sand fescue.

Herbivory can also interact with the effects of pathogens. Bowers and Sacchi (1991) studied the effects of herbivore exclusion in an early successional old field in Virginia. A dominant plant species, *Trifolium pratense* (purple clover, Fabaceae), increased in response to herbivore exclusion. However, these high-density, ungrazed clover populations became severely infected with the fungus *Uromyces trifolii*, which killed many of the plants. In contrast, grazed areas had much less disease. As a result, in the year following the infection, the density of clover plants was much higher in the presence of grazing than in its absence. Other factors affecting the progress of disease in natural populations include plant genotype, microhabitat, and the diversity of the plant community, with the presence of neighbors of other species reducing infection rates (Morrison 1996).

Summary

Plants are the foundation of nearly all terrestrial food webs and, thus, are eaten by a very wide range of animals and other heterotrophic organisms. In addition, plants can be infected by an array of fungi, bacteria, and viruses, many with pathogenic effects. Herbivory—the consumption of plant material—can either kill a plant or reduce its growth and reproduction, depending on what part of the plant is consumed and the life stage at which the herbivory occurs. Consumption can sometimes be beneficial, as when an animal consumes a fruit and disperses rather than destroys the seeds inside. Whether other types of herbivory are ever favorable to plants through the mechanism of overcompensation is disputed among ecologists, but empirical evidence suggests that herbivory is generally harmful to individual plants.

Despite herbivory, the world looks mostly green. We do not yet know to what extent herbivory controls plant population dynamics and community structure in general. It may be that most of the time herbivores are kept at low levels by their predators and so have little overall effect on plants. Alternatively, chronic herbivory may change relative abundances and population distributions. The effects of herbivory on plant populations are obvious in some instances, such as during insect outbreaks. Changes in herbivore populations can also result in changes in community structure, with formerly rare species becoming common. Generalist herbivores are thought to promote species diversity. Amounts of herbivory on leaves vary among biomes, with greater herbivory in tropical than in temperate forests. This observation suggests that the importance of herbivory may differ among communities. Animals can also affect plant community structure indirectly by manipulating the physical environment through activities such as digging or trampling. Pathogens can have effects similar to those of herbivores on plant population and community structure.

Plants defend themselves in a variety of ways. Some plants have tough leaves that resist attack, leaves that are covered with defensive hairs, or stems covered with spines. Other plants have leaves that are low in nutrients or contain substances that make it difficult for herbivores to digest them. To defend themselves against herbivores and pathogens, plants produce toxic substances such as phenolics, alkaloids, terpenes, toxic proteins, protease inhibitors, and cyanogenic compounds. As plants have evolved these defenses, herbivores have evolved countermeasures, although the extent to which one-to-one coevolution occurs is unclear.

Additional Readings

Classic References

Tansley, A. G. and R. S. Adamson. 1925. Studies of the vegetation of the English chalk. *J. Ecol.* 13: 77–223.

Hairston, N. G., F. E. Smith and L. B. Slobodkin. 1960. Community structure, population control, and competition. *Am. Nat.* 44: 21–425.

Rhodes, D. and R. Cates. 1976. Toward a general theory of plant antiherbivore chemistry. *Rec. Adv. Phytochem.* 10: 68–213.

Contemporary Research

Agrawal, A. A. 2000. Benefits and costs of induced plant defense for *Lepidium virginianum* (Brassicaceae). *Ecology* 81: 804–1813.

Augustine, D. J. and D. A. Frank. 2001. Effects of migratory grazers on spatial heterogeneity of soil nitrogen properties in a grassland ecosystem. *Ecology* 82: 3149–3162.

Blossey, B., L. C. Skinner and J. Taylor. 2001. Impact and management of purple loosestrife (*Lythrum salicaria*) in North America. *Biodiversity and Conservation* 10: 1787–1807.

Gutierrez, J. R., P. L. Meserve, S. Herrera, L. C. Contreras and F. M. Jaksic. 1997. Effects of small mammals and vertebrate predators on vegetation in the Chilean semiarid zone. *Oecologia* 109: 398–406.

Additional Resources

Burdon, J. J. 1987. *Diseases and Plant Population Biology.* Cambridge University Press, Cambridge.

Coley, P. D. and J. A. Barone. 1996. Herbivory and plant defenses in tropical forests. *Annu. Rev. Ecol. Syst.* 27: 305–335.

Karban, R. and I. T. Baldwin. 1997. *Induced Responses to Herbivory.* University of Chicago Press, Chicago, IL.

Rosenthal, G. A. and M. R. Berenbaum, eds. 1992. *Herbivores: Their Interactions with Secondary Plant Metabolites*, 2nd ed. Academic Press, San Diego.

12 Community Properties

*A*s we hike up a mountain in the central Rockies, we might start in a forest dominated by ponderosa pine (*Pinus ponderosa*), then soon find ourselves among lodgepole pine (*Pinus contorta*), then move on to spruce-fir forest (*Picea engelmannii* and *Abies lasiocarpa*) and end up in alpine tundra. Each area contains a very different collection of species, yet some species are found in many different areas, and boundaries can be hard to discern. In other environments, such as grasslands with patches of serpentine soils (soils derived from serpentine rock that are nutrient-poor and sometimes high in toxic elements; see Box 15A), differences in species composition between communities can be abrupt and dramatic. Humans also create boundaries that define communities. A vacant lot is a community, as is a park in the middle of an urban landscape.

In previous chapters, we looked at how species interact with one another and their environments; here, we explore the community-scale patterns created by those interactions. What determines the boundaries of a community? Are communities real entities with their own properties, or are they just happenstance collections of individuals and populations? This chapter investigates what a community is, and discusses methods for describing communities. The following chapters will cover some additional processes that create community patterns: disturbance, succession, dominance, and species invasions.

What Is a Community?

A **community** is a group of populations that coexist in space and time and interact with one another directly or indirectly. By "interact" we mean that they affect one another's population dynamics. This definition of "community" includes all plants, animals, fungi, bacteria, and other organisms living in an area. It seems simple enough, but ecologists often use terminology inconsistently, and unfortunately, traditional usage in plant ecology sometimes differs from that in animal ecology. For example, we speak of "plant communities" even though plants are only a subset of the community—we are ignoring the decomposers, herbivores, diseases, pollinators, and many other organisms. Fauth et al. (1996) have discussed some ways through this terminological tangle (Box 12A).

Two closely related older terms that were formerly in wider use in plant ecology have been incorporated recently into some modern vegetation classi-

Box 12A

Communities, Taxa, Guilds, and Functional Groups

A very confusing array of terms has developed to describe communities. Sometimes the same term is used in different ways by different ecologists, while in other cases different terms are given to the same concept. Here we describe a recently proposed scheme to define these terms.

This scheme defines groups of species based on three criteria: geography, phylogeny, and resource use, as shown in the accompanying figure. Geography in this scheme defines **communities**, sets of organisms living in the same place at the same time. **Phylogeny** is the pattern of relationships among species (or higher taxa) based on evolutionary ancestry. Phylogeny defines **taxa**, sets of organisms that share a common ancestor. Resource use defines **guilds**, sets of organisms that use biotic or abiotic resources in a similar way. The term "guild" is taken from animal ecology, but has been used by plant ecologists as well.

Plant ecologists often use the term **functional group** to describe a concept related to the guild. Functional groups can be defined in a variety of ways, depending on the application, but these definitions are all based on a set of traits that identify functionally similar species. The traits used to identify functional groups can be chosen informally or based on formal mathematical algorithms. Ecologists have used the concept of functional groups in a variety of contexts, including studies of the relationship between productivity and diversity (see Chapter 14) and attempts to reduce the number of types of plants that must be taken into account in global climate modeling. Extensive recent reviews of the concept of plant functional groups and ecological applications of this concept are provided by Lavorel and Cramer (1999) and Woodward and Cramer (1996).

Intersections of the sets described by the terms above define more narrow distinctions. The intersection of geography and phylogeny defines **assemblages**, groups of related organisms liv-

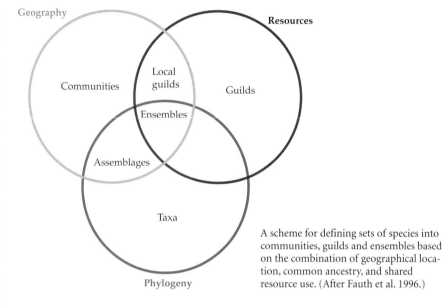

A scheme for defining sets of species into communities, guilds and ensembles based on the combination of geographical location, common ancestry, and shared resource use. (After Fauth et al. 1996.)

ing in the same place. The intersection of geography and resource use defines **local guilds**. The trees in a forest in Ontario are an example of a local guild: they include distantly related species such as sugar maple (*Acer saccharum*), an angiosperm, and eastern hemlock (*Tsuga canadensis*), a conifer. The intersection of all three sets defines **ensembles**. The grass species living together in a prairie are an ensemble. The annual Asteraceae in Australia's Great Sandy Desert are another ensemble.

Plant communities as traditionally defined are made up of a combination of ensembles, local guilds, and assemblages. Typically, terrestrial plant communities are defined as all the vascular plants living in a given space. Most species in this group are primary producers with similar resource requirements. So, for example, all of the grass species in a forest understory are one ensemble of that community. The combination of all understory forbs and graminoids would constitute a local guild, as these species would include more distantly related species of monocots, dicots, and possibly ferns. Some flowering plants are not primary producers, but parasites or saprophytes; however, these species are also includ-

ed in the plant community. The plant community could therefore be considered an assemblage, as it includes species that use different resources. The true community, of course, includes all species (e.g., animals, fungi, bacteria), not just plants. Thus, the traditionally defined plant community is actually a subset of the full community and has properties of ensembles, local guilds, and assemblages.

Monotropa uniflora (Indian pipes, Ericaceae) is an example of a flowering plant that is a saprophyte rather than a primary producer, obtaining its nourishment not by photosynthesis but from dead or decaying organic matter. (Photograph by S. Scheiner.)

fication schemes (see Table 16.3). An **association** was defined as a particular community type, found in many places and with a certain physiognomy and species composition; modern usage is not much different (e.g., Table 16.3 refers to the juniper-sage woodland association). The term **formation** was originally used to denote a regional climax community; modern usage is generally more specific, referring to a physiognomic subtype.

In practice, the boundaries of plant communities are usually defined operationally, based on changes in the abundance of the dominant, or most common, species. Sampling is then confined within those boundaries. A **stand** is a local area, treated as a unit for the purpose of describing vegetation. Typically, a number of stands are used to sample the presence and abundances of species as well as associated environmental variables. Based on data from a number of stands, the community can be characterized.

Only in special cases (e.g., islands, ponds, forest preserves surrounded by suburban development, vacant lots) are community boundaries defined easily. Even then, the movement of organisms and the transport of matter by wind and water make their boundaries fuzzy. Ecologists, therefore, are often of two minds when dealing with communities. On the one hand, we recognize that their boundaries are fuzzy; on the other hand, we often need to define discrete entities for convenience of analysis. Typically, plant ecologists define a community based on the relative uniformity of the vegetation and use their knowledge of species biology to decide when they are moving from one community to another.

Ecologists based in different countries, educated in different historical traditions, tend to view communities in somewhat different ways. In particular, ecologists in continental Europe were historically influenced by the floristic-sociological approach, most extensively developed by Josias Braun-Blanquet (see Chapter 16). This approach emphasizes the discreteness of communities. In contrast, most ecologists in English-speaking countries have been more strongly influenced by the history described in the following paragraphs; as a result, they tend to think of communities as blending continuously into one another. These distinct ways of thinking are becoming less prominent as a result of increased travel and communication among ecologists worldwide.

The History of a Controversy

Within plant ecology, there has been, and still is, a range of views on the nature of communities. The extremes are sometimes labeled the Gleasonian and Clementsian views, named after Henry A. Gleason and Frederic Clements, their first major proponents in the English-speaking world. These two extreme viewpoints differ in the importance they ascribe to biotic versus abiotic factors and predictable versus random processes in shaping community structure. Today, most ecologists hold

a middle ground between these viewpoints, and to a large extent have moved beyond both of them.

The Clementsian view was the majority view among English-speaking plant ecologists during the first half of the twentieth century. Clements saw plant communities as highly organized entities made up of mutually interdependent species. In his view (Clements 1916), communities are **superorganisms**—the analogue of individual organisms—that are born, develop, grow, and eventually die. Succession, in this view, is analogous to the process of development and growth; its trajectory and end point are highly predictable (Clements 1937). Two of the hallmarks of the superorganism concept were the presence of very tight linkages among species within communities and cooperation among the species in a community for the benefit of the function of the entire community.

Frederic and Edith Clements

Even at the height of Clements's influence, many ecologists held more moderate views. This version of Clementsian ecology asserted only that species interactions such as competition, mutualism, and predation are important in determining community structure. Clements himself acknowledged the effects of abiotic factors such as site history and soils in determining community composition. He focused on the idealized nature of communities, and saw them as spatially distinct, with one superorganism complex giving way to another with a very different collection of species. His major focus was on the nature and development of the community as a superorganism, however, rather than on the boundaries between communities. Some ecologists of the day accepted a more moderate version of this view, which admits that communities are not entirely discrete, but still divides them into nonarbitrary groups with recognizable boundaries.

In striking contrast to Clements, Gleason believed that communities are the result of interactions between individual species and the environment (biotic and abiotic factors) in combination with chance historical events. Each species has its own environmental tolerances and thus responds in its own way to environmental conditions (Gleason 1917, 1926). The implication of this belief was that along an environmental gradient, different species would have their boundaries at different places. Not only were communities not tightly linked superorganisms, but defining the collection of species living together in a particular place as a community was an arbitrary human construct.

Henry A. Gleason

According to the Gleasonian view, within the range of environmental conditions a species can tolerate, chance events determine whether a species is actually found in a given location. At the local scale, chance dictates whether a seed happens to get to a particular spot. On larger scales, the chance events of history play a strong role. For example, species in the family Cactaceae (the cacti) are found in the desert communities of the Americas because the family originated in this region, while deserts elsewhere have no cacti (except where they have been recently introduced by humans). Furthermore, the mix of species changes from place to place as one moves across the landscape. The Gleasonian viewpoint posits gradual changes in community composition as opposed to abrupt boundaries between communities unless there are abrupt and large environmental boundaries. A more moderate viewpoint was that some identifiable community types exist, but that these tend to blend into other community types.

Clementsian ecologists were not receptive to Gleason's ideas. While Gleason's work was well known, it failed to influence many plant ecologists until after Clements's death in 1945. Why Clements's views held sway for so long is not fully understood. Clements was said to have an extremely strong personality that could dominate scientific meetings. Given the very small number of plant ecologists active during the first half of the twentieth century, perhaps this alone was sufficient. Gleason, finding little interest in his ideas, abandoned his work in plant ecology by 1930 and spent the rest of his career as a taxonomist.

Not until the 1950s did the separate but almost simultaneous work of John T. Curtis and Robert H. Whittaker convince many ecologists that Gleason's views were largely correct. Curtis and his students mapped the vegetation of Wisconsin and looked at how species' optima and ranges were distributed along environmental gradients (Curtis 1959). The Clementsian theory predicts that species optima and ranges should show distinct groupings, whereas the Gleasonian theory predicts independence of optima and ranges. The researchers found just such independence; each species had a different set of environmental tolerances (Figure 12.1). A key innovation that contributed to this study was the development of ordination, a set of techniques for describing patterns in complex vegetation; we discuss ordination in detail in Chapter 16.

Whittaker (1956) also demonstrated that Gleason was right about the nature of boundaries between communities. One of the most striking patterns one encounters going up an elevational gradient is the dra-

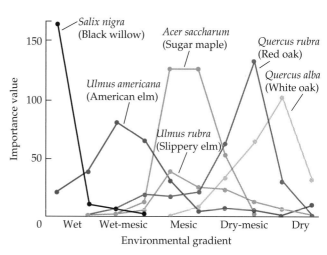

Figure 12.1
Change in the importance of various tree species along a moisture gradient in Wisconsin. Importance was measured as the sum of the relative cover, relative density, and relative frequency of a species in a community. (After Curtis 1959.)

matic turnover from one community type to another as altitude increases. Whittaker realized that if he could demonstrate that even along such a gradient, species turnover was gradual, this would provide very powerful evidence in support of Gleason's ideas and in contradiction to Clements's superorganism model. Whittaker did just that. He showed that forest communities along an elevational gradient in the Great Smoky Mountains of Tennessee changed gradually in species composition without abrupt boundaries (Figure 12.2). He then repeated this work in other areas, including the Siskiyou Mountains of Oregon and the Santa Catalina Mountains in southern Arizona.

Another important line of evidence that strongly affected many ecologists' views of communities was a series of studies conducted since the 1970s on the distributions of plant species during and after the most recent glaciation (see Chapter 21). Many of these studies looked at fossil pollen from lake bottoms. Some of the earliest and most influential of this research was carried out by Margaret Davis. She showed that many species that co-occur today did not always do so during glacial periods; rather, species were distributed among communities in the past in very different combinations from those that are found today (Davis 1981). For example, *Pinus strobus* (eastern white pine, Pinaceae), *Tsuga canadensis* (eastern hemlock, Pinaceae), *Castanea dentata* (American chestnut, Fagaceae), and *Acer* spp. (maples, Sapindaceae; maple pollen cannot be identified to species) are often found together in the eastern deciduous forest of North America today. However, these forest tree species were not always associated in the past. White pine and eastern

Robert H. Whittaker

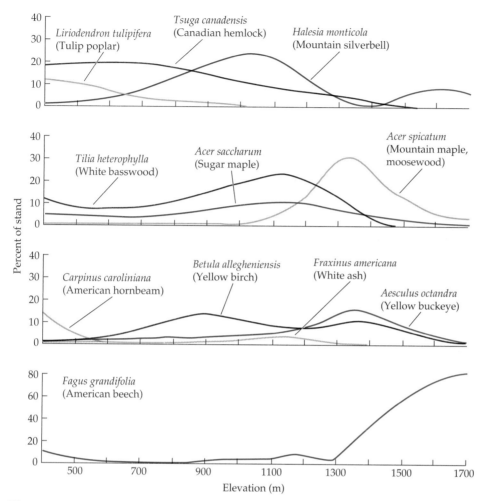

Figure 12.2
Changes in plant species frequencies along an altitudinal gradient in the Great
Smoky Mountains of Tennessee. (After Whittaker 1956.)

hemlock survived the Wisconsin glaciation (about 75,000 to 12,000 years before present) in a region east of the Appalachian Mountains, while during the same period, chestnut and maples were found near the mouth of the Mississippi River (Davis 1981), more than a thousand kilometers to the southwest.

Today most plant ecologists take a middle position between Clements's and Gleason's views, and in many ways have diverged from both of their views. There is wide agreement that species are distributed individualistically, and that community composition typically changes gradually along environmental gradients. Abrupt changes are most likely to be found where there are abrupt changes in the environment. However, abiotic boundaries and community boundaries do not always match. Because of processes such as dispersal from one habitat into another (see Chapter 17), a population may extend partway into an unfavorable environment. Abrupt changes may also reflect past events, such as the edge of a fire or a part of a forest that was

plowed at some time in the past. Thus, current environmental boundaries do not always match past boundaries. Ecologists still disagree as to the relative importance of biotic and abiotic processes and chance events in determining community structure.

Echoes of the controversy between the Clementsian and Gleasonian views of community among English-speaking ecologists of North America and the United Kingdom continue to influence ecologists schooled in those traditions. Related issues were debated among European and Russian ecologists, but not to the extent that they were argued in American and British journals; the ecologists of continental Europe and Russia had different concerns. Heavily influenced by Braun-Blanquet, they focused primarily on systems of community classification. We will return to their traditions in Chapter 16. The key difference is that the European tradition was more concerned with describing patterns than with analyzing processes, mainly sidestepping the arguments described here and in the next chapter.

A Modern Perspective on the Issues in Contention

The primary issues surrounding the nature of plant communities divide roughly into those of pattern and those of process. Underlying these issues is theory: the effort to explain the patterns and processes, which includes a search for the mechanisms responsible (see Chapter 1).

The issues of pattern focus on how species and communities are distributed over the landscape. Are boundaries among communities abrupt or gradual? How predictable are the patterns? Questions of pattern are critical because they set the stage for the rest of the debate. Once patterns are identified, theories can be constructed to explain them. In this chapter we examine ways of measuring patterns within communities; among-community patterns are dealt with in Chapters 16 and 17.

The issues of process focus on what processes actually function in natural communities and which of these are most important in determining the observed patterns. We know of many processes that influence community composition, including the physiological tolerances of species, competition, herbivory, biogeography, historical contingency, and random factors influencing colonization. All can be shown to operate at some times in some places; the question is their relative importance. Do some processes predominate in determining patterns? Is the relative importance of these processes always the same, or are some more important some of the time, or in certain situations, or in certain types of communities? One fundamental issue is whether communities are primarily static, exist at some sort of dynamic equilibrium, or are always changing. We will explore this issue in detail in Chapters 13 and 14. Finally, pattern and process are tied together by theories regarding the nature of communities—such as Clements's superorganism theory—which seek to explain the mechanisms responsible for the patterns and processes that are documented.

An overarching issue is the problem of scale. Recently, ecologists have paid increasing attention to the reality that different patterns and processes may exist and function at different scales. The processes that are important within communities at the scale of meters may differ from the processes that are important across biomes at the scale of thousands of kilometers. Throughout this book, we examine how scale affects both the patterns and the processes one finds, as well as the answers one reaches to questions of the relative importance of different causes and mechanisms.

In Clements's original superorganism theory, the community was an organic entity: a distinct unit with strong emergent properties. Species within the community were tightly linked and interdependent. In contrast, in Gleason's view, any community-level properties were simply the sum of the properties of individual species. Determining the truth of the matter requires documenting patterns (as Curtis, Whittaker, and others did), understanding the processes responsible for creating those patterns, and posing plausible explanations—elucidating the mechanisms causing those patterns and processes.

Whether communities can be considered to have emergent properties depends in part on the relative importance of biotic and abiotic processes in shaping community structure, including how strongly species interact with one another within communities. (We view the issues in somewhat different terms today than either Clements or Gleason viewed them, as we will see in Chapter 13.) **Emergent properties** are those that come about through interactions, such as competition, predation, and mutualism, that occur among the populations in a community (Box 12B). If these processes play a major role in shaping communities, then communities have at least some emergent properties. But if communities are mainly structured by the tolerances of individual species for abiotic factors (such as minimum temperatures) in the environment, then community properties are largely aggregates of the individual species' properties. It is important to note that one can recognize that emergent properties are important without accepting Clements's superorganism view, by discarding the ideas that communities consist of species adapted for one another's benefit and that all or most species in a community are tightly interlinked. Among modern ecologists, for example, E. P. Odum, H. T. Odum, R. V. O'Neill, and their co-workers have strongly emphasized the idea that communities and ecosystems have emergent properties, but none of them embraces the superorganism view (E. P. Odum 1971; H. T. Odum 1983, 1988; O'Neill et al. 1986).

On the whole, ecologists have shifted their attention to working out the mechanisms behind the patterns found within particular communities. This shift from the study of emergent properties to the study of mechanisms is emblematic of the seesawing between reductionist (mechanistic) and holistic (emergent) approaches that has characterized community ecology for the past century. Ecologists in the 1980s and 1990s tended toward reductionist approaches. Now there is some movement back in the direction of holistic studies under the mantle of macroecology (Brown 1995, 1999; Lawton 1999).

Are Communities Real?

As we have seen, the extent to which communities are "real" has been a contentious issue among plant ecologists for much of the twentieth century. The heart of the debate has been philosophical: What types of entities are real, and what types are just mental constructs? It is clear that populations and species are real entities, but are communities real entities as well, or are they merely convenient but arbitrary human inventions? In

Box 12B

A Deeper Look at Some Definitions: Abiotic Factors and Emergent Properties

Ecologists have distinguished between the effects of **abiotic** (nonliving) and **biotic** (living) factors in the environment for a very long time. Typical biotic factors include competition and predation; typical abiotic factors include soil nutrients, microclimate, weather, and general climatic influences. The problem with this terminology is that the distinction between abiotic and biotic factors may be far less clear than we might think. As we emphasize in Chapters 4 and 15, soils are a product of organisms and their interactions with the environment. The nitrogen available to the roots of a plant, for instance, depends on the actions of many different kinds of soil organisms and the interactions of the plant with those organisms. Likewise, the microclimate—and even global climate (see

Chapter 22) is affected by living things. We do not have a good substitute to suggest for the term "abiotic," so while we use it here to mean "things like climate and soils," we recognize that these may have major biotic components.

Another term that bears closer examination is "emergent properties." A central issue in the argument over the nature of communities is the question of whether emergent properties exist. An emergent property is one that is found at a certain level of organization due to properties, structures, and processes that are unique to that level of organization. Emergent properties can be contrasted with properties that are merely aggregates of properties at a lower hierarchical level. For example, the total biomass of all of the plants in a community is merely an aggregate

property of the biomass of each individual: We can determine community biomass by just adding up the biomass of all of the species in the community. While the growth of each individual may depend on the other individuals, community biomass is simply the mass of all individuals; community biomass is not an emergent property.

In contrast, consider canopy photosynthesis: We cannot measure the photosynthetic rate of an entire forest canopy just by measuring the photosynthetic capabilities of the individual plants. Canopy photosynthetic rates depend on how individuals interact in several ways, including shading one another, interfering with wind, and taking up CO_2. Canopy photosynthetic rate is an emergent property of the community.

---

the past, these questions often focused on the problem of boundaries—of identifying where a community begins and ends. Many studies have shown that, except where there are abrupt physical discontinuities, plant communities tend not to have discrete boundaries. At one extreme, this observation might be interpreted as implying that there are no community-level processes worth studying.

This conclusion would be wrong; in fact, the debate is miscast. Instead of focusing on patterns, let's shift the focus to processes and rephrase the question by asking whether community-level processes are important in structuring the living world. We have already discussed several processes responsible for interactions among species: competition (see Chapter 10), herbivory (see Chapter 11), and mutualisms (such as those discussed in Chapter 4). These are all community-level processes that occur among the component parts of communities (e.g., populations). If such processes are significant factors in the structuring of a particular system, then we can regard that system as a community with unique properties. Such an outlook eliminates the need to worry as much about the existence of clear boundaries between communities. We can recognize the existence of communities if it is useful to do so, and ignore

them if it is not. The debate as to whether communities are real entities or imaginary constructs is more than just an academic exercise. For example, The Nature Conservancy (TNC), based in the United States, makes decisions about land acquisition and restrictions on land use based on the classification of the communities present. In collaboration with the Natural Heritage network, TNC has devoted substantial effort to describing and classifying communities under the U.S. Natural Vegetation Classification (USNVC) system (see Table 16.3; Maybury 1999). Classifications of communities are even incorporated into law in a number of places. In southern California, for example, land development is regulated quite differently for "coastal sage scrub" than for "chaparral" communities, although by any definition the two include many of the same species, and different scientists may have used one name or the other to categorize the same place.

Describing Communities

Determining which processes are most important in shaping communities requires that we be able to describe and compare communities. What kinds of properties do plant communities have, and how can they be characterized? One set of community properties encompasses the numbers of species present and their relative abundances: species richness and diversity. A second set encompasses the physical structure of the plant community: its physiognomy.

Species Richness

One way to describe a community is by a list of the species in it. **Species richness** is the number of species on such a list. Because we often have information about the biology and ecology of those species, the list can point to many other types of information, such as the numbers of trees or herbs present, or the numbers of species in different taxonomic groups.

How would one gather such a list? A simple and widely used method is to establish the boundaries of the community and then walk through it, identifying all of the plants and listing them. Such a survey should be done several times during the year because some species may be visible only during a single season. Spring ephemerals, for example, are perennial plants common in temperate deciduous forests whose aboveground parts are present for only one to two months in the spring; during the rest of the year they exist only underground as dormant corms or rhizomes.

While simply finding and listing all species is useful, this method has limitations. Most importantly, if we wish to compare communities, we need comparable samples—otherwise we might find that two communities are "different" simply because we sampled one more intensively than the other. The area sampled can have strong effects on the number of species found. This sort of sampling effect is best dealt with by using plot-based methods, in which a series of sample plots, or **quadrats**, are marked out in a community and a list of species is collected for each. The quadrats may be any shape, such as square, rectangular, or round. They may be nested, contiguous, spaced along a line, placed in a grid, or placed randomly. These different arrangements can be used to ask different types of questions or to control for variation that occurs on different spatial scales (Krebs 1989).

By looking at how the total number of species found increases as quadrats are combined, one can examine the effects of area on species richness. The **species-area curve** (Arrhenius 1921; Gleason 1922; Archibald 1949) describes the increase in the number of species found as the area sampled increases (Figure 12.3A; see also Figure 17.2). This increase comes about for two reasons: first, as more individuals are sampled, the chance of encountering a new species increases; and second, a larger area is more environmentally heterogeneous. For a given community, if it has a relatively uniform environment, the number of new species found for each increment in sampling area decreases. Eventually, very few or no new species are found, and the species-area

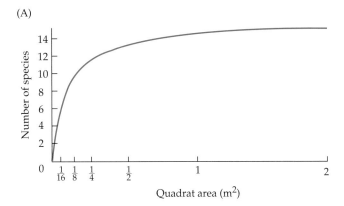

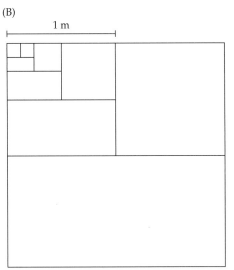

Figure 12.3
(A) A hypothetical species-area curve. (B) A system of nested quadrats used to determine a species-area curve.

curve reaches a plateau. Of course, if the area becomes large enough to encompass new environments, the curve begins to rise again.

The best mathematical model for the shape of the species-area curve has been debated over the years. Three candidates have been proposed: the exponential curve,

$$S = Z \ln(A) + C$$

the power curve,

$$S = CA^Z$$

and the logistic curve,

$$S = \frac{B}{C + A^{-Z}}$$

where S is the number of species, A is the area, and B, C, and Z are constants (Figure 12.4). The three equations are actually members of the same mathematical family, such that the exponential function is a special case of the power function—that is, they are equivalent when

$$C = \frac{Z \ln(A)}{A^Z - 1}$$

and the power function, in turn, is a special case of the logistic function—that is, the power and logistic functions are equivalent when

$$Z = \frac{\ln\left[(B-C)/C^2\right]}{\ln(A)}$$

Which of the three curves best fits a given data set? This depends on the area being sampled and the scale of environmental variation. For small or intermediate-sized areas, a logistic curve often fits bests, while for large areas a power curve is often the best fit.

These three models make different assumptions about the relationship of species number to area. Both the exponential and power functions result in curves that continue to rise indefinitely. The logistic function, on the other hand, eventually reaches a plateau. This distinction is important if one wishes to answer the question, "How many species are in the community?" If the species-area curve rises indefinitely with increasing area, then the answer depends entirely on the area of the community. However, the "area" of a community is often arbitrary, making the question meaningless. Instead, we can rephrase the question as, "How many species are contained in an area of size X?" This value is known as **species density**. If the species-area relationship is best described with a logistic curve, then this question has meaning, because the curve has a definite plateau or asymptote, calculated as B/C. The issue of the shape of the species-area curve will become important when we explore comparisons of diversity among communities in Chapter 16.

An alternative method of estimating species richness is to ask how many species are found in a sample of some number of individuals. For some functional groups, such as trees, in which individuals are easy to identify, this approach can be useful. Mathematical techniques, called **rarefaction**, exist for standardizing estimates among samples that include different numbers of individuals and for estimating the number of very rare species that were missed in the sample (Krebs 1989). These techniques were developed first and are used most often in animal ecology, although they are being used increasingly by plant ecologists. Because animals often move, area-based measures are less useful to animal ecologists. Plant ecologists, on the other hand, tend to emphasize area-based techniques because the clonal nature of many plants can make it difficult to distinguish individuals (see Chapter 8). In plant ecology, **dry weight** (the weight of a sample after it has lost all its moisture and is at a constant weight), or **biomass**, is often used instead of number of individuals.

How is the number of species in a community actually determined? One can use species-area curves to determine the total area required for standardized sampling. By using equal areas in different stands within a particular community, we avoid differences due to sampling intensity. We may also wish to compare different community types in a study using the same methods. For example, we might sample a beech-maple community and an oak-hickory community using the same approach. For very different kinds of plant communities, however, such as grasslands and forests, we might need to use different sampling areas and somewhat different techniques to best determine the characteristics of each.

How large an area should be sampled? We want a sample that is large enough to contain most of the species and which will minimize differences due to random sampling effects, such as where plots happen to be placed. On the other hand, for practical purposes, we do not want the area to be bigger than necessary. With respect to a species-area curve, we want to sample a total area for a given community that is at least as large as the point at which the curve begins to level off at a plateau, or, if there is no plateau, at which the rate of increase with increasing area is very small. We will return to the particular methods used for sampling vegetation shortly.

Diversity, Evenness, and Dominance

Species richness is only one aspect of diversity. Not all species exist in equal numbers: some are rare, some are common but not numerous, and others are very abundant. Imagine two forest stands, both of which contain a total of 100 individuals (or 100 kg of biomass) belonging to 5 different species. In one forest stand, there are 20 individuals of each species. In the other, one species has 60 individuals, while each of the other four species

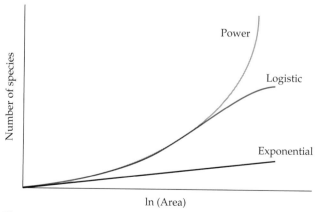

Figure 12.4
Examples of three differently shaped species-area curves: exponential, power, and logistic.

Community A   Community B

Figure 12.5
Samples from two different communities. In the sample on the left all the species have equal numbers of individuals; this sample has greater evenness than the sample on the right.

has 10 individuals. These two samples differ in a property called **evenness** (Figure 12.5). The first, in which the species are represented by the same number of individuals (or the same amount of biomass), is more even, and thus has one of the essential elements of being more diverse than the second. The species diversity of a community depends on both its richness and its evenness: higher species numbers, with the individuals (or biomass) more evenly distributed among them, contribute to higher community diversity. A way to think about evenness as a contributor to diversity is to consider the following thought experiment: Pick two plants at random from a community. Are they members of the same species or different species? In a highly diverse community, it is more likely that they will belong to different species.

In this example, the species that has by far the greatest number of individuals, or biomass, in the second community is the **dominant** species in that community. The first community, with greater evenness, does not have any dominant species. The greater the numerical preponderance of one or a few species, the lower the diversity of the community tends to be. Of course, a community may have one strongly dominant species and a large total number of species, giving it a high diversity value, but this is actually fairly unusual. Other ways of characterizing variation among species in abundance are considered in Chapter 14.

There are various measures and methods for expressing species diversity (Table 12.1). Which one is most appropriate in a given situation depends on what aspects of a community one wishes to highlight; each has different assumptions, advantages, and limitations. Two commonly used measures of species diversity (which combine the effects of species richness and evenness) are the Shannon-Weiner index and the inverse Simpson's index.

The Shannon-Weiner index is computed as

$$H' = \sum_{i=1}^{s} (p_i \ln p_i)$$

where s is the number of species, p_i is the proportion of individuals found in the ith species ($p_i = n_i/N$), n_i is the number of individuals of species i in the sample, and N is the total number of individuals sampled. This index assumes that individuals were sampled from a very large population and that all species are represented in the sample.

Simpson's index measures the chance that two individuals chosen at random from the same community belong to the same species:

$$L = \sum_{i=1}^{s} p_i^2$$

where the summation is over all the species. Both of these indices are commonly transformed so that they range from a minimum of 1 to a maximum of S, the total number of species in the sample, when all species are equally common. For the Shannon-Weiner index, this transformation is the exponential ($e^{H'}$), and for Simpson's index it is the inverse ($1/L = D$). The ratio $J' = H'/\ln S$, called the Shannon evenness index, provides a measure of evenness.

If L is the probability that two randomly chosen individuals are from the same species, then $(1-L)$ is the probability that they are from different species. This number is also known as the Gini coefficient. While the Gini coefficient has been used as a measure of evenness among individuals in a population, it could also be used as a measure of species evenness in a community (Lande 1996).

Which of these indices is more useful? This question has been the topic of much discussion, which has led to the development of numerous ways of correcting for **sampling bias** (departure of the estimated value for the sample from the true value due to systematic, non-chance causes). Lande (1996) showed that Simpson's

Table 12.1 Some species diversity and evenness indices

Index	Usual symbol	Formula	Emphasizes
Species number indices			
Species density	None	$\dfrac{S}{\text{area sampled}}$	
Margalef's index	D_{Mg}	$\dfrac{S-1}{\ln N}$	
Menhinick's index	D_{Mn}	$\dfrac{S}{\sqrt{N}}$	
Proportional abundance indices			
Shannon-Weiner index	H'	$-\sum[p_i \ln(p_i)]$	Rare species
Inverse Simpson's index	D	$1/\sum p_i^2$	Common species
Pielou's index	HP	$\dfrac{\log_2\left(\dfrac{N!}{\sum n_i!}\right)}{N}$	Rare species
Brillouin index	HB	$\dfrac{\ln(N!)-\sum\ln(n_i!)}{N}$	Rare species
McIntosh's U index	U	$\sqrt{\sum n_i^2}$	Rare species
McIntosh's D index	D	$\dfrac{N-\sqrt{\sum n_i^2}}{N-\sqrt{N}}$	Common species
Berger-Parker index	d	$\dfrac{N_{max}}{N}$	Common species
Cuba's index	DC	$S+1-\dfrac{\sum\left|n_i-\dfrac{N}{S}\right|}{2N}$	Common species
Q statistic	Q	$\dfrac{\frac{1}{2}n_{R1}+\sum n_r+\frac{1}{2}n_{R2}}{\log\left(\dfrac{R_2}{R_1}\right)}$	Rare species
Evenness indices			
Shannon evenness	J	$\dfrac{H'}{\ln(S)}$	
Brillouin evenness	EB	$\dfrac{HB}{\frac{1}{N}\ln\left(\dfrac{n!}{[\frac{N}{S}]!^{S-r}\{(\frac{N}{S}+1)!\}^r}\right)}$	
McIntosh evenness	EU	$\dfrac{N-\sqrt{\sum n_i^2}}{N-\dfrac{N}{\sqrt{S}}}$	

Definitions of symbols:

S = the number of species in the sample

p_i = the proportion of individuals in the ith species ($p_i = n_i/N$)

n_i = the number of individuals of species i in the sample

N = the total number of individuals sampled

N_{max} = number of individuals of the most abundant species

n_r = the total number of species with abundance R

R_1 and R_2 are the 25% and 75% quartiles

n_{R1} = the number of individuals in the class where R_1 falls

n_{R2} = the number of individuals in the class where R_2 falls

Note: While these indices were originally defined with regard to numbers of individuals, other measures, such as cover or biomass, can be used. For more information on diversity indices, see Magurran (1988).

index can be estimated without bias, but that estimating the Shannon-Weiner index requires actually knowing the true number of species. The unbiased estimate of D is $(1 - N)/(1 - NL)$. This lack of bias is an advantage for Simpson's index over the Shannon-Weiner index. Which index is preferable in a given study depends in part on what you wish to emphasize: the Shannon-Weiner index is sensitive to changes in the proportions of rare species, while the inverse Simpson's index is sensitive to changes in the proportions of common species (Table 12.2). If one had hypothesized, for example, that some environmental factor were having a negative effect on community diversity over time by disproportionately affecting rare species, reducing their abundance or eliminating them, the Shannon-Weiner index might pick up such effects more readily.

Just documenting that one community has a higher diversity than another community is not enough to conclude that the two communities are truly different. The communities may actually be identical, and the differences in diversity found might be due to chance sampling events. One more piece of information, a measure of the precision of our diversity parameter, is needed before comparisons can be made. Methods for estimating precision are described in the Appendix. A formula to estimate the variance for Simpson's index is given in Lande (1996), and the variance for the Gini coefficient can be estimated with methods in Lande (1996) or Dixon (2001). More detailed information on such topics as sampling effects, effects of quadrat size and shape, and sensitivities to various patterns of abundance can be found in Greig-Smith (1964), Pielou (1975), Magurran (1988), and Krebs (1989).

There are other methods besides the use of indices for quantifying diversity and comparing diversity among communities. These methods include graphical approaches based on comparisons of dominance-diversity curves (discussed more fully in Chapter 14). Other approaches have been used, but they go beyond the scope of this book.

Sampling Methods and Parameters for Describing Community Composition

Several techniques can be used to sample a community. Which to use in a given case will depend on the type of vegetation being sampled and the goal of the survey. Plant ecologists influenced by the Zurich-Montepellier school (the Braun-Blanquet approach; see Chapter 16) typically sample and compare multiple communities by placing a single large sample plot—called a **relevé**—in each stand. The relevé is located subjectively, with an attempt to place it within a uniform patch of vegetation that is representative of the community. How large should the relevé be? Ecologists determine the appropriate relevé size for a given type of community by using a set of nested relevés in one or a few communities, beginning with a small one and increasing the area sampled around it in a series of steps (see Figure 12.3B). As the total area is increased, a species-area curve is constructed until the curve reaches a plateau. That size relevé is then used for the remainder of the communities sampled (if they have similar enough characteristics and structure). Because European botanists carried out many of the early vegetation surveys in the Tropics, often the only available quantitative descriptions of these areas have been done using the relevé method.

Ecologists in English-speaking countries have long used the approach of sampling a given stand or community using some number of smaller quadrats, or sample plots, making the size and number of quadrats the same across stands or communities. These quadrats can be placed either randomly within the community or regularly along a grid. The advantage of using a number of smaller quadrats, rather than one large relevé, is that any local patchiness will even out across the entire sample. This method also provides a measure of local heterogeneity. The issue of regular (systematic) versus random placement of the quadrats has been heatedly debated, and involves considerations beyond the scope of this book (interested readers should consult the methodological references at the end of this chapter).

The size and shape of the quadrats is also an issue. Forests in North America are

Table 12.2 A comparison of three species diversity measures as applied to six communities, each containing five species

	Community					
	1	2	3	4	5	6
Species A	20	30	40	50	60	960
Species B	20	30	30	20	10	10
Species C	20	20	10	10	10	10
Species D	20	10	10	10	10	10
Species E	20	10	10	10	10	10
Sample size	100	100	100	100	100	1000
Species richness	5	5	5	5	5	5
$e^{H'}$	5	2.84	2.07	1.95	1.36	1.04
D	5	4.17	3.57	3.13	2.50	1.08
J	1	0.65	0.45	0.42	0.19	0.02

Note: The exponential Shannon-Weiner index and the inverse Simpson's index differ in how they change as evenness changes from sample 1 to sample 6. The numbers refer to numbers of individuals, but could also represent biomass.

often sampled using 0.1-hectare quadrats. The total area sampled should generally be determined by the species-area curve, as we saw above. As for quadrat shape, square quadrats are typically used, especially for small areas (< 4 m²). Usually the quadrat is marked with a rigid sampling frame, formerly made of wood or aluminum, now typically made of PVC pipe. Larger quadrats can also be set up using posts and measuring tapes in the field, particularly in forests. For the largest areas, circular quadrats may be more efficient. In this case, a post is placed at the center of the sampling area with a rope or tape measure attached to a swivel point at the top. The desired radius is marked on the rope, and a circular area is swept out. This method has the advantages of avoiding the problem of precisely determining 90° corners for a square or rectangular plot and of having a lower area-to-circumference ratio.

Ecological conditions may also dictate the shape of the quadrat. For example, on a steep hillside, if you wanted a given sample to be within a particular elevational contour, a narrow rectangular plot would be most appropriate. A narrow rectangle might also be preferable in a patchy habitat if one wanted to include more patches in the samples. With a narrow rectangle, however, there is a greater chance of miscounting because more plants are near the edge of the quadrat. The shape and placement of quadrats are known to affect the number of species found. Depending on the pattern of environmental heterogeneity, rectangular quadrats may contain more species than do square or circular quadrats of the same size.

A related approach to sampling vegetation is the use of **transects**: very long, narrow sampling areas along which species abundances are determined. Often transects are designed to run either across or parallel to gradients of environmental change, depending on the purpose of the sample. Transects are particularly useful in sampling very large areas. One kind of transect is just a line. In the line-intercept method, a line of a set length (e.g., 100 m) is laid through the community, and the proportion of the length along the line occupied by each intercepted species is determined. In a forest, a tape might be suspended at some height, and the cover of overstory species above the tape and understory species below the tape recorded along the length of the line.

Another method is the belt transect. As with the line-intercept method, a line of a set length is laid through a community. Now, all individuals lying within a belt of some set width on either side of the line (e.g., 0.5 m) are counted (belt transects can also be used to measure cover). A belt transect, in other words, is a long, thin quadrat. A long belt transect may encompass more topographic or soil variation, and thus may include more species, than a square grid for a given total area sampled.

Quadrats and transects can be used to estimate several parameters, based on species' presence or abundance in the samples. **Frequency** is the percentage of quadrats in which a species appears. Such a measure has the advantage that it can be done quickly, even by a single person. In particular, it is not necessary to be able to determine how many individuals of a given species are in a quadrat. The limitation of this method is that it is inaccurate for very rare species or species with clumped distributions and uninformative for very common species (those that appear in every quadrat).

Another quantitative measure is **cover**: the percentage of the ground covered by a given species. Cover may be measured as **basal cover** or **basal area**—the area occupied by the base of a plant, such as a tussock grass or a tree with a definable base—or as **canopy cover**. Cover values can tally to more than 100% if there are several layers of vegetation (e.g., in a forest, trees may cover 70% of the ground, herbs 50%, and shrubs 20%). Cover can be estimated in several ways.

The most precise, but most laborious, method of estimating cover is to map the vegetation. Efficient mapping techniques exist for large individuals—shrubs and trees. Alternatively, for herbaceous vegetation, if it is not too dense and the species are distinctive, a photograph can be taken from above. Areas of cover are then determined using image analysis software. The point-intercept or point-quadrat method for quantifying cover uses a sampling frame with a grid of strings or a frame that is used to drop pins at precise locations. The identity of the plant at each point where the strings cross or the pin is dropped is then determined. The total number of "hits" (points occupied) by a species is an estimate of its cover, given that enough points are sampled. In other words, if a species were found at 12% of the points, its estimated cover would be 12%.

Photographic and point-intercept techniques are intensive and time-consuming, and are best for small sampling areas. For larger areas or less intensive efforts, visual estimates are typically used. Humans are able to discern differences of about 10% by eye, so cover can be estimated to approximately the nearest 10% in this way. Nonequal cover classes can also be used (Table 12.3). The use of cover classes is an established part of relevé sampling.

A further quantitative measure is **density**, the number of individuals of a species per unit area. Density can be measured only for species with distinct individuals (e.g., trees) or where ramets are treated as individuals. For small plants, quadrats are used and all of the individuals are counted. For larger individuals, especially trees, plotless methods, such as the point-centered quarter method, can be used. To implement this method, a series of random points are located in the community,

Table 12.3 Three different systems for estimating cover by categorizing estimates into a limited number of classes

Braun-Blanquet		Domin-Krajina		Daubenmire	
Class	Cover range	Class	Cover range	Class	Cover range
5	75–100	10	100	6	95–100
4	50–75	9	75–99	5	75–95
3	25–50	8	50–75	4	50–75
2	5–25	7	33–50	3	25–50
1	1–5	6	25–33	2	5–25
+	<1	5	10–25	1	0–5
r	<<1	4	5–10		
		3	1–5		
		2	<1		
		1	<<1		
		+	<<<1		

Source: Mueller-Dombois and Ellenberg 1974.

often along a transect line. The area around each of the points is divided into 90° quarters, usually along compass headings. Then, in each quarter, the distance of the nearest individual from the central point is measured. These distance measures are then converted into densities. Other methods, such as nearest individual, nearest neighbor, and random pairs, are variants of this technique. Greig-Smith (1964) and Krebs (1989) discuss the details of using these methods, none of which are perfect. For example, clumping of individuals can result in incorrect estimates of density, depending on the details of how sampling is carried out.

Because different species may be represented more heavily by one or another of the above measures of abundance, it is sometimes useful to combine them in a weighted measure called an **importance value (IV)**. Curtis and McIntosh (1951) first defined the importance value of a species as the sum of the relative cover (i.e., cover of that species divided by total cover of all species present), relative density, and relative frequency of the species in a community. Importance values can also be defined by combining other measurements of relative abundance, depending on what is most relevant in a given study.

Biomass is generally correlated with cover. Cover, however, is limited to two dimensions, whereas biomass is correlated with three-dimensional properties because plant parts fill space. In some ways, it is a better measure of abundance than cover. For many species, biomass is a more useful measure than number of individuals because individuals within a population of the same species can often differ in size by several orders of magnitude (consider a seedling and a full-grown oak tree), and the effect of an individual within its community typically depends on its size (see Chapter 10). Biomass is

usually determined by harvesting samples of the vegetation, but can sometimes be estimated nondestructively. For trees, a typical measure of biomass is diameter at breast height (DBH). DBH is measured at a standard height above the ground, often by measuring circumference using a special tape that converts circumference to diameter units. For some tree species, especially those that are commercially important, the relationship between DBH and total biomass is known. The investigator can also determine this relationship by harvesting a sample of individuals, if necessary.

Physiognomy

Physiognomy is the form, structure, or appearance of a plant community. Knowledge of the physiognomy of a community may tell us about the adaptations of its dominant species to environmental conditions. The Danish ecologist Christen Raunkaier developed one widely used system for describing community physiognomy. He classified species based on their growth form and on the location of their overwintering organs (those that can survive an unfavorable season), such as meristems and underground bulbs (see Figure 8.2). According to this system, a tropical forest is made up almost exclusively of phanerophytes (Table 12.4). A desert community is dominated by shrubs—chamaephytes—and annual herbs. A prairie is dominated by grasses and perennial forbs—hemicryptophytes and cryptophytes. We can use this system both to classify general vegetation types (Raunkaier's original intent) and to begin to make inferences about the environmental factors that shape these communities. The prairie, for example, is dominated by species whose meristems are at or below ground level, suggesting that meristems above ground level are vulnerable to damage. In this case the critical environmen-

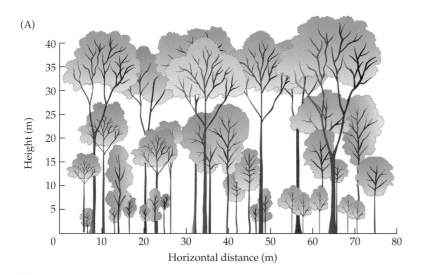

(A)

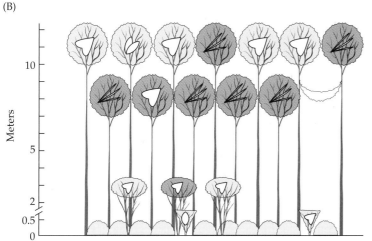

(B)

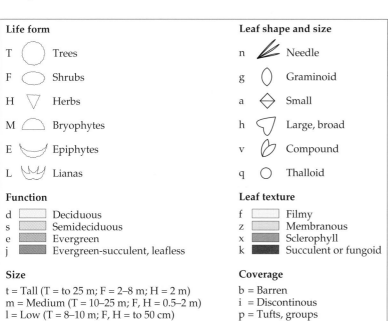

Life form			Leaf shape and size		
T	◯	Trees	n	⫰	Needle
F	⬭	Shrubs	g	◯	Graminoid
H	▽	Herbs	a	◇	Small
M	⌂	Bryophytes	h	▽	Large, broad
E	‿	Epiphytes	v	◊	Compound
L	⨇	Lianas	q	◯	Thalloid

Function			Leaf texture		
d	▭	Deciduous	f	▭	Filmy
s	▭	Semideciduous	z	▭	Membranous
e	▭	Evergreen	x	▭	Sclerophyll
j	▭	Evergreen-succulent, leafless	k	▭	Succulent or fungoid

Size

t = Tall (T = to 25 m; F = 2–8 m; H = 2 m)
m = Medium (T = 10–25 m; F, H = 0.5–2 m)
l = Low (T = 8–10 m; F, H = to 50 cm)

Coverage

b = Barren
i = Discontinous
p = Tufts, groups
c = Continuous

Figure 12.6
(A) A bisect of a tropical rainforest in Trinidad, British West Indies. (After Beard 1946.) (B) A diagrammatic bisect of a temperate woodland. The letters in the key can be used as "shorthand" to provide information about the life form, leaf shape, function, and leaf texture of the species. For example, Tmdh(v)zi would indicate a deciduous tree of medium stature with large, compound membranous leaves and discontinuous coverage. (After Dansereau 1951.)

tal factors are fire and grazing (see Chapters 13 and 19).

An important feature of plant communities is **vertical structure**. A forest community, for example, may consist of canopy trees, understory trees, shrubs, and herbs. A tropical rainforest may have as many as six different vegetation layers, or strata. Grasslands can consist of a mixture of ground-hugging forbs and tall grasses. These properties can be quantified in several ways. The **bisect** is a visual technique consisting of a scale drawing of the vegetation along a transect (Figure 12.6A). A more diagrammatic form of such a drawing is also possible (Figure 12.6B). Data from bisects or other surveys can be used to determine the vertical profile of the community. Along with density, cover, or basal area, the height of each individual is measured. Each species will occupy some range of heights; often seedlings or saplings of canopy tree species are treated separately from adults. By combining species that occupy the same stratum, one can produce a graph that summarizes the distribution of vegetation in different vertical strata (Figure 12.7). These data can also be used to statistically compare vertical structure among sites or community types.

Long-Term Studies

The measurements described above are like snapshots of a community at a single point in time. But ecological processes, such as succession (see Chapter 13) or response to global change (see Chapter 22), can take many years, so these measurements may be inadequate to describe a community. Changes in a community may be gradual, as the environment changes from year to

Table 12.4 Distribution (percentage) of growth forms for various representative communities based on the Raunkaier system[a]

	Phanerophytes	Chamaephytes	Hemicryptophytes	Cryptophytes	Annuals
World average	46	9	26	6	13
Tropical rainforest	96	2	0	2	0
Subtropical forest	65	17	2	5	10
Warm-temperate forest	54	9	24	9	4
Cold-temperate forest	10	17	54	12	7
Tundra	1	22	60	15	2
Mid-temperate mesophytic forest	34	8	33	23	2
Oak woodland	30	23	36	5	6
Dry grassland	1	12	63	10	14
Semidesert	0	59	14	0	27
Desert	0	4	17	6	73

Source: Whittaker 1975, Table 3.2.
[a]These terms are defined in the legend of Figure 8.2.

year, or dramatic, as after a hurricane or major fire. Therefore, a sample taken in a single year may not be very meaningful; in any case, its value is greatly enhanced by other samples taken at other times. If we want to study the effects of environmental variation, we need to sample that variation adequately; even if we are interested only in knowing what happens on average, a single year's data may not be "average."

Some of our most important ecological questions can be answered only with long-term studies (Leigh and Johnston 1994). A good example of such a question is the effect on global warming. The changes predicted to occur in temperature and precipitation will, in some cases, take decades to be manifested (see Chapter 22). From one year to the next, these changes are expected to be small relative to normal year-to-year variation. By monitoring the same communities and individuals over long periods of time, we may be able to most precisely detect the effects of long-term climate change.

Within long-term study areas, quadrats or individuals are permanently marked. Repeated measurements—perhaps of cover, productivity, diversity, phenology, or other population, community, or ecosystem features—can then be taken according to a set schedule, often yearly but possibly at shorter or longer intervals. In such situations, nondestructive sampling methods are usually preferable to destructive ones. If destructive methods must be used, they need to be done in predetermined locations so that, for example, the destructive sampling does not affect other sampling areas in the future.

Ecology is a relatively young science, so most studies that we call "long-term" are less than a few decades old. However, there are several studies that have been

(A) Eastern deciduous forest

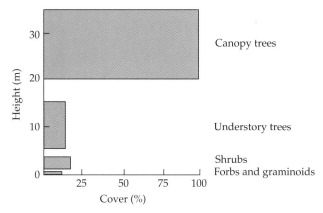

(B) Boreal coniferous forest

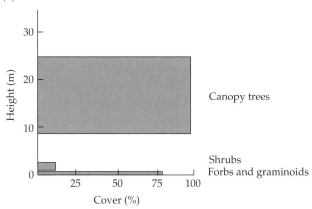

Figure 12.7
Vertical profiles of two different forest types. Each bar represents a different stratum of vegetation. The average height range of a stratum is given by the width of the bar. The total cover by that stratum is given by the length of the bar. (A) Typical eastern deciduous forest. (B) Typical boreal coniferous forest.

The Long-Term Ecological Research Network

The Long-Term Ecological Research (LTER) network is a collaborative effort involving more than 1100 scientists and students, with the goal of investigating ecological processes that operate at long time scales. The network promotes synthesis and comparative research across sites and ecosystems and among related national and international research programs. The National Science Foundation established the program in 1980 to support research on long-term ecological phenomena in the United States. The network now consists of 24 sites representing diverse ecosystems and research emphases, including arctic tundra (Toolik Lake, Alaska), southern deciduous forest (Coweeta, Georgia), old fields (Kellogg Biological Station, Michigan), and the desert-urban interface (Phoenix, Arizona). Besides promoting long-term studies, these sites also encourage studies that are integrated across a range of disciplines: physiological ecology, population biology, community ecology, and ecosystem science.

Global scientific interest in developing long-term ecological research programs is expanding rapidly, reflecting the increased appreciation of their importance in assessing and resolving complex environmental issues. In 1993, the U.S. LTER network hosted a meeting on international networking in long-term ecological research. Representatives of scientific programs and networks that focus on ecological research over long temporal and large spatial scales decided to form the International Long-Term Ecological Research (ILTER) network. They recommended developing a worldwide program to facilitate communication and the sharing of data. As of May 2002, 26 countries had formed national LTER programs and joined the ILTER network. Ten more are actively pursuing the establishment of national networks, and many others have expressed interest in the model.

going on for much longer. The oldest experiment in ecology is the "Park Grass Experiment" of nutrient addition and mowing on pasture communities at the Rothamsted Experimental Station (now the Institute for Crops Research-Rothamsted) in Hertfordshire in the United Kingdom, which as been ongoing since the 1850s. Studies on the Swedish island of Öland, led by Eddy van der Maarel and conducted since the 1960s, have included both descriptions of vegetation dynamics and experimental studies of factors maintaining species diversity. Forest dynamics have been studied at Hubbard Brook Experimental Forest in the White Mountain National Forest, near Woodstock, New Hampshire, since 1963. Hubbard Brook is now part of the Long-Term Ecological Research network that was initiated in 1980 (Box 12C).

Summary

A community is a group of organisms that coexist in space and time and interact with one another. Ecologists have long debated the relative importance of the processes responsible for community structure. One extreme view is that communities are highly predictable entities in which species are tightly interlinked. An opposing view is that communities are chance groupings of species that are shaped by abiotic factors and historical contingency. The current consensus lies between these views. Deterministic and chance events and biotic and abiotic processes are all considered important.

Communities can be described by several measures. The simplest of these is species richness, the number of species in a community. Other indices of species diversity take into account both the number of species and the relative abundance of each species. Relative abundance can be measured as numbers of individuals, frequency, cover, or biomass, using a variety of techniques. Physiognomy, the general form of the vegetation, is another community property that can be measured. These measures are used to describe community patterns in time and space. Describing patterns is a first step in determining which processes are responsible for shaping the plant communities that we see around us.

Additional Readings

Classical References

Clements, F. E. 1916. *Succession*. Carnegie Institution of Washington, Washington, DC.

Curtis, J. T. 1959. *The Vegetation of Wisconsin*. University of Wisconsin Press, Madison, WI.

Gleason, H. A. 1926. The individualistic concept of the plant association. *Bull. Torrey Bot. Club* 53: 7–26.

Mueller-Dombois, D. and H. Ellenberg. 1974. *Aims and Methods of Vegetation Ecology*. John Wiley & Sons, New York.

Contemporary Research

Hoagland, B. W. and S. L. Collins. 1997. Gradient models, gradient analysis, and hierarchical structure in plant communities. *Oikos* 78: 23–30.

Lande, R. 1996. Statistics and partitioning of species diversity, and similarity among multiple communities. *Oikos* 76: 5–13.

Lawton, J. H. 1999. Are there general laws in ecology? *Oikos* 84: 177–192.

Additional Resources

Krebs, C. J. 1989. *Ecological Methodology*. Harper Collins, New York.

Magurran, A. E. 1988. *Ecological Diversity and Its Measurement*. Princeton University Press, Princeton, NJ.

CHAPTER *13* *Disturbance and Succession*

*T*he living world is in a constant state of flux. In previous chapters, we looked at processes of change at the level of the individual (Part I) and the population (Part II). Change also occurs at the level of the community. Communities change as the species present change, as the populations change in number, age structure, or size structure, resulting in changes in physiognomy and ecosystem function.

In Chapter 12 we described a variety of techniques for quantifying community characteristics. In this chapter we explore patterns of community change and their causes. We also explore a variety of other aspects of communities: their continuity in time, their predictability, and whether they reach an equilibrium.

The process of succession has been central to plant ecologists' study of plant communities. **Succession** is directional change in community composition and structure over time. The term denotes changes over periods longer than a single season, although very long-term trends, such as those due to climate change, are not regarded as part of succession. Succession begins when a disturbance—an event that removes part or all of a community—is followed by colonization or regrowth of the disturbed site by plants. Consider the change in the plant community following the abandonment of an agricultural field in North Carolina (Figure 13.1; Oosting 1942; Keever 1950). In the first year, a variety of species are likely to colonize the field. The dominant species are typically short-lived ones—annuals and biennials such as *Erigeron canadense* (daisy fleabane, Asteraceae) and *Gnaphalium purpureum* (purple cudweed, Asteraceae). Over the next few years these species tend to be replaced by herbaceous perennials such as *Andropogon virginicus* (broomsedge, Poaceae) and *Aster ericoides* (white aster, Asteraceae). By 10 years after abandonment, shrubs such as *Rhus radicans* (poison ivy, Anacardiaceae; this species also grows as a vine) and trees such as *Pinus taeda* (loblolly pine, Pinaceae) begin to dominate the field. Eventually, after 150 to 200 years, the pines are replaced by hardwood tree species—*Quercus rubra* (red oak, Fagaceae) and *Carya* spp. (hickory, Juglandaceae). If no subsequent large-scale disturbance occurs, these hardwood species will remain the dominant species.

Discussions about the causes and nature of the successional process are part of the debate about the nature of communities (see Chapter 12). One extreme view is that succession is an orderly and predictable process that is a result of emergent community properties. An alternative view is that succes-

(A)

Figure 13.1
Old-field succession in the Piedmont region of North Carolina. (A) This field, three years after abandonment, is dominated by perennial species, *Andropogon virginicus* (broom-sedge, Poaceae) and *Solidago* sp. (goldenrod, Asteraceae). (Photograph courtesy of N. Christensen.) (B) This field, eight years after abandonment, has been colonized by *Pinus taeda* (loblolly pine, Pinaceae). (C) A mature oak-hickory forest with a canopy of *Quercus alba* (white oak, Fagaceae), *Q. rubra* (red oak), *Carya glabra* (pignut hickory, Juglandaceae), and *C. tomentosa* (mockernut hickory). Note the absence of a ground layer, a characteristic of the infertile soils of this region. (Photographs courtesy of R. Peet.)

(B)

ory, what we have is a complex set of interlocking models (Figure 13.2). These models can be applied to particular systems, providing descriptions of the patterns of succession and the processes responsible.

Theories of the Causes of Succession

Several issues form the core of discussion about the processes responsible for succession. One group of questions flows directly from the sharply divergent views of communities advocated by Clements and Gleason (see Chapter 12). First, what processes are responsible for successional changes? Do emergent properties and interactions among species play a major role? Or is succession the mere unfolding of the life history of each species independent of the others? If species interactions are important, are they mainly in the form of mutualisms, or is competition among species more important?

(C)

Second, how predictable is succession? If two adjacent areas, or the same area at two different times, were to undergo succession, would the same changes occur, leading to the same end point? A **climax** is the hypothetical deterministic end point of a successional sequence. Do plant communities undergoing succession usually reach a stable climax state? To the extent that they do not, is this mainly because extrinsic disturbances (such as storms) prevent it, or because of internal dynamics (such as population cycles) inherent in the communities?

sion is a rather unpredictable series of events that result from interactions between individuals and the abiotic environment. As we will see, succession involves a number of different patterns, mechanisms, and causes. Ecologists have amassed a great deal of information about successional processes. Rather than a single, unified the-

Figure 13.2
A hierarchical framework for a theory of plant community dynamics. Three main processes affect community dynamics: site characteristics, availability of species, and species characteristics. Each of these processes is, in turn, driven by particular interactions and conditions. (After Pickett et al. 1987.)

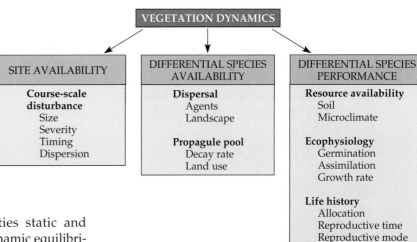

Third, are most plant communities static and unchanging, or are they in a state of dynamic equilibrium, or are they always in a fundamental state of flux? A stable equilibrium implies that they would return to the same structure and composition after small to moderate perturbations, remaining in that state over time. Clements's view was that communities are unchanging. Dynamic equilibrium implies very different processes, but in some ways is a modern substitute for Clements's perspective.

The Clements-Gleason dichotomy, however, does not account for many of the other questions we consider relevant to understanding succession today. For example, it seems obvious that one might ask about the role of competition (see Chapter 10) in successional processes. This question, though, does not fit neatly into the Clements-Gleason dichotomy, as neither of these scientists ascribed an important role to competition in succession. This is especially odd, given that Clements wrote extensively about competition, but (given his superorganism viewpoint) he thought of it in a rather different sense than modern ecologists do. Similarly, neither Clements nor Gleason appears to have thought of herbivory as an important factor affecting plant community structure and function, though they were certainly well aware of it. Thus, while it is useful to begin with the historical differences between Clements's and Gleason's views on succession, it is a mistake to think that those differences are all that one must consider in trying to piece together a contemporary view of successional processes.

As with much in ecology, the answer to all of the questions posed above is that a variety of patterns and processes can be found in different communities and, to some extent, at different times. We will return to these questions after we examine the factors that lead to successional change. Then we will ask, is climax a fixed, immutable state—an end point at which the community remains for all time, barring a major catastrophic change?

The initial ideas about succession as a predictable process leading to a static climax were developed in the 1890s and early 1900s, largely through the work of Cowles (1899, 1911) and Clements (1916). These ideas were a major force in plant ecology in North America (and to some extent in the entire English-speaking world) for the next half century or more. The move away from these ideas took several directions. One influential new perspective on community dynamics came from Alex S. Watt (1947). In his view, a plant community is composed of a mosaic of patches, and the patches are dynamically related to one another (see Chapter 10). Community structure results from a balance between factors that are predictable and those that are unpredictable (what we now might view as factors that drive the community toward equilibrium and unpredictable disturbances that disrupt that tendency). Watt was careful to distinguish this view—that the community is a dynamic mosaic of patches—from that of a successional sequence in which a homogeneous community tends toward a final, determinate end point. He emphasized the ubiquitous patchiness of natural plant communities and stated that the persistence of patches is a fundamental characteristic of plant communities.

Several other early ecologists shared this perspective. R. T. Fisher studied old-growth forests on Pisgah Mountain in New Hampshire, including both hardwood and coniferous forests, over several decades, beginning in 1910. The dynamic and patchlike character of these ancient forests was cogently summed up by Fisher in a preface to a study by his successors (Cline and Spurr 1942, as cited by Dunwiddie et al. 1996):

> *The primeval forests, then, did not consist of stagnant stands of immense trees stretching with little change in composition over vast areas. Large trees were common, it is true, and limited areas did support climax stands (or preclimax stands), but the majority of the stands were in a state of flux resulting from the dynamic action of wind, fire and other forces of nature. The various successional stages brought about, coupled with the effects of elevation, aspect, and other factors of site, made the virgin forest highly variable in composition, density and form.*

Interestingly, a major hurricane on September 22, 1938, uprooted many of the trees in this forest, subsequently changing the vegetation, and ongoing studies suggest that such infrequent, large-scale disturbances are an important factor in forest community dynamics (Foster 1988).

By the 1980s, influenced in particular by work in tropical forests and by broader trends in ecological thinking, many ecologists had begun to view plant communities as mosaics of patches rather than as homogeneous entities. In other words, they ceased trying to average over the entire community and refocused on the smaller scale of gaps and patches (Forman and Godron 1986; Pickett and White 1985). At the same time, ecologists abandoned the search for idealized, static types—such as climax communities—to embrace concepts of variability and dynamic processes as essential to the true nature of communities. In the 1980s, this changing perspective intersected with a very different tradition from Europe. There, the floristic-sociological school was part of the development of landscape ecology. The European floristic-sociological approach focused on measuring static patterns. When wed to the thinking of North American ecologists interested in patch dynamics, the result was a dynamic outlook that helped put successional studies into a wider landscape framework. We discuss these issues in more depth in Chapter 17.

Disturbance

Let's begin by considering the mechanisms that underlie successional change. Complex as the questions listed above may seem, a loose consensus has emerged among ecologists since the early 1980s on at least two crucial points. First, most ecologists now think that most communities do not tend to stable equilibria. Moreover,

Steward T. A. Pickett

different kinds of disturbances over a range of scales clearly play a crucial role in community ecology.

Peter White and Steward Pickett (1985, p. 7) define a **disturbance** as a relatively discrete event in time that causes abrupt change in ecosystem, community, or population structure, and changes resource availability, substrate availability, or the physical environment. Disturbances can result in the removal of a substantial portion of the existing vegetation in an area. They can range from damage to a single plant to the destruction of forests or prairies over thousands of hectares. Sources of disturbance include fire, storms (windstorms, ice storms, tornadoes, etc.), landslides, earthquakes, mudslides, volcanic eruptions, floods, the activities of animals, and disease. Because of the wide diversity of disturbances and their effects, it is almost always more useful to discuss specific kinds of disturbance rather than a single phenomenon.

One way of classifying disturbances is by whether or not they completely remove a community, including any organic soil. These two categories of disturbance lead to two categories of succession. **Primary succession** occurs when plants colonize ground that was not previously vegetated. Examples include the establishment of plant communities on lava fields, earth laid bare by the retreat of a glacier, rock outcrops, sand dunes, and newly formed beaches and sandbars in rivers, or, at a much larger scale, islands newly emerged from the sea. Primary succession also occurs where human activity results in massive soil disturbance, as on mine wastes, roads blasted through rock outcrops, and other sites that are left with undifferentiated soil parent material (Walker 1999). A key aspect of primary succession is that the process begins with the development of soil, which forms as the plant community develops (see Chapter 4). We will take a closer look at primary succession later in this chapter.

Secondary succession occurs when plants colonize ground previously occupied by a living community. In this case, soil already exists, and plant propagules, such as seeds and ramets, are readily available. Forest regrowth following a catastrophic fire is an example of secondary succession, as is the colonization of an abandoned agricultural field, called **old-field succession**. Once soils are well established, there is no fundamental difference between primary and secondary succession, as long as sources of colonizing plants are readily available.

Since disturbances are normal characteristics of all ecosystems, ecologists often discuss the **disturbance regime** of an ecosystem—the characteristics of the dis-

(A)

(B)

(C)

Figure 13.3
(A) A surface fire in a forest in northern lower Michigan, in which only ground-level vegetation is burnt. (B) A crown fire in central New Jersey in 1986 in which entire trees were killed. (C) A crown fire occurred three years earlier, killing most of the mature trees in this spruce-fir forest in the Rocky Mountains. At this time only low-growing herbs and shrubs are found. (Photographs A and C by S. Scheiner; B by D. Burgess.)

turbances occurring in that ecosystem. We can describe disturbance regimes using three characteristics: intensity, size, and frequency. The **intensity** of a disturbance is the amount of change that it causes. A forest fire (Figure 13.3), for example, can be a light **surface fire**, which runs along the soil surface or ground vegetation and destroys only herbaceous plants or low-growing shrubs, or it can be a severe **crown fire** that spreads from the canopy of one tree to another, killing most of the canopy trees as well as most of the shrubs and other vegetation. The **size** (spatial extent) of a disturbance is the amount of area affected. A windstorm, for example, might blow down a single tree, while a hurricane might blow down hundreds or even thousands of trees over an entire section of a forest. The **frequency** of a disturbance—also called the **return interval**—quantifies how often, on average, it occurs in a particular place. These three characteristics are often correlated: small disturbances of low intensity are generally much more frequent than large, intense disturbances. These factors vary, however, depending on the type of community and disturbance. The timing of the disturbance can also be important. For example, a fire that occurs in early spring before plants are actively growing can have very different effects from one in the late summer when they are just beginning to set seed (Biondini et al. 1989; Howe 1995).

Gaps

The size of a disturbance is a major factor in the types of species that can colonize a disturbed site. Disturbances create **gaps** in communities that can be filled by colonizing species. In any given community, gaps of various sizes can exist at any given time, so that succession may be occurring on a variety of scales. For example, mounds made by badgers (*Taxidea taxus*) in North American prairies are 0.2–0.3 square meters in size. Each year about

0.01% of the prairie is turned over by the creation of new mounds (Platt and Weis 1977). The mounds are colonized by nearby plants whose seeds have limited dispersal abilities. In contrast, a massive forest fire that burns thousands of hectares may leave no nearby seed sources. In this case, the initial colonists will be seeds that have traveled long distances by wind or animal or have remained dormant in the soil.

A forest can be thought of as a set of patches of widely varying sizes that have experienced disturbances of different types and intensities. The size of a patch has major effects on species composition, the successional trajectory within that patch, and ecosystem processes. In an experimental study on gaps of different sizes in the southern Appalachian Mountains in North Carolina, solar irradiance was two to four times higher in large than in small gaps, and soil and air temperatures were much higher (Phillips and Shure 1990). Standing crop biomass and aboveground net primary productivity (see Chapter 15) were three to four times greater in the largest than in the smallest gaps. Species richness was also greater in large gaps, and species composition differed among gaps of different sizes. We will return to the question of how gap formation and gap size influence overall community species richness in Chapter 14.

Ecologists' thinking about disturbance has changed substantially over the past few decades. Early ecologists, for example, thought of disturbances as being unusual—or unnatural—occurrences that disrupted the ordinary and orderly processes of a community. In the past quarter century, however, disturbance has been recognized as a natural part of many communities. This change in viewpoint has altered management practices, as we will see in the next section.

Fire

Fire is a major source of disturbance in many communities. Fires can vary tremendously in intensity, size, and frequency. One measurement of fire intensity is the amount of heat transferred per unit area per unit time. The most intense forest fires can release as much as 500,000 kJ/m^2 in a few minutes, which is enough heat to melt an aluminum-block engine. The Mack Lake fire (discussed below) released approximately 3×10^{12} kJ of energy, as much as 90 thunderstorms, or 9 times the energy of the atomic bomb dropped on Hiroshima (Pyne et al. 1996). The speed with which a fire spreads is another aspect of its intensity. Surface fires can move quickly through an area. In a North American short-grass prairie on a very windy day, a fire can move as fast as 22 km/hr—as fast as an Olympic sprinter—for at least short periods. Crown fires have been clocked at 12 km/hr.

Fire frequency is a key determinant of community structure and composition. In mesic prairies, fires can recur as often as every 2 to 3 years. As we will see in Chapter 18, these frequent fires destroy colonizing trees and so maintain the prairie as grassland. Chaparral communities in southern California burn approximately once every 25 years. Average fire return intervals in different community types vary greatly—from every few years to once a century to once every millennium (Table 13.1).

Many plant species have adaptations to fire. One such adaptation is the location of the meristems (see Chapter 8) in a place protected from fire. Prairies, where fires recur frequently, are dominated by grasses and forbs, which have meristems at or below the ground surface. A light, quickly moving surface fire is unlikely to

---

Table 13.1 Examples of fire regimes in various community types in North America

Fire regime	Communities
Natural fires rare or absent	Pacific Northwest coastal forests; wetter regions of eastern deciduous forests; southwestern deserts
Infrequent, low-intensity surface fires with a ±25-year return interval	Most eastern deciduous forests; pinyon-juniper woodlands of the Southwest; some montane meadows in the Rockies and Sierras
Frequent, low-intensity surface fires with a 1- to 25-year return interval combined with high-intensity surface fires with a 200- to 1000-year interval	Sierra mixed coniferous forests; western montane-zone pine forests; southeastern pine forests; prairies of Nebraska and Oklahoma; sawgrass prairie in the Florida Everglades
Infrequent, intense surface fires with a +25-year return interval combined with crown fires with a 100- to 300-year interval	Pine forests and boreal forests in the Great Lakes region; lower-elevation forests of the Rockies; California redwood forests
Frequent, intense surface and/or crown fires with a 25- to 100-year return interval	Most boreal forests; higher-elevation western forests; chaparral from California to Texas
Infrequent crown fires with a return interval >100 years	Northwestern wet coastal montane forests; subalpine forests of the western mountains; rainforests of Hawaii

Source: Davis and Mutch 1994.

harm these meristems, although the tops of the plants may be destroyed. (Having meristems at ground level also protects these plants from damage by grazers, which may have been the primary selective force acting on this adaptation; see Chapter 11.) Several tree and shrub species have the ability to resprout from roots, rhizomes, or buds underneath the bark if the aboveground portions of the plant are killed or severely damaged by fire. Examples from the frequently burned pine barrens of the eastern United States are *Pinus rigida* (pitch pine, Pinaceae) and *Quercus ilicifolia* (scrub oak, Fagaceae). Other species with such adaptations include *Eucalyptus* spp. (Myrtaceae) in Australia, *Populus tremuloides* (quaking aspen, Salicaceae) at high latitudes or high altitudes in North America, and *Adenostoma fasciculatum* (chamise, Rosaceae) in southern California. Other species, such as *Quercus velutina* (black oak) in the eastern United States and *Pinus ponderosa* (ponderosa pine) in the western United States, have very thick bark that protects the tree's cambium (the living, growing part of the trunk just under the bark) during a surface fire. As they grow, these species also tend to shed their lower branches, which

could serve as a "fire ladder," with the result that fire cannot as easily spread to the canopy.

Some species release their seeds from fruits or cones following a fire. Some pines have a trait called **serotiny**, in which they retain their seeds in tightly sealed cones for many years, releasing them only after exposure to fire. The cones are sealed by resin, which melts during a fire, releasing the seeds. Following a fire, these trees can release many years of accumulated seed production. The sealed cones, therefore, serve as an aboveground seed bank. Cleared mineral soil may be required for germination in many serotinous species. Among the pine species with serotiny are *Pinus contorta* (lodgepole pine) in western North America, *P. banksiana* (jack pine) in north-central North America, and some populations of *P. rigida* in the eastern United States (Figure 13.4). Serotiny also occurs in other fire-adapted taxa unrelated to pines, such as many Australian species in the Proteaceae. A related adaptation occurs in a number of annual and short-lived perennial herbs that appear after fires. Some of these plants have a soil seed bank (see Chapter 8) in which seed dormancy is broken by chemical compounds found in ash or smoke.

Pinus palustris (longleaf pine) is a dominant tree in forests of the southeastern United States, where surface fires recur approximately every 3 to 5 years. Young seedlings are especially vulnerable to these fires. *P. palustris* has an unusual life history that appears to be an adaptation to these frequent fires: after the first year or so, the young plant exists as a small tuft of needles grow-

Figure 13.4
Serotiny occurs in some *Pinus rigida* (pitch pine, Pinaceae) populations in the pine barrens of Long Island, NY. (A) Ordinary open cones on a typical, non-serotinous tree. (B) Closed, serotinous cones on a tree belonging to a frequently burned population with a high frequency of serotiny, which is a genetically determined trait. (C) A serotinous cone that has opened as a result of a forest fire, releasing its seeds. (Photographs by J. Gurevitch.)

Figure 13.5
A seedling of *Pinus palustris* (longleaf pine, Pinaceae) looks very much like a clump of grass—hence the term "grass stage." The apical meristem is located out of sight, just below the ground, where it is protected from frequent fires. (Photograph © G. Grant/Photo Researchers, Inc.)

ing at ground level, looking very much like a clump of grass (Figure 13.5). The apical meristem sits just below the ground, where it is undamaged if a low-intensity fire occurs. During this time, the plant grows a large root system and creates a large nutrient store. Finally, when its nutrient store is large enough, it grows very quickly. Within a few years, the tree is big enough, and has bark thick enough, to be quite fire-resistant. Thus, its growth pattern minimizes the number of years that it is vulnerable to fire.

If weather conditions are favorable to fire, the probability of a fire occurring increases with the **fuel load**, the amount of combustible plant material in the community. In prairies, the grasses die back each winter, and highly combustible dead grass builds up. This buildup of material increases the probability of fire with time, as well as increasing the intensity of fire when it finally occurs. Likewise, when exotic grasses invade nongrass-dominated communities, they can contribute to increased fire frequency by creating fuel where it did not previously exist, which in turn appears to facilitate further grass invasion (D'Antonio and Vitousek 1992).

In many *Eucalyptus* forests in Australia, the pine barrens of the eastern United States, and chaparral in California and other parts of the world, similar processes of fuel buildup occur. Here the buildup is of living twigs,

stems, and leaves of the dominant species. In some other community types, the probability of fire does not increase with time. In these communities, weather conditions, rather than fuel availability, limit fire frequency and intensity. In the boreal forests of North America, for example, the chance of a fire occurring in a given stand is roughly independent of the time since the last fire in that stand. There is usually adequate fuel for a fire, but weather is only sometimes conducive to forest fires (Johnson 1992). Interestingly, a roughly constant chance of fire also occurs in stands of California chaparral that have long been unburned: apparently, once the fuel load reaches a critical level, fuel availability no longer strongly affects the chance of fire (Johnson 1992; Keely et al. 1999).

Some plant species are **pyrogenic**—that is, their accumulated leaf or twig litter tends to promote fire more than one would expect based on the mass of the litter alone. Good examples include many *Eucalyptus*, some chaparral shrubs, and possibly some pines. Oils and other flammable chemicals produced by these plants are often of central importance in pyrogenicity. Mutch (1970) proposed that this property might be an evolved adaptation. This proposal remains quite controversial, however, as the conditions required for such a trait to evolve as a direct adaptation to fire may be quite restrictive (see Chapter 6; Kerr et al. 1999). Regardless of whether pyrogenicity is an evolutionary adaptation in itself or a by-product of selection on other traits, pyrogenic species reestablish themselves following a fire either because adults resprout or because the population has a seed bank (which may be located in the soil or in serotinous cones or fruits). Meanwhile, competing species are often killed by the fire. Thus, these pyrogenic plants create an environment that enhances their own persistence (or facilitates their invasion into a new community).

The problem of increasing fuel load with time led to a strenuous debate about how best to manage forests in the United States. For most of the twentieth century, the policy of the U.S. government was to suppress fire as much as possible. As a result, fuel loads and the density of young trees increased. Therefore, when fires did occur, rather than light surface fires, they often became raging crown fires, often killing mature trees and threatening human lives and property. In the 1970s, this management practice began to change for two reasons. First, the problems and dangers associated with accumulating fuel loads were recognized. Second, because of the increased understanding gained by ecologists, U.S. government agencies started to recognize that disturbance and fire are natural parts of ecosystems. In the absence of the natural disturbance regime, many properties of a community can change, including species composition. In the southeastern United States, for example, in the absence of fire, hardwood-dominated forests eventually replace longleaf pine forests.

The new U.S. government management practices include letting naturally occurring forest fires burn themselves out (if they are not threatening human lives or property) and setting **prescribed** or **controlled burns** to reduce accumulated fuel loads (to avert a much bigger, uncontrolled fire). Both of these practices are controversial, however. The extensive wildfires in Yellowstone National Park in the summer of 1988, which began naturally and were allowed to burn, were critical in turning public opinion against a policy of allowing fires to burn. Prescribed burns, too, have become a subject of controversy.

Prescribed burns are planned by foresters and set only after careful evaluation of weather and other conditions. Most have been safe and effective in reducing fuel load and in facilitating regeneration of fire-adapted species such as longleaf pine. They are used routinely to manage forests in many regions of the United States and elsewhere. There have been occasional but spectacular mistakes, however, in which controlled burns were set without the proper precautions and became uncontrolled wildfires. The Mack Lake fire on May 5, 1980, in the Huron National Forest of Michigan began as a prescribed burn, set to help manage the jack pine community. Unfortunately, the fire got out of control. In 12 hours, the fire had burned 10,000 hectares; tragically, it also killed a firefighter and destroyed 44 homes (Pyne et al. 1996). More recently, a prescribed burn in the spring of 2000 near Los Alamos, New Mexico, escaped control, burning thousands of hectares and destroying several dozen homes; fortunately, no lives were lost.

Managing forests and forest fires has become more complex and difficult as suburbanization and the rise in ownership of vacation and retirement homes in scenic areas has led to increasing numbers of homes encroaching into large, previously almost uninhabited forests. This was the case in Los Alamos, as well as in some of the areas devastated by the wildfires (started by lightning) that burned more than 2.5 million hectares in the western United States during the very dry La Niña summer of 2000 (see Chapter 18). Furthermore, incidents such as the Mack Lake and Los Alamos fires have made decisions regarding fire management political and contentious. Many of these discussions are underlain by profound disagreements between environmentalists, logging corporations, and loggers over the extent to which the U.S. government should facilitate the logging of public forests. The logging industry proposes that they be permitted to thin forests to prevent severe fires from occurring. Critics of this proposal say that this is just an excuse to continue extensive logging of public lands, as the type of logging that is actually done removes most of the larger, commercially (and ecologically) valuable trees, leaving the smaller trees that are most vulnerable to fire. An alternative that has been suggested by some environmentalists and foresters is to thin commercially valueless small trees and saplings, which are the most likely to contribute to catastrophic fires, but this proposal has not been enthusiastically embraced by the logging industry because it is not profitable. The results of these controversies have been protracted court battles without any clear resolution at present. Even without a political, economic, or environmental agenda to consider, there is also the problem of determining what a "natural" fire regime for a forest community might be and attempting to get to that state without having a catastrophic wildfire.

Wind

Wind can be another significant source of disturbance. At one end of the scale, wind can blow down a single branch or tree. Such **windthrows** range from blowdowns of branches or large parts of trees to losses of single trees and neighboring groups of trees. Windthrows are important in many tropical forests (Figure 13.6; Brokaw 1985), where trees can be very tall and are often

Figure 13.6
A large *Dipteryx panamensis* (almendro tree, Arecaceae) has created a gap in the tropical forest of La Selva Biological Station in Costa Rica. This photograph was taken approximately one year after the tree fell. (Photograph © G. Dimijian/ Photo Researchers Inc.)

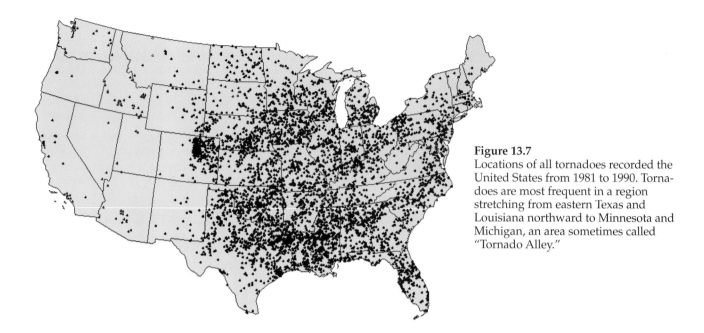

Figure 13.7
Locations of all tornadoes recorded the United States from 1981 to 1990. Tornadoes are most frequent in a region stretching from eastern Texas and Louisiana northward to Minnesota and Michigan, an area sometimes called "Tornado Alley."

connected by vines. As a result, when one tree falls, it often brings down others (Putz 1983). In the MPassa forest of Gabon, Africa, for example, 51% of gaps were caused by single falling trees and were responsible for 38% of the total gap area (Florence 1981). Trees that fell in a domino-like fashion accounted for 14% of gaps and 16% of the total gap area. Once a gap forms, neighboring trees become more susceptible to being blown down. In MPassa, such adjacent treefalls made up 13% of the gaps and 36% of the total gap area. In that forest, the average time between gap formation at any one spot is about 60 years, and the average size of a gap is 3 ha. Treefall in such forests is clearly a very important source of disturbance.

Extremely powerful storms such as hurricanes and tornadoes, while rare, are an important source of wind damage. Hurricanes, typhoons, and cyclones are important in coastal regions. Caribbean hurricanes, for example, pass over any given patch of forest, on average, every 15–20 years. These storms can also be important in temperate regions. In the northeastern United States, hurricanes severe enough to blow down large tracts of forest recur once or twice a century. In 1938 a hurricane blew down 253,000 ha of forest in central New England (the same hurricane that destroyed most of the Pisgah forest mentioned above). Tornadoes are another important source of wind damage, especially in the midwestern U.S. "Tornado Alley" (Figure 13.7). In the forests of northern Wisconsin, catastrophic windthrows create, on average, 52 patches each year, ranging in size from 1 to 3785 ha, with a return interval of 1210 years (Canham and Loucks 1984).

The gaps and patches created by all of these kinds of wind damage are important factors in forest commu-

nity dynamics. Soil erosion by wind can also be a major disturbance factor in more open communities, from arid lands to plowed former grasslands (see Chapter 4).

Water

Water can be an important source of disturbance in both its liquid state, as floods and nonflood erosion (including soil erosion), and in its solid state, as snow and ice. Floods are most important in **riparian** habitats (areas adjacent to streams and rivers) and areas near swamps and bogs. In many such systems, annual floods are continually creating and destroying habitat. In Alaska, for example, the sandbar willow (*Salix exigua*, Salicaceae) grows on sandbars along rivers that are continually being destroyed and re-formed by snowmelt-caused floods. Heavy rainstorms can also lead to landslides. These types of disturbances can initiate primary succession because they tend to create new, previously unvegetated ground. Similarly, snow avalanches are responsible for disturbances in temperate mountain regions. In the Canadian Rockies, avalanches remove 1% of the forests each winter.

Ice storms are important sources of disturbance in many temperate regions. In temperate deciduous forests, such as those of the southeastern United States, they are often responsible for many small-scale disturbances, such as the removal of single branches from trees. Such disturbances create new patches in the forest. Ice storms can also cause large-scale disturbances. In January 1998, a massive ice storm covered large areas of New York and New England in the United States and Ontario, Quebec, and New Brunswick in Canada. Approximately 25% of the trees in the region were damaged moderately to severely in that single storm.

The combination of disturbance by wind and by water can be important in some communities, such as the cove forests of the southern Appalachian Mountains (Runkle 1985). These highly diverse forests occur in sheltered areas, near creeks, at middle elevations. Fire is very rare in these forests; windthrows are the major source of disturbance. Most gaps are caused by the deaths of single trees due to ice storms, lightning, or high winds. The average gap is small, about 31 m^2, although gaps can be as large as 0.15 ha when several trees fall together. On average, a single spot experiences a gap-causing disturbance once every 100 years.

Animals

For the most part, disturbances caused by animals are small and often associated with herbivory (see Chapter 11). Sometimes this herbivory can be quite dramatic, however, as when a herd of elephants strips the foliage from a stand of trees. A single elephant can consume 225 kg of forage per day!

Even small animals such as insects can cause widespread disturbance during periods of population out-

Figure 13.9
The 1980 eruption of Mount St. Helens in Washington state created nearly 16 km^2 of bare ground, leading to primary succession. (Photograph courtesy of U.S. Geological Survey.)

breaks. The gypsy moth (*Lymantria dispar*), for example, was introduced from Europe into Massachusetts in about 1868, in an ill-advised attempt to establish a silk industry in North America by crossing it with the silk moth. It escaped from cultivation and spread rapidly throughout the eastern deciduous forests of North America; today, it can be found as far west as Oregon and as far south as Arkansas. During an outbreak, gypsy moth caterpillars can defoliate large tracts of forest, under some circumstances killing many of the trees (Figure 13.8).

Many other kinds of animal activities can create disturbances (see Chapter 11). Large animals, such as the American bison in North America and the African buffalo in Africa, graze and also create wallows. In the Konza Prairie Research Natural Area (see Chapter 1), bison wallows are important habitats for annual species found nowhere else on the prairie.

Earthquakes and Volcanoes

Seismic events such as earthquakes and volcanic eruptions lead to primary succession. Volcanic lava flows can create new islands in the ocean or bury existing lands. The eruption of Mount St. Helens in Washington on May 18, 1980, created nearly 16 km^2 of bare ground from lava flows (Figure 13.9; we will look at the patterns of succession that followed this eruption below). Seismic events can also have indirect affects by triggering land-

Figure 13.8
Gypsy moths caterpillars (*Lymantria dispar*) defoliated this *Fagus grandifolia* (American beech, Fagaceae) in an infestation that occurred in Connecticut in 1980. Deciduous trees do not always recover from such an attack. (Photograph by J. Bova/Visuals Unlimited.)

slides. In Chile, a series of devastating earthquakes in 1960 set off thousands of debris avalanches, landslides, and mudflows, which covered 25,000 ha, or 2.8%, of the land surface of the province of Valdivia (Veblen and Ashton 1978).

Disease

The spread of disease can also be a source of disturbance. Chestnut blight was introduced to North America in the early twentieth century, leading to the death of chestnut trees (*Castanea dentata*, Fagaceae, see Figure 11.18) over most of the eastern deciduous forest biome. Other examples of pathogen-caused disturbances are described in Chapter 10.

Humans

One of the most important causes of disturbance in natural communities today is human activity (see Chapter 22), including disturbances caused by domestic animals. Sometimes the disturbances we create are accidental or due to casual mischief. For example, approximately 35% of the forest fires in the United States, representing 19% of the area burned, are due to either human carelessness or arson.

Other human disturbances are more purposeful. Commercial logging removes large numbers of trees from U.S. forests annually in a combination of **clear-cutting** (removal of all trees) and **selective logging** (removal of some commercially valuable trees, leaving others). After clear-cutting, timber companies are legally required to replant to replace the trees they have removed. However, because the replanted trees are typically a single economically valuable species that grows quickly, the resulting forests are often dramatically different from the original forest communities. Following clear-cutting and replanting, most forests become **monocultures** (plantations composed of a single tree species), with all of the trees usually of the same age. Furthermore, logging roads can act as corridors for the movement of invasive species, while creating barriers to the movement of other species (see Chapter 17).

In many tropical forests, clear-cut timber harvesting and clearing of land for farming has been extensive. Worldwide, rainforest destruction is currently estimated at 1 ha per second (about equal to two U.S. football fields), 87,000 ha per day (an area larger than New York City), and 32 million ha per year (an area larger than Poland). But not only the forest that is destroyed suffers disturbance. The remaining areas of forest are also affected, as fragmentation of the landscape changes species migration patterns and population sizes (see Chapters 17 and 22).

Grazing by domestic cattle, sheep, and goats has drastically altered the composition of grasslands and shrublands in huge tracts of arid and semiarid coun-

tryside around the world (see Chapter 11), in many cases greatly accelerating soil erosion. Even the trampling associated with recreational hiking and other uses of wilderness areas can be substantial in areas with large numbers of visitors, or in areas that are especially vulnerable, such as deserts and tundra.

Colonization

All successional processes begin with propagules (seeds or other dispersal structures, such as seed clusters). Primary succession differs from secondary succession in the types of propagules involved and in their sources. Generally, in primary succession, all propagules must be seeds or spores carried to the newly bare ground by wind or water. Animal dispersal plays a small role in primary succession because relatively few animals visit unvegetated sites. First to appear are species of mosses and lichens that specialize in colonizing bare ground. These species tend to be slow-growing and to possess unique adaptations that enable them to live under the difficult conditions found at primary successional sites. In addition to the difficulties posed by the lack of organic matter, nutrients, and soil structure, other conditions, such as widely fluctuating temperatures and exposure to high winds, make survival at these sites difficult for most plants.

Peter Grubb

Colonization by dispersing seeds is also important in secondary succession, but the colonizers usually differ from those involved in primary succession. Because soil is already present, many fast-growing species can grow in the newly available patches. Peter Grubb (1977), inspired by Watt's work, defined the **regeneration niche** as the set of environmental requirements necessary for a plant species to reproduce. Grubb proposed that by differing in their regeneration niches, species that were otherwise very similar could coexist.

While newly disturbed areas such as gaps present many challenges to colonizing individuals, they also offer numerous advantages. Light levels are generally high because of the lack of a dominant plant canopy. Soil nutrients may also be high because of both reduced competition and the release of nutrients from decaying plants or plant parts (e.g., leaves from the canopy trees killed in a large blowdown). The colonizing species in secondary succession are often (but not always) very fast growers with small, wind-dispersed seeds. *Taraxacum officinale* (dandelion, Asteraceae), for example, has wind-borne seeds. Often these colonizing species are found at a site for only a year or two following the dis-

(A)

(B)

(C)

Figure 13.10
Disturbance and early succession in a forest in northern lower Michigan. (A) The site was clear-cut and burned in August 1980. (B) Early in June, 1981, the site is covered by *Populus grandidentata* (bigtooth aspen, Salicaceae) about 10 cm tall that has re-sprouted from roots in response to the fire. (C) One year following the fire, in August 1981, the aspens are now about 1 meter tall. (Photographs by S. Scheiner.)

tats have seeds that are able to remain dormant for many years and germinate under specific conditions that follow a disturbance. In particular, the seeds of many short-lived herbs have dormancy that is broken only by exposure to sufficient light, which occurs only after soil disturbance. You may have observed this in a garden: after you turn the soil over, many weedy species seem to appear suddenly. These plants have a soil seed bank and light-induced germination. Other cues can also be important; for example, some seeds germinate in response to the chemicals in ash or smoke left from a fire.

The composition of the soil seed bank can be quite different from that of the plant community growing above it (Oosting and Humphreys 1940; Livingston and Allessio 1968). In an upland oak-pine forest in Michigan (Figure 13.10), for example, *Arenaria serpyllifolia* (sandwort, Caryophyllaceae) was common in forest seed banks, yet was absent above ground. In contrast, *Erigeron canadensis* (fleabane, Asteraceae), which has wind-dispersed seeds, was one of the most common species following a fire, but was absent from the soil seed bank (Scheiner 1988). A gap in this forest might be quickly colonized by both species: *Arenaria* from the seed bank and *Erigeron* by dispersal of wind-borne seeds from other areas.

Vegetative propagules can also be an important source of regrowth following disturbance. Aspens (*Populus tremuloides* and *P. grandidentata*) in northern temperate forests specialize on growing in early successional habitats. These trees can spread by "root suckering"; that is, through the growth of new stems as offshoots from rhizomes. Fire stimulates their tendency to produce new

turbance before being outcompeted by other species (see Figure 13.1).

Soil seed banks (see Chapter 8) can also be an important source of propagules for secondary succession. Some species that specialize on early successional habi-

shoots. Following an experimental burn in northern lower Michigan, there were 24,750 aspen stems per ha, and these shoots grew to an average height of over 1 m in a single summer (see Figure 13.10).

Determining the Nature of Succession

Early ecologists thought of succession as an orderly progression of **seral** (pre-climax) communities leading to a predictable end point—a stable climax state. Beginning in the 1950s, as we saw above, a number of new views about succession emerged. Ecologists are still trying to sort out the patterns and processes of succession.

Interaction between Methodology and Understanding

The ability of scientists to understand nature is affected by the methods they use to study it. This is nowhere more apparent than in the study of succession. Changes in methods for studying succession have affected our understanding of the process; conversely, changes in ecologists' understanding of succession have also fostered changes in methods.

A central difficulty in studying succession is that it is often a very slow process—it can take decades or centuries. Ideally, one would like to be able to observe the entire successional process in a single community, but this is generally not possible. One traditional way to get around this limitation is by studying a **chronosequence**, a set of communities of different ages since disturbance. Users of the chronosequence approach assume that differences among these communities represent what would occur within a single community over time. H. C. Cowles (1899) pioneered and popularized this approach, which was based on his studies of patterns of primary succession in the dunes and forests along the southern shore of Lake Michigan (Figure 13.11). Cowles hypo-

Henry Chandler Cowles

thesized that the dunes were initially colonized and stabilized by plants with extensive root systems, such as the grasses *Ammophila breviligulata* (beach grass) and *Elymus canadensis* (wild rye), and small woody plants such as *Prunus pumila* (sand-cherry, Rosaceae). Once the dune is stabilized by these plants, Cowles argued, larger shrubs, such as *Salix glaucophylloides* (dune-willow, Salicaceae) and *Cornus stolonifera* (red osier, Cornaceae), can colonize it. Eventually, trees such as *Tilia americana* (basswood, Tiliaceae) enter the community, converting it from a thicket to a forest. Finally, under Cowles' scenario, the forest becomes dominated by *Acer saccharum* (sugar maple, Sapindaceae) and *Fagus grandifolia* (beech, Fagaceae).

The chronosequence approach rests upon three important assumptions. First, it assumes that successional processes are highly predictable; that is, that similar communities will all undergo the same successional sequence. Second, it assumes that the climate and other aspects of the environment, as well as the **species pool** (the species available for colonization), will remain the same during the entire chronosequence. We know, however, that this is not always true. For example, new species may migrate into a region, species combinations may change during major periods of climate change (see Chapter 21), or species may go locally extinct. In recent years, especially, invasive species have changed successional trajectories (see Chapter 14). Finally, the chronosequence approach assumes that communities in the same general region are similar enough to be considered part of the same successional sequence. This last assumption

Figure 13.11
A foredune along the shore of Lake Michigan at Big Point Sable, Ludington, Michigan, taken on September 13, 1916. On the left is *Ammophila breviligulata* (beach grass, Poaceae), and on the right is *Prunus pumila* (sand-cherry, Rosaceae). The root systems of these species stabilize the open dunes. (American Environmental Photographs Collection, AEP-MIS136, Department of Special Collections, University of Chicago Library.)

was called into question during the 1950s undermining the use of chronosequeces—an example of how conceptual shifts drive changes in methods.

Two new research approaches were used to directly test the first assumption, that the outcome of succession is highly predictable—an example of how methods sometimes drive conceptual advances. The first of these methods was the use of long-term (or longitudinal) studies, which provided data that were not available to the earliest ecologists. Beginning in the 1920s and 1930s, ecologists established permanent study sites that were monitored for decades or longer. Ecologists have now directly observed long successional sequences and compared them with those predicted by the chronosequence method. In a famous example from North Carolina, Oosting (1942), along with his student Catherine Keever (1950), used the chronosequence method to predict that pine forests would eventually yield to hardwood-dominated communities. This hypothesis was tested using long-term study plots established in the Duke Forest in 1934 and continuously monitored ever since. The methods of observation included tagging every tree with a trunk greater than 1 cm DBH. Based on these studies, Christensen and Peet (1980) determined that Oosting's hypothesized successional sequence was correct. A drawback to long-term studies is that time is still a limiting factor. A single researcher can examine succession over a period of decades, but not centuries. In addition, this method can be used only in situations where sites can be established and monitored without fear of human disturbance, and where there is a dedicated effort to maintain the study over a long period as investigators, funding, and even institutions change.

The second shift in methodology was an increase in the 1970s in the use of manipulative experiments to study community processes. Until then, plant community ecology was mainly an observational discipline, involving the collection and analysis of extensive quantitative data. Now ecologists began to test the explanations proposed for successional processes by applying different experimental treatments to communities and studying their responses. Three of the most extensive of these studies in the United States have been conducted at sites that are part of the Long-Term Ecological Research (LTER) network (see Box 12C). In Chapter 1 we describe one of these sites: the Konza Prairie Research Natural Area in Kansas. A second is Hubbard Brook, New Hampshire, where Frank Bormann and Gene Likens used whole-watershed manipulations to study succession (Bormann and Likens 1979). The third is the Cedar Creek Natural History Area in Minnesota, home to an ongoing study of prairie succession established by David Tilman. The experimental manipulations used at these sites include fertilizing plots with different levels of nitrogen, creating plots of different sizes, and

Figure 13.12
Annual rings show clearly in this cross section of the trunk of a loblolly pine (*Pinus taeda*). (Phograph © G. Grant/Photo Researchers, Inc.)

establishing plots with different sets of species (see Chapter 15). Time is again an important limiting factor in these kinds of experiments. They are most useful for addressing questions over a time span of a few years to over a decade in unusual circumstances. That is why they have mainly been used in communities such as prairies and old fields, where successional change initially proceeds at a relatively fast pace.

Another method that has been used extensively over the years to reconstruct the history of particular communities is **dendrochronology**, the study of tree rings. This method is particularly useful for picking up the study of succession where the other two methods leave off: over a period of centuries. Many temperate-zone trees produce annual growth rings (Figure 13.12). By taking a core from a tree and counting the rings, the tree's age can be determined. Knowing the ages of the trees in a forest allows one to determine when various species became established at that site. Other characteristics of tree rings, such as size, appearance, chemical composition, and isotopic composition, can provide additional information about the climate, soil, and growing conditions in times past. We discuss other methods of determining long-term, nonsuccessional change in plant communities in Chapter 21.

Mechanisms Responsible for Successional Change

Using these various methods, plant ecologists have slowly uncovered successional patterns and the processes responsible for them. The first comprehensive successional theory was that of Clements (1916); it was part of the superorganism concept, his overall theory of the nature of plant communities (see Chapter 12). In Clements's view, some species played the role of preparing a site for the later occupation of other species. Interactions among species were seen as benefiting the functioning of the whole community. Some species existed in the community for a limited time before bowing out and allowing others to take over. Clements believed that these species were somehow designed for this purpose. Today, we recognize that Clements's ideas were profoundly non-Darwinian. He developed them at a time when the Darwinian theory of evolution by natural selection was still a highly debated issue. Today, no scientist would accept the notion of a species existing only for the good of the rest of the community. Mutualistic interactions can be favored by natural selection (Wilson 1980). The evolution of mutualisms is complex, and their role in driving succession is much more limited than envisaged by Clements.

How, then, do populations of successional species interact with one another? To develop a set of testable hypotheses, Connell and Slatyer (1977) proposed three possible mechanisms by which succession might occur.

First, early successional species might **facilitate** the colonization and growth of late successional species. This hypothesis looks at succession from a viewpoint akin to Clements's, albeit within an evolutionary context and in a more mechanistic fashion. A second hypothesis is that a process of **inhibition** drives succession. An early successional species might inhibit colonization by later successional species by monopolizing resources such as light, water, or nutrients. Succession would eventually occur when later species outcompeted earlier species. Finally, a process of **tolerance**, in which species neither help nor hinder colonization by other species, might determine succession.

There is evidence for all three of these mechanisms. Facilitation is best demonstrated during primary succession. Some species are adapted to eking out a living under difficult conditions. Lichens, for example, can grow on bare rock. They do so by slowly dissolving away the rock to gain nutrients, thus beginning the process of soil formation. Another example of facilitation during primary succession occurs along lakeshore dunes, such as those of the Great Lakes in the north-central United States. Species such as the grass *Elymus canadensis* can colonize newly deposited sand, with its very low levels of nutrients and poor water-holding capacity. In areas of active sand dune formation, the fibrous root systems of this grass act to stabilize the sand (see Figure 13.11). Once the sand is stabilized, other species with smaller root systems are able to colonize the site; the establishment of these species is thus facilitated by *Elymus*.

Figure 13.13
In Arizona's Sonoran desert, a *Carnegia gigantea* (saguaro cactus, Cactaceae) grows in the shade of its nurse plant, a *Parkinsonia microphyllum* (palo verde tree, Fabaceae). Once the cactus grows large enough, it no longer requires protection and will displace the tree as *Carnegia* outcompetes *Parkinsonia* for water. (Photograph courtesy of T. Craig.)

Secondary succession in very harsh environments also provides examples of facilitation. In the Sonoran Desert, seedlings of *Parkinsonia microphyllum* trees (palo verde, Fabaceae) survive poorly in the open. Instead, they germinate and grow in the shade of shrubs such as *Ambrosia deltoidea* (Asteraceae). Because of this role in the community, the latter are called nurse plants (see Chapter 10). Similarly, *Carnegia gigantea* (saguaro, Cactaceae) germinate and grow under both *Parkinsonia* and *Ambrosia* (Figure 13.13), so *Parkinsonia* shifts roles to become a nurse plant. The eventual replacement of early successional species by later ones in these instances is not a result of one species voluntarily giving way to another, as Clements suggested. The later species eventually outcompete the early species—in fact, there is good evidence that saguaros eventually kill off their nurse trees. The early successional species persist in the landscape because they are able to colonize and live in ephemeral habitats.

Inhibition and tolerance are the easiest successional processes to demonstrate and observe. Inhibition comes about because of competition for light, water, and nutrients (see Chapter 10). One or more of these resources is always in limited supply in a given community. In a desert, the limiting resource is often water. In very sandy soil, the limiting resource is often nitrogen (see Chapters 4 and 15). In a forest understory, the limiting resource is usually light. Competition for these limiting resources results in reduced growth and reproduction. Thus, competition reduces a species' ability to colonize and grow in a site.

Competition is less intense if the competing species are adapted to somewhat different conditions. For example, some species grow best under high light conditions, while others do best under low light conditions. During forest succession, light levels decrease as trees grow up and the canopy fills in. In this case, species adapted to high light intensities tend to occur early in succession (or in gaps in a mature forest), while species adapted to low light usually occur in older communities. In the temperate forests of the northeastern United States, *Prunus pensylvanicus* (pin cherry, Rosaceae), a species that grows best in high light conditions, occurs in early successional forests. Because of its high light requirement, its seedlings cannot grow up underneath its own canopy. Instead, stands of pin cherry are often colonized by *Acer saccharum*, which can grow under much lower light conditions. Growth under these conditions is actually very slow, and a sugar maple sapling would never become a full-sized tree if it were always growing in the shade of another tree. Rather, the strategy of the sugar maple is one of tolerance. It can survive, growing slowly, under low light conditions. Then, if a canopy tree happens to fall and create an opening, the sugar maple sapling can grow quickly to fill that gap. Because the sugar maple is already at the sapling stage, it can reach the canopy

before any of the other trees that germinate in that gap. So pin cherry inhibits—slows down—the colonization of sugar maple, while sugar maple is able to colonize because of its tolerance of low light.

Another theory points to additional processes that are important in succession. In 1954, Frank Egler described two alternative scenarios for succession (Figure 13.14). The first, known as **relay floristics**, was largely a Clementsian view, in which integrated communities replace one another like runners in a relay race. Egler proposed a second scenario that he thought was more likely: the **initial floristic composition** hypothesis, which emphasized the process of colonization and differences in the life spans of species. Focusing on suc-

(A)

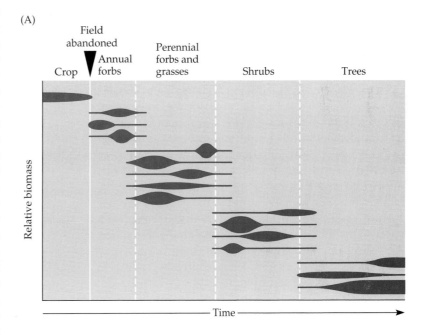

(B)

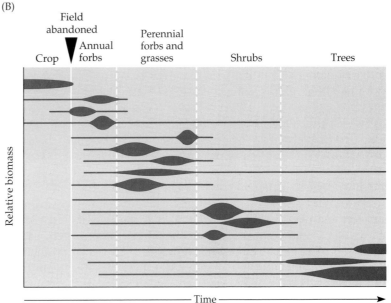

Figure 13.14
F. E. Egler's theories of succession (Egler 1954), stylistically diagrammed for a hypothetical abandoned field in North Carolina. Each line represents a single species of the vegetation type indicated. The thicker the line, the more important the species at a given time. (A) According to relay floristics, groups of species replace one another like runners in a relay race. (B) Egler's modification of relay floristics was the initial floristic composition theory. Here all of the species are present at the beginning of succession, which is simply a process of the unfolding of their various life histories.

cession in abandoned fields in the eastern United States (see Figure 13.1), Egler claimed that all species reach a site early in the successional process. Early succession-al species, such as annuals and perennial herbs, domi-nate the site because they grow quickly. Tree species also arrive early, but do not dominate until much later because they grow more slowly. Thus, in Egler's view, a particular successional sequence is a direct consequence of the initial composition of the plant community. Put differently, Egler's theory was that succession is sim-ply a process of the unfolding of various life histories that proceed at different rates; species neither facilitate nor inhibit other species. In this sense, Egler's theory of initial floristic composition resembles Connell and Slaty-er's tolerance model of succession.

Life history differences do cause shifts in dominance in some communities. *Populus grandidentata* (big-toothed aspen) is an early successional species found in upland forests with poor soils around the North American Great Lakes. It is replaced by a variety of species, including *Quercus rubra* (red oak). This replacement occurs because aspen grows very quickly. Its quick growth means that its wood is not very strong and is susceptible to damage by fungal attack and by high winds. Aspen trees rarely live beyond 60–70 years. In contrast, oaks grow very slowly, producing the strong, dense wood prized by fur-niture makers. Oaks may colonize an aspen stand very quickly, but they will not dominate for decades. In other cases, however, later dominance is clearly the result of a species' inability to colonize an early successional site. For example, *Tilia americana* cannot colonize new sand dunes. It requires other species to stabilize the dunes and begin the process of soil formation.

The Predictability of Succession

One of the most important contributions of the initial floristic composition theory was that it called attention to the process of colonization. While earlier theories of succession saw colonization as an important process, they viewed it as a **deterministic** one—a process with a fixed outcome. Colonization, however, is a process fraught with random elements. These elements include the timing of seed arrival, the number of propagules that arrive at a site, whether those propagules land in a **safe site** (a favorable spot for their germination and growth), and whether climatic conditions are favorable for their establishment when they arrive. Two identical sites might experience different successional sequences due to these chance events. Thus, the predictability of suc-cession is an open question. (Recall that the use of chronosequences to study succession assumes that suc-cession is predictable and repeatable.)

One way of thinking about the predictability of suc-cession is by dividing it into two components: the start-ing point of succession, and the path (or trajectory) that

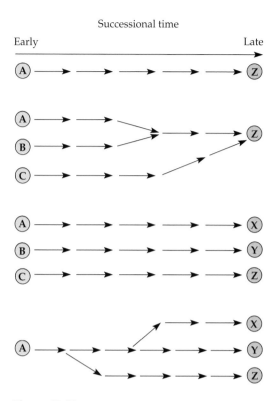

Figure 13.15
Schematic of predictable and unpredictable successional tra-jectories. (1) Early ideas about succession saw all sites in an area as starting from approximately the same starting point and following a single predictable trajectory to the same end point. (2) Different starting points might still lead to the same end point. In this scenario, deterministic processes dominate, and the outcome is predictable. (3) Different start-ing points might lead to various unpredictable end points. (4) Even similar starting points might lead to different unpredictable end points.

succession takes to its end point (Figure 13.15). The start-ing point encompasses the physical conditions follow-ing a disturbance and the propagules that either survive the disturbance or arrive at the site. The early ideas about succession saw all sites in an area as starting from approximately the same point and then following a sin-gle trajectory to the same end point. Once we recognize the possibility of unpredictability, however, three other outcomes are possible. The following descriptions assume that basic conditions (e.g., climate, soil type, slope, aspect) are similar among sites.

First, different starting points might still lead to the same end point. While early successional communities might differ considerably in composition, and while the exact successional sequence might differ, all sites in an area might eventually converge on the same type of com-munity. In this scenario, deterministic processes domi-nate over random ones in the long run. Second, different starting points might lead to various end points. Given different starting points, different successional sequences might unfold. The initial floristic composition hypothesis would fit under this scenario. Here, random processes dominate in setting the stage for a given successional sequence, although deterministic processes take over thereafter. Third, even similar starting points might lead to different end points. In this scenario, random processes dominate during the entire course of succession.

Currently we have limited data indicating which of these three scenarios occurs most often. The best evidence that we have favors the first two scenarios (i.e., that deterministic processes dominate in the long run). A study of tropical rainforest communities in the Amazon basin of eastern Peru, for example, found that species composition tended to converge among sites over time (Terborgh et al. 1996). Similarly, in Oosting's study of old-field succession in North Carolina, fields of similar age had very similar sets of species across many different sites. Other factors can complicate the pathway to convergence, however. In a study of succession on mudflows resulting from the eruption of Mount St. Helens, carried out 16 years after the eruption, an area that was relatively close to an intact forest—a source of many propagules—showed evidence of convergence to the composition of that forest, while an area farther away did not (del Moral 1998). So, dispersal rates and opportunities do affect the successional trajectory. In this case, however, given that the entire successional process takes several centuries, even the more distant site may converge, given enough time. Ecologists will need many more examples from many different types of communities before we can reach a final decision on this issue.

Primary Succession

There are three main reasons why primary succession can be quite different from secondary succession. First, in primary succession, there is generally little or no true soil—the substrate lacks structure and any substantial organic component. Most species are incapable of persisting under these circumstances and can colonize the site only after substantial soil development. Thus, a key issue in many primary successional settings is how the initially colonizing plants affect soil development. A second characteristic of primary succession is that, at least initially, propagules (mainly seeds and spores) that arrive at the site have few or no interactions with resident populations, because there are none. Competition is absent or of minor importance, and herbivory may be drastically reduced. Third, depending on the spatial scale of the disturbance that initiates succession, the physical environment (e.g., temperature at the ground surface) can be much more variable and much more extreme in primary succession than in secondary succession.

Probably the best-studied site of primary succession is at Glacier Bay, Alaska (Cooper 1923). William Cooper set up permanent study plots there in 1916 in areas from which a glacier had retreated fewer than 40 years earlier (Figure 13.16). In this system, the first colonist of bare ground usually is a **cryptogamic crust**, a thin layer of mosses, fungi, various unicellular photosynthetic organisms, lichens, and photosynthetic bacteria (cyanobacteria). The cryptogamic crust limits erosion and helps to retain moisture, dead parts of the crust provide initial organic matter for soil formation, and dispersing seeds may lodge in cracks in the crust. The small shrub *Dryas drummondii* (Rosaceae) is typically the next colonist. A plant with nitrogen-fixing symbionts in its roots, *Dryas* forms dense mats. In some regions of Glacier Bay, *Alnus sinuata* (alder, Betulaceae), another plant with nitrogen-fixing symbionts (see Chapter 4), also forms dense thickets. Not surprisingly, it appears that these early nitrogen fixers facilitate the later colonization of *Populus*

Figure 13.16
A rock-strewn area of bare ground at the base of Franz Josef Glacier in New Zealand. This area has not yet been colonized by any plants. (Photograph courtesy of L. Walker.)

trichocarpa (cottonwood) and *Picea sitchensis* (Sitka spruce, Pinaceae), which do not fix nitrogen. Spruce is much more successful as a colonizer of alder thickets than at earlier successional stages, no doubt partly because of the soil nitrogen, soil organic matter, and presence of mycorrhizae that the alder provides. Some spruce trees are inhibited by, while others are facilitated by, the presence of cryptogamic crusts, *Dryas*, and alder, as well as by other spruce trees (Chapin et al. 1994). Thus, both facilitation and inhibition occur in this system. Different sites at Glacier Bay follow different successional trajectories; in particular, alder does not occur in some areas, mainly due to its limited dispersal ability (Chapin et al. 1994).

Volcanic activity is important in much of the world, and lava flows and ash deposits create new ground on which primary succession occurs. Though volcanic eruptions can be massive disturbances, it is a mistake to think that they always create large areas devoid of life. Studies of succession following the 1980 eruption of Mount St. Helens have shown that most of the initial growth and spread of plant populations was dominated by "biological legacies"—buried roots, small stands of surviving plants, and seeds from nearby slopes (Frenzen et al. 1986; Franklin 1990; Halpern et al. 1990). These foci for the spread of populations remain important in the area today. Around the Kilauea volcano in Hawaii, lava flows often leave small islands of vegetation (kipukas), which are important sources of propagules for the colonization of the lava (Figure 13.17).

The Krakatoa volcanic complex in Indonesia has provided important data on primary succession because of its size and isolation. In 1883, Krakatoa erupted so forcefully that the explosion was heard 3500 km away. Only about a third of the original island remained above sea level after the eruption, but several new islands were formed from the volcanic material ejected. Continued volcanic activity has added to the land area. Early descriptions of Krakatoa following the main eruption say that it was essentially scoured of living things and that colonists arrived slowly from distant sources. It is impossible to judge whether this was truly the case or whether the observers failed to look for the "biological legacies" that have proved to be important in studies of other regions.

Recent research on the Krakatoa archipelago (Schmitt and Whittaker 1996) suggests that limitations on dispersal may be quite important there. Shoreline species recolonized the islands very rapidly, but these were largely water-dispersed species. Species that grow farther inland colonized the islands much more slowly. Wind-dispersed species, especially grasses, composites, and ferns, predominated among the first inland colonists, creating grasslands. Woody plants began colonizing the islands somewhat later; forests dominated the islands by the 1920s. The successional trajectories were similar among the islands, but there were also differences among islands and among sites within islands. These differences seemed to be due to chance differences in dispersal and environmental differences (Schmitt and Whittaker 1996). Continued volcanic activity, resulting in the heavy deposition of ash on some islands, has contributed to increasing differences among the islands over time.

Figure 13.17
A lava field on the island of Hawaii showing a young tree fern *Sadleria cyantheoides* (amaumau fern, Blechnaceae) that has colonized the bare ground. This species is an endemic and common on relatively new lava. (Photograph courtesy of J. Thomson.)

Although (by definition) primary succession occurs in unvegetated areas, dispersal distances may not be great enough to limit the ability of particular species to colonize a disturbed area. At Mount St. Helens and Kilauea, remnants of the prior vegetation are scattered throughout the disturbed sites, where they serve as important sources of propagules. Most other primary succession, such as that on new beaches and sand bars in rivers, occurs on smaller disturbed sites, where dispersal distances may be short.

Climax Revisited

The concept of the climax—the hypothesized static, deterministic end point of succession—once dominated ecologists' thinking about succession. According to this idea, once a community has reached the climax state, it stops changing, unless a disturbance occurs that resets the community back to an earlier seral stage. There are serious difficulties with this concept, which have become increasingly apparent over the years.

The notion of a successional climax was devised by Hult (1885) and developed over the next several decades (Clements 1916). One of its proponents was Frederick Clements. His ideas may have been influenced by classical Greek philosophy, a major part of the intellectual tradition of his day, which emphasized idealized types in nature. Ecologists of the time viewed the natural state of the community as being unchanging and disturbance as an unnatural, external process. As we have seen in this chapter, however, various kinds of disturbances are intrinsic to most communities. As plant ecologists have recognized this fact in recent decades, many have called into question the entire concept of the climax as the end point of succession.

The result was a conceptual shift to considering patterns of stasis and change at different spatial scales. Consider a landscape dotted with many different communities, each containing many different patches. A single small patch may always be in a state of flux as populations change and one species replaces another (as in van der Maarel's carousel model; see Chapter 10). At this scale, change may seem directionless and at the constant mercy of unpredictable factors such as climate and herbivory. In contrast, the community as a whole may be undergoing a slow successional process of gradual replacement of one set of species by another. At this larger scale, disturbances may still occur every century or so, resetting the successional cycle. Finally, the entire landscape may consist of a mosaic of communities at different points along the successional cycle. Although each community is changing, the proportion of the entire landscape at each successional stage remains about the same. Thus, a dynamic equilib-

rium may exist at the scale of the landscape, even though no single community is at equilibrium. We will return to this issue of hierarchies of scales when we discuss landscapes in more detail in Chapter 17.

Today, plant ecologists recognize that communities and landscapes never attain a constant, unchanging state. Rather, communities may attain a dynamic equilibrium, if an equilibrium is reached at all. Long-term climate changes (see Chapter 21), introduction or evolution of new species, and landform changes such as erosion, mountain building, and continental movements mean that the world is always in flux. At best, these changes occur slowly enough that communities or landscapes are at quasi-equilibrium. Thus, one's view of change and equilibrium depends on the scale considered.

Summary

In this chapter, we have explored the process of succession and the implications of that process for our understanding of the nature of communities. Succession starts with disturbance. All communities experience disturbances. Those disturbances can be small (a limb falling from a tree, a badger creating a mound) or large (a forest fire, a volcano erupting). They can occur every year, or once in a millennium. They can result from a multiplicity of factors—fire, wind, rain, snow, animals, disease. After a disturbance, new species may colonize the site. These species then interact with one another so that the community changes over time, resulting in the process of succession. Finally, if disturbances are relatively infrequent, the community may reach a state of dynamic equilibrium. Earlier ecologists thought that a static end point to succession, called the climax, was the usual state of affairs in plant communities. There are now strong reasons to think that even a state of long-term dynamic equilibrium may be exceptional.

In this chapter and the previous ones, we addressed a number of questions about the nature of communities. We began by asking whether communities are entities in their own right, or merely collections of populations that happen to co-occur. We described a number of ways in which species interact. These interactions can be direct or indirect, positive or negative. For example, competition for common nutrients is a direct, negative interaction. In contrast, the stabilization of a sand dune by *Elymus canadensis* allowing colonization by other species is an indirect, positive interaction. These interactions help to shape communities. Succession occurs, in part, because of these interactions. Thus, a community is more than simply the sum of its constituent species. How those species interact is also important in shaping community composition and structure.

Additional Readings

Classic References

Cowles, H. C. 1899. The ecological relations of the vegetation on the sand dunes of Lake Michigan. *Bot. Gaz.* 27:95–117, 167–202, 281–308, 361–391.

Cooper, W. S. 1923. The recent ecological history of Glacier Bay, Alaska. II. The present vegetation cycle. *Ecology* 4:223–246.

Oosting, H. J. 1942. An ecological analysis of the plant communities of Piedmont, North Carolina. *Am. Midl. Nat.* 28:1–126.

Contemporary Research

Chapin, F. S., L. R. Walker, C. L. Fastie, and L. C. Sharman. 1994. Mechanisms of primary succession following deglaciation at Glacier Bay, Alaska. *Ecol. Monogr.* 64:149–175.

Halpern, C. B., P. M. Frenzen, J. E. Means, and J. F. Franklin. 1990. Plant succession in areas of scorched and blown-down forest after the 1980 eruption of Mount St. Helens, Washington. *J. Veg. Sci.* 1:181–195.

Keeley, J. E., C. J. Fotheringham, and M. Morais. 1999. Reexamining fire suppression impacts on brushland fire regimes. *Science* 284:1824–1832.

Phillips, D. L. and D. J. Shure. 1990. Patch-size effects on early succession in southern Appalachian Forests. *Ecology* 71:204–212.

Additional Resources

Johnson, E. A. 1992. *Fire and Vegetation Dynamics*. Cambridge University Press, Cambridge, UK.

Pickett, S. T. A., and P. S. E. White. 1985. *The Ecology of Natural Disturbance and Patch Dynamics*. Academic Press, Orlando, FL.

Pyne, S. J., P. L. Andrews, and R. D. Laven. 1996. *Introduction to Wildland Fire*, 2nd ed. Wiley, New York.

14 Local Abundance, Diversity, and Rarity

*I*n this chapter we address a series of questions that raise fundamental issues in ecology as well as having important implications for applied ecology and conservation. What determines the relative abundances of the species in a community? Are abundant species competitively superior to less common species, or are they more abundant for other reasons? Why are some species rare, others common, and still others invasive? Our focus in this chapter is mainly at the scale of communities; we return to some of the same issues at larger scales in Chapter 20.

The abundances of individuals within species are only part of the story, however. G. Evelyn Hutchinson asked a deceptively simple question in the title of his 1959 paper, "Homage to Santa Rosalia, or why are there so many kinds of animals?" Ecologists are still probing the implications of this general question. What determines the number of species that can coexist in a community? The number of species in a community might be determined largely by deterministic processes such as competition, or by chance processes, such as what species are able to reach the community. How do species interactions affect the ability of a new species to become integrated into the community? It may be that biotic and abiotic factors make a plant community more susceptible or more resistant to invasion by exotic species.

Much effort has gone into looking at how processes that determine local abundances play out within a single community. More recently, there has been a growing recognition that landscape and regional processes can also be very important in determining the relative numbers of individuals per species as well as the number of species in communities. In this chapter we examine the interaction between local and regional processes from a local perspective. In Chapter 17 we will look at landscape-level processes that determine population dynamics and species distributions, and in Chapter 20 we will consider abundance and diversity at regional and global scales.

Dominance

Even a cursory walk through any plant community will reveal that not all species are equally abundant. Typically, at least in temperate communities, most of the individuals (or biomass) in a community will belong to one or a few species, while many other species will be rare. The most conspicuous and abundant species (numerically or in terms of biomass) in a plant community are called the **dominants**. Dominant species are often used to characterize plant communities: a cattail marsh, a pine barren, an oak-hickory forest, or a saguaro "forest" are all characterized by their most conspicuous or abundant plants.

Are Dominant Species Competitively Superior?

The term "dominant" implies competitive superiority. Are numerically dominant species really competitively dominant as well? While plant ecologists have generally assumed that this is the case, few have directly tested that assumption. We have already talked about differing views on the consistency of competitive hierarchies in Chapter 10. Here we examine the extent to which numerical abundance is determined by competitive ability. Random processes, such as which species happen to occupy a site first, can also contribute to the pattern of species abundances at a site. The current pattern of abundances in a community may also be temporary. For example, if a site is undergoing succession following a disturbance (see Chapter 13), one species may be more abundant than another because it tends to dominate during early succession, but may be displaced as time passes.

Timothy Howard and Deborah Goldberg (2001) used field and garden experiments on herbaceous perennials in a Michigan old field to test whether rankings of competitive ability were related to observed species abundances in the field. Competitive ability at the stage of seedling germination and early growth was strongly related to species abundance. The effects of adult competitors on germination and seedling growth were small for the most abundant species, *Poa compressa* (bluegrass, Poaceae), and large for the least abundant species, *Achillea millefolium* (yarrow, Asteraceae). In contrast, the competitive ability or survivorship of adult plants had only a limited relationship to species abundance. Other researchers who have examined the effects of competitive ability at single life history stages (mostly early seedling growth) have generally reported positive relationships between abundance and competitive ranking.

Another approach to testing whether numerical dominants are really competitive dominants is to experimentally remove the dominant species from a community and observe the effects on community structure. Gurevitch (unpublished data) found that removing the dominant species from an old-field community had different effects depending on the species removed. Removal of one of the dominants, *Dactylis glomerata* (orchard grass, Poaceae), had little influence on other species, but removal of another, *Solidago rugosa* (wrinkle-leaved goldenrod, Asteraceae), had dramatic effects. Removal of *S. rugosa* from a plot resulted in an increase in species richness by permitting the persistence of species that had disappeared elsewhere in the old field

in the process of succession and by allowing other species to increase in abundance much more than they did in nonremoval plots. This finding implies that the goldenrod was not only numerically dominant, but also competitively superior.

Other researchers have also reported community responses to the removal of dominant species. For example, Kutiel et al. (2000) found that removal of dominant shrubs on coastal dunes in Israel led to greater abundance of annual species and therefore higher diversity. Removal of the C_3 grass *Lolium perenne* (English rye grass, Poaceae) in a New Zealand grassland resulted in increased species diversity, but removal of C_4 grasses did not have that effect, presumably because no other species in the community had a phenology similar to that of the C_4 grasses (Wardle et al. 1999).

Abundance Curves

One way of describing hierarchies of relative abundance is the use of **abundance curves** (Figure 14.1; also known as dominance-diversity curves). Abundance curves offer an alternative to diversity indices (see Chapter 12) for expressing the number and relative abundances of all the species in a community (or sample, or stand). Because they are a graphical presentation rather than an attempt at condensing everything into a single number, they avoid some of the limitations of diversity indices. Abundance curves can provide an excellent graphical contrast among different kinds of community structures—a curve representing a community with a high degree of dominance by a single species and low species richness will have a very different shape from one representing a highly diverse community without strong dominance, with many common species, and with a high total species number. However, these curves are more useful for descriptive purposes than for quantitative comparisons.

During the 1960s and 1970s it was suggested that abundance curves could serve as an indicator of the processes responsible for community structure, but this effort proved disappointing. Robert MacArthur (1960) proposed that the shape of a curve was created by the way in which species divided up the available energy in a system. Each species was hypothesized to monopolize a random fraction of that energy. This theory is referred to as the broken-stick model because the resulting pattern is the same as would be obtained if one took a stick (representing the available energy) and broke it into pieces at random (Figure 14.1, curve A).

Others demonstrated, however, that one could obtain an identical pattern through a strictly deterministic process. Depending on the assumptions used, curves of any shape can be generated, but such assumptions are extremely difficult to verify. Assume, for example, that the most dominant species can preempt 50% of

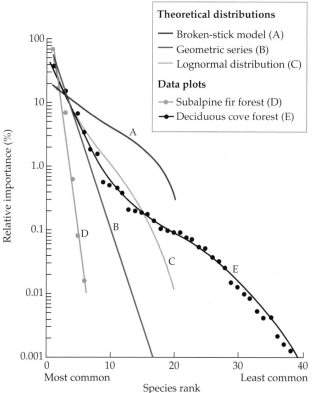

Figure 14.1
Examples of both theoretical and empirical abundance curves. The vertical axis is the relative abundance of a species in a community—that is, the percentage of the community that it represents. Besides numbers of individuals, the relative abundance could be measured in terms of biomass, cover, frequency, or a combination of measures. As is often the case, the vertical axis is shown at a log scale. The horizontal axis arrays the species of the community in rank order, from most common to least common. The most common species has rank 1 (left side), the next most common species has rank 2, and so forth. The green curves represent three theoretical distributions. The black and gray curves represent two sets of empirical data for vascular plant communities in the Great Smoky Mountains of Tennessee—subalpine fir forests and deciduous cove forests. (After Whittaker 1975.)

given dominance pattern by examining the shape of the abundance curve for a given community. Instead, ecologists have shifted their attention to working out the mechanisms causing dominance patterns within different communities.

Rarity and Commonness

Species differ greatly not only in how abundant they are in any one place, but in the number of places in which they are found. Some species have a wide geographic distribution, while others are **endemic** (found in only one place). All species are rare when they first evolve; some go on to become common, while others remain rare. An understanding of rarity and the processes responsible for it is central to the preservation of endangered and threatened species. Endangered and threatened species are defined as being rare, although not necessarily globally rare. A species can be rare and listed as endangered in one state of the United States while being common in another state. In the United States and many other countries, the official listing of a species as endangered triggers a host of laws and regulations concerning how the species and its habitat must be protected. Thus, the topic of rarity is fraught with legal, political, and economic consequences.

The Nature of Rarity

While a plant species may be rare, common, or invasive (see below), these categories are not necessarily fixed characteristics of the species. A remarkable example is *Pinus radiata* (Monterey or radiata pine, Pinaceae) (Lavery and Mead 1998). Once a rare endemic limited to a narrow strip of the California coast, Monterey pine has escaped from timber plantations in the Southern Hemisphere to become a major invasive pest in a number of countries. In its native habitat, the trees are small-statured and gnarled and, thus, are not used as timber. In plantations in the Southern Hemisphere, however, Monterey pine grows rapidly into large, straight trees. Chile, South Africa, New Zealand, and Australia all now have major naturalized populations of this species, and its geographic ranges and population sizes in these countries vastly exceed those in its native habitat. We will return to the factors that may make a species invasive after taking a look at some ecological aspects of rarity and commonness.

Deborah Rabinowitz (1981) identified three different aspects of species distributions: geographic range (wide or narrow), habitat specificity (broad or restricted), and local abundance (somewhere large or everywhere small; Table 14.1). These three aspects result in eight possible combinations of characteristics, and a given species could display any one of those combinations. Only one of the eight possible combinations—

the available energy in the system, that the next most dominant species can preempt 50% of the energy that remains, and so forth. This deterministic process would produce a geometric series (Figure 14.1, curve B). On the other hand, if dominance were determined by many independent factors that affected different species in different ways, the curve would be lognormal (Figure 14.1, curve C).

The problem with deciding among these alternatives (random division, preemption, or many factors) is that all three models (broken stick, geometric, and lognormal) could be fit to any given set of data. Thus, it is not possible to decide which process is responsible for a

Table 14.1 A scheme for describing eight categories of commonness and rarity based on three traits

Geographic range		Wide		Narrow	
Habitat specificity		**Broad**	**Restricted**	**Broad**	**Restricted**
Local abundance	**Somewhere large**	Common	Predictable (habitat specialists)	Unlikely	Endemic
	Everywhere small	Sparse		Unlikely	Rare on all counts

Source: Rabinowitz 1981.

wide range, broad habitat specificity, and large local abundance somewhere—is classified as commonness. The other seven each constitute some form of rarity.

Commonness and rarity are the result of processes that operate at local, regional, and even global scales. Local differences in species abundances may or may not be reflected at larger scales, and these patterns may change through time. Plants that were once common may decline in numbers, while rare species may, over time, become common or even dominant, particularly in new or changed environments. Local and regional patterns may be uncoupled or may be linked in a variety of ways. By chance alone, a species may be introduced to a new site to which it is well adapted and may become highly abundant, as the Monterey pine has done. From there, it may spread and become abundant everywhere in a broad region. In this case, events at a local level spread to a regional scale. Conversely, processes at a regional scale (such as gene flow among sites; see Chapter 5) may result in a species becoming adapted to varying conditions across a wide geographic range. A high average level of adaptation, coupled with metapopulation processes (see Chapter 17), may result in the species reaching high abundances everywhere it occurs. In this manner, regional-scale processes can determine local-scale patterns.

Patterns of Rarity and Commonness

While all possible combinations of local and regional patterns of commonness and rarity are conceivable, what combinations do we actually find in nature? There is often a positive relationship between abundance and geographic range (Figure 14.2): species with large population sizes tend to be widespread, while species with small geographic ranges tend to have small population sizes.

In a similar fashion, species that can live in a wide variety of habitat types tend to have large geographic ranges because in any given area there is likely to be some suitable habitat. While these relationships are the typical ones, exceptions exist. Some species are habitat specialists with small geographic ranges, yet dominate the communities they are found in. An example is the mint *Mentha cervina* (Lamiaceae), a specialist of montane wetlands of the eastern Mediterranean. While it is found in a very limited range of sites in central Spain, where it is found, it exists at high densities.

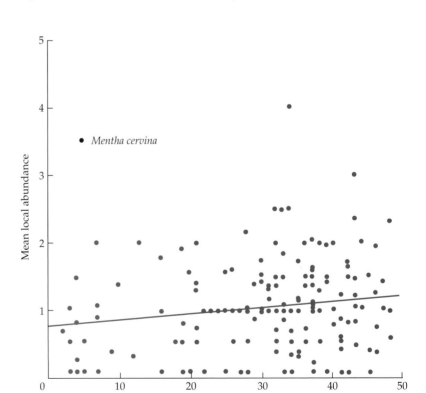

Figure 14.2
Correlation of mean local abundance and geographic range for vascular plants in montane wet meadows in Spain ($r_S = 0.17$, $P < 0.01$). *Mentha cervina* is an exception to the general trend that species with the largest geographic ranges also have the highest abundances where they are present. (Data from Rey Benayas et al. 1999.)

Table 14.2 Percentages of species falling into each of eight commonness and rarity categories in three studies

(A) Vascular plants of the British Isles (*n* = 160)

Geographic range	Wide		Narrow	
Habitat specificity	Broad	Restricted	Broad	Restricted
Local abundance large	36	44	4	9
Local abundance small	1	4	0	2

(B) Vascular plants of montane wet meadows in Spain (*n* = 220)

Geographic range	Wide		Narrow	
Habitat specificity	Broad	Restricted	Broad	Restricted
Local abundance large	14	7	12	13
Local abundance small	2	10	2	39

(C) Trees of a tropical rainforest in Manú National Park, Peru (*n* = 381)

Geographic range	Wide		Narrow	
Habitat specificity	Broad	Restricted	Broad	Restricted
Local abundance large	68	19	0	0
Local abundance small	7	6	0	0

Sources: (A) Rabinowitz et al. 1986 ; (B) Rey Benayas et al. 1999 ; (C) Pitman et al. 1999.

In an analysis of the entire native British flora, Rabinowitz and colleagues (Rabinowitz et al. 1986) found that 36% of species were common, with wide geographic ranges and large local abundances; almost 60% were restricted to particular habitats, while 15% had small geographic ranges and 7% had small maximum population sizes (Table 14.2A). This study was the first to analyze these patterns and is often used as a basis for generalizations about rarity. However, other researchers have found that these patterns vary depending on how one looks at the flora. A study by José Rey Benayas and colleagues considered the entire set of vascular plant species inhabiting a specialized habitat type, wet meadows in the mountains of central Spain (Table 14.2B). In this instance, the most frequent category was species with narrow geographic ranges, restricted habitat specificities, and small local abundances. In another study, Nigel Pitman and colleagues (Pitman et al. 1999) examined tree species in a highly diverse tropical rainforest in Manú National Park and the surrounding region in eastern Peru (Table 14.2C). Most of the tree species in this study were locally common in at least one plot and were found across a wide range of habitats (fewer than 25% were habitat specialists), and all had broad geographic distributions. This pattern contrasts strongly with the Rabinowitz et al. study of the British flora, in which many more species were habitat specialists (Ricklefs 2000). These results may have important implications for the conservation of tropical biodiversity because if most species are widespread, it will be easier to find land to preserve them.

We can also look at rarity from a taxonomic perspective. A study of rarity in the vascular plants of the United States and Canada found that taxonomic groups (genus, family, order, and class) with many species also tended to have higher than average numbers of rare species, while smaller groups had fewer rare species than expected (Schwartz and Simberloff 2001). This pattern would occur if large groups were more likely to produce new species through allopatric speciation (see Chapter 6), in which case, on average, the new species would be isolated, have small range sizes, and hence be more likely to be rare. Conversely, groups with high rates of extinction might lose their rare species first, so that small groups would be left with common species only.

Causes of Rarity and Commonness

Why is a particular species common or rare? The patterns described above suggest that ecological specialization may be an important cause of rarity. Species that are ecological specialists tend to have low abundances and small geographic ranges. This observation raises another question: why are there both ecological specialists and ecological generalists? The traditional explanation is that there are trade-offs that result in commonness or rarity. By specializing, according to this theory, a species can better exploit a particular niche. The theory predicts that local abundance should be negatively correlated with the number of habitats occupied because a specialist should do very well where it is found, but be found in only a few places. However, this prediction is contrary to what is typically observed (see Figure 14.2).

The "environmental control" or "superior organism" theory was proposed and later elaborated by James Brown (1984, 1995) to account for differences among species in abundance. Brown began with the premise that species vary in their abilities to exploit nature. This premise is built on Hutchinson's (1957) concept of the multidimensional niche. A **niche** is the set of all of the conditions and resources that determine the ability of the individuals of a species to survive and reproduce. A **fundamental niche** is the set of conditions and resources that the species could use in the absence of competitors (or, by extension, predators, pathogens, and parasites).

Brown assumed that some species have broad fundamental niches, while others have narrow ones. The center of a species' geographic range, he proposed, is the place where it can exploit the widest range of resource combinations and, therefore, the widest number of habitats. In this region it should be at high abundance. As the distance from this center increases, the conditions favoring the species become less common, because, from its perspective, environmental conditions are deteriorating (since environmental variables tend to be correlated with one another). Therefore, the species' local habitat specificity narrows and its abundance decreases. Eventually, the edge of its geographic range is reached. Those species that begin with a broad niche will have a large geographic range, while those that begin with narrow niches—that is, with restricted environmental tolerances—will be rare and have a small geographic range. This theory has similarities to the centrifugal theory of community organization (see Chapter 10).

A major problem with Brown's theory is that it does not account for the persistence of inferior species. According to the theory, the superior performers should simply overwhelm the inferior ones, eventually replacing them. In all likelihood, both mechanisms (the existence of trade-offs and the superior adaptations of some species) play a role in determining patterns of commonness and rarity.

Evidence relating to Brown's theory is mixed. An analysis of the abundance and occurrence of plant species in 74 landscapes across the globe found support for his theory (Scheiner and Rey Benayas 1994). Two predictions of the theory were confirmed: first, there was a positive correlation between the number of patches a species occupied and its local abundance, and second,

the most frequent class of species in any given landscape was those found at only a single site, especially in middle- and low-latitude areas.

In contrast, Mark Burgman (1989), in a test using the flora of the mallee vegetation of western Australia, did not confirm the major predictions of Brown's theory. This region has a mediterranean climate and highly variable soil conditions. The vegetation of the region is characterized by high species richness, with many endemic species, and is dominated by low shrubs and herbaceous perennials. Burgman's analysis strongly refuted Brown's "environmental control" theory. It found no support for the hypothesis that rare species have smaller fundamental niches than do widespread, abundant species. The low abundances of the rare plants in this study were not caused by limited environmental tolerances, nor was the high abundance of common plants attributable to wide environmental tolerances. Burgman explained these results by saying that even if a species can tolerate a broad range of environmental conditions, if those combinations of conditions and resources are found in only limited geographic areas or patches, those species will be limited as well.

The flaw in Brown's theory, therefore, could be the assumption that the range of variation in environmental conditions is positively related to the geographic extent of the range. Imagine, for example, that a plant species can tolerate a wide range of soil textures, from sandy to loamy soils, but that in the environment in which it is found, there are no sandy soils, a few patches with loamy soils, and large areas with clay-loam and clay soils. Then, even though the species could do well on many different soil types, in that environment it will be restricted to a small number of patches and will be rare. The rare species in Burgman's study appeared to be rare because, although they could tolerate a wide range of environmental conditions, those conditions were found at few sites. In contrast, the conditions tolerated by the common and widespread species were found in many places in Western Australia.

Gaston (1994) and Kunin (1997) list a number of hypotheses proposed to explain what makes a species rare. They suggest that the explanations for the immediate causes of rarity fall into three categories: ecological specialization, lack of dispersal, and historical accident. While various explanations have been proposed for why species become ecologically specialized (Rosenzweig and Lomolino 1997; see also other chapters in Kunin and Gaston 1997), the causes of specialization are not well understood, and the evidence connecting specialization with rarity is limited.

Colleen Kelly (1996; Kelly and Woodward 1996) searched for ecological correlates of commonness and rarity in the floras of Great Britain and Crete. She asked whether there are particular attributes of species that give them wide geographic ranges. She looked at a number of attributes, including growth form, pollination syndrome, and dispersal mechanism. Only growth form (trees were more common than shrubs) and pollination syndrome (wind-pollinated plants were more common than animal-pollinated plants) were found to explain geographic range.

Invasive Species and Community Susceptibility

Ecologists and environmentalists have become alarmed at the tremendous spread of invasive species in the past few decades because their spread is often associated with the displacement of native species. In some cases invasive species have drastically altered community and ecosystem patterns. Many definitions for **invasive species** have been proposed, but the term generally refers to species that are rapidly expanding outside of their native range. Invasive species may be native to the general region, but usually are **exotics** that are far from their native habitats, often from other continents. In most cases they have not dispersed naturally, but have been brought either purposely or inadvertently to a new habitat by people. A large proportion of the plants that have become seriously invasive were deliberately introduced, planted, and cultivated by people (Figure 14.3). Between 2000 and 3000 exotic plant species have been introduced into the United States, most within the last 100 years.

Invasiveness is related to, but not exactly the same as, being **weedy** or **ruderal**. Weeds are generally defined as uncultivated species that proliferate in agricultural settings, interfering with crop production, although the term has been used in a broader context on occasion. Weeds are often exotic species, but may sometimes be natives that thrive in the conditions of high nutrient and light availability and frequent soil disturbance typical of agricultural fields. Ruderal species possess traits that allow them to do well in temporary habitats, including agricultural fields and roadsides, as well as areas subject to frequent disturbances such as rockslides or flash floods. The word "ruderal" derives from "rude," in the sense of "wild" or "rough." While "weedy" and "ruderal" are often used synonymously, the former is typically used in agriculture and the latter is used in ecology.

Ecologists are concerned with a number of questions regarding invasiveness. In this section we ask what makes species successful invaders and what factors make a community susceptible or resistant to invasion by exotic species. In Chapter 22 we return to this topic to look at the ecological effects of invasive species.

Why Do Some Species Become Invasive?

Ecologists have struggled to predict which plants might become invasive for many years. Charles Elton (1958) and Herbert Baker and Ledyard Stebbins (1965) attacked this question four decades ago, tabulating the characteristics and attributes of species known to be success-

(A)

(B)
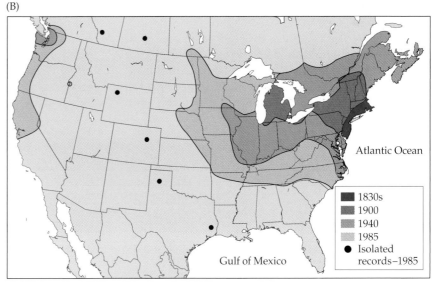

Figure 14.3
(A) *Lythrum salicaria* (purple loosestrife, Lythraceae) is an invasive wetland species that now dominates many marshes and lakes in the northeastern United States. (Photograph courtesy of P. O'Neil.) (B) Purple loosestrife spread across North America over a 150-year period. It may have been introduced into the United States in the 1700s, but it did not begin spreading for a half-century or more. (After Thompson et al. 1987.)

ful invaders. But the attempt to successfully predict invasiveness has been a frustrating exercise. Different plant invaders have contrasting strategies: for example, one might have a very high seed output, while another invests in extensive rhizomatous spread. Most of the data addressing this question are qualitative observations of unrelated organisms; such observations are difficult to compare or synthesize across studies. Furthermore, data on failed invasions are, of course, unavailable, leaving us with only half the picture.

Recently, however, there has been progress in understanding what might make some plant species likely to become successful invaders. Rejmánek and Richardson (1996) quantified the traits of 24 species within a single genus—the pines—12 that are highly invasive and 12 that have been widely planted worldwide but have never become invasive. Because the pines are evolutionarily closely related and are ecologically and physiologically similar in many respects, this is a powerful approach for isolating those characteristics that might be closely linked to invasiveness. The researchers analyzed ten life history traits, and found that three were most closely related to invasiveness: reproduction at younger ages, smaller seeds, and large seed crops produced at shorter intervals. Rejmánek and Richardson then applied this information to 34 other pine species that were not in the original data set, but whose invasive status was known, and accurately predicted which of those species were invasive. They also offered a tentative set of rules for predicting woody plant invasiveness (Table 14.3). Mack et al. (2000) discuss more generally the (usually inadequate) attempts of ecologists to predict invasiveness.

Plant invaders may have the ability to grow rapidly in a new environment for several reasons. They may escape from herbivores and parasites that control them in their native environments (see Chapters 7 and 11). Human disturbances may reduce competition from other plants, allowing invaders to become established and multiply. "Empty niches" may exist in the new communities in which an invader may be able to proliferate. An invasive shrub, for example, might succeed in a forest that has no native shrubs. Finally, invaders may alter ecosystem characteristics in ways that favor their own subsequent increase (see Chapter 22).

Williamson and Fitter (1996) proposed the "tens rule" in an attempt to make (and explain) statistical generalizations about the success of invaders. The tens rule states that 1 in 10 of the plant and animal species brought into a region will escape to appear in the wild, 1 in 10 of those escaped species will become naturalized as a self-sustaining population without cultivation, and of those naturalized species, 1 in 10 will become invasive. Rules of this sort are appealing, but we need to know why (and how well) they hold and when to expect them to fail. Many invasions seem to follow the tens rule, but there are significant exceptions, suggesting that it should be applied with caution.

The process by which a species makes the transition from its first introduction to invasive status consists of several phases (Mack et al. 2000). Many invasive species have a long **lag phase** (Figure 14.4) during which their numbers are low and they are not particularly noticeable. This lag phase may be brief or may last well over a century. Because most introduced species disappear during this lag phase (see the tens rule above), it is difficult to tell during this period which introduced species are headed for extinction and which for invasive status.

Populations of successful invaders increase during this period even though their total numbers are still low. Many invasive plant species are introduced not once, but multiple times, and at multiple sites, during the lag phase. At some point, however, the population begins to grow rapidly and at an accelerating pace in both numbers and area occupied—at this point, the species has

Table 14.3 Tentative rules for predicting which woody seed plants are likely to become invasive

| Z^* | Type | Opportunities for vertebrate dispersal | |
		Absent	Present
> 0	Dry fruits and seed mass > 2 mg	Likely invasive	Very likely invasive
	Dry fruits and seed mass < 2 mg	Likely invasive in wet habitats	
	Fleshy fruits	Unlikely invasive	Very likely invasive
< 0	All	Noninvasive unless dispersed by water	Possibly invasive

Source: Rejmánek and Richardson 1996.
$^*Z = 19.77 - 0.15\sqrt{M} - 3.14\sqrt{J} - 1.21S$, where M = mean seed mass (mg), J = minimum juvenile period (years), and S = mean interval between large seed crops (years). Z is thus a function that integrates several life history traits.

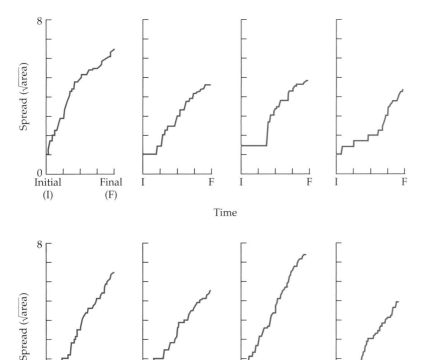

Figure 14.4
Some outcomes for a spatially explicit simulation model of the spread of an invasive species in a new environment. In this model, about half the time, the rate of spread is slow at first, then becomes more rapid as the invading species begins to more fully occupy the new habitat. This pattern corresponds to the lag phase seen in many actual plant invasions that have been documented, in which the rate of spread is unnoticeable at first, sometimes for decades to hundreds of years, but at some point becomes very rapid. (Model results taken from Hastings 1996.)

become invasive. Ultimately the invader reaches new geographic and ecological limits at which its population stabilizes.

What Makes a Community Susceptible to Invasion?

Various predictions have been made regarding the factors that make communities vulnerable to invasion. Some kinds of communities appear to be highly susceptible to invasion by exotic species, while others appear better able to resist invasion. In his classic study reviewing invasions by plants and animals, Elton (1958) hypothesized that invaders were most likely to become established in disturbed and species-poor communities.

While most ecologists continue to believe that disturbance is a critical factor in promoting invasion, there is conflicting evidence on its role, and its effects may differ among communities (Lodge 1993). In a native grassland community in Kansas, for example, Smith and Knapp (1999) showed that increased disturbance (annual burning) strongly decreased invasion by exotic species.

Furthermore, disturbance clearly encompasses more than a single factor; it may reflect changes in factors as diverse as soil surface microclimates, predator-prey relationships, resource availability, and competitive interactions (Orians 1986). Rigorous data appropriate for testing the importance of disturbance, or of its many different effects, remain very limited (Lodge 1993; Rejmánek 1989). In examining the role of disturbance in promoting or inhibiting invasion, elucidating the effects of specific factors (such as soil disturbance, logging, or fire) is a much more satisfactory approach than lumping those factors together into a general category.

Likewise, evidence as to whether species-poor communities are indeed more vulnerable to invasion is limited and contradictory. In a recent review, Levine and D'Antonio (1999) evaluated data and theory linking diversity and biological invasions. They found that while both classic and recent theory predicted that more diverse communities would be more resistant to invasion (e.g., Elton 1958; Tilman 1999), data from experimental and descriptive studies were ambiguous. Some studies supported this predicted relationship and others refuted it.

Two recent studies in grasslands showed negative relationships between vulnerability to invasion and both species and functional group diversity (Tilman et al. 1997; Symstad 2000). Other recent studies, however, have shown the opposite pattern. Thomas Stohlgren and his colleagues (1999) studied the relationship between plant species richness and invasions in the Colorado Rockies and in the Central Great Plains of western North America across a wide range of plant community types (including forests, meadows, and grasslands) and over

(A)

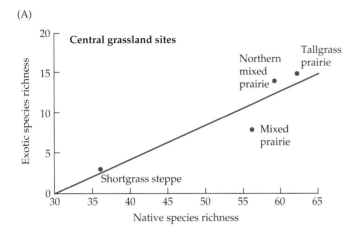

(B)

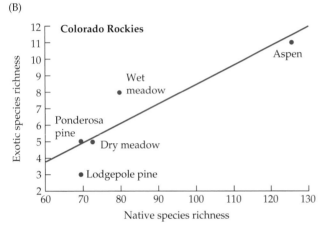

Figure 14.5
The relationship between native species richness and exotic species richness in the central Great Plains grasslands (A) and in the Colorado Rockies (B). The results are based on lists of species in 1000-square-meter plots of each vegetation type. Results of linear regressions (grasslands: $r^2 = 0.85$, $P = 0.078$; Rockies, $r^2 = 0.81$, $P = 0.038$). (After Stohlgren et al. 1999.)

a range of spatial scales. They concluded that exotic species primarily invaded areas of high species richness (Figure 14.5). In a survey of 184 sites distributed globally, Lonsdale (1999) also found a positive relationship (on a landscape scale) between the degree of invasion and native species richness.

The same positive relationship between native species richness and invasion by exotics holds in systems as diverse as the South African fynbos (Higgins et al. 1999), New Zealand mountain beech forests (Wiser et al. 1998), and Great Plains grasslands (Smith and Knapp 1999). In a review of plant invasions globally, Levine (2000) states that, in general, the most diverse plant communities are the most likely to be invaded.

Variation in resource availability has been proposed as an underlying cause for the positive association between native and invasive species diversity. This factor

may be particularly important in systems on soils with low soil nutrient resources, such as soils on glacial outwash (e.g., Long Island, New York) and very old soils (e.g., Australia). A study of Long Island forest communities found that soil characteristics are critical in the facilitation or inhibition of invasion by exotic species (Howard et al., unpublished). There was a strong positive relationship between both nitrogen and calcium levels and plant invasions, and a weaker but also positive relationship between phosphorus levels and invasions. This study also found a strong positive relationship between native species diversity in forests and the degree to which those forests had been invaded by exotic species.

These results echo the pioneering experimental work of Hobbs (1989) in Australia, which showed that disturbance alone did not make a plant community susceptible to invasion, but that disturbance coupled with the addition of soil nutrients greatly facilitated invasion. Clearly, more information is needed before we can definitively answer why plant communities differ so much in their likelihood of becoming invaded by exotic species.

Abundance and Community Structure

As we have just seen, species can be rare or common within a community. They may be long-term or short-term dominants, or they may be invading the community. If we shift our focus from individual species to the entire plant community, we recognize that communities may differ greatly in the number of species present and in the degree to which individual species are dominant. In other words, there are great differences in species richness and diversity among communities. Total biomass also differs greatly among communities, as does **productivity**—the rate of energy flow through a trophic level (see Chapter 15). What are some of the causes of these contrasts among communities? Are variations in these different aspects of communities causally related?

Various hypotheses have been proposed to explain the observed relationships between relative species abundances and the productivity and diversity of communities, and we briefly discuss some of those hypotheses here. We examined some of the causes of species diversity in Chapters 10, 11, and 13. Hypotheses explaining differences in species diversity at local and regional scales have been reviewed by Rosenzweig (1992), Huston (1994), Ricklefs and Schluter (1993), and Tilman (1982). We will examine patterns and possible explanations for differences in species diversity at larger spatial scales in Part V.

Productivity and Diversity

Productivity can affect community structure in a variety of ways. If rates of herbivory and decomposition differ little between communities, a community with higher productivity will have a greater total biomass, or **standing crop** (the total living biomass in an area): individual plants will be larger, or there will be more of them, or both. In most terrestrial ecosystems, this strong positive correlation between standing crop and productivity sets the stage for many important processes that result in structural differences among communities.

Differences in productivity can have dramatic effects on physiognomy and community structure. If higher productivity, for example, gives rise to larger individual plants, the community may have trees instead of shrubs, or shrubs instead of herbaceous perennials. We will examine these differences in more detail in Chapters 18 and 19 when we look at global vegetation patterns. More subtly, higher levels of productivity can lead to larger, more vigorous individuals within a species, or to a change in the particular species with a given growth form occupying a site (for example, a large tree such as eastern white pine, *Pinus strobus*, instead of a small tree such as jack pine, *P. banksiana*, or even the replacement of pine species with hardwoods).

Changes in productivity can also result in changes in the number and diversity of species in a community. The nature and causes of the relationship between productivity and species diversity have been disputed by ecologists for many years (we visit the converse question of the effect of increased diversity on productivity later in this chapter). The nature and shape of the relationship between productivity and diversity in plant communities depends on spatial scale. Here we look at the nature of this relationship at local to regional scales. In Chapter 10 we saw how competition affects this relationship, while in Chapter 20 we examine the patterns observed at larger spatial scales.

Preston (1962) argued that when the total area (or, by extension, the resources) available to support species in a given region is small, then the numbers of individuals of the rarest species will be too low to maintain the existence of that species in that region. Therefore, as the available area (or resources) increases, the number of species will increase.

Wright and his colleagues (reviewed by Wright et al. 1993) extended Preston's theory to explicitly predict that an increase in productivity should lead to an increase in species diversity. They began by assuming that greater productivity reflects a greater amount of energy available to the system, resulting in a greater number of individuals at a site. If a site can support more individuals, then more species can each have the minimum number of individuals necessary to prevent their extinction. Wright et al. allowed, however, that the shape of the relationship between productivity and diversity could be unimodal (a hump-shaped curve), or even negative, under certain circumstances or at smaller scales. If increased productivity resulted in larger individuals, for

example, rather than a greater number of individuals, there might be no change in the number of species, or it might even decrease if the larger individuals crowded out smaller ones. Such declines in individual density and species richness are typically seen after fertilization in old fields and other herbaceous communities (Figure 14.6).

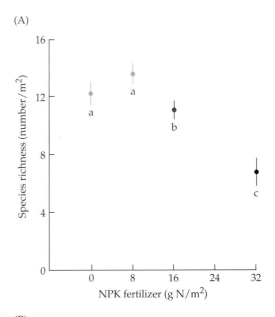

(A)

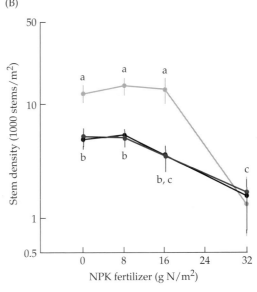

(B)

Figure 14.6
The results of experimental fertilization of 72 plots in an old field in northwestern Pennsylvania. A slow-release NPK fertilizer was applied in May and June, and all aboveground biomass was harvested and measured in September. (A) Species richness declined by about 50% at the highest level of fertilizer treatment. (B) Stem density declined nearly 80%, with the greatest decline in the experimental block with the highest density at the lowest level of fertilizer treatment. Means (bars are ± standard error) with different letters are significantly different ($P < 0.05$). (After Stevens and Carson 1999.)

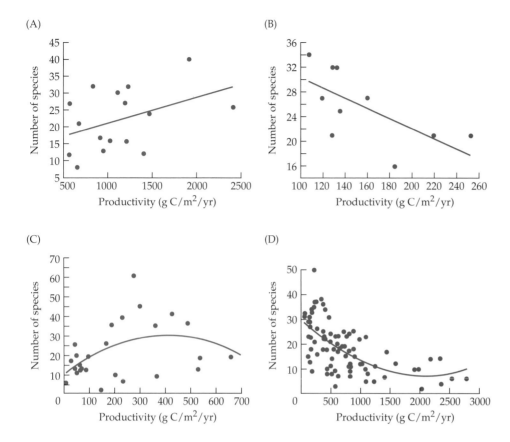

Figure 14.7
The relationship between productivity and species richness can have different shapes in different communities. All relationships are statistically significant. (A) Aboveground yearly productivity versus total number of tree, shrub, and herb species in a deciduous forest in the Great Smoky Mountains. (Data from Whittaker 1966.) (B) Aboveground yearly productivity versus total number of herb species in a grassland in western North Dakota. (Data from Redmann 1975.) (C) Standing crop versus total number of ground-layer species in grasslands and deciduous forests in Estonia. (Data from Zobel and Liira 1997.) (D) Standing crop versus total number of herb species in a fen in the United Kingdom. (Data from Wheeler and Shaw 1991.)

Many authors have asserted that at local and regional scales, there is generally a unimodal, or hump-shaped, relationship between productivity and diversity (e.g., Grime 1973; Whittaker and Niering 1975; Huston 1979; Tilman 1982; Rosenzweig 1992; Huston and DeAngelis 1994). That is, diversity increases as productivity increases up to a maximum at intermediate productivity, but then decreases with further increases in productivity. James Grace (1999) reviewed evidence for this relationship at small spatial scales in herbaceous communities and concluded that the data were consistent with such a unimodal pattern.

Gary Mittelbach and colleagues (2001) conducted a broad review of the literature on the relationship between productivity and diversity. They concluded that while unimodal patterns are found about 40% of the time, there are also many examples of positive patterns, negative patterns, and even U-shaped patterns (Figure 14.7). They emphasized that the shape of the relationship changes as a function of scale. Hump-shaped patterns, for example, are more common in studies that look across communities than within a single community (Figure 14.8). Nearly all studies of the relationship between productivity and diversity are carried out at a single scale, and almost no theories explicitly incorporate scale. The issue of scale and productivity-diversity relationships is further explored in Chapter 20.

Rosenzweig (1971) called the pattern of decline in diversity at high productivity the "paradox of enrichment." This pattern has been found for many different taxa, not just terrestrial plants. Plant ecologists have addressed this paradox experimentally by fertilizing plots and seeing how that treatment affects diversity. All of these experiments (reviewed by Grime 1973; Huston 1979; Tilman 1982) were done in herbaceous plant communities because these communities are likely to respond to treatment within a year or two. In most cases, a decline in diversity was found after fertilization. However, this decline occurs even when very low-productivity sites are raised to intermediate values (Gough et al. 2000). It may be that such short-term declines in diversity are due to changes in competitive dominance among species (see Chapter 10), while an increase in diversity requires the migration of new species into the plot, a longer-term process.

A number of reasons have been suggested for the increases in diversity observed as productivity increases from very low to moderate levels. As mentioned above, one explanation is that more energy or resources can support more individuals, leading to a greater number of species (Wright et al. 1993). Another is that very few species are adapted to survive the difficult conditions that underlie low community productivity (such as low temperatures, low water availability, infertile

(A) **Within communities**

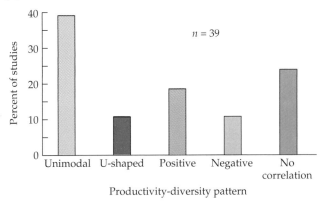

(B) **Across communities**

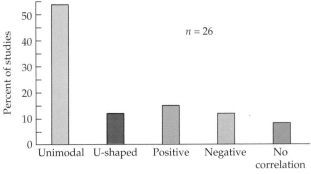

Figure 14.8
Percentage of studies showing various relationships between productivity and plant species richness at two ecological scales: (A) within communities and (B) across communities. (Data from Mittelbach et al. 2001.)

soils, or short growing seasons; Grime 1973). According to this hypothesis, as conditions become more favorable and productivity increases, more species are able to survive because the specialized adaptations necessary to survive in stressful conditions are not required. Yet another explanation is that in nutrient-poor environments, the best competitor for the limiting soil nutrient should dominate, while at intermediate levels of fertility (and productivity), spatial heterogeneity in soil nutrients should lead to the coexistence of multiple species due to niche differences among those species (Newman 1973; Tilman 1982; Tilman and Pacala 1993).

There are also many contradictory explanations for the often-reported decline in diversity at high productivity (see Table 20.4). Grime (1973) proposed that highly competitive species dominate the most favorable environments and suppress other, less competitive species (Grime 1973; see Chapter 10). Tilman (1982; Tilman and Pacala 1993) theorized that in the most fertile and productive environments, individual plants should be much

larger than in other environments. Denslow (1980) suggested that because habitats of intermediate productivity are far more common than those with very low or very high productivity, more species should have evolved to survive in the conditions presented by those intermediate habitats, and far fewer should be adapted to conditions found in environments of very high or very low productivity.

As you might gather, the question of the relationship between productivity and diversity is currently an active area of research. Because more than one process may contribute to the observed patterns, with different processes dominating in different circumstances, resolving their causes is difficult. While it is satisfying to reach a conclusion on such questions, it is the process of working out complex matters such as these that makes ecology an exciting and challenging science.

Niche Differentiation, Environmental Heterogeneity, and Diversity

Why does diversity exist? Another way of asking this question is, given that a positive relationship exists between the number of individuals and the number of species in a community, why is it that adding individuals leads to more species, rather than just more individuals of a single species?

One reason why a single species does not replace all the others in a community by outcompeting them is that different species are best adapted to different conditions. It is not possible to be best at everything simultaneously, and environments are commonly heterogeneous in both space and time. Adaptations to different conditions combined with environmental variation allow species to divide up the environment in various ways. Understory species, for example, are able to tolerate low light conditions, while canopy species are superior performers under conditions of full sunlight. In areas of outcrops and patchy, rocky soils, some species are able to dominate pockets of deeper soil, while others are found in the shallower soils on ridges and among boulders. Some plants are able to tolerate saline soils, or soils with heavy metals, while others are restricted to soils free of these materials.

Temporal heterogeneity may also offer opportunities for specialization. In the Sonoran Desert, which has both a winter and a summer rainy season (see Chapter 18), some annuals are specialized to germinate in response to winter rains in December, January, and February and grow in cool spring temperatures. They flower in March and are dead by May. Others germinate in response to summer rains in July and August, grow in the hot temperatures of summer, flower in September, and die in October. Because these two sets of species have different phenologies, they do not compete with each other. This specialization leads to species coexistence and, thus, increased diversity in the community. Explaining the maintenance of species diversity on the basis of environmental heterogeneity and niche differentiation also may depend on differences in species' competitive abilities under different conditions (see Chapter 10).

Gaps, Disturbance, and Diversity

Disturbances can create gaps in mature communities, particularly in forests, as we saw in Chapter 13. Researchers have proposed two different reasons why these gaps might increase community diversity. The first is that some species might be specialized to take advantage of the different conditions afforded by gaps of different sizes (the niche-partitioning hypothesis). If this were the case, certain species could be maintained in the larger community by being superior at colonizing and dominating small gaps, for example, while others could dominate intermediate or large gaps, and still others would largely be found only in intact forest.

The second reason why gaps might increase total forest diversity is based on chance, or **stochastic** (random) events. According to this second hypothesis, the plants that dominate any particular gap are those that happen to be there when the gap forms, or those with propagules nearby. Species coexistence in the case of niche partitioning would depend on competitive interactions, with different species being competitively superior in different gap types. In the case of stochastic events, competitive exclusion would be prevented because many competitively inferior species would be able to persist by chance, essentially by being in the right place at the right time.

Brokaw and Busing (2000) reviewed the evidence for the role of gaps in maintaining tropical and temperate forest diversity and evaluated the data in support of these two hypotheses. They concluded that gaps are indeed very important in maintaining species diversity in both temperate and tropical forests. They did not find any differences in the role of gaps in temperate and in tropical regions. The evidence pointed to a major role for stochastic events in allowing species coexistence in forests. There are clearly some species that are better than others at dispersing to and growing quickly in gaps (pioneer species). However, while there is some evidence for niche differentiation in different kinds of gaps, this evidence is much weaker than the evidence for the role of stochastic processes, and niche partitioning probably does not play a large role in maintaining diversity in forests. Other studies have agreed with these conclusions for both tropical and temperate forests (e.g., Brown and Jennings 1998; Busing and White 1997), although Schnitzer and Carson (2001) found evidence for niche partitioning in tropical forest gaps.

Gaps are not the only result of disturbance that may

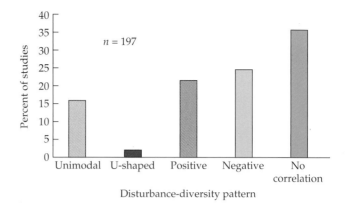

Figure 14.9
Percentage of studies showing various relationships between disturbance and species richness. (Data from Mackey and Currie 2001.)

affect diversity. If disturbances reduce the abundance, distribution, or height and spread of competitive dominants, competitively inferior species may be able to coexist with them. But if disturbances are too frequent or severe, few species will be able to survive. This observation led to the formulation of the **intermediate disturbance hypothesis** (Connell 1978), which states that species diversity should be highest at intermediate levels of disturbance. According to this hypothesis, competitive exclusion reduces species diversity at low levels of disturbance. When disturbances are very frequent, most existing species are destroyed, with insufficient time available for full recolonization before the next disturbance. The intermediate disturbance hypothesis has been tested in many different systems (not only in plant communities) and a conceptual model has been developed and formalized (reviewed by Huston 1994). In a survey of 197 studies, Mackey and Currie (2001) found that the most common pattern was no relationship between disturbance and diversity, while humped, positive, and negative relationships were all frequent (Figure 14.9). Thus, while it is clear that disturbance promotes diversity in some cases, what we do not know is how important it is relative to other factors in determining diversity.

Effects of Increasing Diversity

In the previous section we looked at some of the evidence for the effects of increased productivity on diversity. The direction of causation can also be reversed: some ecologists believe that productivity, as well as other ecosystem properties (see Chapter 15), should be enhanced by increased diversity. ("Diversity," in this context, can refer to either species diversity or diversity of functional groups.) It is easy to imagine how this might occur. Imagine, for example, a forest with no spring ephemerals; no species are growing actively in the early spring. If a spring ephemeral species were introduced into the forest, it would not displace any of the current species; rather, it would occupy an empty niche. It might use up nutrients that another herbaceous species growing later in the year would use. But if light were generally limiting for those species, the effect of the spring ephemeral on the growth of the later species might be small relative to the amount of new biomass it would add to the community.

David Tilman

This idea has been the subject of heated debate among ecologists. Shahid Naeem and his colleagues (Naeem et al. 1996), David Tilman (1996), and others have been strong supporters of the idea that an increase in biodiversity increases productivity as well as affecting other ecosystem properties. The idea has been strongly criticized by David Wardle and other ecologists, however (e.g., Wardle et al. 1997, 2000), for a number of reasons, particularly its proponents' interpretations of experiments and data. Wardle and his colleagues have argued that increased productivity and the other ecosystem properties attributed to increased biodiversity are mostly a result of traits of the particular dominant species and functional groups present, rather than of increased diversity itself. Recently, many of the protagonists in this debate published a joint paper (Loreau et al. 2001), in which they agreed

on what the main issues were and on the types of data needed to settle those issues.

Testing the Effects of Diversity on Ecosystems

The effect of diversity on productivity has been tested experimentally in herbaceous communities (old fields and prairies). One large recent experiment was conducted by a group of plant ecologists in Europe (Hector et al. 1999). They sowed seeds of the same species at two sites each in Germany, Greece, Ireland, Portugal, Switzerland, Sweden, and the United Kingdom. Among plots within each site, they varied the number of species and the number of functional groups. At the end of the growing season, they counted the number of species present in each plot and harvested all of the aboveground biomass to measure productivity. They found that greater diversity was related to greater productivity at all of the sites (Figure 14.10A). Most of this effect could be attributed to differences in the number of functional groups present in a plot (Figure 14.10B). Other experimental tests of the effects of species or functional group diversity on ecosystems have also been carried out (e.g., Tilman 1996; Hooper and Vitousek 1998; see Chapter 15), but the results have been subject to very different interpretations (Wardle et al. 2000; Naeem 2000). As with the effects of productivity on diversity, it is likely that the effects of diversity on productivity will depend in some as yet unknown ways on spatial and temporal scale.

These effects, if substantiated in other systems, could have important implications for conservation efforts. A failure to conserve large tracts of wildlands could lead to a decrease in plant species diversity (see Chapter 17).

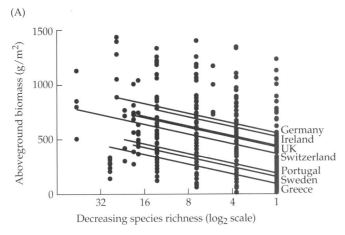

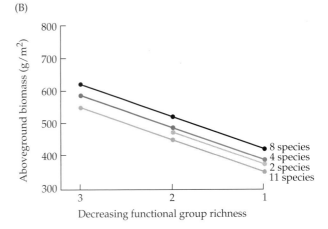

Figure 14.10
The relationship between plant diversity and aboveground productivity in experimental plots at eight sites across Europe. (A) Species richness. The points represent individual plots; the lines are slopes from a series of regressions. (B) Functional group richness. The points are averages over all sites. (After Hector et al. 1999.)

If such a diversity loss led to a decrease in primary productivity, the effects could potentially reverberate through the food chain, leading to extinctions of herbivores and carnivores.

Diversity and Stability

Yet another contentious issue of long standing in ecology is whether community diversity is related to the stability of communities through time. The answer depends on which of several senses of "stability" one has in mind. In mathematics, **dynamic stability** means that a system tends to return to its original state after a small perturbation. Robert May (1973), using a mathematical model, predicted that increased diversity would have the opposite result: after perturbation, a more diverse community should be less likely to return to its original state than a less diverse community. Subsequent theoretical studies have slightly modified this conclusion: the robust theoretical result is that increasing diversity in ecological models is never likely to increase the dynamic stability of a community, but may not decrease it.

More recently, several ecologists have studied a quite different type of stability: they have attempted to determine whether more diverse communities have less variation in productivity from one year to the next. The hypothesis is simple: with more species present, it is more likely that (regardless of environmental fluctuations) in any given year, some species will have a "good" year. In other words, the ability of different species to grow well under different conditions should mean that more diverse communities vary less in productivity.

In a study of prairie communities in Minnesota, Tilman and Downing (1994; Tilman 1996) found that more diverse communities (those with a greater number of species) maintained higher productivity, even during a period of drought (Figure 14.11). Some species were better adapted to low water conditions than others; in a drought year, those species constituted a greater percentage of the total biomass than in other years. Their biomass increased when drought caused other species to grow poorly. Ecologists have debated whether this

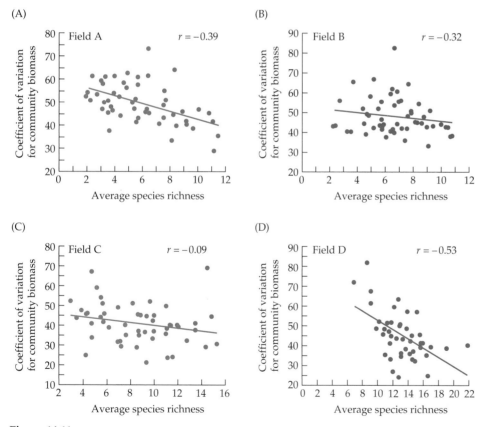

Figure 14.11
Relationship between plant species richness and stability of productivity (variability in total biomass among years) for 4×4 m^2 plots in four fields in a prairie in Minnesota from 1984 to 1994. A very severe drought occurred during this period in 1987–1988. (After Tilman 1996.)

finding represents an important emergent property of communities (Tilman 1997) or a statistical inevitability of sampling larger groups of species (Doak et al. 1998).

Regional Processes

In this chapter we have focused on the role of local processes in determining community structure. However, local diversity, dominance, and abundances might be determined not only by the interactions among species within a site, but also by regional processes. If a species is common throughout a region, for example, then a continual influx of migrants into a site might keep its densities high. We explore this process in more detail when we look at metapopulation models in Chapter 17.

Similarly, the number of species found at a site might depend, in part, on the regional species pool. If we compare the number of tree species in deciduous forests in the eastern United States, Europe, and China, for example, we see a curious result. Although we can find sites with nearly identical climates and soil conditions, we will not find the same numbers of species. Species richness will be highest in China, somewhat lower in the United States, and least in Europe. These local differences reflect differences in the regional species pool. Europe has the fewest tree species because of extinctions during the Pleistocene glaciations. China, on the other hand, has the most species because its temperate zone is adjacent to tropical land areas that have been a source of new species over geological time (Qian and Ricklefs 1999). These regional differences in species richness are a result of nonequilibrium processes that operate over very long time scales. Determining how local and regional patterns and processes are linked is currently one of the grand challenges in ecology (Lawton 1999). We take up these issues in more depth in Chapters 17, 20, and 21.

Summary

In most communities, most species are rare, and only a few are common. The most conspicuous or numerically abundant species in a plant community are called the dominants. The limited data available suggest that numerically dominant species tend to be competitive dominants as well, at least in some communities. Species may be rare in different ways and for different reasons, and a species' status as rare or common may sometimes change dramatically over time. Rare plant species can be limited in terms of geographic distribution, local population size, or range of habitats occupied. A common explanation for rarity, specialization and narrow environmental tolerances, does not appear to be a good general explanation for why species are rare.

Various explanations have also been proposed for the ability of some plant species to become invasive when they are introduced into new environments. The characteristics of species themselves, and of their situation in a new environment, have been used with varying success to predict invasiveness. Characteristics of communities that may make them susceptible to invasion by exotic species have also been proposed and, to a limited extent, tested.

Changes in productivity can alter the richness and diversity of species in a community. One common pattern is a unimodal (hump-shaped) relationship, in which diversity is highest at intermediate productivity and lower at both low and high productivity, but other patterns are also commonly found. Species diversity can also be related to the frequency and severity of disturbance. Gaps created by disturbance are important in maintaining species diversity in temperate and tropical forests. The question of the effects of species diversity on ecosystem properties is still under active investigation and remains highly controversial.

Additional Readings

Classic References

Hutchinson, G. E. 1959. Homage to Santa Rosalia, or why are there so many kinds of animals? *Am. Nat.* 93: 145–159.

Preston, F. W. 1962. The canonical distribution of commonness and rarity. *Ecology* 43: 185–215, 410–432.

Tilman, D. 1982. *Resource Competition and Community Structure*. Princeton University Press, Princeton, NJ.

Contemporary Research

Howard, T. and D. E. Goldberg. 2001. Competitive response hierarchies for different fitness components of herbaceous perennials. *Ecology* 82:979–990.

Kelly, C. K. and F. I. Woodward. 1996. Ecological correlates of plant range size: Taxonomies and phylogenies in the study of plant commonness and rarity in Great Britain. *Phil. Trans. R. Soc. Lond. B* 351: 1261–1269.

Smith, M. D. and A. K. Knapp. 1999. Exotic plant species in a C_4-dominated grassland: Invasibility, disturbance, and community structure. *Oecologia* 120: 605–612.

Additional Resources

Huston, M. A. 1994. *Biological Diversity*. Cambridge University Press, Cambridge.

Rosenzweig, M. L. 1995. *Species Diversity in Space and Time*. Cambridge University Press, Cambridge.

PART **IV** *From Ecosystems to Landscapes*

CHAPTER 15

Ecosystem Processes

The focus here is somewhat different from that of the other chapters in this book. In this chapter we look at plants as critical players in ecosystems, rather than focusing on plants themselves. An **ecosystem** consists of all of the organisms in an area and all of the abiotic materials and energy with which they interact; it is a bounded ecological system. The word *ecosystem* was devised by Sir Arthur Tansley (1935) to include the entire system of living organisms in the context of the physical factors on which they depend and with which they are interconnected.

Tansley also stated that ecosystems could exist over a range of scales—a very contemporary concept. He further stated that "the [eco]systems we isolate mentally are not only included as parts of larger ones, but they also overlap, interlock and interact with one another. The isolation is partly artificial, but is the only possible way in which we can proceed." Ecosystem ecologists ask questions about the roles of plants as conduits for energy and materials in ecosystems and as agents for the transformation of energy and materials into different forms, and about the effects of the supplies and flows of energy and materials on plants.

What determines plant productivity, and why is it so different in different places on Earth? What are the important controls over the flow of energy and materials in ecosystems? What are the most important chemical elements affecting plant growth and decomposition, and how do the relative amounts of these nutrients affect these processes? How do ecosystem processes affect, and how are they affected by, characteristics of communities such as species diversity and plant-animal or plant-microorganism interactions?

We begin with a look at biogeochemical cycles in general, then look at the water cycle as an example. Then we turn to the cycling of carbon in ecosystems and its roles in biogeochemical processes from decomposition to productivity. (We will revisit the global carbon cycle in Chapter 22, when we consider it in the context of climate change.) We then reexamine the nitrogen cycle, first introduced in Chapter 4, and go on to the phosphorus cycle, with a brief look at how some other elements move through the biosphere. Here we begin by looking at how a young plant grows, widening our focus to the ecosystem context from there.

Biogeochemical Cycles: Quantifying Pools and Fluxes

A seed, with its embryo and other associated tissues, contains very limited quantities of energy and materials. Seeds are generally very small—orders of magnitude smaller than the plants into which they will grow. The spore that will develop into a mature fern is even smaller than most seeds. As the new individual grows into a mature plant, it transforms inorganic forms of carbon and other materials into complex organic molecules, and it stores the energy captured from photons of light in the form of chemical bonds in these molecules (see Part I). In terrestrial systems, most of the materials and energy used by heterotrophs—animals, fungi, bacteria, and protists—are transformed from unusable inorganic forms into usable organic materials by plants. (Photosynthetic bacteria, diatoms, green and red algae, and other organisms play a similar role in aquatic systems and sometimes in soils.) The story is actually more complicated than that, however, because some of the materials needed by plants must first be transformed by microorganisms into forms the plants can use (see Chapter 4).

The quantity of material transformed into organic forms by an individual plant can range from fractions of grams to metric tons (1 metric ton = 1000 kg), depending on the size of the plant and the substance being transformed. At the ecosystem scale, this can add up to many metric tons of material. Terrestrial **primary productivity**—the amount of carbon transformed from CO_2 into organic carbon (C) by terrestrial plants per unit area per year—is typically on the order of 5 to 10 metric tons per hectare (far smaller amounts are also fixed by other organisms). On the global scale, fluxes of carbon, for example, are on the order of 10^{15} grams of carbon per year (gigatons).

A major focus of ecosystem ecology is understanding what regulates the **pools** (quantities stored) and **fluxes** (flows) of materials and energy in the various abiotic and biotic components of ecosystems. The pools and fluxes of some major mineral nutrients in a California chaparral system are shown as an example in Table 15.1.

A basic approach that is often used in accounting for the magnitude of pools and fluxes is the mass-balance approach to constructing nutrient budgets. Simply put, the mass-balance approach states that

$$\text{inputs} - \text{outputs} = \Delta \text{ storage}$$

In other words, if we can measure everything that goes into a system and everything that comes out, the difference between the two quantities must be reflected by a change, Δ, in the material stored in the system. The mass-balance approach allows ecosystem ecologists to account for difficult-to-measure quantities by subtrac-

tion from accurate measurements of the other terms (Vitousek and Reiners 1975).

Ecosystem ecologists also seek to make predictions about the future sizes of pools and the magnitude and direction of fluxes. The major fluxes and pools of a substance in a system are collectively called a **cycle**; for example, we might diagram the global water cycle, or at a smaller scale, the nitrogen cycle or phosphorus cycle of a forest ecosystem. Most of the nutrients that are important to plants are cycled biogeochemically—that is, both biological and chemical reactions are involved.

The material contained in a plant is composed mostly of carbon (C) and oxygen (O) (see Table 4.2). The oxygen available to aboveground parts of plants is almost never limiting to terrestrial plant growth (oxygen can, however, be limiting below ground, particularly in waterlogged soils). Carbon, in the form of CO_2, is on average equally available everywhere on Earth (at sea level; it is considerably less available at high altitudes). Yet the production of plant biomass varies enormously over Earth's surface (see Table 19.2). What are the causes of this tremendous variation?

Briefly, primary productivity is determined both by climatic factors (such as moisture, light, temperature, and length of the growing season) and by the supply of nutrients essential for plant growth. We have examined some of these factors in Part I and in Chapter 14. Here we look at the major factors controlling the ability of plants to obtain essential nutrients, as well as the roles of plants as conduits for the fluxes of those nutrients, as agents controlling those fluxes, and as reservoirs for their storage.

Some essential nutrients are needed by plants in relatively large quantities,* but their availability to plants is limited. Nitrogen (N) and phosphorus (P) are in this category; both are major constituents of essential organic molecules, but their supply is limited in most soils. Consequently, the availability of N and P controls primary productivity in many ecosystems. In contrast, in the case of some nutrients that are available in larger quantities, such as sulfur (S) and potassium (K), primary productivity can determine the rate at which they cycle in the ecosystem (Schlesinger 1997). In both cases, living organisms have a major effect on the geochemistry—the pools and fluxes—of these major components of living things.

In contrast, the cycling of elements that are not major constituents of living things, such as sodium (Na) or aluminum (Al), is relatively independent of the actions of living organisms (Schlesinger 1997). It is because plants have a major role in the local, regional, and global cycles of water and some chemical elements that we include

---

*You may wish to review Table 4.2, which summarizes the roles of the various mineral elements necessary for healthy plant growth and nutrition.

Table 15.1 Nutrient cycling in a 22-year-old stand of the chaparral scrub *Ceanothus megacarpus* near Santa Barbara, California

	Biomass	N	P	K	Ca	Mg
Atmospheric input (g/m²/year)						
Deposition	—	0.15	—	0.06	0.19	0.10
Nitrogen fixation	—	0.11	—	—	—	—
Total input	—	0.26	—	0.06	0.19	0.10
Pools (g/m²)						
Foliage	553	8.20	0.38	2.07	4.50	0.98
Live wood	5929	32.60	2.43	13.93	28.99	3.20
Reproductive tissues	81	0.92	0.08	0.47	0.32	0.06
Total live	6563	41.72	2.89	16.47	33.81	4.24
Dead wood	1142	6.28	0.46	2.68	5.58	0.61
Surface litter	2027	20.5	0.6	4.7	26.1	6.7
Annual flux (g/m²/year)						
Requirement for production						
Foliage	553	9.35	0.48	2.81	4.89	1.04
New twigs	120	1.18	0.06	0.62	0.71	0.11
Wood increment	302	1.66	0.12	0.71	1.47	0.16
Reproductive tissues	81	0.92	0.08	1.47	0.32	0.07
Total in production	1056	13.11	0.74	4.61	7.39	1.38
Reabsorption before abscission	—	4.15	0.29	0	0	0
Return to soil						
Litterfall	727	6.65	0.32	2.10	8.01	1.41
Branch mortality	74	0.22	0.01	0.15	0.44	0.02
Throughflow	—	0.19	0	0.94	0.31	0.09
Stemflow	—	0.24	0	0.87	0.78	0.25
Total return	801	7.30	0.33	4.06	9.54	1.77
Uptake (=increment – return)	—	8.96	0.45	4.77	11.01	1.93
Streamwater loss (g/m²/year)	—	0.03	0.01	0.06	0.09	0.06
Comparisons of turnover and flux						
Foliage requirement/total requirement (%)	—	71.3	64.9	61.0	66.2	75.4
Litterfall/total return (%)	—	91.1	97.0	51.7	84.0	79.7
Uptake/total live pool (%)	—	21.4	15.6	29.0	32.6	45.5
Return/uptake (%)	—	81.4	73.3	85.1	86.6	91.7
Reabsorption/requirement (%)	—	31.7	39.0	0	0	0
Surface litter/litterfall (year)	2.8	3.1	1.9	1.2	3.3	4.8

Source: Schlesinger (1997), modified from Gray (1983) and Schlesinger et al. (1982).

this topic in a plant ecology textbook. The large part that plants—and humans—play in global biogeochemistry also has major implications for global change, as we will see in Chapter 22.

Cycles of energy and materials are often illustrated by diagrams that show the fluxes and pools of a single element. While this makes for a convenient presentation, it is important to recognize that the cycles of various materials can be highly interconnected, and that they interact to affect plant growth. The carbon cycle, for example, both depends on and strongly affects the nitrogen and phosphorus cycles (Shaver et al. 1992). These cycles are interconnected by their mutual effects on plant

growth, tissue composition, leaf longevity, litter production, and litter decomposition rates. These interconnections determine the flux and pool sizes of C, N, and P. All three cycles are also simultaneously affected by climatic and local environmental factors such as temperature, precipitation, and light availability.

There are large differences in the magnitudes of pools and fluxes among different materials and ecosystems. Fluxes of water and carbon are vastly greater than fluxes of phosphorus, for example. Pools of many nutrients are much larger in living plants in tropical forests than they are in tundra plants. There are generally huge pools of stored carbon in bog soils, substantial (but far less) stored carbon in grassland soils, and very little stored carbon in desert soils.

The ultimate sources of the different essential nutrients also differ. The atmosphere is the source of N, O, and C, although plants obtain N from the air only indirectly. The atmosphere contains about 78% N and about 21% O. The concentration of CO_2 in the atmosphere, while rising (see Figure 22.2), is surprisingly low—about one-third of one percent on average, or about 370 parts per million by volume (ppmv). Nevertheless, because the volume of the atmosphere is so large, the total pool of carbon in the atmosphere is very large. Large amounts of carbon are also stored as carbonate ions (CO_3^{2-}) in rock and dissolved in seawater; additional carbon is contained in living biomass and in soils. Liquid water is the source of hydrogen (H), and is the ultimate source of the O released to the atmosphere by photosynthesis. Rock weathering is the major source for most of the other elements needed by plants (such as Ca, Mg, K, Fe, and P), while S comes both from atmospheric deposition and rock weathering (Schlesinger 1997). In addition to new inputs from the atmosphere and the weathering of rock, plants rely on ecosystem recycling of elements (Table 15.2).

Before proceeding with a discussion of individual cycles, we must confront the issue of scale (see Chapter 17), which we purposely blur in this chapter. Ecosystem ecologists treat local, regional, and global cycles separately. Our focus here is that point at which biogeochemistry most strongly interacts with plants. We consider aspects of the ecosystem ecology of several elements in other chapters where they are particularly relevant. Water moves over large scales on Earth, and we focus on the pools and fluxes of water at large scales here. Carbon, nitrogen, and phosphorus move through plants and affect plants at a range of scales, from the rhizosphere to the globe, and we consider these elements over a range of scales both in this chapter and elsewhere.

The Global Water Cycle

Terrestrial plants are the only living things (except humans) to have a substantial effect on the global water cycle (Figure 15.1). Most of the world's available water (96.5%) is stored, not surprisingly, in the oceans, with all of the fresh (i.e., non-salty) water in the world's rivers, lakes, ice, atmosphere, and groundwater constituting the remaining few percent. Living things store relatively minuscule amounts of water, but the amount of water that moves through plants by transpiration is large and globally important.

The flux of water through plants can also be important regionally. Plants can contribute a major share of the water in the atmosphere and in regional precipitation, but the effects of plants on local precipitation patterns vary greatly. Precipitation in California, for instance, comes largely from water evaporated over the Pacific Ocean. In contrast, between one-quarter and one-half of the rainfall over the Amazon basin comes from **evapotranspiration** (evaporation and transpiration) from the Amazon forest itself, with the rest coming from outside the region (Salati and Vose 1984; Eltahir and Bras 1994). Cutting large areas of the Amazon forest can therefore result in decreased evapotranspiration, decreased rainfall, and increased temperatures at the ground surface (Lean and Warrilow 1989; Shukla et al. 1990).

Plants can also indirectly affect water fluxes by intercepting precipitation, reducing the impact of rain hitting the ground, and decreasing runoff from the soil. If vegetation is removed—for example, by clear-cutting forests—runoff and soil erosion can be greatly increased, particularly in areas where the terrain is hilly or mountainous. This can sometimes result in flooding and mudslides, further damaging vegetation and threatening human lives and homes. Even in semiarid regions, removal of vegetation can result in reduced precipitation, increased soil warming, and the onset of desertifi-

Table 15.2 Annual nutrient requirements supplied by various sources in the Hubbard Brook Experimental Forest, New Hampshire

	N	P	K	Ca	Mg
Requirement kg/ha/year	115.4	12.3	66.9	62.2	9.5
Percentage supplied by:					
Intersystem inputs					
Atmospheric	18	0	1	4	6
Rock weathering	0	1	11	34	37
Intrasystem transfers					
Reabsorptions	31	28	4	0	2
Detritus turnover (includes return in throughflow and stem flow)	69	67	87	85	87

Sources: Reabsorption data are from Ryan and Bormann (1982). Data for N, K, Ca, and Mg are from Likens and Bormann (1995) and for P from Yanai (1992).

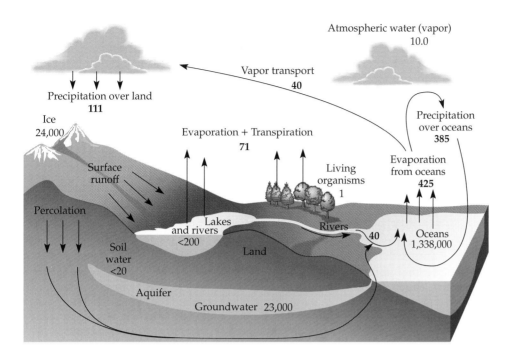

Figure 15.1
The global water cycle. The numbers show the pools (in units of 1000 km^3 of water, shown in regular type) held in various major components of the global ecosystem, and the fluxes among those components (1000 km^3/year, shown in bold-faced type). Notice that the amount of water leaving the land to return to the oceans (40,000 km^3/year) is equal to the amount of water that returns from the oceans to the land as water vapor evaporated from the oceans. The numbers given are approximate, as estimates vary considerably between different authors. (After Schlesinger 1997 and Gleick 1996; data from Lvovitch 1973 and Chahine 1992.)

cation (Schlesinger et al. 1990; Dirmeyer and Shukla 1996).

In intact vegetation, the balance between precipitation, evapotranspiration, and runoff varies greatly among biomes (Table 15.3). In deserts and grasslands, all or most of the water that enters the system as precipitation is lost by evaporation or transpiration, leaving almost nothing for runoff and recharge of groundwater. Both temperate and tropical forests generally lose less water by evapotranspiration than they receive as precipitation, so considerable amounts often run off into rivers and streams and percolate to the groundwater. Over the Earth as a whole, rivers carry about a third of the precipitation that falls over land back to the oceans. While the amount of rainfall over the oceans is greater than that over land (see Figure 15.1), the evaporation of water from the oceans

Table 15.3 Relative importance of pathways leading to the loss of water in a variety of terrestrial ecosystems

Biome	Evaporation (%)	Transpiration (%)	Runoff and groundwater recharge (%)
Tropical rainforest	25.6	48.5	25.9
Tropical rainforest	10	40	50
Tropical rainforest	11	56	32
Temperate forest	13	32	53
Temperate grassland	35	65	0
Temperate grassland	33	67	0
Temperate grassland	55	45	0
Temperate grassland	56	34	10
Desert	28	72	0
Desert	20	80	—
Desert	73	27	—
Desert	65	35	—

Source: Schlesinger (1997).

is also much greater than the evapotranspiration of water over land. The net movement of water vapor from the oceans to the land through the atmosphere is balanced by the runoff of water from the land to the oceans in the form of groundwater and river flow. While only a minuscule amount of water is stored as water vapor in the atmosphere, the flux of water vapor through the atmosphere is tremendous.

Potential evapotranspiration (PET) is the maximum amount of water that would be lost to evapotranspiration in a particular place if water were freely available in the soil and plant cover were 100%. PET depends on the energy available to evaporate water, which depends on the ambient temperature. You may be surprised to know that PET can be much greater per unit area than the amount of water that would be lost by evaporation from an open body of water. This is because the surface area of the roots taking up water is much greater than the surface area of the ground in which the plants are rooted, and because the surface area of the leaves from which water is evaporating is often greater than the surface area of the ground—that is, the **leaf area index (LAI)** is often greater than 1.0.

Actual evapotranspiration (AET) is equal to the amount of water that enters the system in precipitation minus the amount that is lost in runoff and percolation to groundwater (the amount of water stored in living systems is minimal relative to the other terms). In areas with long dry periods, PET is much greater than actual evapotranspiration, but in tropical rainforests PET and AET are approximately equal. Actual evapotranspiration is closely linked to other major ecosystem processes, particularly productivity, decomposition, and other soil processes (Schlesinger 1997). This is not surprising, given the dependence of plant and microbial activity on soil water content and the close functional linkage between carbon assimilation and water loss in plants (see Chapters 2 and 3).

Carbon in Ecosystems

Decomposition and Soil Food Webs

From the shortest-lived microorganism to the longest-lived tree, everything that lives eventually dies. If the organic **detritus**—the dead bodies and waste products of living and once-living organisms—were not decomposed, most ecosystem processes would come to a screeching halt, and needless to say, the piled-up dead bodies and wastes would not be a pretty sight. Sooner or later, much of the carbon and other material and some of the energy released by the process of decomposition are taken up by living organisms and used again. How are materials that were once part of living organisms recycled?

Decomposition transforms dead plants and animals, their shed or removed parts (e.g., leaves, bark, branches, and roots), and animal feces into soil organic matter and ultimately into inorganic nutrients, CO_2, water, and energy. In terrestrial ecosystems, decomposition involves both physical and chemical alteration of the original materials by a complex food web, which exists mostly (but not entirely) on or just under the surface of the soil (Figure 15.2).

Decomposition is largely an aerobic process (one requiring oxygen), and it occurs slowly, if at all, in waterlogged soils or other anaerobic environments. Warm,

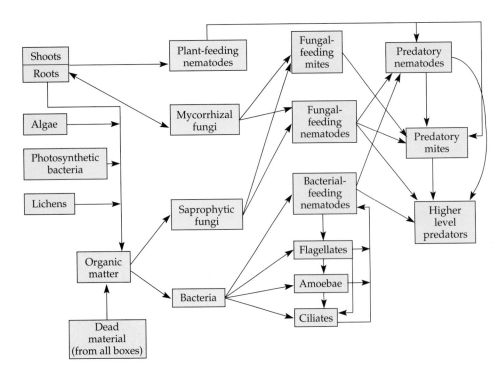

Figure 15.2
Diagram of a soil food web. Autotrophs at the soil surface and upper layers of the soil include plants, with aboveground photosynthetic parts and roots below ground, photosynthetic single-celled protists (including diatoms and green algae), and cyanobacteria (photosynthetic bacteria). Heterotrophs include many different kinds of bacteria, single-celled animals (amoebae, ciliates, flagellates), mycorrhizal and saprophytic (decomposing) fungi, nematodes, soil microarthropods and macroarthropods, including mites, and a variety of vertebrates as well. (After Ingham, unpublished, at http://www.soilfoodweb.com)

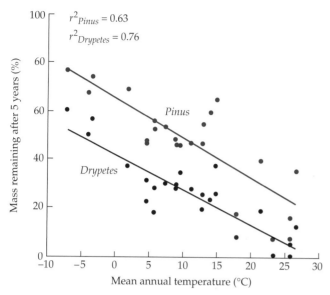

Figure 15.3
Decomposition rates of the roots of two species of trees with highly contrasting properties: *Drypetes glauca* (Euphorbiaceae), a tropical hardwood, and *Pinus elliotii* (slash pine, Pinaceae), a temperate softwood. Mesh bags with root material were buried in 28 sites in North and Central America. The roots of both species decomposed faster at sites with higher annual temperatures. *Drypetes* roots decomposed faster than *Pinus* roots at all sites. (After Gholz et al. 2000.)

(A)

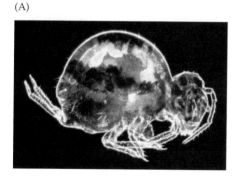

Figure 15.4
Some invertebrates that are important in soil food webs. (A) *Sminthurinus elegans* (a springtail; Collembola). Springtails are very common tiny soil insects that consume fungi. The organism pictured is about 3 mm long. (Photograph courtesy of K. Brocklehurst.) (B) A soil mite (in the Cepheidae; genus and species unknown), from a subtropical rainforest in Queensland, Australia. These tiny (about 2 mm) arachnids consume fungal hyphae and spores, as well as algae, lichens, and bacteria. Soil mites are common worldwide. (Photograph courtesy of D. E. Walter.) (C) A nematode, *Acrobeloides nanus*, showing its front (anterior end). This nematode species is probably one of the most common invertebrates in the world. It lives on all continents, including the subantarctic islands, and in favorable sites may have populations of over 1,000,000 individuals per square meter. Its entire length is about 0.2 mm. (Photograph courtesy of S. Boström.)

(B)

(C)

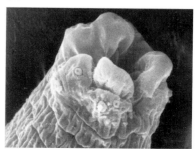

moist conditions favor the highest rates of decomposition. Cold, anaerobic (oxygen-free) conditions inhibit decomposition, resulting in the buildup of organic material. Decomposition rates also depend on other factors, especially the physical and chemical characteristics of the material being decomposed (Figure 15.3).

Scavenger animals begin the process of decomposition of materials of animal origin by consuming animal bodies. These scavengers include large animals (hyenas, vultures, and so forth) as well as insects (such as *Nicrophorus* burying beetles and the Dipteran larvae known as maggots). Dung beetles are important and astonishingly effective consumers of mammal feces in tropical grasslands.

The major source of material for decomposition is plant **litter**, which consists of a variety of materials—from dead roots and shed leaves, needles, or bark, to dead tree trunks. Litter must be broken up into small pieces of organic matter before it can be effectively decomposed by microorganisms. This process of litter fragmentation is largely accomplished by animals. Large animals, from deer and bears to gophers and voles, break litter apart as they search for food in the soil and on its surface.

At an intermediate scale, earthworms, where they are present, ingest and process huge volumes of soil, causing dramatic physical and chemical alteration of the soil and the organic matter contained in it. Termites eat living and dead plant material, particularly decomposition-resistant wood, digesting cellulose by means of mutualistic protists and bacteria that inhabit their hindguts. These insects can be major contributors to tropical ecosystem dynamics.

The smallest multicellular animals in soil food webs are microscopic insects and arachnids known as soil microarthropods, or soil microfauna and mesofauna (Figure 15.4). These organisms are particularly important in the decomposition process, especially in forests. Among them are the Collembola (springtails), which are insects that eat fungi, and mites, which are arachnids that directly consume litter as well as eating soil fungi and bacteria. Nematodes (Figure 15.4C) are spectacularly abundant and critical in soil food webs as herbivores that eat plant roots, carnivores that eat soil animals, and consumers of soil-dwelling bacteria and fungi. Unicellular protists, including ciliates and amoebas, live in the films of soil water between and surrounding soil particles and prey on soil bacteria. By their voracious consumption of soil bacteria and fungi, soil microarthropods, nematodes, and protists are responsible for converting large amounts of microbial nitrogen and phosphorus into forms available to plants, as well as releasing microbial carbon as respired CO_2.

Saprophytic fungi rely on non-living organic material for their carbon and energy. They are the major decomposers of dead leaves and other plant litter. They account for approximately half of the microbial biomass in grassland soils and about 90% in temperate forest soils. Fungi are heterotrophs that secrete powerful digestive enzymes externally, decomposing their substrate.

The fine hyphae of fungi can penetrate plant organs and cells, overcoming the defensive cuticle to gain access to the interior and speeding decomposition. However, both plant structural defenses, such as lignin, and various aspects of plant defensive chemistry, including toxins and pH levels, can inhibit their effectiveness. Such differences in plant chemistry and physical structure contribute to large differences in decomposition rates

among different species (see Figure 15.3) and different plant parts (leaves versus wood, for instance). The extensive multicellular mycelia that make up a fungus can obtain carbon from one source and nitrogen from a different source at some distance, further enhancing the ability of fungi to use dead plant materials as a source of nutrients and energy.

Bacteria play an essential role in soil food webs and nutrient cycling, although they are generally not very important in the initial stages of decomposition in most ecosystems. Bacteria can lyse (rupture) and break down living and dead cells of plants, animals, fungi, and other bacteria. Their presence in the soil is highly variable in space and time; they are found in greatest abundance at warm temperatures and when water and nutrients are available. They are found largely in soil macropores (larger soil pores), in the rhizosphere, and coating aggregates of soil particles (see Chapter 4). They move through the soil passively, carried by water or animals. We will return to the roles of various functional groups of bacteria below.

Productivity

Through photosynthesis, plants and other photosynthetic organisms capture carbon and energy, transform-

Figure 15.5
Global map of estimated NPP (total net primary productivity, above and below ground) for terrestrial plants in metric tons/ha/year of total dry organic matter. (Unpublished data from N. I. Bazilevich, Global Primary Productivity Database, 1993, http://ceos.cnes.fr:8100/cdrom-006/ceos1/casestud/ecoreg/datasets/602/baz.htm)

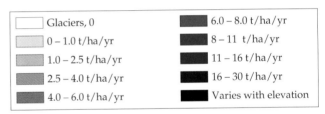

☐ Glaciers, 0	6.0 – 8.0 t/ha/yr
0 – 1.0 t/ha/yr	8 – 11 t/ha/yr
1.0 – 2.5 t/ha/yr	11 – 16 t/ha/yr
2.5 – 4.0 t/ha/yr	16 – 30 t/ha/yr
4.0 – 6.0 t/ha/yr	Varies with elevation

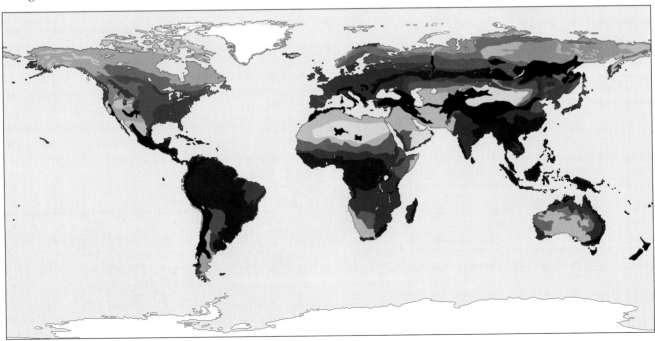

(A)

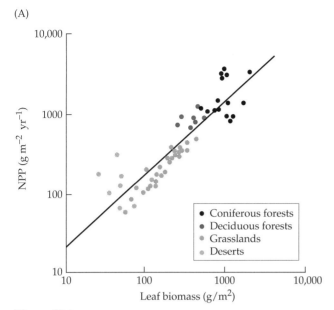

(B)

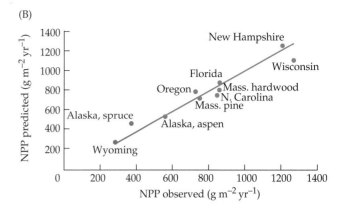

Figure 15.6
(A) Net primary productivity (aboveground, in g/m²/year) across a wide range of ecosystems in North America, from deserts to grasslands to deciduous and coniferous forests. NPP is closely related to leaf biomass. The scale for both axes is logarithmic, so these variables have a very large range. (After Schlesinger 1997) (B) Predicted versus observed NPP (aboveground, in g/m²/year) in ten North American forests. Aboveground NPP in these forests ranges from < 300 g/m²/year to almost 1300 g/m²/year, with a median of about 750 g/m²/year. Predicted values are closely correlated with observed values. (After Waring and Running 1998.)

ing oxidized carbon (CO_2) into reduced forms as organic carbon compounds. **Productivity** is the rate of energy flux through a trophic level in a particular ecosystem. Productivity is often measured in terms of the amount of carbon transferred because the energy we are concerned with is captured together with, and stored in, carbon compounds. Productivity is typically expressed as a rate per unit area (e.g., g C/m²/year).

The total energy (or carbon) fixed by producers in an ecosystem is called **gross photosynthetic production**, or **gross primary production (GPP)**. Since photosynthesizing organisms use some of the total energy fixed, **net primary production (NPP)** is equal to the total energy captured (GPP) minus the losses to respiration by primary producers. The term "primary" indicates that we are concerned with the first trophic level in the system— that is, the capture of solar energy and the fixation of atmospheric CO_2 by autotrophs—in contrast to the productivity of herbivores, carnivores, or detritivores. In practice, GPP and respiratory losses by primary producers are very difficult to measure directly on an ecosystem basis, and most data on productivity are collected as NPP.

NPP varies enormously over Earth's surface (Figure 15.5). The greatest amounts of carbon are fixed in warm, moist ecosystems; much less is fixed in arctic systems, and the least in deserts (see Table 19.2). Across a variety of ecosystems within the North American continent, aboveground NPP varies over more than two orders of magnitude (Figure 15.6A; Table 19.2). Even if we consider only forests within this continent, there can be almost as great a range in aboveground NPP (Figure 15.6B). At a larger scale, across the continental United States, NPP depends on latitude, climate, and elevation.

NPP in terrestrial ecosystems is broadly limited by climate, particularly by inadequate temperatures and insufficient moisture for growth, and by the length of the growing season (Figure 15.7) In addition, productivity can be limited either by a lack of essential nutrients or by an excess of certain other materials (e.g., toxic levels of salt or heavy-metal ions) in the soil. More subtle factors, such as differences in the amount of respiring tissue in forests with predominately young versus old trees, also affect NPP. Climatic or soil factors can influence properties of the vegetation, such as LAI and total leaf biomass, which in turn influence productivity (Figures 15.6A and 15.8).

Productivity can vary greatly even among adjacent sites. Gaius Shaver and F. Stuart Chapin III (1991) compared the relationship between production and biomass among four contrasting types of tundra vegetation in northern Alaska at the Toolik Lake LTER site (see Box 12C). The study sites were chosen to represent extreme examples of the wide range of plant growth forms commonly found in arctic tundra, and included tussock tundra, deciduous shrub-dominated riparian tundra, evergreen heath tundra, and wet sedge tundra (see Figure 19.23). Shaver and Chapin found that the four sites differed greatly in aboveground vascular plant biomass (ranging from 217 to 1877 g/m²), in addition to differing strongly in plant growth form. Aboveground NPP of vascular plants varied over an even greater range (32–305 g/m²/year), and the element content and element

(A)

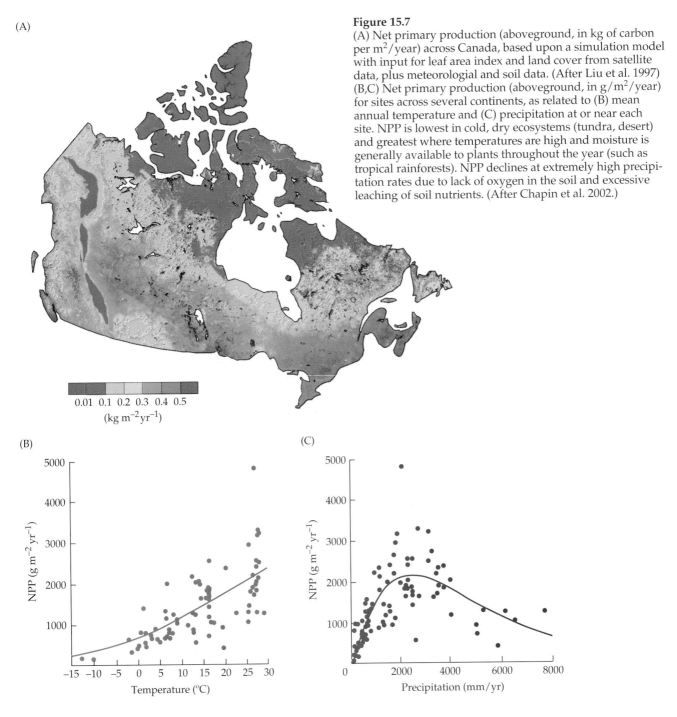

Figure 15.7
(A) Net primary production (aboveground, in kg of carbon per m²/year) across Canada, based upon a simulation model with input for leaf area index and land cover from satellite data, plus meteorologial and soil data. (After Liu et al. 1997) (B,C) Net primary production (aboveground, in g/m²/year) for sites across several continents, as related to (B) mean annual temperature and (C) precipitation at or near each site. NPP is lowest in cold, dry ecosystems (tundra, desert) and greatest where temperatures are high and moisture is generally available to plants throughout the year (such as tropical rainforests). NPP declines at extremely high precipitation rates due to lack of oxygen in the soil and excessive leaching of soil nutrients. (After Chapin et al. 2002.)

0.01 0.1 0.2 0.3 0.4 0.5
(kg m⁻² yr⁻¹)

(B)

(C)

requirements (such as N and P) also varied greatly among the different sites. Among plants with different growth forms there were also dramatic differences in the allocation of carbon and nutrients to the various plant parts, and a wide range in leaf replacement rates and the amount of resources needed to produce each leaf.

Despite these dramatic differences among the four different sites in total biomass, aboveground NPP, and plant nutrient content, Shaver and Chapin found surprisingly little difference among the sites in the relationships between nutrients and NPP, including nutri-

ent cycling (e.g., turnover rates of N), the amount of nutrients required to produce a gram of biomass, and total biomass turnover. **Turnover time**, or mean residence time, is a measure of how rapidly materials move through a system. It can be used to describe systems at any scale—from an individual plant to an ecosystem to the atmosphere. Turnover time is equal to the total mass of a component divided by its flux in or out of the system. The fraction lost each year, $-k$, is equal to 1/turnover time in years, and the amount remaining after t years as a fraction of the starting amount is e^{-kt}.

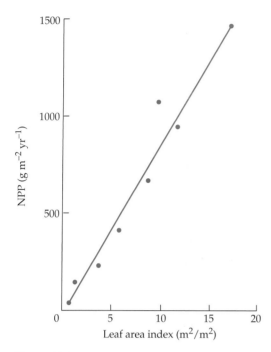

Figure 15.8
Relationship between aboveground net primary production and leaf area index (defined as the total area of one side of all the leaves per unit of ground area) for forests in the northeastern United States. (After Schlesinger 1997.)

The large differences in leaf replacement and allocation patterns that were so evident among different growth forms disappeared at the whole-ecosystem level, because sites or growth forms with rapid leaf replacement had slow stem or rhizome replacement. Thus, in these tundra ecosystems, nutrient storage, particularly in stems and rhizomes, compensated for differences in leaf replacement and resources used for leaf production. The wide variation in leaf replacement and metabolism among the different species and contrasting growth forms was buffered at the ecosystem level by stem processes, which dominated because stems were by far the largest biomass and nutrient pools.

Methods for Estimating Productivity

Measuring productivity and being able to predict its magnitude has great commercial value (for instance, in determining timber production). It is also tremendously important for understanding local ecosystems and as a basis for understanding and predicting components of global carbon balance. While the concept of productivity may seem simple, actually measuring it in practice raises various questions and involves solving some substantive problems.

To begin with, if we wish to know how productivity affects community structure, do we want to know the potential productivity of a site—the maximal amount of carbon that could be fixed given the abiotic conditions

at the site—or are we interested in the actual productivity—the amount that is actually fixed in a given time period by the existing vegetation? The actual productivity will always be less than potential productivity because, for example, not all plant species are equally efficient at fixing carbon, and factors such as competition might limit the growth of some species. In general, we are more concerned with actual productivity, which we focus on here, although making this distinction is important for sorting through theories about how productivity determines species diversity (see Table 20.4).

NPP can be estimated in a variety of ways. In the simplest case, we would start the year with a patch of bare ground. At the end of the year we would measure the **standing biomass** (total aboveground plant mass) by cutting down all of the plants, drying them, and weighing them. In practice, in perennial herbaceous ecosystems, standing biomass is harvested from matched plots at the beginning and at the end of the growing season, with the difference in mass providing an estimate of the amount of material produced. Losses to herbivory and other factors must be estimated to obtain an accurate assessment of NPP. One way to estimate the magnitude of such losses is to exclude herbivores from other matched plots by fencing, insecticides, or other means.

Harvesting plant material to measure productivity has two important limitations. First, harvesting standing biomass is destructive. This method would not be very useful, for instance, in a site that we wanted to study over many years. One way around this limitation is to measure productivity in a number of harvested plots while other measurements are made in adjacent, undisturbed plots.

Second, harvesting only aboveground plant parts ignores belowground productivity, which is very important in many ecosystems. Measuring belowground productivity, however, is difficult. Not only is there a great deal of labor involved in successfully digging up roots, but it creates a substantial disturbance in a plant community under study. Furthermore, much of the root material may be contained in very fine root hairs that are hard to separate from the soil. For these reasons, belowground productivity is often ignored.

The ratio of belowground to aboveground NPP varies greatly among ecosystems. Vogt and colleagues (1982), for example, compared aboveground and belowground NPP in 23-year-old and 180-year-old *Abies amabilis* (silver fir, Pinaceae) forest stands in the Cascade Mountains of Washington. In the young forest, belowground NPP was 45% of the total; in the old forest, the belowground component was 75% of total NPP.

To get around these limitations, ecologists often use indirect measures of productivity. One such approach depends on developing formulas relating changes in plant size to changes in biomass. For many tree species, especially economically important ones, the statistical relationship between average increases in stem diameter and average annual biomass accumulation in stem wood, leaves, branches, and large-diameter roots has been determined by cutting and measuring trees. Such allometric relationships are then used to calculate estimates for NPP and for stores of carbon and nutrients in the different tree components at different stages of forest development. Plant litterfall can be collected to account for that aboveground NPP not retained by the plants. Biomass and productivity are sometimes estimated for shrubs and herbaceous understory plants by similarly relating biomass increases statistically to increases in cover, with the relationships determined initially for harvested plants. These allometric formulas are widely used by forest ecologists and timber companies because extensive data on such relationships are readily available, since measuring tree growth rates is critical for forestry and timber production. While this approach works for well-studied forests, it is not practical for many other ecosystems, because a formula is needed for each species in the community, and obtaining this information is time-consuming and expensive.

A very simple, indirect estimate of productivity can be obtained by extrapolating from observed correlations between AET and productivity. AET has been suggested as a simple, widely available correlate of potential productivity for examining large-scale productivity patterns (Rosenzweig 1968). While AET has been shown to provide fairly good estimates of productivity over a broad range of climates, it is a poor predictor of productivity where precipitation and temperatures are both high. Because it is inadequate as a surrogate for productivity in tropical regions, and because the Tropics account for a large proportion of Earth's total primary productivity, AET is not very useful for predicting global productivity.

A different approach to estimating long-term NPP, based on analyses of soil organic matter, has recently been suggested (Jenkinson et al. 1999). This method has the advantage of providing an estimate of NPP over a much longer period of time than is usually possible with harvest data; it also includes both aboveground and belowground productivity. It estimates NPP by calculating how much organic matter must enter a soil annually to maintain the existing pool of soil organic matter at a steady state. This input of organic matter comes from sources such as plant material, soil fungi, dead animals and heterotrophic bacteria, and carbon fixed by soil autotrophs such as cyanobacteria. The three measurements used are soil mass in each soil layer, the organic carbon content of each soil layer, and the radiocarbon age of each soil layer based on ^{14}C measurements (it is also possible to use data, if available, on the soil microbial biomass).

Along with information on soil clay content and long-term weather information, these data are entered into a simulation model. This method has been tested successfully on tropical grassland and woodland sites. Its limitations include the requirement of a reasonably good knowledge of the site's climate and soil characteristics, the assumption that the ecosystem is at a steady state, as well as the need to account for the amount of carbon removed from the system each year by respiration, burning, logging, or crop harvest. Substantial soil erosion would also invalidate the assumptions of the method. Under the right circumstances, however, this method may prove to be an important complement to other approaches to measuring NPP.

Remote sensing methods (see Chapter 16) can be used to estimate NPP at larger scales than other methods. Images that record the different wavelengths of light reflected by the ground surface can be analyzed to provide an estimate of "greenness," or the amount of chlorophyll present, because chlorophyll absorbs (and reflects) light at specific wavelengths (see Chapter 2 and Box 16A). Since the amount of chlorophyll is related to the amount of live photosynthetic material, this information can be used to estimate live biomass at a site. Successive estimates of biomass can be used to obtain an estimate of productivity.

The **normalized difference vegetation index (NDVI)**, measures standing crop based on the reflectance at all the near-infrared (NIR) and visible (VIS) wavelengths:

$$NDVI = (NIR - VIS)/(NIR + VIS)$$

where the reflectance is integrated over all the relevant wavelengths. Regressions relating NDVI to NPP have been worked out by relating the reflectance values to those from actual harvests of vegetation. NDVI has been widely used, and provides good estimates of relative NPP in many systems (Figure 15.9; Cook et al. 1989). Accuracy can be increased by measuring NDVI at multiple times of the year. If an absolute measure of NPP is needed, it may be necessary to calibrate the measurements using on-the-ground harvests. This measure may be inaccurate in communities with many vegetation layers. New methods are being developed to overcome some of the current limitations.

Another approach to the indirect estimation of productivity is the use of techniques based on micrometeorology that measure concentrations of CO_2 in the air. These techniques have become increasingly important in measuring productivity (and other plant-atmosphere interactions) since the 1980s. The **eddy covariance** (or eddy correlation) method is the most widely used of these methods. This technique measures **net ecosystem exchange (NEE)**, which is the net transfer of CO_2 into or out of an ecosystem .

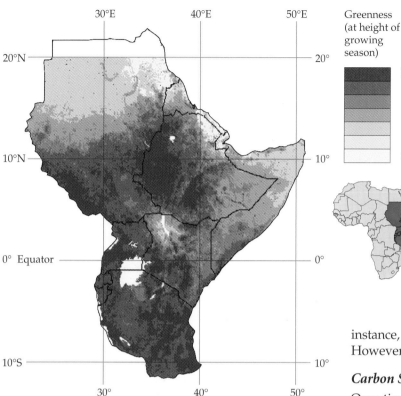

Greenness (at height of growing season)

High

Low

Figure 15.9
Map indicating the maximum NDVI, representing greenness during the height of the main growing season, for the "horn" of east Africa, 1982–1993. These data can be monitored to identify problems in the food production that potentially lead to famine, flood, or other food-insecure conditions in subsaharan Africa. The lowest values are in the northern part of this region and in the tip of the "horn" and other eastern portions. (From U.S. Geological Survey 1996.)

Eddies are parcels of air that move essentially as discrete units, and within which gas concentrations are approximately uniform. Eddies carry CO_2 into and out of the ecosystem, which consists of a complex mix of carbon sources (such as respiring organisms) and sinks (primarily photosynthesizing plants). During the daytime, by far the most important sink for CO_2 is its uptake by the photosynthesizing leaves in the ecosystem. As an eddy comes into contact with a surface, such as a leaf, it slows down because of friction. At that point, CO_2 can diffuse out of the eddy and into the boundary layer of the leaf. When conditions are favorable, there will be a net transfer of CO_2 into the ecosystem from the surrounding air due to photosynthetic uptake by the canopy (that is, gross uptake minus losses due to respiration in the ecosystem).

During the night, there will be a net transfer of CO_2 out of the ecosystem into the atmosphere, due primarily to respiration by the plants and soil organisms in the ecosystem. Using the eddy covariance method requires intensive measurements of local wind speeds and velocities as well as local gas concentrations. It is highly instrument- and computer-intensive; most systems are mounted on either permanent or movable instrument towers in forests, although some have been mounted on aircraft. Because NEE includes all sources and sinks for CO_2 in the ecosystem, it is not a direct estimate of NPP (for instance, it includes respiration by soil microorganisms). However, NEE is in itself a valuable measure.

Carbon Storage

Over time, ecosystems typically accumulate and store carbon. **Net ecosystem production (NEP)** is the net accumulation of carbon per year by an ecosystem (Odum 1969). Our vast underground stores of coal, for example, are the result of geological periods over which NEP was high. NEP is equal to NPP minus total heterotrophic respiration (primarily due to soil organisms). During the course of the growing season, NEP is positive because plant photosynthesis is far greater than heterotrophic respiration. During the season when plants are dormant (winter or dry season) and photosynthesis is minimal, NEP is negative (Figure 15.10).

NEP may also be defined as GPP minus total ecosystem respiration (the sum of all autotrophic and heterotrophic respiration), assuming that no other losses of carbon occur. (Other losses of carbon besides respiration can result from leaching, plant volatile emissions, methane flux, fire, and logging; erosion or deposition and animal movements can add carbon to or subtract carbon from a system as well.) NEP generally depends more on the length of time that an ecosystem has remained undisturbed than on temperature or moisture, and is generally small: undisturbed ecosystems typically show small positive accumulations of carbon each year.

Over the course of ecosystem development following a major disturbance, the biomass of woody tissue in long-lived plants increases. The sum of all organic matter in living vegetation (including the nonliving structural material in wood)—the total standing biomass—may become very high, particularly in older forests, where carbon gradually accumulates in long-lived trees over many

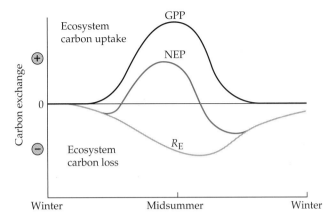

Figure 15.10
Net ecosystem production, NEP, is the net accumulation of carbon per year by the entire ecosystem, including living organisms and soil carbon. NEP increases during the spring to a peak in midsummer, because photosynthesis by plants is greater than total ecosystem respiration. Ecosystem respiration is minimal during the winter. It increases in the spring, causing larger ecosystem carbon losses as soil temperatures warm. The ecosystem reaches a peak of carbon loss in late summer or early autumn. As GPP declines in the autumn, NEP also declines, even though losses due to respiration are also being reduced. (After Chapin et al. 2002.)

centuries. Eventually, the allocation of carbon to new woody tissue is balanced by losses due to the deaths of individuals and plant parts, and the ecosystem no longer gains additional standing biomass. The amount of carbon in the ecosystem may continue to rise after that time as soil organic matter continues to accumulate.

The relative proportions and total amounts of carbon stored in different ecosystem components vary great-ly among ecosystems. Flux rates among the components within ecosystems also vary considerably. For example, in a grassland system, the turnover rate for undecomposed litter was two orders of magnitude greater than that for humic acids in the lower parts of the soil profile, while the amount of carbon stored in the humic acids was much greater than that in the litter (Figure 15.11).

NPP tends to be closely tied to soil microbial respiration because the carbon and energy fixed by plants is the substrate for microbial respiration, and because microbial respiration releases nutrients that plants need in order to capture carbon by photosynthesis (as well as to survive and grow). In some cases, however there is no strong link between NPP and soil microbial activity, and NEP rises or falls accordingly. For example, in peat bogs, the anaerobic soil environment severely limits microbial activity, but does not strongly limit plant productivity, resulting in high carbon accumulation in the soils, and thus high NEP. Disturbances such as fire or logging can result in substantial losses of carbon from an ecosystem. These disturbances remove living plant biomass and sometimes soil organic matter as well, and can alter subsequent NPP in complex ways.

Models of Ecosystem Carbon Cycles

Models have been developed to describe and predict fluxes and pools of carbon from the ecosystem to the global scale. These models typically include the pools of carbon in different fractions in the soil and in vegetation as well as the fluxes of carbon from the atmosphere to plants (NPP), from plants to the soil, and from the ecosystem (plants and soil) back to the atmosphere. These

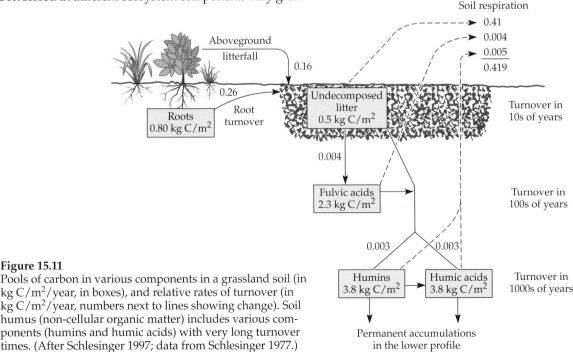

Figure 15.11
Pools of carbon in various components in a grassland soil (in kg C/m²/year, in boxes), and relative rates of turnover (in kg C/m²/year, numbers next to lines showing change). Soil humus (non-cellular organic matter) includes various components (humins and humic acids) with very long turnover times. (After Schlesinger 1997; data from Schlesinger 1977.)

models vary from being highly detailed to very coarse and general, require different inputs of data, and have different particular objectives. Waring and Running (1998) discuss different ecosystem modeling approaches and compare some of the major models.

One example is the CENTURY model (Parton et al. 1988, 1993), which began as a model of carbon storage in agricultural grasslands but has since been extended to model carbon cycling in many other biomes (Figure 15.12). It is a highly detailed model that requires a great deal of information about the system and produces very complex simulations. While there are advantages to this approach, one disadvantage is that, due to the model's complexity, it may be difficult to understand how a particular factor acts to produce a particular result.

The various models in use incorporate different approaches to predicting how environmental factors may affect ecosystem processes, as well as the conse-quences for NPP and other variables. The modeling of carbon fluxes is critical to our understanding of and our ability to predict human effects on the global carbon cycle (see Chapter 22).

Nitrogen and the Nitrogen Cycle at Ecosystem and Global Levels

In striking contrast to the water cycle, living organisms other than plants are important at many junctures in the nitrogen cycle. Water and nitrogen are similar, though, in that they are both commonly limiting to individual plant growth and to community and ecosystem productivity (Figure 15.13). Water and nitrogen cycling are also functionally linked.

Nitrogen has a fairly rapid flux through living organisms coupled to a large global pool with slow turnover—that is, the N_2 in the atmosphere. Nitrogen enters the biosphere from the atmosphere through the process of **nitrogen fixation**. Eventually it is returned to the atmosphere by the process of **denitrification** (and by other means, including fires). The balance between nitrogen fixation and denitrification maintains the concentration of N_2 in the atmosphere at equilibrium (Schlesinger 1997). Nitrogen is recycled among living things at a relatively rapid rate, so primary productivity does not depend solely on nitrogen fixation.

Nitrogen Fixation

Very few organisms—only a relatively small number of prokaryotes—have the ability to fix atmospheric N_2, creating NH_4^+ (ammonium) by breaking the double bond between the two atoms in gaseous N_2. These nitro-

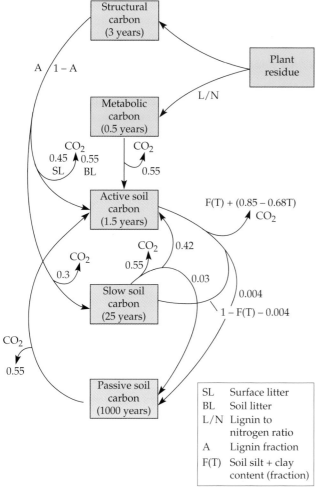

Figure 15.12
The CENTURY model of carbon flow through the biosphere. The fractional numbers indicate the proportion of carbon moving along each pathway. The boxes represent carbon reservoirs, and turnover times are given for each reservoir. (After Parton et al. 1988.)

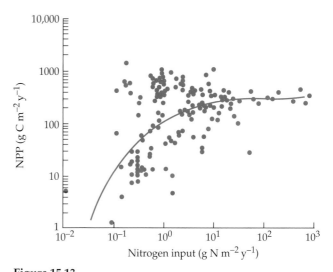

Figure 15.13
Net primary productivity versus annual nitrogen input for a wide range of terrestrial, freshwater, and marine ecosystems. NPP reaches an asymptote at about 10 g N/m²/year. (After Schlesinger 1997.)

Figure 15.14
Traditional rice paddy culture in Mai Chau, Vietnam. Rice is the primary crop and primary export of Vietnam. An important component of the nitrogen economy of paddy agriculture is the symbiosis between the freshwater fern *Azolla* and the nitrogen-fixing cyanobacterium *Anabaena* (see Figure 4.12). (Photograph courtesy of M. Harms.)

gen-fixing microorganisms exist both in free-living forms and in symbiotic relationships with the roots of a few plant species, as described in Chapter 4. The most common nitrogen-fixing symbioses are those between bacteria in the genus *Rhizobium* and members of the Fabaceae (legumes; see Figure 4.13) and that between an actinomycete, *Frankia*, and alder (*Alnus* spp.), *Ceanothus*, and a few other tree and shrub species. Free-living nitrogen fixers are found in soil, in aquatic environments, and in anoxic environments such as mud.

A particularly important group of free-living nitrogen fixers is the cyanobacteria (the microorganisms formerly known as blue-green algae). These photosynthetic bacteria are abundant as free-living organisms in the oceans, in fresh water, and on soil surfaces. Cyanobacteria are also important partners in various nitrogen-fixing symbioses. Lichens are symbiotic associations between a photosynthetic species and a fungal species. The photosynthetic species can be cyanobacteria or members of the Chlorophyta (green algae); lichens containing cyanobacteria can fix nitrogen.

Another nitrogen-fixing symbiosis is that between *Azolla*, a tiny aquatic fern, and *Anabaena* (or other cyanobacteria) in rice paddies and other freshwater tropical systems (Figure 15.14). This *Azolla*-cyanobacterium symbiosis can be a major source of nitrogen in the ecosystems where it occurs, including traditional rice production (where it is deliberately maintained by farmers).

Nitrogen fixation uses a great deal of energy. The symbiotic nitrogen fixers obtain this energy in the form of carbohydrates from their host plants, supplying nitrogen to the plant in return. Free-living nitrogen fixers (other than cyanobacteria) must obtain carbon and energy from other sources, such as soil organic matter. It takes a great deal more energy for a plant to support nitrogen-fixing bacteria than to obtain NH_4^+ from the soil, as long as NH_4^+ is readily available. Because it is less

expensive energetically, plants that are capable of supporting nitrogen-fixing bacteria cease to do so when soil nitrogen is abundant, instead obtaining nitrogen from the soil. This is the case, for example, in agricultural fields fertilized with nitrogen.

Other Sources of Nitrogen Input to Living Organisms

In addition to the atmospheric N_2 that is fixed by prokaryotes, nitrogen that is in suspension in the atmosphere in more available forms can enter terrestrial ecosystems in several different ways. The major forms of N deposition are wet deposition of nitrogen dissolved in rainwater, dry deposition of nitrogen in particles and dust, and fog or cloud deposition of nitrogen dissolved in water droplets onto plant sufaces (Figure 15.15). Nitrogen in a wide variety of forms, including nitrate (NO_3^-), ammonium (NH_4^+), organic N, and gaseous forms (NH_3, HNO_3, or NO_2), can be transferred to plants by deposition.

Much of this atmospheric nitrogen deposition results from human activities, and it has increased enormously in recent decades, with potentially severe environmental consequences (Vitousek et al. 1997). In the eastern United States, for example, atmospheric deposition of N is now 10 times greater than it was before industrialization, while in northern and central Europe it is 20 times greater. The nitrogen enters the atmosphere largely from the burning of fossil fuels, particularly from automobile exhausts and industrial sources. Humans also add about 80 million metric tons of nitrogen per year to the biosphere by fixing atmospheric N for nitrogen fertilizer, which is a large source of the N found in runoff and groundwater (Figure 15.16).

Volatilization of methane and ammonium by domesticated cattle (from flatulence and decomposing manure) is a substantial source of anthropogenic N particularly in some ecosystems, but also globally. Nitrogen deposition by humans is spatially highly variable; it is usually

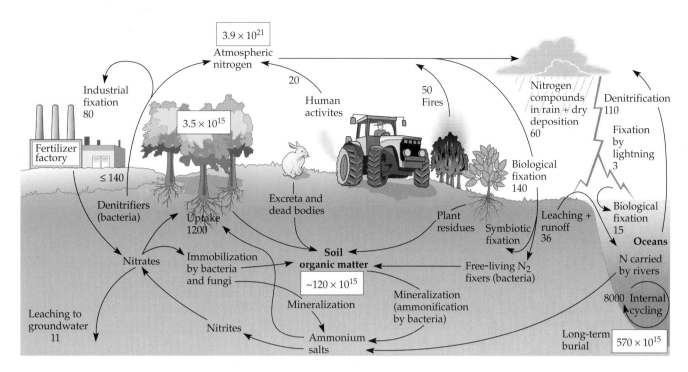

Figure 15.15
The global nitrogen cycle, showing some of the major fluxes (in 10^{12} g N/year, shown in regular type) and pools (shown in boxes; units as indicated). The atmosphere is by far the largest pool of nitrogen. (Data from Schlesinger 1997 and Taiz and Zeiger 1998.)

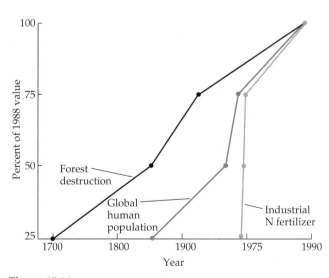

Figure 15.16
Global changes in forest destruction, human population size, and industrial fertilizer production from 1700 to the late 1980s. The chart shows the year in which each variable reached 25%, 50%, and 75% of its value at the end of this time period. Industrial production of nitrogen fertilizer has grown exponentially since the late 1940s. (After Vitousek et al. 1997; data from Kates et al. 1990.)

much greater near industrialized urban centers and where prevailing winds deposit nitrogen from those areas. Not surprisingly, dumping huge amounts of nitrogen on natural systems can have striking effects on ecosystem function. High rates of nitrogen deposition have been shown to have disruptive effects on forests, streams, and coastal waters and estuaries (Aber 1992; Aber et al. 1998) and can lead to general forest decline and dramatic changes in stream ecology (Schlesinger 1997).

Nitrogen Mineralization

Most of the nitrogen used by plants comes from the recycling of N from decaying biomass (see Figure 15.15). The initial stages of decomposition transform these materials into soil organic matter. Further decomposition releases nutrients via a set of processes collectively called **mineralization**, in which microorganisms release carbon as CO_2 and nutrients in inorganic form. The nutrients may then be taken up by plants directly, or they may be taken up after further conversions occur. The process of N mineralization begins with **ammonification**, by which nitrogen in organic compounds, such as proteins and nucleic acids, is released in the form of ammonium ions (NH_4^+). The NH_4^+ ions may be directly taken up by some plants, or may undergo **nitrification** to nitrite (NO_2^-), and nitrate (NO_3^-) (Figure 15.17). Ammonification can be carried out by a variety of soil-dwelling

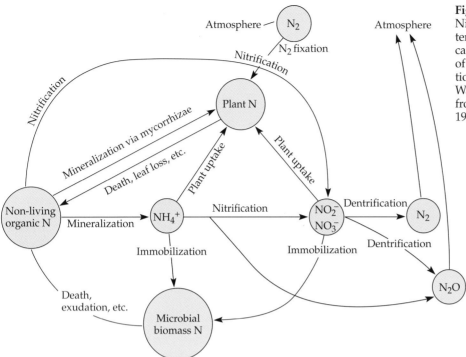

Figure 15.17
Nitrogen cycle within a soil ecosystem, including the soil surface, indicating many of the major components of nitrogen and types of transformations between components. (After Waring and Running 1998, with data from Davidson et al. 1992, Schlesinger 1997, and Drury et al. 1991.)

microorganisms, but nitrification is carried out only by specialized **nitrifying bacteria** (or nitrifiers) and a few other organisms.

In ammonification, heterotrophic microorganisms release N from organic matter by breaking the chemical bonds in carbon compounds to release fully reduced ammonium ions (NH_4^+) in a series of enzymatically controlled steps. Various things can happen to the nitrogen released by these microorganisms, including being used by the same or other microorganisms or by plants. Ammonium ions that are "immobilized" by being taken up by microorganisms (and thus converted back into organic form) can be recycled one or more times through microorganisms before being ultimately released to be taken up by plants, bound to soil minerals or soil organic matter, volatilized to ammonia gas (NH_3) and lost, or oxidized.

Ammonium ions are relatively immobile in the soil and are likely to be taken up only by roots in close proximity. The rate at which ammonification occurs is determined by a variety of environmental factors, particularly temperature and moisture and the chemical characteristics of the soil (including pH and other factors).

In many soils, much of the NH_4^+ produced undergoes further transformation by being oxidized to NO_2^- and NO_3^- through the process of nitrification. Bacteria in the genera *Nitrosomonas* (which convert NH_4^+ to NO_2^-) and *Nitrobacter* (which oxidize NO_2^- to NO_3^-) are the two most important nitrifiers. These bacteria are chemoautotrophs that obtain energy from the oxidation of NH_4^+ or NO_2^- and fix carbon from atmospheric CO_2. They

require oxygen (i.e., they are obligate aerobic organisms). The two groups are generally found together in soils, with the result that high levels of NO_2^- do not often accumulate. There are also a few other organisms that can carry out nitrification. They belong to a variety of heterotrophic groups, including bacteria and fungi. They utilize NH_4^+ to produce NO_3^- or NO_2^-, obtaining energy from the breakdown of carbon compounds in soil organic matter. These organisms may be particularly important in soils that are low in nitrogen.

The most important immediate factor limiting nitrification rates is the amount of available NH_4^+. The factors limiting nitrification therefore include anything that affects the production of ammonium as well as its immobilization, uptake by plants, and capture on clay particles and organic matter. Nitrification rates are also affected by factors that control the population size and physiological activity of nitrifying bacteria, particularly soil water content, but to some extent temperature, pH, and characteristics of the soil and the plant materials. Soil water content is important because very dry soils limit the growth and activity of the bacterial populations, while overly wet and waterlogged soils limit these aerobic microorganisms by excluding oxygen.

Denitrification and Leaching of Nitrogen

Unlike positively charged ammonium ions, negatively charged nitrate ions do not readily bind to soil minerals and organic matter, and are therefore highly mobile in the soil solution. Nitrate ions in the soil solution can be transported some distance to soil roots both by diffusion and by mass flow. Because they are so mobile, they can also be easily leached from the soil into groundwater and runoff, and thus are readily lost. (In some tropical soils, however, anions can be adsorbed onto positively charged aluminum or iron oxides and thus prevented from leaching.) Leaching of nitrate from fertilizer and animal waste can be an important source of water pollution.

Nitrate is also lost from ecosystems by the process of microbial denitrification, which is the reduction of NO_2^- and NO_3^- to various gaseous nitrogen compounds, including N_2 and N_2O. The latter is an important greenhouse gas that can contribute to global warming (see Chapter 22). The composition of the denitrifying bacterial community can differ strikingly between disturbed and undisturbed soils, and these differences can be reflected in very different responses to environmental conditions and in different rates of denitrification and production of N_2O (Cavigelli and Robertson 2000).

Fire can be a major cause of nitrogen loss in some ecosystems. Most of the N released into the atmosphere during a fire comes from the vegetation itself, but substantial amounts can come from the burning of accumulated litter, and very hot fires can burn the soil organic matter in the uppermost part of the profile, thus volatilizing nitrogen from the soil.

Decomposition Rates and Nitrogen Immobilization

As soil organic matter is decomposed by microorganisms, much of the N and P they contain is incorporated into the living microbial biomass (see Figure 15.15) The ratio of C to N in plant tissues ranges from about 25:1 to 150:1, but the ratios are much lower in fungi (between 4:1 and 15:1) and decomposer bacteria (3:1–5:1). As decomposition occurs, soil C:N (as well as C:P) ratios decline as the biomass in decomposing plant material is replaced by biomass in the fungi and bacteria growing on it. When more of the soil N and other nutrients (such as P) become incorporated into living soil microorganisms as organic molecules, these nutrients become less available for uptake by plants. This phenomenon is called biological **immobilization** of the nutrients.

Eventually the microorganisms die, and some of the nutrients again become available for uptake by plants. However, nitrogen immobilization by microorganisms may be a very important short-term limitation on plant nutrient uptake and growth. This effect can be particularly strong when the plant tissues that are decomposing have relatively high C:N ratios or are rich in plant materials (such as lignin) that resist decomposition. High C:N ratios and the presence of lignin and similar materials result in slower decomposition rates and a longer period of N and P immobilization. For instance, evergreen leaves and leaves from plants native to low-nutrient environments typically decompose more slowly than

those from deciduous plants and plants from fertile sites. The leaves of many oak species, which are high in materials that physically and chemically resist decomposition, are incorporated into soil organic matter much more slowly than, for instance, maple leaves from the same site. Leaves of nitrogen-fixing species decompose more quickly than those of co-occurring nonfixers, with consequently less immobilization of soil nutrients. Decomposing animal tissues have much lower C:N ratios, and tend to decompose much more rapidly, than plant tissues.

Plant Uptake of Nitrogen

The balance between NO_3^- and NH_4^+ in soils differs among habitats, largely as a result of environmental effects on nitrification. While most plant species are better able to take up NO_3^-, plants adapted to environments where soil N is largely in the form of NH_4^+ (i.e., where nitrification is inhibited, as in cold, waterlogged soils) may show a preference for uptake of nitrogen in that form. In most plants, both forms of nitrogen are converted in the roots to amino groups (NH_2^-), which are then attached to various organic compounds (NH_4^+ is toxic if it accumulates in plant tissues).

The conversion of NO_3^- to amino groups is energetically expensive. The conversion of NH_4^+ to NH_2^- requires much less energy. However, NO_3^- is transported to plant roots readily by mass flow in the soil solution, but the lower mobility of NH_4^+ in the soil may require greater energy expenditure in the form of root proliferation (Schlesinger 1997).

Nitrogen uptake can be energetically costly for plants for a number of reasons. The uptake of nitrogen as either NO_3^- or NH_4^+ is aided by enzymes that expend energy to transport the ions across cell membranes in the roots. Active root growth and root proliferation into areas of the soil where N is more available also requires energy. In environments where N is less available, plants tend to allocate more biomass to roots, and root:shoot ratios are generally high, compromising the plant's capacity to capture carbon.

Uptake of positively charged ions, such as NH_4^+ and many other nutrients, such as K^+, from the soil could potentially lead to a charge imbalance in plant tissues. To counteract the positive charges, plant roots release hydrogen ions (H^+) into the soil solution, acidifying the soil. When plants take up NO_3^-, they release negatively charged ions such as organic acids and HCO_3^- to maintain their charge balance. The release of these ions in turn affects soil chemistry and the availability of nutrients to plant roots and soil microorganisms. Some ions (e.g., Ca^{2+} and Na^+) are actively excluded by the roots of some plants in environments where their availability is too high.

In some unusual cases, plants obtain nitrogen from sources other than NO_3^- and NH_4^+ in the soil solution.

Figure 15.18
Sarracenia purpurea (purple pitcher plant, Sarraceniaceae) in a bog in northern New York. Bogs are low in nutrients. The pitcher-plant obtains substantial amounts of nitrogen by absorption from detritus captured in its pitcher-shaped leaves. The detritus comes from insects that fall into the leaves, but cannot crawl out because of downward-pointing trichomes (hairs), and drown. Decomposition is carried out by a community of bacteria and invertebrates living in the water that collects in the "pitcher." (Photograph by S. Scheiner.)

"Carnivorous" plants native to extremely nitrogen-poor bogs may obtain nitrogen from decomposing insects (Figure 15.18). Other species may take up nitrogen in the form of amino acids directly from the soil.

Phosphorus in Terrestrial Ecosystems

Although phosphorus is needed for plant growth and function in smaller amounts than other macronutrients (see Table 4.2), it is fairly common for plant growth to be limited by P availability. We considered P availability in the soil and its uptake by plants in Chapter 4. Here we discuss P at the ecosystem and global levels.

Most of the phosphorus that cycles through ecosystems is found in microorganisms, in decomposing plant material, and in soil organic matter—that is, in organic material (Figure 15.19). This is curious, because most of the P in terrestrial systems occurs as minerals in rocks. Inorganic P is often complexed with other soil minerals and is relatively unavailable to plants directly (but see

Chapter 4 for a discussion of how mycorrhizal fungi and other organisms can access this phosphorus).

The original source of most of the phosphorus that enters the biosphere is weathering of rock containing apatite minerals. The minerals in the apatite group are the only minerals common in the lithosphere that con-

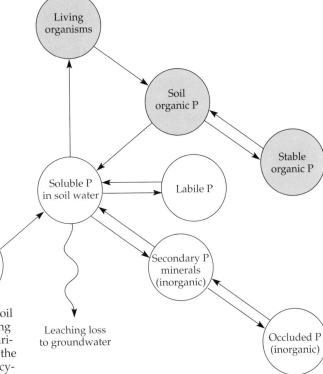

Figure 15.19
Phosphorus from minerals in rock (primary P) dissolves in the soil solution, from which it can be absorbed by plants and other living organisms. Some soluble soil phosphorus is transformed into various inorganic forms or is leached; however, the vast majority of the phosphorus taken up by organisms in an ecosystem has been recycled through other living organisms. Biological transformations are shown in green, geochemical transformations in white. (Modified from Schlesinger 1997, after Smeck 1985.)

tain a substantial amount of phosphorus. As rock weathers and soils form, and as the soils develop and age, much of the phosphorus becomes bound in the interior of iron and aluminum oxide crystals, from which living organisms cannot remove it.

In old soils that have become dominated by iron and aluminum oxides (such as those that are common in Africa, Australia, and other environments, particularly in the Tropics), the lack of available phosphorus is one of the most important environmental factors affecting plants. Inorganic P may also be held in more available forms on the surface of other soil minerals by anion adsorption and other reactions. The complicated soil chemistry involved in these reactions depends on soil pH and other factors. It is therefore generally difficult for scientists to accurately measure the amount of phosphorus available to plants.

Unlike the global cycles of carbon, nitrogen, and water, the global cycle of phosphorus does not have a major atmospheric component (Figure 15.20). Most of the phosphorus that is used by organisms has been recycled in organic form through other living organisms, although its ultimate source is rock weathering. Organically bound phosphorus is released in the form of PO_4^{3-} (orthophosphate) by plant roots, fungi (including those in mycorrhizae as well as free-living fungi), bacteria, and algae. All of these organisms mineralize phosphorus through the action of extracellular phosphatases (enzymes that cleave ester bonds). This process is expensive both energetically and in terms of the nitrogen invested in the phosphatase enzymes. Microorganisms as well as plants can take up the PO_4^{3-} that is released, but if it is not immediately taken up, the orthophosphate does not remain available for long in the soil. Rather, it readily binds to organic particles or to soil minerals, becoming partially or completely unavailable to plants.

Ecosystem Nutrient Cycling and Plant Diversity

Is there any relationship between the characteristics of a plant community, such as diversity, and ecosystem processes such as nitrogen and phosphorus cycling? Recently considerable effort has been made to determine and understand the potential connections between these important ecosystem and community properties (Loreau et al. 2001; Tilman et al. 1997; see Chapter 14).

David Hooper and Peter Vitousek (1998) set out to answer this question experimentally in a California grassland. The grassland was located on serpentine soil (Box 15A). The plants were categorized into four functional groups that differed in phenology (seasonal growth pattern), root:shoot ratio, and other traits relevant to the use and cycling of N and P. The four functional groups were early-season annual forbs (herbaceous plants), late-season annual forbs, nitrogen fixers, and perennial bunchgrasses. Experimental plots contained plants of either a single functional group or combinations of functional groups. The researchers measured pools and fluxes of nitrogen and phosphorus, as well as other ecosystem variables, on the experimental plots.

Hooper and Vitousek found that plant functional group diversity increased the total use of nutrients, as they had hypothesized, because of seasonal differences in plant growth and resource use among the functional groups. They also found, however, that functional group richness did not decrease nitrogen leaching, although much more nitrogen was lost when plants were totally absent than when plants of any of the functional groups were present. Ecosystem nutrient retention was deter-

Figure 15.20
The global phosphorus cycle, showing major fluxes (in 10^{12} g P per year, shown in bold) and pools (in 10^{12} g P). Marine sediments, at approximately 4×10^{21}, are by far the largest pool, and cycling by land plants, at 3000×10^{12} g P per year is the largest flux. (After Schlesinger 1997, with data from Jahnke 1992.)

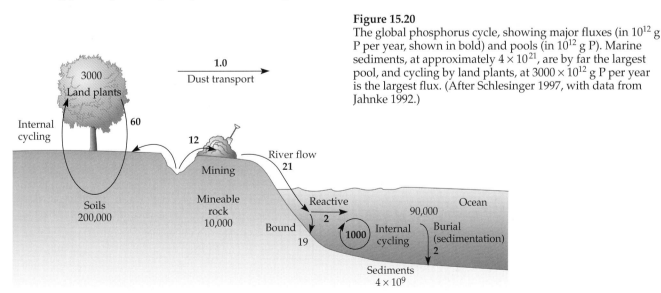

Box 15A

Serpentine Soils

Important ecological research has been conducted over the years on various aspects of the ecology of vegetation on serpentine soils (Kruckeberg 1954; Whittaker 1954; Hooper and Vitousek 1998). Serpentine soils are derived from serpentine rock, which is a magnesium silicate rock of metamorphic origin. These soils are poor in many nutrients—particularly calcium, nitrogen, and phosphorus—and they may have either very low or very high pH. Toxic elements, particularly heavy metals, are also characteristic of serpentine soils.

Areas with serpentine soils occur in many places in the world. Serpentine soils often occur in patches of limited extent, or "islands," surrounded by more ordinary nonserpentine soils. These serpentine islands (sometimes called serpentine barrens) typically support distinctive plant communities that contrast strongly with the vegetation of the surrounding region. For instance, serpentine vegetation in California often contains many showy dicot

species, surrounded by grasslands. Serpentine vegetation is often very high in species diversity and may have many endemic species. The species from surrounding areas are unable to tolerate the toxic, low-nutrient soils, and this excludes them from serpentine patch-

es. In addition, Kruckeberg (1954) showed that the serpentine species appear to be excluded from the surrounding regions by competition (echoing Bradshaw's findings for plants growing on mine tailings, discussed in Chapter 6).

Serpentine grassland in the state of California. (Photograph courtesy of D. Hooper)

mined at least as much by the indirect effects of plants on microbial activity and subsequent nutrient retention by microorganisms as by the indirect effects of plants on microbial activity. The researchers concluded that the identity of the particular plants present explained more about ecosystem nutrient cycling than could be explained by considering the number of functional groups present.

Ecosystem Processes for Some Other Elements

We have examined the major ecosystem processes of primary importance for plants: those involving water, carbon, nitrogen, and phosphorus. Other nutrients, of course, are necessary for plant growth (see Table 4.2). Here we briefly discuss the cycling of two other important elements, sulfur and calcium, both as general examples and because of the influence of human activities on their cycles.

Sulfur

Sulfur is necessary primarily for building the amino acids cysteine and methionine, important constituents of various proteins. Like nitrogen, phosphorus, and carbon, sulfur is cycled biogeochemically in ecosystems. The weathering of rocks with minerals containing sulfur (such as pyrite and gypsum) is one major source of the sulfur that ultimately becomes available to plants. Atmospheric deposition is also an important source of sulfur. A large proportion of the sulfur pool in soils exists as a part of soil organic matter. Inorganic sulfate ions (SO_4^{2-}) are adsorbed onto soil mineral particles, in equilibrium with SO_4^{2-} dissolved in the soil solution. Bacterial mineralization and immobilization of sulfur are analogous to the same processes for nitrogen and phosphorus.

Sulfur is lost from ecosystems in a variety of forms. In addition to the loss of sulfate ions from ecosystems by leaching, many plants release volatile organic sulfur (particularly H_2S), and waterlogged, anaerobic soils can also produce large quantities of sulfur gases by bacterial reduction.

On a global scale, large pools of sulfur are dissolved in seawater. In the atmosphere, sulfur-containing gases are short-lived and are found at low concentrations. However, dust storms, sea spray, and volcanic eruptions can contribute large amounts of particulate sulfur (as well as some sulfur in the form of gases and aerosols) to the atmosphere. Human activities are the largest current source of atmospheric sulfur gas, although recent air pollution controls have decreased the yearly input of sulfur from industrial sources considerably. The SO_2 that enters the atmosphere reacts with water to form sulfuric acid, which, along with nitric acid (from anthropogenic NO and NO_2), is the chief cause of acid precipitation (see Chapter 22).

Calcium

Calcium is abundant in the Earth's crust. It is an important component of limestone and other common rocks such as gypsum ($CaSO_4 \cdot H_2O$) and calcium carbonate ($CaCO_3$). Weathering of these rocks releases calcium ions (Ca^{2+}), which are carried by rivers to the oceans; large pools of calcium ions are found in seawater. Marine invertebrates use this calcium to build calcium carbonate shells; when these organisms die, their remains form thick sediments. Through burial, metamorphosis, and crustal movement, these marine sediments may ultimately be transformed into calcium-bearing terrestrial rock.

Calcium ions readily dissolve in the soil solution and are taken up by plants passively in the **transpiration stream**. Calcium concentrations can vary dramatically among soils, from being apparently limiting to plant growth to being present in excessive amounts. In humid regions with low-pH soils, calcium ions can be heavily leached from the soil. Huntington (2000) studied fluxes of calcium in southeastern U.S. forests, and found that losses of calcium caused by logging and soil leaching often were greater than inputs from atmospheric deposition and rock weathering, leading to severe calcium depletion in the forest soils. His results suggest that calcium depletion could result in widespread problems for southeastern forest trees over the next few decades.

In contrast, in many deserts, calcium can accumulate in the soil. Rainwater dissolves the calcium ions and then evaporates, precipitating and thus concentrating the calcium into either a soft or a cement-hard layer called a **calcic horizon** (also called a "pan" or "caliche"). The excessive amounts of calcium in these calcareous, high-pH soils can be actively excluded by the roots of some plant species.

Within an ecosystem, the calcium contained in plant biomass is returned to the soil through litterfall and sub-

sequent litter decomposition, as well as **throughflow** (rainwater) that passes through the canopy and travels down the surface of tree trunks, carrying dissolved ions leached from the plant surfaces. Some calcium is carried in particulate form in the atmosphere and deposited in rainfall or by dry deposition (Figure 15.21).

The requirement of plants for calcium is generally very high, second only to that for nitrogen (see Table 4.2). Calcium is unusual in that it plays both chemical and structural roles in plant tissues, including regulation of growth and response to stress, stomatal function, cell division and cell wall synthesis, and structural support in both leaves and wood (McLaughlin and Wimmer 1999). Decreasing calcium availability may lead to decreased growth and higher mortality in forest trees, making them more susceptible to pathogens, for example.

Some recent evidence suggests that calcium depletion is occurring in many forests, with negative consequences for forest vitality (Likens et al. 1998), although other researchers have disputed these findings (Yanai et al. 1999). The potential for calcium depletion is a result

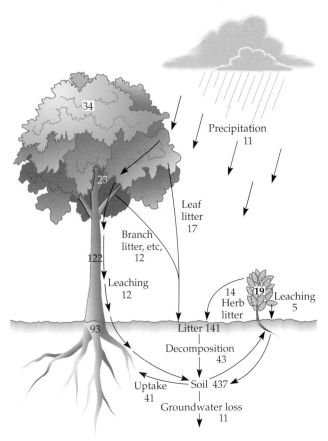

Figure 15.21
Local calcium cycle within a forest ecosystem in the United Kingdom. Pool amounts are given in kg/ha, and annual fluxes (arrows) are shown in kg/ha/year. (After Schlesinger 1997 and Whittaker 1970.)

of several factors, including decreasing amounts of calcium in rainfall and removal of calcium from the ecosystem in biomass (by logging and other human activities). Most important are the direct negative effects of anthropogenic nitrogen deposition (particularly in the form of HNO_3), which results in displacement of calcium ions in the soil, causing them to be leached into stream water and lost.

Summary

Ecosystem ecologists study the flow of energy and materials through ecosystems. In terrestrial ecosystems, plants play the essential role of primary producers: nearly all production of biomass depends on plant photosynthesis. Fluxes of water and carbon, as well as nitrogen and most other nutrients, also depend critically on the plants in an ecosystem, since plants constitute a large proportion of the living biomass, and other organisms acquire carbon and nutrients from food webs that are based mainly on plants. These materials cycle through the biosphere at very different rates. Thus, changes in the amount of plant cover or in the efficiency of uptake by plants have differing effects on each biogeochemical cycle. Similarly, changes in the different biogeochemical cycles have very different effects on plants. Plants and the water cycle affect one another quickly, as changes in regional plant cover can affect humidity and rainfall within years to decades, and changes in moisture can affect plants within days to weeks. Changes in the carbon cycle occur on a larger spatial and longer time scale, because CO_2 is (on average) evenly distributed in the atmosphere, and much carbon is stored as biomass and dissolved in the oceans.

Net primary productivity (NPP) the total energy captured minus respiratory losses by primary producers, is a critical component of many ecosystem processes. Recent technological developments have made it possible to obtain far better measurements of productivity than was possible in the past. Nevertheless, we are still not able to model the carbon cycle with full confidence, although doing so is important for predicting the consequences of deforestation and CO_2 emissions from fossil fuels. The carbon that is captured in NPP accumulates in an ecosystem primarily in living biomass, plant litter, and soil organic matter, and is lost from the ecosystem primarily through soil microbial respiration.

The nitrogen cycle depends strongly on microbial activity. Biologically available nitrogen originates in soils mainly by microbial fixation, and the decomposition of biomass recycles available nitrogen in the soil. Nitrogen is often limiting to plants; increases in the amount of available nitrogen can have major effects on plant growth rates. Use of synthetic fertilizer by humans has roughly doubled the amount of biologically available nitrogen in the world, with serious consequences for both terrestrial and aquatic ecosystems.

The global phosphorus cycle does not have a major atmospheric component. Phosphorus is limiting for plant growth in many ecosystems, particularly those on old soils. Sulfur and calcium are necessary for plant growth; their availability varies widely among different soils. Human activities may be depleting calcium in some ecosystems. Improving our understanding of biogeochemical cycles is critical both for understanding how plants and plant communities change over time and for understanding the environmental challenges presented by anthropogenic change.

Additional Readings

Classic References

Tansley, A. G. 1935. The use and abuse of vegetational concepts and terms. *Ecology* 16: 284–307.

Lindeman, R. L. 1942. The trophic-dynamic aspect of ecology. *Ecology* 23: 399–418.

Odum, E. P. 1960. Organic production and turnover in old field succession. *Ecology* 41: 34–49.

Contemporary Research

Gohlz, H. L., D. A. Wedin, S. M. Smitherman, M. E. Harmon and W. J. Parton. 2000. Long-term dynamics of pine and hardwood litter in contrasting environments: toward a global model of decomposition. *Global Change Biol.* 6:751-765.

Shaver, G. R. and F. S. Chapin III. 1991. Production-biomass relationships and element cycling in contrasting Arctic vegetation types. *Ecol. Monogr.* 61: 1–31.

Vogt, K. A., C. C. Grier, C. E. Meier and R. L. Edmonds. 1982. Mycorrhizal role in net primary production and nutrient cycling in *Abies amabilis* ecosystems in western Washington. *Ecology* 63: 370–380.

Additional Resources

Chapin, F. S., P. Matson and H. A. Mooney. 2002. *Principles of Terrestrial Ecosystem Ecology*. Springer-Verlag, New York.

Schlesinger, W. H. 1997. *Biogeochemistry: An Analysis of Global Change*. 2nd ed. Academic Press, New York.

Waring, R. H. and S. W. Running. 1998. *Forest Ecosystems: Analysis at Multiple Scales*. 2nd ed. Academic Press, New York.

CHAPTER 16 Communities in Landscapes

ommunities exist as components of a larger landscape. The populations that make up a community are linked to populations of the same species in other communities by migration (see Chapter 5). Communities are also linked by the movement of air, water, and nutrients (see Chapter 15). Landscape ecology is the study of these larger-scale relationships among communities. In this and the following chapter we will examine ecological patterns and processes that occur at the level of landscapes. In this chapter we explore methods for examining patterns of variation among communities within a landscape.

How can we quantify differences in community composition and structure? How are species distributed among communities? Can we relate these differences to variation in the environment? These types of questions and analyses are some of the oldest in plant ecology. Comparing communities across a landscape and looking for environmental causes of differences and similarities goes back to the origins of plant ecology at the end of the nineteenth century. While technological advances have enhanced ecologists' ability to collect and analyze complex data to use in making these comparisons, many of the basic underlying questions and approaches have remained the same for the past 100 years.

Comparing Communities

The first question we might want to ask about the communities in a landscape is how different they are from one another. In Chapter 12 we described several ways of characterizing communities. Those characteristics can be compared among two or more communities. The techniques used for such comparisons depend on the nature of the variables being compared. If each community is described by a single characteristic (such as total biomass or mean canopy height), then **univariate** (single dependent variable) statistical procedures are used. In contrast, if each community is described by multiple parameters (such as a species list), then **multivariate** (multiple dependent variable) procedures are needed. The latter methods can be complex and have been the subject of much debate among ecologists.

Table 16.1 Using the Jaccard index to measure the similarity of sites based on presence/absence data

(A) Typical presence/absence data[a]

Species	Sites				
	A	B	C	D	E
1	1	1	0	0	1
2	1	1	1	0	0
3	0	0	1	1	1
4	0	0	1	1	0
5	1	0	1	1	0
6	0	1	0	0	1
7	0	0	1	1	1
8	1	1	0	0	0
9	1	1	1	0	0
10	0	1	1	0	0

(B) Reordered matrix[b]

Species	Sites				
	A	B	E	D	C
8	1	1	0	0	0
1	1	1	1	0	0
9	1	1	0	0	1
2	1	1	0	0	1
6	0	1	1	0	0
10	0	1	0	0	1
5	1	0	0	1	1
3	0	0	1	1	1
7	0	0	1	1	1
4	0	0	0	1	1

(C) Matrix of Jaccard similarity values for the data in (A) or (B)

Site	Sites				
	A	B	C	D	E
A	1.00	0.57	0.33	0.13	0.13
B	0.57	1.00	0.30	0.00	0.25
C	0.33	0.30	1.00	0.57	0.22
D	0.13	0.00	0.57	1.00	0.33
E	0.13	0.25	0.22	0.33	1.00

[a] The presence of a species in a site is indicated by a 1.
[b] A reordered matrix attempts to group sites that share species and species that share sites.

Non-numerical Techniques

While the methods used in plant community ecology have changed considerably in response to the availability of powerful computers, the questions asked have not. One such basic question is, which communities in a landscape are most similar to each other? Later in this chapter we will show how we can quantify the answer to this question. However, we can also address this question in a more qualitative way, using a method favored by European ecologists during the mid-twentieth century.

We might begin, for instance, with a survey of five sites that contain a total of ten species (Table 16.1A). The order of the sites and species in Table 16.1A is arbitrary. (Because this is a hypothetical example, we use numbers for the species, rather than names.) We can then try to change the order of the sites and species so that we group the sites that share the most species and group the species that are most often found together. This reordering is shown in Table 16.1B. The reordering has produced some patterns, with a group of species that are shared by sites A and B appearing in the upper left-hand corner and a group of species that are shared by C and D in the lower right-hand corner, while site E shares species with both groups. Species 9, 2, 10, and 5 are shared by both groups, while some of the species might be considered indicators of each group (8 and 4) since they each appear in just one group. With a much larger table showing many more sites and species, such patterns usually become even more apparent.

Using this sort of reordering method, plant community ecologists tried to discern patterns of commu-

nities across landscapes. Dissatisfaction was expressed with this method, however, because it is partially subjective. For example, site C shares more species with site A than site E shares with site A. So one might rearrange the columns in the table to place sites A and C next to each other. Also, we decided that there were two groups of sites, (A, B) and (C, D), even though there is considerable overlap in the species shared by sites in different groups. To avoid this subjectivity, ecologists developed a variety of quantitative techniques.

Univariate Techniques

Univariate techniques are used whenever a single type of measurement is made, such as biomass per unit area. We might determine biomass in ten quadrats within each of six communities. A typical question would be, "Do these communities differ in their average biomass per square meter?" There are many standard statistical techniques for analyzing such data, such as analysis of variance (ANOVA). A number of statistics textbooks describe these methods in detail (e.g., Sokal and Rohlf 1995; Zar 1999; Scheiner and Gurevitch 2001). Most often, though, we compare communities by measuring more than one variable.

Multivariate Techniques

All multivariate techniques rely on the same basic principles and approach. Each is concerned with comparing how different the members of a group of objects (such as communities) are, based on the values obtained for a set of characteristics measured for all of the objects. The mathematical techniques then arrange those objects in one or more dimensions, based on their differences in the entire set of characteristics taken together.

Let's begin with a simple univariate (one-dimensional) example. Imagine that we wish to compare three plant communities (A, B and C) with mean biomasses of 500 g/m², 725 g/m² and 625 g/m². We can put them in order based on their mean biomass (A C B). Another way of accomplishing the same goal is to first determine the difference between each pair of means (d_{AB} = 225, d_{AC} = 125, d_{BC} = 100). Using this information, we determine that A and B are most different, or that the order is (A C B). Note that $d_{AC} + d_{BC} = d_{AB}$ because there is only one way to put these numbers in order from smallest to largest—they fall on a line. While the second method, using differences, seems more complicated than the first, it is actually what you do when you use ANOVA to determine whether groups differ from one another.

With two descriptive variables, the process becomes a bit more complicated, but the basic principles are the same. Let's assume that we also measured mean foliage height for the three communities and found that the values are A = 3.2 m, B = 3.5 m, and C = 5.6 m. Instead of a single number, or scalar, giving the difference between two communities, we now need an ordered list of numbers, or vector (see Chapter 7), to give the difference. In this case, we can use that old standby, the Pythagorean theorem ($x^2 + y^2 = z^2$), where x and y are the lengths of the sides of a right triangle and z is the length of the hypotenuse. The measure of the difference between each pair of communities is referred to as the **Euclidean distance**. If we let x represent mean biomass and y represent mean foliage height, then for communities A and B, x = 725 – 500, y = 3.5 – 3.2, and z = 225.0002 = d_{AB} (Figure 16.1). Similarly, d_{AC} = 125.0230 and d_{BC} = 100.0220. Now the distances do not sum because the points no longer lie on a straight line. Typically units of measure are not used because the distances are a complex combination of the units of measure for each descriptive variable.

With only two dimensions, it is easy to graph and visualize our data (Figure 16.1). But what if we have many descriptive variables? Information on species presences and abundances constitutes just such a large set of descriptive variables. Think of the two axes in Figure 16.1 as representing the abundances of species 1 and species 2. Now, imagine that we add a third axis projecting out of the page, on which the abundance of species 3 is graphed. We might not be so readily able to imagine a fourth axis for the abundance of species 4. Typical data include, however, not just three or four, but tens to hundreds of species. Clearly, we cannot graph such a multidimensional object. Instead, we resort to various techniques that reduce the problem from n dimensions (n = the number of species) to the two or three dimensions that we can visualize.

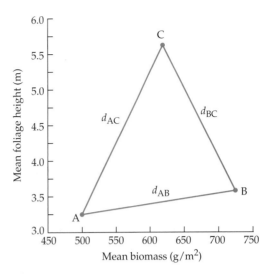

Figure 16.1
Three plant communities (A, B, and C) measured for biomass and foliage height. Euclidean distances between the points are given by d_{AB}, d_{AC}, and d_{BC}.

The first step is to devise a distance measure. If the descriptive variables are related monotonically (meaning that the abundances of species 1 and species 2 always increase together, for instance, and when species 1 increases, species 3 always decreases), and also increase or decrease together linearly (when graphed against one another, the points fall on a straight line), then we can use the multivariate equivalent of the Euclidean distance. This measure is the Pearson product-moment correlation, which is the familiar correlation coefficient (see Appendix A). However, species abundances are usually not related in such a simple fashion.

Consider two species with different soil moisture requirements: species 1 does best in wet soils, while species 2 does best in partly moist soils. Going from very dry soils to partly moist soils, both species would increase in abundance. But going from partly moist soils to wet soils, species 2 would continue to increase in abundance, while species 1 would decline in abundance (Figure 16.2). The descriptive variables would no longer be related in a monotonic fashion. In addition, many of the descriptive variables used are not continuous. Data on the presence and absence of species, for example—the information contained in a species list—are dichotomous, consisting of just 1's and 0's (see Table 16.1A).

For these types of data, ecologists have devised a number of distance metrics called **similarity measures**. Consider first those that are used for presence/absence data. For a pair of sites, we can define four types of species: those that are present in both sites (*a*), those that are present in the first site but absent in the second (*b*), those that are present in the second site and absent in the first (*c*), and those that are absent in both (*d*). This pattern can be represented as follows:

Second site	First site	
	Present	Absent
Present	*a*	*b*
Absent	*c*	*d*

For sites A and B in Table 16.1, $a = 4$, $b = 1$, $c = 2$, and $d = 3$.

The simplest, and oldest, measure of similarity, invented by the French ecologist P. Jaccard, is the Jaccard index (Jaccard 1901). It is the percentage of species contained in two sites that are shared by those sites:

$$S_J = \frac{a}{a+b+c}$$

For sites A and B, $S_J = 4/(4 + 1 + 2) = 0.57$.

Over the years a number of similarity measures have been invented for both presence/absence and abundance data (Table 16.2). While various indices work best in different situations and with different types of data, for presence/absence data the Jaccard index consistently performs the best, or close to the best, in the widest number of situations. Using the Jaccard index, we can now proceed to the second step, measuring the distance between each of our sites (Table 16.1C).

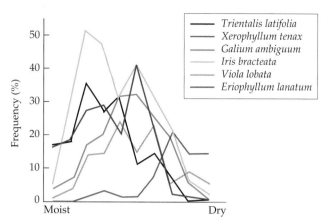

Figure 16.2
Patterns of species abundance along a moisture gradient in the Siskiyou Mountains, Oregon. As seen in this case, changes in the abundances of species along an environmental gradient are unlikely to be related monotonically. Species abundance was measured as the frequency of quadrats within a site in which a species appeared. (Data from Whittaker 1960; after Brown 1984.)

Landscape Patterns

Ordination: Describing Patterns

Ordination is the process of taking information such as that contained in Table 16.1C—points arrayed in an *n*-dimensional space—and reducing it to fewer dimen-

---

Table 16.2 Some similarity measures used by plant ecologists

Index	Formula	Index	Formula
Presence/absence indices		*Abundance indices*	
Jaccard index	$S_J = \dfrac{a}{a+b+c}$	Percentage of similarity	$S_{PS} = \sum \lvert p_{1i} - p_{2i} \rvert$
Sørensen-Dice index	$S_{SD} = \dfrac{2a}{2a+b+c}$	Asymmetrical percentage of similarity	$S_{APS} = \dfrac{\sum(n_{1i} - n_{2i})}{\sum n_{1i} + \sum n_{2i}}$, for $n_{1i} \neq 0$
Simple matching	$S_{SM} = \dfrac{a+d}{a+b+c+d}$	Minimum similarity	$S_{MS} = \sum \min(p_{1i}, p_{2i})$
Ochioi index	$S_O = \dfrac{a}{\sqrt{(a+b)} + \sqrt{(a+c)}}$	Bray-Curtis index	$S_{BC} = \dfrac{\sum \min(n_{1i}, n_{2i})}{\sum n_{1i} + \sum n_{2i}}$
Asymmetrical similarity	$S_{AS} = \dfrac{b}{2a+b}$	Morisita's index	$S_M = \dfrac{2\sum(n_{1i}n_{2i})}{(\lambda_1 + \lambda_2)N_1 N_2}, \lambda_j = \dfrac{\sum n_{ji}(n_{ji}-1)}{N_j(N_j - 1)}$

Definitions of symbols:
a = number of species in both sites
b = number of species in second site only
c = number of species in first site only
d = number of species in neither site

p_{1i} = the proportion of individuals in the *i*th species in sample 1 ($p_{1i} = n_{1i}/N$)
n_{1i} = the number of individuals of species *i* in sample 1
N = the total number of individuals sampled
$\min(x,y)$ = the smaller of the two values

Note: These indices differ in the factors they emphasize (e.g., common versus rare species) and their robustness to deviations in assumptions (Whittaker 1972; Janson and Vegelius 1981; Wolda 1981; Austin and Belbin 1982; McCulloch 1985).

Figure 16.3
The principle of dimension reduction. The flashlight causes the balls, arrayed in three-dimensional space, to cast a two-dimensional shadow on the back wall.

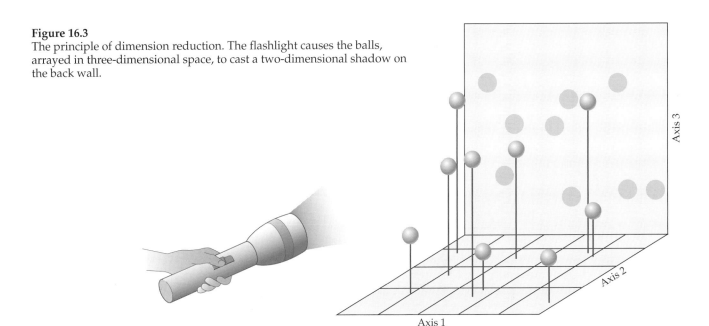

sions; in other words, ordination is just a quantitative re-description of the data. Often the results are shown as a two- or three-dimensional graph. There are many ordination techniques, each of which has had its advocates. They are usually known by acronyms: PO (polar ordination), PCA (principal components analysis), PCoA (principal coordinates analysis), RA (reciprocal averaging), DCA (detrended correspondence analysis), NMDS (nonmetric multidimensional scaling). Each technique uses somewhat different mathematical approaches and has different assumptions, limitations, and advantages. While the techniques differ in their mathematical details, they all rely on the same basic principles, and they all provide broadly similar information. The distinctions between and details of these methods are beyond the scope of this book, but those who wish to explore them further will find the topic covered in depth by Legendre and Legendre (1998).

The process of **dimension reduction**—taking highly multivariate data and collapsing them into a small number of dimensions—can be visualized as follows. Imagine a physical representation of a three-dimensional data set like that shown in Figure 16.3, in which each axis shows the abundance of a species and each ball represents a different site. Now, we shine a flashlight through the model so that a shadow falls on one wall. Each ball (site) is now represented by a point on that wall. We have reduced the data from three dimensions to two. In the process we have lost some information. Two points that were actually far apart in the three-dimensional model may now appear to be close to each other because of the way their shadows fall. We can minimize this problem, though, by carefully choosing the angle at which we shine the flashlight. Most likely the balls are not arrayed

randomly throughout the model. Rather, they are probably arrayed in some more compact shape, such as an elongated ellipse. If we shine the flashlight so that it is at right angles to the long axis of the ellipse, then points that were the farthest apart in the original model are still far apart in the shadow.

Of course, with actual data, the points are arrayed in a much more complicated n-dimensional space, and the "flashlight" is a set of mathematical steps that determine the coordinates of those points in the reduced two- or three-dimensional space. Figure 16.4 shows two ordinations of the data from Table 16.1C using two different techniques; notice how both techniques produce the same overall picture of the relationship between the sites. Since ecologists are usually interested in broad patterns, rather than the exact numerical relationships between sites, the choice of distance metric and ordination technique is often a matter of convenience.

Multivariate statistical methods started to permeate plant community ecology in the mid-1950s, sometimes by importation of techniques being used in other fields and sometimes by independent invention. Perhaps the most influential work was the development of polar ordination by J. Roger Bray in the mid-1950s while he was a student of John Curtis (Bray 1955; Bray and Curtis 1957). Independently, David Goodall (1954) showed how factor analysis (now referred to as principal components analysis) could be applied to community surveys.

Robert Whittaker developed a general approach called **direct gradient analysis**, in which the ecologist chooses environmental axes and orders vegetation samples (stands) along those axes, examining the resulting patterns. Bray and Curtis adopted a different approach,

(A)

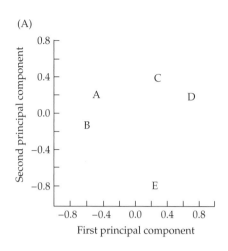

(B)

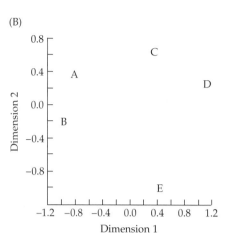

Figure 16.4
Ordinations of the data in Table 16.1C using two different ordination techniques: (A) principal coordinates analysis (PCoA), and (B) nonmetric multidimensional scaling (NMDS). Notice that the two techniques produce the same broad pattern of relationship between the sites.

called **indirect gradient analysis**, in which the stands are ordered by their similarity in species composition and the environmental factors responsible for the resulting patterns are inferred. The latter approach (a form of ordination) has the advantage of being more objective, but it cannot be used to examine patterns of vegetation along particular environmental axes that are known in advance to be of interest.

Ordination and related methods were further refined in the 1960s, 1970s, and 1980s by a number of ecologists, most notably Whittaker's students and collaborators in the United States (e.g., Gauch and Whittaker 1972), Michael Austin in Australia (e.g., Austin 1977), and Cajo ter Braak in the Netherlands (e.g., ter Braak 1986). Over the same period, the development of multivariate techniques in systematics (under the heading of numerical taxonomy; Sneath and Sokal 1973) further influenced their use in ecology. Advances in computer hardware and software at this same time were particularly important in the spread of these techniques. During the 1960s and 1970s, using these methods required writing complex programs, making stacks of punch cards, and gaining access to a mainframe computer.

Today, these techniques are still commonly used, especially for pattern detection and hypothesis generation. A series of communities might be sampled to document relationships involving a particular ecological phenomenon—for example, nitrogen deposition, invasion by exotic species, or damage by a pathogen—with the object of relating the phenomenon to species compositions and environmental variables across communities. The patterns that are revealed can form the basis of experiments to test hypotheses about the mechanisms causing the patterns.

Determining Causes of Patterns

Ordination can describe patterns of species distribution among communities in a landscape. But what are the causes of those patterns? For plant communities, the primary causes of differences in species composition are climatic, topographic, and edaphic factors—the physical factors that determine plant growing conditions (see Part I). A number of techniques exist for determining which of these factors are most important in a given set of communities. All of these techniques use basically the same strategy, looking for correlations between species distributions and environmental variables.

Consider a simple problem, that of determining the causes of variation in the abundance of a particular species across a landscape. One strategy for attacking this problem is to sample a number of sites across the landscape. At each site we measure the abundance of the species as well as a number of environmental variables, such as temperature, precipitation, elevation, slope, aspect, soil texture, and various soil nutrients. (Ideally, the climatic data would be based on long-term averages.) Then we calculate the correlation of each environmental variable with species abundance. This procedure contains two unstated assumptions: (1) that the pattern of species distribution is based on a long-term equilibrium with the environment, and that the current pattern is close to that equilibrium, and (2) that we have measured the correct environmental variables.

Of course, correlation does not demonstrate causation, but a strong correlation provides one piece of supporting evidence for a possible cause of the observed pattern. The method just described has a problem, however: the environmental variables are themselves correlated with one another. Sites at higher elevations, for example, are also likely to have lower temperatures. To account for these correlations, we use a technique called **multiple regression**. This technique determines the regression (Appendix A) of the abundance of our species with each environmental variable while correcting for correlations among the environmental variables themselves. The limitation of this technique is that it cannot correct for very large correlations. For example, if the abundance of our species decreased with increasing ele-

(A)

Figure 16.5
(A) Stands of *Pinus contorta* (lodge-pole pine, Pinaceae) dominate the landscape in the middle elevations of the Canadian Rockies. The cones open in response to fire, and the result is stands like this one near Allison pass. (Photograph courtesy of John Worrall.) (B) Direct gradient analysis of *P. contorta* stands in the Canadian Rockies. The stands are arrayed along two gradients: moisture and elevation. Green numbers indicate stands in Banff National Park. Black numbers indicate stands in Jasper National Park. (After La Roi and Hnatiuk 1980.)

vation and temperature also decreased with increasing elevation, then we could not determine whether the cause of decreasing abundance was elevation or temperature. However, our analysis would have narrowed the possible factors responsible for decreasing abundance, although the possibility would remain that the actual cause was some unmeasured third variable that is also correlated with elevation and temperature.

We could also perform manipulative experiments (e.g., growing the plant in growth chambers at different temperatures) to further narrow the possible causes and to test specific hypotheses. While such experiments are an important source of information, they also have limitations. Because they are often done in highly artificial environments such as a growth chamber or greenhouse, they cannot account for the myriad of abiotic and biotic factors and their interactions in nature. Competitive interactions, for example, might differ at high and low temperatures. Manipulative experiments done in a field setting are technically more challenging, but can provide a more realistic test of the importance of an environmental factor.

We can do the same sort of analysis for whole communities as was just described for a single species. Now, instead of using species abundance, we use position along an ordination axis as our dependent variable. We can see how this works with data from a study of *Pinus contorta* (lodgepole pine, Pinaceae) forests in the Canadian Rockies. Lodgepole pine is the most widely distributed tree in western North America (Figure 16.5A). In Alberta, Canada, where this study took place, it is the dominant species following fire in mid-elevation forests (1000–2000 m). For this study, 63 stands were sampled in Banff and Jasper National Parks; measurements included the number and sizes of all trees, cover of all

(B)

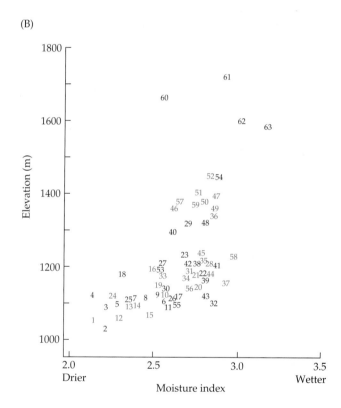

understory species, soil chemistry, soil temperature, soil moisture, elevation, slope, and aspect.

Figure 16.5B shows data from these stands arrayed along two environmental axes, elevation and moisture. Because each axis consists of a single environmental variable, this graph is a form of direct gradient analysis. If each axis consisted of combinations of multiple environmental variables, we would refer to this method as **canonical correspondence analysis** (CCA; ter Braak 1986). In effect, canonical analysis is a hybrid between direct and indirect gradient analysis. Two ordinations are carried out, one on the species data and one on the environmental data, and then the correlations between the axes of the two ordination sets are determined. If we are lucky, the first one or two axes of each set will be highly correlated with each other. An analysis of macrophytes growing on dikes in the Netherlands provides an example of this type of analysis (Figure 16.6). As with ordination, various distance metrics, dimension reduction techniques, and correlation procedures can be used (Ludwig and Reynolds 1985).

Types of Data

We have described analyses that can use either presence/absence data or abundance data. Which type of data is most appropriate? The answer depends on the questions being addressed relative to the scale of variation. Different kinds of variation are expected to occur at different scales. By "scale" we do not mean geographic extent; rather, we mean ecological scale—how different the communities are ecologically. A set of neighboring old fields all abandoned within the past five years would represent small-scale ecological differences. All of the fields would have similar soil conditions and species composition. Differences among the fields would most likely show up as differences in species abundances rather than differences in species presences. In contrast, if those fields were included in an analysis with surrounding forest patches, or with fields abandoned decades ago, then we would have a much larger ecological scale. Now species presences might capture the differences among communities. Geographic extent is not synonymous with ecological scale, but the larger the geographic extent, the more likely that a wider range of ecological conditions will be captured in the analysis.

At larger scales, presence/absence data may be more informative than abundance data because of the signal-to-noise problem. At large scales, we are usually trying to find the one or few causes of differences among very different types of communities. But abundances can vary for a number of reasons having to do with random demographic factors (see Chapters 7 and 14). While this variation may be of interest in relation to other questions, in this instance it acts as "noise" (random variation) in the analysis, obscuring the main pattern. By using presences and absences instead of abundances, we smooth out this noise. However, if we use presence/absence data, it is important that we include as many species as possible, especially rare species. The distribution of each species contains information about the environment and its effects on the plant community. The more species we include in the analysis, the more information we have. There is a redundancy to this information, so that the more species we use, the more confident we can be about our conclusions. A good rule of thumb is that in

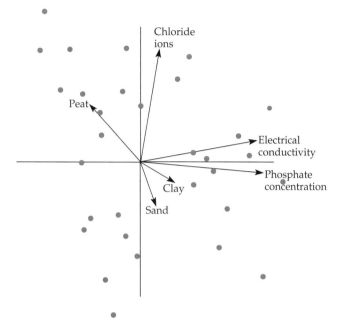

Figure 16.6
Canonical correspondence analysis of macrophytes growing on dikes in the Netherlands. The points show the average position of each species along the vegetation axes. The first vegetation axis is highly correlated with electrical conductivity ($r = 0.83$) and phosphate concentration ($r = 0.86$), while the second vegetation axis is correlated with chloride concentration ($r = 0.86$) and to a lesser extent with the amount of peat ($r = 0.49$) and sand ($r = -0.40$) in the soil. These correlations are indicated by how closely the environmental variables (arrows) line up with the vegetation axes. (After ter Braak 1986.)

any ordination, the number of species used should be at least three or four times the number of sites.

Classification

One use of landscape analysis is the classification of communities into larger groupings (Table 16.3). This classification can then be used for landscape management purposes. For example, if we wished to preserve particular types of communities, our classification scheme could help us choose which places to preserve to meet this goal. Classification techniques are complementary to those of ordination. In effect, what we are doing is taking the pattern revealed by the ordination and drawing circles around subsets of the points. For example, in Figure 16.4, we might put sites A and B into one group, sites C and D into another group, and site E into a third group. We could even construct larger groups from these smaller ones. Just because we have created groupings does not mean that the communities form discrete types; our different groups might gradually blend into one another. Yet, for planning purposes, defining discrete community types would still be useful.

Table 16.3 An example of the classification of a North American plant community

Physiognomic categories
Class . . . . Woodlands
  Subclass . . . . Mainly evergreen woodlands
    Group . . . . Evergreen needle-leaved woodlands
      Subgroup . . . . . . Natural/seminatural
        Formation . . . . . Evergreen coniferous woodlands with rounded crowns
Floristic categories
          Alliance . . . . *Juniperus occidentalis*
            Association . . . . *Juniperus occidentalis/Artemisia tridentata*

Note: This classification follows the National Vegetation Classification system proposed by the Ecological Society of America. The classification uses a dual system in which higher categories are based on physiognomic criteria and finer-level categories are based on floristic criteria.

A variety of methods can be used to create groups. The two major types of classification techniques are monothetic-divisive and polythetic-agglomerative. A **monothetic-divisive analysis** begins by considering all of the sites and dividing them into two groups depending on the presence or absence of a single key species. Each of the new groups is further divided based on new key species contained within each. The process is continued until we have reached the desired number of groups, a set of groups that seems natural and useful, or until division is no longer possible. Thus, "monothetic" refers to the use of a single species as the criterion at each step, and "divisive" refers to the dividing process.

A **polythetic-agglomerative analysis** works in the opposite direction. Using some distance metric (such as the Jaccard index), the two most similar sites are grouped togther. Then the next two most similar sites are grouped together, and so on. The grouping may involve two individual sites, a single site and a previously created group, or two groups. The process stops when all sites have been grouped together. Thus, "polythetic" refers to the use of multiple species in the distance metric, and "agglomerative" refers to the joining process. Various distance metrics and joining algorithms can be used. The result is a tree diagram (Figure 16.7). As with the monothetic-divisive approach, this diagram

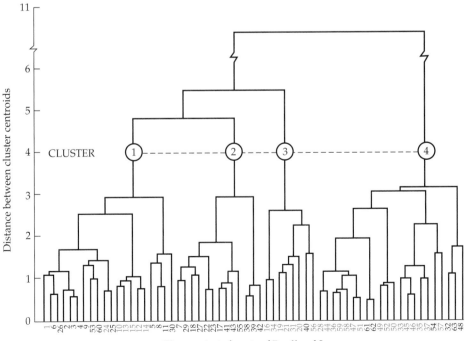

Pinus contorta forests of Banff and Jasper
Polythetic-agglomerative cluster analysis

Figure 16.7
A tree diagram developed by a cluster analysis using a polythetic-agglomerative method for the *Pinus contorta* stands in Figure 16.5. (After La Roi and Hnatiuk 1980.)

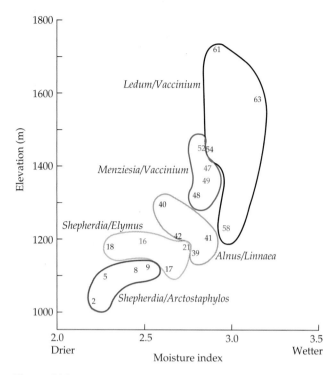

Figure 16.8
Classification of the *Pinus contorta* stands in Jasper National Park graphed in Figure 16.5, using a combination of techniques, including the cluster analysis shown in Figure 16.7 and indicator species analysis. The clusters are named by the dominate species. The number of clusters differs from that in Figure 16.7 based on the additional analyses. (After La Roi and Hnatiuk 1980.)

is used to devise some natural and useful classification scheme (Figure 16.8).

A useful adjunct to classification is the designation of **indicator species**. An ideal indicator species is found in all communities of a given type and not in any other community type. As with other procedures, there are various methods for picking indicator species (Dufrêne and Legendre 1997). The use of indicator species makes classification procedures much easier. An analysis of a landscape might proceed along the following lines. First, one would pick a number of representative sites in the landscape and do a complete species survey of each. Second, one would classify the sites and identify indicator species. Third, one would survey the rest of the landscape, looking just for the indicator species. By just looking for a much smaller number of species, many more sites could be surveyed in the same amount of time and effort. This variety of methods leads to three different approaches to classifying landscapes. The first approach relies on using all species. Polythetic agglomerative methods are typical of this approach and are a hallmark of the floristic-sociological school of central Europe. The second approach relies on using dominant or indicator species. Monothetic-divisive methods are typical of this

approach and have been used primarily by Russian plant ecologists. The third approach relies on physiognomy rather than species identity. In this approach, all broad-leaved deciduous forests would be classified together, even if they shared no species in common. It is also possible to combine these approaches. The current National Vegetation Classification scheme for classification of North American vegetation (Table 16.3) combines aspects of the second and third approaches.

Which approach is most appropriate depends on the scale of classification and its purpose. At very broad continental or global scales, physiognomy-based classifications are the most useful because we are often looking for commonalities among communities that transcend species identities. Our discussion of global biomes in Chapter 19 uses this type of system. In contrast, at local scales, we often are interested in how closely-related species sort themselves into communities, so either of the other approaches would be useful. We may, on the other hand, wish to emphasize continuity among samples. If that is the case, ordination is most appropriate.

Remote sensing is the process of collecting data about an entity of interest without being in direct contact with that entity. It is especially useful for gathering and analyzing ecological information on a large scale, as in classifying vegetation over large areas. Data collection is done using airplanes or satellites, as in the case of the popular Landsat Thematic Mapper. A landscape is captured in an image that is divided into a grid. The most common grid size is $30 \times 30 \ m^2$, although with newer technology $1 \times 1 \ m^2$ grids are now available. Each grid square is called a pixel, like the pixels of a computer image. The data consist of the spectral (light wave) properties of each pixel (Box 16A). Different satellite- or airplane-borne cameras record different numbers of wavelengths, from three or four to hundreds.

Classification using remote sensing proceeds by grouping pixels with similar spectral properties into a single category. There are two basic approaches. **Unsupervised classification** separates pixels into a user-defined number of classes based on statistical similarities between the pixels. This method is similar to the classification methods described previously. In this case, however, instead of each pixel (site) being characterized by the presence or abundance of a set of species, each pixel is characterized by the presence or intensity of a set of light waves. In contrast, **supervised classification** requires the groups to be defined explicitly by the user. Specific locations in the image (training regions) are characterized using ground-based knowledge. For instance, one begins by surveying a series of sites, which are then classified into groups using the techniques described above. Those sites are also mapped using a Global Positioning System (GPS) and located on the remotely sensed image. These sites then serve as the

Box 16A

Differentiating Vegetation Based on Spectral Quality

Remote sensing has been widely used to discriminate and map vegetation cover and community types since the 1960s. Discrimination is based on the different responses of different plant species to the various regions of the electromagnetic spectrum (microwave, visible, infrared, and radar) and to different bands within those regions. Thus, the 520 to 600 nm (green) band within the visible light region matches the green reflectance peak of vegeta-tion, which is useful for assessing plant vigor. The 630 to 690 nm (red) band matches a chlorophyll absorption band that is important for distinguishing vegetation types. The 1550 to 1750 nm (reflected infrared) band indicates the moisture content of soil and vegetation and also provides a good contrast between vegetation types. For example, Rey Benayas and Pope (1995) were able to distinguish six different tropical for-est types by differences in the moisture content of their leaves, photosynthetic activity, standing dead biomass, and senescent woody biomass. The radar wavelengths (0.1 to 70 cm) provide information on vegetation density, architecture, and flooding. For exam-ple, Pope et al. (1994) distinguished six marsh types with different species composition, standing biomass, and flooding status using radar bands of 68, 24, and 5.7 cm.

training regions. It is best to have multiple training regions per group in order to account for the inevitable variation between sites. Following either type of classi-fication, accuracy is judged using ground-truthing oper-ations in which randomly selected sites are surveyed to see if their actual species composition matches that predicted by the remote sensing classification.

Views on Continuous versus Discrete Landscapes

Do communities tend to blend into one another, or do they form discrete types? One's answer to this question can be influenced by how one analyzes landscape data—by ordination or by classification. Ordination assumes that landscapes are continuous, while classification assumes that they are not. The dominance of ordination as an analytic technique among North American aca-demic ecologists is a reflection of the current consensus among those ecologists that landscapes are mostly con-tinuous and are likely to be dominated by one or two very strong gradients. Applied ecologists in North America often use classification methods to categorize vegetation for practical reasons, however.

Most classification systems and techniques were orig-inally devised by European and Australian ecologists (Mueller-Dombois and Ellenberg 1974). The Zurich-Mont-pellier or floristic-sociological approach was developed by the Swiss ecologist Josias Braun-Blanquet (1932). While these ecologists recognized some continuity, this approach emphasizes discontinuity. Northern Europeans, primari-ly in Sweden, took an even stronger position toward dis-continuity in the early part of the twentieth century. More recently they have blended their methods with the floris-tic-sociological approach. Russian ecologists participated in many of these same debates. They were most strongly influenced by ecologists in northern Europe, and had sim-ilar views on discontinuity.

Today science is becoming increasingly global in scope and outlook, and it is becoming increasingly dif-ficult to assign particular viewpoints to particular geo-graphic areas. For example, as exemplified by European journals such as *Oikos, Journal of Vegetation Science*, and *Plant Ecology*, one can find articles representing all tra-ditions being published by plant ecologists from all over the world.

Landscape Diversity

Imagine yourself walking through a forest. You stop and count the number of species around you. You walk a bit farther and stop again. Again you count the number of species you find. You are measuring species diversity (see Chapter 12). But now you ask an additional ques-tion: How different is the forest from one place to the next? Did you mainly find the same species as you walked along, or different ones? Are communities com-posed of equal proportions of common and rare species, or do some communities contain many more rare species than others? Answering these questions both informs us about the processes responsible for structuring land-scapes and provides critical information for deciding how many and what kinds of communities we need to preserve.

Differentiation Diversity

Any ecological unit—be it a single quadrat, a commu-nity, a landscape, or a biome—has several measurable properties (Table 16.4). **Inventory diversity** is the species diversity found within the unit. **Differentiation diver-sity** is the way in which the species are grouped into sub-units. Differentiation diversity is most commonly stud-ied at the landscape level and is often simply referred to as β diversity.

Table 16.4 Definitions of some diversity concepts

Inventory diversity

Measures: Species density; Shannon-Weiner index;
Simpson's index

Spatial scale:

1. Point diversity: The diversity of a single small or microhabitat sample within a community that is regarded as homogeneous
2. Alpha (α) diversity: The diversity of a sample representing a single community
3. Gamma (γ) diversity: The diversity of a landscape or a set of samples that includes more than one community
4. Epsilon (ε) diversity: The diversity of a broader geographic area including different landscapes

Differentiation diversity

Measures: Mean similarity; percentage of species richness; turnover

Spatial scale:

1. Beta (β) diversity: The difference in community composition between communities along an environmental gradient or among communities in a landscape
2. Delta (δ) diversity: The difference in community composition between communities between geographic regions

Pattern diversity

Measures: Mosaic diversity; nestedness

Spatial scale: No explicit terminology

Sources: Whittaker 1977 and Scheiner 1992.

In a study in Spain, landscapes that were found in transition zones were more species-rich—had a greater inventory diversity—than landscapes outside those zones (see Figure 20.5). In other words, these transition zones contained a mingling of species from two different biogeographic floras. This mingling could potentially be of two types: either a patchwork of different types of communities from each zone, or a mixture of species from both zones within single communities. Measures of differentiation diversity tell us which one of these is the actual pattern.

The differentiation diversity of a set of sites can be measured in several ways. One measure is mean similarity—for example, the mean of the values in Table 16.1C. This measure indicates whether the communities in the landscape all tend to contain the same species (a high mean similarity) or whether they tend to contain different species (a low mean similarity). In Spain, the landscapes in each transition zone had the same mean similarity as landscapes outside the zone, indicating that species from the two biogeographic floras were mingling within single communities (Rey Benayas and Scheiner, 2002).

A second measure of differentiation diversity is the percentage of total species richness:

$$\left(\frac{S}{S_{i\bullet}}\right)-1$$

where S is the total number of species found and $S_{i\bullet}$ is the mean species richness of each sample.

A third measure is the **turnover** of species along a gradient—the average number of species that appear and disappear as you pass from one community to the next along a gradient. Turnover can be estimated with abundance data or presence/absence data. To do so, it is essential that the samples can be arrayed along a single gradient. Turnover is often calculated in conjunction with an ordination of the samples.

Pattern Diversity

A third aspect of diversity is the arrangement of subunits within an ecological unit, referred to as **pattern diversity**. These patterns can be of three types: spatial, temporal, and compositional. **Spatial pattern diversity** is the arrangement of subunits in physical space (Turner 1989). The measurement of spatial patterns often begins with a map created from an aerial photograph or by satellite remote sensing. From this map we can calculate metrics such as patch size, nearest neighbor probability (the probability that two types of habitat will be adjacent), and fractal dimension (a measure of the complexity of a spatial pattern). These aspects of spatial pattern are of particular importance for habitat preservation and the design of nature reserves. We explore these measures in more detail in Chapter 17.

Temporal pattern diversity is the arrangement of subunits in time. The same techniques that are used to measure spatial pattern diversity and compositional pattern diversity can be used, with the restriction that all subunits must be a single spatially defined area measured at different times. **Compositional pattern diversity** is the arrangement of subunits in the mathematical space defined by the site-species composition matrix. Two metrics have been developed for quantifying compositional pattern diversity: mosaic diversity (Istock and Scheiner 1987; Scheiner 1992) and nestedness (Patterson and Atmar 1986). **Mosaic diversity** is a measure of landscape complexity due to variation in species richness among communities and variation in commonness and rarity among species within the landscape. A high value for mosaic diversity indicates a landscape with many environmental gradients and strong differentiation among communities, from those that are rich in common species to those that are rich in rare species. In contrast, a simple landscape is one dominated by a few species and controlled by one or a few environmental gradients.

Table 16.5 Two examples of nested landscapes

(1) Species	A	B	C	D	E	F	G	H	I	(2) Species	A	B	C	D	E	F	G	H	I
1	1	1	1	1	1	1	1	1	1	1	1	1	1	1	1	0	0	0	0
2	1	1	1	1	1	1	1	1	0	2	1	1	1	1	0	0	0	0	0
3	1	1	1	1	1	1	0	0	0	3	1	1	1	1	0	0	0	0	0
4	1	1	1	1	1	0	0	0	0	4	1	1	1	0	0	0	0	0	0
5	1	1	1	0	0	0	0	0	0	5	1	1	0	0	0	1	1	1	1
6	1	1	0	0	0	0	0	0	0	6	1	1	0	0	0	1	1	1	0
7	1	0	0	0	0	0	0	0	0	7	1	1	0	0	0	1	1	0	0
8	1	0	0	0	0	0	0	0	0	8	1	0	0	0	0	0	0	0	0

China is more species-rich than the United States because eastern Asia is rich in older lineages (Latham and Ricklefs 1993; Qian and Ricklefs 1999; see Chapter 20). Is the greater species richness in eastern Asia due to a few communities that are especially rich in rare species? Perhaps the species from the old lineages tend to concentrate in just a few unusual community types. However, a comparison of alpine tundras in eastern Asia and western North America found that mosaic diversity was the same in the two areas, indicating that rare species were distributed in similar ways in both regions (Qian et al. 1999).

Nestedness is the tendency of communities to be subsets of other communities. To put it another way, when communities are highly nested, a rare species will be found only in communities that also contain all of the more widespread species. The pattern of nestedness is critical in deciding how many communities are necessary to protect in order to preserve the most species (see Chapter 17).

Two examples of nestedness are given in Table 16.5. In the first example (Table 16.5A), site A contains all eight species, site B contains six of the eight species contained in site A, and so forth, to site I, which contains just one species, which is found in all sites. In the second example (Table 16.5B), site A again contains all eight species and site B contains seven of those eight. But now sites C–E form one nested subset of the landscape, while sites F–I form another.

Nestedness is a frequent condition of animal communities, especially communities on islands or in island-like habitats, such as mountaintops (Wright et al. 1998). However, plant communities within a landscape are generally not nested. Instead, plant species tend to replace one another along gradients. Plant communities that contain all the species in a landscape, like site A in our example, rarely exist except when the species are drawn from a very small spatial and ecological scale.

Summary

A plant community exists as part of a set of communities that make up a landscape. To study landscapes, we need methods and measures for comparing communities and arraying them within this larger context. Communities can be compared by a variety of measures. Properties of communities such as biomass or productivity can be compared using standard univariate and multivariate statistical techniques, such as analysis of variance. Using information on species presence or abundance, various measures of similarity are possible.

This information can be further utilized to examine patterns among sets of communities. Ordination is a method for arranging these sets in continuous space. These arrangements can then be used to look for correlations between vegetation patterns and environmental variables. Several classification methods can be used to sort these sets into groups, which is useful for management purposes.

Diversity is another property of landscapes. Most diversity studies are concerned with inventory diversity: the number of species present and their relative abundances. Variation among communities can also be measured by differentiation diversity and pattern diversity. All of these measures of communities within landscapes provide additional ways of looking at the world. Using these measures of pattern, we can study the processes that shape the world's vegetation.

Additional Readings

Classic References

Bray, J. R. and J. T. Curtis. 1957. An ordination of the upland forest communities of southern Wisconsin. *Ecol. Monogr.* 27: 325–349.

Greig-Smith, P. 1980. The development of numerical classification and ordination. *Vegetatio* 42: 1–9.

Whittaker, R. H. 1972. Evolution and measurement of species diversity. *Taxon* 21: 213–251.

Contemporary Research

Jackson, D. A. and K. M. Somers. 1991. Putting things in order: the ups and downs of detrended correspondence analysis. *Am. Nat.* 137: 704–712.

ter Braak, C. T. F. 1986. Canonical correspondence analysis: a new eigenvector technique for multivariate direct gradient analysis. *Ecology* 1167–1179.

Wartenberg, D., S. Ferson and F. J. Rohlf. 1987. Putting things in order: a critique of detrended correspondence analysis. *Am. Nat.* 129: 434–448.

Additional Resources

Legendre, L. and P. Legendre. 1998. *Numerical Ecology.* 2nd English ed. *Developments in Environmental Modelling,* Vol. 20. Elsevier, Amsterdam.

Ludwig, J. A. and J. F. Reynolds. 1988. *Statistical Ecology.* John Wiley & Sons, New York.

McCune, B. and M. J. Mefford. *PC-ORD: Multivariate Analysis of Ecological Data.* MjM Software Design, Gleneden Beach, OR.

17 *Landscape Ecology*

*I*n the last chapter, we looked at communities embedded within landscapes, as well as patterns of species composition and diversity among communities. Here we shift our perspective to consider the landscape as a whole. Science involves the search for patterns and for the processes that cause those patterns (see Chapter 1). What might seem to be a third activity—prediction—depends on having identified the processes that explain the observed patterns. These two aspects of science are especially apparent in landscape ecology.

Landscape ecology is the study of the spatial distributions of individuals, populations, and communities and the causes and consequences of those spatial patterns. Landscape ecologists have a particular interest in spatial scale, and have brought this topic to the forefront of contemporary ecological thinking.

While the roots of landscape ecology go back to the origins of ecology, it gained its current identity only in the 1980s. The term "landscape ecology" was coined by Carl Troll, a German geographer, in 1939. Until 1980, the field was mostly confined to Europe, where it was most closely allied with the floristic-sociological approach of Braun-Blanquet and emphasized static patterns (Naveh and Lieberman 1994). In the early 1980s, Richard T. T. Forman (1995) and others brought landscape ecology to North America, modifying the European tradition by adding dynamics and scale. The theoretical roots of some of these efforts go back to the 1960s, to island biogeography theory and metapopulation theory. The coalescence of landscape ecology into its modern form has also relied on contemporary developments in the field of geography. Other threads that have contributed to the emergence of landscape ecology are the development of spatially explicit statistics and modeling and technological advances in Geographic Information Systems (GIS) and remote sensing.

Because of the relative youth of landscape ecology, the field is still very much in flux. New perspectives, methodology, and conceptual advances continue to develop. As a result, links between theory and data are still limited; questions have been posed that do not have answers, and relationships have been hypothesized but not yet tested. While such ambiguities can be frustrating for students, the study of landscape ecology provides an opportunity to grapple with a scientific field at a very dynamic stage in its development.

Many ecological phenomena are best studied at the landscape level. For example, low-intensity fires can be quite patchy on a local scale: severely damaged trees may be found near unscathed trees. As burn intensities increase, however, this local patchiness is greatly reduced (neighboring trees tend to experience similar burn damage), but a new spatial pattern emerges because intense fires tend to "spot": flying embers ignite material at a distance from the initial fire. The spatial arrangements of individual trees, groups of trees, lakes and rivers, and topography all combine to determine fire spread and intensity. Thus, we can gain unique insights into fire ecology and the resulting responses of plant populations by studying how the elements in a landscape are arranged.

Landscape ecology is concerned with questions such as, How are individuals, species, and communities distributed within a landscape? How many types of patches are there in a landscape? What are the sizes, shapes, and distributions of those patches? How do these patterns affect the movement of individuals, materials, and energy among patches and across the landscape? What processes are responsible for which patterns? Do the processes occur at the level of the landscape (e.g., migration) or within communities (e.g., competition)? How do the patterns we find, and the interpretation of those patterns, differ as scale changes? The answers to these questions have important implications for conservation and landscape management. The design of nature reserves, for example, is sometimes explicitly based on landscape ecology theory. Realistically, political and economic realities often intrude on scientific considerations in the establishment and management of nature reserves; nevertheless, it is important to know what one is aiming for in making management decisions. We return to issues of conservation and management at the end of this chapter.

The name "landscape ecology" can be misleading—landscape ecology does not necessarily concern large (to humans) spatial scales, but rather scales that are large relative to the species or phenomena of interest. For example, a study in landscape ecology concerned the movements of beetles over a 10-m^2 patch (Johnson and Milne 1992). Many studies of landscape ecology are carried out at the scale of a **watershed**—an area that includes a stream or river and all of the land that drains into it. While the focus of interest for many landscape ecologists may be at the scale of watersheds or other whole landscape elements, others are most interested in how ecological phenomena are affected by changes in scale, or in the effects of spatial patterns on ecological properties regardless of spatial scale.

Spatial Patterns

There are two general ways to study patterns on a landscape level. One, a **spatially explicit approach**, depends on determining the particular spatial arrangement of species and landscape elements (e.g., habitat patches, farms, roads). The other, often called the **mean field approach**, focuses on describing average parameter values.

To see this distinction, consider a species-area curve (see Chapter 12). Imagine that we do the following: On the landscape shown in Figure 17.1, we overlay a grid of 10×10 m squares, and we compile a list of every vascular plant species growing in each square. We can now build a species-area curve in two ways (Figure 17.2). Using the first, spatially explicit approach, we start in one spot. Imagine that this spot is the bottom left corner

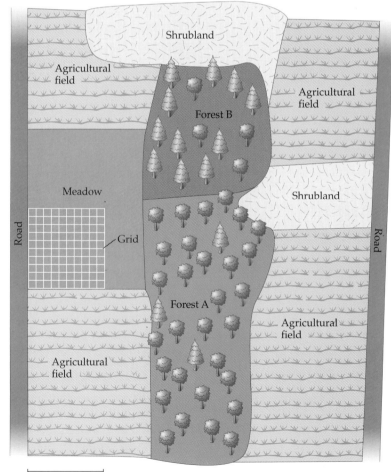

Figure 17.1
Diagram of a landscape showing a patchwork of meadows, forests, and shrublands. For sampling purposes we can lay down a 100×100 m grid that arbitrarily divides the area into smaller plots.

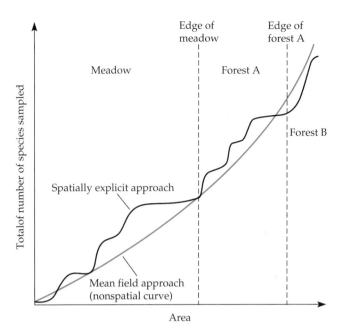

Figure 17.2
Two species-area curves. One curve is from a spatially explicit tally, which forms a stairstep pattern. Within a community, the number of species found as area increases will level off. When a community boundary is crossed, however, the number of species found will once again rise rapidly. The other curve, from a mass field analysis that is not spatially explicit, forms a smoothly rising pattern.

of the meadow. Our first data point is the number of species in the 10×10 m square in that corner. Next, we expand our square to 20×20 m and again count the total number of species. This number will include all of the species in the original square plus any new species in the larger square. We repeat this operation for a 30×30 m square. Initially, the number of species will rise rapidly. But as more and more of the meadow is captured in our ever-growing plot, the number of new species will drop off. At some point, it is likely that we will even stop finding new species, and the curve will level off. If we keep going, however, our plot will eventually run into the forest abutting the meadow. Now, suddenly, we will find a whole new suite of species, such as trees and forest-floor herbs. The species-area curve will again rise rapidly with increasing area until, once again, it levels off as most of the forest gets sampled. This stairstep pattern will be repeated each time we cross a new community boundary.

In contrast, we can build our species-area curve using the mean field approach, which ignores the spatial arrangement of the squares. First, we calculate the average number of species in all of the 10×10 m squares. Next, we take all possible combinations of two squares, determine the total number of species in each pair of squares, and again calculate the average. (While we

would not want to do this by hand, it can be done in a few seconds on a typical desktop computer.) We continue this process with all sets of three squares, four squares, and so forth. The result is a smoothly rising curve. Any discontinuities in the environment that would have produced a stairstep effect with the spatially explicit approach are, instead, averaged out.

Which of these approaches is preferable? The answer depends on the questions one wishes to address. The pattern produced by each of these species-area curves can provide important information about both landscape patterns and the processes that account for them. Next, we explore these patterns in more detail.

Defining Patches

The basic unit we use to examine spatial patterns is the patch. A **patch** is any specified area, either an area defined arbitrarily or, often, an area that is in some way relatively homogeneous or internally consistent. A patch can be defined by naturally occurring edges (e.g., the boundary between a meadow and a forest), but need not be. For example, each of our 10×10 m squares in the exercise above could be a patch. However, if we start with such artificially defined patches, we might want to aggregate them into "natural" units. Sometimes this is done based on **compositional similarity**—the extent to which adjacent patches share a similar set of species at similar frequencies (see Chapter 16). Of course, adjacent patches usually share at least some species, so the cutoff point for when patches should be joined or not is somewhat arbitrary. Often the decision is aimed at ending up with a manageable and interpretable number of patch types.

Remote sensing images have become a popular method for sampling large areas quickly. In this case, the minimum patch size is determined by the resolution of the remote sensing equipment. Typically this is 30×30 m, although newly available technology is pushing resolution down to 10×10 m, and even 1×1 m. Our aggregation of patches is again based on their similarity, but now similarity is determined by spectral quality rather than by shared species (see Box 16A). Because spectral quality is an indirect measure of community composition, it limits our view of landscape structure.

Aggregating patches and deciding on the number of patch types is, therefore, as much an art as a science. Often it depends on the ability of measurement instruments to distinguish between different patch types. With remote sensing, for example, the indirect measures we are using need to be related to the actual vegetation through a process called **ground-truthing**: one goes out to the field and determines empirically what the remote sensing devices are recording. Two communities with very different species compositions may be indistinguishable in spectral quality. Thus, if remote sensing were being used to classify a large region, one would be forced to lump those two communities into one patch type.

Because landscape ecology is built on the study of patches, it is, inevitably, built on a categorical view of the world. Categorical versus continuous views of nature have been a regular source of controversy in plant ecology. In Chapter 12 we explore the history of this controversy, beginning with the disagreements of Clements and Gleason. It is somewhat ironic that nearly 50 years after the Gleasonian continuum was declared triumphant, at least among English-speaking ecologists, a categorical framework has once more emerged to dominate landscape-level studies.

Quantifying Patch Characteristics and Interrelationships

Once we have defined our patches, we can measure some of their attributes. For individual patches, we can measure sizes and shapes. A typical measurement of shape is the perimeter-to-area ratio, which indicates how much edge a patch has relative to its interior. Edges are important ecologically for many reasons. In particular, they are the focal areas for processes such as the movement of individuals or materials between the outside and inside of a patch of forest.

One approach to quantifying the shapes used in landscape ecology is to calculate the fractal dimension of the landscape elements. A **fractal dimension** is a fractional dimension—for example, something in between a one-dimensional line and a two-dimensional surface, or in between a two-dimensional surface and a three-dimensional area. For example, as a line (which has one dimension) becomes increasingly convoluted, it becomes space-filling, finally turning into a surface with two dimensions. The fractal dimension is thus a measure of the complexity or degree of convolution of a line (such as the perimeter of a patch or the coastline of an island) or a surface. Fractal geometry can also be used to quantify patterns of branching and fragmentation.

Imagine that you have a very large map of a coastline showing very fine details. You have to measure the length of the coastline using a ruler with no markings. Thus, the smallest distance that you can measure is the length of the ruler. A small ruler will pick up more of the ins and outs of a coastline than a large ruler, so that using the smaller ruler will result in a longer measurement. As the coastline or the perimeter of a patch becomes increasingly convoluted, the effect of ruler size becomes greater. The magnitude of the ruler size effect is the fractal dimension. Pennycuick and Kline (1986) used just such an approach to quantifying fractal dimension to compare bald eagle nest density on two Alaskan islands that differed in the complexity of their coastlines. This approach might be useful for various problems in plant ecology as well.

Bruce Milne (1992) connected the concepts of fractal dimension and scaling explicitly by quantifying the degree of habitat fragmentation in a landscape. Milne quantified the fractal dimension of a New Mexico grassland and showed that the habitat was patchy. Beyond that simple result, he showed that the patchiness was scale-dependent. Scale dependence implies that species might perceive the landscape differently. A beetle that lived in grass clumps might perceive a very disconnected landscape, while a grazing pronghorn might perceive that same landscape as one large patch. Some areas of the landscape showed high connectivity at three different scales, so here the beetle and the pronghorn might perceive similar landscapes. Other areas consisted of isolated patches that might attract species that disperse readily, but then settle within a single patch (e.g., small rodents).

As researchers become concerned with going beyond the individual patch, they can examine the arrangement of patches in the landscape. Using a different approach than the one taken by Milne, researchers may ask, are patches mostly aggregated into like types or not? A checkerboard pattern, for example, represents the maximal interdispersion of two patch types in a landscape (Figure 17.3A). The maximal aggregation of two patch types would have all the patches of each type grouped into two large clumps (Figure 17.3B). A measure of this aggregation is connectedness. If there are more than two patch types, then other measures of context come into play. For example, for a landscape with three patch types, a patch of type A could be surrounded entirely by patches of type B, or entirely by patches of type C, or by half of each. Clearly, such second-order patterns can get very complicated with a large number of patch types. We can quantify these patterns as the **nearest neighbor probability**, the chance that two patch types will be adjacent.

Scale

Definitions and Concepts

When we study an ecological pattern, we must establish the scale at which we are conducting our analysis. Before we examine the implications of this seemingly straightforward statement, however, we must point out a potentially confusing difference in how people in different fields discuss scale. Ecologists generally equate scale with the physical dimensions of the phenomenon being studied: a process that occurs over 10 km is at a larger scale than one that occurs over 10 cm. We follow this practice in this textbook. Some landscape ecologists refer, instead, to "coarse scale" and "fine scale" for larger and smaller physical areas, respectively (Forman and Godron 1986). However, cartographers and remote sens-

(A)   (B)   (C)

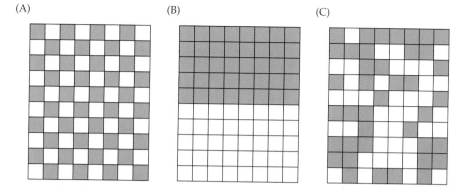

Figure 17.3
Three different patterns of patch aggregation in a landscape with two patch types. (A) Maximal dispersion of patches. (B) Minimal dispersion of patches. (C) Random dispersion of patches.

ing researchers use exactly the opposite terminology, in which scale is the relative size at which an object of a standard size appears in a map or photo. Thus, a world map has a much smaller scale than a regional map (that is, a meter is much smaller on a world map). It is therefore important to be explicit as to how one is using the term "scale" (e.g., Turner and Gardner 1991), although some ecologists have advocated abandoning the term entirely, replacing it with its components (Csillag et al. 2000).

Scale is composed of several components, two of which are grain and extent. **Grain** is the size of the primary unit used in a study—for example, a 10×10 m square or a 10 km^2 area in which species are counted. While the grain is usually the smallest distinguishable unit, or the finest level of spatial (or temporal) detail that can be resolved in a given data set, for some types of data it is possible to extrapolate to a smaller grain; in contrast, data aggregation will result in a larger grain. **Extent** is the total range over which a pattern is examined. The extent of a study might be across a local region of 100 km^2 or across the entire globe. For example, a research project might have a grain of 1-ha squares and an extent of 10,000 ha in central Africa, which together define the scale of the analysis.

Some researchers recognize a third component of scale, **focus** or **resolution**. These terms refer to the sampling level or the area represented by each data point. Consider an analysis of diversity in a 1-km^2 area that is divided into 1-ha squares. Those 1-ha squares become the focus of our sampling. In each square we randomly place 10 1-m^2 quadrats and record the species contained in each quadrat. These data can be analyzed in two ways: using the average species richness of the 10 quadrats in each 1-ha square (in which case the grain equals 1 m^2), or using the total number of species in the 10 quadrats in each 1-ha square (in which case the grain equals 10 m^2). In both cases, though, the focus is 1 ha, because the 1-ha square is the basic unit of analysis. Changing the grain, extent, or focus of a study can result in fundamentally different patterns (Figure 17.4; see also Figure 20.11).

Consideration of scale is central to understanding ecological processes. The same process may function differently at different scales. For example, the ecological process of herbivory is generally a very local phenomenon. However, the effects of an elephant herd might be manifested at the landscape level, and those of a locust swarm at the regional level. Population dynamics can be a local phenomenon, but if metapopulation dynamics is involved, then it becomes a regional phenomenon as well. The evolutionary process of extinction can occur at a local, regional, or global scale; its implications at these different scales can be very different.

Jonathan Levine (2000) found that at different scales, different processes explained the success of plant invasions. In northern California, *Carex nudata* (Cyperaceae), a tussock-forming sedge, forms clumps along river edges. A tussock (clump) creates a microhabitat that can shelter up to 20 species of angiosperms and bryophytes during winter flooding. Across a 7-km stretch of river, Levine found that the higher the species richness of a tussock, the more likely it was to be invaded by the non-native species *Agrostis stolonifera* (creeping bent grass, Poaceae), *Plantago major* (common plantain, Plantaginaceae), and *Cirsium arvense* (Canada thistle, Asteraceae). This pattern was the result of processes working in opposite directions at smaller and larger scales. Very small-scale processes—those occurring within a single tussock—were investigated by an experimental manipulation of the number of native species on individual tussocks and by the introduction of seeds of the invaders into the experimental tussocks. At this small scale, as species richness increased, the germination, survival, and growth of the invaders declined. In contrast, at the larger scale of the 7-km stretch of the river, Levine found that seeds of all species (both natives and invaders) moved with the flow of the river, and so both native diversity and frequency of invasion increased downstream. Thus, although native diversity contributed to community resistance to invasion at very small scales, at larger scales, other factors that changed with diversity (here, seed supply of invaders) were more important,

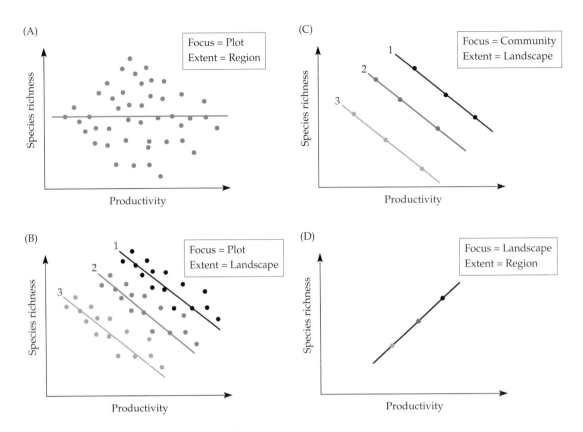

(A) Focus = Plot Extent = Region

(C) Focus = Community Extent = Landscape

(B) Focus = Plot Extent = Landscape

(D) Focus = Landscape Extent = Region

Figure 17.4
An illustration of how changing the focus and extent of a study can alter the results—in this case, the relationship between species richness and productivity. Each diagram represents a region. In this region there are three landscapes, with three communities in each landscape and five plots in each community. In all cases, the grain of the study remains the plot. Species richness is measured as the total number of species in each plot. (A) Each point represents one plot, so the focus is the plot, and the extent is the entire region. An analysis of the relationship between species richness and productivity results in a slope of zero, or no relationship. (B) Each point represents one plot, so the focus is the plot. Each landscape is analyzed separately, so the extent is the landscape. Analyses of the relationship between species richness and productivity for each landscape result in a negative slope. (C) Each point represents the average of the plots in a community, so the focus is the community. Each landscape is analyzed separately, so the extent is the landscape. Analyses of the relationship between species richness and productivity for each landscape result in a negative slope. (D) Each point represents the average of the plots in a landscape, so the focus is the landscape. The extent is the entire region, and the relationship between species richness and productivity is positive. (From Scheiner et al. 2000.)

creating a positive relationship between invader and native diversity.

Interestingly, there was even a suggestion of the potential importance of factors operating at much larger scales. At the beginning of the study in 1998, the river chosen for the experiment was one of the few whose banks were not flooded by high-water conditions, and thus had one of the few stretches of *Carex* tussocks available for study. In the following year, water levels dropped severely, threatening the survival of the *Carex* tussocks on this stretch of river, while they recovered elsewhere in the region. This variation in local flooding is related to a 3- to 7-year climatic cycle, the El Niño Southern Oscillation, that affects weather conditions across the Pacific (see Chapter 18). How the temporal variation in stream flow affected invasions is not known.

However, such landscape-scale variations in flooding, across temporal changes in weather happening at the scale of thousands of kilometers, could potentially influence regional invasion success by creating greater opportunities for invasion as tussocks become disturbed by flooding followed by drought. Thus, the process of invasion could be influenced in this system by processes occurring over a range of spatial and temporal scales, sometimes acting in opposition to one another.

Hypotheses that seek to explain large-scale patterns may be based on processes that operate at smaller scales than, or at the same scale as, the phenomena they seek to explain. If the underlying process on which the explanation rests operates at a smaller scale, then the hypothesis must either include an explanation for how local processes scale up to larger areas or assume that large-scale patterns are simply the sum of local-scale patterns. Scaling up and down from one set of patterns to another is a major current focus of research in many different areas of plant ecology, from the modeling of carbon fluxes (leaf-canopy-ecosystem-global) to determining the causes of diversity patterns.

Spatial and Ecological Scale

How do we specify (and think about) scale? It is common for people to define scale in an exclusively spatial manner, but we can contrast this approach with a consideration of scale in an ecological sense. This latter meaning of scale is related to the concept of hierarchy in ecology. Using a strictly spatial definition, ecologists conventionally speak of local, landscape/regional, and continental/global scales, and they use area to distinguish these scales—for instance, local scales are often thought of as being up to 10^2 km^2, landscape/regional scales as being between about 10^2 km^2 to 10^8 km^2, and continental/global scales as being 10^8 km^2 and larger. It is usually important, however, to define scale in terms of the particular ecological processes involved. For example, 10^2 km^2 may be about the scale at which local pollen dispersal occurs for pines or some other plants with wind-dispersed pollen, but it is much too large a scale for pollen dispersal by animals such as ants, beetles, or Australian possums.

Ecologically, we often recognize three basic scales: within a community, across communities, and across biomes. It is easiest to appreciate the distinctions among these scales by starting with biomes, the largest scale. Biomes are defined by distinct differences in physiognomy over large regions (e.g., grasslands, shrublands, forests; see Chapters 18 and 19). Patterns that cross biome boundaries involve variation in the physical sizes, morphologies, and ecological niches of the dominant species—for example, shrubs versus trees. Within biomes, moving across communities may result in more subtle, yet equally important changes—for example, from a forest dominated by fast-growing hickory (*Carya*) trees to one dominated by slower-growing oaks (*Quercus*). Even within a single community, such as the meadow in Figure 17.1, there may be variation in factors such as soil conditions, land use history, microhabitat, and herbivore activity.

These two different approaches to thinking of scale—spatial and ecological—are not entirely independent. The farther we move in space, the more likely it is that we will cross community or biome boundaries. But we may also cross community boundaries over very short distances if the environment changes rapidly. Moving away from the edge of a stream or lake, for example, can result in abrupt changes in the amount of water in the soil or the amount of disturbance due to wave action (Wilson and Keddy 1986; see Figure 10.21). In arid mon-

tane regions, by moving a distance of just a few kilometers up the side of a mountain, one can travel from desert to temperate forest to boreal forest. Conversely, traveling across the taiga of Russia, one finds very similar forest communities stretching across many hundreds of kilometers.

Quantifying Aspects of Spatial Pattern and Scale

Given the complexity of the mosaic of environmental factors and biotic interactions that determine spatial patterns at different scales, how can we hope to describe or quantify these patterns? Ecologists have developed both spatially explicit and nonspatial methods of doing this. (By "spatially explicit," we mean approaches that take the particular location of each data point into account.) Methods that are not spatially explicit—ordination and some statistical approaches to quantifying spatial association—are older, and we discuss some of them in Chapter 16.

Recently, a spatially explicit graphical technique, **Geographic Information Systems (GIS)**, has become central to landscape ecology. GIS is a method for depicting the relationships between different kinds of information ("layers" in GIS terminology) at particular locations. For example, we might have several types of information that relate to each location in the landscape (Figure 17.5). One set of information might be the species composition of each patch. For each patch, we might also have information on soil type, nutrient levels, water table depth, slope, and the last time the patch was logged or burned. Our information could also include socioeconomic information such as land ownership, taxes on that piece of land, the number of people residing there, and the legal classification determining how the land can be used. GIS allows us to map these variables singly and in any combination to get a visual sense of potential patterns of correlation among them, so that we may pose hypotheses regarding causality. GIS by itself does not allow one to quantify these relationships or statistically test hypotheses about them.

Spatially based relationships can be quantified using spatial statistics. Many advances in the field of spatial statistics have been made since the late 1980s (Legendre and Fortin 1989). These advances include the use of geostatistics, taken from the field of mining geology, and techniques for studying spatial autocorrelation, taken from the field of geography (Liebhold et al. 1993; Liebhold and Gurevitch 2002). Older statistical techniques in ecology were developed to analyze and model spatial variation in ecological data (Pielou 1977; Greig-Smith 1983). These methods were useful in distinguishing certain kinds of spatial patterns, but did not take the actual spatial location of the data into account, and had numerous shortcomings. The newer approaches have contributed substantially to our current understanding

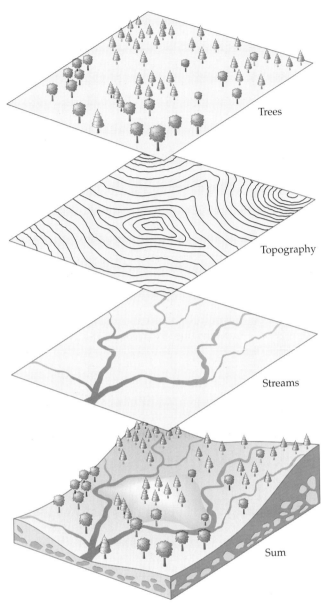

Figure 17.5
GIS is used to examine the relationships among multiple types of information, which are arranged in layers.

chy defines a scale. Ecological processes occur within patches, while links among patches are created by the movement of materials and energy. The pattern of linkages defines the next layer in the hierarchy and determines how processes propagate up the hierarchy. For example, gas exchange by a single leaf can be combined with that of the other leaves on a plant to determine the dynamics of gases around the entire canopy of a tree, which in turn can be used to model an entire forest stand.

We have discussed a number of the issues involved in describing and studying patterns in nature. As you have doubtless concluded, there is no single way to study or even describe such patterns (Levin 1992). Viewing similar questions from different vantage points (such as the spatial versus ecological scales discussed above) or at different scales can lead to very different answers. This may be frustrating if you are expecting simple truths, but as we have seen, considering information at more than one scale, while difficult, can also lead to new insights and understanding of the underlying biology. We now turn to some approaches to studying the processes responsible for some of the types of patterns we have just examined.

Toward a Theoretical Basis for Landscape Patterns: Island Biogeography Theory

Our exploration of landscape processes starts with **archipelagos**, chains of islands. Robert MacArthur and Edward O. Wilson's **theory of island biogeography** (1967) was developed to explain patterns of species presence and absence on islands. That theory was important in the development of much of today's landscape ecology.

Although the initial description of the theory applied it to oceanic islands, many types of communities have island-like properties. An inclusive definition of an **island** is any area that is suitable for the survival and

Edward O. Wilson

reproduction of a species and is surrounded by unsuitable habitat. A landscape made up of isolated woodlots consists of a set of islands from the point of view of an herb that lives only in woodland understories. Mountaintops are islands for alpine species. Plants can be specialized for life on particular types of soil or terrain, such as serpentine barrens (see Box 15A) or rock outcrops, which also function as islands. The theory of island biogeography therefore has applications far wider than oceanic islands.

The basic idea of the theory is quite simple: The number of species on an island is determined by the rate

of the spatial aspects of organisms' responses as individuals, populations, and metapopulations (summarized by Dale 2000; Legendre and Legendre 1998). Curiously, spatial statistics have rarely been used by ecologists in conjunction with GIS. Just as we combine graphical and statistical analyses with other types of data, together GIS and spatial statistics constitute a remarkably powerful set of tools.

Hierarchical patch dynamics is a new approach to examining the effects of scale (Wu and Loucks 1995). In this approach, a patch is seen as existing in a hierarchy of successively larger patches. Each level of this hierar-

of immigration of new species to the island and the rate of local extinction on the island (Figure 17.6). The rate of arrival of new individuals from the mainland determines the immigration rate—the rate at which new species arrive. (This immigration rate differs from that used in other contexts where it refers to the rate at which individuals arrive.) As the number of species on the island increases, the immigration rate declines, because an arriving individual is increasingly likely to be a member of a species already on the island. The relationship between species number and extinction rate is a bit more complicated. Assume that the island can hold a fixed total number of individuals. As more species take up residence on the island, the potential population size of each gets smaller. The chance that a species will go locally extinct—disappear from the island—increases as its population size gets smaller. Thus, the rate of extinction increases as the number of species rises. If we make the further assumption (as MacArthur and Wilson did) that these processes are roughly constant over long periods, then the number of species on the island eventually reaches a steady state, given by the value of the horizontal axis where the immigration and extinction curves intersect.

A key insight of MacArthur and Wilson was that larger islands can hold more individuals and, consequently, extinction rates on larger islands are lower. Moreover, islands closer to the mainland have higher immigration rates because it is more likely that individuals will arrive there from the mainland. Using their assumption that the islands had reached an equilibrium number of species, MacArthur and Wilson predicted that the greatest number of species would be found on large islands near mainlands (the source of immigrant species), and that the smallest number would be found on small islands far from mainlands (Figure 17.6C,D).

Real archipelagos, of course, are more complex: among other things, immigrants can arrive not just from the mainland, but also from other islands. A more important limitation of island biogeography theory is that it treats all species as if they were interchangeable. Clearly, this is not the case. Even very similar species have distinctive ecologies. MacArthur and Wilson acknowledged this limitation, although many ecologists subsequently trying to apply their theory did not. The equilibrium assumptions needed to predict species number are unlikely to hold, and empirical studies have provided little support for the theory's predictions. Despite these limitations, the theory is useful. It provides a starting point for examining patterns and gives us a way of thinking about the movement of individuals in a landscape.

Metapopulation Theory

The theory of **metapopulations**, originally developed by Richard Levins (1969) and subsequently expanded

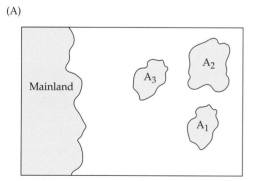

(A)

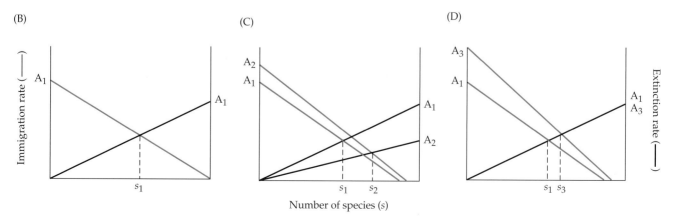

(B)

(C)

(D)

Figure 17.6
(A) Mainland and archipelago. Islands A_1 and A_2 are the same distance from the mainland, while A_1 and A_3 are the same size. (B) The equilibrium number of species on island A_1 is determined by the balance between the immigration and extinction rates. (C) The larger island (A_2) will have a higher immigration rate and a lower extinction rate, resulting in a greater number of species at equilibrium. (D) The nearer island (A_3) will have a higher immigration rate but the same extinction rate, resulting in a greater number of species at equilibrium.

Box 17A

Metapopulation Models

Metapopulation models have the general form dp/dt = immigration rate – extinction rate where p is the percentage of patches occupied by a given species, $0 \leq p \leq 1$, and dp/dt indicates the rate of change in occupation over time. Typically we are interested in solving this equation for p, which predicts the frequency of occupied patches in a landscape. Obviously, a species' biology sets its immigration and extinction rates.

It is important to realize that in these models, immigration and extinction rates may not be independent of each other, and either or both may depend on the frequency of occupied patches, p. It is easy to see how the immigration rate might depend on p: the number of seeds arriving in a patch could depend on the number of occupied patches.

The extinction rate might also depend on p for the same reason: as the number of seeds being dispersed increases (with increasing p), there could be a reduced chance of local extinction. This phenomenon is called the **rescue effect** (Brown and Kodric-Brown 1977) because new immigrants "rescue" the local population from extinction.

If both the immigration rate and the extinction rate are functions of p, the model takes the form

$$dp/dt = ip(1 - p) - ep(1 - p)$$

That is, the immigration rate is a function of the probability of immigration, i, times the fraction of available sources of immigrants, p, times the fraction of available patches, $1 - p$. The extinction rate is a function of the probability of extinction, e, times the fraction of patches with populations that can go extinct, p, times one minus the fraction of sites that can provide immigrants, thereby reducing the probability of local extinction, $1 - p$.

If both extinction and immigration rates are independent of the fraction of occupied patches, the model reduces to

$$dp/dt = i(1 - p) - ep$$

Gotelli (1991) refers to occupancy-independent immigration as the **propagule rain** because immigration provides a steady "rain" of new individuals to each population. All four possible combinations of occupancy-dependent and occupancy-independent immigration and extinction rates are conceivable and represent the extremes of a more general metapopulation model.

---

upon by Illka Hanski (1982, 1999), was another important stimulus for landscape ecology. In some respects, metapopulation theory is similar to the theory of island biogeography, but applied to populations rather than sets of species. Begin by thinking of an archipelago, but one far from a continent. The population size of a species on any given island is a result of local population dynamics plus immigration minus emigration (this is the same as Equation 7.1). A population on a given island may go extinct, but new immigrants from other islands may repopulate that island. The equilibrium between immigration and extinction determines the total size of the metapopulation—the number of surviving populations in the archipelago and their average size. Like the theory of island biogeography, metapopulation theory applies, of course, to many situations other than oceanic islands. The ideas of Levins were undoubtedly influenced by those of the population geneticist Sewall Wright, who envisaged a very similar process as part of his shifting balance theory of evolution (Wright 1931, 1968). We discuss the details of metapopulation models in Box 17A.

Metapopulation theory can be used to look at both population and community patterns. The theory predicts the distribution of population sizes for a species across a set of islands or patches, and how it changes over time. If one is willing to assume that species have

equivalent migration and extinction probabilities, the theory also predicts the distribution of population sizes among species within a community and the frequency of occurrence of species among communities. It is possible to create more complex versions of the theory by relaxing the assumptions of species and patch equality. Susan Harrison recently coined the term "metacommunity dynamics" for the extension of metapopulation theory to explain community patterns (Harrison 1999).

One important conclusion that comes from extending population dynamics to a landscape is that not all local populations need be self-sustaining. A **sink population** may be continually on the road to extinction, but may be maintained by constant immigration from a **source population**. The idea of source and sink populations is important because it emphasizes that local populations may often not be at equilibrium. Thus, testing theories based on equilibrium assumptions may be misleading. Island biogeography theory, for example, makes predictions based on equilibrium assumptions. During the 1960s, especially in response to island biogeography theory, equilibrium was often the dominant assumption among ecologists. Since the 1980s, the pendulum has swung in the opposite direction, and nonequilibrium is generally assumed, at least at a local level. A metapopulation perspective moves the equilibrium assumption up one level to the entire landscape. While individual

Table 17.1 Predictions about the correlation between species frequency and abundance and the shape of the frequency distribution from four different versions of the metapopulation model

| Model | | Predictions | | |
Immigration	Extinction	Correlation	Distribution	Source
Dependent	Dependent	Positive	Bimodal	Hanski 1982
Dependent	Independent	Zero	Single interior mode	Levins 1969
Independent	Dependent	Positive	Single interior mode	Gotelli 1991
Independent	Independent	Zero	Single interior mode	Gotelli 1991

Note: The models are classified by the dependence of immigration and extinction rates on the fraction of sites that are occupied (see text for details).

populations are not at equilibrium, the entire metapopulation is often assumed to be at equilibrium, though this assumption has not been tested.

Metapopulation Patterns

How well does metapopulation theory work at predicting landscape patterns for plant communities? The results are mixed. The key issue is whether plant populations have migration and extinction rates high enough for metapopulation processes to be a significant factor in determining species distributions and population sizes. The various versions of metapopulation models make explicit predictions about two aspects of species distributions (Table 17.1). First, if local extinction rates are dependent on the frequency of occupied patches, then those species occurring in a large number of patches will also have large populations when present in a patch—a positive correlation between occupancy and abundance. If local extinction rates are independent of the frequency of occupied patches, the correlation will be zero. Evidence from many studies shows that in most plant species, occupancy and abundance are positively correlated (see Figure 14.5; Brown 1984; Scheiner and Rey Benayas 1994). Thus, at least some versions of the metapopulation models are consistent with available data.

The second prediction of metapopulation theory is not so well borne out in plant studies. If both immigration and extinction rates depend on the frequency of

occupied patches, then patch occupancy by a species should have a bimodal distribution (Figure 17.7A). In other words, species should tend to be either very common or very rare, although they may switch over time. Hanski (1982) coined the terms "core species" for those species that are common in all communities in a landscape and "satellite species" for the rare ones. This version of the theory is often called the core-satellite hypothesis. Some ecologists occasionally confuse the core-satellite hypothesis with the idea of source and sink populations; they are quite distinct from each other in that the former idea concerns characteristics of entire communities. The bimodal patch occupancy distribution of the core-satellite hypothesis becomes unimodal if either the immigration or extinction rate is occupancy-independent. The mode is located somewhere in the middle of the distribution (Figure 17.7B)—that is, typical species will be somewhat common, with the exact values depending on immigration and extinction rates.

One series of studies of tallgrass prairie communities in Kansas (Gotelli and Simberloff 1987; Collins and Glenn 1990, 1991) found evidence of bimodal patch occupancy distributions for surveys done at a scale of 1 km² or less (Figure 17.8). However, a survey of 74 landscapes ranging in scale from 0.5 to 740,000 km² (mean 755 km²) found no evidence of bimodal distributions (Scheiner and Rey Benayas 1994). Thus, at medium to large scales, the data for terrestrial plants are not consistent with the hypothesis that metapopulation process-

(A)

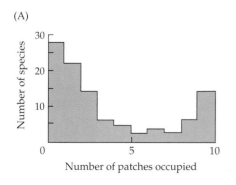

(B)

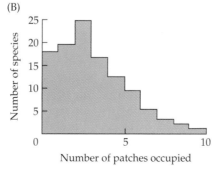

Figure 17.7
Two possible distributions of species frequencies. (A) Both immigration and extinction rates are dependent on patch occupancy rates. (B) Only one of the rates, or neither, is dependent on patch occupancy rates. (After Collins and Glenn 1991.)

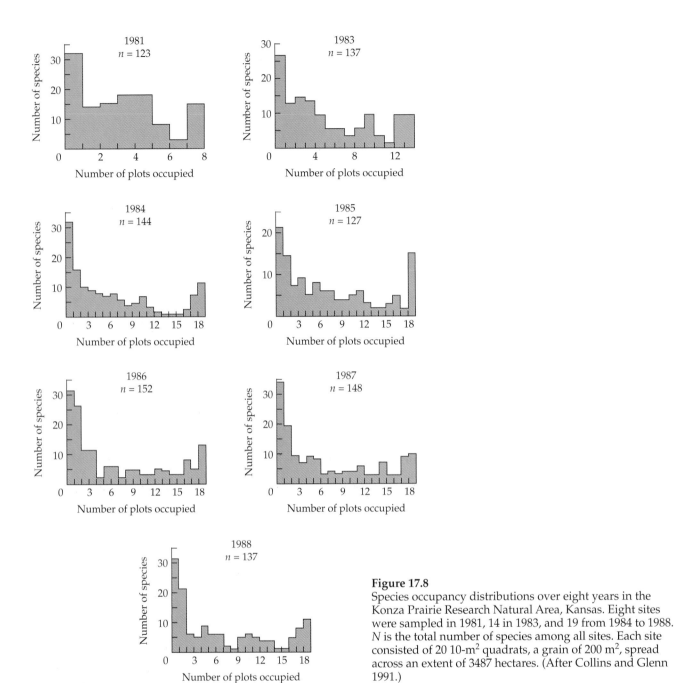

Figure 17.8
Species occupancy distributions over eight years in the
Konza Prairie Research Natural Area, Kansas. Eight sites
were sampled in 1981, 14 in 1983, and 19 from 1984 to 1988.
N is the total number of species among all sites. Each site
consisted of 20 10-m² quadrats, a grain of 200 m², spread
across an extent of 3487 hectares. (After Collins and Glenn
1991.)

es are important, but they may be important at smaller
scales of approximately 1 km² or less.

Why is it that metapopulation theory does not do a
good job of predicting plant species distributions at large
spatial scales? There are two reasons, one mathematical and one biological. The mathematical reason is simple: to make the mathematics analytically tractable, simple models need to assume that the landscape has no
spatial structure—that is, that all patches are equally
close to one another. If dispersal rates are very high, this
is a reasonable approximation, but few organisms, especially plants, have such high dispersal rates. This diffi-

culty is easy to overcome, though, because spatial structure can be incorporated into computer simulation models (e.g., Holt 1997).

The biological reason is more critical (and you may
have already realized what it is): the models work only
if immigration and extinction rates are high enough to
influence local population dynamics. Immigration rates
must be large to generate a positive correlation between
occupancy and abundance. Consider a set of populations, each of which varies most of the time between 40
and 200 individuals. If 5 to 6 individuals enter each population in any generation—a low immigration rate—

(A)

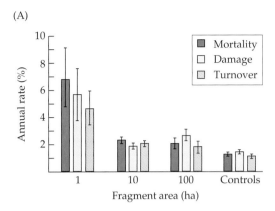

(B)

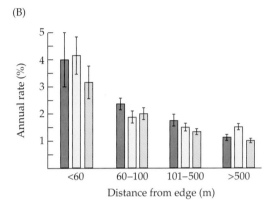

Figure 17.11
Annualized rates (mean ± 1 standard error) of forest dynamic parameters of all trees in each fragment for the Biological Dynamics of Forest Fragments Project. Turnover is the rate at which trees are being replaced. (A) Effects of fragment area (size of fragment). (B) Effects of distance from the edge of the fragment. (After Laurance et al. 1998.)

side of several forests into the forests' interiors. Windborne seeds, including those of invasive non-native species, entered the forests in substantial numbers, an important but not unexpected finding. Long-undisturbed edges with denser, "branchier" vegetation, however, formed far better seed barriers than recently cut edges. Old edges not only blocked far more seeds from entering the forest, but also kept those seeds that made it into the forest much closer to the edge, protecting the interior better. These results imply that recent increases in habitat fragmentation, creating many new edges, will exacerbate the spread of invasive species into the interior of these habitats.

Habitat fragmentation can change not only the sizes, but also the genetic diversity of local populations. Small populations—those in small habitat fragments—are likely to lose genetic diversity through the process of genetic drift (see Chapter 5). In turn, reduced genetic diversity (as well as small population size itself) may increase the chance of extinction because the population will be unable to evolve in response to changes in the environment. Because the process of fragmentation may itself cause environmental change, this problem becomes even more important. For instance, in small fragments of the BDFFP, individuals of the Brazilian canopy tree *Corythophora alta* (Lecythidaceae) are all very similar genetically, in contrast to a lower level of genetic relatedness in large fragments (Hamilton 1999). Because seed and pollen movement are limited, inbreeding is likely to increase, leading to further loss of genetic diversity.

Edges, Connectivity, and Nestedness

Edge effects—systematic differences between areas inside the edges of habitat patches and the interiors of those patches—can also be important to plants. Wind speed, for example, is greater at the edge of a forest than in the interior, and the plants there are exposed to more desiccation. If winds are strong enough, trees on the edges may have more limbs broken off or may have a greater chance of being knocked down. The effect of wind has been particularly important in the BDFFP plots. Near fragment edges, the rates of tree damage and mortality were three times higher than inside intact forests (Figure 17.11B). Other edge effects include differences in available light (and therefore in the magnitude of diurnal temperature changes) and humidity between edge and interior, and in some cases, differences in CO_2 concentrations. There may also be differences in the presence or activity of herbivores, pollinators, and seed dispersers between edge and interior. Thus, patch shape—especially the ratio of perimeter to total area—is important to consider.

In recent years, ecologists have begun to turn their attention to several other aspects of landscape geometry, such as the distances between patches and the degree to which patches are connected to one another. **Connectivity** is the degree to which a landscape facilitates or impedes the movement of organisms (Tischendorf and Fahrig 2000). It is a function both of the physical arrangement of patches and of the behaviors of organisms in response to that pattern. Different species may judge the connectivity of a landscape very differently. Much attention has been devoted to the idea of connecting habitat patches with "corridors," mainly in the context of animal conservation. Dispersal among habitat patches, even if it occurs only occasionally, can be an important means by which populations are preserved in the long run, as isolated populations are more likely to go extinct.

Much less attention has been given to connectivity in plant ecology, probably because most attention in conservation as a whole has been paid to animals, especially vertebrates. An exception is the work of Janez Pirnat (2000), who used GIS and vegetation sampling to study connectivity between forest fragments in Ljubljana,

Slovenia. Here, forest patches existed as urban forests and parks, as suburban forests around the fringe of the metropolitan region, and as connecting forest corridors and fragments. Road building and development disrupted these connections, threatening the dispersal of forest plant species, particularly endangering the maintenance of diversity in the urban forests that are population sinks for various tree species. Similarly, in many landscapes affected by natural or anthropogenic fragmentation, from large lava flows or fires to suburban development, the availability of foci of dispersal—patches of plants that can serve as sources of seeds for dispersal into the sink regions—can be of critical importance in the maintenance and recovery of plant populations.

The nestedness of a set of communities—the degree to which communities tend to be subsets of other communities (see Table 16.4)—is an important landscape-level issue in conservation biology. In a perfectly non-nested set of communities, each community is composed of a unique set of species; for example, the sets of species growing in a forest and in a neighboring lake do not overlap. By contrast, in a perfectly nested set of communities, the smallest patches have only species contained in the larger patches. In a perfectly nested system, one could thus preserve all the species diversity by preserving just the largest patch, while in a perfectly non-nested system one would need to preserve every patch.

A simple application of island biogeography theory predicts that the small patches will contain random samples of the entire set of species, with different species appearing in different patches. But fragmented landscapes have a tendency to become more nested over time: small patches often contain the same few species because those species are the hardiest or the best dispersers. Animal communities tend to be highly nested (Wright et al. 1998), although this is less true of plant communities. The reason is that local environmental effects play a major role in determining plant community composition, and patches tend to differ in important environmental parameters. In the BDFFP, immigration of new individuals into the fragments consisted mainly of species found in disturbed areas, rather than the forest-interior species previously found there. Thus, there was a profound shift in the species composition of the fragments, which was not simply a random sampling of the previous species.

Summary

Landscape ecology is the study of the spatial distribution of individuals, populations, and communities and the causes and consequences of those spatial patterns. There are several ways to describe spatial patterns. Most such measures deal with the sizes, shapes, and arrangements of patches. Other measures, such as some types of species-area curves, are not spatially explicit. These measures highlight the importance of scale in ecological processes. Scale has several components—grain, extent, and focus—and can have both spatial and ecological hierarchies. How we perceive a pattern depends on the scale at which we analyze that pattern. Our ability to measure patterns and examine scale effects has dramatically increased in the past two decades with advances in remote sensing methods and the development of new analysis techniques such as GIS and spatial statistics.

More significantly, processes can operate differently as scale changes. Movement—whether of individuals or of materials such as nutrients—is a key ecological process in landscapes. The movement of individuals affects population dynamics and community structure. The dispersal of individuals is a critical component of two important theories—island biogeography theory and metapopulation theory—that form the basis of landscape ecology. From these theories have come other theoretical advances, such as hierarchical patch dynamics. At present, however, these theories have had only limited success in explaining the landscape patterns of plant communities.

An understanding of landscape dynamics and their consequences is important in conservation biology. Nature reserves can be designed to maximize the number of species preserved. Island biogeography theory was at the center of a debate about whether it is better to create one large reserve or many small reserves. Landscape fragmentation leads to decreases in connectivity among patches and increases in the amount of habitat edges. These changes in landscape structure can have important, and often negative, effects on community structure and composition, including the loss of species and decreases in the genetic diversity of populations.

Additional Readings

Classic References

Hanski, I. 1982. Dynamics of regional distribution: The core and satellite hypothesis. *Oikos* 38: 210–221.

Levins, R. 1969. Some demographic and genetic consequences of environmental heterogeneity for biological control. *Bull. Entomol. Soc. Am.* 15: 237–240.

MacArthur, R. H. and E. O. Wilson. 1967. *The Theory of Island Biogeography*. Princeton University Press, Princeton, NJ.

Simberloff, D. S. and E. O. Wilson. 1970. Experimental zoogeography of islands: A two-year record of colonization. *Ecology* 51: 934–937.

Current Research

Harrison, S. 1999. Local and regional diversity in a patchy landscape: Native, alien, and endemic herbs on serpentine. *Ecology* 80: 70–80.

Johnson, A. R. and B. N. Milne. 1992. Diffusion in fractal landscapes: Simulations and experimental studies of tenebrionid beetle movements. *Ecology* 73: 1968–1983.

Tischendorf, L. and L. Fahrig. 2000. On the usage and measurement of landscape connectivity. *Oikos* 90: 7–19.

Additional Resources

Gotelli, N. J. 2001. *A Primer of Ecology*, 3rd ed. Sinauer Associates, Sunderland, MA.

Hanski, I. 1999. *Metapopulation Ecology*. Oxford University Press, New York.

Naveh, Z. and A. S. Lieberman. 1994. *Landscape Ecology: Theory and Application*, 2nd ed. Springer-Verlag, New York.

Dale, M. R. T. 2000. *Spatial Pattern Analysis in Plant Ecology*. Cambridge University Press, Cambridge.

PART **V** Global Patterns
and Processes

CHAPTER *18* Climate and Physiognomy

*I*f you were to travel any considerable distance, either across a continent or across the globe, you would see strikingly different plant communities. In some places you would see tall trees with broad leaves. Other places would have no trees and dense grass cover, while still others would be home mainly to shrubs. Some places would have a great diversity of plants, while others would have few species. These patterns of variation, both in overall plant diversity and in **physiognomy**—the general form of the vegetation—are the subject of Part V of this book. In this chapter, we consider the climatic factors that control vegetation patterns and how they lead to variation in vegetation form. Subsequent chapters will look at large-scale patterns of vegetation form and species diversity across the globe. Finally (in Chapters 21 and 22), we will examine patterns through geological time and look at some predictions about what may happen in the future as the global climate changes.

Climate and Weather

Any attempt to determine the causes of large-scale vegetation patterns must start with climate. **Climate** refers to the long-term distribution (such as means and variability; see the Appendix) of the weather in a given area (e.g., London generally has cool temperatures and a lot of rain), while **weather** refers to the immediate or short-term conditions (e.g., this week or this month it might be unusually warm and dry in London). Although weather has profound effects on the function, growth, and survival of plants, it is climate that determines the general type of vegetation in an area (e.g., broad-leaved deciduous forest versus desert) and influences large-scale patterns of diversity. In this chapter we describe the patterns of climatic variation around the world (with special attention to North America), explain the mechanisms responsible for those patterns, and look at how climatic variation translates into variation in plant form.

Our understanding of the connection between climate and vegetation was forged during the nineteenth century, spurred by the research of the German naturalist Alexander von Humboldt. He was the first to systematize information on the effects of altitude and air pressure on patterns of temperature and precipitation. He also was the first to codify our understanding of how coastal climates differ from those inland. The first map of mean monthly world temperatures was published in 1848. Two decades later, in 1866, the first world vegetation map was produced. Over the next several decades, a number of naturalists developed classifications of plant communities, noting the relationships between different types of communities (such as forests and grasslands) and the climate at different latitudes and altitudes.

Our understanding of weather has depended heavily on the ability of researchers to gather large numbers of measurements over a wide geographic area. Prior to the mid-eighteenth century, no one had realized that weather moves in predictable ways across the globe—news traveled far more slowly than weather (McIlveen 1992). On October 21, 1743, Benjamin Franklin attempted to observe a lunar eclipse in Philadelphia, but was prevented from seeing it by a storm. Later, he was surprised to learn that the eclipse was visible in Boston and that the storm arrived there the following day. By contacting people living between the two cities, he was able to reconstruct the movement of the storm. It was the development of communication technology, however, that really changed the sciences of meteorology and climatology. The invention of the telegraph in 1835 (with the first message transmitted in 1844) made it possible to organize large numbers of people to observe and forecast the movement of storms. The advent of airplanes and the military technology developed in World War II further spurred the development of weather station networks and made it possible to sample weather conditions high in the atmosphere. Today, we gather information about the weather with a combination of satellites, ground-based stations, and upper-atmosphere probes. The ready accessibility of powerful computers makes it possible to analyze these enormous data sets, and both climatologists and meteorologists are actively developing more sophisticated models of climate and weather (see Box 22A). These efforts are especially important in our attempts to predict changes in weather and climate due to the effects of global warming and to understand how different factors influence those changes.

The two primary components of a region's climate are temperature and moisture. We first turn to the long-term mean (average) temperature and precipitation and the effects of these average conditions on vegetation patterns. Variation around these means is also important, and we look at two aspects of this variation: predictable changes (e.g., it is usually warmer in the summer than in the winter) and departure from average conditions (e.g., in a particular year it may be unusually dry in the Amazon basin). The time scale of variation is also important, with temperature and precipitation varying on scales from a single day, a week, a season, a year, a decade, up to cycles that take hundreds of thousands of years to complete. Here we investigate these patterns of variation and how they shape plant communities.

Temperature

If you sit outside on a sunny day, you can feel the sun's radiant energy on your skin. Radiant energy from the sun, and variation in how much radiant energy reaches different parts of Earth's surface, is the primary cause of variation in temperature, the first of the two major climatic factors. **Heat** is a measure of the total kinetic energy—the energy of motion—of the molecules in a substance, while **temperature** is a measure of the average kinetic energy of those molecules; see Chapter 3. Radiant energy from the sun heats objects (such as plants or the ground) directly. This energy is primarily in the shortwave part of the electromagnetic spectrum (wavelengths < 700 nm) and includes visible light. As objects are heated by sunlight, they emit longwave radiant energy (wavelengths >700 nm, including infrared radiation) in proportion to the fourth power of temperature (see Chapter 3), becoming a secondary source of heating for nearby objects. The warming of physical objects by sunlight, and their slow emittance (i.e., the storage and later release) of that heat energy over time as longwave radiation, creates lags in temperature change, which we will explore in more detail below. The transport of that heat energy from one place to another by atmospheric and ocean currents—a process called convection—is also a key element in global climates.

The temperature at any spot on Earth's surface is determined primarily by the amount of radiant energy it receives from the sun. That amount is determined, in turn, largely by the angle of Earth's surface in that spot relative to the sun's rays, as well as by atmospheric conditions. Consider a spot at sea level on the equator at noon during the vernal or autumnal equinox, when the sun is directly overhead. At the top of the atmosphere at that spot, Earth receives approximately 340 W/m^2/year, or 2 langleys (ly) of energy per minute (1 ly = 1 cal/cm^2) as shortwave radiation. If we consider the amount of energy striking the top of Earth's atmosphere to be 100%, we can trace what happens to that energy (Figure 18.1). Almost a third of it returns to space as shortwave radiation, reflected by Earth's surface (6%) and by clouds (17%) and scattered by the atmosphere (8%). Almost half (46%) is absorbed by Earth's surface. The rest of the energy entering the atmosphere is absorbed by clouds (4%) and by the "greenhouse gases" (19%) in the atmosphere. (We will return to the absorption of energy by greenhouse gases shortly and will discuss it at greater length in Chapter 22.)

Earth's surface emits energy in the longwave part of the spectrum, reradiating 15% of the initial total energy that reaches the upper atmosphere and transferring an additional 24% as latent heat loss (see Chapter 3) and 7% as convection to the atmosphere and clouds. The **greenhouse effect** results when of greenhouse gases in the

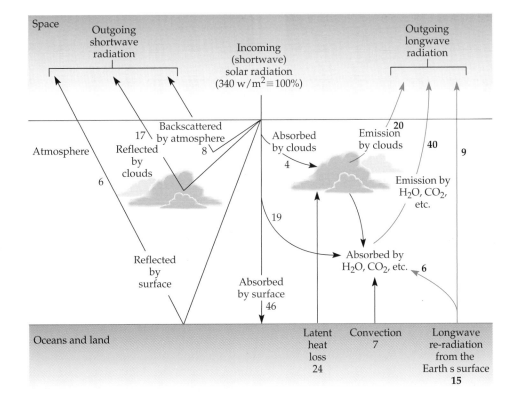

Figure 18.1
The radiant energy balance of the Earth, expressed as the percentage of the energy reaching the top of the atmosphere (340 W/m^2/year, or 100%). This energy is reflected by the surface and by clouds and scattered by the atmosphere, so that about 31% returns to space. Of the remainder, 46% is absorbed by the surface and 23% by the atmosphere, warming both. Longwave radiation (infrared, shown by boldfaced numbers) is emitted by the surface, warming the atmosphere further as well as being lost to space. Convection and latent heat exchange also transfer heat energy from the surface to the atmosphere. The total amount of energy absorbed by the surface, atmosphere, and clouds (as shortwave radiation) and the total amount that leaves the planet (as longwave radiant energy, latent heat loss, and convection) are in balance with each other. The total amount of energy entering the top of the atmosphere (as shortwave radiation) is exactly equal to the sum of the outgoing shortwave radiation (from reflection and scattering) plus the outgoing longwave radiation, which together sum to 100%. (After Schlesinger 1997.)

atmosphere reabsorb longwave radiation emitted by the Earth's surface (see Figure 22.3). Secondary heating of the Earth's surface then occurs by reradiation from the atmosphere.

This reabsorption and reradiation is critical for life on Earth. Without the greenhouse effect Earth would be bitterly cold everywhere, all of the time. Secondary heating by longwave energy from the atmosphere accounts for two-thirds of the total radiant energy received at the surface. The total outgoing longwave radiation from Earth back to space is about 69% of the energy received initially, coming from Earth's surface, the atmosphere, and clouds.

Incoming energy must balance outgoing energy. Temporary imbalances will result in Earth warming or cooling, but eventually a new equilibrium will be reached as Earth's emission of longwave radiation changes according to its new temperature. Thus, the shortwave radiant energy absorbed by the atmosphere and Earth's surface (46% + 23%) is balanced by the longwave energy (69%) emitted.

Short-Term Variation in Radiation and Temperature

The combination of shortwave and longwave radiation determine the ambient temperature at Earth's surface. As you travel away from the conditions described above, of maximal solar energy input at the equator at noon, several factors act to determine the amount of incoming radiation and the subsequent temperature. Most important is the angle of the sun, which determines the amount of incoming shortwave radiation. Over the course of the day, Earth rotates on its axis, and the sun's rays hit the surface at different angles. The solar **angle of incidence** is the angle that a ray of sunlight makes with a line perpendicular to the surface. The smaller the angle of incidence, the closer to 0° from the horizontal, the larger the

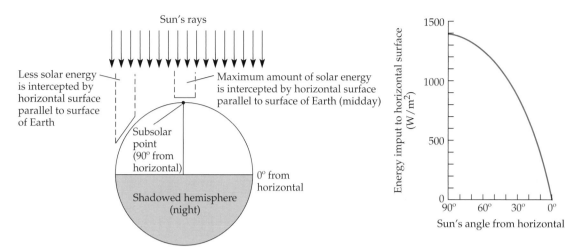

Figure 18.2
Differences in angle and amount of solar radiation reaching the Earth's surface. At the equinox, when neither hemisphere points toward or away from the sun, the diagrams shows the amount of solar radiation received at a single spot over the course of the day. The same differences could apply to the angles and amounts of solar radiation reaching different latitudes at different times of the year (see Figure 18.3B).

area over which a given amount of solar energy is spread. Consequently, when the incoming sunlight is at a steep angle, any given square meter of surface receives less energy than it would if the sun were at a more direct angle overhead (Figure 18.2), Also, at steeper angles, the incoming solar energy must travel a longer distance through the atmosphere, and so more energy is absorbed by the atmosphere and less reaches Earth's surface.

Similarly, as one moves north or south of the equator, the angle of the sun, even at noon, decreases. The actual angle depends on the time of year and how far one is from the equator. Earth's axis is tilted relative to the plane in which it revolves around the sun. Seasons are created by the differences in solar angle, and in the consequent amount of incoming solar energy, over the year-long progression of Earth around the sun (Figure 18.3A). The Northern and Southern Hemispheres are tilted toward or away from the sun at the opposite extremes of Earth's trajectory around the sun, at the solstices in December and June. At the equinoxes, in September and March, Earth's axis is parallel to its plane of revolution, neither pole is tilted toward or away from the sun, and day and night length are equal everywhere on Earth. When the sun is more directly overhead in the Northern or Southern Hemisphere, it is summer in that hemisphere. In the other hemisphere, steep solar angles result in decreased radiant energy input, and it is winter. This change in solar angle with the seasons is greatest at the poles and least at the equator. As a result, over the course of a year at extreme latitudes, there are large changes in radiation input. In contrast, equatorial regions have

nearly constant solar energy inputs throughout the year (Figure 18.3B). These effects combine to produce distinct patterns of temperature variation across the globe (Figure 18.4).

Atmospheric, ground, and oceanic temperatures have a feature called **lag**—a delay in effect—that determines some of the daily and yearly patterns of temperature variation. It is warmer in the late afternoon, even when the sun is low in the sky, than when the sun is at a comparable angle in the morning. The coldest month of the winter in the Northern Hemisphere is not December (when the sun is lowest and the days are shortest), but January; conversely, the hottest summer months are July and August, not June (when the days are longest and the sun most directly overhead). The cause of lag is the storage of heat by the ground, the ocean and other large bodies of water, and the atmosphere itself.

Imagine a place somewhere at a temperate latitude at the winter solstice. Incoming solar radiation is at a minimum. As Earth revolves around the sun and the year progresses, day length increases and the sun is more directly overhead. The ground absorbs solar energy and begins to warm. The ground has a certain **specific heat**—the amount of heat energy that is required to warm 1 gram of it by 1°C. Water has a higher specific heat than the ground, so the ocean takes longer to warm up than nearby land. Air has a far lower specific heat than either land or water, and warms up more readily than they do. Even as our imaginary spot on Earth passes the summer solstice and incoming radiation begins to decline, the longwave output from the now warmed ground and bodies of water nearby continue to warm the air, keeping air temperatures high.

Eventually, as solar input declines, the reverse pattern occurs: air, ground, and finally water become cooler. A high specific heat results in slow cooling, just as it causes slow warming. Coastal environments, influenced by the adjacent bodies of water, consequently cool and warm more slowly than those in the interior of conti-

(A)

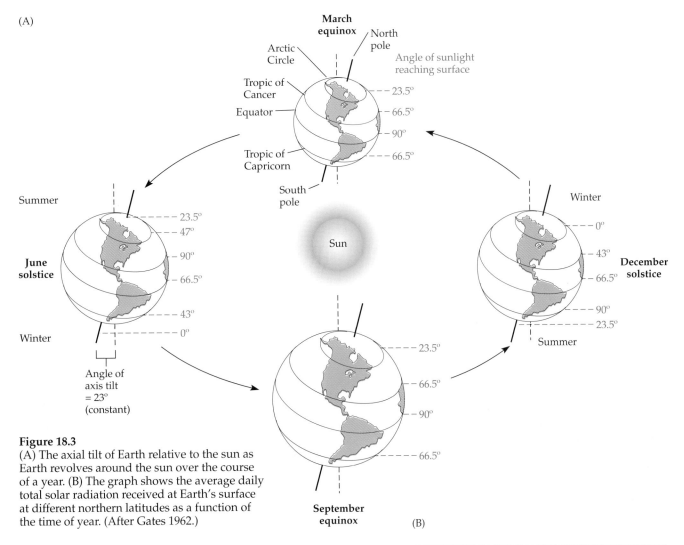

Figure 18.3
(A) The axial tilt of Earth relative to the sun as Earth revolves around the sun over the course of a year. (B) The graph shows the average daily total solar radiation received at Earth's surface at different northern latitudes as a function of the time of year. (After Gates 1962.)

(B)

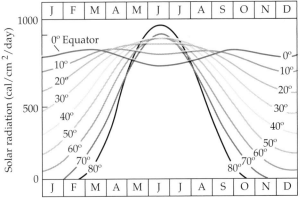

nents, and thus experience more moderate temperatures. The water ameliorates very hot temperatures as cooler offshore breezes convectively reduce heat on land. In autumn, the water releases heat gradually, slowing temperature declines on land by providing warming breezes. These breezes may feel cold to a person, but they are warm relative to the ground and the air over it. One subtle implication of these lags for plants is that times of maximum or minimum solar radiation are not the times of highest or lowest temperatures, either daily or seasonally.

The magnitude of daily variation in temperature—the range from nighttime to daytime temperatures—depends in part on local atmospheric conditions, particularly the amount of moisture in the air. At night, there is no incoming shortwave radiation. We do not immediately freeze after sunset, however, because the surrounding air acts somewhat like a blanket, holding heat and releasing it as longwave radiation all night. Moist air can hold more heat than dry air, as we will see below. A place like Tokyo, adjacent to the Pacific Ocean with

typically humid air, usually has nighttime temperatures that are only a little cooler than daytime temperatures. In contrast, Tucson, Arizona, with very dry air, may have dramatic swings, with nighttime temperatures 20°C or more cooler than those during the day. As altitude increases, the density of the air decreases—the blanket is thinner. As a result, areas at high elevations, such as

(A)

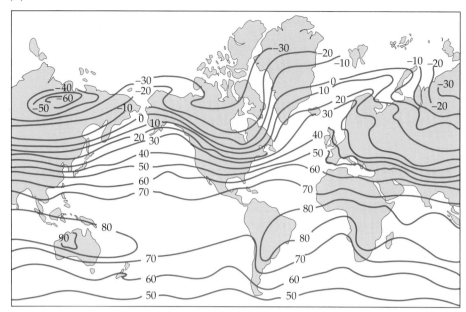

(B)

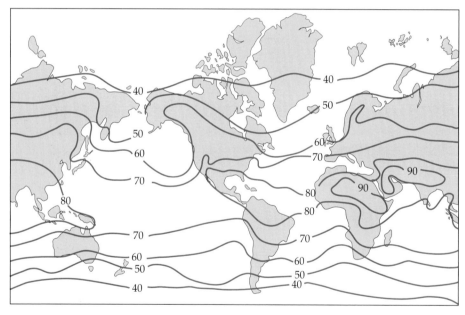

Figure 18.4
Mean temperatures (Farenheit) across the globe in (A) January and (B) July.
(After Rumney 1968.)

mountaintops, also tend to have large swings in temperature from day to night. Plants must be adapted to tolerate the temperature ranges, as well as the average conditions, where they live. Elevation and latitude jointly affect climate in interesting ways. As indicated above, equatorial regions receive nearly constant amounts of radiation throughout the year. But there are high-elevation regions at the equator, such as the Ecuadorian Andes in South America and the Kenyan highlands in

Africa. These spots can experience daily temperature swings approaching those experienced between summer and winter at mid-latitudes, although obviously they also receive much more radiation during the day.

Long-Term Cycles

Besides the daily and yearly variations in solar radiation, there are also longer-term cycles. One of these is the 11-year sunspot cycle. Over this cycle, the amount of solar radiation reaching Earth varies up to 0.1%. A 22- to 23-year double sunspot cycle is also known, and is related to reversals of the solar magnetic field. These cycles probably do not directly affect plant growth or local temperatures. Rather, they affect atmospheric circulation and precipitation patterns, resulting in indirect effects on plants due to changes in weather patterns. The linkage between the sunspot cycle and climate was first studied by the astronomer Andrew E. Douglass, who used the long-term record of good and bad years for plant growth recorded in tree rings. In doing so, he invented the field of dendrochronology (see Chapter 13).

At very long time scales, there are several different kinds of changes in Earth's orbit around the sun that influence climate, collectively called **Croll-Milankovič effects** (Figure 18.5). The longest of these orbital cycles

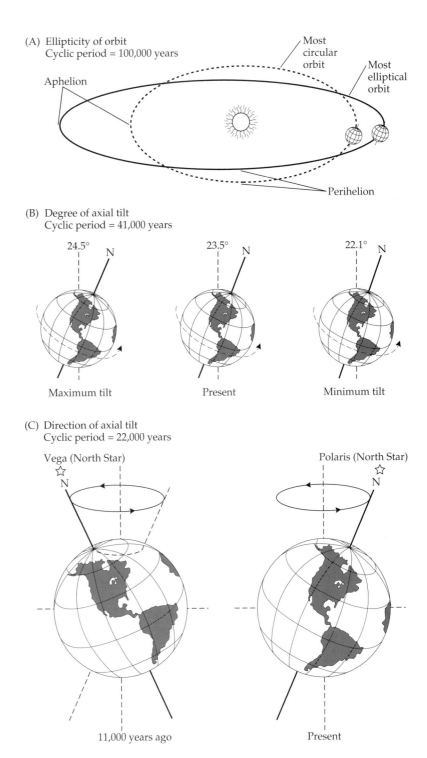

Figure 18.5
Croll-Milankovič cycles are periodic changes in (A) the ellipticity of Earth's orbit, (B) the degree of axial tilt, and (C) the direction of axial tilt. Each affects the amount of solar radiation received at different points on Earth's surface at different times of the year. These cycles interact to determine long-term climatic patterns. (After Gates 1993.)

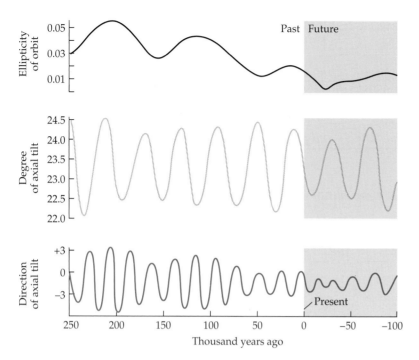

Figure 18.6
Past and predicted Croll-Milankovič cycles. Sometimes the cycles can reinforce each other; for example, when both ellipticity and the degree of axial tilt are large. At other times they can cancel each other. The precentage of axial tilt is the amount of wobble. (After Gates 1993.)

is a 100,000-year cycle of change in the shape of Earth's orbit around the sun, from nearly circular to a more elliptical trajectory. Currently the orbit is almost circular, so that the difference between Earth's nearest point to the sun and its farthest is only 3.5%. When the orbit becomes more elliptical, this difference between the shortest and longest distance of Earth from the sun may be as great as 30%. Clearly, such changes in orbital shape will affect the range of seasonal variation at temperate and polar latitudes, because the amount of incoming solar radiation will change.

The next longest pattern of variation is a 41,000-year cycle in the degree of tilt of Earth's axis from 22.1° to 24.5°; currently, the tilt is 23.5°. These changes in the axial tilt will greatly affect the range of seasonal variation, because changes in the angle of the sun will change the amount of solar energy input at any given spot in temperate and polar latitudes.

Finally, there is a 22,000-year wobble of Earth's axis. This wobble is similar to what you would see on a top that is beginning to slow down, so that its stem no longer points straight up but is beginning to circle around. The result of this wobble is to change whether the Northern or the Southern Hemisphere is pointing toward the sun during the time of Earth's nearest approach to the sun. Earth is now a little bit closer to the sun at the winter solstice in the Northern Hemisphere (the summer solstice in the Southern Hemisphere) than at the opposite point in its orbit. Changes in the direction of the axial tilt will affect the intensity of seasonal variation; for example,

Northern Hemisphere winters currently are milder than Southern Hemisphere winters.

All three of these cycles thus modify seasonality, and they can work together or in opposition to magnify or minimize seasonal variation (Figure 18.6). While the effects may be small, they are sufficient to cause large-scale and long-term changes in Earth's climate and may be responsible for the cycles of ice ages that Earth has experienced over the last million years (see Figure 21.8). We look at the most recent of these ice ages and its aftermath in Chapter 21.

Recently there has been extensive debate about global warming, an overall rise in global temperatures. Part of the debate stems from a genuine scientific question: is the recent warming of the climate likely to be caused by human addition of greenhouse gases to the atmosphere, or does it reflect natural cyclic changes and perhaps some random variation in the weather? Without considering the actual data, either hypothesis (or both) could be plausible. Are the warmer temperatures we are experiencing part of a long-term trend, or are they merely a short-term blip that will soon be reversed? If this warming is part of a trend, to what extent is it due to human actions? Certainly, it is very difficult to separate natural variation in weather and climate from anthropogenic changes. In any case, one may well wonder what the answers to these questions mean for the future: Is the observed warming a harbinger of an increasingly warm climate? We investigate this issue in depth in Chapter 22.

Precipitation

Besides temperature, the other major aspect of climate is moisture: **precipitation**, the amount and pattern of rainfall and snowfall, and **relative humidity**, the water vapor in the air as a percentage of the amount in fully saturated air at the same temperature. Precipitation clearly has a major determining effect on the type and amount of vegetation found in a region. Humidity levels are also important ecologically because fog may be a critical source of moisture for plants in certain habitats (fog occurs only at saturating humidity) and, more generally, because humid air holds more heat than dry air. Water molecules in the atmosphere are very effective at absorbing longwave radiation and have a high specific heat. Therefore, humid air acts as a reservoir of stored heat. Clouds can also absorb and reflect incoming solar energy. A number of factors determine how precipitation and water vapor are distributed over the globe and during the year.

Global Patterns

Warm air is less dense than cold air, and therefore rises, like a piece of wood floating on water. That is why hot air balloons go up, and why it is warmer near the ceiling of a room than at the floor. Warm air also has a greater capacity for holding moisture than does cooler air. You can see this directly by observing moisture condensing on a cold beverage glass on a warm, humid day. As the air near the surface of the glass cools, its capacity to hold moisture is reduced, so moisture condenses as liquid on the cool surface of the glass.

As explained previously, solar radiation is greatest at the equator, and the air there is warmed. This warming creates a pattern of rising air in a belt around the equator. Warm tropical air masses pick up moisture from the land surface, both from transpiration by plants and by evaporation from the ground, and in large amounts from the oceans by evaporation. These warm, moist masses of rising air establish a gigantic three-dimensional conveyor belt for moisture and energy.

This pattern of air movement has two outcomes: large amounts of precipitation in equatorial regions, and air flowing toward the poles at high altitudes. The air rising over the equator begins to cool as it reaches higher altitudes. Why? Air pressure decreases with altitude because as a given packet of air rises, there is less atmosphere above it pushing down on it. The rising warm air therefore expands, because gases at lower pressures expand to fill up a greater volume. Expanding air cools as energy is used to push the air molecules farther apart. The cooler air now cannot hold as much moisture as it did, and some of the water vapor it contained condenses. Clouds form, and eventually the clouds become saturated with moisture, which falls as rain. Equatorial regions consequently tend to have high rainfall.

Above those clouds, the conveyor belt of air continues to move. Moisture continues to be wrung out of the air as this rising, cooling, and condensation process continues. Rising air cools at the **adiabatic lapse rate** when no heat is exchanged with surrounding parcels of air. Air saturated with moisture cools at 5.4°C/1000 m, the saturated (or moist) adiabatic lapse rate, while dry air cools more rapidly, at the dry adiabatic lapse rate of 9.8°C/1000 m. The reason these rates differ is that as air saturated with moisture rises and cools, the water that condenses out of the air releases heat to the air—the opposite of what happens when a surface is cooled by evaporating water (latent heat loss; see Chapter 3).

Eventually, air can no longer rise any higher in the atmosphere. But the continual addition of new air from below means that this air has to go somewhere. The rising mass of air at the equator is thus forced to move away from the equator northward and southward toward the poles. As this air moves poleward, it slowly cools further because it is receiving less radiant energy. By the time it has reached about 30° latitude, it has cooled sufficiently to become denser and begin sinking. The falling air becomes warmer, through the reverse of the process that occurred at the equator. Already dry, this rewarmed air has a large capacity to hold moisture, but contains little water vapor. These falling air masses absorb any available moisture, and release little of it, creating intensely arid regions where they meet Earth's surface. This explains why the great deserts of the world are largely found in two bands, at 30° N latitude (the Sahara, Arabian, and Sonoran) and 30° S latitude (the Atacama of Chile, the Kalahari of southern Africa, and the Great Sandy Desert and Great Victoria Desert of Australia).

Finally, once the air nears the ground, it must move either toward the poles or toward the equator. The Coriolis effect (Box 18A) causes air to veer to the right (relative to the direction in which it is moving) in the Northern Hemisphere and to the left in the Southern Hemisphere. For the air moving from 30° N and 30° S toward the equator, this effect creates the **trade winds** in both hemispheres, which move from northeast to southwest in the Northern Hemisphere and from southeast to northwest in the Southern Hemisphere. Air masses moving poleward from 30° N also veer to the right, while air moving poleward from 30° S veers to the left. This pattern creates the prevailing **westerlies** (winds moving from west to east) in both hemispheres. Early travelers in sailing ships going between Europe and North America took advantage of these patterns, allowing the trade winds to carry their ships southwestward across the Atlantic on their journey out and returning to Europe

Box 18A

The Coriolis Effect

You may have heard the persistent myth that water goes down toilets and drains in a clockwise direction in the Southern Hemisphere, but spins counterclockwise in the Northern Hemisphere. This myth is false (as we know, having personally performed the experiment), but the basis of the claim—the **Coriolis effect**—is true. In fact, the general pattern of circulation in the oceans is clockwise in the Northern Hemisphere but counterclockwise in the Southern Hemisphere. Understanding the Coriolis effect is fundamental to understanding many patterns in global climate and weather, including the reasons for the general patterns of atmospheric and oceanic circulation.

One way of summarizing the Coriolis effect is this: Objects moving toward or away from the equator in the Northern Hemisphere are deflected toward their right, and in the Southern Hemisphere toward their left. If you were in Paris and you fired a missile due north, it would land somewhat to the northeast. If you fired the missile due south, it would land somewhat to the southwest. The reverse would be true in the Southern Hemisphere: a missile fired due north from Sydney would land to the northwest, and one fired due south would land in the ocean, to the southeast. Moving air masses do the same thing: if they are in the Southern Hemisphere, they tend to be deflected to their left, and if they are in the Northern Hemisphere, they tend to be deflected to their right.

Why does this strange phenomenon occur? Odd though it sounds at first, the surface of Earth does not rotate at the same speed everywhere. If you stand on the equator, in one day your body will have traveled much farther in space than if you were standing on the Arctic Circle—this must be so if you are to arrive back at the same place in 24 hours, because there is a lot farther to go at the equator than at the Arctic Circle. An object, whether it be a missile or an air mass, moving toward the equator from 30° N initially has the same rotational speed as the ground where it is "launched;" as Earth rotates on its axis, the object moves along with Earth at the same speed as the ground surface. As this object is shot or forced southward toward the equator, it retains its movement in the direction of Earth's rotation, but as it gets closer to the equator it is moving at a speed *slower than* Earth's surface beneath it. Relative to the surface, it will be deflected to its right (toward the southwest). An air mass originating at the equator and moving in the opposite direction, toward the North Pole, will have a rotational speed *faster than* that of the surface of Earth farther north, so it, too, will veer to the right, being deflecting to the northeast. The analogous pattern is found in the Southern Hemisphere, but there objects are deflected to the left.

Another way to think of the Coriolis effect is to ask whether objects are speeded up or slowed down relative to Earth's surface, considering the direction in which Earth is rotating. Earth rotates counterclockwise if you are looking down at it from the North Pole, or clockwise if you are looking up at it from the South Pole. Objects moving toward the equator in either hemisphere veer in a direction opposite to that of Earth's rotation because they are relatively slower than the ground at their destination, while those moving toward the poles are deflected in the same direction as the Earth's rotation because they are speeded up relative to the ground surface.

This means that air masses moving from 30° latitude toward the equator are deflected toward the southwest in the Northern Hemisphere and toward the northwest in the Southern Hemisphere—these are the trade winds. Air masses moving from 30° latitude toward the poles are deflected toward the northeast in the Northern Hemisphere and toward the southeast in the Southern Hemisphere—these are the westerlies. In turn, these wind patterns are the major reason that ocean currents move clockwise in the Northern Hemisphere and counterclockwise in the Southern Hemisphere.

---

with the westerlies farther north. These large-scale patterns of moving air masses have critical influences on world climates because they convey moisture and energy over great distances, influencing conditions in the lands they cross.

The three-dimensional conveyor belts of air that rise at the equator, traverse a distance, fall, and then move back again along Earth's surface are known as **Hadley cells** (Figure 18.7A). A similar (but weaker) process operates in polar regions, with air rising at about 60° latitude and sinking at the poles; polar regions also tend to be very dry. Between these two sets of Hadley cells in the upper atmosphere is the **jet stream**, a river of air that moves at high speeds from west to east. The Northern Hemisphere jet stream, especially its precise position at any given time, is one of the primary factors determining short-term weather patterns at mid-latitudes. The overall patterns of global air circulation are summarized in Figure 18.7B.

The movement of air masses at the Earth's surface is a major factor in determining surface oceanic currents (Figure 18.8), which in turn play a major role in terrestrial climates. Because of the high specific heat of water, these currents are even more effective conveyors of heat energy than the moving air masses. They bring water warmed or cooled far away to the coasts of all of the con-

We close by returning to the story of toilets and drains: Why don't they do what is so widely claimed? It's a matter of scale: water moving poleward across your sink cannot possibly be deflected very far by the Earth's rotation—unless you have a truly enormous sink!

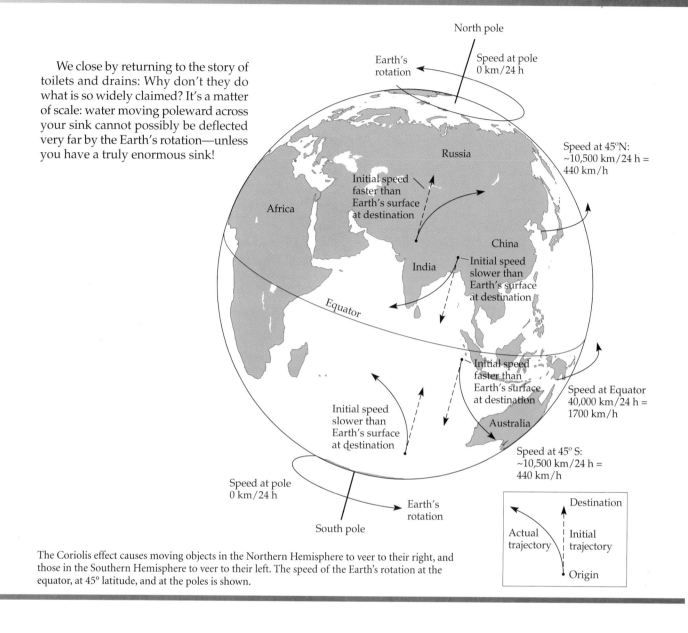

The Coriolis effect causes moving objects in the Northern Hemisphere to veer to their right, and those in the Southern Hemisphere to veer to their left. The speed of the Earth's rotation at the equator, at 45° latitude, and at the poles is shown.

tinents north of Antarctica and to most islands, adding or removing heat energy wherever they travel, and thus have a major influence on local temperature and moisture regimes.

The prevailing winds described above push masses of water, causing currents that can act like rivers of water with their own distinct temperatures, salinities, and nutrient concentrations. These oceanic currents move through the surrounding ocean waters. Like the winds pushing them, these currents are deflected to the right in the Northern Hemisphere and to the left in the Southern Hemisphere by the Coriolis effect.

Unlike air masses, currents cannot rise above the surface, so when they are blocked by continents, they are forced to go either north or south (Figure 18.8). This effect creates **gyres**, enormous circles of water, that move clockwise in the Northern Hemisphere and counterclockwise in the Southern Hemisphere. Major gyres exist in the North and South Atlantic, North and South Pacific, and Indian Oceans. The gyres carry heat energy to and away from the areas they traverse. The Gulf Stream in the North Atlantic, for example, brings tropical waters northeastward, warming the island of Bermuda and resulting in a far warmer climate in Ireland, Great Britain, and other parts of northern and western Europe than they would have based on their latitudes alone. The

(A)

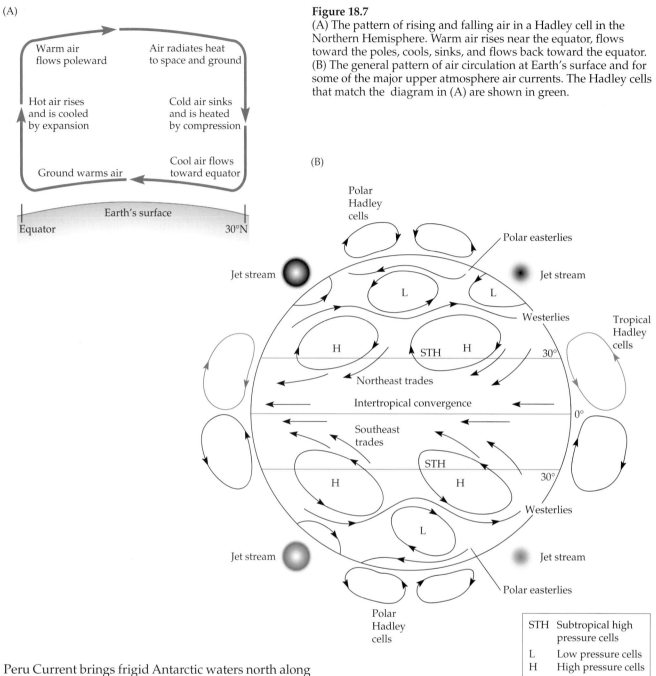

Figure 18.7
(A) The pattern of rising and falling air in a Hadley cell in the Northern Hemisphere. Warm air rises near the equator, flows toward the poles, cools, sinks, and flows back toward the equator. (B) The general pattern of air circulation at Earth's surface and for some of the major upper atmosphere air currents. The Hadley cells that match the diagram in (A) are shown in green.

Peru Current brings frigid Antarctic waters north along the western coast of South America, causing a cooling and drying effect there. We will consider some of the effects of these currents when we look at the El Niño Southern Oscillation and related phenomena later in this chapter.

Continental-Scale Patterns

Precipitation patterns at continental and local scales are strongly affected by two factors in addition to those just described: the distribution of mountain ranges and proximity to large bodies of water. When air moving over Earth's surface meets a mountain range, it has nowhere to go but up. This rising air, as explained pre-

viously, cools and loses moisture. The windward sides of mountains thus tend to be areas of high precipitation. Eventually the air is pushed over the tops of the mountains and falls down the other side. The falling air warms again, and (like the descending air at 30° latitude) tends to pick up and hold any remaining moisture. Thus, the leeward sides of mountains tend to be areas of low precipitation. These areas are known as **rain shadows** (Figure 18.9). This process of pushing air over mountains also explains why mountaintops tend to be windy places.

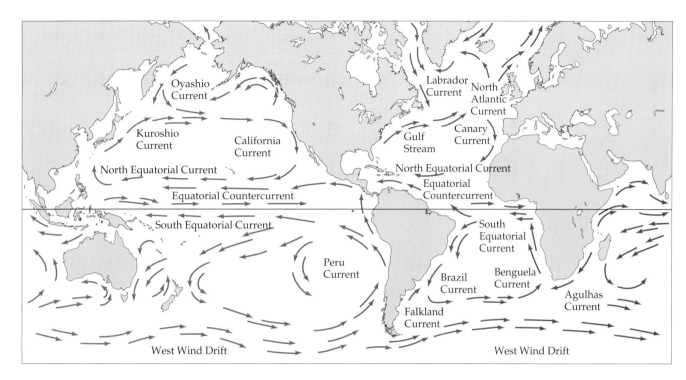

Figure 18.8
The world's major oceanic currents.

Rain shadows occur in many parts of the world and at various scales. One of the driest areas of the world, the Gobi Desert of central Asia, is in the rain shadow of the Himalayas. The Pampas grasslands of Argentina, at mid-latitudes in South America, are in the rain shadow east of the Andes caused by the prevailing westerlies. Farther north in South America, the air masses moving to the northwest across the Atlantic, laden with moisture, cross the eastern Amazon basin and are forced to rise when they reach the Andes at their eastern edge. This pattern results in very high rainfall on the eastern side of the Andes at latitudes north of 30° S, and extreme aridity on the western side of the Andes.

Even a single mountain can cast a rain shadow. For example, Mauna Kea on the island of Hawaii rises 4200 meters above sea level. The eastern face of the mountain receives as much as 750 cm of rain a year. Just 50 km away on the western face, less than 50 cm of rain falls in a year (Figure 18.10). This pattern is repeated on islands in the Caribbean with sufficiently high mountains—for example, the island of Puerto Rico has a rainforest on one side of its mountains and semi-arid conditions on the other side.

Land masses can be affected by their proximity to oceans or other large bodies of water in several ways. Air passing over the water can pick up moisture and deposit it on land as rain or snow, or the land can tend to lose moisture to the body of water, depending on relative air and water temperatures. One example of the first phenomenon is the "lake effect" around the Great Lakes of North America: winter storms tend to deposit much more snow as one moves eastward through this

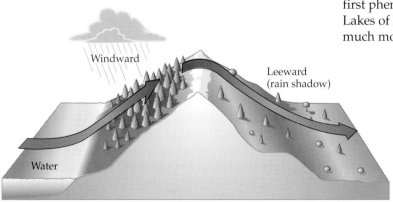

Figure 18.9
Rain shadows are caused when warm, moist air is forced upward and over a mountain or mountain range, depositing high levels of precipitation on the windward side. The rain shadow is created where dry air descends on the leeward side of the mountain.

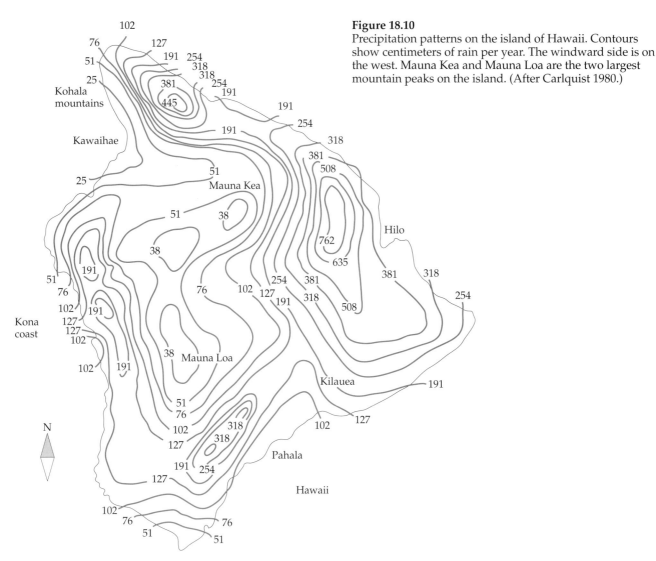

Figure 18.10
Precipitation patterns on the island of Hawaii. Contours show centimeters of rain per year. The windward side is on the west. Mauna Kea and Mauna Loa are the two largest mountain peaks on the island. (After Carlquist 1980.)

region because they gain moisture from the lakes. That is why Buffalo, New York (at the eastern edge of Lake Erie) is locally famous for its winter blizzards, while nearby Toronto, Ontario (on the north shore of Lake Ontario, and thus out of the way of most storms off the lakes) receives few blizzards.

At a larger scale, when air masses cooled by cold oceanic currents travel to lower latitudes and meet warm land masses, the air warms as it passes over the land, absorbing moisture and preventing precipitation. If this air veers offshore, it can even tend to lose moisture to the ocean, compounding the drying effect on the land. This phenomenon occurs to some degree on the western coasts of all continents, all of which have cool currents offshore. The coastal Atacama Desert in Chile, for example, is also affected by the cold Peru Current, and as a consequence it is exceptionally arid. Some areas in that desert have been rain-free for stretches of ten years,

although just offshore there are cool, foggy air masses. Coastal southern California also has limited rainfall (though its vegetation is chaparral rather than desert) because the cold California Current prevents oceanic air masses from delivering rain during much of the year. In winter, the land cools enough to allow rainfall there some of the time. Farther north, however, the land is much cooler, and the cool, moist prevailing westerlies dump large amounts of precipitation, creating temperate rainforests in Oregon, Washington, and British Columbia. Even there, much of the rain falls in winter when the ground is coldest.

Air circulation patterns and geographic features interact to produce continental-scale patterns of precipitation (Figure 18.11). In North America, for example, the major mountain ranges (Appalachians, Rockies, and Sierra Nevada) run in a north-south direction. The primary direction of air flow at mid-latitudes, as we have just seen, is west to east. These factors combine to produce a distinct pattern of precipitation across the continent.

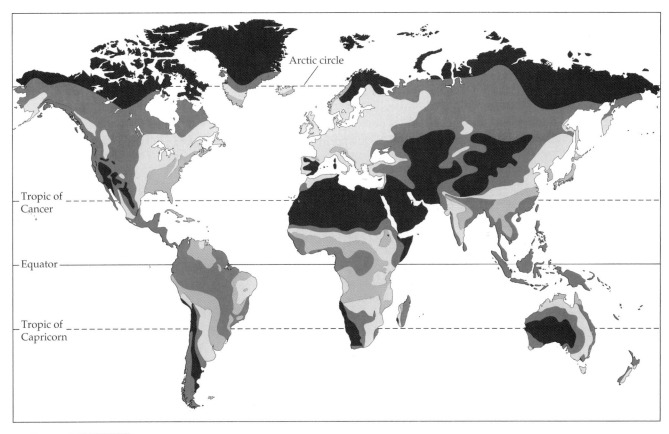

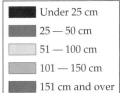

■	Under 25 cm
■	25 — 50 cm
░	51 — 100 cm
▒	101 — 150 cm
▓	151 cm and over

Figure 18.11
Global patterns of mean annual precipitation. (After Rumney 1968.)

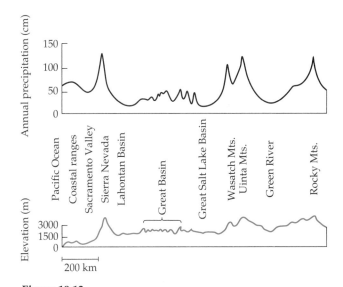

Figure 18.12
Transects showing the patterns of elevation and precipitation across western North America at 40° latitude (from the coast of northern California, through Nevada and Utah, and ending near Boulder, Colorado). (After U.S.D.A. Yearbook 1941.)

Along the West Coast, from northern California to southern Alaska, there is a region of high precipitation where moisture-laden air coming in from the Pacific runs into mountain ranges (and the ground is cool enough to allow precipitation, at least in the cooler months). To the east of these mountains it is dry. The Intermountain Desert of Nevada, for example, is in the rain shadow of California's Sierra Nevada (Figure 18.12). A second area of relatively high precipitation occurs on the western side of the Rockies. To the east of the Rockies there is another rain shadow. Precipitation levels gradually increase toward the Atlantic seaboard. The increased moisture in the Midwest and east comes primarily from moist air moving up from the Gulf of Mexico (Figure 18.13). The Appalachian Mountains are not high enough to create a rain shadow.

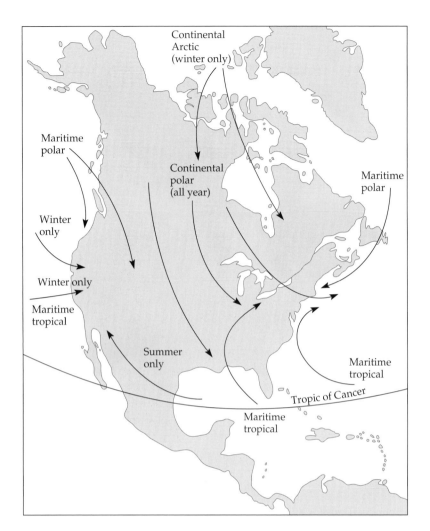

Figure 18.13
Principal air masses and their direction of movement at different seasons in North America. (After Rumney 1968.)

Seasonal Variation in Precipitation

Variation in precipitation, as in temperature, occurs on several different time scales. On a scale of days, precipitation varies as weather systems move across the globe. These weather systems are caused by the movement of high and low pressure areas. Precipitation usually occurs at the leading edge of high pressure systems. On average, these systems take several days to move through an area, creating alternating wet/dry cycles. Weekly and seasonal variation in larger-scale phenomena, such as the position of the jet stream, causes variability on a different spatial and temporal scale.

The most important scales of variation in precipitation are seasonal and yearly. Annual variation in temperature helps to drive variation in precipitation. Several different kinds of annual patterns of precipitation exist. Along western continental margins there are distinct seasonal patterns, with rainy winters and dry summers (Figure 18.14A). Where temperatures are warm, this pattern is often referred to as a **mediterranean** cli-

mate, as it is typical of areas around the Mediterranean Sea. In North America, mediterranean climates are found in coastal southern and central California; other regions that exhibit this pattern include western Australia, southern Chile, and the southwestern tip of Africa. **Continental** climates are found in the interior of large continents at mid-latitudes, as in the North American Midwest and central Russia. Precipitation here is also highly seasonal, with the greatest amount in the summer, although there is generally some precipitation year round (Figure 18.14B). **Maritime** climates are often found along the eastern edges of continents. Here precipitation is more uniform throughout the year, although it comes as rain in the summer and snow in the winter (Figure 18.14C). In equatorial regions, temperatures are high throughout the year. Precipitation in the Tropics may show one of three patterns: relatively constant precipitation throughout the year, a distinct wet season and dry season, or two wet/dry cycles in a year (Figures 18.14D–F; see also Figure 19.6).

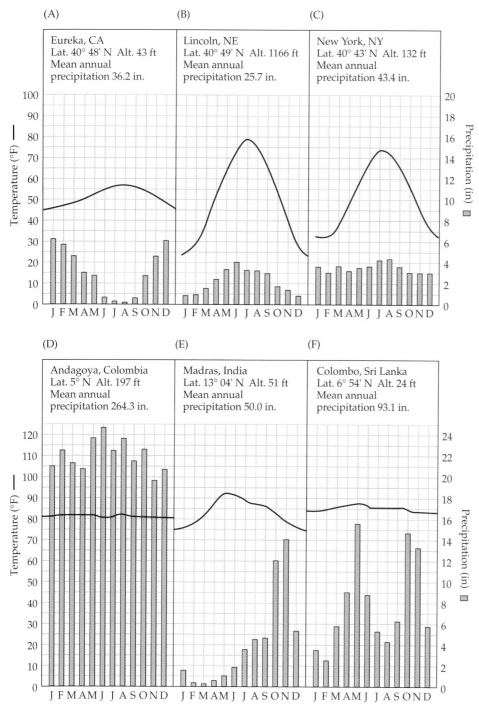

Figure 18.14
Average monthly temperature and precipitation for various locations across the globe. (A–C) Patterns for three locations at approximately 40° latitude in North America. (A) Eureka, California, near the Pacific Coast, has a mediterranean climate. (B) Lincoln, Nebraska, at midcontinent in the Great Plains, has a continental climate. (C) New York City, on the Atlantic Ocean, has a maritime climate. (D–F) Three tropical locations with different patterns of precipitation. (D) Andagoya, Colombia, in South America, has constant precipitation year-round. (E) Madras, India, in southern Asia, has a rainy and a dry season. (F) Colombo, Sri Lanka, has two distinct rainy seasons, separated by drier periods. (After Rumney 1968.)

Figure 18.15
Pacific Ocean events (A) during a normal year and (B) during an El Niño year. Events just above and below the ocean surface, in the deeper waters below the surface, and in the upper atmosphere, as well as the consequences for rainfall patterns, are shown.

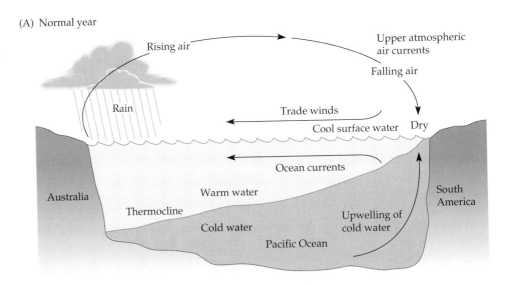

(A) Normal year

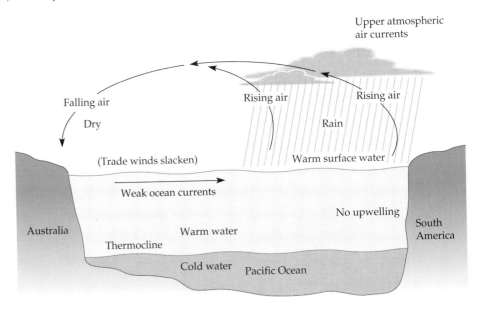

(B) El Niño year

The El Niño Southern Oscillation

Besides these weekly and yearly cycles in precipitation, other cycles also exist. Of these, the most important is a 3–7 year cycle called the **El Niño Southern Oscillation (ENSO)**. Narrowly defined, the ENSO determines weather patterns in a band from 20° N to 20° S across the Pacific Ocean. However, ENSO is closely interconnected with other global atmospheric and oceanic effects. Together, ENSO events and their associated effects drive some of the most dramatic global fluctuations in weather, although the strength of the events, and the magnitude of their effects, can vary substantially.

ENSO effects occur over a very large geographic scale, influencing temperatures and rainfall from Africa to Australia, across the Pacific, and all the way across North and South America, affecting by some estimates about 75% of the globe (among the few places that are *not* much affected are northern Europe and northern Asia, although parts of southern Asia are strongly affected). Remarkably, climatologists began to fully recognize and study this phenomenon only during the record El Niño year of 1982–1983. The study of ENSO and the role of the Pacific Ocean in this cycle has transformed our understanding of world weather patterns, and climatologists are still refining their understanding of all of its effects. These effects may also be changing due to global warming (see Chapter 22).

El Niño (literally "the boy") is named for the Spanish term for the infant Jesus, because its effects were initially noticed off the coast of Peru beginning around Christmas. El Niño is only one half of the ENSO; it alternates with La Niña ("the girl") years, during which conditions are reversed.

The heart El Niño is a cycle of tropical Pacific oceanic and atmospheric conditions. During an ordinary year, the trade winds in both hemispheres blow west across the Pacific, pushing strong ocean currents along in the same direction (Figure 18.15A). The water being pushed away from the western coasts of North and South America is replaced with very cold, nutrient-rich water dragged up from the ocean bottom. This upwelling of cold bottom water in the eastern Pacific supports extremely productive marine food webs as well as affecting global climates. Strong trade winds push warm surface waters into the far western Pacific north of Australia. Warm water has increased evaporation, bringing rainfall to the islands of the western Pacific and to eastern Australia and Asia.

When an El Niño year commences, the trade winds slacken (Figure 18.15B). As a result, upwelling in the eastern Pacific ceases, with huge negative repercussions for fish populations, marine mammals, seabirds, and other organisms that are ultimately dependent on the nutrient-rich bottom water. Because this cold water never makes it to the surface, warm surface water now spreads across the tropical Pacific from east to west. Surface temperatures in the central Pacific Ocean become as much as 5°C warmer than normal, leading to drastic decreases in rainfall in the western and central Pacific.

As an example, the very severe El Niño year of 1997–1998 (Figure 18.16) resulted in extreme drought in Australia, Indonesia, and southern Africa.

Warm, moist air now rises above the eastern Pacific, causing heavy rainfall in ordinarily dry regions such as southern California, northern Chile, and northward to the coasts of Peru and Ecuador. This rain can result in severe flooding and massive landslides. Finally, a complex change in high-altitude winds 12 km over the Pacific forces a jet of air to move eastward, past Central America, and then across the Atlantic to Africa, depressing hurricane activity in the Atlantic and causing other disturbances in normal weather patterns. These changes in upper-level atmospheric winds in the southern Pacific also force changes in the jet stream in northern mid-latitudes, spreading El Niño's effects on weather to a far greater portion of the globe.

Eventually, these conditions reverse. During La Niña years, at the opposite part of the ENSO, strong trade winds push warm surface waters into the far western Pacific north of Australia. Evaporation is intensified in the eastern Pacific, further strengthening the trade winds. Meanwhile, the movement of surface waters eastward causes a strong upwelling of cool, deep waters in the western Pacific along the coast of South America. The Equatorial Countercurrent, which flows counter to the prevailing winds, is relatively weak. The effects on weather patterns are essentially the opposite of those in an El Niño year. At some point, the positive feedback that maintains this system reverses, and El Niño conditions return again.

ENSO is a succession of alternating events. We do not completely understand what triggers the changes between El Niño and La Niña conditions, but the ENSO clearly depends on a complex interplay between ocean

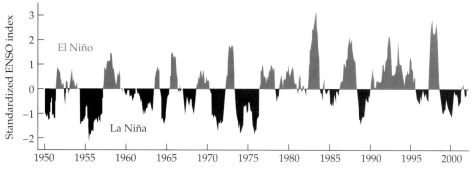

Figure 18.16
Historical record of intensity of El Niño (above the midline of the graph) and La Niña events (below the midline) from 1950 to 2001, based on a standardized multivariate index that includes ocean surface temperatures, surface air temperatures, wind characteristics, and other variables. (After NOAA-CIRES Climate Diagnostic Center, http://www.cdc.noaa.gov)

currents and winds. Ocean waters are always exchanging energy (as heat) with the atmosphere, although they hold far more heat, and move heat around the globe more slowly, than do atmospheric currents. This exchange of heat energy with the ocean drives the movement of winds, and the winds drive ocean currents.

How do these complex changes in atmospheric and oceanic interactions affect weather and organisms in different places? In North America, El Niño events may be linked to milder winters along the U.S.-Canada border, increased winter storms along the California coast, floods in the southeastern United States, increased snowfall in the southwestern mountains, and greatly decreased hurricane activity in the Atlantic. La Niña also can have strong effects on weather patterns in North America, including increases in reported tornado strength and frequency in the Midwest, stronger and more frequent hurricanes in the Atlantic, and drought and forest fires in the Southwest.

Globally, the El Niño of 1982–1983 caused droughts in Africa, Australia, and South America. Coral reefs off Costa Rica, Panama, Colombia, and the Galápagos Islands suffered drastic losses (50%–97% die-offs) because of the warmer water and loss of nutrients. In that same event, 85% of the seabirds on the Peruvian

coast died or abandoned their nests; the same fate befell almost all of the 17 million seabirds living on tiny Christmas Island in the Pacific.

The El Niño of 1997–1998 was even more extreme, with the resulting drought exacerbating forest fires in Indonesia and Malaysia and harming the already threatened rainforests there; the rainforests of the Amazon basin also suffered from extreme drought. In many places, the unusual severity of this particular El Niño, coupled with other problems facing these tropical forests (such as logging and fragmentation), combined to cause much greater damage than any of these factors would have by themselves (Wuethrich 2000). Other areas affected by the drought included Australia and southern Africa, while severe flooding occurred in southern California and in the coastal deserts of northern Chile and Ecuador.

These cycles also affect plant population dynamics. Because the U.S. Southwest normally has low amounts of precipitation, this region provides a sensitive context for studying these effects. The ENSO effects both precipitation and temperature, as can be seen by looking at the climate record for the past century at Las Cruces, New Mexico (Figure 18.17). Periods of dry winters combined with wet summers in this area are asso-

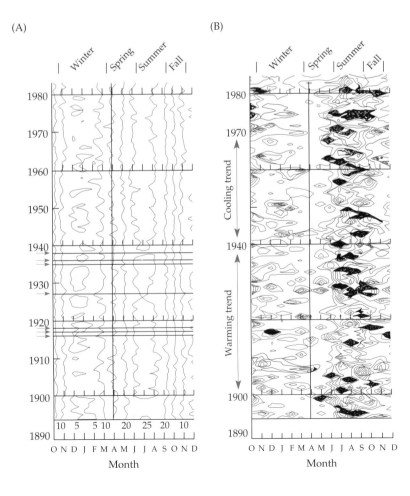

Figure 18.17
Patterns of temperature and precipitation for the past century at Las Cruces, New Mexico. (A) Contour map of average monthly temperature (contour interval 5°C). Colder winters are indicated by larger areas within the 5° contour, such as the period from 1928 to 1934. Warmer summers are indicated by larger areas within the 25° contour, such as the period from 1927 to 1937. Arrows followed by solid lines indicate years with large numbers of seedlings of *Bouteloua eriopoda* (black grama grass, Poaceae). (B) Contour map of average monthly precipitation (contour interval 10 mm). Wet periods are indicated by the dark triangular-shaped areas, such as the period from 1965 to 1975. (After Neilson 1986.)

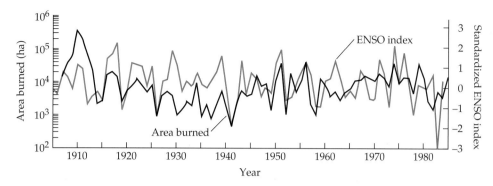

Figure 18.18
Annual area burned in Arizona and New Mexico (black line) and an index of the intensity of the El Niño Southern Oscillation (ENSO) (green line) from 1905 to 1985. (After Swetnam and Betancourt 1990.)

ciated with range increases of the native *Bouteloua eriopoda* (black grama grass, Poaceae; Neilson 1986). Periods with dry spring weather are associated with large-scale forest fires (Figure 18.18).

Short-term oscillations such as ENSO are embedded in larger, long-term phenomena. Over the past 5000 years there may have been periods of more frequent El Niño events from 4800 to 3600 years before present (B.P.), around 1000 B.P., and after 500 B.P. These periods also appear to be associated with increased flooding in the southwestern United States (Ely et al. 1993).

Predictability and Long-Term Change

In this chapter we have described patterns of variation in temperature and precipitation. These patterns are generally regular, with some predictability. Random effects act to introduce unpredictability into these patterns. The degree of unpredictability is one determinant of the types of plant communities found at different locations (Figure 18.19). Uncertainty is also an important determinant of some kinds of adaptations. For example, predictable variation is more likely to favor adaptation by phenotypic plasticity (see Chapter 5), while unpredictable variation is more likely to favor adaptation by a jack-of-all-trades strategy.

The distinction between random and predictable patterns of change depends on the life span of the plant relative to the rate at which the climate changes. The ENSO, for example, varies enough in its details (e.g., intensity and length) that for an annual plant it is effectively random, while for a *Sequoia sempervirens* (redwood, Cupressaceae) that lives for centuries, it forms a predictable cyclic pattern.

At the longest time scale—tens of millions of years—continental drift and tectonic activity play a major role in changing climate. Mountain building, for example, creates rain shadows. The spread of grasslands across the middle of North America is associated with the rise

of the Rocky Mountains during the Eocene 34 million years ago. As continents drift across the face of the Earth, their climates change. For the past 100 million years, Australia has been moving northward; as a result, its climate has changed from mainly temperate and polar to mainly tropical and subtropical. Changes in atmospheric CO_2 concentrations, both natural and anthropogenic, also have had large effects on global climate in the past as well as currently. We discuss these long-term patterns in detail in Chapters 21 and 22.

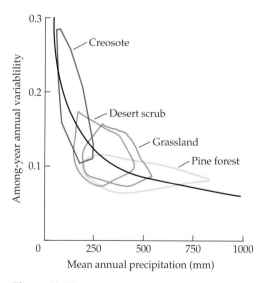

Figure 18.19
Association between the variation and predictability of rainfall and the presence of different types of plant communities in Arizona. Desert areas dominated by creosote bush (lower Sonoran Desert) or desert scrub (upper Sonoran Desert) have low amounts of precipitation and high year-to-year variation. As precipitation amounts increase and variation decreases, plant communities become dominated by grasses and then pine forest. (After Davidowitz 2002.)

Plant Physiognomy across the Globe

A continent is a complex mosaic of climates and vegetation forms. Across a continent, this variation falls into regular patterns. Here we introduce several of these broad patterns. The first of these is a north-south gradient in vegetation form due to temperature. A second major pattern is variation from west to east in response to changes in average precipitation. Such patterns exist on all continents because of the latitudinal differences in solar radiation and the movement of weather systems described earlier in this chapter. While we concentrate here on North America, and while each continent is unique, the principles behind these patterns can be applied to the other continents as well. This variation will be addressed in greater detail when we describe the world's biomes in the next chapter.

Forests

Forests are communities dominated by trees whose leaves touch each other, resulting in a closed canopy. If you were to travel southward in eastern North America from northern Canada to the Carolinas, Georgia, and northern Florida, you would pass through a variety of different-looking forests. Forest composition is affected by both climate and soil; to focus on the effects of climate, we confine our discussion to the relatively rich soils found inland (away from the nutrient-poor coastal plain).

Immediately to the south of the tree line you would find taiga or boreal forest, which is dominated by coniferous, evergreen needle-leaved trees. Next you would enter a region dominated by broad-leaved trees—deciduous angiosperms that drop their leaves in winter—although conifers would still be present. Finally, you would come upon broad-leaved evergreen forests dominated by angiosperms that retain their leaves all year. In summary, there is a transition from evergreen forests to deciduous forests and back to evergreen forests, although the type of evergreen tree changes from needle-leaved to broad-leaved, and from gymnosperms to angiosperms. These changes in tree type with latitude are largely driven by differences in seasonal temperatures. Of course, the description of this imaginary trip is very general; for the moment, we are ignoring many other details to concentrate on temperature effects. Similar patterns are found, for example, in eastern Asia as one travels from Korea and northeastern Russia to the south of China and beyond to Southeast Asia.

To understand why these changes occur, let's reverse our trip. Beginning in southeastern North America, growing seasons are longest in the broad-leaved evergreen forests found along the Atlantic and Gulf coasts. These areas are classified as **subtropical** because they rarely experience temperatures below freezing. Tropical

and subtropical forests can be dominated by either deciduous or evergreen angiosperms.

The critical climatic factor determining which one occurs in a given area is seasonality of precipitation. Areas that receive rain throughout the year are evergreen, while areas with pronounced dry periods are deciduous. As always, there are exceptions to this general pattern. For example, many forests in the southeastern United States are dominated by a particular genus of needle-leaved evergreen gymnosperms: *Pinus* (pines). These pine forests are typically found on soils that are very low in nutrients and are subject to frequent fires. (The effects of soil nutrients on leaf life span are considered in the context of plant strategies in Chapter 9.)

Moving north, the transition from broad-leaved evergreen to broad-leaved deciduous forests depends on the likelihood of freezing temperatures. Broad-leaved evergreen trees are vulnerable to frost through two of its effects. First, leaf tissues will die if kept frozen. Second, the weight of ice or snow on leaves will cause them to be torn off or, worse, will break branches and even tree trunks. A very unusual October snowstorm in 1997 in the Midwest and Great Plains of the United States, which occurred before the leaves had fallen, caused major damage to deciduous trees as branches and even trunks snapped under the weight of the snow-burdened canopies. Tall, narrow conifers such as spruces (*Picea*) and firs (*Abies*) shed snow far more effectively. So it is no surprise, then, that the boundary of broad-leaved evergreen trees corresponds with that of freezing temperatures (Figure 18.20).

Farther north, the transition between broad-leaved deciduous and needle-leaved evergreen forests is related to growing season length. An evergreen tree can take advantage of short periods of favorable conditions as long as there is enough unfrozen moisture in the soil to support transpiration (a major limitation for these trees). Conifers are able to avoid leaf damage due to freezing, by increasing the osmotic concentration of intracellular water (thereby lowering its freezing point). Ice crystals form only between the cells (which does little damage to membranes); this acts to draw water out of cells, which further increases the osmotic concentration inside the cells.

In contrast, a deciduous tree requires time to produce new leaves in the spring before it can begin photosynthesizing. When the new leaves are emerging, they are particularly vulnerable to injury from freezing. Thus, leaf emergence does not occur until the chance of a late frost is very small. Following leaf emergence, the growing season must be long enough for the tree to accumulate enough carbon for its maintenance, growth, and reproduction. Thus, deciduous trees are able to survive only in regions with a relatively long growing season.

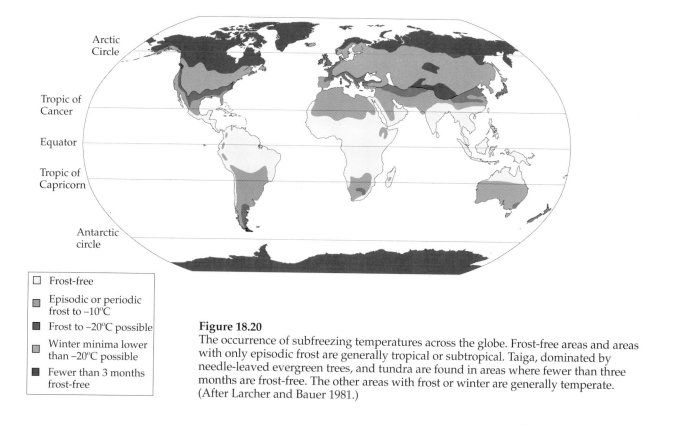

Figure 18.20
The occurrence of subfreezing temperatures across the globe. Frost-free areas and areas with only episodic frost are generally tropical or subtropical. Taiga, dominated by needle-leaved evergreen trees, and tundra are found in areas where fewer than three months are frost-free. The other areas with frost or winter are generally temperate. (After Larcher and Bauer 1981.)

Legend:
- Frost-free
- Episodic or periodic frost to –10°C
- Frost to –20°C possible
- Winter minima lower than –20°C possible
- Fewer than 3 months frost-free

The slow leafing out of deciduous trees in the spring has another effect on the form of deciduous forests: These communities are much more complex in form and higher in species diversity than boreal forests. Because the canopy is deciduous, there are times of year when large amounts of light penetrate to the forest floor. Consequently, there can be up to four layers of vegetation in deciduous forests: herbaceous species along the ground, small shrubs, small trees and large shrubs, and canopy trees. In the spring, the forests turn green from the ground up, with each layer leafing out slightly later. This ground-up effect occurs because the air is warmer near the ground, especially early in the season before the canopy leafs out and reduces convective cooling. The diversity of species and form in deciduous forests, therefore, depends on seasonal variation in temperature.

The evergreen needle-leaved trees that dominate the northern coniferous forests pay a cost for their ability to withstand freezing temperatures: they grow much more slowly than broad-leaved deciduous trees. Once the growing season is sufficiently long, deciduous trees outcompete needle-leaved evergreen trees. The border between the deciduous forest and the needle-leaved evergreen forest corresponds to a temperature line north of which the minimum temperature during the winter falls below –40°C (Arris and Eagleson 1989; compare Figures 18.20 and 19.1). Above this temperature and

below 0°C, plant tissues can supercool, meaning that they can cool without ice formation. It is the formation of ice crystals in plant and animal tissues that is the primary source of damage and injury from freezing.

Tree Line

Temperature has a direct effect on the types of vegetation that can grow in a region. One of the most striking examples of this effect is the alpine **tree line**, an often abrupt boundary where trees are replaced by low-growing vegetation (Figure 18.21). Tree line is most obvious in high mountain ranges, such as the Sierra Nevada and Rockies in North America, the Alps in Europe, the Andes in South America, or the Himalayas in Asia, all of which have forests that stop below the summit. If you were to hike up such mountains, you would reach a point where the forest would thin and the trees would get shorter. At some point, the trees would disappear entirely. A similar sight would greet you at the Arctic or Antarctic tree line on a hike toward the North Pole in North America or Eurasia, or toward the southern tip of South America.

Tree lines are caused by a complex of several factors, including the limits to supercooling mentioned above and others detailed in Chapter 19. In addition, wood will form only when temperatures are above 10°C. Microenvironmental differences in temperature can lead to inter-

Figure 18.21
Tree line at about 1200 m in Rocky Mountain National Park, Colorado. (Photograph by S. Scheiner.)

esting effects, such as the formation of **krummholz** (Figure 18.22), a peculiar tree form found at tree line on mountains in different parts of the world.

Above tree line on Mount Washington, New Hampshire—the highest peak in eastern North America—you might notice an odd-looking plant. It looks like a horizontal woody vine growing along the ground (Figure 18.23). If you examine it closely, however, you will find that it is *Picea mariana* (black spruce, Pinaceae), the dominant tree species in the nearby forest. Seeds taken from these plants, if planted farther down the mountain, will grow into tall trees! The krummholz form of the plant comes about because the temperature right at ground level is just a little bit warmer than the air above it—just enough above 10°C for wood formation (Teeri 1969). The warmer temperatures near the ground are caused by reduced convective cooling and longwave radiation by the ground of the energy it absorbs from sunlight.

Grasslands and Woodlands

Traveling from west to east, starting in the rain shadow on the eastern face of the Rocky Mountains and traversing the center of the United States, you will see another gradual transition in the physiognomy of the vegetation. Your journey begins in short-grass prairie, in the midst of low-growing clumps of grasses, with bare patches interspersed between the clumps. As you travel eastward, the vegetation becomes taller and thicker as you reach the midgrass prairie. Here you may notice fewer grasses growing in clumps and more rhizomatous grasses, as well as many dicots.

By the time you have reached Iowa, you are in tallgrass prairie. In addition to taller vegetation, these communities have greater biomass; they are also more productive and more diverse. Besides the obvious grasses, there are many dicot forbs whose stature matches that of the grasses. This shift in plant stature and diversity closely follows a pattern of increasing rainfall.

As you continue eastward, more and more trees are evident. In the areas dominated by grasslands, trees are clustered along streams. Farther east, they begin to dot the countryside, becoming more dense. These areas are **woodlands**, dominated by trees, but without a closed canopy. Eventually, at about the Illinois/Indiana border, you reach forests.

This transition from grassland to woodland to forest is a function of changes in both the amount and seasonality of precipitation. Annual rainfall, for example, is about 80 cm both in the middle of Kansas and to the northeast in the middle of Minnesota, yet the former is in the center of the tallgrass prairie, while the latter is forested. The difference is that in Kansas, the rainfall is much more seasonal, with more of it occurring in spring and fall, the temperature is warmer, and the chance of drought is greater.

This seasonality of precipitation and propensity for drought leads to another key factor: fire (see Chapter 13). Under natural circumstances, prairies experience fires

Figure 18.22
Krummholz vegetation at treeline in Rocky Mountain National Park, Colorado. Note the twisted trunk of the tree. (Photograph by S. Scheiner.)

Figure 18.23
Picea mariana (black spruce, Pinaceae) growing in a "vine" form above tree line on Mount Washington, New Hampshire, the highest peak in eastern North America. (Photograph courtesy of J. Teeri.)

every 3–5 years, depending on climatic and landform conditions. Each year dry grasses and other plant material accumulate until there is enough fuel to sustain a conflagration, and lightning (or human activity) ignites a fire. Prairie plants are adapted to these frequent fires. Their vulnerable meristems are located at or below ground level. A fast-moving fire will burn off the tops of the plants, but they quickly resprout. Trees, however, have their meristems at the tips of their branches. Seedlings and young saplings are particularly vulnerable to fire. Grasslands begin to give way to woodlands where the frequency of fire drops low enough that some trees have sufficient time between fires to grow tall enough to become resistant to fire. Even so, these woodlands are dominated by species that have thick bark and are tolerant of fire. The woodlands become closed-canopy forests where precipitation levels are high enough that fire frequency becomes low.

Shrublands and Deserts

Shrublands are another widespread physiognomic form dominated by low-growing woody plants with multiple stems. The locations of these communities are determined by a combination of temperature and precipitation effects. They are primarily found in dry to very dry regions. Areas with mediterranean climates, for example, contain a mixture of shrublands and woodlands. Which type of vegetation predominates in an area

depends on disturbance. Fire often favors shrubs, which can resprout readily after being burned, over trees. Annuals are also very common in shrublands and are especially prevalent in the years immediately following a fire. Other disturbances can also favor shrubs over trees. Today grazing by domesticated animals is the most important factor spreading and maintaining shrublands in subtropical regions of the world.

The other major climatic regions with extensive shrublands are **deserts,** which are regions where potential evapotranspiration exceeds actual evapotranspiration. The amount of precipitation and its seasonality are among the most important factors in determining what kinds of plants thrive in very dry environments (Figure 18.24). Summer and winter precipitation represent very different sources of water for plants. At summer temperatures, rain only wets the soil surface, often for only a few days. At winter temperatures, precipitation can infiltrate deeper soil layers and remain available there throughout the year.

Wet winters with dry summers favor deep-rooted, long-lived plants with large leaf areas, such as trees. These plants use almost no summer rainwater, even when it is available. As the amount of winter water declines, plants with greater shoot biomass relative to the amount of root biomass, such as shrubs, are favored. As winter water further decreases, summer water becomes increasingly important, favoring plants with shallower root systems and even less root material, such as herbaceous annuals and perennials. These plants are capable of extracting surface water quickly, before it is lost to evaporation.

When there is no winter water stored deep in the soil, only succulent plants such as cacti, which store water internally and lose it slowly, can survive between summer rain events. The more often it rains in summer, the less need to store water, and the community becomes increasingly more herbaceous. If wet summers are accompanied by moderately dry winters, herbaceous, shallow-rooted plants, such as grasses or summer annuals, are favored.

These effects of the amount and seasonality of precipitation in deserts explain the differences in the physiognomy of the hot deserts of western North America: the Mojave, the Sonoran, and the Chihuahuan. The Mojave Desert (mainly in California) gets almost exclusively winter precipitation. The Chihuahuan Desert to the southeast (mainly in the Mexican states of Chihuahua and Coahuila) gets almost exclusively summer precipitation, whereas the Sonoran Desert between the two in Arizona and Sonora gets both summer and winter precipitation. As a result, while all three regions have many shrubs, the Mojave Desert is dominated by low-growing shrubs (many of which are deciduous) and has a very large number of annual plant species, the Chihuahuan Desert is dominated by evergreen shrubs and grasses, and the Sonoran Desert has a mixture of many growth forms. In all of these deserts, however, several growth forms coexist; no one growth form can preempt all of the available water.

Summary

The climate of an area is determined by the mean, variability, and seasonality of temperature and precipitation. Mean temperature is determined by the amount of incoming solar radiation, which varies daily, seasonally, yearly, and over the course of centuries. At the equator solar radiation inputs are generally high and even throughout the year; in contrast, solar radiation inputs at high latitudes vary greatly over the year. At much longer time scales—tens to hundreds of thousands of years— changes in Earth's orbit around the sun and in the tilt of Earth's axis also create variation in solar radiation.

Patterns of precipitation are caused by differential heating and cooling of the atmosphere at different lati-

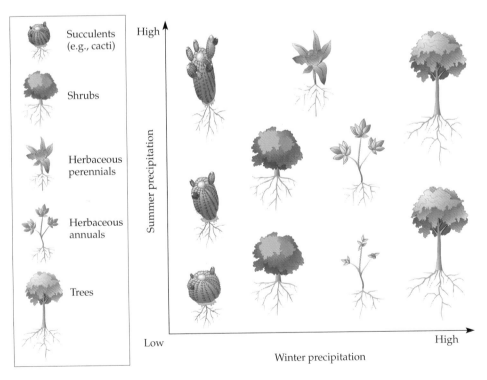

Figure 18.24
Plant types favored by various combinations of winter and summer precipitation.
(After Schwinning and Ehleringer 2001.)

tudes. At the equator, warm rising air creates a region of high precipitation. That air tends to descend at 30° north and south of the equator, creating regions of low precipitation there. A second area of low precipitation is found near the poles. In addition, the prevailing winds push masses of water, creating ocean currents that carry heat energy to and away from the areas they traverse.

A second pattern is caused by the flow of air across continents and its interaction with mountain ranges. Areas on the upwind (western) sides of mountain ranges tend to be wet, while the downwind (eastern) sides of mountain ranges tend to be dry. The north-south–running mountain ranges of North America create a distinct pattern of alternating wet and dry areas across the continent. Variation in precipitation occurs on time scales of days, seasons, years, and decades. This variation also contains an element of unpredictability that is especially important in extreme climates.

It is the combination of the mean climate of a region, the variability of that climate, and the predictability of that variation that determines the types of plants that will grow there. Trees can grow only where temperatures exceed 10°C for sufficient periods of time. Warm temperatures and long growing seasons favor broad-leaved deciduous trees over needle-leaved evergreen trees. Trees require high amounts of rainfall, however. Areas of lower rainfall are dominated by shrubs or grasses. Succulents are favored in areas with unpredictable summer rainfall. The border between woodlands, shrublands, and grasslands is often controlled by rates of disturbance, especially the frequency of fire. Fire frequency, in turn, is controlled by patterns of temperature and precipitation. Thus, an understanding of climatic patterns across the globe provides the basis for understanding patterns of vegetation. We explore those patterns in more detail in the next chapter.

Additional Readings

Classic References

Beard, J. S. 1955. The classification of tropical American vegetation-types. *Ecology* 36: 89–100.

Gates, D. 1962. *Energy Exchange in the Biosphere.* Harper & Row, New York.

Holdridge, L. R. 1947. Determination of world plant formations from simple climatic data. *Science* 105: 367–368.

Mather, L. R. and G. A. Yoshioka. 1968. The role of climate in the distribution of vegetation. *Ann. Assoc. Am. Geog.* 58: 29–41.

Contemporary References

Ji, J. 1995. A climate-vegetation interaction model: simulating physical and biological processes at the surface. *J. Biogeog.* 22: 445–451.

Neilson, R. P. 1986. High-resolution climatic analysis and southwest biogeography. *Science* 232: 27–34.

Schwinning, S. and J. R. Ehleringer. 2001. Water-use tradeoffs and optimal adaptations to pulse-driven arid ecosystems. *J. Ecol.* 89: 464–480.

Additional Resources

Heinrich, W. 1973. *Vegetation of the Earth in Relation to Climate and the Eco-Physiological Conditions.* Springer-Verlag, New York.

Rumney, G. R. 1968. *Climatology and the World's Climates.* Macmillan, New York.

CHAPTER *19* *Biomes*

Our planet has a complex pattern of climates, which in turn has a major role in creating the complex pattern of vegetation and community types we find on Earth. In the previous chapter, we looked at climate and how it affects plant physiognomy. Here we look at the resulting vegetation patterns in more detail. Ecologists divide these large-scale patterns into units called **biomes**: major biogeographic regions that differ from one another in the structure of their vegetation and in their dominant plant species (Clements 1916). Biomes represent the largest scale at which ecologists classify vegetation. In this chapter, we examine these broad-scale patterns, look briefly at the world's major biomes, and briefly consider some of the ways in which human activities affect them.

Categorizing Vegetation

Classification of the world's vegetation into biomes (Figure 19.1; Table 19.1) provides useful categories that describe major aspects of vegetation, such as its structure, function, and adaptations. Not only do biomes tell us a great deal about what kinds of plants might be found growing in a given area, they also indicate something about the types of animals and other organisms likely to be living there, and about the major environmental constraints on living things. Similar climates in different parts of the world contain similar biomes (Figure 19.2), although their details differ. For example, temperate evergreen forests are similar in being dominated by trees that stay green all year, but differ in being dominated by needle-leaved conifers in North America, by broad-leaved beeches in South America, and by broad-leaved eucalypts in Australia.

Biomes are defined by the physiognomy of the dominant or most obvious plants. Thus, we recognize different types of forests (e.g., evergreen or deciduous, broad-leaved or needle-leaved), shrublands, and grasslands. However, we must not lose sight of the trees for the forest: there may be a tremendous amount of variation within a given biome. In her study of communities within a single biome, the temperate deciduous forest of eastern North America, E. Lucy Braun recognized 12 different major sub-biomes and a number of minor variants (Braun 1950). Within a biome, there may be patches of vegetation that appear "not to belong," such as riparian forests found along streams in grasslands or deserts. The occurrence of these patches reminds us that climate is only one factor in determining vegetation. Local variation in soils and topog-

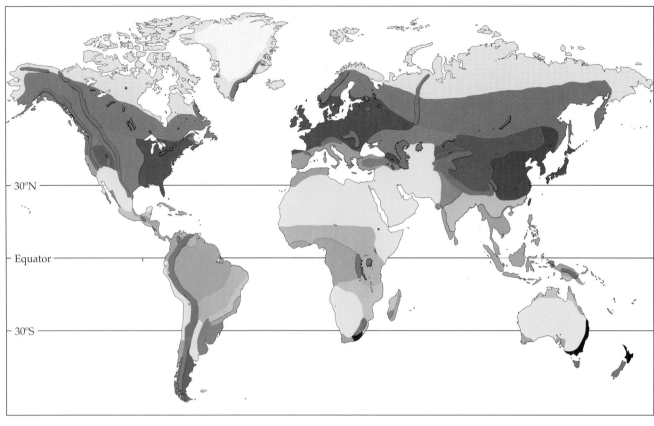

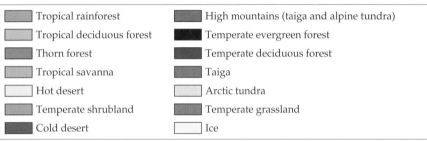

Tropical rainforest	High mountains (taiga and alpine tundra)
Tropical deciduous forest	Temperate evergreen forest
Thorn forest	Temperate deciduous forest
Tropical savanna	Taiga
Hot desert	Arctic tundra
Temperate shrubland	Temperate grassland
Cold desert	Ice

Figure 19.1
Major biomes of the world.

E. Lucy Braun

raphy, especially as the latter affects microclimate, can influence the type of vegetation found in an area. In turn, the animals living in a given biome are determined by the vegetation there, as well as by climate and other factors.

Biomes are useful, if somewhat arbitrary, descriptive classifications, rather than quantitative or objective categories. Boundaries between biomes may represent the range limits of the dominant species, whereas other species may have ranges that span those boundaries. While we draw sharp boundaries on a map, in reality the boundaries of biomes are often fuzzy. Similarly, the categories that we define ignore the fact that a particular place might not

fit easily into any given biome. Furthermore, different scientists might include a place in different biomes. As the definitions of biomes are somewhat arbitrary, they also differ in detail among scientists. Some use fewer, broader groupings, whereas others use a more detailed breakdown of vegetation types. The definitions of biomes given here may not exactly match those given in other texts. However, there is a broad consensus on the rough categories and approximate definitions of biomes.

Our use of "temperate rainforest," for example, may not be identical to someone else's use of that term, but it will be close. If you were to search the scientific literature, you would find many different descriptions of places called rainforests. There is, however, no single, rigorous definition of the term—it is, in fact, used to refer to forests occurring in places where rain is abundant and there is no extended dry season. Tropical rainforests occur where there is substantial year-round rain; tem-

Table 19.1 Major biomes of the world: dominant growth forms and general climatic conditions[a]

Biome	Dominant growth form	Angiosperms or gymnosperms dominant or common	Temperature	Moisture
Tropical rainforest	Broad-leaved evergreen trees	Angiosperms	Hot	Wet
Tropical montane forest	Broad-leaved evergreen trees	Angiosperms	Mild	Wet
Tropical deciduous forest	Broad-leaved deciduous and semi-evergreen trees	Angiosperms	Hot	Seasonally dry
Thorn forest	Broad-leaved deciduous trees	Angiosperms	Hot	Dry
Tropical woodland	Broad-leaved semi-evergreen trees and grasses	Angiosperms	Hot	Moderate
Temperate deciduous forest	Broad-leaved deciduous trees	Angiosperms	Seasonally cold	Moderate
Temperate rainforest	Needle-leaved evergreen trees	Gymnosperms	Seasonally cool	Wet
Temperate evergreen forest	Needle-leaved or broad-leaved evergreen trees	Gymnosperms or angiosperms	Various	Various
Temperate woodland	Needle-leaved evergreen or broad-leaved deciduous trees and grasses	Both	Mild	Moderate
Taiga	Needle-leaved evergreen trees	Gymnosperms	Cold	Moderate
Temperate shrubland	Evergreen shrubs, annual forbs	Angiosperms	Mild	Moderate
Temperate grassland	Perennial grasses	Angiosperms	Seasonal	Moderate
Tropical savanna	Perennial grasses	Angiosperms	Hot	Moderate
Hot desert	Shrubs, succulents, annual and perennial grasses, annual forbs	Angiosperms	Hot	Dry
Cold desert	Shrubs	Angiosperms	Mild	Dry
Alpine shrubland	Deciduous shrubs	Angiosperms	Cold	Moderate
Alpine grassland	Perennial grasses	Angiosperms	Cold	Moderate
Tundra	Perennial grasses, sedges, shrubs, forbs	Angiosperms	Cold	Moderate

[a]Other factors, especially seasonality, are also important.

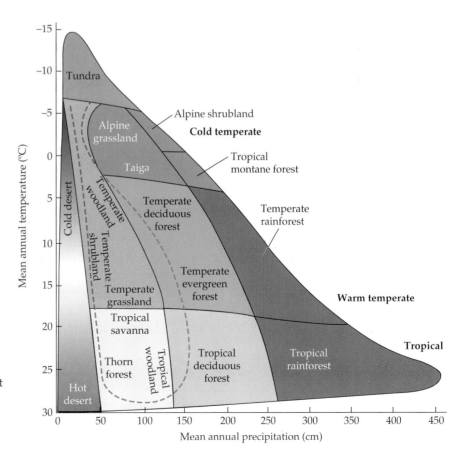

Figure 19.2
The distribution of biomes is determined by climate, especially annual temperature and precipitation. In regions within the dashed lines, other factors—such as fire, grazing, and seasonality of precipitation—strongly affect which biome is present. Climate can also interact with factors such as soil type to determine biome distributions. (After Whittaker 1975.)

Table 19.2 Primary productivity of selected major biome types

Biome	Area (× 10⁶ km²) Amount	Area % of total terrestrial area	Net primary productivity per unit area (g/m²/yr) Normal range	Net primary productivity per unit area (g/m²/yr) Mean	Global net primary productivity (10¹² g/yr) Amount	Global net primary productivity % of total terrestrial productivity	Biomass per unit area (kg/m) Normal range	Biomass per unit area (kg/m) Mean	Total global biomass (10¹² g) Amount	Total global biomass % of total terrestrial biomass
Tropical rainforest	17.0	11.4	1000–3500	2200	37.4	32.5	6–80	45	765	41.7
Tropical deciduous forest	7.5	5.0	1000–2500	1600	12.0	10.4	6–60	35	260	14.2
Temperate deciduous forest	7.0	4.7	600–2500	1200	8.4	7.3	6–60	30	210	11.4
Temperate evergreen forest	5.0	3.4	600–2500	1300	6.5	5.6	6–200	35	175	9.5
Woodland and shrubland	8.5	5.7	250–1200	700	6.0	5.2	2–20	6	50	2.7
Taiga	12.0	8.1	400–2000	800	9.6	8.3	6–40	20	240	13.1
Temperate grassland	9.0	6.0	200–1500	600	5.4	4.7	0.2–5	1.6	14	0.8
Tropical savanna	15.0	10.1	200–2000	900	13.5	11.7	0.2–15	4	60	3.3
Desert	18.0	12.1	10–250	90	1.6	1.4	0.1–4	0.7	13	0.7
Tundra	8.0	5.4	10–400	140	1.1	1.0	0.1–3	0.6	5	0.3
Extreme desert, rock, sand, and ice	24.0	16.1	0–10	3	0.07	0.1	0–0.2	0.02	0.5	0.03
Cultivated land	14.0	9.4	100–3500	650	9.1	7.9	0.4–12	1	14	0.8
Swamp and marsh	2.0	1.3	800–1500	2000	4.0	3.5	3–50	15	30	1.6
Lake and stream	2.0	1.3	100–1500	250	0.5	0.4	0–0.1	0.02	0.05	0.003
Total terrestrial	149.0			773	115			12.3	1837	
Total world	510.0			333	170			3.6	1841	

Source: Whittaker and Likens 1975.

perate rainforests tend to occur where there is heavy winter rain and lighter summer rain but considerable summer fog. The term "rainforest" is thus useful as a broad description, but not as a rigorously defined category or unique entity.

While we have emphasized that, scientifically, the boundaries between biomes (or communities) are arbitrary, this is not necessarily true under the law. In Australia, for example, areas designated "temperate rainforest" are legally afforded a greater measure of protection than other areas. Some eucalypt forests in very rainy places exist above an understory that is floristically a rainforest and would replace the eucalyptus canopy in about 300–500 years in the absence of fire. However, because these forests are fire-dependent, some states do not legally considered them rainforests. If the legal Australian definition were applied in North America, the redwood forests of the Pacific coast would not be classified as temperate rainforests—a categorization with which some North American scientists have agreed.

In Chapter 12, we discussed a controversy that centered on the contrasting views of Frederic Clements and Henry Gleason about the nature of communities. Clements saw communities as highly predictable entities primarily controlled by large-scale climatic patterns. In contrast, Gleason saw communities as unpredictable and variable and subject to the vagaries of dispersal, microclimatic conditions, and individualistic species distributions. One way to resolve this conflict is to recognize that each scientist emphasized a different spatial scale. Clements was focusing on patterns at the level of biomes. At that scale, we can see regular patterns and draw boundaries on our maps. Gleason was most concerned with patterns at local levels. At that scale, the boundaries blur, and we mostly see local variability. Both perspectives are valid, but each tells us different things about the world. In the following sections, we will look at vegetation patterns from a large-scale perspective as we survey the major biomes and see what distinguishes them from one another.

Moist Tropical Forests

Tropical Rainforest

Tropical rainforests are among the most diverse and productive biomes on the planet. Although they make up only 11% of Earth's land area, they account for over 30% of terrestrial net primary production (Table 19.2) and contain perhaps half of all living species. Tree species diversity can be as high as 300 species per 0.1 ha. Animal diversity is also high, especially that of insects and other invertebrates.

Tropical rainforests are found within a band extending from 10°N to 10°S latitude at elevations below 1 km, in areas that get more than 250 cm of rainfall each year, with some rain falling in all months. The most extensive rainforests are found in the Amazon basin of South America, the equatorial regions of western Africa, and parts of southeastern Asia. As precipitation becomes more seasonal, this biome grades into tropical semi-evergreen forests dominated by broad-leaved trees, some of which retain their leaves year-round and others that lose their leaves seasonally.

Tropical rainforests are characteristically multilayered, with two, three, or even four layers of understory trees (Figure 19.3). Tallest are the **emergent trees**, which are widely spaced and often more than 40 m in height, with umbrella-shaped canopies that extend above the general canopy of the forest. Below this is a closed canopy of trees, typically 30–40 m high. Light is readily available at the top of this layer, but greatly reduced below it. Under the closed canopy, there may be another layer of trees less than 20 m in height. Under these trees, there is often a shrub/sapling layer, and finally a sparse ground layer.

The floors of tropical rainforests can be very dark, with less than 1% of above-canopy light levels. While Hollywood movies may portray tropical rainforests as jungles with a tangle of vines and understory plants, the understory is generally sparsely vegetated with highly specialized plants capable of tolerating low light levels

(A)

(B)

Figure 19.3
(A) The interior of a tropical rainforest in La Selva, Costa Rica. (Photograph courtesy of E. Orians.) (B) Epiphytes are symbiotic commensals (see Box 4.1) that grow on the surfaces of other plants and are common in most tropical rainforests. Numerous bromeliads (Bromeliaceae) can be seen on the trunks of several of the trees in this forest in Puerto Rico. (Photograph courtesy of J. Thomson.)

(see Chapter 2). The interior of a rainforest was compared by early European explorers to a dimly lit, soaring green cathedral with an expansive space reaching to a high, vaulted roof. "Jungles" are the consequences of disturbed, high-light environments on the margins of the forest, along roads, rivers, and in gaps, or where logging has thinned the canopy.

Because of this vertical light distribution within the rainforest, much of its productivity is concentrated high in the canopy. Plant life forms that are not present in most temperate forests greatly contribute to the diversity of tropical forests. These plants include many species of **lianas** (woody vines rooted in the ground) and **epiphytes** (plants that grow on other plants but are not rooted in the ground; Figure 19.3B), including many species of ferns, orchids, and bromeliads. Lianas and epiphytes are common in tropical forests because of the very low light levels near the ground. An alternative strategy to being a tree is to grow high up on a tree to gain access to light.

Most epiphytes and lianas have **commensal** relationships (the host neither benefits nor is harmed) with the trees on which they grow unless the burden of epiphytes and lianas becomes too large, weakening the tree and possibly pulling it down. Some species are not so benign, however. Strangler figs (*Ficus* spp., Moraceae) begin as a number of separate individual plants that send roots downward to the forest floor, develop thick trunks and canopies, and gradually envelop and kill the host tree (see Figure 10.7A). The separate individual plants comprising the strangler ultimately merge to become a functional individual tree. Thus, this plant has a combined strategy of beginning life high in the canopy as an epiphyte, becoming rooted in the ground, which allows more growth, and eventually becoming a tree and outcompeting other species for light.

Moist tropical forests typically have highly infertile soils. In tropical rainforests, warmth and abundant moisture contribute to an extraordinary amount of respiration in the soil—not only by roots, but also by many kinds of soil dwellers, particularly decomposing organisms. When CO_2 is dissolved in water, it produces carbonic acid (that is why natural rainwater is always slightly acidic). The huge amount of soil respiration in tropical rainforests releases large amounts of CO_2, much of which goes into solution in the soil. The net effect is that nutrient ions, which bind to soil particles (especially clay) less tightly than hydrogen ions bind to soil particles, tend to enter the soil solution. As a result, they are either taken up almost instantaneously by plants or washed away by the high rainfall (see Chapter 4).

If the soils are so poor, how is it that tropical forests are so productive? Factors contributing to their high productivity are year-round water availability, warm temperatures, and duration of daylight varying little throughout the year. As for the mineral nutrients needed for growth, there are large quantities of nutrients in these forests, but they are mainly found in the living biomass rather than in the soil. Because decomposition is extremely rapid in these environments, nutrients cycle very rapidly. Like plants in many areas where nutrients are limiting, those in tropical forests are very good at acquiring and retaining nutrients. Particularly important in this regard are mycorrhizal associations (see Chapter 4).

Not only is a large proportion of the world's biodiversity located in the Tropics; a large proportion of its human population is located there as well. The greatest growth in human populations is also found in the Tropics, along with poverty and diseases associated with poverty and poor living conditions. All of these factors can drive the rapid obliteration of natural communities by people. Although until recently much of the human population in the Tropics was rural, it is becoming increasingly urban. Lagos, Nigeria, for example, with about 13 million people, is one of the fastest-growing cities in the world. Unfortunately, this urbanization does not mean that there is less pressure on natural habitats; in many cases, the opposite is true. Economic, population, and political pressures have contributed to the destruction of a large proportion of the world's rainforests (Table 19.3; see Chapter 22). Exploitation comes not only from within the tropical countries themselves, but also in great measure from individuals and corporations based in the United States, Europe, and Japan. Today all tropical rainforests are threatened by logging, and even more by the clearing of land for a range of purposes, including subsistence slash-and-burn agriculture and corporate-owned cattle ranching (see Chapter 22).

Logging and the clearing of land for cattle grazing or human habitation can have numerous effects on rainforests. Because most of the nutrients are tied up in biomass, logging, burning, and agriculture remove substantial amounts of nutrients from the system. The remaining soils are often very nutrient-poor, and they can quickly achieve a cementlike texture or become severely eroded. These changes can make agriculture impossible in a very short time—in other words, the purposes for which the forest was removed are not sus-

Table 19.3 Global tropical rainforest loss

Region	Area ($\times 10^6$ km^2)		% lost
	Original	Remaining	
Central/South America	8.0	5.8	28
Asia/Pacific Islands	4.4	2.2	39
Africa	3.6	1.2	66

tainable. Removal of rainforest from large enough areas can drastically affect regional rainfall levels. When the forest is present, rainfall is quickly taken up and tran- spired back into the atmosphere. When the forest trees are no longer there, the rainfall runs off, which leads to increasing erosion, particularly for forests on slopes. Even when the destruction is not so dramatic, as in selec- tively logged forests, non-native plant species can invade disturbed forests, threatening the regeneration of native species.

While habitat destruction is the key factor threaten- ing tropical biodiversity, other human activities, from the illicit trade in rare species such as orchids and par- rots brought into the United States and Europe, to the hunting of chimpanzees and other primates for the del- icacy called "bush meat" in African cities, also endanger animal and plant species within these forests. These forms of forest destruction are not particularly new; wealthy nineteenth-century orchid collectors in England and Germany destroyed entire tropical forests to assure that their competitors could not collect orchids from them. However, the scale of recent damage dwarfs that of the past, and, of course, the destruction is cumulative.

Tropical Montane Forest

Tropical montane forests—also called elfin forests or cloud forests at the highest elevations—are higher-ele- vation neighbors of rainforests (Figure 19.4). They are called cloud forests because they are often swathed in clouds, and much of the moisture available to plants comes from deposition of condensation. They are cool- er than tropical rain forests because they are at higher elevations. The trees are typically shorter, and there are many more epiphytes festooning the branches of trees.

Overall diversity is lower than in rainforests, however, because there are fewer lianas. These forests are found at middle elevations in the mountains of Africa, South America, Central America, and New Guinea.

Seasonal Tropical Forests and Woodlands

When people think about the ecology of tropical regions, they typically think about rainforests, especially when they think about species diversity and conservation. However, many tropical regions are considerably drier than rainforests, and many of these forests are also rich in species. The seasonal tropical forest biome, with its associated species, is being lost very rapidly. In fact, these forests are disappearing—or have already disap- peared—even more rapidly than tropical rainforests, as they are often on soils that are much better for farming. This biome includes a range of types that grade from deciduous forest to thorn forest and woodlands.

Several geographic factors determine the locations of dry tropical forests and woodlands. First, they tend to occur relatively close to the Tropics of Cancer and Capricorn, rather then near the equator. This pattern is a result of the shifting throughout the year of the **intertropical convergence**—the latitude at which the trade winds from the Northern and Southern Hemi- spheres tend to converge (see Figure 18.7), causing a large amount of rainfall. The intertropical convergence is usually near each of these latitudes only once a year, whereas it passes over the equator twice a year. As a result, rainfall is highly seasonal, whereas closer to the equator, rainfall is more evenly spread throughout the year (see Figure 18.14). Second, dry forests and wood- lands tend to occur in the rain shadows of mountain

Figure 19.4
A tropical montane forest in the Santa Elena Cloud Forest Reserve, Costa Rica. (Photo- graph © G. Dimijian/Photo Researchers, Inc.)

Figure 19.5
The top view of a tropical deciduous
forest in Palo Verde National Park,
Costa Rica was taken during the rainy
season; the bottom photograph is the
same view shown in the dry season,
when many of the trees are leafless.
(Photograph courtesy of D. L. Stone.)

ranges (see Figure 18.9) and on the
western slopes of continents. For
example, the eastern (Caribbean)
side of the mountain range in Costa
Rica is tropical rainforest, whereas
the western (Pacific Ocean) side is
dry tropical forest.

Tropical Deciduous Forest

Tropical deciduous forests, also
called tropical seasonal forests,
include a range of community
types, from wholly deciduous to
semi-evergreen (Figure 19.5). The
trees are **drought-deciduous**, meaning that they lose their leaves during the dry season. These communities are found in tropical areas that
have pronounced wet and dry periods, and were once especially extensive in India and Southeast Asia,
also occurring in Central and South
America and elsewhere. Depending
on the latitude, there might be a single wet and a single
dry period, or two of each (Figure 19.6). Although most
photosynthetic activity and plant growth occurs during
the wet season, the dry season is often when flowering
takes place. There may be advantages to flowering in the
dry season for both animal-pollinated and wind-pollinated plants. The lack of leaves makes the flowers more
conspicuous to pollinators, and the pollen and nectar
rewards may be one of the few food sources available to
them. Similarly, wind-dispersed pollen can travel with
fewer obstructions.

    Tropical deciduous forests are similar in overall form
and structure to rainforests, with a species richness lower
than but approaching that of tropical rainforests in some

Figure 19.6
The length and number of rainy seasons depends on latitude. Near the equator there are two wet and dry seasons
each year, while farther north and south there is a single
rainy season in the summer and a dry season in the winter.
The timing of the seasons is reversed north and south of the
equator.

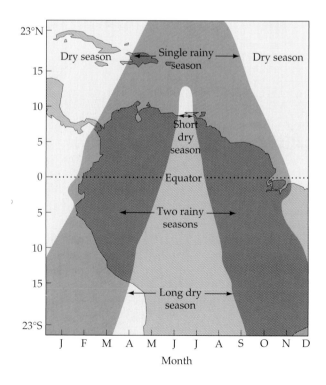

Figure 19.7
A thorn forest on the island of Madagascar, off the east coast of Africa. This photo was taken after heavy rains in December. From August through November—the dry season—the trees are parched and brown. (Photograph by N. Garbutt/ Indri Images.)

Central America, and in Asia (Myanmar, India, and Thailand), continental Africa, and Madagascar. They have pronounced wet and dry seasons, and the total yearly rainfall may be half that of tropical deciduous forests. Species of *Acacia* and other woody members of this part of the pea family (Fabaceae) often dominate these communities. These plants have leaves that are divided into many small leaflets, which helps to minimize water loss. Diversity is considerably lower than in tropical rainforests and deciduous forests, but can still be quite high compared with that in temperate systems. As in tropical seasonal forests, the trees are typically drought-deciduous. Succulents are also common, especially in drier areas. As the climate becomes even drier, these communities grade into either desert scrublands or tropical savannas.

Tropical Woodland

Forests are dominated by trees whose canopies touch each other. In contrast, the trees in woodlands are spaced apart, although the density of those trees can vary from closely spaced to sparse. Tropical woodlands are found in areas with strongly seasonal climates, like tropical deciduous forests, but which are drier or have less favorable soils. They occupy areas of Brazil (Figure 19.8) and interior southern Africa (miombo), and also occur in northern South America, the West Indies, northern Australia, and Myanmar and other parts of Southeast Asia.

areas. Localities with lower amounts of rainfall generally have shorter canopies and fewer species. Human populations are often very high in these biomes, and they face many of the same issues and problems as tropical rainforests.

Thorn Forest

Tropical areas that are even drier support thorn forests (also called thornwoods), named for the menacing armament of many of the plant species there (Figure 19.7). They are common from South America to Mexico, including those on limestone soils in the West Indies and

Figure 19.8
The vegetation on Brazil's Sete Cidades National Park is an example of tropical woodland. (Photograph © J. Jangoux/Photo Researchers, Inc.)

The trees in tropical woodlands range in height from 3 to 7 m. Most trees and shrubs have large, tough, semi-evergreen leaves and thick, fire-adapted bark. Many of the trees are members of the Fabaceae, as in thorn forests. Palms are also common, whereas succulents and spiny plants are rare. African woodlands—as well as savannas—are rich in large mammal species, including herds of elephants, giraffes, wildebeest, and their predators, such as lions and hyenas.

Temperate Deciduous Forest

Temperate deciduous forests are found in three disjunct areas of the Northern Hemisphere: eastern North America, eastern Asia, and Europe (Figure 19.9). These forests grow on young soils formed since the most recent glaciation; the lack of such forests in the Southern Hemisphere is probably a result of the absence of such soils there. Temperate deciduous forests are found in areas with continental or maritime climates with warm, moist summers and cold, sometimes snowy winters, with a range of 50–250 cm rainfall annually, distributed through the year. Diversity and productivity are generally moderate in these forests, but span a fairly wide range. In North America and Asia, this biome is known for the brilliant red, orange, and gold colors of its leaves in autumn, before they are shed for the winter. The dominant trees are 18–30 m tall. Growing seasons range from very short in the northernmost and high-altitude parts of this biome to very long in its southern portions. The genera of the dominant trees common to temperate deciduous forests on all three continents include *Quercus* (oak), *Acer* (maple), *Fagus* (beech), *Castanea* (chestnut), *Ulmus* (elm),

Tilia (basswood or linden), *Juglans* (walnut), and *Liquidambar* (sweet gum). Different species of these genera occur on each continent. The diversity of local forest types on all continents may be very great, depending on the details of topography, climate, soils, and disturbance regimes. In North America, for instance, the eastern deciduous forest consists of many different forest subtypes (Braun 1950; Greller 1988).

At higher latitudes and elevations, temperate deciduous forest sometimes grades into mixed deciduous-evergreen forest. An example is the hemlock-white pine-northern hardwood forest in the northern United States (especially around the Great Lakes) and extending south in the upper elevations of the Appalachian Mountains. The dominant trees in this forest are a mixture of evergreen needle-leaved and deciduous broad-leaved species. Elsewhere, patches of pine-dominated communities can be found where the soil is especially shallow or infertile or there is a high frequency of fire. If such areas are large enough, they may be separately classified as temperate evergreen forest.

Tree species diversity in temperate deciduous forests is generally moderate, with 5–30 dominant tree species in any one place. However, the temperate deciduous forests of the southeastern United States are a center of plant diversity, with the highest tree species diversity in North America as well as a very high diversity of ferns and deciduous and shrubby angiosperms. For example, there are over 120 species of trees and over 3000 plant species in the southern Appalachian region. These forests escaped glaciation, and some areas have been continuously vegetated for as long as 200 million years. The dominant genus here is *Quercus* (oak, Fagaceae). Other common canopy trees include numerous species of *Carya* (hickory, Juglandaceae), *Acer* (maple, Sapindaceae), *Aesculus* (buckeye or horse chestnut, Sapindaceae), and *Magnolia* (Magnoliaceae), as well as *Tilia americana* (basswood, Tiliaceae), *Liriodendron tulipifera* (tulip poplar, Magnoliaceae), and *Betula lenta* (black birch, Betulaceae).

Common in the understory are trees such as *Cornus florida* (flower-

Figure 19.9
A temperate deciduous forest in Michigan, here dominated by American beech (*Fagus grandifolia*, Fagaceae) and sugar maple (*Acer Saccharum*, Sapindaceae). (Photograph by S. Scheiner.)

Figure 19.10
An old-growth temperate rainforest in Olympic National Park, Washington state. (Photograph © G. Ranalli/ Photo Researchers, Inc.)

Other Temperate Forests and Woodlands

Temperate Rainforest

Substantial temperate rainforests are located in southern Chile and along the northwest coast of North America (Figure 19.10). Smaller ones occur on the western coast of New Zealand, the southeastern and southwestern coasts of mainland Australian, and the island of Tasmania. Fragments remain in Norway; in other parts of Europe, such as Ireland, these forests were completely destroyed several centuries ago. These regions are all similar climatically: under maritime influences, temperatures stay cool, and there is abundant winter rainfall and summer fog. Productivity and decomposition are extremely high in some of these forests, due largely to the mild temperatures and extended rainy period, which result in a very long growing season. In contrast to tropical rainforests, however, species diversity here can be quite low. In North America, these forests are dominated by needle-leaved conifers. In contrast, temperate rainforests are dominated by broad-leaved trees, such as *Nothofagus* (Fagaceae), in Chile, New Zealand, and Australia, although conifers such as those in the family Podocarpaceae also play an important role in some Southern Hemisphere forests.

Temperate rainforests contain some of the tallest trees in the world, including the giant redwoods (*Sequoia sempervirens,* Cupressaceae) of California and some of the eucalypts of Australia, with heights exceeding 100 m. Not all forests of this type contain trees of such enormous stature, however; the coastal Sitka spruce forests of Alaska have trees similar in height to those of temperate deciduous forests. Some of these forests (including the redwood forests) depend on occasional fires for regeneration.

Temperate rainforests are threatened more by their economic value than by human population growth because most of them are not near centers of human population. Because of their enormous standing crop of wood, many of these forests on the western coast of Canada and the United States have been clear-cut for lumber, resulting in high rates of erosion and losses of

ing dogwood, Cornaceae) and large shrubs such as *Kalmia latifolia* (mountain laurel, Ericaceae) and *Rhododendron* spp. (evergreen rhododendrons and deciduous azaleas, Ericaceae). In wetter areas such as coves or stream banks, *Tsuga canadensis* (eastern hemlock, Pinaceae) is found, while pines are common on drier ridges. The formerly dominant *Castanea dentata* (chestnut, Fagaceae; see Figure 11.18) was devastated by chestnut blight during the early part of the twentieth century and has largely been replaced by various oak species in its former habitats.

Deciduous forests are threatened by a variety of factors. Some of the world's largest cities (such as London and New York) occur on land that was once temperate deciduous forest, although large areas of this biome still exist. In Europe, air pollution, including the effects of acid precipitation and nitrogen deposition, has been a major cause of physiological stress in forest trees and general forest decline (see Chapter 22). Worldwide, logging has been a major cause of forest destruction for hundreds of years, as has development for farmland or residential use. In some places the forests have been destroyed, while in others, such as the eastern United States, secondary forests have regenerated over large regions. In some areas of the world (such as parts of Japan and Mexico), the only forests to escape cutting occur in places that are inaccessible or very steep. Road building and expanding suburbanization are renewing threats to many secondary forests in North America. Invasion by exotic plant species is a threat to this and many other biomes worldwide as well (see Chapter 22).

Figure 19.11
This privately owned temperate rainforest near Port Angeles in Washington state is being "clear cut." All the trees will be felled for commercial lumber. (Photograph © C. Larsen/Photo Researchers, Inc.)

species diversity. Because of the high potential growth rates for trees, some of these areas have been replanted with tree monocultures for lumbering, with the result being low-diversity tree plantations in place of natural old-growth forest. The remaining stands of old-growth forest have been the focus of extended and bitter fights between environmentalists and logging interests. Old-growth temperate rainforests in North America are disappearing more rapidly than tropical rainforests, almost entirely as a result of commercial lumbering operations (Figure 19.11).

Temperate Evergreen Forest

Temperate evergreen forest is a catch-all category that includes forests of very different types. All of them are located in areas that have pronounced seasonal climates and are drier than areas with deciduous forests, due to either low rainfall or soils that do not retain moisture well (Figure 19.12). These forests generally experience relatively mild temperatures but have strong wet/dry cycles (e.g., in the wetter parts of the Mediterranean region, southern California, and southern Australia), or have greater temperature ranges and drier continental climates (e.g., in the mountains of the western United States). Temperate evergreen forests often grade, at their drier margins, into temperate woodlands or shrub-dominated systems such as chaparral. The trees can be broad-leaved or needle-leaved, but their leaves are retained throughout the year. Trees with **sclerophyllous** (tough, evergreen, relatively small, roundish and thick) leaves are common, especially in areas with milder temperatures. In cooler regions, needle-leaved forests often predominate, including the pine and fir forests at mid-altitudes in the Sierra Nevada of California and the mixed needle-leaved and broad-leaved forests of conifer and beech in Chile and New Zealand. The latter forests, depending on the relative proportions of needle-leaved

Figure 19.12
In their natural state, temperate evergreen forests are one of the most endangered habitats in the world. Shown here is Hartwick Pines State Park in central Michigan. The last remaining stand of old-growth white pine forest in Michigan, it is home to the largest white pine tree (*Pinus strobus*, Pinaceae) in the world. (Photograph by S. Scheiner.)

evergreen trees and broad-leaved deciduous trees, might be classified with temperate deciduous forests.

Needle-leaved forests are found in areas of the eastern United States and western Europe where drought due to sandy soil, low soil fertility, or shallow soil on rocky outcrops favors pines over deciduous trees. Examples are the "pine barrens" and "piney woods" of the eastern and southern coastal plains of the United States. These forests are dominated by pitch pine (*Pinus rigida*) in parts of the eastern coastal plain and on rock outcrops along the East Coast, and by longleaf pine (*P. palustris*) and slash pine (*P. elliottii*) from Virginia to East Texas, along with various tree and shrub oaks (*Quercus* spp.) and low-growing ericaceous shrubs. All of these forests experience frequent fires, and to varying extents depend on fire for their maintenance and regeneration. Many of the species in these forests have numerous adaptations to fire (see Chapter 13). Productivity and decomposition rates are typically low in these forests. Diversity can vary extensively; while many have low diversity, longleaf pine forests include some of the most species-rich systems in North America.

In the southeastern United States, pine-dominated forests are currently more extensive than they were several hundred years ago. Areas that were originally broad-leaved forests were cleared for farming, which depleted the soil of nutrients. When the farms were abandoned during episodic convergences of bad economics and bad weather, from the nineteenth century to the Great Depression of the 1930s, the forests returned, but they are now dominated by pines and differ sub-

stantially from the original forests. Today, because various pine species can be grown quickly for pulp, many pine forests are maintained in this state by logging and forestry practices, with none of the natural diversity or structure remaining. Periodic fires in unmanaged areas may also help to maintain pine dominance. The remaining natural pinelands, especially those dominated by *Pinus palustris* are among the most endangered ecosystems in the United States. Many of these areas are now being urbanized.

Temperate Woodland

Temperate woodlands, like their tropical counterparts, occur on the drier margins of forests and often form transition zones between forests and either grasslands or shrublands. Diversity and productivity are lower in woodlands than in deciduous forests. Temperate woodlands range from nearly continuous tree canopies to very open, with scattered trees (Figure 19.13). The tree species can range from needle-leaved evergreens to broad-leaved sclerophylls to broad-leaved deciduous trees with heights from 3 to 7 m. In North America, coniferous woodlands are extensive in the area between the Rocky Mountains and the Sierra Nevada, where they are dominated by several pinyon pine (*Pinus*) and juniper (*Juniperus*) species. Oak (*Quercus*) and oak-pine woodlands occur from Washington to the Central Valley of California and across areas of the southwestern United States and northern Mexico, east to Texas. They are also common around the Mediterranean Sea (where they are again oak-dominated) and in South America, Africa, and Australia.

Figure 19.13
A pinyon-juniper woodland in northern Arizona. (Photograph by S. Scheiner.)

Taiga

The taiga is a broad swath of primarily coniferous forest that covers an enormous region of the circumpolar subarctic. Similar vegetation is found in many subalpine regions globally (Figure 19.14). The name "taiga" comes from the Russian word for boreal forest, which is another name for this biome. The taiga is one of the largest of the world's biomes. It encompasses approximately 12 million km², occupying 8% of the Earth's land surface, makes up about a third of the world's forested area, and stretches from Canada and Alaska to Scandinavia, Russia, China, Mongolia, and parts of Japan and the Korean peninsula. In North America, boreal forests can also be found at high elevations in the Appalachian Mountains, the Rocky Mountains, and the Sierra Nevada; they are found in similar mountainous regions across Eurasia. At their northern or upper elevation margins, these forests open up into woodlands, **muskeg** (northern bogs), or wind-swept shrublands.

Taiga climates are largely continental, with long, bitterly cold winters (up to six months below freezing) and short, warm summers (with only 50 to 100 frost-free days). Precipitation is low (38–50 cm/year), but the climate is humid due to very low evaporation rates. Large parts of this region were glaciated during the last Ice Age. Recurring disturbances are an important feature of the landscape and include forest fires, treefalls due to storms, and large areas defoliated by insect attack. Productivity and decomposition rates tend to be very low, and soils are generally very acidic.

Species diversity and productivity in boreal forests are low, and these forests are generally dominated by just a few species of coniferous trees. Four tree genera dominate much of the circumpolar taiga: *Picea* (spruce), *Abies* (fir), *Pinus* (pine), and *Larix* (larch or tamarack). (Larches are one of the few deciduous conifer species.) The understory of a boreal forest is typically shady, with few species. Broad-leaved deciduous trees and shrubs are common dominants in both primary and secondary succession, with *Alnus* (alder), *Betula* (birch), and *Populus* (aspen) particularly common. In much of Canada and Alaska, for example, the dominant species are *Picea glauca* (white spruce) and *Picea mariana* (black spruce), with lesser numbers of *Abies balsamea* (balsam fir), *Larix laricina* (larch), *Pinus banksiana* (jack pine), *Pinus contorta* (lodgepole pine), *Populus tremuloides* (quaking aspen), *Populus balsamifera* (balsam popular), and *Betula papyrifera* (paper birch).

Because low decomposition rates lead to the accumulation of peat (see Chapter 15), the taiga is a "net carbon sink," playing a critical role in the global carbon balance by absorbing about 700 million tons of carbon per year, about equal to the annual emissions of China. Large areas of the taiga remain as old growth, or at least are intact and relatively undisturbed. The taiga of Russia and Canada constitutes more than half the world's remaining old-growth forest. Because human populations in the taiga are low, the primary threat to these forests is commercial lumbering for timber, pulp, and paper. In Sweden, for example, large-scale commercial logging for over a century and replanting with monocultures of timber species has transformed many boreal forests into low-diversity timber plantations. Climate change, mining, and oil and gas drilling also threaten many taiga ecosystems.

Figure 19.14
Taiga in James Bay, Quebec, Canada. (Photograph © F. Lepine/Earth Scenes.)

Temperate Shrubland

Temperate shrublands are found in regions with mediterranean climates: cool, wet winters and warm to hot, dry summers. They occur from 30° to 40° latitude on the western coasts of continents, adjacent to cold ocean currents offshore. These regions are highly isolated from one another on the different continents, resulting in many endemic species. Names for this biome differ around the world: garrigue and maquis around the

Figure 19.15
Temperate shrubland (chaparral) in California. Chaparral vegetation such as this is subject to periodic fires. (Photograph courtesy of J. Thomson.)

Mediterranean Sea, fynbos in the Cape region of southern Africa, matorral in central Chile, kwongan in southwestern Australia, and chaparral on the southern coast of California. Species diversity ranges from low to exceptionally high, most notably in the fynbos (see Box 20A). All of these different types of temperate shrubland are dominated by sclerophyllous evergreen shrubs (Figures 19.15 and Figure 19.16). Many members of the shrub flora are highly aromatic, including many members of the mint (Lamiaceae) family (e.g., sage, rosemary, thyme, and oregano). Many contain highly flammable oils.

Total annual precipitation in temperate shrublands is 40–100 cm/year, but precipitation is highly concentrated in a few months of the cool season. Temperatures are cool to hot, moderated by proximity to the ocean and by fog caused by the cold ocean currents. There is a short, predictable growing season when soil moisture and temperatures are sufficient for growth; most plant growth and flowering occurs from late winter to spring. Many of the plants are adapted to drought. Fires are frequent, as the residents of California suburbs encroaching into this biome have been dismayed to discover. In addition to the dominant shrubs, annuals are very common, especially in the years immediately following a fire.

Temperate shrublands have been affected by humans for millennia by both livestock grazing and the use of fire. The Mediterranean region itself, as we know from classical Greek literature, was formerly forested with evergreen oaks, pines, cedars, wild carob, and wild olive. Many scientists believe that the shrublands of California, likewise, became much more extensive as a result of deliberate burning by Native Americans as a management tool, as well as livestock grazing by Span-

ish settlers (Woodward 1997). There is wide variation among continents in the current human population size in and the nature of the human effects on shrublands. Human populations are not large or rapidly growing in most shrubland regions, although in California, suburbanization and development coupled with rapid population increases are major threats to chaparral. The shrubland of Australia has a low human population, but in some places has been cleared for agriculture. The fynbos region is particularly threatened by invasive plant and animal species.

Grasslands

Temperate Grassland

Temperate grasslands (Figure 19.17) are known by many names around the world: prairie (North America), steppe (vast parts of north-central Eurasia), veldt (South Africa), pampas (South America), and puszta (Hungary). They are generally found in moderately dry continental climates with cold, dry winters and warm to hot summers with higher rainfall. Grasslands are the largest of the North American biomes; at one time, they made up 266 million ha (35%) of the United States. The grasses can be short, especially in drier areas, or tall. *Andropogon gerardii* (big bluestem), a dominant species of the tallgrass prairie of central North America, can grow as tall as 2.5 m. The tallest grasses and greatest productivity are in the eastern most part of the North American biome, with stature and productivity declining to the west.

The invention of the steel plow by John Deere in 1837 resulted in the conversion of almost all of the tall-

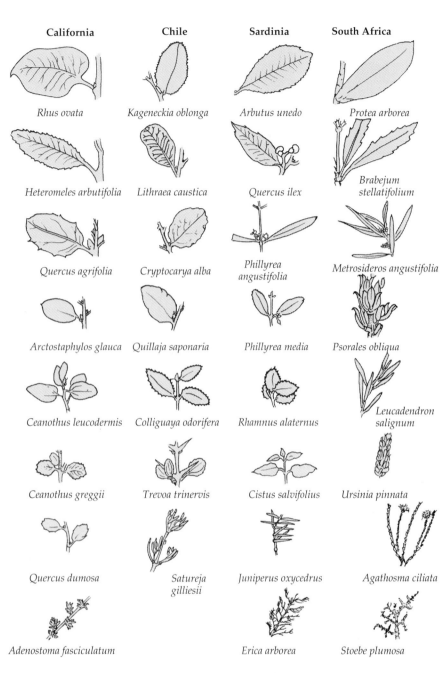

California	Chile	Sardinia	South Africa

Rhus ovata — *Kageneckia oblonga* — *Arbutus unedo* — *Protea arborea*

Heteromeles arbutifolia — *Lithraea caustica* — *Quercus ilex* — *Brabejum stellatifolium*

Quercus agrifolia — *Cryptocarya alba* — *Phillyrea angustifolia* — *Metrosideros angustifolia*

Arctostaphylos glauca — *Quillaja saponaria* — *Phillyrea media* — *Psorales obliqua*

Ceanothus leucodermis — *Colliguaya odorifera* — *Rhamnus alaternus* — *Leucadendron salignum*

Ceanothus greggii — *Trevoa trinervis* — *Cistus salvifolius* — *Ursinia pinnata*

Quercus dumosa — *Satureja gilliesii* — *Juniperus oxycedrus* — *Agathosma ciliata*

Adenostoma fasciculatum — *Erica arborea* — *Stoebe plumosa*

Figure 19.16
Examples of leaf morphology in four widely separated areas with temperate shrubland biomes. The leaves illustrated are from southern California in North America (chaparral), northern Chile in South America (matorral), the island of Sardinia in the Mediterranean Sea, and South Africa (fynbos). Leaves in temperate shrublands tend to be sclerophyllous: tough, evergreen, relatively small, rounded, and thick. (After Cody and Mooney 1978.)

grass prairie, and much of the mixed prairie to its west, to cropland. Today only a few prairie remnants are preserved, and the only common tallgrass species that you are likely to see is maize (*Zea mays*, which does not occur at all in nature, but was created by human selection and breeding in pre-Spanish Mesoamerica). In Illinois, for example, less than 1% of the original prairie still exists. The grasslands that have been converted to croplands for growing grain worldwide are often described as "the breadbasket" of their particular country.

Rainfall is strongly seasonal in most temperate grasslands, leading to seasonal drought. In some cases, the total precipitation would be enough to support trees if it were more evenly distributed throughout the year.

In many grasslands, fire is the key factor in the exclusion of trees and shrubs. All of the U.S. prairie regions east of the Mississippi River are fire-maintained; evidence suggests that Native Americans used fire to maintain these grasslands in areas that would otherwise be temperate deciduous forest.

Human effects on temperate grasslands are extensive. Fire exclusion has led to the invasion of many grasslands by woody plants. Human hunting of large native grazers, such as bison in North America, has also favored woody plants such as shrubs that are vulnerable to grazing. High densities of domestic cattle have also had major effects. One major difference between bison and cattle is that while bison were always on the move, allowing the grass-

Figure 19.17
Tallgrass prairie in Nebraska, in the center of the United
States. It is easy to see why this biome is sometimes poetical-
ly referred to as "a sea of grass." (Photograph by R. and J.
Pollack/Biological Photo Service.)

es to recover, cattle are kept permanently in one place
for long periods. The introduction and spread of invasive
non-native plant species in North America has also resulted in dramatic changes in grasslands at the community,
ecosystem, and landscape levels.

Enormous regions of temperate grassland have been
converted to irrigated cropland. Such intensive agriculture is often not sustainable over the long term. For
example, much of the North American Great Plains is
irrigated with water from the Oglala aquifer that underlies the southern part of the region. This water will not
be available forever; the aquifer is being used up much
faster than it is being recharged. Over many years, irrigation leads to salinization of soil because the water
leaves behind concentrated mineral deposits when it
evaporates. Moreover, the costs of pumping the water
increase over time as the water table drops. Irrigated
agriculture has already become untenable in parts of the
Great Plains, and the problems being experienced there
will expand to other regions in the future. Meanwhile,
the native grassland is gone and cannot replace the irrigated cropland—and the change has occurred within
just two or three generations. The efforts of restoration
ecologists to restore grassland ecosystems have met with
a range of results, from what optimistically appear to be
long-term successful restorations to complete failures.
Successful restoration of prairies and other grasslands

to their original condition is costly, labor- and time-inten-
sive, and depends on a range of conditions; it is therefore unlikely at the present time to provide workable
solutions for anything other than very limited areas. On
the other hand, the return of ecosystem structure and
function, albeit to a very different community, is much
easier to accomplish.

Anthropogenic disturbances in grasslands can have
interactive effects, as seen in the disappearance of the valley grasslands of California. The Central Valley of California was once a vast perennial grassland, with extensive wetland areas and fertile riverbeds. Spanish explorers
wrote that, in wet years, it was impossible to ride a horse
across the Sacramento Valley due to the density and
height of the vegetation, as well as the widespread wetlands. This vegetation is now entirely gone, having been
replaced by short Mediterranean annual grasses (and
some exotic forbs). Conversion to irrigated cropland,
while important, was not the main reason for this change.
Interactions between drought, high densities of non-
native domestic cattle and sheep, and the introduction
of non-native plants all contributed to the demise of the
perennial grasslands. The growth form of the native grasses in clumps provides open spaces that allow the establishment of invading plants. The native grasses also do
poorly under heavy or close grazing. When the United
States took possession of California from Mexico, and
grazers were introduced in the 1850s and 1860s, the balance began to tip toward the successful incursion of non-
native plant species. The fact that the area was then
undergoing a prolonged drought also played a role in giving the invasive annuals an advantage.

Figure 19.18
The Masai Mara Reserve of Kenya is part of the great East
African savanna known as the Serengeti. (Photograph ©
A. Jones/Photo Researchers, Inc.)

Tropical Savanna

Tropical savannas are perennial grasslands that may have
an open tree canopy or scattered shrubs. "Savanna" is a
Hispanicized variation of a Native American word for
"plains." Like temperate grasslands, savannas occur in
areas with distinct dry and wet seasons that would be
seasonal forests or woodlands if soil conditions or disturbances did not limit the establishment of canopy trees.
Rainfall is 75–125 cm per year, with a dry winter season
that lasts at least five months. Disturbances are caused
largely by grazing or frequent fires, or occasionally by
other factors such as soil saturation (due to impermeable
soils). The most extensive tropical savannas are found in
Africa and Brazil (cerrado); others are found in Central
America, Australia, and Southeast Asia.

Savannas are dominated by tall perennial grasses
1–2 m high, typically interspersed with drought- and
fire-resistant trees or shrubs. Palm, eucalyptus, and acacia savannas are examples of this vegetation type, each
named for its scattered overstory trees. Savannas have
a wide range of characteristics. Most are fairly high in
species diversity. Decomposition rates are lower than
those in tropical forests, but are still generally very high,
and soils are quite variable. The acacia savannas of eastern and southern Africa are a mosaic of local communities created and maintained by fire and grazing, and
are preserved in Kenya, Tanzania, Zimbabwe, Botswana,
Namibia, and South Africa. The famous Serengeti Plains
of Tanzania are acacia savanna growing on nutrient-rich
volcanic sands, although most savanna vegetation

occurs on nutrient-poor soils. The
llanos of the Orinoco basin of
Venezuela and Columbia are grass
savannas maintained by the annual flooding of the Orinoco River,
which results in long periods of
standing water that prevent the
establishment of trees. The cerrado
of Brazil has an open canopy of
short trees with a distinctive twisted form; its acidic, heavily leached
soils are high in toxic aluminum
ions and low in plant nutrients. This
region has many endemic species,
and the overall plant species diversity approaches that of tropical rainforest. The pine savannas of Central
America occur on dry, low-nutrient
quartz sands. The savannas of
Southeast Asia are generally con-
sidered to be created and maintained by humans through
frequent fires.

Grazing can combine with fire to establish or maintain savanna vegetation. In East Africa, elephants debark
trees and, more dramatically, knock them over, opening
woodlands to invasion by grasses. The presence of grasses increases the attractiveness of the area to other grazing animals, such as antelopes and zebras, which eat and
trample tree seedlings and prevent woodland regeneration. Only thorny trees and shrubs can become established in this environment, leading to the creation of an
acacia savanna (Figure 19.18).

The world's greatest diversity of ungulates (hoofed
mammals) is found on the savannas of Africa, including
buffalo, giraffes, eland, impalas, gazelles, oryx, gerenuk,
wildebeest, plains zebra, rhinoceroses, elephants, and
warthogs. The vast herds of herbivores support carnivores that include lions, leopards, cheetahs, jackals, and
hyenas. In contrast, few Neotropical (Central and South
American) mammals or birds are restricted to savannas,
although capybaras (a very large, semiaquatic rodent)
are typical of llanos, and anteaters are common in South
American savannas and are endemic to the Neotropics.

Termites are especially notable residents of all tropical savannas; these detritivores are important in soil formation. The termite-eating aardvarks and pangolins that
live in the savannas are among the oddest mammals
on Earth.

Like many other biomes, tropical savannas are
threatened by human population growth. In Africa the
greatest challenge of preserving the large herbivores is
to find ways in which they can coexist with humans.
Beyond the problems of poaching, elephants, for example, are at risk when they wander off the nature reserves
and humans shoot them to prevent them from destroy-

ing vital food crops. Conservation plans that include economic incentives and participation in decision making by the people living in the region, and not just by outside conservation experts, are likely to be the most successful strategies for preserving these and other natural systems.

Deserts

Hot Desert

Deserts and semi-deserts occupy large areas of western North America, Australia, Asia, and Africa. Deserts occur where potential evapotranspiration exceeds actual evapotranspiration—in other words, where water can be lost to the atmosphere at a greater rate than it is normally made available to plants. Typically this means that desert regions receive less than 25 cm of precipitation per year, but such numbers can be deceptive; for example, the coastal regions of southern California receive less rainfall than this, but they are not deserts because cold offshore currents act to lower temperatures and increase relative humidity. Thus, San Diego, which has a mediterranean climate, typically receives less annual rainfall than Tucson, in the Sonoran Desert.

Hot deserts (also called "warm deserts") are found in two belts centered at 30° N and S latitudes. The hot desert biome includes the Mojave, Sonoran, and Chihuahuan deserts of North America, the Sahara, Kalahari, and Namib deserts of Africa, the Arabian Desert of Asia, the Atacama Desert of South America, and the deserts of Australia (Figure 19.19). These areas all have pro-

nounced seasonal wet/dry cycles, although some of the driest deserts may experience periods of many years without any measurable rainfall.

Precipitation in the desert typically comes at one time period during the year. In North America, the Mojave Desert receives almost all of its rain in winter, while the Chihuahuan Desert receives mainly summer rains. The Sonoran Desert is unique among the deserts of the world in that it has two periods of rainfall, a winter and a summer rainy season, with very dry autumn and spring seasons separating them. As a result, the Sonoran is the greenest of the world's deserts, with the highest standing biomass and productivity, and is one of the most floristically diverse.

As a rule, the lower the average annual rainfall, the greater the variance in both the amount and timing of annual rainfall. The "rainy season" in most deserts is a time when rain is more likely, although it is not uncommon to have very little rain in any given rainy season. Consequently, desert organisms must be adapted not only to limited moisture and high temperatures, but also to the lack of predictability in rainfall (see Chapter 3). At one extreme of adaptation, annual plants spend most of their lives as desiccation-resistant seeds, living only briefly as photosynthetic organisms when conditions are most favorable. At the other extreme, succulents and evergreen shrubs have photosynthetic tissue that is always exposed to potential desiccation. Drought-deciduous shrubs are intermediate, with little moisture-losing tissue exposed during the long dry periods. In part because of the unpredictability of rainfall, succulents are less abundant or absent in the driest deserts; they are characteristic of deserts with relatively greater and more predictable rainfall. In Death Valley, California, which is extremely hot and dry, annual plants are the dominant growth form. The other common growth form in Death Valley is drought-deciduous shrubs. In contrast, succulent cacti, agaves, and yuccas are abundant and diverse in the Sonoran Desert.

Because annual species predominate in hot deserts, their seeds are usually rela-

Figure 19.19
The Sonoran Desert outside Phoenix, Arizona. (Photograph by S. Scheiner.)

tively abundant in the soil, and consequently there are many **granivorous** (seed-eating) animals, particularly rodents, ants, and birds. Southern Arizona, for instance, has the highest diversity of rodents in the world, and almost the highest diversity of seed-eating ants. Although the seeds of desert annuals are abundant, they do not usually disperse well. Many desert plant species have adaptations that limit their ability to disperse. Areas from which plants have been cleared often are not recolonized for decades or even centuries.

Deserts are important areas for agriculture, grazing, and mineral exploitation, and they are becoming increasingly important for recreation as well. For almost all of human history and prehistory, human populations in deserts have been very low. In the last few decades, however, there have been rapid increases in human populations in some desert areas. The explosive growth of desert cities, such as Phoenix, Arizona, has meant that human activities are having massive effects on natural systems in such areas. Underground aquifers are being depleted by urban dwellers and by agricultural use, and surface waters (such as the Colorado River in the southwestern United States) face very heavy demands. Irrigated agriculture can lead to soil salinization in deserts, just as it does in grasslands. This process is thought to have caused the collapse of several ancient civilizations in the American Southwest. Environmentally damaging mining and oil extraction are additional sources of severe disturbance to biotic systems in many deserts. Overgrazing, by both livestock and feral ungulates, has widespread effects in many deserts. Heavy grazing disturbs the soil, increasing evapotranspiration rates by reducing cover. Use of recreational off-road vehicles, common across the American desert landscape (as well as in other regions), is highly destructive to the fragile vegetation. Because the vegetation is responsible for holding the soil in place, the loss of vegetation through grazing and other human activities leads to soil loss and severe erosion.

Cold Desert

Cold deserts (also called "cool deserts") are found in rain shadow areas in the centers of continents, the largest being the Kazakho-Dzungarian and Gobi deserts of central Asia and the Intermountain or Great Basin desert of North America (Figure 19.20). Unlike hot deserts, cold deserts experience freezing temperatures for part of the year. All deserts can be quite hot in the summer, however. Precipitation is generally concentrated in the winter. Cold deserts are dominated by shrubs and grasses and have few of the succulents found in some hot deserts, as most succulents are vulnerable to damage by ice crystal formation. Cold deserts also have few annual plants because moisture is available mainly as snowmelt, precisely when temperatures are too cool for rapid growth. There may also be historical reasons for the dearth of annuals. In the Great Basin of North America, for example, there are few native annual species, but the successful invasion of the Old World annual grass *Bromus tectorum* is evidence that annuals can do well there. Most of the anthropogenic threats to hot deserts also threaten cold deserts.

Figure 19.20
Sagebrush (*Artemesia,* Asteraceae) dominates this cold desert landscape near Mono Lake, California, part of the North American Great Basin. (Photograph by E. Ely/ Biological Photo Service.)

Figure 19.21
Alpine shrubland on Mt. Washington in the state of New Hampshire. (Photograph © M. P. Gadomski/Photo Researchers, Inc.)

Alpine and Arctic Vegetation

Grassland and Shrubland

Alpine shrublands extend upward from tree line on high mountains, and are often interspersed with alpine grasslands (Figure 19.21). The vegetation in these areas can include both what we typically think of as shrubs—multi-stemmed woody plants of short stature—and bizarre-looking plants that have a rosette form that can grow 3–5 m tall (see Figure 6.12). Alpine shrublands are found on all of the continents except Australia and Antarctica. In South America, they constitute a large part of the high-altitude vegetation type called paramo. **Heath** is the name for a general vegetation type that is dominated by low-growing ericaceous shrubs, and in some parts of the world alpine shrublands consist to a large extent of heath vegetation. In the northern temperate zone, the shrubs are typically *Salix* (willow) and *Betula* (birch).

Above tree line in mountainous areas or in areas with cold air drainage, alpine grasslands are the major community type (Figure 19.22). These grasslands are often dominated by **sedges**, plants in the family Cyperaceae that are closely related to and similar in growth form to grasses. These communities can also contain scattered shrubs, dwarf trees, and small, herbaceous perennial plants that are shaped like little cushions, thus grading into alpine shrublands and tundra.

Figure 19.22
Alpine grassland in Rocky Mountain National Park, Colorado. (Photograph by S. Scheiner.)

Temperate and tropical alpine areas differ in some important ways. Temperate alpine regions have long, cold winters, during which the ground is frozen and deep snow covers much of the vegetation for long periods. Snowmelt generally marks the start of spring growth. In contrast, tropical alpine regions, especially in very high mountains, have a strong daily temperature cycle, as if it were winter every night and summer every afternoon. Growth continues year-round, and the soil does not freeze (at least not deeply or for long periods of time). All alpine regions have very high solar energy input because the atmosphere is thinner than at sea level, so there is less to filter (i.e., absorb and scatter) incoming sunlight. The sunlight is particularly intense in tropical alpine regions because the angle of incoming solar radiation is most direct over the Tropics. The resulting high levels of ultraviolet radiation can damage biochemical processes in leaves (and can cause severe sunburn in humans).

The total area occupied by alpine vegetation of all types is low. Precipitation is generally limited in these areas, and productivity and species diversity are also low. The harsh climate, steep slopes, and low productivity of alpine regions have combined to keep human population sizes low from prehistory to the present, although several distinct cultures, such as those of the Incan and Tibetan peoples, have developed and thrived in high mountain regions. Even small human populations, however, can still have significant effects on alpine habitats. The cold temperatures mean than many processes, such as plant growth, soil formation, and decomposition, are very slow. As a result, human activities such as mining and grazing (the primary form of agriculture in these regions) can have very long-lasting effects.

Tundra

Tundra is found in areas where temperatures are too cold for tree growth. There are vast areas of arctic tundra in northern North America and Eurasia. Because there is little land in the Southern Hemisphere at high latitudes, there is little tundra there—only a few areas on the Antarctic Peninsula and neighboring islands are referred to as tundra. Conditions at the tops of high mountains, including wind and extreme cold, create alpine tundra.

The arctic tundra has long, dark, and often bitterly cold winters and short, cool summers with up to 24 hours of sunlight. The soil thaws each summer to a depth of only 25–30 cm, resulting in **permafrost**, a permanently frozen layer of soil, which severely limits root growth. Trees are excluded from the arctic tundra by a combination of factors, including permafrost and extreme cold. Temperatures below 40°C kill almost all living aboveground plant tissue, preventing long-lived plants such as trees from becoming established. Another factor lim-

iting tree growth is that emergent vegetation above the snow cover in winter would experience not only extreme cold, but also high winds, especially in alpine areas. Tundra soils are generally young and poorly differentiated (often inceptisols; see Table 4.1) because most arctic tundra sites were glaciated in the recent past. Litter decomposition is generally slow due to the cold temperatures and consequently low microbial activity and, secondarily, the anaerobic conditions in many places.

Alpine and arctic tundra have similarities and differences. Both have short growing seasons, and alpine tundra (at least in temperate areas, although not in the Tropics) is subject to severe cold in winter. In both types of tundra, most precipitation is in the form of snow, so that there are long periods during which plants are unable to take up moisture; in many settings (such as the alpine tundra at mid-latitudes in western North America), drought is common during the summer. Arctic and alpine tundra differ greatly in other factors, including day length, atmospheric pressures, partial pressures of gasses such as CO_2, exposure to ultraviolet radiation, and soil conditions.

Many genera and species found in Northern Hemisphere alpine tundra are also found in the arctic tundra. *Oxyria digyna* (mountain sorrel, Polygonaceae) is a plant found in the Arctic and in alpine areas of the western and eastern United States and Europe. A comparative study of arctic and alpine populations of this plant concluded that they are each adapted to specific local conditions. Compared with the arctic plants, the alpine plants had a lower leaf chlorophyll content, a higher respiratory rate, and a lower light saturation point. Alpine tundra species are often evolved from local lowland species. In the high Andes, for example, there are many species of high-altitude bromeliads, a group we normally associate with tropical forests.

The growth forms of tundra plants include low shrubs, taller shrubs alongside rivers, herbaceous perennials, grasses and sedges, mosses, and lichens. Different plant communities exist within the tundra, depending on topography, soils, and other factors. Tussock tundra is dominated by grasses and sedges (notably by the sedge *Eriophorum vaginatum*), deciduous shrubs, and evergreen shrubs in roughly equal abundance. Riparian bottomlands are dominated by relatively tall (1–4 m) willows and other shrubs. Wet sedge tundra, dominated by low-statured (< 20–30 cm) rhizomatous sedges, covers flat areas with restricted drainage, forming a broad expanse on the Arctic coastal plain of Alaska as well as in other areas. Soils typically thaw to only 25–30 cm here, and the freeze-thaw cycle often creates a characteristic ice wedge-shaped pattern in the soil. The soils of wet sedge tundra have very high organic matter content and may be covered with water during the summer because the frozen soil prevents drainage.

Plant species diversity in the tundra is very low. Many genera (such as *Salix*) and even some species (e.g., *Dryas octopetala*, *Eriophorum vaginatum*, and *Oxyria digyna*) are common across the circumpolar Arctic (Figure 19.23), and similar communities are found over wide ranges. Animal population densities and species richness are also low. However, the Arctic provides critical habitat both for species that are characteristic of this region, such as lemmings, and for others, such as wolves, that have been reduced to low numbers elsewhere. Brown bears, grizzly bears, polar bears, snowshoe hare, lynx, moose, and caribou (called reindeer in northern Europe) are some of the other large mammals that are notable components of the arctic fauna, and marine species such as walrus and various seal and whale species are native to, or rely heavily on, arctic habitats. The arctic tundra serves as critical nesting area for many bird species, particularly those associated with wetlands.

Human population sizes and densities in the tundra are extremely low (under 1 person/km^2), although humans have lived in, traveled through, and exploited arctic environments since prehistoric times. In spite of these low population sizes, humans have managed to pollute arctic environments in disproportion to our numbers, and the cold temperatures mean that the pollution persists for a long time. Small soil disturbances can last a very long time. In the arctic tundra, disturbance may result in the melting of permafrost, which leads to subsidence and soil compaction. Perhaps the most threatening of all ecological effects is human-caused global warming, which is predicted to be greatest in polar regions (see Chapter 22).

Summary

Biomes are large regions that differ from one another in the structure of their vegetation. They are strongly determined by climate—temperature, rainfall, and seasonality—and, to a lesser extent, by soil properties. While the boundaries of biomes are defined arbitrarily, the concept of the biome remains a useful one so long as it is clear that it is a descriptive categorization. There is a broad consensus among scientists on the classification of biomes.

Biomes are classified based on whether they are dominated by trees (forests and woodlands), shrubs, or grasses and other herbaceous species. Forests are further classified by whether the trees are evergreen or deciduous, broad-leaved or needle-leaved. Where conditions become too dry for trees, grasses predominate; grasslands can be tropical, temperate, alpine, or arctic. Shrubs dominate under a variety of conditions, including those that are too hot or dry for grasses, those with very seasonal precipitation, in some very cold regions, or as a secondary result of overgrazing or fire suppression.

All biomes are heavily affected by humans. Some, such as temperate deciduous forests, can recover relatively rapidly from human disturbance because many of their species disperse readily and grow rapidly. Others, such as hot deserts, recover poorly because their plants disperse poorly and grow slowly. Full recovery in still others, such as tropical forests, is difficult because of soil destruction and species extinctions in these highly diverse communities. Recovery of all damaged or degraded biomes is limited by continued encroachment, habitat destruction, and global climate change.

Figure 19.23
The vegetation of this Arctic tundra along Glacier Bay in southern Alaska is dominated by species of *Dryas* (Rosaceae). (Photograph courtesy of L. Walker.)

Additional Readings

Classic References

Braun, E. L. 1950. *Deciduous Forests of Eastern North America*. Blakiston, Philadelphia, PA.

Rübel, E. 1936. Plant communities of the world. In *Essays in Geobotany in Honor of William Albert Setchel*, T. H. Goodspeed (ed.), 263–290. University of California Press, Berkeley, CA.

Schimper, A. F. W. 1903. *Plant-Geography upon a Physiological Basis*. Clarendon Press, Oxford.

Tansley, A. G. 1939. *The British Islands and Their Vegetation*. Cambridge University Press, Cambridge.

Contemporary Research

Brais, S., C. Camire, Y. Bergeron and D. Pare. 1995. Changes in nutrient availability and forest floor characteristics in relation to stand age and forest composition in the southern part of the boreal forest of northwestern Quebec. *Forest Ecol. Mgmt.* 76: 181–189.

Pastor, J., Y. Cohen and R. Moen. 1999. Generation of spatial patterns in boreal forest landscapes. *Ecosystems* 2: 439–450.

Pitman, N. C. A., J. Terborgh, M. R. Silman and P. Nuñez V. 1999. Tree species distribution in an upper Amazonian forest. *Ecology* 80: 1651–1661.

Additional Resources

Bowden, C. 1993. *The Secret Forest*. University of New Mexico Press, Albuquerque.

Brown, J. H. and M. V. Lomolino. 1998. *Biogeography*. 2nd ed. Sinauer Associates, Sunderland, MA.

Davis, M. B. 1996. *Eastern Old-Growth Forests*. Island Press, Washington, DC.

Reisner, M. 1993. *Cadillac Desert: The American West and Its Disappearing Water*. Penguin Books, New York.

CHAPTER 20 *Regional and Global Diversity*

About a quarter of a million plant species are currently known worldwide, and there may yet be many undescribed species. Strikingly, some parts of the world have a great many more species than others. Some tropical forests can have as many as 1500 species of flowering plants (including 750 species of trees) in 1000 hectares—in Indonesian Borneo, researchers recorded more than 400 tree species in a 0.75 km² plot! In contrast, the northern regions of Canada and Russia may have only a few dozen species spread across hundreds of square kilometers. Why do some regions of the world have so many more plant species than others?

In this chapter, we describe regional and global patterns of diversity and discuss progress in understanding their causes. Local-scale patterns of diversity were discussed in Chapter 14. Here we look at patterns at two other scales: at the largest scale (across continents and the entire globe), and at intermediate scales (across communities within a region). At each scale, we examine some of the hypotheses that have been proposed to explain the patterns observed. Some of these hypotheses have now been ruled out, others explain the observed patterns at least in part, and still others are being vigorously debated by ecologists. No one explanation can account for all of the observed patterns of species diversity. Rather, ecologists seek to assess the relative importance of the various factors contributing to biodiversity in different places and over time. In Chapters 18 and 19 we saw how one of those factors—climate—determines patterns of community physiognomy and the distribution of biomes. Here we see how climate, among other factors, influences biological diversity.

Besides being an intrinsically interesting subject, understanding patterns of diversity is important for conservation. An ever greater proportion of the planet is occupied by human-dominated environments such as farms, homes, and industrial areas. Our domination of the landscape leaves wildlands more diminished every year, and alters those that remain. Old-growth forests and other relatively untouched plant communities are shrinking. With these changes in land use and in the characteristics of the remaining natural habitats comes an accelerating rate of species extinction (see Chapter 22).

If we wish to preserve species and community diversity, we need to understand current patterns of diversity and the processes that drive these patterns. Because plants are at the base of almost all terrestrial food webs (the only exceptions being anaerobic food webs), the diversity of animals, fungi, and microbes is often linked to plant diversity. Thus, the issues discussed in this chapter have practical as well as fundamental implications.

Large-Scale Patterns of Diversity

Scientists first became aware of large-scale patterns in species diversity as a result of the voyages of exploration and colonization undertaken by Europeans in the eighteenth and nineteenth centuries. Their ships often carried botanists and zoologists as part of the crew, the most famous example being the voyage of the *Beagle* (1831–1836), on which Charles Darwin served as ship's naturalist in. These naturalists made records of the plants and animals they found, brought back specimens, and added them to the growing catalog of described species. In the eighteenth century, Carl Linnaeus named over 9000 terrestrial plant species; today there are approximately 250,000 named plant species.

From these records, it soon became apparent that tropical regions were typically very rich in species, polar regions were very species-poor, and temperate regions had intermediate numbers of species. For example, Brazil has over 56,000 named plant species, the United States has about 18,000, while Canada has about 4200. Species richness in tropical regions is probably underestimated because many of these areas have still not been thoroughly explored. Even in North America, which has been studied by botanists for hundreds of years, new plant species are being named at a rate of about 60 per year. These plants are occasionally found right under one's feet—for instance, in 1997, a new species, the mustard *Lesquerella vicina* (Brassicaceae), was discovered by the botanist James Reveal growing in his neighbor's yard in Montrose, Colorado.

An explanation for the general latitudinal gradient in species richness may be the oldest major ecological hypothesis (Hawkins 2001). Between 1799 and 1804, Baron Alexander von Humboldt traveled through Mexico, Central America, and northwestern South America. He subsequently published a series of essays under the title *Ansichten der Natur* ("Views of Nature"), in which he described a global gradient in species diversity. He postulated that this pattern was due to differences in climate, specifically winter temperatures and the effects of freezing.

Besides the latitudinal gradient, many other patterns of diversity have been observed at regional and global scales. The regional diversity of a particular taxonomic group may be very similar to or extremely different from the diversity of that group in similar habitats in other parts of the globe (Schluter and Ricklefs 1993). Comparisons among Northern Hemisphere regions with similar conditions show that there are more plant species in east-

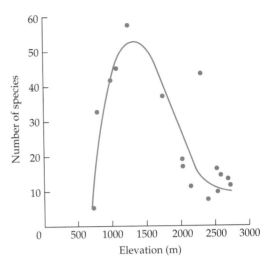

Figure 20.1
Plant species richness along an elevational gradient in the Santa Catalina Mountains in southern Arizona. Both the low- and high-elevation communities—the Sonoran Desert and boreal forest, respectively—are in extreme environments, resulting in a peak in diversity at intermediate elevations. (After Brown 1988; data from Whittaker and Niering 1975.)

ern Asia than in Europe, while North America has an intermediate number. In high mountain ranges, there are typically more species at mid-elevations than elsewhere (Figure 20.1). Within continents, there are longitudinal gradients of diversity; in the boreal forest of North America, species richness within sites is higher in central Canada than it is in either western or eastern regions (Figure 20.2). In the temperate zone, between approximately 35° and 60° latitude, there is a peak of within-community species diversity (Figure 20.3). In this chapter we explore some of the hypotheses that have proposed to explain these patterns of terrestrial plant species diversity.

Levels of Explanation

How have ecologists attempted to explain the causes of variation in species diversity along geographic or environmental gradients? The processes that we use to explain a pattern change with the temporal and spatial scale of that pattern. At short to medium time scales—up to centuries or millennia—ecological processes dominate. The physiological and ecological properties of each species result in species sorting among communities within a region. Over longer time periods, these properties, together with the history of each species, help determine the distributions of species at regional and continental scales. At even longer time scales, ecological and historical factors affect the distributions of species across the globe. At the longest time scales—from millennia to hundreds of thousands or millions of years—

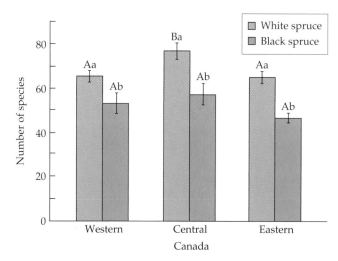

Figure 20.2
Number of species (mean ± 1 standard error) in 30 × 30 m² plots in boreal forests of Canada. Species richness is higher in the central region, especially in white spruce communities. Bars with different capital letters indicate statistically significant differences among regions within community types. Bars with different small letters indicate statistically significant differences within regions among community types. (Data from Qian et al. 1998.)

allows it to occupy a new habitat. At regional scales, these processes of adaptation can lead to the formation of new species. If the new species is different enough, it may coexist in the same habitat or landscape with the species from which it evolved, increasing species richness in that landscape. At continental to global scales, as new species appear, they may migrate to new regions, so that species richness increases over an entire set of regions.

Some proposed explanations for patterns of species diversity assume that species distributions are in equilibrium with current ecological conditions, while other explanations do not make that assumption. Whether a given explanation assumes that diversity is at equilibrium or not depends, in part, on scale. Within a single community, for example, disturbance can increase diversity (see Chapter 14); this is a nonequilibrium explanation for diversity. But at a larger scale, across communities in a landscape or region, disturbance in particular communities balanced by migration among communities may determine diversity (see Chapter 17). This is an equilibrium explanation for diversity, although the equilibrium is a very dynamic one.

evolutionary processes (adaptation, isolation, speciation, and extinction) predominate. Some processes, including range expansions and contractions and extinctions of populations and species, may happen over long periods of time or very quickly. Humans and their effects on the environment have greatly accelerated these processes for some species (see Chapter 22).

The mix of processes governing diversity will also differ depending on spatial scale. At local scales—within communities—interactions among species are of primary importance in determining species richness. These interactions include competition, herbivory, and mutualisms (see Chapters 10 and 11). As our scale expands to the level of the landscape or sets of communities, metapopulation processes may become increasingly important. These processes include migration among communities and extinction within communities (see Chapter 17). Finally, at continental to global scales, migration and extinction in response to climate change become important. We discuss one example, migration following the most recent glacial retreat, below and in Chapter 21.

Evolutionary processes also operate at all of these scales, although over longer time spans. At small spatial scales, local adaptation can occur (see Chapter 6). In response to competition with other species, a population may evolve to occupy a different niche, thereby reducing competition and increasing local diversity. Similarly, a population may evolve a defense against an herbivore that

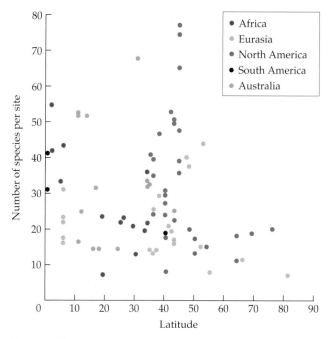

Figure 20.3
A latitudinal pattern in vascular plant species richness. Each point represents the mean number of species per site averaged across sites in a particular landscape. Species richness of tropical landscapes is underestimated in this analysis. Latitude is given in absolute degrees so that landscapes in the Northern and Southern Hemispheres are superimposed. The latitudinal gradient is very weak because many factors affect large-scale patterns of diversity (see text). (After Scheiner and Rey-Benayas 1994.)

Explanations for Latitudinal Gradients

A natural tendency in the face of the strong patterns of latitudinal diversity is to seek simple, unifying explanations for those patterns. If, for example, we can account for the temperate-tropical diversity gradient on a number of continents with a single explanation, we are likely to have a deeper understanding than if we try to explain the patterns on different continents separately by invoking details of the floras and histories of each continent. Several theories have been put forward to explain latitudinal diversity patterns (Table 20.1), some of which we delve into here.

A useful starting point for evaluating hypotheses is to build a **null model**. Used throughout the sciences, null models describe the patterns that would be observed if only random processes were operating. One such null model is the bounded ranges hypothesis (Colwell and Hurtt 1994; Willig and Lyons 1998). Developed as a possible explanation for the latitudinal gradient of animal diversity, this hypothesis uses a simple geometric argument. Species ranges are bounded by the North and South Poles; otherwise, species can appear any-

where. Random placement of species on a uniform globe would result in many ranges overlapping along the equator, thus creating a peak in diversity in tropical regions and decreases in diversity toward the poles.

Because the bounded ranges model includes no ecological processes—only geometry—it serves as a useful baseline against which to compare observed patterns. For example, the model was found to account for some, but not all, of the latitudinal pattern of diversity in New World bats (Lyons and Willig 1999). Actual species diversity was higher in the Tropics and lower at higher latitudes than predicted by the model. Other biological processes, therefore, are needed to explain the entire pattern. Thus, a null model provides a useful starting point for discovering where interesting patterns lie and what about those patterns requires explanation. No analysis of this kind has yet been done with plants.

A related idea depends on the observation that species' range sizes tend to increase with latitude and elevation (Rapoport 1982; Stevens 1992). This pattern is known as Rapoport's rule. Because temperate-zone species have larger ranges than tropical species, temperate species' ranges are more likely to extend into the

Table 20.1 Theories explaining the latitudinal gradient in species richness

Theory	Mechanism	References
Ecological and evolutionary theory		
Available energy	Increased energy supply leads to greater resource availability and thus larger population sizes. With more resources available, more species can coexist. Larger populations have higher speciation rates and lower probabilities of extinction.	1, 2
Evolutionary theories		
Speciation rate and temperature	The Tropics have higher rates of speciation due to warmer temperatures, which decrease generation times, increase mutation rates, and increase the speed of selection.	3
Area and latitude	There is more land area in the Tropics than in the temperate zones. Larger areas support more species due to higher rates of speciation and lower rates of extinction.	4, 5
Time	There are more species in the Tropics because the communities there are older.	6
Climatic stability	The Tropics have not been subjected to the large climatic fluctuations experienced by temperate and polar regions during recent glaciations, so more species have evolved or survived in the Tropics.	7
Ecological theories		
Bounded ranges	Peaks in species richness near the equator arise from a random association between the size and placement of species ranges.	8, 9
Rapoport's rule and species spillover	Relatively low climatic variability at low latitudes leads to narrower species ranges. Species at higher latitudes are thus more likely to spill over into the Tropics than vice versa, inflating species richness in the Tropics.	10
Spatial heterogeneity	The Tropics contain more different kinds of habitats than other regions.	
Predation	Higher levels of herbivory in the Tropics prevent competitive exclusion.	
Disturbance	Diversity is highest in regions with frequent disturbances, which is true of the Tropics.	

References: 1, Wright 1983; 2, Wright et al. 1993; 3, Rhodes 1992; 4, Rosenzweig 1995; 5, Terborgh 1973; 6, Fischer 1960; 7, Sanders 1968; 8, Willig and Lyons 1998; 9, Colwell and Hurtt 1994; 10, Stevens 1989.

Tropics than tropical species' ranges are to extend into the temperate zones. The result will be more species in the Tropics—species that live only in the Tropics plus those that have overlapping ranges from neighboring temperate regions. Fine (2001) tested this hypothesis using data for North American trees. He found that few tropical species are also found in extratropical areas, and concluded that Rapoport's rule cannot explain the latitudinal diversity gradient of plants.

Another simple explanation for latitudinal diversity patterns, especially for the most extreme environments, is differences in opportunities for growth. At very high elevations and latitudes, conditions are so unfavorable that few species are physiologically capable of persisting there. At the summit of the Mauna Kea volcano in Hawaii (4700 m elevation), there is one vascular plant species present, a grass, with only a few individuals in an entire hectare. The climate, with very cold air and little precipitation, means that the soil is poorly developed. Not surprisingly, there are few organisms of any kind in this environment, while plant species richness is very high on the lower slopes of the mountain, especially on the wet side of the island (see Figure 18.10).

An analogous argument explains why species diversity is so low in cold, arid Patagonia, Argentina, in southernmost South America, and so much greater in the Amazon basin near the equator, and why diversity is lower in the Canadian tundra than in the rainforests of Central America. Unfortunately, this simple explanation does little to explain the rest of the latitudinal (or elevational) diversity gradient. More complex explanations are needed.

One major nonequilibrium hypothesis for the increase in diversity toward the equator is that communities in the Tropics are older than those closer to the poles. If communities in the Tropics have remained intact for a longer period of time, then the species in those communities should have had more opportunity to diversify into many species. Given enough time, it has been proposed, there might be more speciation in temperate and polar communities, or more migration of species from low to high latitudes, and the latitudinal gradient might slowly disappear. A related idea is that tropical communities have experienced fewer extinctions because they have experienced fewer climatic fluctuations, and in particular have never been glaciated.

Explanations such as these that invoke time and climatic stability have a number of serious weaknesses. For example, there is much evidence that there has been sufficient time for plant species to migrate into temperate and polar regions (see Chapter 21). In addition, a great deal of evidence makes it clear that tropical communities were strongly affected by recent glaciations, even though they were not covered by glaciers themselves. Similarly, the existence of elevational gradients, for which migrational lag cannot be the explanation, weakens this hypothesis.

In contrast to the notion that climatic stability in the Tropics has led to increased diversity, some ecologists have proposed the opposite argument, suggesting that speciation and species coexistence are most likely to occur with more frequent disturbances. One reason to believe that such a relationship might exist is that disturbance may reduce the chance of competitive exclusion. However, there is no evidence that tropical communities differ from temperate ones in frequency of disturbance. In addition, as originally formulated, this hypothesis was unclear as to the scale of the disturbance effects involved. That is, it might explain differences in local diversity, but not the large difference in species diversity between Brazil and the United States.

Disturbance might also increase diversity through its evolutionary effects. For example, speciation rates might have been greater in the Tropics because of the repeated contractions and expansions of the tropical rainforest during repeated glaciations. Geographic isolation, which would have been heightened by these shifts, is known to promote speciation (see Chapter 6). However, both temperate and tropical species should have been subjected to similar kinds of isolation during recurrent periods of glaciation and warming. The distribution of species in northern boreal and polar regions show evidence for this pattern of speciation (Figure 20.4). In Chapter 21, we will look at general patterns of speciation and diversity as affected by long-term climate changes.

An old and influential equilibrium explanation for the latitudinal diversity gradient is the available energy, or productivity, hypothesis (Preston 1962). The notion is simple: increasing productivity means that more energy is available for growth and reproduction and that more individuals can therefore live in the same amount of area (see Chapter 14). Because a sustainable population requires some minimum number of individuals (see Chapter 7), more individuals in an area means that they can be divided up among more species, resulting in higher species richness.

On an evolutionary time scale, larger populations reduce the chance of extinction, again resulting in higher species richness. Analysis of the latitudinal pattern shown in Figure 20.3 revealed that productivity differences were associated with about 25% of the total variation in species richness and accounted for most of the latitudinal pattern (Scheiner and Rey Benayas 1994). However, while productivity is an important explanation—indeed, it may be the most important single factor in the latitudinal gradient—it is certainly not the only factor causing the gradient. The effects of productivity on diversity also vary with scale, as we will see later in this chapter.

Figure 20.4
Distribution of the *Lathyrus maritimus* (Fabaceae) species complex around the Northern Hemisphere polar region. Populations of this plant in different regions differ morphologically, particularly in whether and where leaves have hairs. The different morphological types may be different species or incipient species. The circumpolar distribution of this species complex is typical of many polar and boreal species and genera.

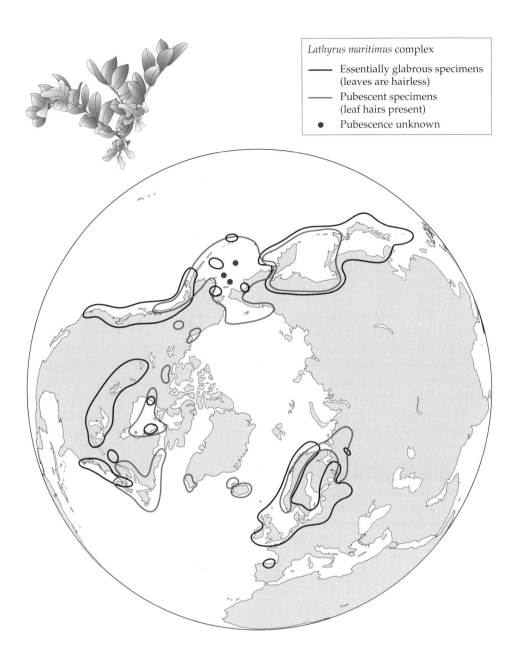

Lathyrus maritimus complex

—— Essentially glabrous specimens (leaves are hairless)

—— Pubescent specimens (leaf hairs present)

● Pubescence unknown

A simple and currently controversial equilibrium theory is often called the area hypothesis. It proposes that the Tropics are more species-rich because the total tropical land area is greater than the total temperate land area (Rosenzweig 1995). This hypothesis contains both ecological and evolutionary components. Ecologically, one can reason that as area increases, spatial heterogeneity also increases. Because the Tropics encompasses more area, it contains more habitat types, and therefore more species. Furthermore, over evolutionary time, a larger area affords greater opportunities for geographic isolation, and therefore promotes greater rates of allopatric speciation. Michael Rosenzweig and Klaus Rohde have conducted a vigorous debate over this issue (Rohde 1992; Rosenzweig 1995; Rohde 1997; Rosenzweig

and Sandlin 1997; Rohde 1998). While the two ecologists agree that area has some effect, they have been unable to agree on how important the area effect is relative to other factors such as productivity. While it is relatively easy to identify which factors are important in patterns of diversity, assigning exact levels of importance is much more difficult.

Continental Differences

If we compare similar regions in eastern Asia, North America, and Europe, we find that eastern Asia has the most plant species, while Europe has the fewest. This pattern holds over a range of taxa. For instance, if we look at the number of tree species across all plant families, eastern Asia has the greatest number of species, North America has an intermediate number, and Europe has the fewest. These patterns are also robust across higher taxonomic levels: the same relationships hold for orders, families, and genera of trees (Latham and Ricklefs 1993).

What can explain these patterns? The first explanation relies on differences in species extinctions. A key difference between Europe and the other two regions has to do with an accident of topography. The Alps, the main mountain range in Europe, runs in an east-west direction, while most or all of the principal ranges in eastern Asia and North America run north-south. During periods of glaciation, climate zones are pushed southward. At the height of the most recent glaciation in North America, for example, the edge of the ice sheet was in Wisconsin and boreal forests extended as far south as the Gulf Coast (see Chapter 21). As climate zones moved southward, the plants followed them, and the southern portions of the United States provided refugia for many species.

In Europe, however, the Alps created a barrier to this southward migration. Species that were adapted to warmer conditions were unable to migrate over the tops of the mountains to escape from the cold, and went extinct. As the glaciers retreated, the species in North America and Asia that had retreated southward migrated northward again. But in Europe, those species had disappeared and were not available to recolonize the newly ice-free temperate areas. The end result of these events, repeated during each of the Pleistocene glaciation and warming periods, is the greatly reduced species diversity of Europe (Table 20.2). North-temperate and boreal communities in North America contain about 50% more vascular plant species than equivalent European ones (approximately 18,000 versus 12,000: Niemelä and Mattson 1996). Fossil evidence supports this explanation: European forests before the Pleistocene had much greater species diversity than they do now (Latham and Ricklefs 1993). This explanation depends on historical contingency: the pattern of diversity is the result of a particular sequence of historical events in a particular place, not of a general or fundamental process that explains all such patterns.

The difference between the floras of North America (particularly the eastern part of the continent) and eastern Asia is also thought to be the result of a particular history, but one operating on a longer time scale and depending on rates of evolution—the rise of orders, families, and species—rather than on extinctions (Latham and Ricklefs 1993; Qian and Ricklefs 1999). China and the United States are approximately the same size and contain a similar range of ecological conditions. Yet China contains 60% more species of vascular plants (29,188 vs. 17,997). Much of this difference is caused by the fact that plant families in China tend to have more species than the same families in the United States.

Hong Qian and Robert Ricklefs (1999) hypothesize that these differences in diversity are largely a result of continental-scale geographic relationships. The southern United States is separated from tropical regions by water to the east and south and by very dry conditions to the west. In contrast, southern China is directly connected with tropical and subtropical regions. The temperate regions, Qian and Ricklefs hypothesize, were repeatedly invaded by species from adjacent subtropical and tropical areas. These species then underwent extensive adaptive radiation in the temperate regions. What makes this argument especially compelling is the finding that the greatest differences in diversity between the two continents are found in older taxonomic groups with strong tropical affinities. Europe also suffers from the same limitation as North America in this regard because the Mediterranean Sea and the arid expanse of North Africa create a barrier between it and tropical and subtropical floras. Thus, a happenstance of geography appears to have been largely responsible for differences in species evolution over very long time periods as well as extinction rates over a shorter ecological time scale.

Table 20.2 The number of woody species of various genera that went extinct in Europe during the Pleistocene but are still found in North America

Genera	Number of species
Gymnosperms	
Chamaecyparis	3
Sequoia	1
Taxodium	2
Thuja	2
Torreya	2
Tsuga	4
Angiosperms	
Asimina	3
Carya	11
Diospyros	2
Lindera	2
Liquidambar	1
Liriodendron	1
Magnolia	8
Morus	2
Nyssa	3
Persea	1
Robinia	4
Sabal	3
Sapindus	2
Sassafras	1

Source: Niemelä and Mattson 1996.

Other Geographic Patterns

Species Diversity and Patterns of Overlap

There is often an increase in diversity in **transition zones**, areas of contact between communities, regions, or biomes. This increase in diversity occurs because the transition zone contains species from the areas on either side (Shmida and Wilson 1985). For example, the mid-latitude peak in terrestrial plant species richness in North America (see Figure 20.3) is associated with the area around the Great Lakes in the biogeographic region called the hemlock-white pine-northern hardwoods. This region is a transition zone between the boreal forest to the north and the eastern deciduous forest to the south.

A similar pattern is found in the Iberian Peninsula, where there are two transition zones (Figure 20.5A). One transition zone is found along the Atlantic coast and the Pyrenees mountain range, where the Eurosiberian flora abuts that of the Mediterranean region. The other is found in the southeast, in the region adjacent to Africa. These two areas have more species than others in Iberia because species ranges tend to overlap there (Figure 20.5B).

Within the boreal forest of North American there is also a longitudinal pattern of species richness, as mentioned above (Qian et al. 1998) (see Figure 20.3). The differences in species composition found along this longitudinal transect are apparently related to migration following the most recent glaciation (see Chapter 21). The higher species richness in the middle of the continent may be due to an overlap of species ranges, caused by the different pathways from the east and west taken by plants expanding their ranges northward from refuges in the southeast and southwest after the glacial retreat. This explanation is a longitudinal version of the bounded range hypothesis.

While the transition zone effect provides part of an explanation for the mid-latitude peak in species richness in North America, an additional explanation can be found by looking at patterns of seasonality. The mid-latitude landscapes with the greatest species richness have highly seasonal continental climates. This seasonality allows for the coexistence of species with different life history patterns. In many temperate deciduous forests, for example, there is a suite of plants that are active only in early spring. During this time, it is warm enough for plants to grow close to the ground, but too cool for growth at greater heights. Consequently, herbaceous plants in the understory can use this period for growth before the trees leaf out and cut off their access to light. Seasonality creates this temporal niche. Note that this argument is analogous to the argument that tropical diversity is high due to spatial heterogeneity; in this case, however, the heterogeneity creates opportunities in time, rather than across space.

(A)

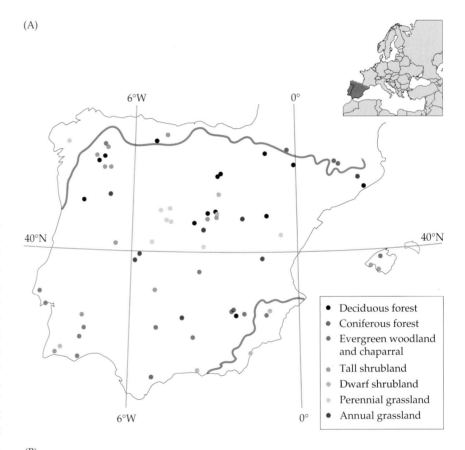

- • Deciduous forest
- • Coniferous forest
- • Evergreen woodland and chaparral
- • Tall shrubland
- • Dwarf shrubland
- • Perennial grassland
- • Annual grassland

(B)

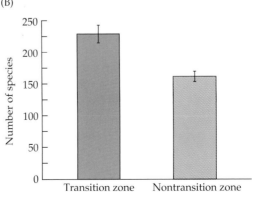

Figure 20.5
(A) Map of Iberia, showing location of the two transition zones (areas along green lines). Each point is the center of a large sampled landscape. (B) Landscapes in the transition zones (areas along the green lines) contain more species (mean ± 1 standard error) than those elsewhere in Iberia. (Data from Rey Benayas and Scheiner 2002.)

Figure 20.6
Tree species richness as measured in 2.5° × 2.5° quadrats across North America. The contour lines connect points with the same approximate number of species per quadrat. (After Currie and Paquin 1987.)

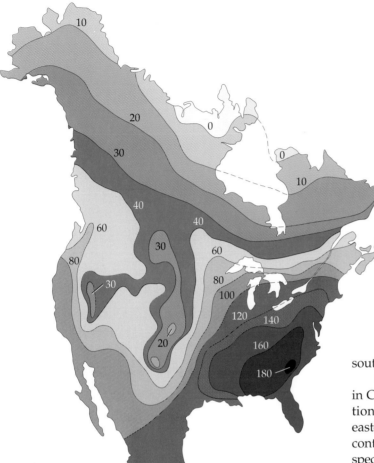

As with global patterns, different continental patterns of diversity exist for different subsets of plants. For trees in North America, the highest number of species is found in the southeastern United States, with steady decreases to the north—because of decreasing tempera-

tures—and west—because of decreasing precipitation (Figure 20.6). This pattern contrasts with the pattern for the entire flora, which shows a peak around the Great Lakes (see Figure 20.3). The mid-latitude peak does not exist for trees, presumably because seasonal niche partitioning occurs among species with different growth forms, and so would not be expected to hold for trees, which belong more or less to cies richness can differ between the entire plant community and a single functional group.

Endemism, Centers of Diversification, and Isolation

All of the patterns we have discussed so far involve gradual changes in species diversity over large spatial scales. Diversity can also vary in patchy patterns, in which some regions have many species compared with adjacent regions. Often these differences are a result of evolutionary processes. For example, high species diversity in a region may occur because a particular lineage had a large evolutionary radiation in that region. Cacti, for example, are especially diverse in central and southern Mexico because many species originated there.

Geographic isolation promotes speciation, as we saw in Chapter 6. Islands are an obvious case of such isolation. Madagascar is a very large island off the southeastern coast of Africa that became separated from that continent 90 million years ago. During that time, many species evolved in isolation there. Madagascar is now one of the most floristically diverse places on Earth, with approximately 10,000 native plant species, about 80% of which are **endemic** (found nowhere else). Similarly, the Pacific islands of Hawaii, New Zealand, and New Caledonia have all been isolated for millions of years, and all have very high species richness and very high proportions of endemic plants (Table 20.3).

Table 20.3 Diversity of plant genera and species and percentage of endemism in the native flora on some isolated islands and island groups

Island	Area (km²)	Total genera	Endemic genera	Total species	Endemic species	Endemism (%)
Cuba	114,914	1308	62	5900	2700	46
Hispaniola	77,914	1281	35	5000	1800	36
Jamaica	10,991	1150	4	3247	735	23
Puerto Rico	8,897	885	2	2809	332	12
Galápagos	7,900	250	7	701	175	25
Hawaii	16,600	253	31	970	883	91
New Zealand	268,000	393	39	1996	1618	81
New Caledonia	17,000	787	108	3256	2474	76

Source: Gentry 1986.

Even continental areas may have endemic species. Rare endemic plants may be restricted to unusual habitats, such as serpentine barrens (see Box 15A). In some cases a species may be endemic to an area because it is the product of a recent speciation event and has not yet spread. Another possibility is that species may be prevented from spreading from their place of origin. If speciation rates are high in a region and species cannot spread, that region will become species-rich and have a large number of endemic species

The general problem of explaining variation in speciation rates is still one of the greatest challenges in evolutionary biology. Other than geographic isolation, there is little agreement about the mechanisms of speciation. Factors that have been hypothesized to cause high levels of speciation or endemism in plants include protection from herbivores, low productivity, the existence of many local topographic barriers, habitat diversity, a low frequency of major disturbances, and isolation. The case of the astounding levels of plant diversity found in the fynbos of the Cape region of South Africa presents many of these issues and the difficulties in explaining them (Box 20A). Such centers of diversification are also important in conservation, as we will see in Chapter 22.

Relationships between Regional and Local Diversity

Most studies of the causes of diversity have focused either on local patterns of diversity (see Chapter 14) or on regional or global patterns, with few studies seeking to connect the two scales. Recently ecologists have started to make this connection, although the roles of large-scale and small-scale processes and how they interact to determine local and regional diversity are still not well understood.

One question we can ask is whether local diversity is set by the number of species available in the regional species pool, or whether it is determined by processes acting locally. We can test these two hypotheses by sampling a series of regions, measuring both the total number of species in each region and the number of species in a small subset of the region, and then plotting local species richness as a function of regional species richness (Figure 20.7A). If local species richness is determined by the size of the regional species pool, then we should see a straight-line relationship, in which a larger regional species pool leads to more species locally. On the other hand, if local processes, such as competition, limit the local number of species, the relationship should be concave; at some point, adding more species to the regional pool should have little or no effect on local species richness. In an analysis of plant communities in Estonia, Meelis Pärtel and colleagues (1996) found that local diversity was a function of regional diversity, with

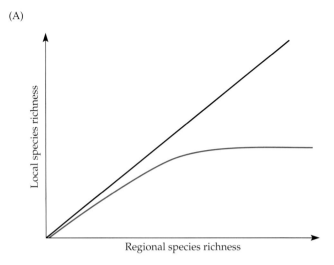

(A)

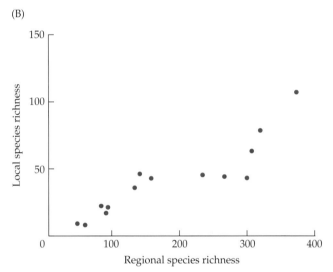

(B)

Figure 20.7
(A) Two possible relationships between the size of the regional species pool and local species richness. If local richness depends only on the size of the regional pool, then local richness should increase with regional richness (black line). If local richness depends in addition on local processes, then at some point, adding species to the regional pool should not result in any more species locally (green line). (B) The relationship between regional and local species richness for 14 plant communities in Estonia, including grasslands, shrublands, and forests. (After Pärtel et al. 1996.)

no leveling off of the relationship at the highest regional diversity (Figure 20.7B). As of yet, few other studies have quantified this relationship, and a general answer to this question awaits more research.

Some ecologists are attempting to examine the mechanisms that link local and regional diversity. Jonathan Shurin and Emily Allen (2001) modeled the conditions under which local species interactions (competition and predation) contrast with regional patterns in determining species richness at both local and regional scales. Their model included both deterministic, equilibrium processes at the local scale and nonequilibrium, stochastic processes at the regional scale. They found that predation expands the range of environments in which competitors can co-exist at a regional level, promoting higher levels of regional diversity by preventing competitive exclusion in some locations (see Chapter 11). The effect of these interactions on local diversity was

dependent on the regional context and on the abilities of species to disperse among sites (see Chapter 17). Thus, there is a reciprocal relationship between local and regional scales: feedbacks between processes at the two scales result in interactions such as predation leading to increased diversity at both scales. As with any mechanism shown to be plausible through modeling, the current challenge is to determine whether this mechanism is important in nature. Currently, little consensus exists

Table 20.4 Theories explaining the hump-shaped productivity-diversity relationship observed at regional scales

Hypothesis	Mechanism	References
Evolutionary theories		
Available habitat	More species have evolved to tolerate more common habitats, and the most common habitats are of intermediate productivity.	1, 2
Speciation rates and productivity	The difference between rates of speciation and extinction is maximal at intermediate productivities.	3
Rate of evolution	The rate of evolution is maximal at intermediate productivities.	3
Adaptive trade-offs	More species can evolve tolerance to intermediate conditions.	3
Ecological theories		
Competition for resources and resource heterogeneity	Species compete for resources, and more species coexist where resource heterogeneity is greater. Effective resource heterogeneity first increases with productivity and then decreases due to light limitation.	4,5
Transport-limited competition	Consumption of nutrients by plants and sufficiently high nutrient supply: nutrient diffusion ratios generate heterogeneity in soil resources and allow the coexistence of competing species. Increased soil fertility allows for the coexistence of more species until light limitation reduces species richness.	6
Predation (herbivory) and resource competition	Species compete for resources, and there is a trade-off between competitive ability and vulnerability to predators. Predator density increases with productivity, resulting in a shift to more predator-resistant taxa at higher productivities. Spatial or temporal variation in resources allows coexistence of multiple species at low to moderate productivities, but at high productivities only the most predator-resistant species persist.	7
Intertaxonomic competition	A given taxon can exploit only a portion of the productivity range and is then replaced by a competing taxon. Species richness within a taxon increases and then decreases with productivity.	2, 8, 9
Dynamical instability	Increased productivity reduces stability in multitrophic-level interactions.	10, 11
Disturbance and competitive exclusion	Population growth rates are positively correlated with productivity. Disturbance acts to reduce population density and slow the time to competitive exclusion. Therefore, at low productivities, species diversity is low because most species' growth rates are not high enough for them to tolerate disturbance. At high productivities, species diversity is low because of competitive exclusion. Species richness is thus maximal at intermediate productivities.	12, 13
"Hump-backed model," including environmental stress, competitive exclusion, and disturbance	Plant species diversity is low in low-productivity habitats due to environmental stress. In high-productivity habitats, dominant species competitively exclude other species, and species diversity is low. At moderate productivities, species density is high because moderate levels of stress and disturbance suppress competitive dominants.	14
Predator (herbivore)-victim ratios	As productivity increases, predators absorb much more than a proportional share and reduce the diversity of consumers.	8, 15
Temporal covariance	Temporal variation in resources allows the coexistence of competing species. If temporal variation correlates with productivity, then a hump-shaped productivity-diversity relationship may result.	8, 16

Note: General discussions of these theories can be found in Tilman and Pacala 1993; Huston 1994; Abrams 1995; Rosenzweig 1995; and Leibold 1998.
References: 1, Denslow 1980; 2, Rosenzweig and Abramsky 1993; 3, VanderMeulen et al. 2001; 4, Tilman 1982, 1988; 5, Abrams 1988; 6, Huston and DeAngelis 1994; 7, Leibold 1999; 8, Rosenzweig 1995; 9, Tilman and Pacala 1993; 10, Rosenzweig 1971; 11, Wollkind 1976; 12, Huston and Smith 1987; 13, Huston 1979, 1994; 14, Grime 1973, 1979; 15, Oksanen et al. 1981; 16, Chesson and Huntley 1988.

Box 20A

The Fynbos and the Cape Region of Africa

You may never have heard of the fynbos, but it deserves to be far better known. The fynbos is a small region in the southernmost part of Africa that has some of the highest levels of plant diversity on Earth. Why should anyone care about this tiny area with its weird plants and its animals with odd names such as grysbok, klipspringer, and Cape dainty frog? We should care because, among other reasons, it is a priceless resource for advancing our understanding of biodiversity, evolution, and ecology.

The vegetation of the fynbos is so distinctive that a few scientists have elevated it to the status of a unique biome. Its distinctiveness was recognized in the early twentieth century by the German botanist Adolph Engler, who classified it (including the surrounding plant communities in the Cape region of Africa; see Figure 22.14) as the smallest of the world's six great floristic divisions.* The fynbos is home to about 8000 species, some 5000 of which are endemic. It has an extremely high concentration of plant species, with 1300 species per square kilometer and up to 121 species per square meter. The latter statistic points to a truly remarkable feature of the fynbos: its high diversity at small to medium spatial scales.

Derived from the Dutch words *fijn bosch* ("fine bush"), fynbos (pronounced roughly "fane boss") is named for the narrow-leaved, grasslike plants and sclerophyllous evergreen shrubs with fine, small or rolled leaves that are dominant there (see Figure 19.16). It has a mediterranean climate, with cool, wet winters and hot, dry summers. Like other mediterranean vegetation, the fynbos is adapted to frequent fires (which occur predominantly in late summer). The soils are infertile, acidic, sandy, and very ancient. The topogra-

A lowland fynbos landscape. The fynbos, a region of about 46,000 km² (about the size of the states of Maryland and Massachusetts together), boasts a mediterranean climate and landscapes that range from the heathlike lands shown above to craggy montane regions. The vegetation is highly endemic, with many species found only in this tiny region. (Photograph courtesy of E. Orians.)

phy is diverse, with sandstone mountains, steep, dissected valleys, limestone outcrops, shale flats, and coastal plains.

Fynbos is defined by the presence of plants in the Restionaceae (Cape reeds). Restios are somewhat grasslike in form and replace grasses in the fynbos. The Ericaceae (the heaths—generally low-growing, shrubby plants typically found on acid soils) is another important fynbos plant family; the genus *Erica* reaches a peak of diversity here, with about 600 species, whereas there are fewer than 30 species in the entire rest of the world. Other characteristic plants of the fynbos are members of the Proteaceae and the Asteraceae; Asteraceae is represented by over 1000 fynbos species, more than half of them endemic. There are a large number of plants in the Iridaceae with particularly beautiful flowers, including many that are grown horticulturally. Unlike vegetation in most other places, the fynbos typically has no one dominant species or group of species.

Various distinctive local plant communities occur as part of the fynbos

and the surrounding Cape region, with wonderful names such as the Succulent Karoo and Nama Karoo, the Mountain Fynbos, and Renosterveld (small leafed shrublands). Animals are far less diverse in the Cape region than are plants, although there are rare and endemic species in a number of animal taxa, including fishes, reptiles, mammals, birds, and many insect groups. Most dramatic of all may be the highly specialized animal-plant interactions on which many fynbos plant species depend for reproduction and seed dispersal. *Protea acaulis* and many other proteas are totally dependent on mice for pollination. Their flowers are held near the ground and produce a strong odor that mimics yeast. Mice drink the stinky nectar, carrying pollen on their noses to the next plant. Ants are very important seed dispersers for many fynbos plants, such as those in the genus *Callistemon* (bottlebrush, Myrtaceae). The seeds of these plants are attached to special structures called elaiosomes, which offer a food reward to the ants for carrying the seeds off.

*The six floral kingdoms categorized by Engler are the Cape, Antarctic, Australian, Neotropical (Western Hemisphere tropical regions), Paleotropical (African and Asian tropical regions), and Holarctic (the entire Northern Hemisphere north of the Tropics).

The great diversity of the fynbos is a geologically recent phenomenon. Until some 5 million years ago, the Cape region was covered with subtropical forests. As the climate became cooler and drier, mediterranean shrublands spread into the Cape from the north. Much of the diversification of the flora of the fynbos occurred during the large climatic fluctuations of the past 2 million years, as glacial periods alternated with warm periods.

Why is this tiny region home to so many plant species? How did they evolve, and how do they coexist? Why is the fynbos so rich in species when most other mediterranean shrublands are low in plant diversity? The poor, acidic soils are hypothesized to be an important factor in the evolution of this astounding diversity—but the poor, acidic soils of the pine barrens of North America are supposed to explain the very low plant species diversity there!

The soils of the fynbos are very old, in contrast to the very young soils of the pine barrens, but soil age is not reliably correlated with diversity. Another hypothesis is that the highly dissected and varied terrain of the Cape region has created many ecological opportunities for specialization. Evolution of ecological specialization can occur over short distances if selection is sufficiently strong, even if some gene flow still occurs. Whether this phenomenon accounts for part of the diversity of the fynbos flora is unknown. Future study of the fynbos may hold the key to many mysteries critical to our understanding of plant evolution.

Tragically, the fynbos faces serious threats. Some 1700 plant species are threatened with extinction in the Cape region. Many fynbos species are highly localized, and there are also many very local centers of species diversification and endemism. The city of Cape Town is located on two such centers of endemism, and its expansion threatens several hundred species. Agriculture and the spread of fertilized commercial crops threatens lowland fynbos. An-

(A)

(A) Endemic fynbos species include the mutualistic species *Protea magnifica* (queen protea, Proteaceae) and its pollinator, *Promerops cafer*, the Cape sugarbird. (B) This tiny *Protea* is one of the ground-growing species dependent upon mice for pollination. (Photographs courtesy of C. Oertel.)

(B)

thropogenic changes in fire regimes threaten species throughout the fynbos.

Most threatening to the fynbos flora, however, is the invasion of exotic species. The silky hakea, *Hakea sericea*, is a fire-adapted member of the Proteaceae from southeastern Australia that was brought to South Africa as a hedge plant and for firewood around 1830. It soon escaped cultivation and has become a major threat to fynbos vegetation. Clumps of this shrub or small tree develop into dense, impenetrable thickets that eliminate native species, alter fire regimes, and deplete soil water reserves. Other devastating invaders include *Pinus radiata* (Monterey or radiata pine, Pinaceae; see Chapter 14) from coastal California, other pine species from Europe, and *Acacia cyclops* (Fabaceae) and *Acacia saligna* from Aus-

tralia. An invasive, highly aggressive ant species from South America, *Iridomyrmex humilis* (the Argentine ant), is eliminating the native ant species on which many fynbos plants depend for seed dispersal, threatening the regeneration of many plant species.

Finally, global climate change is especially threatening to the survival of highly restricted ecosystems such as those of the Cape region and to species with very limited distributions. The fynbos is located at the very southern edge of the African continent; if the climate warms and species are driven poleward, there is nowhere for them to go but the salty deep.

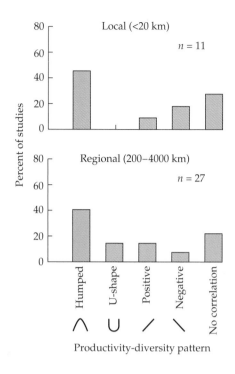

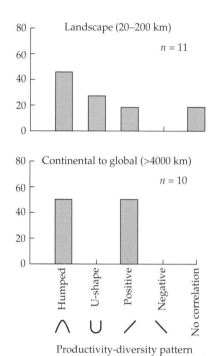

Figure 20.8
Percentage of studies showing various relationships (humped, U-shaped, negative, positive) between productivity and vascular plant species richness at different geographic scales. Geographic scale is based on the farthest distance between sites in a data set. Sample sizes (*n*) refer to the number of data sets. (After Mittelbach et al. 2001.)

on the relative importance of various mechanisms in different systems, or the ways in which they may commonly interact (Caswell and Cohen 1993; Lawton 1999).

Productivity and Scale

We can also look at patterns of diversity in relationship to environmental factors. Ecologists agree that productivity is an important determinant of species richness. Beyond that agreement, however, lies a much murkier realm, consisting of a myriad of processes operating at a variety of scales. The relationship between species richness and productivity is complex; it can be positive, negative, hump-shaped, or U-shaped (see Figure 14.7). To make matters even more complicated, the shape of the relationship often varies with scale (Figure 20.8). Many theories have been proposed to explain these patterns (Table 20.4).

The effects of scale on productivity-diversity relationships were recently explored by Scheiner and Jones (unpublished), using data on the vegetation of Wisconsin originally collected by John Curtis and his students in the 1950s (Curtis 1959; Figure 20.9). For the state as a whole, the relationship was U-shaped (Figure 20.10), with the lowest diversity at intermediate productivities—a rather unusual pattern (see Figures 14.7 and

John Curtis

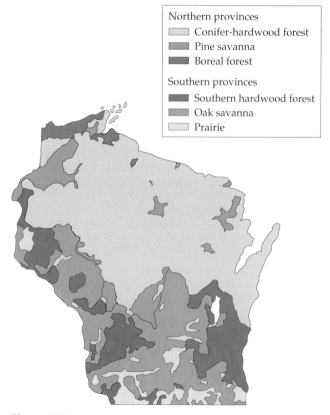

Northern provinces
  Conifer-hardwood forest
  Pine savanna
  Boreal forest
Southern provinces
  Southern hardwood forest
  Oak savanna
  Prairie

Figure 20.9
The state of Wisconsin, showing the ecological divisions and provinces defined by Curtis. (After Curtis 1959).

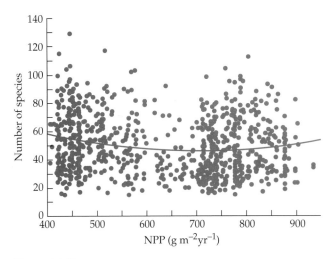

Figure 20.10
The relationship between terrestrial vascular plant species richness and net primary productivity for 901 plots across the state of Wisconsin. (Data from Scheiner and Jones, unpublished.)

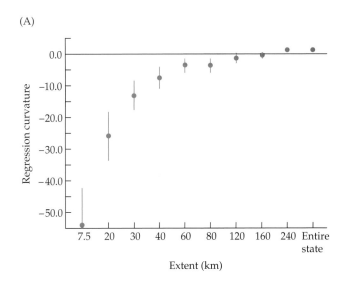

(A)

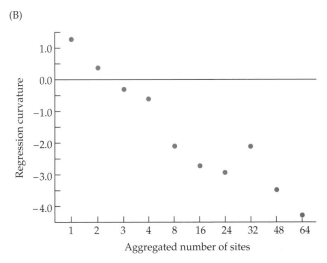

(B)

Figure 20.11
Effects of changing (A) the extent and (B) the grain of the data on the relationship between plant species richness and productivity across the state of Wisconsin. The graph shows the average amount of regression curvature for 1000 randomly chosen samples. Negative curvature indicates a hump-shaped relationship; positive curvature indicates a U-shaped relationship. Error bars show the 95% confidence interval. (Data from Scheiner and Jones, unpublished.)

20.8). The relationship is statistically significant despite the obviously large amount of variation around the regression line. The variation is due to a combination of error in estimating productivity as well as other, unaccounted for factors that also affect diversity. Such variation from multiple causes is common with ecological data.

One way to examine the effects of scale is by considering how the relationship between productivity and diversity changes when different components of scale are varied (see Chapter 17 for a discussion of scale and its components). First, Scheiner and Jones grouped their data points so as to vary geographic extent (the total area examined). If very small areas (7.5×7.5 km^2) were sampled, the relationship between productivity and diversity was hump-shaped—that is, diversity was highest at intermediate levels of productivity (Figure 20.11A). Increasing the area sampled changed this relationship. At intermediate areas, there was a negative relationship between productivity and diversity, while at the largest area—the entire state—the relationship was U-shaped.

A second analysis examined the effects of changing grain size (the size of the sampling unit represented by each data point) (Figure 20.11B). Grain size was manipulated by combining data from adjacent sites, beginning with two sites, then three sites, and so forth. At the smallest grain sizes the relationship between productivity and diversity was U-shaped; at intermediate grain sizes the relationship became negative, while at the largest grain size the relationship was hump-shaped (Figure 20.11B). What these analyses show is that even for a single set of sites there is no unique relationship between productivity and diversity. The relationship depends on the scale of the analysis, with different components of scale having different effects.

A critical factor responsible for the effects of scale on diversity-productivity patterns in Wisconsin is that the state sits at the intersection of three major biomes, with eastern deciduous forest to the southeast, hemlock-white pine-northern hardwood forest to the east, and tallgrass prairie to the west. Nonlinear (hump-shaped or U-shaped) patterns are more often associated with scales that span more than one community type or biome (see Figure 14.8), which is what was found for these data.

Summary

The explanation of large-scale patterns of plant diversity—which regions have many species and which have few—has been an ongoing effort of ecologists since before ecology developed as a distinct discipline. Today these issues have taken on a new urgency as global change and habitat destruction have increased the need to preserve areas of high diversity. We need to know why species are where they are if we are to predict the effects of global change and preserve as many species as possible.

One of the oldest known and best established pattern is a latitudinal gradient in diversity: highest in the Tropics, intermediate in temperate zones, and low in polar regions. Similar gradients are found from low to high elevations. There are also longitudinal gradients, such as the peak in diversity in the middle of Canada.

The effects of productivity—greater productivity allows the coexistance of more species—is likely the primary cause for large-scale patterns of species diversity.

It is clear, however, that it is not the sole explanation. Many different hypotheses have been proposed to explain these patterns. Current research is attempting to penetrate this thicket of ideas. While we have been able to discard some hypotheses, others remain.

One likely explanation for the current lack of resolution on this problem is that no single process accounts for all of the observed patterns. Rather, different processes are important in different circumstances and at different scales. Other important processes include the ecological and evolutionary histories of particular regions, and interactions between local and regional processes that determine diversity.

Our job now is to try to find rules defining those circumstances and to determine when and how two or more processes may function together. Investigations of how patterns of diversity change as a function of scale is one important piece of this puzzle. The rigorous evaluation of the hypotheses presented in this chapter is an important first step in all of these efforts.

Additional Readings

Classic References

Curtis, J. T. 1959. *The Vegetation of Wisconsin*. University of Wisconsin Press, Madison.

MacArthur, R. H. 1965. Patterns of species diversity. *Biol. Rev.* 40: 510–533.

Pianka, E. R. 1966. Latitudinal gradients in species diversity: A review of concepts. *Am. Nat.* 100: 65–75.

Contemporary Research

Colwell, R. K. and G. C. Hurtt. 1994. Nonbiological gradients in species richness and a spurious Rapoport effect. *Am. Nat.* 144: 570–595.

Currie, D. J. and V. Paquin. 1987. Large-scale biogeographical patterns of species richness of trees. *Nature* 329: 326–327.

Mittelbach, G. G., C. F. Steiner, S. M. Scheiner, K. L. Gross, H. L. Reynolds, R. B. Waide, M. R. Willig, S. I. Dodson and L. Gough 2001. What is the observed relationship between species richness and productivity? *Ecology* 82: 2381–2396.

Additional Resources

Huston, M.A. 1994. *Biological Diversity: The Coexistence of Species in Changing Landscapes*. Cambridge University Press, Cambridge.

Rosenzweig, M. L. 1995. *Species Diversity in Space and Time*. Cambridge University Press, Cambridge.

Ricklefs, R. E. and D. Schluter (eds.). 1993. *Species Diversity in Ecological Communities: Historical and Geographical Perspectives*. University of Chicago Press, Chicago, IL.

CHAPTER 21 *Paleoecology*

In previous chapters we have looked at global patterns of vegetation and diversity and at how those patterns have resulted from ecological and evolutionary processes. But the world is constantly changing, and the picture presented in those chapters is just a "snapshot" of today's world. The world was a very different place at various times in the past. Today, if you were to stand at a spot in rural northern Illinois, you would see cornfields all around you. Just 150 years ago, the scene would have been one of rolling prairie. Earlier—20,000 years ago—you would have been standing in arctic tundra, with sunlight glinting off the massive glaciers to your north. Go even farther back, and you might have been standing in the middle of a dense forest, or in a swamp dominated by tree ferns.

In this chapter, we consider some of these long-term changes in plant communities, the scientific methods used for understanding them, and some of their implications. Knowledge of the past both informs our understanding of the present and provides critical data for the management of today's ecosystems (Swetnam et al. 1999). We cannot, of course, cover the entire history of Earth, or even of terrestrial plants, in one chapter. The ecology of many earlier periods is still poorly understood. The field of **paleoecology**—the study of historical ecology—depends heavily on inferences and analogies with current species and communities. Our understanding of the very distant past is especially tenuous because of a dearth of fossils from those times, and because many ancient species have no modern analogues. Our knowledge of the past, however, is increasing rapidly. The development of new methods, such as the use of carbon isotope signatures of C_3 and C_4 plants and the creation of extensive computerized databases of the fossil record, have made paleoecology an active and lively field of research.

In this chapter we first survey several geological eras and periods (Figure 21.1), especially those that represent significant times of change for plant communities. Then we look in detail at the past 20,000 years. Because of the amount of detail involved, we focus on North America. The history of this recent period is now well known and provides special insight into current plant communities and biomes. And what of the future? That is the subject of the next and final chapter of this book.

Figure 21.1
The geological time scale (numbers are in millions of years). Although the Cenozoic era was traditionally split into the Tertiary and Quaternary periods, a newer system now divides it into two periods of more equal duration, the Paleogene and the Neogene. (After Stanley 1987.)

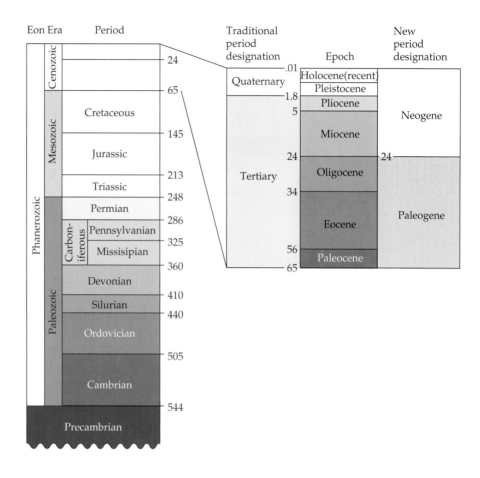

The Paleozoic Era

Fossil evidence tells us that plants first invaded land in the Paleozoic era, during the late Ordovician through Silurian periods. In Chapter 3 we considered some of the evolutionary challenges they faced in accomplishing this major transition. Land plants are descended from the Charophyta, a group of branched, filamentous green algae. Today many green algae grow along seashores, where they are subjected to daily drying as the tides rise and fall. Undoubtedly these were the conditions under which land plants evolved.

The invasion of plants drastically changed terrestrial environments, setting the stage for two major (and separate) animal invasions (by arthropods and by vertebrates) as well as for all subsequent plant evolution.

During the Silurian and Devonian periods, the terrestrial landscape was completely transformed. This period of time saw a steady increase in the morphological and ecological complexity of both plants and animals and the consequent dramatic radiation of major phyla. At the beginning of this period, the only vegetation consisted of ground-hugging mats in very damp areas, dominated by plants related to liverworts (Figure 21.2). Based on fossil evidence, we can infer that these early plants were mechanically supported by turgor

pressure, that their gametophytes produced both male and female reproductive organs like those of modern bryophytes and ferns, and that they had, at most, only rudimentary roots (Bateman et al. 1998). The forms of their sporophytes were simple, but at least one species,

Figure 21.2
Marchantia is a liverwort, a nonvascular plant. The 3-centimeter-tall structures in the center are reproductive organs of the female (left) and male (right) gametophytes. Early terrestrial plants may have had a similar form. (Photograph by S. Scheiner.)

Figure 21.3
A forest in what is now Colorado as it might have looked 300 million years ago. A shallow sea lies to the east, and rivers choked with gravel flow from the mountains to the sea. The banks of the rivers are colonized by forests of scale trees (*Sigillaria*, Lycopodophyta) and giant horsetails. The trees form a very minimal canopy with little shade. They have green trunks, and even their surface roots are green; the whole tree can photosynthesize. Forests of primitive conifers grow in the higher terrain of the foothills and mountains. (Painting by Jan Vriesen, courtesy of Denver Museum of Nature and Science.)

ity probably triggered an "arms race." There is generally little advantage to a plant in becoming very tall unless it is likely to be shaded by its neighbors (see Figure 6.6). If one species in a community is taller than all of the others, it will outcompete its neighbors for light. Natural selection will then favor individuals of those other species that can grow even taller. This process continues until other constraints balance selection for height. Selection for increased height, as well as other selective forces, resulted in the evolution of wood, which in turn enabled the great diversity of forms of trees and shrubs to evolve. The evolution of roots, in contrast, was probably driven by a combination of abiotic factors as well as competition for nutrients and water. As a result of these adaptations, plants were able to spread from wet areas, eventually spreading over nearly the entire surface of Earth that is not covered by water or ice.

As plants diversified, biotic interactions probably came to play an increasingly important role in plant evolution. Many modern ecological interactions evolved

Rhynia gwynne-vaughanii, had already evolved the ability to spread by clonal growth.

By the end of the Devonian, soils had developed in many habitats, and tall forests rose above primordial swamps dominated by giant lycopods up to 40 meters tall (Figure 21.3). As root systems evolved, land plants diversified and spread to mesic and drier uplands. By the end of the period, many major vascular plant phyla—Psilotophyta (whiskferns), Lycopodophyta (clubmosses; Figure 21.4), Sphenophyta (horsetails) and Pteridophyta (ferns)—had all evolved, as had the seed plants (Bateman et al. 1998). Today the first three of these groups grow only as small understory plants. Many ferns also grow as small plants, but tree ferns are large plants that grow in a number of tropical and subtropical habitats. A consistent trend during this period was the increasing dominance of the sporophyte (diploid) generation. As the sporophyte became larger and longer-lived, the gametophyte became smaller and shorter-lived.

We can speculate about the selective processes that created this diversification. The evolution of vascular tissue allowed plants to increase in height, and this abil-

Figure 21.4
Lycopodium obscurum (ground pine, Lycopodiaceae) is a clubmoss, a vascular plant approximately 10 centimeters tall. It has a stem, but no true roots or leaves. (Photograph by S. Scheiner.)

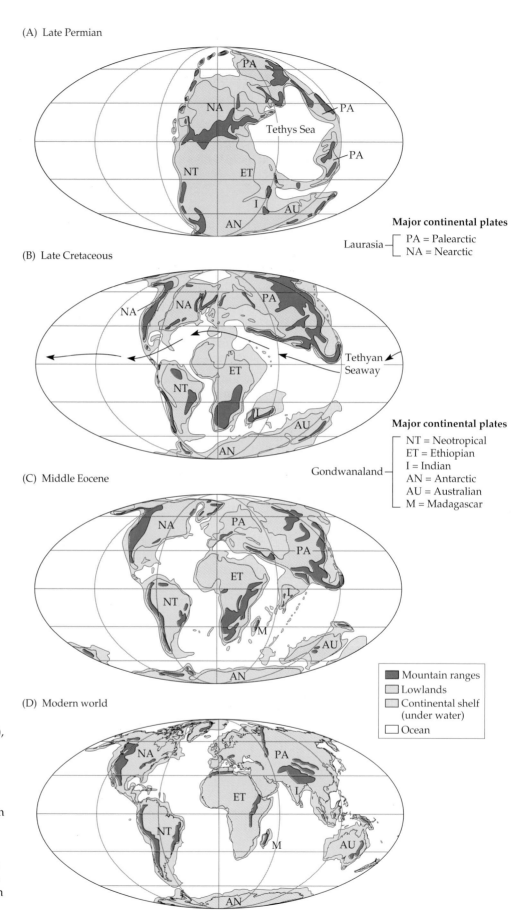

(A) Late Permian

Major continental plates

Laurasia — PA = Palearctic
NA = Nearctic

(B) Late Cretaceous

Major continental plates

Gondwanaland —
NT = Neotropical
ET = Ethiopian
I = Indian
AN = Antarctic
AU = Australian
M = Madagascar

(C) Middle Eocene

■ Mountain ranges
Lowlands
Continental shelf (under water)
□ Ocean

(D) Modern world

Figure 21.5
The positions of the continents over geological time: (A) Late Permian (255 million years ago), when the land masses were joined together into one super-continent, Pangaea. (B) Middle Cretaceous (94 million years ago), when the northern continent, Laurasia, and the southern continent, Gondwana, had begun to break apart. (C) Middle Eocene (50 million years ago), when the continents were taking on their current configurations. (D) Today. (After Brown and Lomolino 1998.)

very early, including mycorrhizal symbioses and herbivory during the Devonian and animal pollination and seed dispersal during the Carboniferous and Permian periods.

The Carboniferous period is so named because of the enormous amounts of fossil carbon—oil and coal—that were deposited that time. Coal deposits were generated mainly from the remains of wetland plants, including ferns and other spore-producing plants, as well as early gymnosperms (DiMichele et al. 2001). The organic sources of oil are less well understood; they appear to be primarily marine autotrophic and heterotrophic plankton, but may also include wetland plants. As climates changed in the late Carboniferous—mainly by becoming drier—seed plants began to dominate the landscape. This change in the vegetation was rapid, resulting in a nearly complete change in the species present at any given location, with upland plants rapidly replacing those of wet lowlands.

The Mesozoic Era

The Dominance of Gymnosperms

At the beginning of the Mesozoic era, all of the continents were joined together into one supercontinent, **Pangaea**. They have been carried apart and reshuffled by plate tectonics since that time (Figure 21.5). The coalescence of Pangaea, along with increases in atmospheric CO_2 and temperature, led to widespread continental climates, greater seasonality, and widespread aridity. Any productivity benefits of the high CO_2 levels may have been more than counterbalanced by the combination of high temperatures (from the greenhouse effect) and low or seasonal rainfall.

By the late Jurassic period, 150 million years ago, Pangaea had separated into two large continents, northern Laurasia and southern Gondwana. These continents, in turn, would break up into our current continents by the end of the Cretaceous period. The Atlantic Ocean was coming into being by the early Cretaceous, first with the separation of North America and Eurasia about 130 million years ago, and then with the separation of South America and Africa about 90 million years ago. As a result of these continental movements, areas that were once inland became coastal. Coastal areas are both wetter and less seasonal than inland regions (see Chapter 18). Thus, overall conditions for plant growth tended to improve throughout the Jurassic and Cretaceous.

We can consider the history of land plants as the successive dominance of three major groups. The pteridophytes (ferns) and pteridosperms (seed ferns) dominated the flora during the late Paleozoic, when climates were warm and wet. Despite the similarity in their names, ferns and seed ferns are not closely related; **seed ferns** are an extinct group that superficially resembled ferns, but reproduced through seeds rather than free living gametophytes.

With increasing aridity and **continentality**—influence of land masses on weather, in this case due to the formation of Pangaea—during the first half of the Mesozoic, the flora became dominated by conifers, cycadophytes, and other gymnosperms, and seed ferns gradually went extinct. Nonflowering families, genera, and species from the Mesozoic which persist today include: *Osmunda* (a fern genus), *Ginkgo biloba* (ginkgo, Ginkgoaceae), *Sequoia* (a conifer genus that today includes the giant sequoia of California), and *Araucaria* (a conifer genus of the Southern Hemisphere that today includes several important forest trees, such as *A. excelsa*, the Norfolk Island pine of the South Pacific, and *A. araucana*, the monkey puzzle tree of the southern Andes of South America). The angiosperms came to dominate the terrestrial fauna during the Cretaceous.

The early Cretaceous was a time of warm temperatures and very little difference among the seasons. CO_2 levels were somewhere between 3 and 4 times higher than they are today (Figure 21.6), contributing to warmer temperatures through the greenhouse effect (see Chapter 22). The warm climate and ready availability of CO_2 made the early Cretaceous a period generally favorable to plant growth. By the late Cretaceous, however, these favorable conditions had gradually declined, with climates becoming much more seasonal and larger differences in temperature existing between the equator and the poles.

The plant communities of the Mesozoic were similar to those of today in some ways. There were distinct biomes dominated by trees, shrubs, or herbs. However, the trees were predominately conifers; today, tropical forests and many temperate forests are dominated by angiosperms, with coniferous forests mostly found in temperate or boreal regions. Savannas are a particularly arresting example of how ancient biomes differ from those of today. Today's savannas are dominated by grasses along with scattered angiosperm trees. In the Mesozoic, there were no grasses; instead, ferns provided most of the ground cover, while the trees were largely gymnosperms.

If we consider the modern distribution of ferns, we might think of them as a largely tropical group, as they are most abundant and diverse in the Tropics. Interestingly, most of today's fern families originated in moist temperate regions (above 30° N and S latitude) during the Mesozoic (Skog 2001). During this time, ferns were noticeably absent from the Tropics. Long-distant dispersal of spores allowed fern families to spread widely in temperate regions, so that during much of the Mesozoic, many families were found in both the Northern and Southern hemispheres. Only later did they spread into the Tropics.

(A)

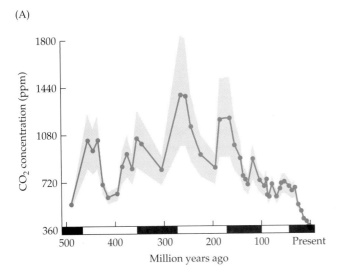

(B)

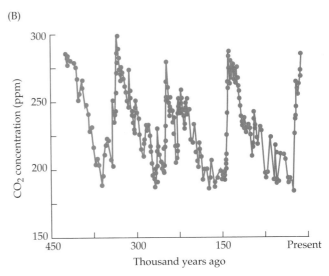

Figure 21.6
(A) Estimated concentrations of atmospheric CO_2 over the past 500 million years. The shaded areas above and below the solid line are upper and lower limits of the estimate. The black bars indicate periods when Earth's climate was relatively cool; white areas are periods of relative warmth. CO_2 levels have mostly decreased for the past 175 million years. (After Rothman 2002.) (B) Estimated concentrations of atmospheric CO_2 over the past 450,000 years, measured from trapped air in ice cores from the Antarctic. Over this period CO_2 levels have varied, but there has not been a trend either up or down. (After Barnola et al. 1999.)

The Breakup of Pangaea and the Rise of the Angiosperms

The rise of the angiosperms was a critical event in the history of plants. A few tantalizing fossils suggest an origin as early as the Triassic for this group, but that interpretation is controversial and not well supported by current evidence. What is certain is that by the middle to late Cretaceous, angiosperms were spreading throughout the world. This group now accounts for approximately 75% of all land plant species and a level of plant diversity not seen previously on Earth. The separation of the continents during the Cretaceous not only contributed to climate change, but also helped to accelerate speciation rates in both plants and animals through geographic isolation (see Chapter 6). In this world of newly divided continents the angiosperms spread, proliferated, and diversified. Their rise created a profound change in the structure of communities. First, the nature of animal communities went through a dramatic transformation. Insects began to diversify; today their species numbers vastly overwhelm those of any other group of animals. In some cases, such as the diversification of lepidopterans (butterflies and moths), the timing of insect radiations can be linked directly to the evolutionary radiation of angiosperms.

The evolution of the flower—a key feature of angiosperms—opened possibilities for more specialized relationships between plants and pollinators. The pollination of earlier vascular plant taxa depended mainly on the wind (as in conifers) or on beetles (as in cycads), which mainly consumed pollen and only occasionally transferred some of it to other plants. With the evolutionary diversification of flowers came nectar rewards, and diversification in flower shapes, colors, and scents increased the reliability of insects in moving pollen to receptive stigmas on other flowers of the same species. Today some of the best-known examples of mutualisms are those of flowers and their pollinators. The evolution of fruits similarly opened new possibilities for interactions between plants and their animal seed dispersers (and seed predators).

Another important difference between angiosperms and the groups that preceded them is that angiosperms include many herbaceous and small woody species that often have faster growth and life histories. In conifers, seed maturation can take from six months to three years, while almost all angiosperms can mature seeds in just weeks to months. More generally, angiosperms encompass an extremely wide diversity of plant architectures, life histories, and reproductive modes.

At the end of the Cretaceous, tropical forests were found in what is now southeastern North America. These forests had open canopies with relatively tall conifers similar to *Sequoia* and *Metasequoia* (Taxodiaceae), with an understory consisting of a variety of angiosperms, including rosette palms. At this time, the middle of North America was covered by a vast shallow sea. The areas west of the sea were covered by a broad-leaved evergreen community with a variety of species like those in the southeast. Temperate forests, dominated by broad-leaved evergreen conifers, cycads, and ginkgos, covered what is today the northwestern United States and southwestern Canada. In those forests, the angiosperms were generally deciduous species growing along streams and in other disturbed habitats. The far northwestern areas of North America—the area above 60° N latitude—were covered by broad-leaved deciduous forests with a mixture of conifers (such as Taxodiaceae and *Gingko*) and angiosperms, including species related to birches (Betulaceae) and elms (Ulmaceae). The milder climates of the Cretaceous allowed these broad-leaved communities to exist where now there is taiga and tundra.

The Cretaceous-Tertiary (K-T) Boundary

The Mesozoic era came to an end 65 million years ago. There is strong evidence that a large asteroid (10 ± 4 kilometers in diameter) slammed into what is now the Caribbean Sea near the current Yucatán Peninsula. The asteroid struck from the southeast, throwing vast amounts of debris into the atmosphere to the north. This debris eventually settled to the ground, forming a distinct geological layer marking the transition between the Cretaceous and Tertiary periods, known as the K-T boundary.

How this asteroid impact led to massive global change—including the extinction of the dinosaurs—is still an area of active research and controversy. Under one hypothesis, the debris thrown up by the impact may have been sufficiently dense to darken the skies for long enough that photosynthesis was largely shut off. Other evidence suggests that the impact radiated enough heat to start massive and widespread fires that lasted for several years. The debris remaining in the atmosphere from the impact and the fires is likely to have led to a drastic cooling in the world's temperatures for a period of years to decades. It has also been hypothesized that the disruption in the global carbon cycle caused by the impact led to major oscillations in global climates for the next million years. It is now generally accepted that this event is tied to the extinction of the dinosaurs, along with 70% of the marine species that existed at the time.

In North America, the flora of the Paleocene epoch contained many fewer species than the flora of the late Cretaceous. In the southern part of the continent, almost 80% of existing plant species became extinct, while polar forests had a 25% extinction rate. Even as far away from the impact as New Zealand, a recent study of fossil pollen and spores found that a diverse flora consisting of gymnosperms, angiosperms, and ferns was abruptly replaced with one consisting of a small number of fern species (Vajda et al. 2001). Recovery occurred only gradually over more than a million years, with the eventual evolution of many new species, particularly angiosperms.

The Cenozoic Era

In the early Eocene epoch, global temperatures began to fall, following falling global CO_2 levels. In North America,

the Rocky Mountains and Sierra Nevada continued rising, creating a mid-continent rain shadow (see Chapter 18). Along with continued mountain building in the Himalayas, these geological changes led to a cooling in the global climate. The cooling was strongest in the Southern Hemisphere: the South Pole was covered by ice during the Oligocene epoch, but the North Pole was not. This cooling may have been reinforced by another effect. Massive mountain building during this period resulted in the weathering of large amounts of rock: chemical reactions during weathering remove CO_2 from the atmosphere (Ruddiman and Kutzback 1989). Temperatures remained high enough, though, that throughout the Miocene epoch the vegetation was tropical even at high latitudes.

These changes in climate had important ecological consequences. By the Oligocene epoch, broad-leaved evergreen forests in the middle of North America had thinned out to tropical woodlands and savannas. Grasses, which originated in the Paleocene, began to spread widely during the Miocene. Grass eaters evolved from ancestors that ate leaves from trees or shrubs. Most notably, horses evolved in North America during this time. (Eventually horses migrated to Eurasia over the land bridge between Alaska and Siberia, and went extinct in North America 10,000 years ago. Today's wild herds in North America are descendants of horses brought by the Spanish conquistadores just 500 years ago.)

The drop in CO_2 levels may have had a second important effect on plant ecology beyond the change in temperature. Studies of the isotopic signatures of fossil plants show that C_4 angiosperms first evolved from C_3 ancestors during the Oligocene. C_4 plants are more efficient than C_3 plants at CO_2 uptake (see Chapter 2), especially at low CO_2 concentrations. During the late Miocene, tropical and subtropical grasslands dominated by C_4 grasses spread widely, especially between 7 and 8 million years ago. Their spread has often been explained as a response to the drop in CO_2 concentrations. However, the drop in CO_2 had already become substantial by the early Miocene, some 16 million years earlier. So, the delay in the spread of C_4 grasslands is unexplained. Researchers have suggested that climate change, rather than changes in global CO_2 concentration, might be the most important factor in the rise of C_4 grasslands (Pagani et al. 1999).

We can look at changes in the distribution of C_3 and C_4 grasses in response to much more recent climate changes in order to understand what might have happened during the Miocene. Studies of fossils from 10,000 years ago in two lake beds in Mexico and Guatemala support the hypothesis that climatic drying was responsible for the local spread of C_4 grasses (Huang et al. 2001). Atmospheric CO_2 concentrations were the same at both sites, and thus can be ruled out as a cause of differences in the vegetation. At the Mexican site, C_3 grasses spread as the local climate become rainier at the end of the most recent glaciation; at the Guatemalan site, C_4 grasses spread as the local climate become drier. Studies of grasslands in southern Africa also suggest that the relative proportion of C_3 to C_4 plants varies with climate (Scott 2002).

How will the Earth's vegetation change in response to the current rise in global CO_2 levels? Will it favor the spread of C_3 plants over C_4 plants (see Chapter 22)? It seems likely that the answer will depend on how changes in the abundance of CO_2 affect regional climates.

Paleoecology Methods

How do we know so much about the history of plant communities? Studies of the past use a variety of techniques. Of primary importance is the study of fossils. From these remains of leaves, stems, flowers, and other plant parts, we can reconstruct past communities and climates (Prentice et al. 1991).

The sources of fossils depend on the habitat where they were formed. Because most organic material decomposes quickly, **anoxic**—oxygen-free—conditions are ideal for its preservation. In nonmarine habitats, these conditions are most often found at the bottoms of lakes and bogs. Sediments formed in lakes and bogs are the primary source of **macrofossils**, such as leaves, flowers, stems, and seeds. For information from hundreds of thousands or millions of years ago, we rely on hardrock fossils, in which the organic material has been replaced by nonorganic material. Material from more recent periods may still be organic. Bogs, which have highly acidic waters that slows decay, are especially useful in this regard.

Fossils can be retrieved from lakes and bogs by removing a core from the sediments at the bottom. A long tube or pipe, open on the bottom, is pushed down into the sediments, and the core is then carefully drawn up. Thin slices of the core are examined; fossils, if present, are identified, and the slices are aged using isotopic dating.

In addition to macrofossils, these cores yield pollen grains (Figure 21.7) and other microfossils. **Palynology** is the study of spores and pollen. Wind-borne pollen often lands in a bog or lake and settles to the bottom. Because pollen can travel long distances, cores from lake or bog sediments provide a profile of the community over a wide area around the body of water. In contrast, macrofossils usually provide a picture only of the plants growing immediately adjacent to the area studied.

Palynology has strengths and weaknesses as a method of studying the past. One strength is that pollen is readily preserved, and large amounts of it can usually be found. A weakness is that not all species are present or equally represented, or even identifiable. Only the pollen of wind-pollinated species makes it into lakes and bogs in quantity; animal-borne pollen is largely missing. Thus, sediment core samples are biased toward trees—because most temperate-zone trees are wind-pollinated—and wind-pollinated herbaceous groups, especially grasses and sedges.

The level to which pollen can be identified varies among taxa. Pollen from some trees can be identified to species. In many genera, however, only subgenera can be distinguished. For example, among pines, the pollen of *Pinus resinosa* (red pine, Pinaceae) and *P. banksiana* (jack pine) cannot be told apart, although they can be differentiated from that of *P. strobus* (white pine). In some cases, such as the grasses, pollen can be identified only to the family. Despite these limitations, palynology has been key to reconstructing communities and migration patterns in temperate regions.

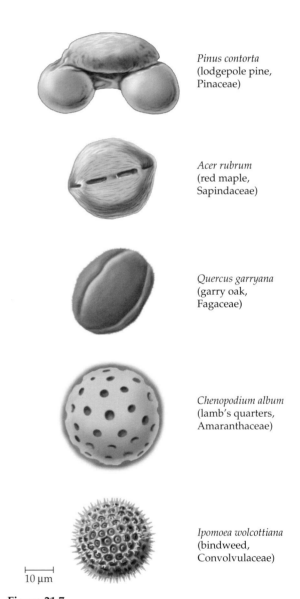

Pinus contorta
(lodgepole pine,
Pinaceae)

Acer rubrum
(red maple,
Sapindaceae)

Quercus garryana
(garry oak,
Fagaceae)

Chenopodium album
(lamb's quarters,
Amaranthaceae)

Ipomoea wolcottiana
(bindweed,
Convolvulaceae)

10 μm

Figure 21.7
Pollen grains preserved in sediments can be used to identify plant species present at some time in the past. (After Pielou 1991.)

Another way in which plant material can be preserved is by drying. This means of preservation has provided macrofossils in arid regions of North America where packrats (*Neotoma*) live. As their name implies, these rodents accumulate large amounts of plant material, which they store in their dens. They urinate on the plant material, caking it with mineral salts. Urination provides substantial evaporative cooling for the den and, incidentally, provides excellent conditions for preservation of the plant material. The accumulations of packrats are referred to as **middens**. A given den site may be used for centuries by successive generations of packrats, with its middens growing to several meters across. A fossilized midden may be preserved for thousands of years in a dry climate. Samples from packrat middens can be radiocarbon-dated and the plant material identified, allowing researchers to reconstruct changes in the plant community surrounding the den site (Cole 1985; Betancourt et al. 1990). In Arabia and Africa, the rock hyrax creates a similar structure, called a hyraceum.

The Recent Past

We now leave the distant past and jump ahead to a point just 20,000 years before the present, when much of the Northern Hemisphere was covered by glaciers. Over the previous 2 million years, during the Pleistocene epoch, Earth had experienced a period of major glaciations. Starting about 850,000 years ago, there were gradual periods of glacial buildup, followed by abrupt switches to warmer weather and short periods of glacial retreat (called interglacials) (Figure 21.8). Each of these periods of alternating cold and warm weather lasted about 100,000 years. These cycles were caused mainly by changes in the tilt and precession of Earth's orbit (see Chapter 18). The various glacial advances and retreats are known by various names. In North America, the most recent glacial advance is called the **Wisconsin glaciation**, named for one of the locations of its southernmost extent (Figure 21.9). Because each glacial advance destroyed most of the traces of previous ones, we focus here on the Wisconsin glaciation, for which the best data exist.

The last warming period began about 12,000 years ago. Since that time, the average global temperature rose, peaked about 7000 years ago, and then declined (Figure 21.8B,C). If not for human activities bringing on a new, rapid period of global warming, we would likely be seeing the growth of glaciers in the next few thousand years.

Figure 21.8
Mean global temperatures over geological time for various periods of time: (A) the past 850,000 years; (B) the past 140,000 years; (C) the past 10,000 years; (D) the past 1,000 years. (After Gates 1993.)

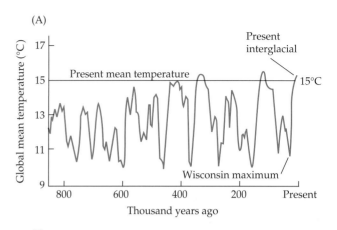

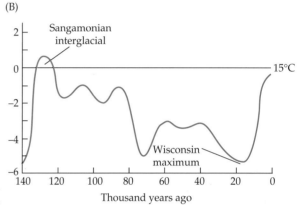

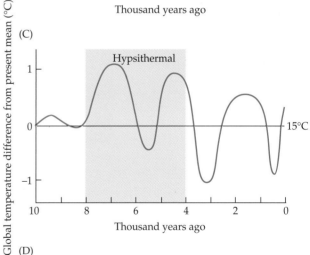

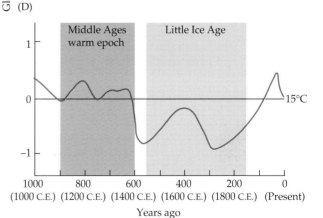

What has all of this climate change meant for plants? As glaciers advanced and retreated, plants in northern latitudes migrated across continents, following moving climate zones. In the Tropics, these climate changes resulted in the contraction of forests and the expansion of grasslands. Our study of these changes in plant distributions has led to new understanding of the causes of diversity patterns across the globe, as well as new insights into the nature of plant communities.

At the Glacial Maximum

At the maximum of the Wisconsin glaciation, large continental glaciers extended as far south as 40° N latitude in North America and 50° N in Eurasia, while smaller glaciers were found in mountainous regions such as the Rocky Mountains in North America, the Himalayas in Asia, and the Alps in Europe. Few glaciers formed in the Southern Hemisphere, excepting Antarctica, because there is little continental land area at high latitudes, although some glaciers existed in the Andes mountain range of South America. Regional climatic conditions maintained some glacier-free regions even at high latitudes. For example, some coastal regions remained free of glaciers in Canada and Alaska.

With so much water tied up in glaciers, sea levels were lower than today's by 85 to 130 meters, resulting in coastal areas that were substantially broader than at present. Of particular importance to plants were the

(A)

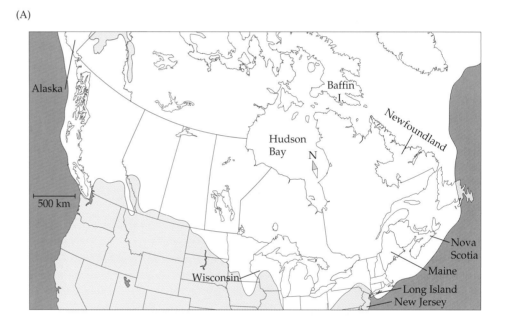

(B)

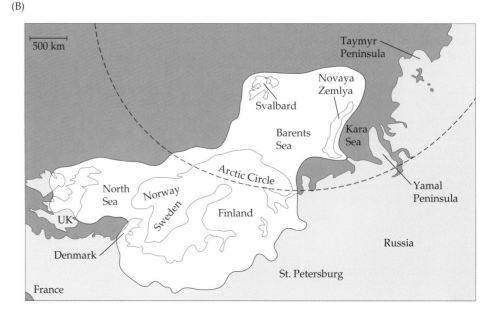

Figure 21.9
The greatest extent of continental glaciers during the most recent glaciation in (A) North America and (B) Eurasia. (After Mayewski 1981 and Siegert et al. 2002.)

southeastern coast of North America, which served as a refuge for species pushed southward by the glaciers, and the Bering land bridge connecting Alaska with Siberia, which was an important migration route for plants and animals between North America and Eurasia. (It was by this route that humans probably first reached North America.)

South of the glaciers, and in other ice-free areas such as eastern Siberia, were vast stretches of tundra and **muskeg**—a grassy bog habitat with scattered, stunted conifers. The tree species of the taiga and temperate forests that today cover these areas were then found far to the south. In southeastern North America, along the coastal plain, there were forests of deciduous trees such as *Quercus* (oak), *Carya* (hickory), *Fagus* (beech), and *Acer* (maple). Along the southern reaches of the Mississippi River Valley were coniferous forests dominated by species such as *Picea* (spruce), *Thuja* (cedar), *Abies* (fir), and *Larix* (larch).

The discovery of these distributions addresses an important question in plant community ecology: whether communities are highly coordinated units or simply collections of coexisting species (see Chapter 12). The assemblages of species that existed 20,000 years ago, at the most recent glacial maximum, have no modern equivalent. For example, *Tsuga canadensis* (eastern hemlock, Pinaceae) and *Acer saccharum* (sugar maple, Sapindaceae) are codominants in forests around the Great Lakes today. Yet during the glacial maximum, the former tree species was found mainly along the Mississippi River, while the latter was found several hundred kilometers to the west along the Gulf Coast and parts of inland Texas. In the late 1970s and early 1980s, palynological studies of past species distributions and the mapping of northern migrations in response to climatic warming demonstrated that the species compositions of plant communities have been reshuffled many times. These data demonstrate that communities are not highly coordinated units, as well as showing that communities rarely reach an equilibrium of either species composition or population structure (Davis 1981).

Studies of the refuges where plant species survived during the glacial maximum also point to important causes for contemporary differences between North American and European plant communities. As mentioned in Chapter 20, north-temperate and boreal communities in North America contain approximately 50% more vascular plant species than equivalent European ones. This difference is due to the many species in Europe that went extinct during the Pleistocene (see Table 20.2), but which survived in North America because they were able to reach refuges.

These global climate changes also affected the Tropics, even though those regions were not covered by glaciers. Most tropical studies have focused on events in the Amazon basin. While the exact details of what happened there are still under active study, researchers agree that the climate in the Amazon basin at the glacial maximum was cooler than at present. Until recently, it was generally thought that the climate was markedly drier as well—in fact, precipitation was thought to be the most important factor in determining tropical plant distributions during that time (Graham 1999). However, a recent review suggests that the evidence for decreased rainfall at the glacial maximum is equivocal (Burnham and Graham 1999).

Previously, almost all paleoecologists believed that cycles of glaciation and warming created, alternatively, highly connected forests and a mixture of forest fragments separated by savannas and grasslands. Some ecologists have hypothesized that these changes in forest coverage in the Tropics contributed to the high species diversity in tropical regions (see Chapter 20) Current data do not support this hypothesis. Comparisons between taxonomic groups that should have been highly affected by this fragmentation (such as terrestrial mammals) and groups less likely to be affected (such as birds) have failed to find differences in the numbers of species that evolved. Some recent studies suggest that forests in the Amazon basin may not have become so severely fragmented after all. Studies of pollen deposited at the mouth of the Amazon suggest that throughout the last 50,000 years, forests have always dominated the region (Haberle 1999; Colinvaux and De Oliveira 2001). A recent study suggests an alternative mechanism that might have fragmented the Amazon basin, however: increases in sea level during the interglacials may have broken the area into two major islands and a number of smaller archipelagos (Nores 1999).

An important event for North American plant communities was the arrival of humans, possibly as early as 30,000 years ago, but certainly by 12,000 to 15,000 years ago. This migration is linked to a major extinction about 10,000 years ago of most of the large mammal species in the New World (Martin and Klein 1984). Similar mass extinctions of large animals followed the arrival of humans in Australia and Madagascar Such extinctions can create pitfalls for present-day plant ecologists. Today, if you travel the western deserts of the United States, you find many shrubs with sharp thorns and spines. Yet, except for introduced species such as cattle, horses, and donkeys, there are almost no large browsers, except for bighorn sheep, pronghorn, and mule deer. Only after studying the fossil record would you realize that these plants evolved during a time when North America had many large native browsers, such as horses, camels, elephants, and giant ground sloths, all of which are now extinct.

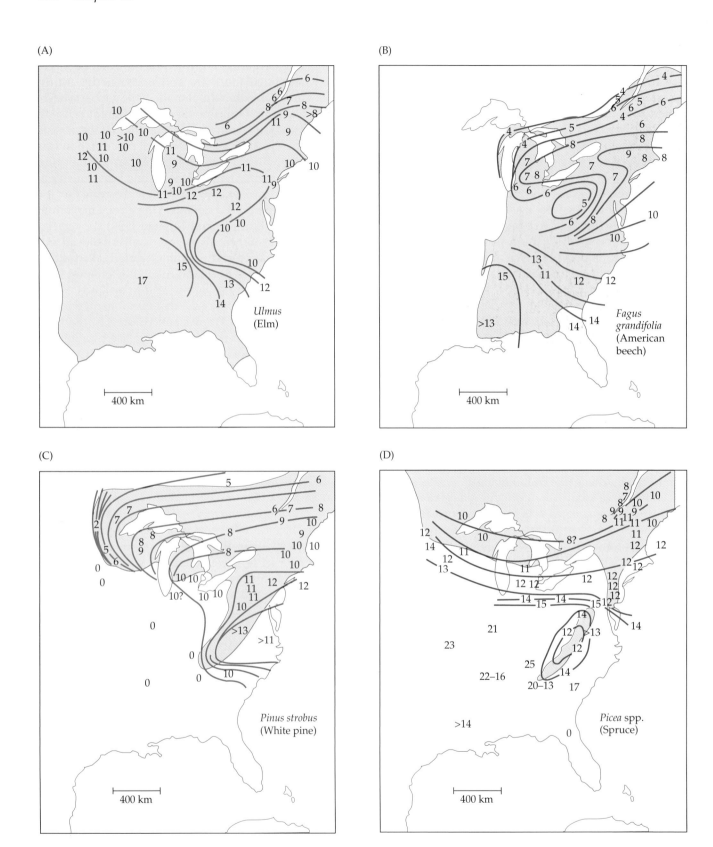

Figure 21.10
Examples of migrations patterns for trees in North America based on pollen records. Counters indicate the time of arrival (thousands of years ago). The green area shows the current distribution. In some cases we can identify the pollen to species, and in others only to genus. (A) *Ulmus* spp. (elm, Ulmaceae). (B) *Fagus grandifolia* (American beech, Fagaceae). (C) *Pinus strobus* (white pine, Pinaceae). (D) *Picea* spp. (spruce, Pinaceae). (After Davis 1983.)

Glacial Retreat

Beginning about 20,000 years ago, the climate warmed and the glaciers retreated, slowly at first, then more rapidly starting about 14,000 years ago. For several thousand years, as the ice melted, large lakes (much bigger than the current Great Lakes) dominated the interior of North America. Areas that had been under ice became tundra and then forest or grassland.

During this warming period, trees migrated northward. However, each tree species was unique in its migration pattern (Figure 21.10). Many of these patterns were deciphered by Margaret Davis, using pollen samples gathered from lakes and bogs throughout eastern North America (Figure 21.11). Among the fastest-moving species was *Pinus strobus* (white pine, Pinaceae), which migrated at an average rate of 400 meters per year. In contrast, *Ulmus* (elm, Ulmaceae) moved at a rate of 250 meters per year, while slowpoke *Castanea dentata* (American chestnut, Fagaceae) moved a mere 100 meters per year (Table 21.1).

One might imagine that wind-dispersed and animal-dispersed species would have migrated at about equal rates, and that among animals, birds would have moved seeds faster than mammals. Yet Table 21.1 clearly shows that wind-dispersed seeds generally traveled much faster than animal-dispersed seeds. The exception, *Quercus* (oaks), is dispersed mainly by small mammals in eastern North America. The bird-dispersed species

Margaret Davis

(e.g., *Fagus* and *Castanea*) were actually the slowest to migrate northward.

The migration of a species is not a steady progression. For example, the migration of *Fagus grandifolia* (American beech, Fagaceae) over the last 7000 years has included both periods of rapid expansion and periods of relative stability (Woods and Davis 1989). During this period, beech migrated northward and westward from the Lower Peninsula of Michigan into the Upper Peninsula and Wisconsin. Some of the delays are attributed to the barrier presented by Lake Michigan; long-distance dispersal by blue jays or passenger pigeons may have been responsible for breaching this barrier. Other starts and stops may have been responses to climatic fluctuations. Beech arrived in Wisconsin on the western shore of Lake Michigan approximately 6000 years ago, but did not spread farther west and north for another 1500–2000 years, probably due to cold winter temperatures.

Migration depends on the successful establishment of seedlings. When we look closely at the frontier of a migration, we find a patchwork of populations, with establishment and spread highly dependent on local conditions. Davis et al. (1998) studied the spread of east-

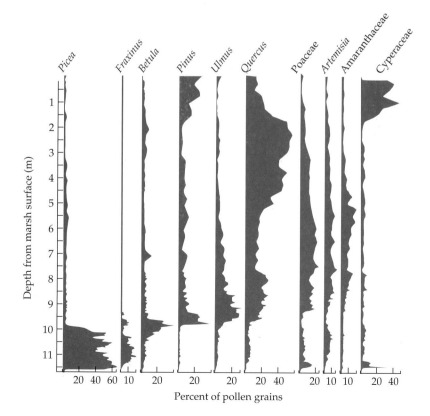

Figure 21.11
Example of a pollen diagram for Kirchner Marsh, Minnesota. The deepest samples are the oldest. About 20,000 years ago, this area was dominated by *Picea* (spruce) and *Fraxinus* (ash). These were followed by a succession of species: *Betula* (birch), *Pinus* (pine), *Ulmus* (elm), and *Quercus* (oak). The subsequent decrease in oaks and increase in grasses (Poaceae) indicates the existence of a woodland vegetation. Currently the area is again dominated by pines mixed with oaks, but now with a sedge (Cyperaceae) understory. (After Grimm 1988.)

Table 21.1 Average rates of northward range expansion for trees in eastern North America following the most recent glacial retreat

Species	Rate (m/yr)	Dispersal agent
Pinus banksiana/resinosa	400	Wind
Pinus strobus	300–350	Wind
Quercus spp.	350	Animals
Picea spp.	250	Wind
Larix laricina	250	Wind
Ulmus spp.	250	Wind
Tsuga canadensis	200–250	Wind
Carya spp.	200–250	Animals
Abies balsamifera	200	Wind
Acer spp.	200	Wind
Fagus grandifolia	200	Animals
Castanea dentata	100	Animals

Source: Davis 1981.

ern hemlock into an area in the western Upper Peninsula of Michigan. Four communities dominated by white pine were invaded by eastern hemlock about 3000 years ago over a period of 800 years. Over the next 2000 years, climate changes caused the water table to rise and the soil to become wetter. Eastern hemlock became the dominant species, and white pine disappeared from all but one site. In contrast, four other nearby sites that were initially dominated by sugar maple and oak were never invaded by eastern hemlock, possibly because the greater availability of soil nitrogen and higher light levels in the spring under the deciduous canopy allowed sugar maple seedlings to outcompete hemlock. Thus, the exact details of migration patterns may be highly contingent on local conditions and on which species happen to arrive first.

Climate Fluctuations in the Recent Past

Superimposed on the long-term glacial-interglacial climate swings are shorter climate fluctuations (see Figure 21.8D). The period from 1500 to 1850, known as the "Little Ice Age," was a time of low solar activity, cold temperatures, and glacial advances (interrupted by relatively warm interludes in 1540–1590 and 1770–1800). These shorter-term fluctuations also leave their imprint on plant communities and often help explain puzzling aspects of current species distributions.

Grassy **balds**—treeless mountain peaks—in the southern Appalachian Mountains are odd in that they do not occur where you might expect them, on the tallest mountains. Rather, they are found on mountains of intermediate height, but not on all such mountains. Closer inspection shows that the tree line on these mountains occurs at about 1500 meters, the elevation where deciduous forest gives way to coniferous forest

on nearby, taller mountains. The initial hypothesis was that the soil on these mountains was not suitable for coniferous species. But no differences from nearby mountains could be found, and coniferous trees planted on the balds grew well.

The mystery was solved once the effects of climate change and migration patterns were considered. As the climate warms and cools, climate zones move up and down mountainsides, and plant species follow this movement. During the most recent glaciation, the balds were covered with coniferous forests. As the climate warmed, the lower edge of the ranges of the conifers moved upward until it was higher than the tops of the mountains. At this point, the mountains were entirely covered with deciduous forests. Then the climate cooled again. The upper edge of the species ranges of the deciduous trees moved downward, below the tops of the mountains. Now, however, there were no coniferous trees left on the mountains to replace them, leaving the tops bald—dominated by grasses and forbs. On higher mountains the coniferous species were still present, so they simply migrated back down. Given enough time, long-distance migration might return conifers to these mountaintops.

This pattern of vertical migration was even more important in the American Southwest. This region has many high mountains, plateaus, and other vertical features, so that during the Pleistocene much of the plant migration consisted of elevational changes. During periods of maximum glaciation, the climate was cooler and wetter, and pine forests and pinyon-juniper woodlands spread out over the landscape. Large lakes dotted the landscape in what is now the Great Basin desert. Their remnants can be seen today as the Great Salt Lake and the Bonneville Salt Flats. During interglacial periods (like the present), the low-elevation areas became hotter and drier, and desert shrubland became dominant in these areas.

Studies of these patterns in the Grand Canyon in Arizona have disclosed an important phenomenon, called **vegetation lag**. Such lags occur when changes in the vegetation happen more slowly than the rate at which the climate changes. Using evidence from packrat middens, Kenneth Cole (1985) found that as the climate warmed, species disappeared locally as the lower edges of their ranges moved upward. However, these species were not immediately replaced; it took centuries for new species to become established (Figure 21.12). The studies of migration in the Great Lakes region described previously also found evidence of vegetation lag, with species taking 1000 to 2000 years to reach and then dominate areas after the climate had become suitable for them. The studies of the Grand Canyon vegetation also reinforce the notion that plant species migrate individualistically, with plant communities changing through time as species combinations changed (Cole 1990).

Plant species may also undergo evolutionary adaptation in response to climate change. Rapid climate change—of the kind now occuring in response to rising concentrations of atmospheric CO_2—imposes very strong natural selection on populations. Studies of tree species point to the conclusion that adaptation and migration occur simultaneously (Davis and Shaw 2001); they are not alternative responses to climate change. Will plant populations be able to meet the challenge of today's changing climates through some combination of adaptation and migration? Evolutionary responses are limited by the amount of heritable vari-

ation with a population (see Chapter 5). Migration is limited both by dispersal biology (for example, wind can move pine seeds only so far) and by anthropogenic changes in land use that present barriers to species movement. Together, these factors suggest that rates of extinction are likely to increase as a consequence of current global climate change.

Summary

The plant communities that we see today are not static entities; they are the current state of a dynamic sys-

(A)

(B)

Figure 21.12
An area near the southern rim of the Grand Canyon in Arizona at 1500 m elevation. (A) Today, this area is a desert shrubland of *Ephedra viridis* (Mormon tea, Ephedraceae), *Coleogyne ramosissima* (blackbrush, Rosaceae), and *Opuntia engelmannii* (prickly pear cactus, Cactaceae), with a few juniper (*Juniperus* sp., Cupressaceae) along the cliff tops. (B) Reconstructed image of the same site as it looked 20,000 years ago, based on plant fossils from packrat middens. The bottom slopes support a woodland of juniper, *Artemisia tridentata* (sagebrush, Asteraceae), and *Atriplex canescens* (shadscale, Amaranthaceae). In the shady alcoves of the cliffs grow *Pseudotsuga menziesii* (Douglas fir, Pinaceae) and *Abies concolor* (white fir, Pinaceae), with a few *Pinus flexilis* (limber pine, Pinaceae) along the cliff tops. (Cole 1985; photograph and reconstruction courtesy of K. Cole.)

tem that changes in a complex fashion. These changes occur on both ecological and evolutionary time scales. Unless we recognize this, we can be seriously misled about the causes of ecological patterns. Among plant ecologists, appreciation for this dynamic has grown during the past century. Today we have a much better understanding of the origins of modern plant communities, although much still remains to be discovered, especially about the Paleozoic and Mesozoic eras.

The Paleozoic era witnessed the establishment of plants on land and the origins of many of the basic patterns of plant and community diversity. The early Mesozoic era saw the dominance of the first seed plants, the seed ferns and the gymnosperms. An explosion of new forms and patterns occurred in the Cretaceous with the diversification of the angiosperms.

During the Cenozoic, angiosperms continued to evolve, and more recognizably modern species and communities appeared. These changes were driven by environmental change, caused primarily by continental drift, mountain building, and variations in Earth's orbit. Equally important were evolving biotic interactions such as competition, herbivory, and pollination. The evolution of these interactions created feedback effects, spurring further change.

Today's temperate plant communities were shaped mainly by patterns of glaciation during the past 2 million years. The last glacial retreat began about 12,000 years ago, with temperatures first rising, peaking at about 7000 years ago, and then dropping. Along with these climate changes have come migrations of species, resulting in shifting mosaics of species and communities across the landscape. Today these changes are happening at an even more dizzying pace, but they are now driven primarily by human-caused environmental changes—the topic of the next chapter.

Additional Readings

Classic References

Davis, M. B. 1981. Quaternary history and the stability of forest communities. In *Forest Succession Concepts and Applications*, D. C. West, H. H. Shugart and D. B. Botkin (eds.), 132–153. Springer-Verlag, New York. Contemporary Research kill?

Krasilov, V.A. 1975. *Paleoecology of Terrestrial Plants: Basic Principles and Techniques.* Wiley, New York.

Contemporary References

Davis, M. B., R. R. Calcote, S. Sugita and H. Takahara. 1998. Patchy invasion and the origin of a hemlock-hardwood forest mosaic. *Ecology* 79: 2641–2659.

Jacobs, B. F., J. D. Kingston and L. L. Jacobs. 1999. The origin of grass-dominated ecosystems. *Ann. Missouri Bot. Gard.* 86: 590–643.

Skog, J. E. 2001. Biogeography of Mesozoic leptosporangiate ferns related to extant ferns. *Brittonia* 53: 236–269.

Swetnam, T. W., C. D. Allen and J. L. Betancourt. 1999. Applied historical ecology: Using the past to manage for the future. *Ecol. Appl.* 9: 1189–1206.

Additional Resources

Graham, A. 1999. *Late Cretaceous and Cenozoic History of North American Vegetation.* Oxford University Press, New York.

Niklas, K. 1997. *The Evolutionary Biology of Plants.* University of Chicago Press, Chicago.

Pielou, E. C. 1991. *After the Ice Age.* University of Chicago Press, Chicago.

Betancourt, J. L., T. R. V. Devender and P. S. Martin. (eds.). 1990. *Packrat Middens: The Last 40,000 Years of Biotic Change.* University of Arizona Press, Tucson.

CHAPTER 22 *Global Change: Humans and Plants*

*E*arth is going through one of the most rapid periods of change in its history, and the primary cause of the ongoing changes is humans. The global carbon cycle is central to these changes.

One of the major consequences of alterations in the carbon cycle has been a global rise in temperatures. Some of the other likely changes in weather include shifts in rainfall patterns, so that some areas will become wetter and others drier; changes in the seasonality of rainfall in many locations; and increases in rainfall intensity during storms. These changes will affect most living organisms.

The global carbon cycle is intimately intertwined with plant ecology. Plants both drive and respond to the carbon cycle. As conditions for plant growth change over the face of the globe, dramatic changes—some predictable, but others unpredictable—will affect agricultural systems and natural ecosystems in all biomes. Plant and animal distributions will shift, natural communities will have different species compositions than they have today, some species will become extinct, and some natural communities will decline or vanish altogether. Humans and human institutions (such as national economies) will almost certainly be affected by these changes as well, directly by changing weather patterns and rising sea levels and indirectly by alterations in agricultural productivity (for better or worse, depending on the location).

While shifting weather patterns are the best known of its effects, anthropogenic global change encompasses far more than warmer temperatures. Other global changes caused by humans over the past 150 years include greatly altered patterns of land use and a large and continuing decline in the diversity of plants and animals living on Earth. To adequately address the role of plants, and consequences for plants, in this global drama, we must digress in places from our focus on plants to look more closely at the actions of people.

Carbon and Plant-Atmosphere Interactions

The Global Carbon Cycle

The uptake of carbon by plants was introduced in Chapter 2, and we examined in detail how carbon moves through ecosystems in Chapter 15. To recap briefly, plants take up carbon in the form of CO_2 from the atmosphere and incorporate it into organic compounds using energy from sunlight. Carbon moves through ecosystem food webs via consumption of plants and plant parts and by the actions of decomposers. Organic carbon accumulates in ecosystems for a period of time both in living organisms and as soil organic matter. Carbon returns to the atmosphere primarily through the respiration of soil organisms. We now widen our focus to a global scale, examining the pools and fluxes of carbon in terrestrial environments, the atmosphere, and the oceans (Figure 22.1).

By far the largest active pool of carbon on or near the Earth's surface is that dissolved in the oceans. (This does not include the substantial pools of carbon contained in rock, oil, and natural gas, which are naturally largely inactive.) Most of the active marine carbon is found in deep ocean waters, and the exchange of this carbon with the atmosphere occurs slowly, with a turnover time of about 350 years. Surface waters exchange CO_2 with the atmosphere much more quickly. It takes about 11 years, on average, for a molecule of CO_2 in the atmosphere to be dissolved at the ocean surface and then released again. Slightly more CO_2 is dissolved every year than is released, so the oceans currently act as a net sink for carbon.

Another substantial pool of carbon is found in soil organic matter. A major flux of CO_2 into the atmosphere comes from the respiration of soil organisms, primarily microorganisms (Schlesinger 1997). The atmosphere and

living plants contain the two other major active pools of CO_2, with somewhat more carbon stored in the atmosphere than in plants. (The organic matter stored in living animals is minuscule because the total biomass of animals is small relative to the global carbon budget.)

Gross primary production—the total flux of CO_2 into living terrestrial plants—removes the greatest amount of carbon from the atmosphere yearly. About half of this carbon is returned to the atmosphere through the respiration of terrestrial plants, so that terrestrial global NPP (net primary production: see Chapter 15) is about 60×10^{15} g C/year. Marine phytoplankton—particularly cyanobacteria—exchange CO_2 directly with that dissolved in surface ocean water and only indirectly with the atmosphere. NPP in the oceans is estimated to be about $35–50 \times 10^{15}$ g C/year, or somewhat less than that of terrestrial plants. Most of the carbon fixed by marine phytoplankton is taken up by heterotrophic bacteria and zooplankton in surface waters. Much of the CO_2 taken up by the bacteria is quickly transformed to inorganic forms again and released into the water, while the zooplankton form the basis for marine food webs. Only a tiny proportion of the NPP eventually sinks to the ocean bottom. The total pool of carbon in marine organisms at any one time is very small, although the turnover rate of carbon in this pool is very rapid.

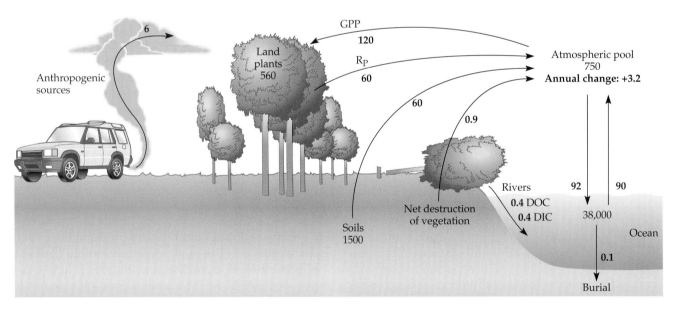

Figure 22.1
The global carbon cycle. The numbers show the pools (in units of 10^{15} g C) of carbon in various major compartments and the fluxes (in units of 10^{15} g C/year, shown in boldfaced type) of carbon moving between portions of the land, oceans, and atmosphere. Notice that land plants take up about twice as much carbon annually as they release, and that land plants store almost as much carbon as the atmos-

phere. Earth contains about 10^{23} g carbon; most of this carbon is buried in sedimentary rocks. Only about 10^{18} g carbon is contained in active pools, as shown here, with the largest of these as marine dissolved inorganic carbon. DOC, dissolved organic carbon; DIC, dissolved inorganic carbon; GPP, gross primary production; R_P, plant respiration. (After Schimel et al. 1995 and Schlesinger 1997.)

(A)

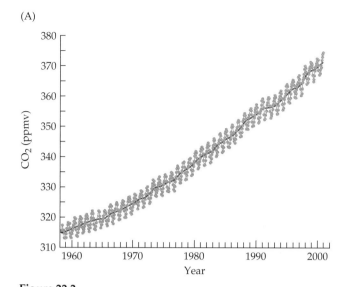

(B)

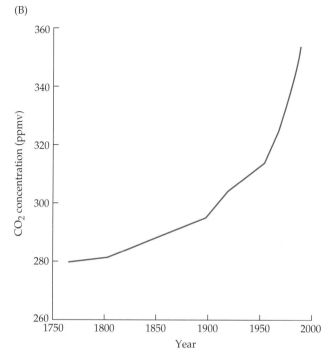

Figure 22.2
(A) Recent increases in atmospheric CO_2 (in parts per million by volume, ppmv) recorded at the Mauna Loa observatory in Hawaii. This site was chosen because it is not subject to the variation in CO_2 levels that would be found close to sources of anthropogenic emissions, but the results are similar to those found at many other sites. The actual monthly values are shown by data points; the wiggling line is a result of seasonal variation due to the decrease when Northern Hemisphere terrestrial vegetation is most actively photosynthesizing in summer and the increase when it is least actively photosynthesizing in winter. The long-term trend is shown by the colored line, which has been statistically smoothed based on the monthly values. Data for 1958 to 1974 are from the Scripps Institution of Oceanography, and for May 1974 to 2001 from the U.S. National Oceanic and Atmospheric Administration. (Keeling and Whorf 2001.) (B) The trend in CO_2 concentrations over the past two centuries, determined from bubbles of gas in ice cores from Antarctica. (After Vitousek 1992.)

These fluxes of CO_2 are natural, although they can be affected by human activities. Two additional important fluxes in the global carbon cycle, however, are entirely anthropogenic, at least on their current scales. These two fluxes, both of which add CO_2 to the atmosphere, result from the cutting and burning of large areas of forests and from the extraction and combustion of fossil fuels, including oil, natural gas, and coal. The immediate consequence of these additional fluxes is that atmospheric concentrations of CO_2 are rising. Before the Industrial Revolution of the nineteenth century, CO_2 was 280 ppmv (parts per million by volume) of the atmosphere. The concentration of CO_2 is now about 370 ppmv on average at Earth's surface and, if we continue our current practices, could reach twice preindustrial levels—560 ppmv—by 2065 (Figure 22.2). These higher levels of CO_2 will have direct effects on plants as well as indirect effects mediated by changes in climate. After exam-

ining the direct effects of increasing CO_2 levels, we discuss the effects they may have on global climates. We then examine the sources of this anthropogenic atmospheric carbon.

Direct Effects of Increasing CO_2 on Plants

Plants and other photosynthetic organisms use atmospheric CO_2 in photosynthesis, and the carbon in the atmosphere is thus the basis for most food webs on Earth. For plants that use C_3 photosynthesis—that is, the majority of photosynthesizers—the concentration of CO_2 in the atmosphere is suboptimal (see Chapter 2). It is reasonable to assume, therefore, that rising atmospheric CO_2 levels might act as a "fertilizer" to stimulate plant productivity. Elevated CO_2 levels can also increase the efficiency of water and nutrient use (at least for C_3 plants) and alter patterns of carbon allocation within the plant body. Because different plants respond very differently to enhanced CO_2 levels, global changes in atmospheric CO_2 concentrations might change intra- and interspecific competitive interactions among plants. Changes in competitive abilities, as well as any other responses that lead to differences in relative performance among individuals, can change the relative reproductive output of individuals, thus altering the future genetic structure of plant populations. Elevated CO_2 can also affect leaf chemistry, alter local ecosystem carbon cycles, and change nitrogen cycling and other ecosystem-level processes. Some ecologists have predicted that although there will be large positive responses to higher CO_2 levels initially (due to the "fertilizer" effect), this will be followed by adjustment and reduced responses over time, over a range of scales from individual plants to ecosystems. The details of these changing physiological and biogeochemical responses to rising CO_2 levels are interesting, but beyond the scope of this book.

It has been predicted that C_3 plants should grow more vigorously under elevated CO_2, while C_4 plants should not (see Chapter 2), and that C_4 plants may therefore be at a greater competitive disadvantage in an enhanced-CO_2 world. Differences among species in growth and other responses to enhanced CO_2 would be likely to change community composition, as some species would do better and others relatively worse. Wand et al. (1999) carried out a quantitative synthesis of the results of published studies on the responses of uncultivated grasses to elevated CO_2. The researchers showed that across all studies, both C_3 and C_4 plants had higher photosynthetic rates on average in response to elevated CO_2 (33% and 25%, respectively). Total biomass increased for both C_3 and C_4 grasses (by 44% and 33%, respectively), and leaf water use efficiency was also greater for both grass types. These findings were surprising, and show that the presumed competitive disadvantage of C_4 species at higher CO_2 levels may not exist.

In an analysis of the effects of elevated CO_2 on trees (all with C_3 photosynthesis), Curtis and Wang (1998) found that both photosynthesis and total growth increased substantially when CO_2 levels were elevated to twice the current atmospheric levels, although stressful conditions such as low soil nutrient levels reduced these effects of enhanced CO_2. They also found no evidence that trees eventually would acclimate to high CO_2, reducing its effect over time, as some have predicted. It is probably safe to say, therefore, that many plants would have positive direct responses to higher atmospheric CO_2 levels, including higher photosynthetic rates, greater growth, and reduced water loss, if nothing else changed. However, the indirect effects of the climate changes that are likely to occur due to elevated CO_2 will probably dwarf these direct effects.

Anthropogenic Global Climate Change

The Greenhouse Effect

Carbon dioxide molecules in the atmosphere are highly effective at absorbing infrared (or longwave) radiation (see Chapter 18). Visible (short) wavelengths of sunlight pass through the atmosphere without being absorbed by it, warming the Earth's surface. Earth's surface reradiates some of that energy in the longwave part of the spectrum (see Chapter 3). Some of that reradiated energy is absorbed and trapped by the atmosphere, particularly by water and CO_2 molecules (Figure 22.3). This recaptured energy warms the atmosphere, which then further warms the surface. The longwave energy that is not trapped by the atmosphere escapes to space.

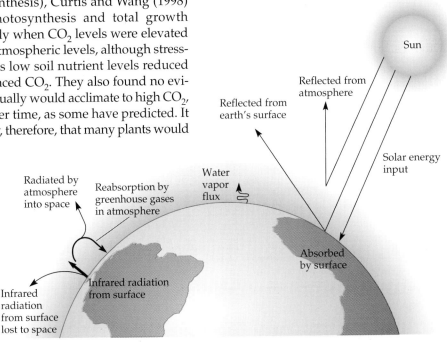

Figure 22.3
The greenhouse effect. About 70% of the incoming short-wave radiant energy from the sun is absorbed by Earth's surface; the remainder is reflected by the surface and the atmosphere. About half of the absorbed energy heats the Earth's surface. Some of that energy is radiated by the surface as longwave (primarily infrared) radiant energy. Most of that radiated energy is reabsorbed by atmospheric gases, which then reradiate it back to further warm the Earth's surface. A small portion of the longwave energy escapes from the atmosphere to space. The other half of the energy absorbed by the Earth's surface is used in the evaporation of water, which adds water vapor to the atmosphere and initially cools the surface. When that water vapor condenses to form clouds, the energy removed from the surface is transferred to the atmosphere, driving the creation of storms and releasing precipitation.

The carbon dioxide in the atmosphere is a natural part of the carbon cycle. Why, then, does it threaten to change global climates? The problem is not that there are CO_2 molecules in the atmosphere, but that there has been a tremendous rise in their concentrations. The greater the number of CO_2 molecules in the atmosphere, the more infrared radiation is trapped, and the more heat is retained at the Earth's surface. CO_2 is called a **greenhouse gas** because, like the glass in a greenhouse, it lets visible light energy enter the atmosphere but prevents longwave radiation from escaping, thereby warming the Earth's surface. The **greenhouse effect** is the warming of the Earth's surface and atmosphere due to the retention of heat by the greenhouse gases in the atmosphere. The greenhouse effect is natural and is necessary for the continuation of life on Earth—without it, the average temperature of the Earth's surface would be –18°C (0°F) instead of its actual 14°C (57°F). However, most popular usage of the term "greenhouse effect" refers to the rise in temperatures resulting from the increases in atmospheric CO_2 and other greenhouse gases caused by human activities. It is the increase in the greenhouse effect that is leading to global climate change.

Water vapor is the most predominant and effective greenhouse gas, followed by CO_2. Both occur in the atmosphere naturally. Because it is produced in the largest quantities (after water vapor), carbon dioxide is the most important greenhouse gas currently increasing due to human activities, but it is unfortunately not the only one. A variety of other gases also contribute to global warming. (Anthropogenic effects on water vapor are complex and not fully understood; they are not thought to be a major contributor to the increase in the greenhouse effect.) In addition to CO_2 and water vapor, four of the most important greenhouse gases are methane (CH_4), nitrous oxide (N_2O), ozone (O_3), and chlorinated fluorocarbons (CFCs). The first three occur naturally, but humans have greatly increased their concentrations in the atmosphere.

Anthropogenic methane (Figure 22.4) is released in large quantities by leakage during the drilling and transportation of natural gas, by anaerobic prokaryotes in cattle rumens, and in decomposing trash in landfills and dumps. Increases in nitrous oxide are largely a by-product of commercial agriculture, as it is released by the microbial degradation of nitrogen-containing fertilizers; nylon production is another substantial source of N_2O. Ozone, O_3, is produced in the lower atmosphere (the troposphere) through a series of chemical reactions where air pollution is high, particularly as volatile hydrocarbons and carbon monoxide in the presence of high NO. Some ozone is also produced naturally from compounds emitted by vegetation and by forest fires. CFCs are artificial compounds used primarily as refrigerants and propellants. While recent international treaties have dramatically reduced their production, they are so stable chemically that the molecules already in the air will persist for a very long time.

While humans produce far smaller amounts of these four compounds than of CO_2, these other greenhouse gases are much more potent absorbers of infrared radiation. Methane traps more than 20 times the amount of heat per molecule than carbon dioxide does, while nitrous oxide absorbs 270 times more. Carbon dioxide emissions are responsible for about two-thirds of the anthropogenic intensification of the greenhouse effect, with most of the remainder due to CH_4 (about 19%), CFCs (10%), and N_2O (6%).

(A)

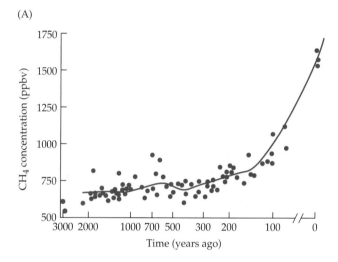

(B)

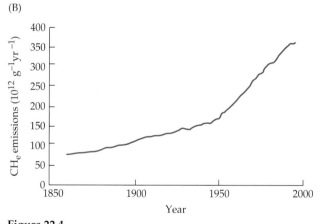

Figure 22.4
Increases in atmospheric concentrations of methane, a major greenhouse gas. (A) Atmospheric methane (CH_4) levels (in parts per billion by volume, ppbv) during the past 3,000 years (note that the x-axis expresses time on a logarithmic scale). Concentrations were measured from air bubbles trapped in ice cores from Greenland and Antarctica, or from historic and current air samples. (After Schlesinger 1997; Cicerone and Oremland 1988.) (B) Global anthropogenic emissions of methane (CH_4, in 10^{12} g/year) from all sources from 1860 to 1994. (After Stern and Kaufmann 1998.)

Many people, incidentally, confuse the intensification of the greenhouse effect with the problem of the ozone hole in the atmosphere. They are two different—although related—problems. The ozone hole is the loss of parts of the protective layer of ozone high in the stratosphere (the part of the atmosphere from 10 to 50 km above the surface), caused primarily by a class of artificial compounds called halogenated hydrocarbons (including CFCs and related chemicals). Ordinarily the stratospheric ozone layer absorbs much of the high-energy ultraviolet radiation coming from the sun, preventing it from reaching the Earth's surface. Halogenated hydrocarbons destroy ozone in the stratosphere, allowing more ultraviolet radiation to reach the surface; this effect has been particularly severe at high latitudes in the southern hemisphere. The increase in ultraviolet radiation poses a risk of damage to many organisms, including humans. This phenomenon contrasts with the intensification of the greenhouse effect, which is caused by a buildup of heat-trapping gases in the atmosphere, with consequent increases in global temperatures. The ozone hole is connected to the greenhouse effect and the problem of global warming specifically because CFCs act both to destroy stratospheric ozone and as greenhouse gasses, and more generally because both are anthropogenic changes in Earth's atmosphere.

Global Climate Change: Evidence

The evidence is clear that the concentrations of CO_2, methane, and certain other greenhouse gasses have been increasing in the atmosphere. Predictions based on physics say that these increases will result in a warmer Earth. But what is the evidence, if any, that Earth is actually warming, or that any other climate changes have resulted from the rising concentrations of greenhouse gasses? Weather and climate are complex phenomena, and natural variation can confound and mask general trends (see Chapter 18), so sifting out the effects of any particular factor can be difficult.

That is why it has taken so long to nail down the effects of CO_2 on past and future climate change. By now, however, the weight of evidence is so great that there is little scientific doubt remaining that the Earth is rapidly becoming warmer even if the exact magnitude is still uncertain. We now take a brief look at some of the different lines of scientific evidence that have been amassed on anthropogenically caused global climate change.

Over the past hundred years, the average temperature of the Earth's surface has increased by 0.6°C (± 0.2°C). Almost half of this change (0.2–0.3°C) has occurred over the last 25 years, and temperature increases may now be accelerating (Figure 22.5). Since 1980, the Earth has experienced 17 of the 18 hottest years of the

Figure 22.5
Temperature increases over the past 120 years, expressed as differences between annual values and long-term mean temperatures (°C) for the entire Earth (terrestrial and oceanic), the world's oceans, and the world's land surfaces. Global temperatures over the past 20 years were consistently above the long-term averages, and most of the hottest years on record occurred recently. Temperature increases have been greater on land than in the oceans, although temperatures are consistently increasing for both. The long-term mean temperatures were calculated from data recorded at thousands of worldwide terrestrial and oceanic observation sites for the entire period of the data, and by interpolation for areas that were not monitored. (Unpublished data of T. C. Peterson, U.S. National Climatic Data Center/NOAA, October 2001.)

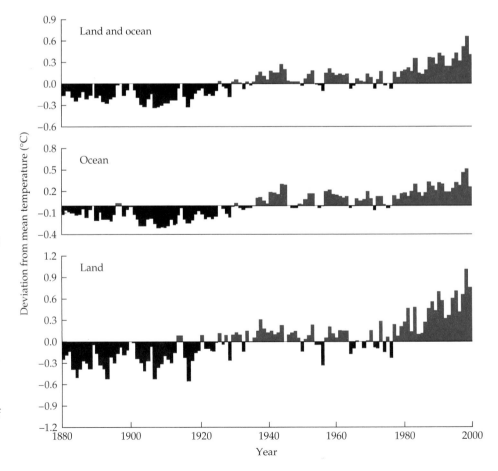

twentieth century. Some regions have experienced much greater warming, while others have not warmed as much, or have even cooled. The greatest warming has occurred in northern latitudes (between 40° N and 70° N across Eurasia and North America). Summer temperatures in the Northern Hemisphere in recent decades have been the warmest in the past thousand years. Nighttime temperatures are increasing more than daytime temperatures, leading to reduced day/night differences in many places.

In addition to direct measurements of increasing temperatures, indirect data also show evidence of general global warming. Glaciers are retreating, snow cover is decreasing and melting earlier in the spring, and Arctic sea ice has declined both in thickness and in the area it covers. Permafrost in the Arctic is melting. In the Antarctic, the rate at which the ice shelves that ring much of the continent are breaking up appears to be increasing. There is evidence of both surface melting and warming below floating ice, and average air and water temperatures around the coastlines of Antarctica are becoming higher as well. Sea levels globally have already risen between 10 and 20 centimeters in the past hundred years, largely due to the expansion of water when it warms.

Temperature is not the only aspect of global climates that is changing. The frequency and intensity of El Niño Southern Oscillation events (see Chapter 18) also appears to have increased over the past few decades, with an abrupt increase in the mid-1970s that has persisted since then. ENSO events have been occurring for at least hundreds of years (and maybe far longer). However, it is hypothesized that recent global warming, by pumping heat energy into the surface of the oceans, will intensify their effects. Although some scientists have predicted that a consequence of global warming will be a global increase in extreme weather, there is presently no evidence for general increases in extreme events or weather variability across the entire globe.

At a regional scale, however, there has been an increase both in the variability of weather conditions and in extreme events. It has not been firmly demonstrated that there is a tie between these ENSO-related increases and global warming, but there are some reasons to believe that there may be a connection between the two. There is evidence suggesting that heavy rainfall events have become more severe regionally, particularly at higher latitudes in the Northern Hemisphere. Parts of Africa and Asia have suffered from more frequent and severe droughts, associated in part with shifts in ENSO patterns. ENSO is linked to other regional circulation phenomena that may interact with global warming to further change climates in complex ways that are not yet fully understood.

How long have we known about the greenhouse effect, changes in it as a result of increasing atmospheric CO_2 levels, and global warming? Probably the first scientist who published the prediction that CO_2 from the combustion of fossil fuels would lead to global climate change was the Swedish chemist Svante Arrhenius in 1896. It was first confirmed that CO_2 concentrations in the atmosphere were actually rising by the 1930s, and by the 1950s there were precise measurements of the magnitude of the increase. Most scientists had accepted the reality of global warming by the 1990s. A small number of scientists do not accept the generally held conclusion that anthropogenic rises in CO_2 concentration are responsible for global temperature increases. There is, however, agreement that adding greenhouse gases to the atmosphere must ultimately warm it, and a large majority of scientists today believe the evidence is strong enough to link recent warming to anthropogenic changes in the concentrations of greenhouse gases.

Global Climate Change: Predictions

We have just taken a look at some of the actual changes in the Earth's climate that are associated with the intensified greenhouse effect. What kinds of changes are predicted for the future? Climatic scientists use a variety of direct and indirect measurements to document climate changes. Predictions for what future climates might be like are made by constructing and refining complex computer simulation models (Box 22A).

The popular reaction to evidence for global warming sometimes seems to be that some warm winter days in temperate regions would actually be pleasant and welcome. What could be wrong with getting to play ball or wash your car outside in the middle of winter? To answer that question, we need to consider more closely what global climate change may entail.

A variety of changes are predicted to result from the greenhouse effect. Averaged across the entire globe, temperatures are likely to rise by 1.4–5.8°C or more over the next hundred years. To get an idea of how large such a change would be, consider that during the most recent glacial maximum 20,000 years ago, temperatures were only about 9°C cooler globally than they are today (see Figure 21.8). Or consider that the average temperature over the year is only 3°C higher in Milan, Italy, than in London, England, and 4°C higher in Charlotte, North Carolina, than in New York City. Clearly, what might seem to be a small average temperature increase can mean a very different climate.

But these temperature increases will not be evenly spread across the globe. Changes in atmospheric and oceanic circulation patterns are forecast to affect temperature, rainfall, and other climatic factors differently at different locations. Land areas will probably warm much more than oceans, and higher latitudes in the Northern Hemisphere will warm much more than other places, particularly in the winter. Summer warming is

Box 22A

Modeling Climate

How do scientists know what the values for temperatures, precipitation, and other weather and climatic variables were in the past, particularly the distant past? There are both direct records of these variables—measurements that have been taken and recorded around the world—as well as indirect records of various kinds.

Reasonably good instrumental records have been kept in many places for temperature, rainfall and snowfall, humidity, and air pressure at Earth's surface for at least a hundred years, but some locations are missing data entirely or for earlier years. For recent years, data from satellites and high-atmosphere weather balloons exist. There are also excellent recent data for global ocean temperatures.

Physical, biological, and chemical indicators offer insight into climate and weather further back in time (see Chapter 21). Some of these indicators include measurements of tree rings from trees that are hundreds or thousands of years old, sediments at the bottom of lakes and oceans, air trapped in ancient ice cores, and annual density bands in long-lived corals.

How do scientists make predictions about future climates? Atmospheric scientists use complex computer simulation models, called General Circulation Models or Global Climate Models (GCMs), that incorporate existing weather and climate data, as well as what is known about the major processes that determine climate, to predict climate change. These climatological models have been constructed using largely physical rather than biological data, in contrast to the ecological models discussed in Chapter 15, although they can incorporate some effects of vegetation.

Once a model is constructed, changes in some factors—such as CO_2 levels in the atmosphere—can be introduced to see how the climate might be affected in different parts of the globe. Many scientists must work together to construct a GCM, and only a small number of major GCMs exist. These models incorporate large amounts of known data and make predictions about future temperatures, rainfall, snow cover, sea level, and other climatic factors.

Recent research in this area, for example, has included efforts to incorporate more accurate satellite data on Earth's radiation balance into GCMs (Wielicki et al. 2002). These attempts have revealed greater variation in radiation fluxes between Earth's surface and the atmosphere than was previously thought to exist, particularly in the Tropics. Differences in cloudiness appear to account for this variation; clouds both increase reflection of incoming solar radiation (a cooling effect), and reduce infrared radiation to space (a warming effect). A complex set of feedbacks between Earth's surface and the atmosphere can affect the degree of cloudiness.

The predictions based on GCMs are uncertain to various extents, due to limitations in what is known as well as uncertainty about what human actions will be taken (for example, whether CO_2 emissions will be reduced). However, these models have been continually refined and improved, and they are becoming increasingly accurate and detailed. They offer a reasonable set of predictions that will help us to anticipate changes in global climate.

---

expected to be particularly great in central and northern Asia. More hot days and periods of extended intense heat are predicted over all land areas, and average humidity is expected to increase. Cold waves will become fewer and less intense. Glaciers and ice caps will continue their dramatic retreats.

Large changes in precipitation levels and storm severity are also predicted. Large increases in yearly rainfall are expected at high latitudes across the Northern Hemisphere. Winter precipitation is predicted to increase in tropical Africa and at mid-latitudes in the Northern Hemisphere, and summer rainfall is predicted to increase in southern and eastern Asia. Rainfall will decrease in Australia, Central America, and southern Africa, particularly in winter. Rainfall patterns will shift, becoming more variable from year to year almost everywhere, with increased flooding in some years and drought in others. Extreme rainfall events will become

more frequent and more intense (more rainfall per event) over most land areas. Mid-continental areas will become drier, with major implications for the world's "breadbasket" regions (see Chapter 19). Both mean and peak rainfall during tropical cyclones are expected to increase a great deal, although the frequency of cyclones is not predicted to be greater. The effects on mid-latitude storm intensities are still uncertain.

Sea levels are predicted to rise between 11 and 77 centimeters globally, due largely to the continued thermal expansion of water and, to a lesser extent, to melting glaciers. If this rise occurs as predicted, it will result in widespread coastal flooding. The consequences of this flooding will be felt from low-lying island nations such as the Maldives to highly populated coastal regions from Bangladesh to China to Florida. Most coastal cities—which contain a large proportion of the world's population—will also be threatened.

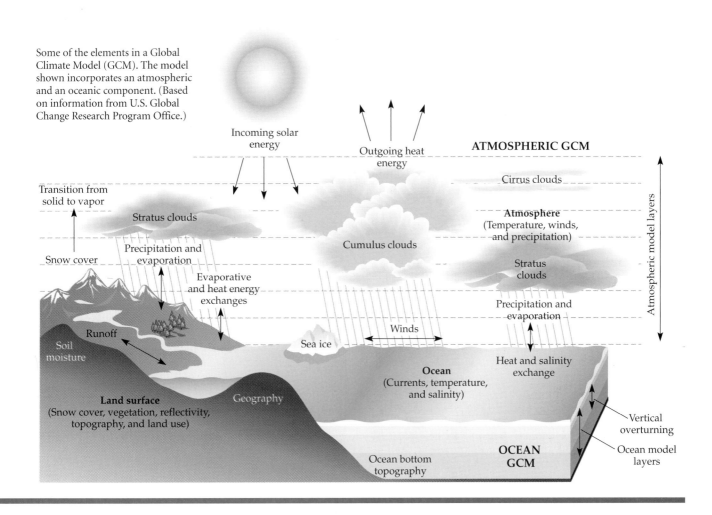

Some of the elements in a Global Climate Model (GCM). The model shown incorporates an atmospheric and an oceanic component. (Based on information from U.S. Global Change Research Program Office.)

Biotic Consequences of Climate Change

The consequences of regional and global changes in climate are likely to have large, though only partially predictable, effects on plants. The details of these changes are not yet known, but parts of the bigger picture are beginning to come into focus. Some species, ecosystems, and vegetation types will benefit from these changes and increase, while others will lose ground, declining or disappearing (Figure 22.6). While there are models that make specific predictions about particular regions, we will focus here on the general nature of the changes we can expect.

Individual plant physiology, growth, reproduction, and allocation are all likely to be affected by changing climates. Interactions of plants with other species—competitors, pollinators, herbivores, and pathogens—will probably change as well. For example, plants that are dependent on highly specialized pollinators or dis-

persers will clearly suffer if environmental changes reduce or eliminate the animals they depend on. Climate change is predicted to alter ecosystem cycling for many major nutrients. Soil respiration and nitrogen mineralization rates, for example, have been shown to increase substantially when ecosystems are warmed experimentally. Some models of climate change predict increases in fire frequency and severity in many ecosystems.

The distributions of many plant species are expected to change, resulting both in shifts in the locations of plant communities and in different combinations of species growing together in the communities of the future. The increases in productivity resulting from higher CO_2 levels may be amplified in some ecosystems by warmer climates and longer growing seasons. Other ecosystems will be affected by increased drought, which will cancel out the positive effects of higher CO_2 levels on photosynthesis and growth. Changes in temperatures

Figure 22.6
(A) Current distribution of the world's biomes, as derived from current climatic data using the MAPSS model (Mapped Atmosphere Plant Soil System). (B) Predicted altered distribution of biomes 75 to 100 years in the future, based on one set of scenarios for global CO_2 increases. According to this model, the early stages of global warming will lead to increases in the productivity and density of forests worldwide, as increased carbon dioxide levels act as a fertilizer. Continued elevated temperatures, however, will strain water resources, in time producing drought-induced stress and broad-scale die-offs of vegetation, with associated increases in wildfires. (Based on models by Neilson 1995 and data from http://www.fs.fed.us/pnw/corvallis/mdr/mapss/)

(A) Current biome distribution

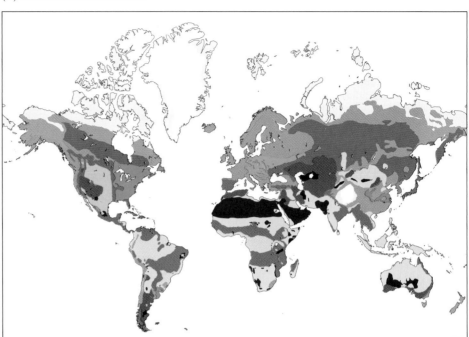

(B) Predicted distribution, 2070–2099

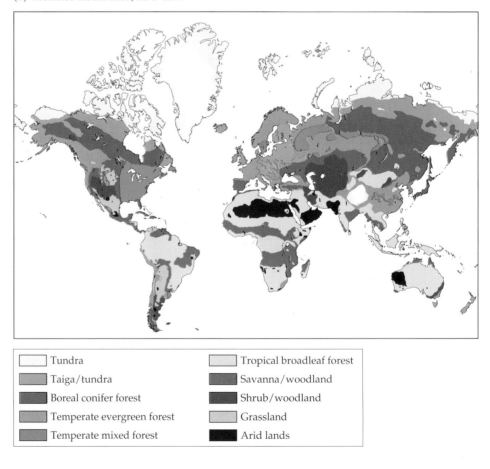

Tundra		Tropical broadleaf forest	
Taiga/tundra		Savanna/woodland	
Boreal conifer forest		Shrub/woodland	
Temperate evergreen forest		Grassland	
Temperate mixed forest		Arid lands	

and rainfall regimes are expected to have major consequences for agricultural regions. Some highly productive areas are predicted to become unproductive, while some regions that are currently marginal agriculturally will become highly productive. For example, there are expected to be declines in the agricultural productivity of the U.S. Midwest and increases in that of the grain belt of central Canada. The crops that are usually grown in particular areas may change substantially, with grain belts shifting and frost-sensitive crops being grown at higher latitudes.

As we have seen, atmospheric CO_2 levels have fluctuated widely over the course of Earth's history (see Figure 21.6), and global and regional climates have fluctuated widely as well (see Figure 21.8). Biomes and the species inhabiting them have changed dramatically over the eons. Many plant species have moved as climate has changed. After the most recent glaciation in North America, for example, species moved northward to recolonize areas that became deglaciated as temperatures warmed (see Figure 21.10). The biggest difference between the global changes of the past and those occurring now is that the current changes are happening at much faster rates—far greater, ecologists predict, than the rates at which many species can migrate or adapt. Many of the past changes were on a regional or continental scale, rather than the current global scale, so refuges may be scarcer today: all species everywhere are more likely to experience at least some of the effects of climate change.

Habitat reduction and fragmentation are likely to make it more difficult for species to migrate in response to climate change. Parks and preserves that currently protect endangered populations, species, or communities may become ineffective refuges: as climatic conditions change, such refuges may no longer be as suitable for those organisms or systems. If species are unable to reach new favorable habitat, or if habitat in newly suitable environments is not protected, those species will disappear. Species will almost certainly not move together, but will migrate (or fail to migrate) individually (see Chapter 21). As a result, many vegetation and community types may disappear and be replaced with new species combinations having new properties.

Arctic and boreal forest species, facing the greatest degree of climate change, may find themselves with no place to go as warming shrinks the extent of suitable habitat. Alpine species forced to move to higher altitudes may face similar problems. Many plant species are reported to be moving to higher elevations in mountainous areas, including trees that are becoming established above current tree lines. In Europe, current plant species richness on 30 peaks in the Alps has increased

by 70% over the oldest accurate historical records, apparently as a result of colonization by lower-elevation species; at the same time, glaciers there have lost 30–40% of their surface area. Likewise, butterflies, birds, and other species that interact closely with plants have been moving upward in elevation and poleward (reviewed by Hughes 2000).

Reductions in available habitat as elevation increases, coupled with species being forced upward by warming, are likely to result in at least regional extinctions of some alpine species. Changes in the vegetation of the San Francisco Mountains in northern Arizona, for example, were modeled by Francis (1999). Currently, this vegetation varies from sage shrublands at the base of the mountains, through *Pinus ponderosa* (Ponderosa pine, Pinaceae) forests at mid-elevations, to spruce-fir forests, to tundra at the summit. The predicted effects of climate change include a substantial spread of the sage shrublands, the movement of ponderosa pine forests to higher elevations, and the loss of spruce-fir forests and tundra (Figure 22.7).

Range shifts, range contractions and expansions, and changes in phenological patterns are likely to occur, as changes take place not only in average temperatures, but in the length of growing seasons, time of snowmelt or even the existence of snowpack, the duration of drought and extreme heat events, fire regimes, and soil moisture availability at different times of the year. Furthermore, the ecology of organisms at all levels, from populations to ecosystems, will be affected by changes in many of these factors at the same time. Due to the combined effects of multiple stresses, species currently threatened or endangered are more likely to go extinct, and additional species are likely to become endangered.

Finally, changes in land forms are likely to have substantial effects on vegetation at the landscape and regional levels and on the global persistence of particular groups of species dependent on affected habitats. Sea level rises and increased rainfall have already caused many seacoasts to experience increased flooding, storm damage, and erosion (Figure 22.8). These processes are likely to accelerate over the coming decades. Freshwater marshes close to coastal areas may be affected by saltwater incursion. Highly diverse and productive coastal ecosystems, such as salt marshes and mangrove forests, will be subjected to negative and sometimes disastrous effects (including erosion and submersion). Small islands close to sea level may disappear entirely as oceans rise, taking their endemic and rare plant and animal species with them (as well as the homes of the people living there). Even far from coastlines, increases in storm intensity, sometimes coupled with the effects of deforestation, are expected to result in major disturbances to plant

(A)

Figure 22.7
(A) The San Francisco Peaks are a volcanic range in northern Arizona. The highest point is at 3850 meters elevation. (Photograph by S. Scheiner.) (B) Current and predicted vegetation patterns on the San Francisco Mountains. Currently, a small area of alpine tundra exists at the highest elevation, with spruce-fir forests at slightly lower elevations. Intermediate elevations are dominated by ponderosa pine forests. The lower elevations are covered by sage shrubland and pinyon-juniper woodland. The predicted changes are based on a climate change model (see Box 22A). Rising temperatures will result in the spread of the sage shrublands, the movement of ponderosa pine forests to higher elevations, and the loss of spruce-fir forests and tundra. (After Francis 1999.)

(B) Current vegetation   Predicted vegetation

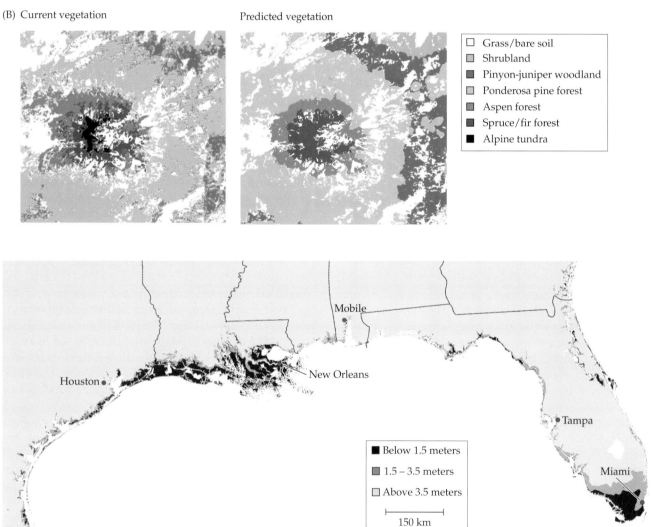

☐ Grass/bare soil
▨ Shrubland
▨ Pinyon-juniper woodland
▨ Ponderosa pine forest
▨ Aspen forest
▨ Spruce/fir forest
■ Alpine tundra

■ Below 1.5 meters
▨ 1.5 – 3.5 meters
☐ Above 3.5 meters

⊢———⊣
150 km

Figure 22.8
Lands vulnerable to sea level rise along the Gulf Coast of the southern United States. The map shows land below the 1.5-m and 3.5-m contours (elevations above sea level in 1929), which are vulnerable under different models of future climate change. A rise in sea level sufficient to inundate lands at the 1.5-m contour at high tides appears to be most likely to occur in the next 120 years, but has a 1% chance of occurring in the next 60 years. Sea level has risen about 20 cen-

timeters since 1929. Approximately 58,000 km² of land along the Atlantic and Gulf coasts lie below the 1.5-m contour. Those above the 3.5-m contour would be vulnerable after a period of several centuries with a greater sea level rise. Rising sea levels will threaten, drastically alter, or destroy many coastal ecosystems, such as highly productive salt marshes in many areas, and the Florida Everglades in the southern part of Florida. (After Titus and Richman 2001.)

communities by causing flooding in river valleys and landslides, mudslides, and avalanches on steep slopes.

Anthropogenic Effects on the Global Carbon Cycle

The initially pleasant image you may have had of global warming meaning milder days in winter may begin to look less appealing upon seeing more of the details of the changes that may be in store. Where does the anthropogenic carbon responsible for these changes originate? We mentioned earlier that increases in atmospheric CO_2 are caused mainly by two factors: **deforestation** (the destruction of forests) and the combustion of fossil fuels (Figure 22.9). Where are these things happening, and who is causing them? To fully investigate these questions, we would need to discuss not only the science involved, but also the social, political, and economic bases for anthropogenic CO_2 emissions. These topics can be treated to only a limited extent in a plant ecology textbook. Here we mention some of the factors that may be involved in order to suggest some of the possible directions you may want to take in your future thinking and reading on this issue.

Deforestation

While trees naturally accumulate carbon in wood and release it when they eventually die, human activities are dramatically accelerating the rate at which this is happening. Cutting down a tree affects atmospheric CO_2 levels in two ways. First, if the tree is not replaced with another tree, the CO_2 it would have taken up will remain in the atmosphere rather than being fixed. Second, if the tree is burned or allowed to decompose, much of the carbon accumulated over its long life—which can be decades or centuries for canopy trees—is released to the atmosphere suddenly or over a relatively short time. (Herbaceous plants, in contrast, rapidly exchange carbon with the soil and the atmosphere.)

If trees are logged to build things that remain intact for a long time (a good violin, perhaps, or a building that lasts a long time, rather than a stack of bulk mail solicitations), then some of the carbon in the wood will be preserved in the structure for decades or even centuries, rather than being immediately released to the atmosphere (although the carbon in roots, branches, and leaves that remain when the trunk is harvested will decompose quickly, even when the wood is used in something long-lasting). If trees are logged to make newspaper, disposable chopsticks, or other short-lived items, the postdisposal fate of those items (whether they are burned, stored in a landfill, or recycled) will determine the fate of that carbon. The fate of the organic carbon in the forest soil will depend on a number of factors, including what happens to the land after the forest is logged.

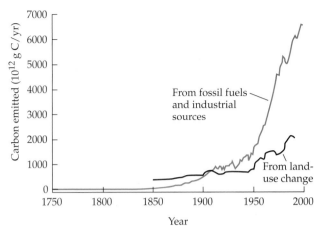

Figure 22.9
Carbon emissions to the atmosphere (in 10^{12}g C/year) from fossil fuel combustion and other industrial sources (green line) from 1750 present, and carbon flux to the atmosphere from land-use change (primarily deforestation; black line) from 1850 to the present. (Data from Houghton and Hackler 2001; Marland et al. 2001.)

If new trees are planted where a forest is logged, the new trees will accumulate carbon and net ecosystem production (see Chapter 15) is likely to be positive. If the forest is replaced by buildings and pavement, that obviously will not happen. Removing forests and replacing them with agricultural lands, tree plantations, parking lots, or barren wastelands has enormous ecological implications beyond the effects on the carbon cycle, and we will briefly consider some of these implications toward the end of this chapter.

Two factors act directly to counter the emission of CO_2 to the atmosphere caused by forest destruction: forest regeneration and higher photosynthetic rates due to the "fertilization" effect of higher CO_2 concentrations in the atmosphere. Forest regeneration has, in some cases, been dramatic. Large parts of the northeastern United States, for example, were heavily logged and cleared for farming in the eighteenth and nineteenth centuries. The abandonment of these farms led to the regeneration of large areas of secondary forest (see Figure 13.1), sufficient to act as a substantial carbon sink. However, these lands have recently once again become sources rather than sinks for atmospheric CO_2, both because many of these forest ecosystems have reached maturity and their rate of carbon accumulation is declining (see Chapter 15), and because of renewed forest clearing for both lumber and suburban development.

Cutting trees for firewood and burning trees to clear land have been anthropogenic sources of carbon input to the atmosphere throughout human history. These activities were sufficiently limited before about 1800 that they had little or no measurable effect on atmospheric CO_2 concentrations (see Figure 21.6B). That is no longer the case, however, and the effects of these activities are particularly severe for tropical forests. Today, about 1.1×10^{15} g C is emitted to the atmosphere from the burning of tropical forests each year, and almost twice that amount is released from all tropical deforestation activities. A gross total of 3.4×10^{15} g C is produced by tropical forest destruction if decomposition of uncombusted material is included (Fearnside 2000). The total net emission of carbon to the atmosphere resulting from tropical land use changes (particularly destruction of forests) is equal to approximately 29% of the total global anthropogenic emissions from all fossil fuel combustion and nontropical land use changes combined (Fearnside 2000). Tropical forest clear-cutting has been carried out both by people living in tropical countries and by corporations owned by people in developed countries. Corporations based in Japan, Europe, and North America are all active in tropical deforestation.

Not all deforestation has occurred in the Tropics, however. In many cases, developed countries have cleared large areas of their own forested land. An example is the temperate rainforests of the northwestern U.S. and British Columbia in Canada, which have experienced large-scale deforestation and clear-cut logging in the past 20 years (see Figure 19.11). (Where these old-growth forests are replanted or undergo secondary succession, the young trees will once again serve as sinks for CO_2, although the removal of old-growth forests results in other ecological effects.) One important reason why forest destruction is concentrated in developing countries today is that forests in the wealthy industrial countries have already been altered or destroyed. Both Europe and Japan, for example, have very few areas that are not heavily affected by human use. In Japan, virtually all lowland areas are either farmed or occupied by human structures; on the mountainsides, there are extensive forests, but most of these are regularly logged. In the Swiss Alps, livestock are grazed even on some of the highest peaks. In Spain, the highest elevations contain some of the greatest plant species diversity, not because those areas are inherently more species-rich, but because more productive lowland areas are heavily used for agriculture.

Fossil Fuel Combustion

The burning of fossil fuels provides the second major anthropogenic contribution to atmospheric CO_2. The carbon contained in fossil fuels accumulated over eons, as it was taken up from an ancient atmosphere by organisms now long extinct, over the course of millions of years. Vast amounts of carbon in the form of fossil fuels are being taken from deep in the earth and released in a geological blink of an eye. The extractable pool—the portion feasible to get at—of carbon contained in Earth's fossil fuels is about 4×10^{18} g C, which is about one-tenth of the total active pool of carbon at or near the Earth's surface. The combustion of fossil fuels releases approximately 6×10^{15} g C per year. A little over half of this carbon remains in the atmosphere, adding about 1.5 ppmv CO_2 each year to the atmosphere. Some of the rest of the CO_2 from fossil fuels enters the oceans, while some remains unaccounted for. Not until the middle of the 20th century did the carbon released to the atmosphere by the combustion of fossil fuels become greater than the carbon released by the cutting or burning of forests, but now it represents about 70% of all anthropogenic CO_2 released.

CO_2 emissions differ enormously among the countries of the world, whether we look at the total amount released into the atmosphere by each country or consider how much, on average, each person in the country is responsible for producing (Figure 22.10). The highest total emissions per country come from the United States, but China is a remarkably close second. U.S. emissions have been growing rapidly for the past hundred years, while those from China have been growing very steeply since the late 1960s (although the most recent data for China suggest a potential slowdown). The United States and China maintain these high emissions by very different means. The United States has a fairly large population and very high rates of energy use per person. China has an enormous population, but the amount of energy used on average per person is quite modest (Figure 22.10C).

The top ten countries in total fossil fuel carbon emissions as of 1998, from the highest in descending order, are the United States, China, the Russian Federation, Japan, India, Germany, the United Kingdom, Canada, Italy, and Mexico. This list contains a mix of developed and developing countries in North America, Europe, and Asia. Most industrialized countries in Europe have shown dramatic increases in energy efficiency and, consequently, both decreased per capita emissions (such as those for Germany and Poland; Figure 22.10C and D) and decreased total emissions (Figure 22.10A and B). Modest per capita increases in countries with large and rapidly growing populations, such as India and Brazil, over the past 15 years (Figure 22.10D) have led to steep increases in those countries' total emissions (Figure 22.10B). On the other hand, one could argue that, from the standpoint of reducing human suffering, some countries actually produce too little CO_2 from fossil fuels. The tiny and erratic emissions of countries such as Congo are a reflection of disaster, war, deprivation, and human misery. For other countries, low emissions may reflect a tiny population, particularly one that is largely non-

(A)

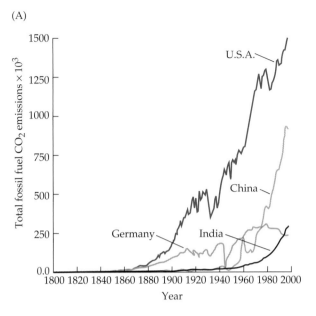

(B)

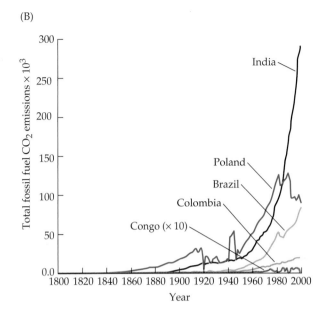

(C)

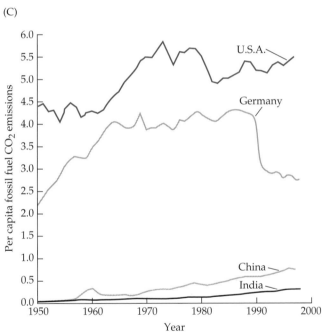

(D)

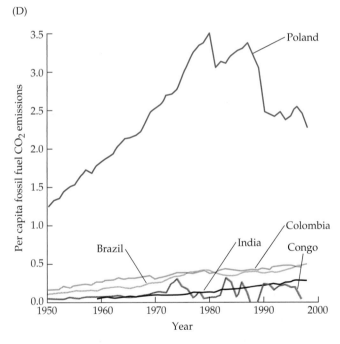

Figure 22.10
CO_2 emissions resulting from the combustion of fossil fuels. (A, B) Total emissions (in 10^{12} g C/year) for various countries from 1800 to 1998. Note that the scales on the y-axes of the two graphs are very different because the total amounts differ by almost an order of magnitude; the data for India are included in both graphs to show these scale differences. The values for Congo were multiplied by 10 so that they would be visible. (C, D) Per capita emissions (in 10^6 g C/year) for various countries from 1950–1998. Note that the scales on the y-axes of the two graphs are different; data for India are included in both graphs to show these scale differences. (Data for Germany include combined totals for the former German Democratic Republic and Federal Republic of Germany between 1946 and 1990.) (Data from Marland et al. 2001.)

industrialized, with few automobiles and trucks, little industry, and largely nonmechanized agriculture.

Thus, the anthropogenic contribution to atmospheric CO_2 is highly unequal. Almost any way one measures them, U.S. emissions are disproportionately large. The combination of a fairly high per capita emission rate with a fairly large population makes U.S. emissions the largest source of anthropogenic carbon input to the atmosphere. The United States has contributed about 30% of the anthropogenic increase in atmospheric CO_2 to date, and currently adds 25% of the carbon emitted per year. While only 5% of the world's popula-

Box 22B

Daily Human Activities and CO_2 Generation

Every time you breathe you release CO_2 into the atmosphere. The amount varies from person to person, of course, depending on an individual's weight, activity level, fat:muscle ratio, and other factors; but on average, in a year each person respires 365 kg of CO_2.

But human respiration is not really adding to atmospheric CO_2. As with all animals, the carbon respired comes from carbon removed from the atmos-

phere a short time before by living plants. Human use of fossil fuels, however, increases atmospheric releases of CO_2.

The amount of CO_2 removed from the atmosphere by one mature tree annually is about 9 kg; based on this figure, it takes 1,430 trees to use up the CO_2 produced by one sport utility vehicle of the largest class (e.g., Lincoln Navigator).

The table on the right gives approximate amounts of CO_2 released into the atmosphere by some common human activities. All of the numbers are general approximations, made using various assumptions; the amount of CO_2 actually produced varies as conditions change. The purpose of these numbers is not to provide exact figures, but to make general comparisons among different activities.

---

tion lives in the United States, it accounts for 25% of the global daily usage of petroleum (about 3 billion of the more than 12 billion liters used globally per day), and about 25% of the world's energy consumption.

The current rate of CO_2 emission in the United States is 5.43 metric tons of carbon per year per person. Can this be accounted for by factors such as the high standard of living in the United States, the need for home heating during the winter in much of the country, or transportation needs in the large geographic area it encompasses? Not readily. The U.S. per capita emission rate is far higher, for example, than that in Sweden, a colder country with a comparable standard of living. Sweden's average annual emission rate is 1.50 metric tons of CO_2 per person, and, as in many European countries, this rate has declined steadily (from a peak of 3.13 in 1970), while emissions in the United States, after a modest drop during the 1970s, otherwise continued to climb (Figure 22.10C). U.S. emissions are also higher than those in Australia (4.88) and Argentina (1.03), which, like the United States, have populations spread over large geographic areas.

Average emission rates per person must be interpreted cautiously, however, for many reasons—for example, the fuel used by ships and planes visiting Aruba, a popular tourist destination, is considered to be part of the per capita fossil fuel use by the people of that small island country, leading to a figure deceptively similar to that of the United States.

One argument in defense of the U.S. CO_2 emission rate is that although that country is responsible for a large proportion of the world's anthropogenic carbon emissions, it also provides the greatest number of goods and services to world markets. But the European Union produces almost as great an amount of goods and services while consuming substantially less (16%) of the

world's energy. Furthermore, U.S. rates of fossil fuel consumption per unit of Gross National Product (a measure of economic productivity) per person are very high.

What is producing all of that anthropogenic carbon? Americans use almost 1.3 billion liters of petroleum *per day* in their personal vehicles. Americans now purchase more light trucks (pickup trucks, sport utility vehicles, and minivans) than cars for personal transportation, and these light trucks on average use far more petroleum than cars do with consequently greater CO_2 emissions (Box 22B). Transportation of all sorts (commercial as well as individual) is responsible for about a quarter to a third of the total U.S. emissions. The rest come from industrial and commercial sources (about half of the total) and residences (about a fifth of the total). Box 22B lists some common activities and their resulting fossil fuel carbon emissions.

Different fossil fuels release very different amounts of carbon for each unit of energy generated—coal, for example, produces much more atmospheric carbon for the same amount of energy yield than natural gas. Some sources of electricity other than fossil fuels and the burning of plant material produce no carbon when generating energy: nuclear power, hydroelectric power, solar power, and wind and geothermal energy all release no carbon. (They may cause other environmental problems, however.)

It is easy to see from Box 22B that even modest increases in per capita fossil fuel use can have dramatic consequences for a country's total emissions if the country has a large population. If everyone in China kept just one more light bulb on for 16 hours a day, for example, 364 billion (or 364×10^9) kg more carbon would be added to the atmosphere per year. If every family in India and China had a refrigerator powered by fossil fuel-generated electricity—surely a modest and reasonable

Approximate amounts of CO_2 produced per year by various human activities

Activity	Annual CO_2 production
Production of 100 kilowatt hours of electricity	
Using natural gas	64 kg (141 lbs)
Using oil	84 kg (186 lbs)
Using coal	104 kg (230 lbs)
Using hydroelectric power	0 kg (0 lbs)
Common human activities and appliances[a]	
60 W incandescent light bulb (16 hours a day, 365 days)	364 kg (803 lbs)
Pentium III computer and 17-inch color monitor (on 24 hours a day, no "energy saver" features)	8,232 kg (18,148 lbs)
Home heating/hot water using natural gas (based on average usage per U.S. home in U.S.; varies with ambient and indoor temperature, etc.)	4,113 kg (9,068 lbs)
Home heating/hot water using fuel oil (based on average usage per home in the northeastern U.S.)	9,525 kg (21,000 lbs)
Color TV (on 8 hours a day, which is average for U.S. households)	848 kg (1,869 lbs)
Refrigerator (average size, 10 years old)	939 kg (2,070 lbs)
Refrigerator (0.6 m^3/20 cu ft, new, energy-efficient)	522 kg (1,150 lbs)
Clothes dryer (52 uses)	261 kg (575 lbs)
Toaster (365 uses)	20 kg (44 lbs)
Single round trip, New York to San Francisco on commercial airline	330 kg (728 lbs)
Vehicle emissions: Conventional gasoline engines[b]	
Small car (4 cylinders, 1.7 L engine; e.g., Honda Civic, Toyota Corolla)	5,171 kg (11,400 lbs)
Family vehicle (6 cylinders, 3 L engine; e.g. Ford Taurus wagon)	8,074 kg (17,800 lbs)
Minivan (6 cylinders, 3.3 L engine; e.g., Dodge Caravan)	8,528 kg (18,800 lbs)
Full-size van (8 cylinders, 5.4 L engine; e.g., Ford Econoline)	11,521 kg (25,400 lbs)
Sports car (8 cylinders, 5.7 L engine; e.g., Chevrolet Camaro)	8,255 kg (18,200 lbs)
Midsize SUVs (sport utiltiy vehicle; 4WD, 6 cylinders, 4 L engine; e.g., Ford Explorer)	9,344 kg (20,600 lbs)
Largest SUVs (4WD, 8 cylinders, 5.4 L engine; e.g., Lincoln Navigator)	12,973 kg (28,600 lbs)
Vehicle emissions: Electric vehicles[b,c]	
Ultra-compact (Ford Th!nk)	4082–5008 kg (9,000–11,040 lbs)
Midsize SUV (Ford Explorer; U.S. Postal Service vehicles)	7,080–8450 kg (15,600–18,630 lbs)
Vehicle emissions: Compressed natural gas vehicles[b]	
Small car (Honda Civic, 4 cylinders, 2.2 L engine)	4,536 kg (10,000 lbs.)
Midsize car (Toyota Camry, 4 cylinders, 2.2 L engine)	5,715 kg (12,600 lbs)
Luxury car (Ford Crown Victoria, 8 cylinders, 4.6 L engine)	7,983 kg (17,600 lbs)
Full-size van (Ford Econoline E250, 8 cylinders, 5.4 L)	10,977 kg (24,200 lbs)

[a]Electricity is assumed to be generated by a coal-fired power plant, unless otherwise noted.
[b]Assumptions: Based on 45% highway driving, 55% city driving, 15,000 annual miles (24,000 km), and automatic transmission.
[c]Values assume electricity generated from coal or natural gas. Assumptions for CO_2 equivalents are taken from the U.S. DOE GREET (Greenhouse Gases, Regulated Emissions, and Energy Use in Transportation) model, a "wells to wheels" model that includes energy used in extracting the fuel (available on the Web from the U.S. Department of Energy), where electricity generated comes from liquid and compressed natural gas (methane); several estimates are also provided where electricity is generated from less efficient coal, but for those estimates only CO_2 directly expended, and not CO_2 equivalents, are included.

increase in the standard of living—well, you can project the consequences (easily estimated; see Box 22B).

Without question, there would be economic and other costs to changing our current patterns of fossil fuel consumption. But what are the alternatives? The consequences of our current actions—or lack of actions—may be far more complex and expensive than many have anticipated.

Acid Precipitation and Nitrogen Deposition

Two other human-induced phenomena affecting plants are acid precipitation and nitrogen deposition. Like rising CO_2 levels, these anthropogenic changes in biogeochemical cycles affect plants, but they have effects mainly on a regional scale. However, because they affect so many different regions, and because their effects are so profound, we mention them here. Acid precipitation has been a cause for environmental concern for decades, while nitrogen deposition has gained widespread attention only in recent years. The two are closely related to each other.

Acid precipitation refers to precipitation that is more acidic than normal. Normal precipitation is slightly acidic because CO_2 dissolved in water droplets in the atmosphere forms carbonic acid. Acid precipitation is caused primarily by two common components of air pollution, SO_2 and NO_x (which includes NO and NO_2), that are produced by industrial and automotive emissions. These gases undergo chemical reactions in the atmosphere to form sulfuric acid (H_2SO_4) and nitric acid (HNO_3). In solution, these acids produce highly acidified (very low pH) rain and snow.

Acid precipitation harms vegetation directly by damaging foliage and other plant parts. Indirect effects caused by acidification of the soil and surface water have even greater negative consequences for plants. The release of ions harmful to plant growth in acidified soil may cause toxic reactions. More broadly, as soils become more acidic, leaching of cations can reduce nutrient availability or cause imbalances in the ratios of different nutrients, with detrimental effects for many plant species (see Chapter 4). Die-offs of *Acer saccharum* (sugar maple, Sapindaceae) in Quebec, Canada, have been attributed to deficiencies of soil K and Mg caused by acid rain. Forest declines in central Europe have been attributed to acid rain depleting levels of soil Mg relative to N.

Plants can also absorb SO_2 and NO_2 gases directly from the atmosphere. In Germany, damage to conifer foliage seems to be due to direct uptake of anthropogenic N, which results in leaching of Mg and other cations within foliage tissues. SO_2 and NO_2 can also be deposited in particulate form (dryfall) on leaves, soil, and other surfaces and then later react with rainfall to produce acids. High levels of NH_4^+ and NO_3^- in fog may also directly harm trees.

While recent efforts to curb air pollution have been quite successful in limiting SO_2, nitrogen-containing pollution has increased, becoming a greater problem. Nitrogen from anthropogenic sources enters the atmosphere in various forms, primarily NO_x, but also N_2O_5, HNO_3, and NH_3. Humans have doubled the rate at which nitrogen enters the terrestrial nitrogen cycle each year since the 1850s—and the rate continues to rise (Figure 22.11). This additional nitrogen has strong effects on ecosystems at local to regional scales.

Anthropogenic nitrogen comes from a wide variety of sources: automobile exhaust, industrial manufacture, power plants, and agricultural fertilizers and animal wastes. This broad range of sources is one of the main reasons it is much more difficult to control nitrogen pollution than sulfur pollution. Sulfur is produced by far fewer kinds of sources, primarily the combustion of coal and other fossil fuels (particularly by power plants) and some industries. However, the largest contribution of anthropogenic N to the terrestrial nitrogen cycle comes from the industrial fixation of atmospheric N for agricultural fertilizer (currently 80×10^{12} g/year). Eventually, substantial quantities of this N are volatilized or transferred as particulate matter (dust and other small particles) into the atmosphere as NO_x. The use of commercial N fertilizers is increasing rapidly: the amount of N in chemical fertilizers applied from 1980 to 1990, for

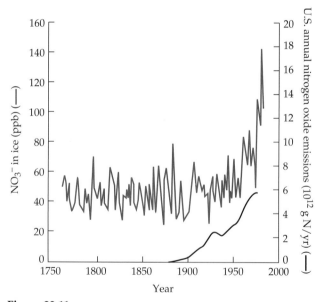

Figure 22.11
Concentrations of nitrate (NO_3^-) deposited in ice in Greenland (in ppb of ice) from the late eighteenth century to the late twentieth century (green line), and annual emissions of nitrogen oxide (NO_x in 10^{12} g/year) by fossil fuel combustion in the United States (black line). (After Schlesinger 1997 and Mayewski et al. 1990.)

instance, was at least as great as the amount applied before that in all of human history.

In addition to its contribution to acid rain, nitrogen pollution returns to the ground as both **wet deposition** and **dry deposition**—dissolved in water and as particulate matter. Current nitrogen inputs in the northeastern United States and central Europe, which suffer from the greatest nitrogen deposition, are 10–50 kg N/ha/year, which is between 5 and 20 times natural levels. Nitrogen is limiting to plant growth in many communities (see Chapter 4), so extra nitrogen in the form of nitrogen deposition initially acts as a fertilizer, leading to increased growth. But excess nitrogen deposition can exceed the ability of the system to use the nitrogen, leading eventually to **nitrogen saturation** of the system. Its effects are nonlinear: negative effects may not be manifested as nitrogen continues to accumulate until a critical threshold is reached. At that point, the effects of additional N rapidly become detrimental to many plants, soil bacteria, and fungi. These effects include leaching of nutrients (such as calcium) from the soil as a result of increased acidity, loss of fine roots, steep declines in mycorrhizal fungi and consequent reductions in plant uptake of phosphorus, and increased leaching of N into ground and surface water due to increased bacterial nitrification and reduced plant uptake.

Ecosystem responses to nitrogen deposition are uneven. Different systems vary greatly in their responses, depending on soils, plant species composition, climate, and other factors. Areas in the United States that show evidence of nitrogen saturation and consequent declines in forest ecosystems include the Catskill Mountains north of New York City, the Great Smoky Mountains in Tennessee, high-elevation forests in the eastern Colorado Rockies, and the San Gabriel and San Bernardino Mountains outside of Los Angeles.

Nitrogen deposition can also reduce plant tolerance of other stresses, including frost, periods of drought, excessive heat, and herbivory. Changes in insect herbivory patterns in forests and other plant community types can be sensitive indicators of nitrogen deposition effects. Changes in community composition, with the disappearance of sensitive species, have been attributed to the effects of nitrogen deposition in various parts of Europe. Mosses and lichens appear to be especially vulnerable to nitrogen deposition, often showing signs of stress first. In the United Kingdom, sensitive *Sphagnum* moss species have declined in the Pennine Mountains due to the toxic effects of NH_4^+ and NO_3^- deposition, and the loss of many species of mosses and lichens in Cumbria since the 1960s also appears to be a consequence of nitrogen deposition.

Changes in community composition, and the losses of some species, may also be the result of changes in competitive interactions due to higher nitrogen input. Experiments in bogs at four sites in Western Europe showed that increases in nitrogen deposition reduced the growth of the dominant *Sphagnum* moss species. This reduction was a result of shading by increased cover of taller vascular plants, and led to changes in community composition (Berendse et al. 2001). Similarly, the widespread replacement of *Calluna vulgaris* (heather, Ericaceae) in East Anglia, U.K., and in other locations by dense stands of grass seems to be due to the indirect effects of nitrogen deposition on competitive interrelationships in heathlands. Interestingly, grazing may reduce nitrogen-induced competitive replacements by grasses in species-rich vegetation in the Netherlands (Kooijman and Smit 2001).

It is difficult to separate the effects of nitrogen deposition on species losses from the effects of other environmental factors that have also changed over the same time period. Simply observing nitrogen deposition and species losses occurring together does not mean that nitrogen deposition necessarily caused the declines. However, experiments have provided evidence of a direct link between atmospheric nitrogen deposition and changes in species composition in heathlands and calcareous grasslands in the Netherlands. Aerts and Berendse (1988) used field experiments to demonstrate that increased nitrogen input in amounts similar to that of anthropogenic deposition led to declining cover of *Erica tetralix* (Ericaceae; a dwarf shrub, formerly dominant) in wet heathlands and increasing cover of *Molinia caerulea* (Poaceae; a perennial grass replacing *Erica*). Changes in species composition in these communities have been attributed to the effects of increased nitrogen (from anthropogenic deposition) on competitive interactions in these naturally low-fertility habitats (Aerts 1999; Aerts et al. 1990). Together with other evidence, studies such as these have implicated atmospheric N deposition in changes in community composition and losses of plant biodiversity in the Netherlands, United Kingdom, and elsewhere in Western Europe.

Declining Global Biodiversity and Its Causes

About 1,500,000 living species of all kinds have been scientifically described and named worldwide, including about 250,000 plants. It is estimated that there are at least 5 million living species that have not yet been described, and there may be many times that number. Only a handful of species have been well studied—primarily those that are economically important (domesticated species, diseases, and pests). Even among most of these "well-studied" species, little is known of their ecology. The evolutionary relationships, genetics, physiology, and other aspects of most nonhuman organisms are almost entirely unknown. At the very least, there is a wealth of understanding to be gained from greater knowledge of these living things and the communities with which they interact.

Biodiversity, short for "biological diversity," refers to all of the populations, species, and communities in a defined area (which may be local, regional, continental, or include the entire globe) and includes the range of genetic variation in these living things. In contrast to the more specific term "species diversity," the term "biodiversity" was coined to emphasize the many complex kinds of variation that exist within and among organisms at different levels of organization. We can speak, for instance, of the biodiversity of major human crop species, which is currently greatly threatened. As a small number of commercial varieties of domestic crops have been widely planted everywhere on Earth, many local varieties have been abandoned. These local varieties, cultivated and selected sometimes over thousands of years for geographically specific desirable traits, may possess genes that enable them to withstand insects, drought, and other stresses—genes that are missing in the more productive but potentially vulnerable commercial strains. If we fail to maintain these ancient crop races, these invaluable genetic storehouses will be lost forever.

A range of anthropogenic factors currently threatens biodiversity from local to global scales. It has been difficult to accurately assess current rates of outright species extinction except in some very limited cases. It is even more difficult to accurately quantify species' declines, and impossible to obtain precise numbers on overall rates of loss of genetic or other biodiversity. Not surprisingly, we have better data for large, attractive, obvious, and useful organisms than for others—we have much better population data for grizzly bears and sugar maples, for example, than for mycorrhizal fungi, even though mycorrhizal species are critical to the existence of many plants (see Chapter 4). One of the difficulties is that most people would not be aware of or recognize the decline of important mycorrhizal fungi. Likewise, declines or even extinctions of many nondomesticated plant species would be likely to go unnoticed in many parts of the world.

Habitat Fragmentation and Loss

Threats to global, regional, and local biodiversity include a range of factors. While the specific threats to biodiversity differ among biomes (Sala et al. 2000), the most important factor threatening plants and plant communities globally is changes in land use, which are resulting in the destruction, degradation, and fragmentation of habitats (Figure 22.12). Besides being entirely destroyed, habitats can be degraded by many factors, including losses of critical species, damage to the soil (such as erosion or nitrogen deposition), overgrazing, and changes in disturbance regimes caused by humans and their domesticated animals. Fragmentation of natural communities by roads, agriculture, or human settlements is changing many natural habitats worldwide

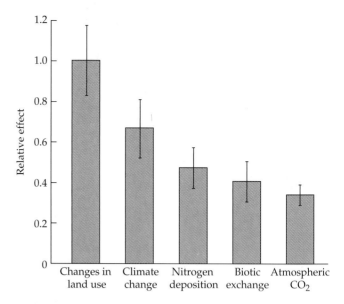

Figure 22.12
Average estimated relative effects of the most important current threats to global biodiversity, projected for the next 100 years. Land use changes (such as habitat loss and deforestation) are predicted to have the greatest harmful effects on biodiversity, followed by climate change, nitrogen deposition, the incursion of invasive exotic species and other biotic exchanges, and direct effects of rising atmospheric CO_2 levels. (After Sala et al. 2000.)

(see Chapter 17). Figure 22.13 is a graphic illustration of forest loss and fragmentation in Central America. Skole and Tucker (1993) quantified changes in forest fragmentation in the Brazilian Amazon over a ten-year period (1978–1988) using satellite images. By the end of that period, almost three times more forested area consisted of forest fragments—patches of forest less than 100 km^2, surrounded by cleared land—than at the beginning. The size and shape of cleared patches also changed dramatically during that period due to transformations in the economy and human demographics. Clearings changed from smaller and more irregular patches to larger and more continuous areas, with negative implications for the viability of the forest ecosystems (Peralta and Mather 2000).

These changes are important because habitat fragmentation can have many different negative effects (Laurance 2000a; see Chapter 17). For example, fragmentation can change forest microclimates as wind, light levels, temperatures, and runoff increase at the edges of the forest, while humidity decreases. Because of the diminished area of the forest environment and the altered conditions within it, deep-forest species can suffer reduced survival and reproduction, while populations of species adapted to edge conditions, such as many invasive plants, can grow. Increases in herbivory and disease and declines or extinctions of some native

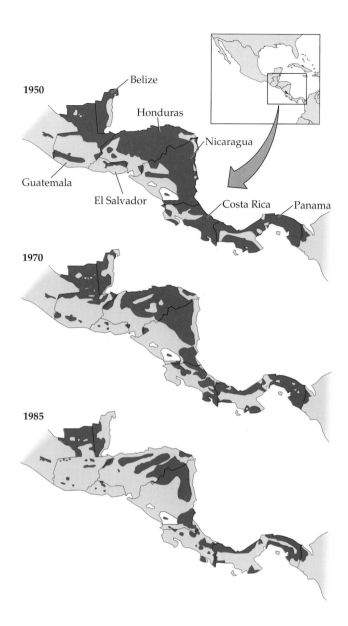

1950

Belize

Honduras

Nicaragua

Guatemala

El Salvador

Costa Rica

Panama

1970

1985

Figure 22.13
Forest fragmentation in Central America between 1950 and 1985. Deforestation and fragmentation have increased in the years since 1985, although some forests are now protected in national parks, most notably in Costa Rica. Not only has the total area of intact forest declined dramatically, but the remaining forest fragments are small, often isolated from one another, and may experience large edge effects. Similar patterns are repeated for forests in many other parts of the world, and over a range of scales.

lion km^2 (1.4% of Earth's dry land surface); originally, these communities covered an area of 17.4 million km^2. Together they have lost 88% of their original vegetation cover. Twenty of these hotspots are in tropical forests or mediterranean climate regions. Nine of them are on oceanic islands or archipelagos, where long isolation has led to the evolution of endemic species. Many of the others are isolated in other ways.

Not only rare species and rare communities are threatened, however. Less diverse but unique communities, sometimes with unusual organisms, including serpentine grasslands (see Box 15A), parts of the North American deserts, and some salt marshes, have also been subjected to various threats not only to their extent, but to their continued existence. In some cases, even originally very common species and ecosystems have been reduced to the point of becoming endangered, or have disappeared altogether. Climate change and human land uses have dramatically increased the extent of newly created depauperate deserts (with little vegetation of any kind, and with few uniquely adapted desert species) at the expense of fragile grasslands and shrublands. In the Sahel region of Africa, south of the Sahara, vast areas of once productive grasslands supporting large populations of people and animals have become shifting sands. In Lagan in central China (near the headwaters of the Yellow River), extensive grasslands have long supported humans and their grazing animals. They have now become barren wastelands supporting little vegetation, with recurrent sandstorms and vanishing rivers, lakes, and pastures, apparently due to a combination of climate change and intensive land use (particularly overgrazing as a result of human population pressure).

Tropical forests are among the best-known threatened biomes. About half of the species on Earth live in tropical rainforests, even though they now encompass only about 7% of terrestrial land cover. Many of these species are found only in very limited areas. For example, organisms in the seasonal tropical forests of the large island of Madagascar, off the east coast of southern Africa, have evolved in isolation for 121 million years, leading to some of the world's highest levels of plant diversity and endemism. Madagascar has an estimated 10,000 plant species, of which about 80% are endemic.

plant species may result from the increased incursion of edge effects into forests and other habitats.

Many habitats around the world are threatened by human activities. When these are rare or unique local habitats, reducing or destroying them may harm the only populations of a particular species in a region, or may harm the only remaining population of that species anywhere on Earth. Unique organisms and communities are found in many locations, and human encroachment is, in one way or another, threatening many of these places. Twenty-five **biodiversity hotspots**—threatened areas with very high species diversity—have been identified worldwide. These areas together contain 133,000 plant species, or about *half* of the world's plant species (Figure 22.14, Table 22.1). The five most species-rich of these hotspots support 20% of all plant species. Taken together, the 25 hotspots currently occupy 2.1 mil-

Table 22.1 "Hotspots" of biodiversity

Location	Original extent of primary vegetation (km²)	Remaining primary vegetation (km²) (% of original extent)	Area protected (km²) (% of hotspot)	Number of plant species	Number of endemic species (% of global plants[b])
Tropical Andes	1,258,000	314,500 (25.0)	79,687 (25.3)	45,000	20,000 (6.7%)
Mesoamerica	1,155,000	231,000 (20.0)	138,437 (59.9)	24,000	5,000 (1.7%)
Caribbean	263,500	29,840 (11.3)	29,840 (100.0)	12,000	5,000 (2.3%)
Brazil's Atlantic forest	1,227,600	91,930 (7.5)	33,084 (35.9)	20,000	7,000 (2.7%)
Choco/Darien/western Ecuador	260,600	63,000 (24.2)	16,471 (26.1)	9,000	2,250 (0.8%)
Brazil's Cerrado	1,783,200	356,630 (20.0)	22,000 (6.2)	10,000	4,400 (1.5%)
Central Chile	300,000	90,000 (30.0)	9,167 (10.2)	3,429	1,605 (0.5%)
California Floristic Province	324,000	80,000 (24.7)	31,443 (39.3)	4,426	2,125 (0.7%)
Madagascar[a]	594,150	59,038 (9.9)	11,548 (19.6)	12,000	9,704 (3.2%)
Eastern Arc and costal forests of Tanzania/Kenya	30,000	2,000 (6.7)	2,000 (100.0)	4,000	1,500 (0.5%)
Western African forests	1,265,000	126,500 (10.0)	20,324 (16.1)	9,000	2,250 (0.8%)
Cape Floristic Province	74,000	18,000 (24.3)	14,060 (78.1)	8,200	5,682 (1.9%)
Succulent Karoo	112,000	30,000 (26.8)	2,352 (7.8)	4,849	1,940 (0.6%)
Mediterranean basin	2,362,000	110,000 (4.7)	42,123 (38.3)	25,000	13,000 (4.3%)
Caucasus	500,000	50,000 (10.0)	14,050 (28.1)	6,300	1,600 (0.5%)
Sundaland	1,600,000	125,000 (7.8)	90,000 (72.0)	25,000	15,000 (5.0%)
Wallacea	347,000	52,020 (15.0)	20,415 (39.2)	10,000	1,500 (0.5%)
Philippines	300,800	9,023 (3.0)	3,910 (43.3)	7,620	5,832 (1.9%)
Indo-Burma	2,060,000	100,000 (4.9)	100,000 (100.0)	13,500	7,000 (2.3%)
South-central China	800,000	64,000 (8.0)	16,562 (25.9)	12,000	3,500 (1.2%)
Western Ghats/Sri Lanka	182,500	12,450 (6.8)	12,450 (100.0)	4,780	2,180 (0.7%)
Southwest Australia	309,850	33,336 (10.8)	33,336 (100.0)	5,469	4,331 (1.4%)
New Caledonia	18,600	5,200 (28.0)	526.7 (10.1)	3,332	2,551 (0.9%)
New Zealand	270,500	59,400 (22.0)	52,068 (87.7)	2,300	1,865 (0.6%)
Polynesia/Micronesia	46,000	10,024 (21.8)	4,913 (49.0)	6,557	3,334 (1.1%)
Totals	17,444,300	2,122,891 (12.2)	800,767 (37.7)	*	133,140 (44%)

[a]Madagascar includes the nearby islands of Mauritius, Reunion, Seychelles, and Comoros.
[b]The total number of plant species globally is here estimated to be ~300,000.
*These totals cannot be summed due to overlapping between hotspots.

Madagascar also has some of the world's most highly threatened habitats. Although forest destruction began many years ago, it accelerated greatly in the twentieth century. Between 1960 and 1990, slightly less than half the forested areas of Madagascar were destroyed. Today, less than 10% of the original (pre-eighteenth-century) forests remain.

Deforestation is occurring in tropical rainforest habitats worldwide, and is threatening or leading to the extinction of many species. Other forests, including highly diverse dry tropical forests, are also severely threatened. Climatic effects sometimes interact with other factors to intensify damage to tropical forests and other biomes. For instance, the 1997–1998 El Niño event caused uniquely ferocious and damaging fires in the rainforests of the island of Borneo, exacerbating the damage already done by extensive logging and land clear-

ing. Some scientists believe that global warming may have intensified these climatic effects.

It is difficult to get precise measurements of how much deforestation has already occurred, or of current regional rates of deforestation. Satellite imaging has greatly aided our ability to track deforestation, particularly in combination with new methods for image analysis. Estimated rates of deforestation differ among regions. Deforestation and logging have been accelerating in the Amazon in the 1990s (Figure 22.15). Similar rates of deforestation are occurring in tropical Africa (1.3–3.7 million ha/year: Boahene 1998). While the rainforests of Southeast Asia (Cambodia, Indonesia, Laos, Malaysia, Myanmar, Thailand, and Vietnam) covered a much smaller area in the decade from the mid-1970s to the mid-1980s, almost as much rainforest there was lost each year as in the Brazilian Amazon. Vast areas of

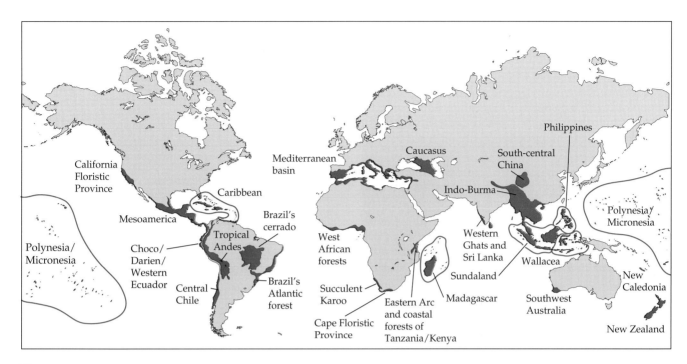

Figure 22.14
Twenty-five regions of the world identified as biodiversity hotspots have exceptional densities of species diversity and endemism. The hotspots are located within the colored regions, but may make up only parts of those regions. (After Myers et al. 2000.)

ancient forest were converted to agriculture or cut for timber. Similar rates of forest loss continue there to the present day. If current rates of deforestation continue, the world's tropical forests will disappear within 100 years, existing only in tiny, isolated fragments (see Figure 22.13).

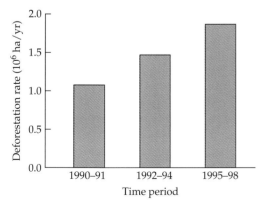

Figure 22.15
Deforestation rates in the Brazilian Amazon have accelerated during the 1990s. Forest losses are expressed in millions of hectares a year. Deforestation rates in other parts of the Amazon basin have also been increasing. (Data from Laurance 2000b.)

Invasive Species and Other Threats to Biodiversity

Other major global threats to biodiversity include overexploitation of species by hunting, fishing, or collecting and the spread of invasive exotic species. Overexploitation more obviously threatens animals than plants, but plants are not immune to this threat. First, logging and removal for other reasons have greatly reduced the populations of some plant species. Several plant taxa, particularly orchids and cacti, are threatened by amateur collectors (Orlean 2000). While collecting and selling plants taken from natural populations is illegal in many areas, steep prices and minimal fines conspire to maximize these depredations. Second, many plant species are threatened indirectly by overexploitation of other species. Killing major predators, for example, can lead to large rises in herbivore populations, resulting in overgrazing or overbrowsing. The extensive loss of understory vegetation in forests all over the northeastern United States in recent years is due in large part to a tremendous rise in populations of white-tailed deer. The rise in deer populations is in turn the result of a number of factors, from suburbanization patterns that fragment forests and increase edge habitats, which favor the deer, to declines in predators and hunting (hunters are unwelcome in suburban backyards). The rise in deer numbers threatens forest regeneration because deer browsing kills tree saplings.

The spread of exotic and invasive species is increasingly threatening ecosystems all over the world (see Chapter 14). This phenomenon has serious consequences both for humans and for the preservation of native species and natural communities (Lodge 1993). Human activities such as deliberate and accidental species introductions, forest fragmentation, logging, and other disturbances have all been factors in the invasion and spread of non-native species. Dukes and Mooney (1999) raise the provocative suggestion that global climate change, including nitrogen deposition and rising atmospheric CO_2, may preferentially enhance the success of invaders.

Exotic plants have invaded forests in many parts of the world; these species may have substantial effects on the retention and restoration of native species. New Zealand, for example, now has at least as many naturalized exotic plants as native species (Allan 1961; Webb et al. 1988). These introduced species have modified the vegetation of large areas in New Zealand. While native species have been not been driven to extinction there by the exotic species, the native plants appear to have been locally displaced and diminished in abundance by the invaders in many cases.

Grasslands, deserts, and other plant communities have also been greatly altered by invasive species. Invasive plant species can decrease the diversity of native plants and disrupt ecosystem processes, altering nitrogen cycling, fire regimes, and other fundamental characteristics of ecosystems (Higgins et al. 1999; Vitousek et al. 1996; Mack and D'Antonio 2001). In addition, biological invasions by exotic plants can exact enormous economic costs, including decreases in timber growth rates and forest value. These ecological and economic impacts can be so great that some ecologists have urged recognition of biological invasions as one of the most important forms of global change affecting ecosystems today.

Human Populations and Land Use Patterns

The number of people in the world has grown dramatically since the Industrial Revolution, and is continuing to grow rapidly (Figure 22.16). The consequences of global increases in the human population have been a cause for alarm, debate, and controversy for many years—as exemplified by the wide range of reactions to the much-disputed book by Paul Ehrlich, *The Population Bomb* (1968). Some of the predictions made regarding human population growth have been seriously flawed and alarmist, but attempts at outright dismissal of the concerns about its consequences have also been highly flawed. Much of the disagreement over and interest in the implications of human population growth has concerned the effects it might have on food supplies, and more generally on human societies. Less popular attention has focused on the possible environmental effects of growing human populations, although both environmentalists and social scientists have expressed concern about these effects.

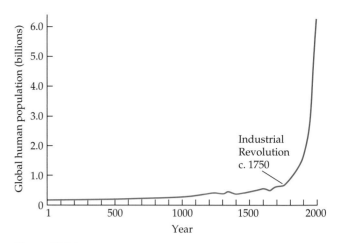

Figure 22.16
Worldwide human population growth, estimated over the past 2000 years.

Recent growth in human populations has been much greater in developing than in industrialized regions, and tends to be highest in the Tropics, where a large proportion of the people are living at or below subsistence levels. Population growth has slowed in many of the developed, industrialized countries. As a result of these population growth patterns, increasingly greater proportions (and absolute numbers) of the world's people live in developing countries, and an increasingly large proportion of the world's people are children who live in poverty in those countries (Figure 22.17).

This rapid human population growth could have large environmental consequences (such as increased resource use, habitat destruction, and species extinctions) as well as economic and political effects and implications for global stability. But how much of the ongoing deforestation and other damage to natural communities can we actually ascribe to human population densities and population growth? How accurate is it to assume that increasing human populations directly affect forests or any other plant communities? Human societies are complex systems, and it is not surprising that human populations affect the environment in indirect and multifactorial ways. It is an oversimplification to assume that human population size is directly correlated with damage to the environment. Instead, both local and international economic and political forces interact with population size to influence changes in land use (Lambin et al. 2001).

Social scientists who seek to characterize and quantify the interactions between humans and their environments disagree about the extent to which different factors, including population size, are responsible for environmental problems such as deforestation. On the one hand, there is solid evidence that across a wide number of developing countries, population growth results

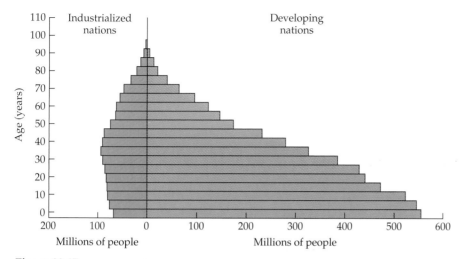

Figure 22.17
Global age pyramid. The total number of people, in millions, in each age category, from birth to 110 years, in industrialized and in developing nations. The total number of people in developing nations is far greater than in industrialized nations. The distribution of ages is also different, with relatively more children and more people at younger ages in the developing countries. The "baby boom" of people now in their thirties to early fifties is clearly evident as a bulge at those ages in the industrialized nations, but not in the developing nations. The number of very old people (over 90) is very small for both groups of nations; the maximum human age (not clearly visible here) is not different for the two groups. (Data from United Nations records.)

in greater deforestation (Ehrhardt-Martinez 1998). This has also been demonstrated at a global scale, although the negative relationship between human populations and declining forest area may have lessened in recent years, at least in some countries (Mather and Needle 2000).

In studies focused at the scale of particular countries or regions, there have been contradictory results. In two studies of the effects of human population density on deforestation in the Amazon basin of Brazil, its effects were at best indirect. Pfaff (1999) found that in the Brazilian Amazon, population density did not directly determine tropical forest clearing. Rather, many factors were implicated, from land characteristics and distance to markets to government development projects. However, Pfaff also found that while a static measure of population size was not a good predictor of deforestation, the first settlers in a region had a far greater effect on forest destruction than later immigrants. He concluded that the timing and distribution of population growth were more important than its absolute magnitude in determining its impact.

In the second study, Parayil and Tong (1998) found that population size and growth were not directly responsible for deforestation in the Amazon basin. Instead, multiple agents were involved, operating at different scales. Major agents responsible for deforestation included large-scale clearing of forests to create cattle ranches and logging of forests for lumber (both gener-

ally by large multinational corporations). The third major agent of forest destruction was the very small-scale clearing of forests for farming (slash-and-burn agriculture) by long-term residents as well as by impoverished immigrants to the region from other parts of the country. Thus, Parayil and Tong found that factors other than population growth were the major direct contributors to plant community destruction at a regional scale.

Patterns of forest loss and fragmentation are also changing (Laurance 2000b). In past decades, many of the largest-scale forest losses in the Amazon basin occurred at its southern and eastern edges. The development of major new highways, however, is now bringing settlers and loggers directly into the heart of the central Amazon forest, leading to great increases in fragmentation and forest loss (Figure 22.18; Laurance et al. 2001). This observation reinforces Pfaff's argument that the distribution of human populations is at least as important as the absolute number of people in a country. The opening of new roads in forest areas also was found to be the most important factor in deforestation in the rainforest of Cameroon, in central Africa (Brown and Ekoko 2001). In that study, multiple agents (from the government to corporate loggers to local village interests and needs) and causes of deforestation acted synergistically to the detriment of the natural systems.

Patterns that are broadly similar in their complexity, but different in important details, have been found in human-environment interactions in the United States.

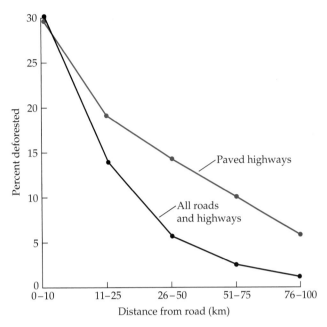

Figure 22.18
Effect of proximity to roads on deforestation in the Brazilian Amazon. The percentage of the original closed-canopy forest that was destroyed by 1992 is shown as related to distance from paved highways and from all roads and highways. Forests close to any roads, particularly paved highways, are far more likely to be lost than are those farther away. (After Laurance et al. 2001.)

In the past half-century, two distinct patterns of changing residential land use have had large effects on the reduction and fragmentation of wildlands. Both are, on the surface, seemingly innocuous, and stem in part from the desire of people to get out of the cities and live in closer contact with nature. The first of these is the rise of suburbanization—the establishment of vast low-density areas surrounding metropolitan centers, particularly in North America. The consequence has been a shift in population from cities to "metropolitan regions" sprawled over thousands of square kilometers. This shift has changed land that was once agricultural or occupied by natural communities into a mix of homes on large plots of land, with commercial development at some distance. The second, more recent change has been the incursion of vacation and retirement homes into wilderness areas, particularly in coastal, montane, and desert areas (Bartlett et al. 2000). While this land use does not consume as much land as suburban development, it can cause fragmentation of natural areas, and results in changes in fire regimes, erosion, and incursions of invasive plant and pest species.

Local planning and policy decisions that create suburban development patterns, in contrast to traditional towns or cities, have other unintended consequences in addition to the direct destruction of natural communities and conversion to human-dominated landscapes. They lead directly to far greater per capita fuel consumption, because suburban residents have no choice but to commute long distances to work, to shop, to attend school, and for social and recreational opportunities. The most fuel-efficient vehicle driven by a commuter with a 100-mile round trip to work, or by a suburban parent who racks up 25,000 miles per year in everyday activities, will have far more impact than even a very large sports utility vehicle driven very few miles by a person living in a compact neighborhood where it is possible to walk or drive short distances for most activities. Recent efforts to build traditional towns and neighborhoods once again offer the opportunity to reverse some of these effects (Langdon 1997).

A Ray of Hope?

Is all gloom and doom? Is the world going to hell in a handbasket? There are clearly reasons to think so. However, there are also people working mightily to change the statement so often heard that "all this will happen if the current rates of [fill in your favorite bad thing] remain unchanged." Perhaps some of these things *can* be changed. Human behavior is not immutable, and if current practices are altered, some of the consequences of our actions can also be changed.

Although there are many factors acting to continue our current momentum, it is certainly not impossible, for instance, to begin to bring carbon emissions in the United States into line with the kinds of reductions that have been accomplished in Western Europe. If we are concerned about the environmental damage being caused by unrestrained suburban sprawl in North America, there are workable, popular, and economically feasible alternatives to current environmentally costly land development patterns (for example, see Duany and Platt-Ziederbeck 2001). People are working in many countries to reduce carbon emissions, stop deforestation and habitat fragmentation, and restore damaged or destroyed ecosystems. We have the option of supporting efforts to protect species and habitats locally and around the world, and of changing our own habits and practices to reduce our personal impact on the environment. We also have the option of acting politically to alter current patterns of environmental destruction. The consequences of doing nothing should be obvious.

Summary

Earth is experiencing one of the most rapid periods of change in its history. The global carbon cycle is central to many of these changes. Anthropogenic increases in atmospheric CO_2 and other greenhouse gases are expected to cause global climate changes that include warmer temperatures, changes in precipitation patterns, and possibly changes in other climatic factors such as storm intensity and frequency. The increase in atmospheric greenhouse gasses is caused by a number of factors, chief among them fossil fuel combustion and deforestation. There is now strong evidence that many of the predicted climate changes are already occurring. Predictions for the future, taking current trends into account, are for warmer climates everywhere, with the greatest changes in polar regions. Rainfall patterns will shift, with some areas becoming wetter and others drier, and both flooding and drought will increase.

Other anthropogenic global changes include habitat destruction and fragmentation, nitrogen deposition, and the spread of exotic invasive species. Nitrogen deposition and habitat decline and fragmentation will result in the decline or extinction of sensitive species. Together, these changes will result in major shifts in species' distributions, changes in plant community composition and ecosystem productivity, and declining biodiversity.

Additional Readings

Classic References

Arrhenius, S. 1918. *The Destinies of the Stars*. G. P. Putnam's Sons, New York.

Hardin, G. 1968. The tragedy of the commons. *Science* 162:1243–1248.

Woodwell, G. M. 1970. Effects of pollution on the structure and physiology of ecosystems. *Science* 168: 429–433.

Contemporary Research

Aber, J. D. 1992. Nitrogen cycling and nitrogen saturation in temperate forest ecosystems. *Trends Ecol. Evol.* 7: 220–223.

Hughes, L. 2000. Biological consequences of global warming: is the signal already apparent? *Trends Ecol. Evol.* 15: 56–61.

Mack, R. N., D. Simberloff, W. Mark Lonsdale, H. Evans, M. Clout and F. A. Bazzaz. 2000. Biotic invasions: Causes, epidemiology, global consequences and control. *Ecol. Appl.* 10: 680–710.

Jefferies, R. L. and J. L. Maron. 1997. The embarrassment of riches: Atmospheric deposition of nitrogen and community and ecosystem processes. *Trends Ecol. Evol.* 12: 74–78.

Additional Resources

Langdon, P. 1997. *A Better Place to Live: Reshaping the American Suburb*. University of Massachusetts Press, Amherst.

Naeem, S., F. S. Chapin III, R. Costanza, P. R. Ehrlich, F. B. Golley, D. U. Hooper, J. H. Lawton, R. V. O'Neill, H. A. Mooney, O. E. Sala, A. J. Symstad and D. Tilman. 1999. *Biodiversity and ecosystem functioning: Maintaining natural life support processes*. Issues in Ecology No. 4. Ecological Society of America, Washington DC.

Vitousek, P. M, H. A. Mooney, J. Lubchenco and J. M. Melillo. 1997. Human domination of Earth's ecosystems. *Science* 277: 494–499.

A Statistics Primer

tatistics are important tools for scientists. They are used in two ways: to provide descriptions of data and to test hypotheses. We provide a brief introduction to these uses of statistics here. Much more thorough treatments can be found in any number of basic statistics books (e.g., Siegel 1956; Snedecor and Cochran 1989; Sokal and Rohlf 1995; Zar 1999). Other books provide more advanced statistics for ecologists (e.g., Digby and Kempton 1987; Ludwig and Reynolds 1988; Hairston 1989; Manly 1992; Shipley 2000; Scheiner and Gurevitch 2001).

Data Description

A collection of data is just a pile of numbers. To make that pile useful, we need some way of simplifying it, which is usually done by describing the data with a few basic parameters. The first thing we usually want to describe is the central tendency of the data, which measures the center, or middle, of the data. Two very common measures of central tendency are the **median**—the value of the point at the exact middle of all the points, or the 50th percentile—and the **mean** ($\overline{X}$), also called the average (Figure 1A). The mean is described by the equation

$$\overline{X} = \frac{\displaystyle\sum_{i=1}^{N} X_i}{N}$$

where X_i is the ith observation (where i is a way of keeping track of the observations, as we count them all) and N is the total number of observations. It is always very important to indicate the number of observations—the **sample size** (N) of the data, without which most other statistics are difficult to interpret.

The second basic parameter measures the variation in the data. It tells us whether the observations are tightly grouped around the central tendency or are spread out over a large set of values. There are several measures of variation that are commonly used. One of the most basic of these is the **range**, which is the difference between the largest observation and the smallest observation. Two other common measures are the **variance** (s^2) and its square root, the **standard deviation** (s). The variance is calculated as the sum of the squares of the

(A)

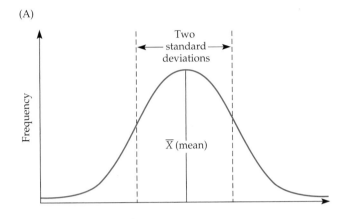

(B)

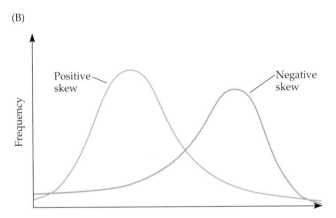

Figure 1
Three frequency distributions of data. (A) The average value of a distribution is the mean. For a normal (bell) curve, two-thirds of the width of the distribution is two standard deviations. This distribution is symmetrical, so the skew is zero. (B) Two asymmetrical distributions, one with a long tail to the left (negative skew) and one with a long tail to the right (positive skew).

deviations of each observation from the mean, divided by the sample size to give something like an average deviation:

$$s^2 = \frac{\sum_{i=1}^{N}(X_i - \overline{X})^2}{(N-1)}$$

The standard deviation is usually reported because its units are the same as those of the mean, while the variance has squared units.

Another basic parameter that is sometimes reported is the **skew**, a measure of the symmetry of the data (Figure 1B). Knowing the shape of the frequency distribution of the data is important, because the mean for a normal (bell) curve tells you a lot about the values in the data, but the mean for a highly skewed distribution can be very misleading and is not a good indication of the "middle" or central tendency of the data. The formula for skew can be found in any basic statistics textbook.

All of the previous descriptors are for single variables. We may also have two variables and wish to describe how they are related to each other. If one variable increases, does the other variable increase, decrease, or not change? This relationship is measured by the **correlation** between the two variables (*r*). The correlation is a dimensionless number (that is, it has no units) between –1 and 1, and is useful for comparing relationships of data with very different units or magnitudes. An example would be the relationship between seed size and stem diameter in redwood trees and in *Arabidopsis thaliana*. A correlation of 1 between seed size and stem diameter in redwoods would tell us that, for the observed data, an increase in seed size is always accompanied by an increase in stem diameter; that is, there is a positively correlation between seed size and stem size. If we graphed the two variables against each other, the points would fall on a straight line with a positive slope (Figure 2A). A correlation of –1 would mean that the points would fall on a straight line with a negative slope (Figure 2B). Correlations between these extremes mean that the points do not all fall on a line, and a correlation of 0 would mean that there is no line we could draw to accurately summarize the relationship between the two variables.

A closely related quantity that retains the units of measurement is the slope of the line describing the relationship between the two variables, also called the **regression**, symbolized as *b*. The regression provides somewhat different information than the correlation: the correlation tells us how closely two variables are related, while the regression tells us what the relationship is. For example, knowing that the data show a correlation of 0.8 between seed size and stem diameter, we might then estimate the regression in order to know how many units of stem diameter accompany each unit increase in seed size.

Estimating Accuracy

Often we wish to go beyond the data that we have in hand—the sample—and make inferences about the population that the data came from. A parameter measured from a sample (e.g., the mean) is an estimate of the true value of that parameter. By random chance, however, the sample will differ from the true value. If one knows how chance affects the sampling process, then one can compute a measure of the **precision** of the estimate—how close the estimate comes to the true value. Two related measures of precision are the **confidence interval** and the **standard error**. If we say that the 95% confidence interval for the mean height of full-grown maple trees in a forest is (3.56 m, 6.42 m), we mean there is a 95% probability that the true value of the mean height lies within

(A)

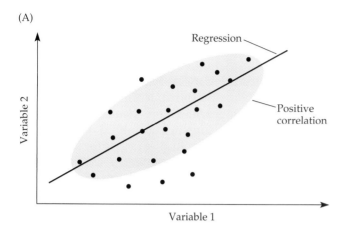

(B)

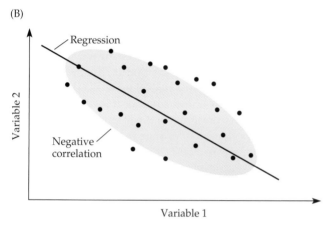

Figure 2
A correlation measures the strength of the relationship
between two variables. A correlation that is close to 1 or –1
indicates a strong relationship, while a value that is close to
0 indicates a weak relationship. The relationship can be (A)
positive or (B) negative. A related measure is the regression,
or slope of the line.

that interval. For certain parameters (e.g., the mean of a
sample), this sampling distribution is understood, and
formulas are available for calculating precision. The standard error of a parameter is a measure of the expected
variation in the distribution of that parameter and is used
to calculate the confidence interval.

However, the distributions for many ecological
indices, and indeed, for a great deal of ecological data,
are not known. Instead, we can use alternative statistical procedures for estimating precision. Two of these
procedures are the **jackknife** and the **bootstrap**. The
name "jackknife" comes from the notion of an all-pur-
pose tool like a pocket knife; the name "bootstrap"
comes from the notion of "pulling yourself up by your
own bootstraps." Here we present the basic concepts
behind these techniques; for details on how to implement them, see Dixon 2001.

To calculate a jackknifed standard error of a parameter, do the following. From your data sample, create a

subsample that consists of all of the data except for the
first observation. Then, calculate the parameter from this
sample and save that number (known as a pseudovalue). Repeat, but now delete the second observation.
Keep doing this until you have deleted each observation
in turn. Now you have a set of N pseudovalues, where
N was the size of your original sample. The standard
error of these pseudovalues is then a measure of the precision of the parameter.

The bootstrap is conceptually similar. From the original data, randomly draw a sample of size N with
replacement. That is, some values may be drawn more
than once, while others might not be drawn at all. Calculate the parameter and save the pseudovalue. Repeat
this many times, typically 1000 times or more. Again, the
distribution of bootstrapped values provides a measure
of the precision of the original parameter. This distribution can be used to calculate both a standard error and
a confidence interval.

But is this "cheating"? Why can we use a single sample over and over again? We assume that our sample is
representative of the entire population from which it was
drawn. This assumption makes the jackknife and bootstrap more reliable as our sample size increases. Both the
jackknife and the bootstrap give standard errors and
confidence intervals that correspond closely to their statistical definitions, the result that we would obtain if we
repeatedly sampled from the same population. For
example, in calculating a standard error of the mean of
a sample, the bootstrap simulates the definition of a standard error: if we repeated the experiment many times,
what would be the dispersion of the estimates of the
mean?

The estimates produced by both the jackknife and
the bootstrap can be biased, but simple bias corrections
are available (Dixon 2001). The two techniques give the
same answers for very large sample sizes. Before the
advent of powerful, cheap computers, the jackknife was
computationally easier; now, one can do either on a PC
in a matter of minutes. It is important to keep in mind
that both procedures assume that the sample is a reasonable representation of the original population from
whence it came. An unrepresentative sample can never
be made any better. For this reason, any estimate based
on a small sample should always be viewed with caution.

Using and Reporting Statistics

Statistics are just one tool among many that scientists
use to gain knowledge. To be used well, these tools must
be used thoughtfully, not applied in an automatic fashion. While there is often no single "right" way to use statistics, there are many wrong ways. Which statistical procedure is best depends on several factors, including how

well the data meet the differing statistical assumptions of the various procedures. Most importantly, choosing the statistical procedure depends on what question is being asked. Procedures differ in the types of questions they are best equipped to address.

When using statistics to test a hypothesis, we often focus on the probability (P) that the null hypothesis is false (e.g., that two means are not different from each other). Somewhere along the line, a value of $P < 0.05$ became a magic number: we reject the hypothesis if the probability of observing a parameter of a certain value by chance alone is less than 5%, or do not reject it (that is, we accept it) if the probability is greater than that. While some sort of objective criterion is necessary so that one does not simply alter one's expectations to meet the results (analogous to throwing darts against the wall and then drawing a target around them), it is important to keep in mind the ultimate goal of any scientific investigation: deciding which hypotheses about the universe are correct. Ultimately, these are yes or no decisions. What the statistical test should indicate is "Yes; the hypothesis is almost certainly true," "No; the hypothesis is almost certainly false," or "Maybe; another experiment needs to be performed." Thus, the P values are only guidelines, and results that lie in the region $0.011 < P < 0.099$ (to pick two arbitrary values) should be judged with caution.

Equally important to using statistics correctly is reporting what you did clearly and completely. Probably the most common statistical sin committed is not describing procedures or results adequately in publications (Fowler 1990; Gurevitch et al. 1992). When reporting a parameter, you should always give the sample size (N) and some measure of the spread in the data, or the accuracy of that parameter (e.g., the standard error or confidence interval around a mean). It is important to report explicitly what was done. A scientific publication should permit a reader to draw an independent conclusion about the hypothesis being tested. While the ecological or other biological assumptions behind an experiment are usually explicitly laid out, statistical assumptions are often not addressed, or worse, the exact statistical procedure is not specified. Thus, when a conclusion is reached that the hypothesis has been falsified, the reader does not know whether it is due to errors in the ecological assumptions of the hypothesis (that is, the scientific hypothesis is false) or the statistical assumptions of the analysis.

Photo Credits

The authors thank and acknowledge the following sources for providing the "yearbook" photographs of scientists that appear as follows:

Dwight Billings, p. 39 courtesy of Duke University Archives

John Harper, p. 96 courtesy of J. Harper

A. D. Bradshaw, p. 102 courtesy of George Robinson

Jose Sarukhán, p. 118 courtesy of Jose Sarukhán

Mary Willson, p. 158 courtesy of Mary Willson

Carol Augspurger, p. 162 courtesy of Carol Augspurger

Eddy van der Maarel, p. 200 courtesy of Opulus Press AB, Länna, Sweden

Arthur Tansley, p. 220 courtesy of the American Environmental Photographs Collection, Special Collections of the University of Chicago

Frederic and Edith Clements, p. 237 courtesy of The American Heritage Center, University of Wyoming

Henry A. Gleason, p. 237 courtesy of The LuEsther T. Mertz Library of The New York Botanical Garden, Bronx

Robert H. Whittaker, p. 238 courtesy of Thomas Wentworth

Steward T. A. Pickett, p. 256 photo by Jill Cadwallader, courtesy of Steward Pickett

Peter Grubb, p. 264 courtesy of David Coomes

H. C. Cowles, p. 266 courtesy of the American Environmental Photographs Collection, Special Collections of the University of Chicago

David Tilman, p. 289 photo by Nancy Larson, courtesy of David Tilman

Edward O. Wilson, p. 340 photo by Jon Chase, courtesy of E. O. Wilson

E. Lucy Braun, p. 382 courtesy of Ronald Stuckey

John Curtis, p. 418 courtesy of the University of Wisconsin Arboretum, Madison

Margaret Davis, p. 433 courtesy of Margaret Davis

In addition, the authors thank the following colleagues who provided the slides and/or electronic images that now enhance the book.

Wayne Armstrong, Palomar College

Robin Chazdon, University of Connecticut, Storrs

Kenneth Cole, U.S. Geological Survey

Scott Collins, National Science Foundation

Timothy Craig, University of Minnesota, Duluth

Peter Curtis, Ohio State University

Thomas Eisner, Cornell University

David Hooper, Western Washington University

Peter Jorgensen, Missouri Botanical Garden, St. Louis

Walter Judd, University of Florida

Susan Kalisz, University of Pittsburgh

Colleen Kelly, University of Southampton, U.K.

Alan Knapp, Kansas State University

Brian Lange, Northern Illinois University

Svata Louda, University of Nebraska, Lincoln

Jeffry Mitton, University of Colorado

Robert Peet, University of North Carolina, Chapel Hill

Pamela O'Neil, University of New Orleans

Elizabeth Orians, Seattle, Washington

Chris Ray, University of California, Davis

Mark Rees, Imperial College of Science, Technology and Medicine, U.K.

Robert Robichaux, University of Arizona

Kayla Scheiner, Arlington, Virginia

Johanna Schmitt, Brown University

Susan Schwinning, University of Arizona

Allison Snow, Ohio State University

Maureen Stanton, University of California, Davis

James Teeri, University of Michigan, Ann Arbor

John Thomson, University of Toronto

Raymond Turner, United States Geological Survey (ret.)

Lawrence Walker, University of Nevada, Las Vegas

Kathy Whitley, University of South Florida

John Worrall, University of British Columbia

Glossary

Abiotic nonliving

Absolute competition index (ACI) a measure of the intensity of competition (see also, log response ratio and relative competition index)

Abundance curve a graph describing a hierarchy of relative abundance among the species in a community

Acclimation a potentially reversible adjustment to environmental conditions

Acid precipitation precipitation that is more acidic than normal

Actual evapotranspiration (AET) the amount of water that enters the system in precipitation, minus the amount that is lost in runoff and to percolation to groundwater

Adaptive trait a trait which increases the probability of an organism to survive, to reproduce, or to leave descendants

Additive design an experimental test of the effects of competition by altering the total density of neighbors while keeping the density of the target species constant

Additive series an experimental test of the effects of competition by altering both the total density and frequency of neighbors

Adiabatic lapse rate the rate of cooling of rising air

Adsorb to attract and hold to a surface

Adventitious roots roots that originate from above-ground tissue

Aerenchyma aerated plant tissues

Aerobic requiring oxygen

Agamospermy the production of seeds without fertilization or meiosis

Age-based methods used for the study of populations requiring only information on the population's age structure

Age-structured a population described by the relative frequency of individuals by age class

Alkaloid a small compound that contains an aromatic ring and nitrogen

Allelopathy the chemical inhibition of one organism by another

Allopatric speciation speciation occurring when a geographically isolated population becomes a new species

Allopolyploidy polyploidy due to the union of gametes from individuals of different species

Alpha diversity inventory diversity within a single community

Alternation of generations a reproductive cycle in which a haploid multicellular organism gives rise sexually to a diploid organism, while the meiotic products of the diploid organism grow directly by meiosis into a haploid organism

Ammonification the release of nitrogen from organic compounds in the form of ammonium ions (NH_4^+)

Amphistomatous having stomata on both sides of the leaf

Androdioecy some plants in a population have only staminate flowers while others have either perfect flowers, or a mixture of both staminate and pistillate flowers

Andromonoecy the occurance of both staminate and perfect flowers on the same individuals

Aneuploidy changes in the genome of a species due to the gain and loss of individual chromosomes

Angle of incidence the angle that a ray of sunlight makes with a line perpendicular to the surface

Annual a plant in which the life cycle is completed in one year or less

Anoxic without oxygen; anaerobic

Anther terminal portion of a stamen containing pollen in pollen sacs

Apical dominance the prevention of growth of other meristems by the apical meristem

Apical meristem a group of undifferentiated plant cells at the tip of a growing stem or branch

Apomixis reproduction by maturing seeds in the absence of fertilization

Apparent competition   density-dependent negative interactions between species that appear to be due to competition for resources, but are actually due to a shared predator or herbivore

Aquatic plants   those that grow submerged in water

Archipelago   a string of islands

Architecture (of a plant)   the arrangement of regenerating parts of the plant

Arithmetic mean   the sum of *n* numbers, divided by *n*; the average

Assemblage   a group of related organisms living in the same place

Association   a particular community type, found in many places and with a certain physiognomy and species composition

Assortative mating   the mating of individuals with similar phenotypes in greater frequency than by chance alone

Asymmetric competition   competition in which one individual has very large negative effects on its neighbors, but not the reverse

Autecology   the ecological study of individuals

Autopolyploidy   polyploidy due to the union of gametes from individuals of the same species

Average ($\overline{X}$)   the sum of the observations divided by the number of observations; the arthmetic mean

Axillary meristem   a group of undifferentiated plant cells at a leaf or stem axil

Bald   a treeless mountain peak

Basal area   the area occupied by the base of a plant such as a tussock grass or a tree with a definable base

Basal cover   the fraction of area occupied by the base of a plant such as a tussock grass or a tree with a definable base

Beneficial mineral element   minerals that either are essential for only some plant species, or are not essential, but stimulate growth

Beta diversity   differences in community composition along an environmental gradient or among communities in a landscape

Biennial   a semelparous plant that flowers after two or more years

Biodiversity   all of the populations, species, and communities in a defined area

Biodiversity hotspot   threatened areas with very high species diversity

Biological control   the deliberate use of one species to control another undesirable species

Biological immobilization   the incorporation of mineral nutrients into soil microorganisms as organic molecules

Biological species concept   a group of actually or potentially interbreeding organisms that are completely or nearly completely genetically isolated from other such groups so that each group is an evolutionarily independent lineage

Biomass   mass of living organisms

Biome   a major biogeographical region that differs from others in the structure of its vegetation and dominant plant species

Biotic   living

Bisect   a scale drawing of the vegetation along a line transect

Bootstrap   a statistical technique to estimate a sampling distribution by resampling observations randomly

Boundary layer   a blanket of relatively still air that surrounds all objects in air or water

Broad-sense heritability   heritability due only to total genetic variation

Browser   an animal that eats leaves from trees or shrubs

Bulb   an underground rosette stem that store nutrients

Bulbil   a tiny bulblike organ vegetatively produced in an inflorescence or leaf axil

C_3 photosynthesis   a form of photosynthesis in which CO_2 is captured by RuBP carboxylase/oxygenase and the first stable product is a three-carbon compound

C_3 plant   a plant species that uses C_3 photosynthesis

C_4 photosynthesis   a form of photosynthesis in which CO_2 is captured by PEP carboxylase and the first stable product is a four-carbon compound

C_4 plant   a plant species that uses C_4 photosynthesis

Calcic horizon   a cement-hard layer of soil formed by a concentration of calcium found in many deserts

Calvin cycle   the biochemical reactions by which carbon from CO_2 is incorporated into organic compounds

CAM photosynthesis   see Crassulacean Acid Metabolism

CAM plant   a plant species that uses CAM photosynthesis

Cambium   a zone of meristematic cells between primary xylem and phloem

Canonical analysis   correlation analysis of two multivariate data sets; in ecology the correlation of a matrix of community similarities with a matrix of environmental relationships

Canonical correspondence analysis   an ordination method based on the correlation of a matrix of community similarities with a matrix of environmental relationships in which the axes are restricted to be linear combinations of explanatory variables

Canopy cover   the amount of ground covered by the highest level of vegetation of a community

Capillary action   the pull of water through a narrow tube by the attraction of the water molecules to the charged particle surfaces or each other

Capitulum   the specialized inflorescence of plants in the Asteraceae; a head

Carbon fixation   the part of photosynthesis involved in the acquisition of carbon from the air

Carboxylation   a chemical reaction in which a CO_2 molecule is added to another molecule

Cation exchange capacity (**CEC**)   a measure of the total ability of the soil colloids to adsorb cations

Cellular respiration   a process by which organic compounds are broken down to release energy through the use of oxygen and the release of CO_2

Cellular respiration   the breakdown of organic compounds to release energy

Census   a count of individuals

Chamaephyte   a shrub with buds less than 25 centimeters above the ground

Characteristic dimension   the effective width of a leaf with respect to fluxes of energy and mass

Characteristic value   an eigenvalue

Characteristic vector   an eigenvector

Chasmogamous   referring to flowers that are open and potentially cross-pollinated

Chronic herbivory   herbivory that occurs over long time periods

Chronosequence   a set of communities of different ages since disturbance, assumed to represent a single community over time

Clayey soil   soil with at least 35% clay particles

Clear-cutting   the removal of all trees

Cleistogamous   referring to flowers that are closed and self-pollinated

Climate   the long-term average and variability of the weather in a given area

Climax   the hypothetical equilibrium endpoint of a successional sequence

Cline   a gradient in genetic composition

Clonal   consisting of potentially independent ramets

Clonal fragmentation   a process in which pieces of a plant break off and are capable of rooting to form new independent plants

Clone   a genetic duplicate

Cohort   a group of individuals in a population that germinated, reached a particular size, or entered a study at about the same time

Collenchyma   support tissue made of living elongated cells with walls thickened irregularly

Colloid   substance in which one or more materials in a finely divided state are suspended or dispersed through a second material

Commensal   a species interaction in which one species benefits and the other neither benefits or is harmed

Common garden experiment   growing plants from diverse locations in a single garden to determine whether differences among populations are due to genetic differences

Community   a group of populations that coexist in space and time and interact with each other directly or indirectly

Compensation   a hypothesized response to herbivory resulting in no net difference between grazed and ungrazed individuals

Competition   a reduction in fitness due to shared use of a resource in limited supply

Competitive effect   the ability of an individual to affect neighbors through resource use

Competitive response   the response of neighbors to resource use by an individual

Compositional pattern diversity   the arrangement of subunits in the mathematical space defined by an object-attribute matrix

Compositional similarity   the extent to which adjacent patches share a similar set of species at similar frequencies

Conduction   the direct transfer of heat energy from the molecules of a warmer object to the molecules of a cooler object

Confidence interval   a numerical interval within which a given parameter has a certain probability of residing

Connectivity   the degree to which a landscape facilitates or impedes movement of organisms

Constitutive defense   a defense that is present in a plant regardless of herbivore damage

Continental climate   the climate of the interior of large continents at mid-latitudes with seasonal changes in precipitation and temperature

Controlled burn   a fire that is planned and purposely set (also see Prescribed burn)

Convection   heat transport by a packet of air or water moving as a unit

Convergent evolution   the independent evolution of similar features in unrelated taxa

Coriolis effect   the tendency of an object moving toward or away from the equator to be deflected, relative to the direction in which it is moving, toward the right in the Northern Hemisphere and toward the left in the Southern Hemisphere

Cork cambium   a layer of meristematic cells under the bark responsible for secondary growth

Correlation (*r*)   a measure of the strength of the relationship between two variables (also see Regression)

Correlational selection   when individuals with particular combinations of trait values have the highest fitness in a population

Cosexual   plants that function as males and females more or less simultaneously

Cover   the percentage of the ground covered by the canopy of a given species

Crassulacean Acid Metabolism (**CAM**)   a form of photosynthesis in which CO_2 is captured by PEP carboxylase and the capture of light energy is temporally separated from the uptake of CO_2

Croll-Milankovič effects   long-time scale changes in the Earth's orbit around the sun

Crown fire   the burning of the tree canopy

Cryptic dioecy   sporophytes that produce only female or male gametophytes even when apparently capable of producing both types

Cryptic species   species that look identical and yet are reproductively isolated

Cryptogamic crust   a thin crust made of mosses, fungi, algae, lichens, and cyanobacteria

Cryptophyte   a perennial herb with its perennating organ below the ground surface

Cuticle   a waxy, nonliving covering over the exposed epidermal cells of a leaf

Cycle   the major fluxes and pools of a substance in a system

Deciduous   the loss of leaves on a seasonal basis

Deforestation   the destruction of forests

Degree-days   the sum of temperatures experienced over a specified time period

Demographic stochasticity   variation in vital rates because of random variation in individuals' survival and reproduction

Dendrochronology   the study of tree rings

Denitrification   the reduction of NO_2^- and NO_3^- to gaseous nitrogen

Density   the number of individuals of a species per unit area

Density-dependent   population processes that vary as a function of population density

Density-independent   population processes that vary as a function of extrinsic factors

Desert   a region where potential evapotranspiration exceeds actual evapotranspiration

Deterministic   having a fixed outcome

Detritus   debris; organic detritus is the dead bodies and waste products of living and once-living organisms

Diameter at breast height (**DBH**)   the diameter of a tree trunk measured at 1.37 m from the ground

Diaspore   the dispersal unit of a plant including the seed and any additional structures

Differentiation diversity   the diversity of a sample based on the way the species are grouped into subunits

Dimension reduction   the process of taking highly multivariate data and collapsing them into a small number of dimensions

Dioecy (dioecious, adj.)   among the plants in a population, at least some individuals have only pistillate or staminate flowers

Direct gradient analysis   analysis of species distributions or community similarities along a single environmental gradient

Directional selection   individuals with the most extreme trait values have the highest fitness in a population

Disassortative mating   the mating of individuals with dissimilar phenotypes is greater in frequency than by chance alone

Dissected   in reference to leaves, a blade divided into small but connected parts

Disturbance   a relatively discrete event in time that causes abrupt change in ecosystem, community, or population structure, and changes resources, substrate availability, or the physical environment

Disturbance regime   the characteristics of disturbances occurring in that ecosystem, generally described by intensity, size, and frequency

Dominance   the expression of an allele depends on the properties of the other allele at that locus

Dominant eigenvalue   the largest eigenvalue

Dominant species   the most conspicuous and numerically abundant species in a community

Drought avoidance   a strategy of growth during periods of sufficient rainfall by plants living in areas with periods of prolonged drought

Drought deciduous   the dropping of leaves in response to low water conditions

Drought tolerance   the ability to live and grow even when experiencing very low water potential

Dry deposition   molecules deposited on a surface as particulate matter

Dry weight   the weight of a sample after it has lost all its moisture and is at a constant weight

Dynamic stability   the tendency for a system to return to its original state after a small perturbation

Ecology   the study of the relationships between living organisms and their environments

Ecosystem   all of the organisms in an area and all of the abiotic materials and energy with which they interact

Ecotype   populations of a species from different habitats or locations that possess a similar set of genetically-based adaptions

Ectendomycorrhizae   mycorrhizae with most characteristics of ectomycorrhizae, but whose hyphae penetrate the cells of the host plant

Ectomycorrhizae (**ECM**)   mycorrhizae that form a sheath around the root, whose hyphae penetrate the intercellular spaces of the root cortex but not the root cells

Edaphic   referring to characteristics of the soil

Eddy   a parcel of air that moves as a discrete unit

Eddy covariance   a method for measureing net ecosystem exchange

Edge effect   a systematic difference between the edge of a habitat patch and its interior

Efficiency of light interception   the proportion of incident light intercepted by a plant canopy

Eigenvalue   a number that, when used to multiply a population vector, gives the same result as multiplication by the transition matrix; given a matrix **A**, if $\mathbf{A}x = \lambda x$, λ is an eigenvalue of **A** and x is the associated (right) eigenvector

Eigenvector   a nonzero vector that, when multiplied by an associated eigenvalue, gives the same result as multiplying by the transition matrix; see Eigenvalue. The eigenvector associated with the dominant eigenvalue gives the stable age- or stage-distribution of the population.

El Niño Southern Oscillation (**ENSO**)   a 3–7 year oscillation in the Earth's global climate

Elaiosome   a lipid body attached to a seed that serves as a reward to a seed disperser

Elasticity   the extent to which a proportional change in a survivorship or fecundity rate causes a proportional change in the population growth rate

Eluviation   the removal from soil layers of soil material in suspension or in solution

Embryo sac   megagametophyte; a cluster of haploid cells in the ovule of flowering plants in which fertilization of the egg and development of the embryo occurs

Emergent property a property that occurs at one level of organization due to properties and processes that are unique to that level of organization

Emergent tree a tree with a canopy that extends above the general canopy of the forest

Emissivity how efficient a body is at emitting energy

Endemic species a species found only in a single, limited area

Endomycorrhizae mycorrhizae whose hyphae penetrate the cells of the root cortex but do not form a sheath around the root

Endosperm triploid nutritive tissue for the embryo of most angiosperms that arises through double fertilization

Energy balance the amount of energy absorbed, emitted, and stored

Ensemble a group of related organisms living in the same place that use biotic or abiotic resources in a similar way

Environmental stochasticity variation in vital rates due to environmental factors that affect all the individuals in a class or population in roughly the same way

Epiphyte a plant that grows upon another plant and is not rooted in the ground

Epistasis the expression of an allele depends on the properties of alleles at other loci

Equilibrium-based model a model that assumes that the system reaches a stable endpoint

Ericoid mycorrhiza a symbiosis between a fungus and the roots of a plant in the family Ericaceae

Errors of development small, random differences in when and how genes are expressed

Essential mineral nutrients those nutrients that plants need to live and grow

Euclidean distance the shortest distance between two points in a flat plane

Evapotranspiration water transferred to the atmosphere in evaporation plus transpiration

Evenness the extent to which the species in a community are equally abundant

Exchangeable base a cation that contributes to making the soil more alkaline

Exotic species a species that is outside its native range

Experiment a test of an hypothesis

Extent a component of scale indicating the total range of a set of data

Extinction probability the percent of replicate populations expected to go extinct in a given time period

Facilitation a process by which early successional species increase the survival of later successional species

Facultative existing only under some conditions

Fecundity the potential reproductive capacity of an organism measured as the number of gametes or propagules

Female fitness fitness through maturing seeds

Fertility the number of viable offspring produced per unit time, the birth rate

Fibrous root systems roots forming a dense network

Field capacity the point at which a soil is completely saturated with water

Floral bract a specialized leaf below a flower

Flux rate of flow

Flux equation a model of flow rates of the general form: flux = (conductance) × (driving force)

Focus a component of scale indicating the sampling level or area represented by each data point (also see Resolution)

Forb a broadleaved herbaceous plant

Forest an area dominated by trees in which the canopies touch each other

Formation a community defined by its physiognomy

Fractal dimension a measure of the complexity or degree of convolution of a line or surface

Frequency (of a disturbance) how often, on average, a disturbance occurs in a particular place (also see Return interval)

Frequency (of a species in a community) the percentage of quadrats in which a species appears

Frequency-dependent selection when the fitness of a genotype depends on whether it is common or rare in a population

Frugivory fruit-eating

Fruit the matured ovary or ovaries of one or more flowers and associated structures

Fuel load the amount of combustible plant material in an area

Functional ecology the study of how the biochemistry and physiology of an individual determines its response to its environment

Functional group a group of organisms that use biotic or abiotic resources in a similar way (also see Guild)

Fundamental niche the environmental range in which a species is physiologically capable of growing

Fynbos a small region in the southernmost part of Africa with a high plant diversity, especially of endemic species

Gametophyte the haploid generation of plants that produces gametes

Gametophytic self-incompatibility self-incompatibility due to the pollen grain's haploid genotype

Gamma diveristy inventory diversity across landscape or a set of samples including more than one community

Gap an opening in a forest canopy

Gene-environment covariation a nonrandom relationship between genes and the environment

Genet a genetic individual, the product of a single seed

Genetic drift changes in gene frequencies due to random sampling effects

Genetic response that part of natural selection consisting of the change in the population across generations

Genome an individual's entire DNA sequence

Genotype an individual's DNA sequence

Genotype-environment interaction   differences in genetic expression as a function of the environment

Geographic Information Systems (GIS)   a method for mapping different kinds of information for the same region

Geometric mean   the *n*th root of the product of *n* numbers

Grain (in landscape ecology)   a component of scale indicating the size of the primary unit in a set of data

Grain (in reproduction)   the fruit of a member of the grass family (Poaceae); a pollen grain is a single pollen individual

Grana stack   in chloroplasts, a group of disc-shaped, flattened vesicles bearing photosynthetic pigments

Granivory   seed-eating

Gravitational potential   water potential due to the pull of gravity

Grazer   an herbivore that eats grass and other ground-growing plants

Greenhouse effect   the re-absorption of longwave radiation emitted by the earth's surface by greenhouse gasses

Greenhouse gas   an atmospheric gas that contributes to the greenhouse effect

Gross photosynthesis   the total amount of CO_2 captured by an individual

Gross photosynthetic production (GPP)   the total energy (or carbon) fixed by primary producers

Gross primary production (GPP)   the total energy (or carbon) fixed by primary producers

Ground-truthing   testing the reliability of remote-sensing data by ground-based measurements

Groundwater   underground water that has percolated to a zone that is saturated

Group selection   variation in group characteristics causes differences in fitness among members of different groups

Guard cell   the two cells on either side of a stomate that determine the size of the stomatal opening

Guild   a group of organisms that use biotic or abiotic resources in a similar way (also see Functional group)

Gynodioecy   some plants in a population have only pistillate flowers while others have either perfect flowers, or a mixture of both staminate and pistillate flowers

Gynomonoecy   the occurance of both pistillate and perfect flowers on the same individuals

Gyre   a circular or spiral motion

Habitat   the kind of environment a population or species generally inhabits, or an area distinguished by its community

Hadley cell   a large cycle of air rising, traversing a distance, falling, and then moving back again along the earth's surface

Halophyte   a plant that lives in saline soil

Hartig net   a complex of mycelia that grow between root cortical cells in ectomycorrhizae

Heat   a measure of the total kinetic energy of the molecules in a substance

Heath   a vegetation type dominated by low-growing shrubs in the Ericaceae

Hemicryptophyte   a perennial herb with buds at the ground surface

Hemiparasite   a plant that facultatively taps into the roots of other plants to gain resources

Herbivory   the consumption of all or part of a living plant

Heritability (h^2)   the resemblance among relatives due to the sharing of genes, the amount of phenotypic variation in a population that is due to genetic variation

Heritable   having a genetic basis

Hermaphrodite   a plant that produces male and female gametes simultaneously

Heterostyly   a population with flowers that have more than one length of styles and stigmas

Hierarchical patch dynamics   an approach to examining the effects of scale in which patches exist in a hierarchy of successively larger patches that are linked by the movement of materials and energy

Horizon   characteristic layer of a soil

Hybrid swarm   a group of species that hybridize frequently

Hybridization   interbreeding between individuals of different species

Hydraulic lift   plants take up water in deep soil horizons and then lose it to the soil at upper horizons

Hygrophyte   a plant that lives in permanently moist soil

Hypothesis   a possible explanation for a particular observation

Hyraceum   the accumulation of large amounts of plant material from around a rock hyrax den

Igneous   of volcanic origin

Illuviation   the process in which soil material is precipitated (from solution) or deposited (from suspension) into a soil layer from an overlying layer

Immobilization   the incorporation of nutrients into soil microorganisms, making them less available for uptake by plants

Importance value   the sum of the relative cover, relative density, and relative frequency of a species in a community

Inbreeding   mating among relatives

Inbreeding depression   a decrease in fitness due to inbreeding

Inclusive fitness   total fitness of an allele of all copies in a group, usually siblings

Indicator species   a species used to place a community into a predetermined group

Indirect gradient analysis   a form of community analysis in which the stands are ordered by their similarity in species composition and the environmental factors responsible for the resulting patterns are inferred

Individual-based model   a model consisting of simulations in which the properties of individual organisms are explicitly modeled

Induced defense   a response to herbivory that protects the plant whether or not it harms the herbivore

Induced resistance   a response to herbivory that harms the herbivore whether or not it protects the plant

Induced response   a defense that is present in a plant elicited by an attack by a herbivore

Inflorescence   an aggregation of flowers

Inhibition   a process by which early successional species decrease the survival of later successional species

Initial floristic composition   a theory of succession that emphasized the process of colonization and differences in the life spans of species

Intensity (of a disturbance)   the amount of change caused by a disturbance

Intercalary meristem   a group of undifferentiated plant cells between stem nodes

Intermediate disturbance hypothesis   an hypothesis that species diversity will be highest at intermediate levels of disturbance.

Internode   the stem segment between nodes

Intertropical convergence   the region where the trade winds tend to converge from the northern and southern hemispheres

Intracellular conductance   for a given concentration gradient, the rate at which a gas moves through leaf cells and cell walls

Invasive species   a species that is rapidly expanding outside of its native range

Inventory diversity   diversity of a sample based on species number or evenness

Island   an area that is suitable for the survival and reproduction of a species, surrounded by unsuitable habitat

Isobilateral   a leaf with a distinctive, symmetrical internal architecture with palisade tissue on both upper and lower sides of the leaf

Iteroparous   reproducing in more than one bout in a lifetime

Jackknife   a statistical technique to estimate the bias in a variance by deleting observations sequentially

Jet stream   a river of air in the upper atmosphere that moves at high speeds from west to east

Kin selection   a form of group selection in which the groups are composed of closely related individuals

Krummholz   stunted forest or scrub in windswept alpine regions close to tree line

***K*-selection**   hypothesized natural selection at high population density for traits that increase the carrying capacity

Lag   a delay in an effect

Lag phase   a period during early population growth of an invasive species when numbers are low

Lamellae   see thylakoid membranes

Landscape ecology   the study of the spatial distribution of individuals, populations, and communities, and its causes and consequences

Latent heat exchange   the transfer of energy in the process of converting water between the liquid and vapor states

Latent heat loss   the loss of energy through the process of converting water between the liquid and vapor states

Latex   milky sap

Leach   to lose from a surface soil through water drainage

Leading eigenvalue   the largest eigenvalue

Leaf area index (**LAI**)   the area of photosynthetic surface per unit of ground area

Leaf axil   the place where a leaf and stem join

Leaf conductance   for a given concentration gradient, the rate at which a gas moves into or out of a leaf

Leaf production efficiency   the net gain in plant weight per unit leaf area

Leaf turnover   the rate at which leaves are replaced

Left eigenvector   a vector y such that, when it premultiplies a matrix $\mathbf{A}$, gives the same result as if it is multiplied by an associated eigenvalue; $y\,\mathbf{A} = \lambda\,y$. The left eigenvector associated with the dominant eigenvalue gives the reproductive values.

Liana   a woody vine rooted in the ground

Life cycle graph   a drawing showing the size-, age-, or stage-classes of a population with arrows indicating transitions among classes

Life history   the schedule of birth, mortality, and growth of individuals in a population

Life table   an estimated list of the mortality, cumulative survival, and related rates of a cohort

Light compensation point   the light level at which photosynthetic gain exactly matches respiratory loss

Light reaction   the capture of light energy during photosynthesis

Light saturation (A_{sat})   the light level at which the maximum photosynthetic rate is reached

Lignin   a phenol that impregnates secondary cell walls providing strength

Limiting resource   a resource that determines the growth or reproduction of the plants in a population

Litter   recently fallen plant material which is only partially decomposed

Loamy soil   soil with approximately equal proportions of sand, silt, and clay particles

Local guild   a group of organisms living in the same place that use the same resources

Loess   fine, unconsolidated wind-blown sediment

Log response ratio (**LRR**)   a measure of the intensity of competition (see also, Absolute competition index and Relative competition index)

Longwave radiation   radiation at wavelengths greater than 700 nanometers

Macrofossil a large fossil such as leaves, flowers, stems, and seeds

Macronutrient an element that is needed by plants in large quantities

Manipulative experiment a test of an hypothesis based on a deliberate change in the physical world

Mantle a dense network of fungal mycelia surrounding plant roots, found in ectomycorrhizae

Maritime climate the climate of the eastern edges of continents with uniform precipitation throughout the year

Mass-balance approach a method for used by ecosystem ecologists to account for system storage by accounting for all inputs and outputs

Masting the large and erratic variation among years in the size of the seed crop that is synchronized among most of the plants in a population

Mating system the biological factors that govern who can mate with whom

Matric potential water potential due to the cohesive force that binds water to physical objects

Mean ($\overline{X}$) the sum of the observations divided by the number of observations; the average)

Mean field approach analysis that uses average parameter values

Mean field model a model in which all individuals in the population are assumed to have the same properties

Median the middlemost number in an ordered set; the 50th percentile

Mediterranean the climate characteristic of the western edges of continents with rainy winters and dry summers

Meiosis two successive divisions of a cell resulting in the production of haploid cells, each of which contains one of each pair of the homologous chromosomes of the parent cell; meiosis in plants results in spores

Meristem a group of undifferentiated plant cells

Mesic moist

Mesophyll the photosynthetic tissue between the upper and lower epidermis of a leaf

Mesophyll conductance for a given concentration gradient, the rate at which a gas moves through leaf mesophyll cells and cell walls

Mesophyte a plant that lives in moderately moist soils

Meta-analysis the statistical synthesis of the results of independent experiments

Metabolic energy energy used by plants for maintenance, growth and reproduction

Metamorphic rocks changed by the action of great pressures and temperatures deep underground

Metapopulation a group of populations linked by migration and extinction

Microfossil a small fossil such as a pollen grain

Microhabitat the conditions in the immediate surroundings of an individual organism

Micronutrient element for growth needed in very small quantities

Midden the accumulation of large amounts of plant material on a packrat den

Migration the movement of organisms or propagules

Mineralization the processes by which microorganisms release carbon as CO_2 and nutrients in inorganic form from decaying biomass

Minimum viable population the smallest size necessary to give a population a probability x of surviving N years

Mixture a community composed of two or more species

Model an abstraction or simplification that expresses structures or relationships

Modular structure a system of repeated units (nodes, lateral organs, and internodes)

Mollicute tiny, wall-less bacteria

Monocarpic reproducing in a single bout in a life time; a botanical term for semelparous

Monoculture a community composed of a single species

Monoecy the occurrence of both staminate flowers and pistillate flowers on the same individual

Monothetic-divisive a classification process based on dividing one group into two groups depending on the presence or absence of a single key species

Morphology the form or structure of an organism

Mosaic diversity a measure of landscape complexity due to variation in species richness among communities and variation in commonness and rarity among species within a region

Multiple regression the regression of a single dependent variable simultaneously on multiple independent variables

Multivariate consisting of more than one variable

Muskeg a grassy bog habitat with scattered, stunted conifers

Mutation a change in a DNA sequence

Mutualism a symbiosis in which both organisms benefit

Mycorrhiza (pl. mycorrhizae) a symbiosis between a fungus and the roots of a terrestrial plant

Narrow-sense heritability heritability due only to additive genetic variation

Natural experiment a test of an hypothesis based on a change in the physical world caused by some natural occurrence

Natural selection the process by which individuals with different phenotypic characteristics leave different number of offspring (also see Sexual selection)

Negative assortative mating mating of individuals with dissimilar phenotypes occurs at greater frequency than by chance alone

Nestedness the tendency of communities to be subsets of other communities

Net ecosystem exchange (NEE) the net transfer of CO_2 into or out of an ecosystem

Net ecosystem production (NEP) the net accumulation of carbon by an ecosystem over the course of succession

Net primary production (NPP) the total energy (or carbon) fixed by primary producers minus respiratory losses

Net reproductive rate (R_0) the expected number of offspring that an individual will have over its entire life

Niche the range of ecological conditions in which a species will grow

Nitrification the oxidization of ammonia to nitrate

Nitrifying bacteria specialized bacteria that carry our nitrification

Nitrogen fixation the process by which nitrogen enters the biosphere from the atmosphere

Nitrogen mineralization the conversion of nitrogen from organic to inorganic form

Nitrogen saturation nitrogen deposition into a system exceeds use

Nitrogen use efficiency maximum photosynthetic rate per gram of nitrogen in the leaf

Node a point of proliferation for the development of leaves or flowers

Normalized difference vegetation index (NDVI) relative reflectance of vegetation in the near infrared and visible wavelengths

Null model a model that describes how data would look if only random processes were operating

Nurse plant a plant that increases the survival of another plant, especially in very harsh environments such as a desert

Obligate necessary

Observational experiment a test of an hypothesis based on the systematic study of natural variation

Old-field succession succession in abandoned agricultural fields

Orchidaceous mycorrhizae a symbiosis between a fungus and the roots of a plant in the family Orchidaceae

Ordination the process of taking points arrayed in an *n*-dimensional space and reducing them to fewer dimensions

Organic matter the decaying and decomposed material in soil that comes from living things

Osmoregulation the regulation by an organism of its water potential

Osmotic potential water potential due to solutes dissolved in the water

Outcrossing the avoidance of mating among relatives

Ovary the ovule-bearing region of a pistil

Overcompensation a hypothesized response to herbivory by growing more than plants not eaten

Ovule in seed plants, the sporophytic structure containing the female gametophyte

Paleoecology the study of historical ecology

Palynology the study of spores and pollen

Parapatric speciation speciation occurring when a new species arises in a population adjacent to other populations of the ancestral species

Parent material the upper layers of the heterogeneous mass that is left over after the action of weathering and other forces on rock

Patch a small area

Pattern relationships between pieces or entities of the natural world

Pattern diversity the relative arrangement of subunits within an ecological unit

PEP carboxylase the enzyme responsible for the initial capture of CO_2 in C_4 photosynthesis

Percentage base saturation the proportion of the cation exchange capacity that is occupied by bases

Perennial a plant in which the life cycle is completed in more than one year

Perfect blackbody radiator an object that absorbs all the radiation that falls on it, and emits all of the energy possible for an object at its temperature

Perfect flower a flower containing both functional stamens and functional stigmas

Perianth the combination of the petals and sepals of a flower

Permafrost a permanently frozen layer of soil

Permanent wilting point when plants can no longer can extract water from the soil and exist in a permanently wilted condition

pH the negative log of the concentration of H^+ ions

Phanerophyte a tree or tall shrub with buds at least 25 centimeters above the ground

Phenol a chemical consisting of an aromatic ring with an attached hydroxyl group

Phenology the timing of life history events, such as initiation of growth, flowering, and dormancy

Phenotype all of the physical attributes of an organism

Phenotypic plasticity differences in the characteristics of a genotype caused by environmental influences

Phenotypic selection that part of natural selection consisting of the combination of phenotypic variation and fitness differences

Phloem plugging a defense against infection in which the phloem clogs up preventing the spread of the infectious agent

Photon a discrete packet of energy that travels at the speed of light

Photorespiration the consumption of O_2 and release CO_2 in plants which depends on light

Photosynthesis the biochemical processes by which an organism captures energy from sunlight and fixes carbon from the atmosphere

Photosynthetic induction the necessary start-up time for plants to reach maximum photosynthesis after exposure to bright light

Photosynthetic photon flux density (PPFD) the amount of useful light energy

Photosynthetically active radiation (**PAR**)   the wavelengths of light that can be used in photosynthesis

Photosystem   an array of chlorophyll and accessory pigments that capture light energy

Phreatophyte   a plant with roots that extend to the water table

Phylogeny   the pattern of relationships among species (or higher taxa) based upon evolutionary ancestry

Physiognomy   the form, structure, or appearance of a plant community

Physiological ecology   the study of the physiological mechanisms which underlie whole-individual responses to the environment

Phytoalexin   a secondary compound that acts as a defense against pathogens

Phytochrome   a bluish pigment that plants use to sense light of specific wavelengths

Pistillate flower   a flower with functional pistils but not functional stamens

Plantlet   small plants created vegetatively

Plasmodesmata   thin strands of living tissue

Pollen   microgametophyte; a microspore containing a mature or immature male gametophyte

Pollination syndrome   a particular combination of floral traits that are associated with particular pollinators

Polycarpic   reproducing in more than one bout in a life time

Polyploidy   a change in chromosome number due to the duplication of the entire set of chromosomes

Polythetic-agglomerative   a classification process based on linking groups into a hierarchy based on similarity of objects or groups of objects

Pool   the stored quantity of a nutrient or element

Population dynamics   changes in the number of individuals alive, genetic composition, or age structure of a population

Population ecology   the study of population growth, composition, and spatial dispersion

Population structure   the relative frequency of individuals in each size-class, age-class, or stage-class

Population vector   the vector of number of individuals of each size-, age-, or stage-classes of a population

Pore space   the total portion of the soil occupied by air and water

Pore width   the size of a stomatal opening

Porosity   in soil, the total volume of pores, and their size, shape and arrangement between and within aggregates

Potential evapotranspiration (**PET**)   the maximum amount of water that would be lost in evapotranspiration in a particular place, if water is freely available in the soil and plant cover is 100%

Precipitation   the amount and pattern of rain- and snowfall

Precision   how close the estimate comes to the true value

Predictive germination   seed germinate when conditions are likely to be favorable

Prescribed burn   a fire which is planned and purposely set (also see Controlled burn)

Pressure potential   water potential due to the hydrostatic or pneumatic pressure in the system

Primary growth   the process of increasing length in shoots or roots

Primary metabolite   a compound necessary for the basic functioning of a plant

Primary productivity   the amount of carbon transformed from CO_2 to organic carbon per unit area in a specified time period

Primary research   gathering information or finding out facts not known before

Primary succession   succession when plants colonize ground that was not previously vegetated

Productivity   the amount of carbon or energy transformed from one trophic level to the next per unit area in a specified time period

Propagule   a seed or other dispersal structure, such as a seed cluster

Propagule rain   in metapopulation models, occupancy-independent immigration

Proteoid roots   a dense clusters of short, fine lateral rootlets that produce organic acids or other chelating agents

Pubescence   mat of short hairs on a leaf surface

Pyrogenic   promoting fire

Quadrat   a sampled area

Radiant energy   energy transferred from one object to another by photons

Rain shadow   an area on the leeward side of a mountain that tends to have low precipitation

Ramet   a potentially physiologically-independent unit of a genet

Range (in statistics)   the difference between the largest observation and the smallest observation

Rarefaction   methods for estimating the total number of species in an area based on a sample of individuals

Realized fecundity   the chance of surviving to maturity times the fecundity of survivors

Realized niche   the environmental range within which a species is found

Regeneration niche   the set of environmental requirements necessary for the germination and establishment of a plant species

Regression (*b*)   a measure of the relationship between two variables (also see correlation)

Relative abundance curve   a graph of species abundance of rank-order abundance versus abundance, often log (abundance)

Relative competition index (**RCI**)   a measure of the intensity of competition (see also, absolute competition index and log response ratio)

Relative efficiency index (REI) a measure of the difference in relative growth of two species grown in competition with each other

Relative growth rate (RGR) a measure of growth in terms of increase in weight per unit time with respect to initial weight

Relative humidity the water vapor in the air as a percent of the amount in fully saturated air at the same temperature

Relay floristics a theory of succession that emphasized the facilitation of early successional species and the replacement of an entire community by another

Relevé a single, large sample plot

Remote sensing the process of collecting data about an entity of interest without being in direct contact with that entity

Reproductive effort the fraction of resources allocated to reproduction at a given age or stage

Reproductive isolation a condition in which two populations are unable to share genes do to barriers to mating or inviability of offspring

Reproductive value the contribution in offspring an average individual now in stage x will make to the next generation before it dies

Rescue effect in metapopulation models, the dependence of population extinction on the frequency of occupied patches

Resistance the inverse of conductance

Resolution a component of scale indicating the sampling level or area represented by each data point (also see Focus)

Response surface experiment an experimental test of the effects of competition by altering both the total density and frequency of neighbors

Return interval the average time between disturbances in a particular place

Rhizome an underground horizontal stem growing near the soil surface

Rhizosphere the microbe-rich environment immediately surrounding plant roots

Richness see Species richness

Riparian areas adjacent to streams and rivers

Root-to-shoot ratio ratio of root mass to aboveground tissue mass

Rosette plant a plant with short or unnoticeable internodes

r-selection hypothesized natural selection at low population density for traits that increase the intrinsic rate of population growth

Rubisco RuBP carboxylase/oxygenase; the enzyme responsible for the initial capture of CO_2 in C_3 photosynthesis or the Calvin cycle

Ruderal species a species that grows well in temporary habitats and areas subject to frequent disturbance (also see Weed)

Rumen a specialized stomach compartment of some mammalian grazers

Safe site a favorable spot for seed germination and establishment

Sample size (N) the number of observations

Sampling bias the departure of the estimated value for the sample from the true value due to causes other than chance

Sandy soil soil with more than 50% sand particles

Saprophyte a plant or fungus that acquires carbon from organic detritus

Scalar a single number

Scale the spatial or ecological context of a set of data, measured as grain, extent and focus

Scale dependence a change in a pattern or process with the scale of measurement

Scarification the abrasion of seeds to promote germinate

Scientific method a process for obtaining knowledge of the natural world consisting of observations, descriptions, quantifications, posing hypotheses, testing those hypotheses using experiments, and verification, rejection, or revision of the hypotheses, followed by retesting of the new or modified hypotheses

Sclereid a sclerenchyma cell with thick, lignified, much-pitted walls

Sclerenchyma a plant tissue containing elongated fiber cells

Sclerophyllous leaves that are tough, evergreen, relatively small, roundish and thick

Secondary chemical a compound that serves one of a wide variety of functions outside a plant's basic metabolism, including defense and attraction of pollinators

Secondary growth the process of increasing girth by producing woody tissues

Secondary metabolite see Secondary chemical

Secondary research gathering data or confirming facts that are already known

Secondary succession succession when plants colonize ground previously occupied by a living community

Sedge a grass-like plant in the Cyperaceae

Sedimentary the deposition and recementation of material derived from other rock

Seed an embryonic sporophyte embedded in a female gametophyte and covered with one or more integuments derived from the maternal sporophyte

Seed bank the seeds buried in the soil, can refer to either a single species or an entire community

Seed fern an extinct group of vascular plants that superficially resembled ferns, but reproduced through seeds rather than spores

Seed predator herbivore that consumes seeds or grains; granivore

Selective logging the removal of only some trees

Self-incompatibility the prevention of self-mating or matings between certain individuals due to a genetically-based recognition system

Self-pruning the shedding of branches by a shrub or tree

Self-thinning the process of density-dependent mortality

Semelparous reproducing in a single bout in a life time

Senescence decline in vigor with age

Sensible heat heat which can be measured

Sensible heat exchange energy exchange which results in a change in temperature

Sensitivity the extent to which a change in a transition rate causes a change in the long-term growth rate

Sequential hermaphroditism beginning life as one gender and then switching to the other

Seral community a plant community that is part of a successional sequence

Serotiny the retention of seeds in cones or fruits that are then released after exposure to heat, normally fire

Serpentine soil soil derived from magnesium silicate rock of metamorphic origin that is low in nutrients and may be high in toxic elements

Sessile attached to a surface, immobile

Sexual selection the process by which individuals with different characteristics leave different number of offspring due to differences in access to mates and fertility (also see Natural selection)

Shoot the stem and leaves of a plant

Shortwave radiation radiation at wavelengths less than 700 nanometers

Shrubland a plant community dominated by low-growing woody plants with multiple stems

Silty soil soil with more than 50% silt particles

Similarity the extent to which two communities share the same species based on presence or abundance

Sink population a population that is maintained by immigration

Size (of a disturbance) the amount of area affected by a disturbance

Size hierarchy a highly unequal size distribution among individuals in a population or community

Skew a measure of the symmetry of a set of data

Soil horizon a characteristic layer of the soil

Soil order the broadest category in a system of soil classification

Soil profile the sequence of horizons that characterizes a soil

Soil solution the water in the soil with its associated dissolved minerals

Soil structure the physical arrangement of soil particles into larger clusters

Soil texture the relative proportions of sand, silt, and clay particles

Soil type a designation determine by the texture of the surface layer

Solar tracking the tracking of the sun across the sky during the day by a leaf

Somatic mutation a mutation in a non-gamete-producing cell

Source population a population that supplies migrants to other populations

Spatial autocorrelation resemblance as a function of distance

Spatial pattern diversity the arrangement of subunits in physical space

Spatially explicit approach analysis that uses the particular spatial arrangement of objects

Speciation the process of origination of new species

Species a groups of organisms given the rank of species; the basic unit of biological classification (cf. biological species concept, taxonomic species concept)

Species density the number of species per unit area

Species pool the species in a region available to colonize a site

Species richness the number of species

Species-area curve a mathematical function showing how the number of species increases as the area sampled increases

Specific heat the amount of heat energy required to warm 1 gram by 1°C

Specific leaf area leaf area per gram of dry weight

Specific leaf area leaf area per unit leaf weight

Specific leaf weight leaf dry weight per unit leaf area

Spiroplasma a helical, motile mollicute

Sporophyte the diploid generation of plants that produces spores

Sporophytic self-incompatibility self-incompatibility due to the pollen parent's diploid genotype

Spring annual a plant that germinates and grows in the spring, reproduces in the late spring or early summer, then dies

Spring ephemeral an herbaceous perennial that leafs out, grows, reproduces, and dies back in the spring

Stabilizing selection individuals with intermediate trait values have the highest fitness in a population

Stable growth rate the rate of population growth achieved when the proportion of individuals in each age- or stage-class ceases to change

Stable stage distribution the population structure at which the proportion of individuals in each stage-class stays constant each generation

Stage-based methods used for the study of populations requiring only information on the population's stage structure

Stage-structured a population in which individuals are classified by their size or morphology

Stamen the pollen-bearing organ or microsporophyll of a flowering plant

Staminate flower a flower with functional stamens but not functional pistils

Stand a local area, treated as a unit for the purpose of describing vegetation

Standard deviation (s) the square root of the mean of the square of the deviations of each observation from the mean (also see Variance)

Standard error a measure of the variation in the distribution of a parameter

Standing biomass the total aboveground plant mass in an area

Standing crop the total of living and dead biomass in an area that is not laying on the ground

Static life table a life table built on the assumption that a population has reached a stable age distribution and is no longer growing

Stephan-Boltzmann equation an equation describing the radiant energy emitted by the surface of an object

Stipule an appendage or bract situated at either side of a leaf axil

Stochastic random

Stolon a stem that runs along the ground

Stomatal conductance a measure of how readily water and CO_2 move into or out of stomata

Strategy in evolution, a set of coordinated adaptive traits

Stroma thylakoid a membrane in a chloroplast that connects grana stacks

Structural coloration color caused by the way light bounces off a physical surface with particular optical properties

Substitutive design an experimental test of the relative strength of intraspecific versus interspecific competition by altering the frequencies of two hypothesized competitors while keeping the total density constant

Subtropical an area that rarely experiences temperatures below freezing

Succession directional change in community composition and structure through time

Succulent a plant that stores large amounts of water in leaves, stems or other tissues

Sucker a shoot that arises from a bud on the near-surface roots of a woody plant

Superorganism a theory that communities are analogous to individuals and are born, develop, grow, and eventually die

Supervised classification a classification process that uses remote sensing data which requires that groups be defined explicitly by the user

Surface fire a fire that runs along the soil surface or ground vegetation

Survivorship the chance of surviving

Symbiosis the relationship between two intimately interacting organisms

Sympatric speciation formation of two species from a single population in one location

Synecology the ecological study of whole communities

Tannin a phenol that reduces the digestibility of plant tissues

Taproot a root that extends deep into the soil, and may also be thickened and capable of storing food

Taxon (pl. taxa) a group of organisms that share a common ancestor

Taxonomic species concept a monophyletic group that shares a set of unique traits

Temperature the average random kinetic energy of the molecules in a substance

Temporal pattern diversity the arrangement of subunits in time

Terpene compounds composed of multiple units of isoprene (C_5H_8)

Tetraploid having twice the usual number of chromosomes (4N)

Theory a broad, comprehensive explanation of a large body of information

−3/2 Thinning law a hypothesis that density-dependent mortality results in a consistent relationship between population number and mass of −3/2 on a log-log scale

Throughflow precipitation that passes through the canopy and travels down the surface of tree trunks

Thylakoid membrane a double membrane in a chloroplast that bears photosynthetic pigments consisting of grana stacks alternating with stroma thylakoids

Tiller the shoot of a plant springing from the root

Tolerance a process by which early successional species neither increase nor decrease the survival of later successional species

Tracheid the water-conducting tissue in earlier-evolved vascular plants consisting of elongated cells connected to one another by thinner parts of the end walls through which water diffuses

Trade winds winds moving from 30° N and 30° S toward the equator, from northeast to southwest in the Northern Hemisphere and from southeast to northwest in the Southern Hemisphere

Trade-off increasing one feature of an organism necessitates decreasing another feature

Transect a long line along which samples are taken

Transition zone an area of contact between two communities, regions, or biomes

Transpiration stream the water that passes through a plant, entering at the roots and exiting at the leaves

Tree line an often abrupt demarcation where trees are replaced by low-growing vegetation

Trichome a single-celled plant hair

Tristyly having styles of three different lengths in different individuals in a population; see heterostyly

Tundra an area at high altitude or high latitude dominated by herbaceous plants because the temperature is too cold for tree growth

Turgid firm and fully hydrated

Turnover (of species along a gradient) the average number of species that appear and disappear as you pass from one community to the next along a gradient

Turnover rate (of nutrients or biomass) the total mass of a component divided by its flux in or out of a system

Univariate concerning a single variable

Unsupervised classification a classification process that uses remote sensing data which separates pixels into a user-defined number of classes based on statistical similarities between the pixels (also see polythetic-agglomerative)

Variance (s^2) the mean of the square of the deviations of each observation from the mean (also see standard deviation)

Vascular cambium a large, tubular meristem just inside the stem, responsible for secondary growth of xylem and phloem

Vector an ordered list of numbers

Vegetation lag a change in the vegetation that happens more slowly than the rate at which the climate changes

Vegetative reproduction increase by growth

Vertical life table a life table built on the assumption that a population has reached a stable age distribution and is no longer growing

Vertical structure the pattern of canopy trees, understory trees, shrubs, and herbs in a community

Vesicular-arbuscular mycorrhizae (**VAM**) a group of endomycorrhizae

Vital rates the reproductive rates and chances of surviving of individuals in a population

Water potential the difference in the potential energy between pure water and the water in some system

Water use efficiency the grams of carbon fixed in photosynthesis per gram of water lost in transpiration

Watershed an area that includes a stream or river and all of the land that drains into it

Weather the immediate or short-term climatic conditions

Weed a species growing where it is not wanted; uncultivated species that proliferate in agricultural settings (also see ruderal species)

Westerlies winds moving from west to east from 30° N and 30° S toward the poles

Wet deposition molecules deposited on a surface dissolved in water

Wilting coefficient the soil moisture content at which a plant reaches its permanent wilting point

Windthrow blowdown of a branch, part of tree, whole tree, or group of trees

Winter annual a plant that germinates in the fall, grows through the winter, and reproduces in the spring, then dies

Wisconsin glaciation the most recent glacial advance in North America named for one of the locations of its southernmost extent

Wood secondary xylem

Woodland an area dominated by trees that are spaced apart, varying from closely spaced to sparse

Xeric dry

Xerophyte a plant that live in regions with frequent or extended droughts

Xylem the tissue that carries water throughout a vascular plant

Xylem vessel element a long water-conducting tube found in some angiosperms

Literature Cited

Aber, J. D. 1992. Nitrogen cycling and nitrogen saturation in temperate forest ecosystems. *Trends Ecol. Evol.* 7: 220–223.

Aber, J. D., W. McDowell, K. Nadelhoffer, A. Magill, G. Berntson, M. Kamekea, S. McNulty, W. Currie, L. Rustad and I. Fernandez. 1998. Nitrogen saturation in temperate forest ecosystems: hypotheses revisited. *Bioscience* 48: 921–934.

Abraham, K. F. and R. L. Jefferies. 1997. High goose populations: causes, impacts and implications. In *Arctic Ecosystems in Peril: Report of the Arctic Goose Habitat Working Group*, B. J. Batt (ed.), pp. 7–72. U.S. Fish and Wildlife Service, Washington, DC.

Abrams, P. A. 1988. Resource productivity-consumer species diversity: simple models of competition in spatially heterogeneous environments. *Ecology* 69: 1418–1433.

Abrams, P. A. 1995. Monotonic or unimodal diversity-productivity gradients: what does competition theory predict? *Ecology* 76: 2019–2027.

Adler, L. S., R. Karban and S. Y. Strauss. 2001. Direct and indirect effects of alkaloids on plant fitness via herbivory and pollination. *Ecology* 82: 2032–2044.

Aerts, R. 1995. The advantages of being evergreen. *Trends Ecol. Evol.* 10: 402–407.

Aerts, R. 1999. Interspecific competition in natural plant communities: mechanisms, trade-offs and plant-soil feedbacks. *J. Exper. Bot.* 50: 29–37.

Aerts, R. and F. Berendse. 1988. The effect of increased nutrient availability on vegetation dynamics in wet heathlands. *Vegetatio* 76: 63–69.

Aerts, R., F. Berendse, H. de Caluwe and M. Schmitz. 1990. Competition in heathland along an experimental gradient of nutrient availability. *Oikos* 57: 310–318.

Agrawal, A. A. 2000. Benefits and costs of induced plant defense for *Lepidium virginianum* (Brassicaceae). *Ecology* 81: 1804–1813.

Allan, H. H. 1961. *Flora of New Zealand*, Vol. 1. Government Printer, Wellington, New Zealand.

Allen, T. F. H., G. Mitman and T. W. Hoekstra. 1993. Synthesis mid-century: J. T. Curtis and the continuum concept. In *John T. Curtis Fifty years of Wisconsin plant ecology*, J. S. Fralish, R. P. McIntosh and O. L. Loucks (eds.), pp. 123–143. Wisconsin Academy of Sciences, Arts & Letters, Madison, WI.

Angiosperm Phylogeny Group. 1998. An ordinal classification for the families of flowering plants. *Ann. Missouri Bot. Gard.* 85: 531–553.

Antonovics, J. 1968. Evolution in closely adjacent plant populations. V. Evolution of self-fertility. *Heredity* 23: 219–238.

Antonovics, J. and A. D. Bradshaw. 1970. Evolution in closely adjacent plant populations. VIII. Clinal patterns at a mine boundary. *Heredity* 25: 349–362.

Antonovics, J. and N. L. Fowler. 1985. Analysis of frequency and density effects on growth in mixtures of *Salvia splendens* and *Linum grandiflorum* using hexagonal fan designs. *J. Ecol.* 73: 219–234.

Archibald, E. E. A. 1949. The specific character of plant communities. II. A quantitative approach. *J. Ecol.* 37: 260–274.

Armesto, J. J. and S. T. A. Pickett. 1985. Experiments on disturbance in old-field plant communities: Impact on species richness and abundance. *Ecology* 66: 230–240.

Armstrong, R. A. and R. McGehee. 1980. Competitive exclusion. *Am. Nat.* 115: 151–170.

Arrhenius, O. 1921. Species and area. *J. Ecol.* 9: 95–99.

Arrhenius, S. 1918. *The Destinies of the Stars.* G. P. Putnam's Sons, New York.

Arris, L. L. and P. S. Eagleson. 1989. Evidence of a physiological basis for the boreal-deciduous forest ecotone in North America. *Vegetatio* 82: 55–58.

Augspurger, C. K. 1984. Seedling survival of tropical tree species: interactions of dispersal distance, light-gaps, and pathogens. *Ecology* 65: 1705–1712.

Augspurger, C. K. and S. E. Franson. 1987. Wind dispersal of artificial fruits varying in mass, area, and morphology. *Ecology* 68: 27–42.

Augustine, D. J. and D. A. Frank. 2001. Effects of migratory grazers on spatial heterogeneity of soil nitrogen properties in a grassland ecosystem. *Ecology* 82: 3149–3162.

Austin, M. P. 1977. Use of ordination and other multivariate descriptive methods to study succession. *Vegetatio* 35: 165–175.

Austin, M. P. and L. Belbin. 1982. A new approach to the species classification problem in floristic analysis. *Aust. J. Ecol.* 7: 75–89.

Baker, H. G. 1972. Seed mass in relation to environmental conditions in California. *Ecology* 53: 997–1010.

Baker, H. G. and G. L. Stebbins. 1965. *The Genetics of Colonizing Species.* Academic Press, New York.

Baldwin, I. T. 1988. The alkaloidal responses of wild tobacco to real and simulated herbivory. *Oecologia* 77: 378–381.

Baldwin, I. T. 1991. Damage-induced alkaloids in wild tobacco. In *Phytochemical Induction by Herbivores*, D. W. Tallamy and M. J. Raupp (eds.), pp. 47–69. John Wiley, New York.

Balzter, H., P. W. Braun and W. Kohler. 1998. Cellular automata models for vegetation dynamics. *Ecol. Model.* 107: 113–125.

Banovetz, S. J. and S. M. Scheiner. 1994. Effects of seed size on the seed ecology of *Coreopsis lanceolata*. *Am. Midl. Nat.* 131: 65–74.

Barbour, M. G., J. H. Burke and W. D. Pitts. 1987. *Terrestrial Plant Ecology*, 2nd ed. Benjamin/Cummings Publishing Company, Menlo Park, CA.

Barnola, J. M., D. Raynaud, C. Lorius and N. I. Barkov. 1999. Historical CO_2 record from the Vostok ice core. Trends Online: A Compendium of Data on Global Change. Carbon Dioxide Information Analysis Center, Oak Ridge National Laboratory, U.S. Department of Energy. [http://cdiac.esd.ornl.gov/trends/co2/vostok.htm].

Barrett, S. C. H. and L. D. Harder. 1996. Ecology and evolution of plant mating. *Trends Ecol. Evol.* 11: 73–79.

Barrett, S. C. H., C. G. E. Eckert and B. C. Husband. 1993. Evolutionary processes in aquatic plant populations. *Aquatic Botany* 44: 105–145.

Bartholemew, B. 1970. Bare zones between California shrub and grassland communities: the role of animals. *Science* 170: 1210–1212.

Bartlett, J. G., D. M. Mageean and R. J. O'Conner. 2000. Residential expansion as a continental threat to U.S. Coastal Ecosystems. *Population and Environment* 21: 429–468.

Bateman, R. M., P. R. Crane, W. A. DiMichele, P. R. Kenrick, N. P. Rowe, T. Speck and W. E. Stein. 1998. Early ecology of land plants: phylogeny, physiology, and ecology of the primary terrestrial radiation. *Annu. Rev. Ecol. Syst.* 29: 263–292.

Batty, A. L., K. W. Dixon, M. Brundrett and K. Sivasithamparam. 2001. Constraints to symbiotic germination of terrestrial orchid seed in a Mediterranean bushland. *New Phytol.* 152: 511–520.

Beard, J. S. 1946. The mora forests in Trinidad, British West Indies. *J. Ecol.* 33: 173–192.

Behrensmeyer, A. K., J. D. Damuth, W. A. DiMichele, R. Potts, H.-D. Sues and S. L. Wing. 1992. *Terrestrial Ecosystems Through Time: Evolutionary Paleoecology of Terrestrial Plants and Animals*. University of Chicago Press, Chicago, IL.

Belsky, A. J. 1986. Does herbivory benefit plants? A review of the evidence. *Am. Nat.* 127: 870–892.

Belsky, A. J., W. P. Carson, C. L. Jensen and G. A. Fox. 1993. Overcompensation by plants: herbivore optimization or red herring? *Evol. Ecol.* 7: 109–121.

Berendse, F., N. Van Breemen, H. Rydin, A. Buttler, M. Heijmans, M. R. Hoosbeek, J. A. Lee, E. Mitchell, T. Saarinen, H. Vasander and B. Wallen. 2001. Raised atmospheric CO_2 levels and increased N deposition cause shifts in plant species composition and production in *Sphagnum* bogs. *Global Change Biol.* 7: 591–598.

Betancourt, J. L., T. R. Van Devender and P. S. Martin. 1990. *Packrat Middens: The Last 40,000 Years of Biotic Change*. The University of Arizona Press, Tucson, AZ.

Bewley, J. D. and M. Black. 1985. *Seeds Physiology of Development and Germination*. Plenum, New York.

Bierregaard, R. O., Jr., T. E. Lovejoy, V. Kapos, A. A. dos Santos and R. W. Hutchings. 1992. The biological dynamics of tropical forest fragments. *Bioscience* 42: 859–866.

Bierzychudek, P. 1982. The demography of Jack-in-the-pulpit, a forest perennial that changes sex. *Ecol. Monogr.* 53: 335–351.

Bigger, D. S. and M. A. Marvier. 1998. How different would a world without herbivory be? A search for generality in ecology. *Integrative Biology* 1: 60–67.

Billings, W. D. and H. A. Mooney. 1968. The ecology of arctic and alpine plants. *Biol. Rev.* 43: 481–529.

Biondini, M. E., A. A. Steuter and C. E. Grygiel. 1989. Seasonal fire effects on the diversity patterns, spatial distribution and community structure of forbs in the Northern Mixed Prairie, USA. *Vegetatio* 85: 21–31.

Björkman, O. 1968. Further studies on differentiation of photosynthetic properties in sun and shade ecotypes of *Solidago virgaurea*. *Physiol. Plant.* 21: 84–99.

Björkman, O. and P. Holmgren. 1966. Photosynthetic adaptation to light intensity in plants native to shaded and exposed habitats. *Physiol. Plant.* 19: 854–859.

Boahene, K. 1998. The challenge of deforestation in tropical Africa: reflections on its principal causes, consequences and solutions. *Land Degrad. Devel.* 9: 247–258.

Bonal, D. and J.-M. Guehl. 2001. Contrasting patterns of leaf water potential and gas exchange responses to drought in seedlings of tropical rainforest species. *Func. Ecol.* 15: 490–496.

Borchert, R. 1994. Soil and stem water storage determine phenology and distribution of tropical dry forest trees. *Ecology* 75: 1437–1449.

Bormann, F. H. and G. E. Likens. 1979. *Pattern and Process in a Forested Ecosystem*. Springer-Verlag, New York.

Bowden, C. 1993. *The Secret Forest*. University of New Mexico Press, Albuquerque, NM.

Bowers, M. A. and C. F. Sacchi. 1991. Fungal mediation of a plant herbivore interaction in an early successional plant community. *Ecology* 72: 1032–1037.

Boyce, M. S. 1984. Restitution of r- and K-selection as a model of density-dependent natural selection. *Annu. Rev. Ecol. Syst.* 15: 427–447.

Bradshaw, A. D. 1965. Evolutionary significance of phenotypic plasticity in plants. *Adv. Genet.* 13: 115–155.

Braun, E. L. 1950. *Deciduous Forests of Eastern North America*. Blakiston Co., Philadelphia, PA.

Braun-Blanquet, J. 1932. *Plant Sociology*. McGraw-Hill, New York.

Bray, J. R. 1955. The savanna vegetation of Wisconsin and an application of the concepts order and complexity to the filed of ecology. Ph.D. Dissertation. University of Wisconsin, Madison, WI.

Bray, J. R. and J. T. Curtis. 1957. An ordination of the upland forest communities of southern Wisconsin. *Ecol. Monogr.* 27: 325–349.

Breck, S. W. and S. H. Jenkins. 1997. Use of an ecotone to test the effects of soil and desert rodents on the distribution of Indian ricegrass. *Ecography* 20: 253–263.

Brokaw, N. V. L. 1985. Treefalls, regrowth, and community structure in tropical forests. In *The Ecology of Natural Disturbance and Patch Dynamics*, S. T. A. Pickett and P. S. White (eds.), pp. 53–69. Academic Press, Orlando, FL.

Brokaw, N. and R. T. Busing. 2000. Niche versus chance and tree diversity in forest gaps. *Trends Ecol. Evol.* 15: 183–188.

Brown, B. A. and M. T. Clegg. 1984. Influence of flower color polymorphism on genetic transmission in a natural population of the common morning glory, *Ipomoea purpurea*. *Evolution* 38: 796–803.

Brown, J. H. 1984. On the relationship between abundance and distribution of species. *Am. Nat.* 124: 255–279.

Brown, J. H. 1988. Species diversity. In *Analytical Biogeography*, A. Myers and R. S. Giller (eds.), pp. 57–89. Chapman and Hall, London.

Brown, J. H. 1995. *Macroecology*. University of Chicago Press, Chicago, IL.

Brown, J. H. 1999. Macroecology: progress and prospect. *Oikos* 87: 3–14.

Brown, J. H. and A. Kodric-Brown. 1977. Turnover rates in insular biogeography: effect of immigration on extinction. *Ecology* 58: 445–449.

Brown, J. H. and M. V. Lomolino. 1998. *Biogeography*, 2nd ed. Sinauer Associates, Inc., Sunderland, MA.

Brown, J. H. and J. C. Munger. 1985. Experimental manipulation of a desert rodent community: food addition and species removal. *Ecology* 66: 1545–1563.

Brown, J. H., T. J. Valone and C. G. Curtin. 1997. Reorganization of an arid ecosystem in response to recent climate change. *Proc. Nat. Acad. Sci.* 94: 9729–9733.

Brown, J. S. and D. L. Venable. 1991. Life history evolution of seed-bank annuals in response to seed predation. *Evol. Ecol.* 5: 12–29.

Brown, K. and F. Ekoko. 2001. Forest encounters: synergy among agents of forest change in Southern Cameroon. *Soc. Nat. Resources* 14: 269–290.

Brown, N. D. and S. Jennings. 1998. Gap-size niche differentiation by tropical rainforest trees: a testable hypothesis or a broken-down bandwagon? In *Dynamics of Tropical Communities*, D. M. Newbery and N. D. Brown (eds.), pp. 79–94. Blackwell Science, Oxford.

Brown, V. K. and A. C. Gange. 1992. Secondary plant succession: how is it modified by insect herbivory? *Vegetatio* 101: 3–13.

Brundrett, M. 1991. Mycorrhizas in natural ecosystems. *Adv. Ecol. Res.* 21: 171–313.

Bryant, J., F. S. Chapin III and D. Klein. 1983. Carbon/nutrient balance of boreal plants in relation to vertebrate herbivory. *Oikos* 40: 357–368.

Buckman, H. O. and N. C. Brady. 1969. *The Nature and Properties of Soils*, 7th ed. Macmillan, New York.

Burgman, M. A. 1989. The habitat volumes of scarce and ubiquitous plants: a test of the model of environmental control. *Am. Nat.* 113: 228–239.

Burley, N. and M. F. Willson. 1983. *Mate Choice in Plants*. Princeton University Press, Princeton, NJ.

Burnham, R. J. and A. Graham. 1999. The history of neotropical vegetation: new developments and status. *Ann. Missouri Bot. Gard.* 86: 546–589.

Busing, R. T. and P. S. White. 1997. Species diversity and small-scale disturbance in an old-growth temperate forest: a consideration of gap partitioning concepts. *Oikos* 78: 562–568.

Cadenasso, M. L. and S. T. A. Pickett. 2001. Effects of edge structure on the flux of species into forest interiors. *Cons. Biol.* 15: 91–97.

Caley, M. J. and D. Schluter. 1997. The relationship between local and regional diversity. *Ecology* 78: 70–80.

Callaway, R. M. and E. T. Aschehoug. 2000. Invasive plants versus their new and old neighbors: a mechanism for exotic invasion. *Science* 290: 521–523.

Campbell, G. S. 1977. *An Introduction to Environmental Biophysics*. Springer-Verlag, New York.

Campbell, I. D., K. McDonald, M. D. Flannigan and J. Kringayark. 1999. Long-distance transport of pollen into the Arctic. *Nature* 399: 29–30.

Canham, C. D. and O. L. Loucks. 1984. Catastrophic windthrow in the presettlement forest of Wisconsin. *Ecology* 65: 803–809.

Carlquist, S. 1980. *Hawaii: A Natural History*, 2nd ed. Pacific Tropical Botanical Garden, Honolulu, Hawaii.

Carr, D. E. and M. R. Dudash. 1996. Inbreeding depression in two species of *Mimulus* (Scrophulariaceae) with contrasting mating systems. *Am. J. Bot.* 83: 586–593.

Caswell, H. 2001. *Matrix Population Models*, 2nd ed. Sinauer Associates, Inc, Sunderland, MA.

Caswell, H. and J. E. Cohen. 1993. Local and regional regulation of species-area relations: a patch-occupancy model. In *Species Diversity in Ecological Communities*, R. E. Ricklefs and D. Schluter (eds.), pp. 99–107. University of Chicago Press, Chicago, IL.

Caughley, G. and J. H. Lawton. 1981. Plant-herbivore systems. In *Theoretical Ecology: Principles and Applications*, 2nd ed., R. M. May (ed.), pp. 132–166. Blackwell Scientific, Oxford.

Cavieres, L. A., A. Penaloza, C. Papic and M. Tambutti. 1998. Nurse effect of *Laretia acaulis* (Umbelliferae) in the High Andes of Central Chile. *Revista Chilena De Historia Natural* 71: 337–347.

Cavigelli, M. A. and G. P. Robertson. 2000. The functional significance of denitrifier community composition in a terrestrial ecosystem. *Ecology* 81: 1402–1414.

Cermák, J., J. Jeník, J. Kucera and V. Zidék. 1984. Xylem water-flow in a crack willow tree (*Salix fragilis* L) in relation to diurnal changes of environment. *Oecologia* 64: 145–151.

Chahine, M. T. 1992. The hydrologic cycle and its influence on climate. *Nature* 359: 373–380.

Chapin, F. S., L. R. Walker, C. L. Fastie and L. C. Sharman. 1994. Mechanisms of primary succession following deglaciation at Glacier Bay, Alaska. *Ecol. Monogr.* 64: 149–175.

Chapin, F. S., III. 1989. The cost of tundra plant structures: Evaluation of concepts and currencies. *Am. Nat.* 133: 1–19.

Chapin, F. S., III, P. A. Matson and H. A. Mooney. 2002. *Principles of Terrestrial Ecosystem Ecology*. Springer-Verlag, New York.

Charlesworth, B. and D. Charlesworth. 1978. A model for the evolution of dioecy and gynodioecy. *Am. Nat.* 112: 975–997.

Charnov, E. L. and W. M. Schaffer. 1973. Life history consequences of natural selection: Cole's result revisited. *Am. Nat.* 107: 791–793.

Chazdon, R. L. 1985. Leaf display, canopy structure, and light interception of two understory palm species. *Am. J. Bot.* 72: 1493–1502.

Chazdon, R. L. and R. W. Pearcy. 1991. The importance of sunflecks for forest understory plants. *Bioscience* 41: 760–766.

Chesson, P. L. and N. Huntly. 1988. Community consequences of life history traits in a variable environment. *Ann. Zool. Fenn.* 25: 5–16.

Clark, J. S. 1998. Why trees migrate so fast: Confronting theory with dispersal biology and the paleo record. *Am. Nat.* 152: 204–224.

Clausen, J. D., D. Keck and W. M. Hiesey. 1940. *Experimental studies on the nature of species. I. Effects of varied environments on Western North American plants.* Carnegie Institute of Washington, Washington, DC.

Clausen, J. D., D. Keck and W. M. Hiesey. 1948. *Experimental studies on the nature of species. III. Environmental responses of climatic races of* Achillea. Carnegie Institute of Washington, Washington, DC.

Clauss, M. J. and D. L. Venable. 2000. Seed germination in desert annuals: an empirical test of adaptive bet-hedging. *Am. Nat.* 155: 168–186.

Clements, F. E. 1916. *Plant Succession.* Carnegie Institute of Washington, Washington, DC.

Clements, F. E. 1937. Nature and structure of the climax. *J. Ecol.* 24: 252–284.

Clements, R. E., J. E. Weaver and H. C. Hanson. 1929. Competition in cultivated crops. *Carnegie Inst. Landscape and Urban Planning* 398: 202–233.

Cline, A. C. and S. H. Spurr. 1942. The virgin upland forest of central New England. A study of old growth stands in the Pisgah mountain section of southwestern New Hampshire. Harvard Forest Bulletin, 21.

Clough, J. M., J. A. Teeri and R. S. Alberte. 1979. Photosynthetic adaptation of *Solanum dulcamara* L. to sun and shade environments. I. A comparison of sun and shade populations. *Oecologia* 38: 13–21.

Cochran, M. E. and S. Ellner. 1992. Simple methods for calculating age-based life history parameters for stage-structured populations. *Ecol. Monogr.* 62: 345–364.

Cody, M. 1966. A general theory of clutch size. *Evolution* 20: 174–184.

Cody, M. L. and H. A. Mooney. 1978. Convergence versus nonconvergence in Mediterranean-climate ecosystems. *Annu. Rev. Ecol. Syst.* 9: 265–321.

Cole, K. L. 1985. Past rates of change, species richness, and a model of vegetational inertia in the Grand Canyon, Arizona. *Am. Nat.* 125: 289–303.

Cole, K. L. 1990. Reconstruction of past desert vegetation along the Colorado River using packrat middens. *Palaeogeogr. Palaeoclim. Palaeoecol.* 76: 349–366.

Cole, L. C. 1954. The population consequences of life-history phenomena. *Quart. Rev. Biol.* 29: 103–137.

Coley, P. D. and J. A. Barone. 1996. Herbivory and plant defenses in tropical forests. *Annu. Rev. Ecol. Syst.* 27: 305–335.

Colinvaux, P. A. and P. E. De Oliveira. 2001. Amazon plant diversity and climate through the Cenozoic. *Palaeogeogr. Palaeoclim. Palaeoecol.* 166: 51–63.

Collins, S. L. and S. M. Glenn. 1990. A hierarchical analysis of species' abundance patterns in grassland vegetation. *Am. Nat.* 135: 633–648.

Collins, S. L. and S. M. Glenn. 1991. Importance of spatial and temporal dynamics in species regional abundance and distribution. *Ecology* 72: 654–664.

Colwell, R. K. and G. C. Hurtt. 1994. Nonbiological gradients in species richness and a spurious Rapoport effect. *Am. Nat.* 144: 570–595.

Connell, J. H. 1978. Diversity in tropical rainforests and coral reefs. *Science* 199: 1302–1310.

Connell, J. H. and R. O. Slatyer. 1977. Mechanisms of succession in natural communities and their role in community stability and organization. *Am. Nat.* 111: 1119–1144.

Connolly, J. 1986. On difficulties with replacement-series methodology in mixture experiments. *J. Appl. Ecol.* 23: 125–137.

Connolly, J. 1987. On the use of response models in mixture experiments. *Oecologia* 72: 95–103.

Connolly, J. 1997. Substitutive experiments and the evidence for competitive hierarchies in plant communities. *Oikos* 80: 179–182.

Cook, E. A., L. R. Iverson and R. L. Graham. 1989. Estimating forest productivity with thematic mapper and biogeographical data. *Remote Sens. Envir.* 28: 131–141.

Cooper, W. S. 1923. The recent ecological history of Glacier Bay, Alaska. II. The present vegetation cycle. *Ecology* 4: 223–246.

Cowles, H. C. 1899. The ecological relations of the vegetation on the sand dunes of Lake Michigan. *Bot. Gaz.* 27: 95–117, 167–202, 281–308, 361–391.

Cowles, H. C. 1911. The causes of vegetative cycles. *Bot. Gaz.* 51: 161–183.

Cruden, R. W. 1977. Pollen-ovule ratios: a conservative indicator of the breeding systems in flowering plants. *Evolution* 31: 32–46.

Csillag, F., Marie-J. Fortin and J. L. Dungan. 2000. On the limits and extensions of the definition of scale. *Bull. Ecol. Soc. Amer.* 81: 230–232.

Currie, D. J. and V. Paquin. 1987. Large-scale biogeographical patterns of species richness of trees. *Nature* 329: 326–327.

Curtis, J. T. 1959. *The Vegetation of Wisconsin.* The Univerity of Wisconsin Press, Madison, WI.

Curtis, J. T. and R. P. McIntosh. 1951. An upland forest continuum in the prairie-forest border region of Wisconsin. *Ecology* 32: 476–498.

Curtis, P. S. and X. Z. Want. 1998. A meta-analysis of elevated CO_2 effects on woody plant mass, form and physiology. *Oecologia* 113: 299–313.

Dale, M. R. T. 2000. *Spatial Pattern Analysis in Plant Ecology.* Cambridge University Press, Cambridge.

Dansereau, P. 1951. Description and recording of vegetation upon a structural basis. *Ecology* 32: 172–229.

D'Antonio, C. M. and P. M. Vitousek. 1992. Biological invasions by exotic grasses, the grass/fire cycle, and global change. *Anu. Rev. Ecol. Syst.* 23: 63–87.

Darwin, C. 1876. *The Effects of Cross and Self Fertilisation in the Vegetable Kingdom.* John Murray, London.

Darwin, C. 1877. *The Various Contrivances by which Orchids are Fertilized.* John Murray, London.

Davidowitz, G. 2002. Does precipitation variability increases from mesic to xeric biomes? *Global Ecol. Biogeogr.* 11: 143–154.

Davidson, C. B., K. W. Gottschalk and J. E. Johnson. 1999. Tree mortality following defoliation by the European gypsy moth (*Lymantria dispar* L.) in the United States: a review. *Forest Sci.* 45: 74–84.

Davidson, E. A., S. C. Hart and M. K. Firestone. 1992. Internal cycling of nitrate in soils of a mature coniferous forest. *Ecology* 73: 1148–1156.

Davis, K. M. and R. W. Mutch. 1994. Applying ecological principles to manage wildland fire. In: Fire in Ecosystem Management, training course (National Advanced Resources Technology Center).

Davis, M. B. 1981. Quaternary history and the stability of forest communities. In *Forest Suc-*

cession Concepts and Applications, D. C. West, H. H. Shugart and D. B. Botkin (eds.), pp. 132–153. Springer-Verlag, New York.

Davis, M. B. 1996. *Eastern Old-growth Forests.* Island Press, Washington, DC.

Davis, M. B. and R. G. Shaw. 2001. Range shifts and adaptive responses to Quaternary climate change. *Science* 292: 673–679.

Davis, M. B., R. R. Calcote, S. Sugita and H. Takahara. 1998. Patchy invasion and the origin of a hemlock-hardwood forest mosaic. *Ecology* 79: 2641–2659.

del Moral, R. 1998. Early succession on lahars spawned by Mount St. Helens. *Am. J. Bot.* 85: 820–828.

Denslow, J. S. 1980. Gap partitioning among tropical rainforest trees. *Biotropica (Supplement)* 12: 47–55.

Detling, J. K., M. I. Dyer, C. Procter-Gregg and D. A. Winn. 1980. Plant-herbivore interactions: examination of potential effects of buffalo saliva on regrowth of *Bouteloua gracilis* (H. B. K.) Lag. *Oecologia* 45: 26–31.

de Wit, C. T. 1960. On competition. *Versl. Landbouw. Onderz.* 66: 1–82.

Digby, P. G. N. and R. A. Kempton. 1987. *Multivariate Analysis of Ecological Communities.* Chapman and Hall, London.

DiMichele, W. A., H. W. Pfefferkorn and R. A. Gastaldo. 2001. Response of Late Carboniferous and Early Permian plant communities to climate change. *Annu. Rev. Earth Plant. Sci.* 29: 461–487.

Dirmeyer, P. A. and J. Shukla. 1996. The effect on regional and global climate of expansion of the world's deserts. *Quart. J. Roy. Meteor. Soc. B* 122: 451–482.

Dirzo, R. and J. L. Harper. 1982a. Experimental studies on slug-plant interactions: 3. Differences in the acceptability of individual plants of *Trifolium repens* to slugs and snails. *J. Ecol.* 70: 101–117.

Dirzo, R. and J. L. Harper. 1982b. Experimental studies on slug-plant interactions. 4. The performance of cyanogenic and acyanogenic morphs of *Trifolium repens* in the field. *J. Ecol.* 70: 119–138.

Dixon, P. M. 2001. The bootstrap and the jackknife: describing the precision of ecological indices. In *Design and Analysis of Ecological Experiments*, 2nd ed., S. M. Scheiner and J. Gurevitch (eds.), pp. 267–288. Oxford University Press, New York.

Doak, D. F., D. Bigger, E. K. Harding, M. A. Marvier, R. E. O'Malley and D. Thomson. 1998. The statistical inevitability of stability-diversity relationships in community ecology. *Am. Nat.* 151: 264–276.

Donald, C. M. 1951. Competition among pasture plants. I. Intra-specific competition among annual pasture plants. *Aust. J. Agric. Res.* 2: 355–376.

Drury, C. F., R. P. Voroney and E. G. Beauchamp. 1991. Availability of NH_4^+-N to microorganisms and the soil internal N cycle. *Soil Biol. Biochem.* 23: 165–169.

Dudley, S. A. 1996. Differing selection on plant physiological traits in response to environmental water availability: a test of adaptive hypotheses. *Evolution* 50: 92–102.

Dudley, S. A. and J. Schmitt. 1995. Genetic differentiation in morphological responses to simulated foliage shade between populations of *Impatiens capensis* from open and woodland sites. *Func. Ecol.* 9: 655–666.

Dudley, S. A. and J. Schmitt. 1996. Testing the adaptive plasticity hypothesis: density-dependent selection on manipulated stem length in *Impatiens capensis. Am. Nat.* 147: 445–465.

Dufrêne, M. and P. Legendre. 1997. Species assemblages and indicator species: the need for a flexible asymmetrical approach. *Ecol. Monogr.* 67: 345–366.

Dukes, J. S. and H. A. Mooney. 1999. Does global change increase the success of biological invaders? *Trends Ecol. Evol.* 14: 135–139.

Dunwiddie, P., D. Foster, D. Leopold and R. T. Leverett. 1996. Old-growth forests of southern New England, New York, and Pennsylvania. In *Eastern Old-Growth Forests: Prospects for Rediscovery and Recovery*, M. D. Davis (ed.), pp. 126–143. Island Press, Washington, DC.

Easterling, M. R., S. P. Ellner and P. M. Dixon. 2000. Size-specific sensitivity: Applying a new structured population model. *Ecology* 81: 694–708.

Eckert, C. G., D. Manicacci and S. C. H. Barrett. 1996. Frequency-dependent selection on morph ratios in tristylous *Lythrum salicaria* (Lythraceae). *Heredity* 77: 581–588.

Egler, F. E. 1954. Vegetation science concepts I. Initial floristic composition. A factor in old-field vegetation development. *Vegetatio* 4: 412–417.

Ehleringer, J. 1985. Annuals and perennials of warm deserts. In *Physiological Ecology of North American Plant Communities*, B. F. Chabot and H. A. Mooney (eds.), pp. 162–180. Chapman and Hall, New York.

Ehleringer, J. and I. Forseth. 1980. Solar tracking by plants. *Science* 210: 1094–1098.

Ehleringer, J. R. and R. K. Monson. 1993. Evolutionary and ecological aspects of photosynthetic pathway variation. *Annu. Rev. Ecol. Syst.* 24: 411–439.

Ehleringer, J., O. Björkman and H. A. Mooney. 1976. Leaf pubescence: effects on absorptance and photosynthesis in a desert shrub. *Science* 192: 376–377.

Ehrhardt-Martinez, K. 1998. Social determinants of deforestation in developing countries: A cross-national study. *Soc. Forces* 77: 567–586.

Ehrlich, P. R. and L. C. Birch. 1967. The "balance of nature" and "population control". *Am. Nat.* 101: 97–107.

Eltahir, E. A. B. and R. L. Bras. 1994. Sensitivity of regional climate to deforestation in the Amazon Basin. *Adv. Water Res.* 17: 101–115.

Elton, C. S. 1958. *The Ecology of Invasions by Animals and Plants*. Chapman and Hall, New York.

Ely, L. L., Y. Enzel, V. R. Baker and D. R. Cayan. 1993. A 5000-year record of extreme floods and climate change in the southwestern United States. *Science* 262: 410–412.

Emlen, J. M. 1984. *Population Biology: The coevolution of Population Dynamics and Behavior*. Macmillan, New York.

Endler, J. A. 1986. *Natural Selection in the Wild*. Princeton University Press, Princeton, NJ.

Ennos, R. A. and M. T. Clegg. 1982. Effect of population substructuring on estimates of outcrossing rate in plant populations. *Heredity* 48: 283–292.

Epperson, B. K. and M. T. Clegg. 1986. Spatial-autocorrelation analysis of flower color polymorphisms within substructured populations of morning glory (*Ipomoea purpurea*). *Am. Nat.* 128: 840–858.

Esau, K. 1977. *Anatomy of Seed Plants*. John Wiley, New York.

Etherington, J. R. 1982. *Environment and Plant Ecology*, 2nd ed. John Wiley and Sons, New York.

Faegri, K. and L. van der Pijl. 1979. *The Principles of Pollination Ecology*, 3rd ed. Pergammon, New York.

Fahrig, L. 1998. When does fragmentation of breeding habitat affect population survival? *Ecol. Model.* 105: 273–292.

Fauth, J. E., J. Bernado, M. Camara, W. J. Resetarits Jr., J. Van Buskirk and S. A. McCollum. 1996. Simplifying the jargon of community ecology: a conceptual approach. *Am. Nat.* 147: 282–286.

Fearnside, P. M. 2000. Global warming and tropical land-use change: Greenhouse gas emissions from biomass burning, decomposition and soils in forest conversion, shifting cultivation and secondary vegetation. *Climatic Change* 46: 115–158.

Feder, M. E., A. F. Bennett, W. W. Burggren and R. B. Huey. 1987. *New Directions in Ecological Physiology*. Cambridge University Press, Cambridge.

Fine, P. V. 2001. An evaluation of the geographic area hypothesis using a latitudinal gradient in Noth American tree diversity. *Evol. Ecol. Res.* 3: 413–428.

Firbank, L. G. and A. R. Watkinson. 1985. On the analysis of competition within 2-species mixtures of plants. *J. Appl. Ecol.* 22: 503–517.

Firbank, L. G. and A. R. Watkinson. 1987. On the analysis of competition at the level of the individual plant. *Oecologia* 71: 308–317.

Firbank, L. G. and A. R. Watkinson. 1990. On the effects of competition: from monocultures to mixtures. In *Perspectives on Plant Competition*, J. B. Grace and D. Tilman (eds.), pp. 165–192. Academic Press, New York.

Fitter, A. H. and R. K. M. Hay. 1981. *Environmental Physiology of Plants*. Academic Press, London.

Fitzhugh, R. D., C. T. Driscoll, P. M. Groffman, G. L. Tierney, T. J. Fahey and J. P. Hardy. 2001. Effects of soil freezing disturbance on soil solution nitrogen, phosphorus, and carbon chemistry in a northern hardwood ecosystem. *Biogeochemistry* 56: 215–238.

Florence, J. 1981. Chablis et sylvigenèse dans une forêt dense humid sempervirente du Gabon. Ph.D. Dissertation. Université Louis Pasteur de Strasbourg, Strasbourg, France.

Flores-Martinez, A., E. Ezcurra and S. Sanchez-Colon. 1998. Water availability and the competitive effect of a columnar cactus on its nurse plant. *Acta Oecol.* 19: 1–8.

Forbes, R. D. 1930. *Timber Growing and Logging and Turpentining Practices in the Southern Pine Region*. Technical Bulletin No. 204, U.S. Department of Agriculture.

Ford, E. D. 1975. Competition and stand structure in some even-aged plant monocultures. *J. Ecol.* 63: 311–333.

Forman, R. T. T. 1995. *Land Mosaics*. Cambridge University Press, Cambridge.

Forman, R. T. T. and M. Godron. 1986. *Landscape Ecology*. Wiley, New York.

Forseth, I. and J. Ehleringer. 1982. Ecophysiology of two solar tracking desert winter annuals. 2. Leaf movements, water relations and microclimate. *Oecologia* 54: 41–49.

Foster, D. R. 1988. Disturbance history, community organization and vegetation dynamics of the old-growth Pisgah forest, southwestern New Hampshire, U.S.A. *J. Ecol.* 76: 105–134.

Fowler, N. 1990. The 10 most common statistical errors. *Bull. Ecol. Soc. Amer.* 71: 161–164.

Fowler, N. L. 1995. Density-dependent demography in two grasses: a 5-year study. *Ecology* 76: 2145–2164.

Fox, G. A. 1989. Consequences of flowering-time variation in a desert annual: adaptation and history. *Ecology* 70: 1294–1306.

Fox, G. A. 1990a. Components of flowering time variation in a desert annual. *Evolution* 44: 1404–1423.

Fox, G. A. 1990b. Drought and the evolution of flowering time in desert annuals. *Am. J. Bot.* 77: 1508–1518.

Fox, G. A. 1992. The evolution of life history traits in desert annuals: adaptation and constraint. *Evol. Trends Plants* 6: 25–31.

Fox, G. A. and J. Gurevitch. 2000. Population numbers count: tools for near-term demographic analysis. *Am. Nat.* 156: 242–255.

Francis, J. M. 1999. A multiscale analysis and model of vegetation change in a semiarid landscape. Ph.D. Dissertation. Arizona State University, Tempe, AZ.

Franklin, J. F. 1990. Biological legacies: a critical management concept from Mt. St. Helens. In *Transactions of the Fifty-fifth North American Wildlife and Natural Resources Conference, March 16–21, 1990, Denver, CO*, R. E. McCabe (ed.), pp. 215–219. Wildlife Management Institute, Washington, DC.

Frenzen, P. M., J. E. Means, J. F. Franklin, C. W. Kiilsgaard, W. A. McKee and F. J. Swanson. 1986. Five years of plant succession on eleven major surface types affected by the 1980 eruptions of Mount St. Helens, Washington. In *Mount St. Helens: Five Years Later*, S. A. C. Keller (ed.), Eastern Washington University Press, Cheney, WA.

Galen, C. and M. L. Stanton. 1991. Consequences of emergence phenology for reproductive success in *Ranunculus adoneus* (Ranunculaceae). *Am. J. Bot.* 78: 978–988.

Gaston, K. J. 1994. *Rarity*. Chapman & Hall, New York.

Gates, D. M. 1962. *Energy Exchange in the Biosphere*. Harper and Row, New York.

Gates, D. M. 1993. *Climate Change and Its Biological Consequences*. Sinauer Associates, Sunderland, MA.

Gauch, H. G., Jr. 1982. *Multivariate Analysis in Community Ecology*. Cambridge University Press, London.

Gauch, H. G., Jr. and R. H. Whittaker. 1972. Comparison of ordination techniques. *Ecology* 53: 868–875.

Gentry, A. H. 1986. Endemism in tropical versus temperate plant communities. In *Conservation Biology: The Science of Scarcity and Diversity*, M. Soulé (ed.), pp. 153–181. Sinauer Associates, Inc, Sunderland, MA.

Gerdol, R., L. Brancaleoni, R. Marchesini and L. Bragazza. 2002. Nutrient and carbon relations in subalpine dwarf shrubs after neighbour removal or fertilization in northern Italy. *Oecologia* 130: 476–483.

Gholz, H. L., D. A. Wedin, S. M. Smitherman, M. E. Harmon and W. J. Parton. 2000. Long-term dynamics of pine and hardwood litter in contrasting environments: toward a global model of decomposition. *Global Change Biol.* 6: 751–765.

Gibson, D. J., J. Connolly, D. C. Hartnett and J. D. Weidenhamer. 1999. Designs for greenhouse studies of interactions between plants. *J. Ecol.* 87: 1–16.

Gilpin, M. and I. Hanski. 1991. *Metapopulation Dynamics: Empirical and Theoretical Investigations*. Academic Press, London.

Gleason, H. A. 1917. The structure and development of the plant association. *Bull. Torrey Bot. Club* 43: 463–481.

Gleason, H. A. 1922. On the relation between species and area. *Ecology* 3: 158–162.

Gleason, H. A. 1926. The individualistic concept of the plant association. *Bull. Torrey Bot. Club* 53: 7–26.

Gleick, P. H. 1996. Water resources. In *Encyclopedia of Climate and Weather*, S. H. Schneider (ed.), pp. 817–823. Oxford University Press, New York.

Goldberg, D. E. 1990. Components of resource competition in plant communities. In *Perspectives on Plant Competition*, J. B. Grace and D. Tilman (eds.), pp. 27–49. Academic Press, New York.

Goldberg, D. E. and A. M. Barton. 1992. Patterns and consequences of interspecific competition in natural communities: a review of field experiments with plants. *Am. Nat.* 139: 771–801.

Goldberg, D. E. and K. Landa. 1991. Competitive effect and response: hierarchies and correlated traits in the early stages of competition. *J. Ecol.* 79: 1013–1030.

Goldberg, D. and A. Novoplansky. 1997. On the relative importance of competition in unproductive environments. *J. Ecol.* 85: 409–418.

Goldberg, D. E. and S. M. Scheiner. 2001. ANOVA and ANCOVA: Field competition experiments. In *Design and Analysis of Ecological Experiments*, 2nd ed., S. M. Scheiner and J. Gurevitch (eds.), pp. 77–98. Oxford University Press, New York.

Goldberg, D. E. and R. M. Turner. 1986. Vegetation change and plant demography in permanent plots in the Sonoran Desert. *Ecology* 67: 695–712.

Goldberg, D. E. and P. A. Werner. 1983. The equivalence of competitors in plant communities: A null hypothesis and a field experimental approach. *Am. J. Bot.* 70: 1098–1104.

Goldberg, D. E., T. Rajaniemi, J. Gurevitch and A. Stewart-Oaten. 1999. Empirical approaches to quantifying interaction intensity: competition and facilitation along productivity gradients. *Ecology* 80: 1118–1131.

Goodall, D. W. 1954. Objective methods for the classification of vegetation. III. an essay in the use of factor analysis. *Aust. J. Bot.* 2: 304–324.

Goodman, D. 1982. Optimal life histories, optimal notation, and the value of reproductive value. *Am. Nat.* 119: 803–823.

Gotelli, N. J. 1991. Metapopulation models: the rescue effect, the propagule rain, and the core-satellite hypothesis. *Am. Nat.* 138: 768–776.

Gotelli, N. J. 2001. *A Primer of Ecology*, 3rd ed. Sinauer Associates, Inc., Sunderland, MA.

Gotelli, N. J. and D. Simberloff. 1987. The distribution and abundance of tallgrass prairie plants: a test of the core-satellite hypothesis. *Am. Nat.* 130: 18–35.

Gough, L., C. W. Osenberg, K. L. Gross and S. L. Collins. 2000. Fertilization effects on species density and primary productivity in herbaceous plant communities. *Oikos* 89: 428–439.

Gould, S. J. and R. Lewontin. 1979. The spandrels of San Marco and the Panglossian paradigm. *Proc. Royal Soc. London B* 205: 581–598.

Grace, J. B. 1991. A clarification of the debate between Grime and Tilman. *Func. Ecol.* 5: 585–587.

Grace, J. B. 1995. On the measurement of plant competition intensity. *Ecology* 76: 305–308.

Grace, J. B. 1999. The factors controlling species density in herbaceous plant communities: an assessment. *Pers. Plant Ecol. Evol. Syst.* 2: 1–28.

Grace, J. B. and D. Tilman. 1990. *Perspectives on Plant Competition*. Academic Press, New York.

Grace, J. B., J. Keough and G. R. Guntenspergen. 1992. Size bias in traditional analyses of substitutive competition experiments. *Oecologia* 90: 429–434.

Gray, J. T. 1983. Nutrient use by evergreen and deciduous shrubs in southern California. I. Community nutrient cycling and nutrient-use efficiency. *J. Ecol.* 71: 21–41.

Greene, D. F. and E. A. Johnson. 1995. Long-distance wind dispersal of tree seeds. *Can. J. Bot.* 73: 1036–1045.

Greig-Smith, P. 1964. *Quantitative Plant Ecology*. Butterworths, London.

Greig-Smith, P. 1980. The development of numerical classification and ordination. *Vegetatio* 42: 1–9.

Greig-Smith, P. 1983. *Quantitative Plant Ecology*, 3rd ed. University of California Press, Berkeley, CA.

Greller, A. M. 1988. Deciduous Forest. In *North American Terrestrial Vegetation*, M. G. Barbour and W. D. Billings (eds.), pp. 287–316. Cambridge University Press, Cambridge.

Grime, J. P. 1973. Competitive exclusion in herbaceous vegetation. *Nature* 242: 344–347.

Grime, J. P. 1977. Evidence for the existence of three primary strategies in plants and its relevance to ecological and evolutionary theory. *Am. Nat.* 11: 1169–1194.

Grime, J. P. 1979. *Plant Strategies and Vegetation Processes.* John Wiley and Sons, New York.

Grinnell, J. 1917. The niche-relationships of the California thrasher. *The Auk* 34: 427–433.

Groom, P. K. and B. B. Lamont. 1997. Fruit-seed relations in Hakea: serotinous species invest more dry matter in predispersal seed protection. *Aust. J. Ecol.* 22: 352–355.

Grover, J. P. 1997. *Resource Competition.* Chapman and Hall, London.

Grubb, P. J. 1977. The maintenance of species richness in plant communities: the importance of the regeneration niche. *Biol. Rev.* 52: 107–145.

Guo, Q. F. and J. H. Brown. 1996. Temporal fluctuations and experimental effects in desert plant communities. *Oecologia* 107: 568–577.

Gurevitch, J. 1986. Competition and the local distribution of the grass *Stipa neomexicana*. *Ecology* 67: 46–57.

Gurevitch, J. 1988. Variation in leaf dissection and leaf energy budgets among poulations of *Achillea* from an altitudinal gradient. *Am. J. Bot.* 75: 1298–1306.

Gurevitch, J. 1992. Sources of variation in leaf shape among two populations of *Achillea lanulosa*. *Genetics* 130: 385–394.

Gurevitch, J. and P. H. Schuepp. 1990. Boundary layer properties of highly dissected leaves: an investigation using an electrochemical fluid tunnel. *Plant Cell Environ.* 13: 783–792.

Gurevitch, J., P. Willson, J. L. Stone, P. Teese and R. J. Stoutenburgh. 1990. Competition among old-field perennials at different levels of soil fertility and available space. *J. Ecol.* 78: 727–744.

Gurevitch, J., L. L. Morrow, A. Wallace and J. S. Walsh. 1992. A meta-analysis of field experiments on competition. *Am. Nat.* 140: 539–572.

Gurevitch, J., P. Curtis and M. H. Jones. 2001. Meta-analysis in ecology. *Adv. Ecol. Res.* 32: 199–247.

Gustafsson, A. 1946. Apomixis in higher plants. I. The mechanism of apomixis. *Lunds Univ. Arssskr.* 42: 1–68.

Gustafsson, A. 1947a. Apomixis in higher plants. II. The causal aspects of apomixis. *Lunds Univ. Arssskr.* 43: 69–180.

Gustafsson, A. 1947b. Apomixis in higher plants. III. Biotype and species formation. *Lunds Univ. Arssskr.* 43: 181–370.

Gutiérrez, J. R., P. L. Meserve, S. Herrera, L. C. Contreras and F. M. Jaksic. 1997. Effects of small mammals and vertebrate predators on vegetation in the Chilean semiarid zone. *Oecologia* 109: 398–406.

Haberle, S. G. 1999. Late Quaternary vegetation and climate change in the Amazon basin based on a 50,000 year pollen record from the Amazon fan, ODP site 932. *Quatern. Res.* 51: 27–38.

Hairston, N. G., F. E. Smith and L. B. Slobodkin. 1960. Community structure, population control, and competition. *Am. Nat.* 94: 421–425.

Hairston, N. G., Jr. and N. G. Hairston Sr. 1993. Cause-effect relationships in energy flow, trophic structure, and interspecific interactions. *Am. Nat.* 142: 379–411.

Hairston, N. G., Sr. 1989. *Ecological Experiments: Purpose, Design, and Execution.* Cambridge University Press, Cambridge.

Hallé, F., R. A. A. Oldeman and P. B. Tomlinson. 1978. *Tropical Trees and Forests: An Architectural Analysis.* Springer-Verlag, Berlin.

Halpern, C. B., P. M. Frenzen, J. E. Means and J. F. Franklin. 1990. Plant succession in areas of scorched and blown-down forest after the 1980 eruption of Mount St.Helens, Washington. *J. Veg. Sci.* 1: 181–195.

Hamilton, M. B. 1999. Tropical tree gene flow and seed dispersal. *Nature* 401: 129–130.

Handel, S. N. 1983. Contrasting gene flow patterns and genetic subdivision in adjacent populations of *Cucumis sativus* (Cucurbitaceae). *Evolution* 37: 760–771.

Hanley, M. E. and B. B. Lamont. 2000. Herbivory, serotiny and seedling defence in Western Australian Proteaceae. *Oecologia* 126: 409–417.

Hanski, I. 1982. Dynamics of regional distribution: the core and satellite hypothesis. *Oikos* 38: 210–221.

Hanski, I. 1999. *Metapopulation Ecology.* Oxford University Press, New York.

Hardin, G. 1968. The tragedy of the commons. *Science* 162: 1243–1248.

Harper, J. L. 1977. *Population Biology of Plants.* Academic Press, London.

Harrison, S. 1999. Local and regional diversity in a patchy landscape: native, alien, and endemic herbs on serpentine. *Ecology* 80: 70–80.

Hartshorn, G. S. 1975. A matrix model of tree population dynamics. In *Tropical Ecological Systems*, F. B. Golley and E. Medin (eds.), pp. 41–51. Springer-Verlag, New York.

Hastings, A. 1996. Models of spatial spread: is the theory complete? *Ecology* 77: 1675–1679.

Hastings, J. R. and R. M. Turner. 1965. *The Changing Mile.* University of Arizona Press, Tucson, AZ.

Hawkes, C. V. and J. J. Sullivan. 2001. The impact of herbivory on plants in different resource conditions: a meta-analysis. *Ecology* 82: 2045–2085.

Hazlett, D. L. 1992. Leaf area development of four plant communities in the Colorado steppe. *Am. Midl. Nat.* 127: 276–289.

Hector, A., B. Schmid, C. Beierkuhnlein, M. C. Caldeira, M. Diemer, P. G. Dimitrakopoulos, J. A. Finn, H. Freitas, P. S. Giller, J. Good, R. Harris, P. Högberg, K. Huss-Danell, J. Joshi, A. Jumpponen, C. Körner, P. W. Leadley, M. Loreau, A. Minns, C. P. H. Mulder, G. O'Donovan, S. J. Otway, J. S. Pereira, A. Prinz, D. J. Read, M. Scherer-Lorenzen, E.-D. Schulze, A.-S. D. Siamantziouras, E. M. Spehn, A. C. Terry, A. Y. Troumbis, F. I. Woodward, S. Yachi and J. H. Lawton. 1999. Plant diversity and productivity experiments in European grasslands. *Science* 286: 1123–1127.

Hedges, L. V., J. Gurevitch and P. Curtis. 1999. The meta-analysis of response ratios in experimental ecology. *Ecology* 80: 1150–1156.

Henneman, M. L. and J. Memmott. 2001. Infiltration of a Hawaiian community by introduced biological control agents. *Science* 293: 1314–1316.

Herben, T. and F. Krahulec. 1990. Competitive hierarchies, reversals of rank order and the deWit approach: are they compatible. *Oikos* 58: 254–256.

Herrera, C. M. 1982a. Defense of ripe fruits from pests: its significance in relation to pest-disperser interactions. *Am. Nat.* 120: 218–241.

Herrera, C. M. 1982b. Some comments on Stiles' paper on bird-disseminated fruits. *Am. Nat.* 120: 819–822.

Herrera, C. M., P. Jordano, J. Guitian and A. Traveset. 1998. Annual variability in seed production by woody plants and the masting concept: reassessment of principles and relationship to pollination and seed dispersal. *Am. Nat.* 152: 576–594.

Hett, J. M. and O. L. Loucks. 1976. Age structure models of balsam fir and eastern hemlock. *J. Ecol.* 64: 1029–1044.

Hickman, J. C. and L. F. Pitelka. 1975. Dry weight indicates energy allocation in ecological strategy analysis of plants. *Oecologia* 221: 117–121.

Higgins, S. I., D. M. Richardson, R. M. Cowling and T. H. Trinder-Smith. 1999. Predicting the landscape-scale distribution of alien plants and their threat to plant diversity. *Cons. Biol.* 13: 303–313.

Hobbs, R. J. 1989. The nature and effects of disturbance relative to invasions. In *Biological Invasions: A Global Perspective*, J. A. Drake, H. A. Mooney, F. di Castri, R. H. Groves, F. J. Kruger, M. Rejmánek and M. Williamson (eds.), pp. 389–403. John Wiley & Sons, Chichester, UK.

Holbrook, N. M. and F. E. Putz. 1996. From epiphyte to tree: differences in leaf structure and leaf water relations associated with the transition in growth form in eight species of hemiepiphytes. *Plant Cell Environ.* 19: 631–642.

Holt, R. D. 1977. Predation, apparent competition, and structure of prey communities. *Theor. Pop. Biol.* 12: 197–229.

Holt, R. D. 1997. From metapopulation dynamics to community structure: some consequences of spatial heterogeneity. In *Metapopulation Biology: Ecology, Genetics, and Evolution*, I. Hanski and M. E. Gilpin (eds.), pp. 149–165. Academic Press, San Diego, CA.

Hooper, D. U. and P. M. Vitousek. 1998. Effects of plant composition and diversity on nutrient cycling. *Ecol. Monogr.* 68: 121–149.

Horn, H. S. 1971. *The Adaptive Geometry of Trees*. Princeton University Press, Princeton, NJ.

Horvitz, C. C. and D. W. Schemske. 1990. Spatiotemporal variation in insect mutualists of a neotropical herb: the myth of tropical stability. *Ecology* 71: 1085–1097.

Houghton, R. A. and J. L. Hackler. 2001. Carbon Flux to the Atmosphere from Land-Use Changes: 1850 to 1990. ORNL/CDIAC–131, NDP-050/R1. Carbon Dioxide Information Analysis Center, U.S. Department of Energy, Oak Ridge National Laboratory, Oak Ridge, TN.

Houle, D. 1991. Genetic covariance of fitness correlates: what genetic correlations are made of and why it matters. *Evolution* 45: 630–648.

Howard, T. and D. E. Goldberg. 2001. Competitive response hierarchies for germination, growth, and survival and their influence on abundance. *Ecology* 82: 979–990.

Howe, H. F. 1995. Succession and fire season in experimental prairie plantings. *Ecology* 76: 1917–1925.

Howe, H. F. and L. C. Westley. 1997. Ecology of pollination and seed dispersal. In *Plant Ecology*, M. J. Crawley (ed.), pp. 262–283. Blackwell Science, Oxford.

Huang, Y., F. A. Street-Perrott, S. E. Metcalfe, M. Brenner, M. Moreland and K. H. Freeman. 2001. Climate change as the dominant control on glacial-interglacial variations in C_3 and C_4 plant abundance. *Science* 293: 1647–1651.

Hughes, L. 2000. Biological consequences of global warming: is the signal already apparent? *Trends Ecol. Evol.* 15: 56–61.

Hult, R. 1885. Blekinges vegetation. Ett bidrag till växtformationernas utvecklingshistorie. *Medd. Soc. Fenn.* 12: 161.

Huntington, T. G. 2000. The potential for calcium depletion in forest ecosystems of southeastern United States: review and analysis. *Global Biogeo. Cycles* 14: 623–638.

Huntly, N. J. 1987. Effects of refuging consumers (pikas: *Ochotona princeps*) on subalpine vegetation. *Ecology* 68: 274–283.

Husband, B. C. and D. W. Schemske. 1996. Evolution of the magnitude and timing of inbreeding depression in plants. *Evolution* 50: 54–70.

Huston, M. 1979. A general hypothesis of species diversity. *Am. Nat.* 113: 81–101.

Huston, M. A. 1994. *Biological Diversity: the coexistence of species in changing landscapes*. Cambridge University Press, Cambridge.

Huston, M. A. and D. L. DeAngelis. 1994. Competition and coexistence: the effects of resource transport and supply rates. *Am. Nat.* 144: 954–977.

Huston, M. and T. Smith. 1987. Plant succession: life history and competition. *Am. Nat.* 130: 168–198.

Hutchinson, G. E. 1957. Concluding remarks. *Cold Spring Harbor Symp. Quant. Biol.* 22: 415–427.

Hutchinson, G. E. 1959. Homage to Santa Rosalia, or why are there so many kinds of animals? *Am. Nat.* 93: 145–159.

Hutchinson, G. E. 1965. *The Ecological Theater and the Evolutionary Play*. Yale University Press, New Haven, CT.

Huxman, T. E. and S. D. Smith. 2001. Photosynthesis in an invasive grass and native forb at elevated CO_2 during an El Niño year in the Mojave Desert. *Oecologia* 128: 193–201.

Inouye, R. S. and W. M. Schaffer. 1981. On the meaning of ratio (de Wit) diagrams in plant ecology. *Ecology* 62: 1679–1681.

Istock, C. A. and S. M. Scheiner. 1987. Affinities and high-order diversity within landscape mosaics. *Evol. Ecol.* 1: 11–29.

Jaccard, P. 1901. Distribution de la flore alpine dans le Bassin des Dranes et dans quelques régions voisines. *Bull. Soc. Vaud. Sci. Nat.* 37: 241–272.

Jackson, D. A. and K. M. Somers. 1991. Putting things in order: the ups and downs of detrended correspondence analysis. *Am. Nat.* 137: 704–712.

Jahnke, R. A. 1992. The phosphorus cycle. In *Global Biogeochemical Cycles*, S. S. Butcher, R. J. Charleson, G. Orians and G. V. Wolfe (eds.), pp. 301–315. Academic Press, London.

Jain, S. K. and A. D. Bradshaw. 1966. Evolutionary divergence among adjacent plant populations. I. The evidence and its theoretical analysis. *Heredity* 21: 407–441.

Janson, S. and J. Vegelius. 1981. Measures of ecological association. *Oecologia* 49: 371–376.

Jefferies, R. L. and J. L. Maron. 1997. The embarrassment of riches: Atmospheric deposition of nitrogen and community and ecosystem processes. *Trends Ecol. Evol.* 12: 74–78.

Jeffrey, D. W. 1987. *Soil-Plant Relationships: An Ecological Approach*. Croom Helm, London and Timber Press, Portland, OR.

Jenkinson, D. S., J. Meredith, J. I. Kinyamario, G. P. Warren, M. T. F. Wong, D. D. Harkness, R. Bol and K. Coleman. 1999. Estimating net primary production from measurements made on soil organic matter. *Ecology* 80: 2762–2773.

Jetz, W. and C. Rahbek. 2001. Geometric constraints explain much of the species richness in African birds. *Proc. Nat. Acad. Sci.* 98: 5661–5666.

Johnson, A. R. and B. N. Milne. 1992. Diffusion in fractal landscapes: simulations and experimental studies of Tenebrionid beetle movements. *Ecology* 73: 1968–1983.

Johnson, D. M. and P. Stiling. 1998. Rate of spread of *Cactoblastis cactorum* (Lepidoptera: Pyralidae) Berg, an exotic *Opuntia*-feeding moth, in Florida. *Florida Entomol.* 81: 12–22.

Johnson, E. A. 1992. *Fire and Vegetation Dynamics*. Cambridge University Press, Cambridge.

Jordan, R. A. and J. M. Hartman. 1996. Effects of canopy opening on recruitment in *Clethra alnifolia* L. (Clethraceae) populations in central New Jersey wetland forests. *Bull. Torrey Bot. Club* 124: 286–294.

Kalisz, S. and M. A. McPeek. 1992. Demography of an age-structured annual: resampled projection matrices, elasticity analyses, and seed bank effects. *Ecology* 73: 1082–1093.

Kalisz, S. and M. A. McPeek. 1993. Extinction dynamics, population growth and seed banks. *Oecologia* 95: 314–320.

Karban, R. and I. T. Baldwin. 1997. *Induced Responses to Herbivory*. University of Chicago Press, Chicago, IL.

Kates, R. W., B. L. Turner and W. C. Clark. 1990. The great transformation. In *The Earth as Transformed by Human Action*, B. L. Turner, W. C. Clark, R. W. Kates, J. F. Richards, J. T. Matherw and W. B. Meyer (eds.), pp. 1–17. Cambridge University Press, Cambridge.

Kausik, S. B. 1939. Pollination and its influences on the behavior of the pistillate flower in *Vallisneria spiralis*. *Am. J. Bot.* 26: 207–211.

Kays, S. and J. L. Harper. 1974. The regulation of plant and tiller density in a grass sward. *J. Ecol.* 62: 97–105.

Kearns, C. A. and D. W. Inouye. 1993. *Techniques for Pollination Biologists*. University Press of Colorado, Niwot CO.

Keddy, P. A. 1990. Competitive hierarchies and centrifugal organization in plant communities. In *Perspectives on Plant Competition*, J. B. Grace and D. Tilman (eds.), pp. 265–289. Academic Press, Orlando, FL.

Keddy, P. A. 2001. *Competition*. Kluwer Academic, Boston, MA.

Keddy, P. A., L. H. Fraser and I. C. Wisheu. 1998. A comparative approach to examine competitive response of 48 wetland plant species. *J. Veg. Sci.* 9: 777–786.

Keeley, J. E. and C. J. Fotheringham. 1997. Trace gas emissions and smoke-induced seed germination. *Science* 276: 1248–1250.

Keeley, J. E., C. J. Fotheringham and M. Morais. 1999. Reexamining fire suppression impacts on brushland fire regimes. *Science* 284: 1824–1832.

Keeling, C. D. and T. P. Whorf. 2001. Atmospheric CO_2 records from sites in the SIO air sampling network. Trends Online: A Compendium of Data on Global Change. Carbon Dioxide Information Analysis Center, Oak Ridge National Laboratory, U.S. Department of Energy. [http://cdiac.esd.ornl.gov/trends/co2/sio-mlo.htm].

Keever, C. 1950. Causes of succession on old fields of the Piedmont, North Carolina. *Ecol. Monogr.* 20: 229–250.

Kelly, C. K. 1992. Resource choice in *Cuscuta europea*. *Proc. Nat. Acad. Sci.* 89: 12194–12197.

Kelly, C. K. 1996. Identifying plant functional types using floristic data bases: ecological correlates of plant range size. *J. Veg. Sci.* 7: 417–424.

Kelly, C. K. and F. I. Woodward. 1996. Ecological correlates of plant range size: taxonomies and phylogenies in the study of plant commonness and rarity in Great Britain. *Phil. Trans. Roy. Soc. London B* 351: 1261–1269.

Kelly, D., D. E. Hart and R. B. Allen. 2001. Evaluating the wind pollination benefits of masting. *Ecology* 82: 117–126.

Kelly, J. K. 1996. Kin selection in the annual plant *Impatiens capensis*. *Am. Nat.* 147: 899–918.

Kerr, B., D. W. Schwilk, A. Bergman and M. W. Feldman. 1999. Rekindling an old flame: a haploid model for the evolution and impact of flammability in resprouting plants. *Evol. Ecol. Res.* 1: 807–833.

Kevan, P. G. 1978. Floral coloration, its colorimetric analysis, and significance in anthecology. In *The Pollination of Flowers by Insects*, A. J. Richards (ed.), pp. 51–78. Academic Press, London.

Kira, T., H. Ogawa and K. Sinozaki. 1953. Intraspecific competition among higher plants. 1. Competition-density-yield interrelationships in regularly dispersed populations. *J. Inst. Polytech. Osaka City Univ. D* 4: 1–16.

Kitajima, K. and C. K. Augspurger. 1989. Seed and seedling ecology of a monocarpic tropical tree, *Tachigalia versicolor*. *Ecology* 70: 1102–1114.

Knapp, A. K., J. M. Briggs, D. C. Hartnett and S. L. Collins. 1998. *Grassland Dynamics: Long-Term Ecological Research in Tallgrass Prairie*. Oxford University Press, New York.

Kooijman, A. M. and A. Smit. 2001. Grazing as a measure to reduce nutrient availability and plant productivity in acid dune grasslands and pine forests in The Netherlands. *Ecol. Engineer.* 17: 63–77.

Kramer, P. J. 1983. *Water Relations of Plants*. Academic Press, New York.

Kramer, P. J. and J. S. Boyer. 1995. *Water Relations of Plants and Soils*. Academic Press, New York.

Krebs, C. J. 1989. *Ecological Methodology*. HarperCollins, New York.

Kruckeberg, A. R. 1954. The ecology of serpentine soils. III. Plant species in relation to serpentine soils. *Ecology* 35: 267–274.

Kuittinen, H., M. J. Sillanpää and O. Savolainen. 1997. Genetic basis of adaptation: flowering time in *Arabidopsis thaliana*. *Theor. Appl. Genet.* 95: 573–583.

Kunin, W. E. 1997. Introduction: on the causes and consequences of rare-common differences. In *Causes and consequences of rare-common differences*, W. E. Kunin and K. J. Gaston (eds.), pp. 3–11. Chapman and Hall, London.

Kunin, W. E. and K. J. Gaston. 1997. *The Biology of Rarity. Causes and consequences of rare-common differences*. Chapman and Hall, London.

Kutiel, P., Y. Peled and E. Geffen. 2000. The effect of removing shrub cover on annual plants and small mammals in a coastal sand dune ecosystem. *Biol. Conserv.* 94: 235–242.

Lambers, H., F. S. Chapin III and T. L. Pons. 1998. *Plant Physiological Ecology*. Springer-Verlag, New York.

Lambin, E. F., B. L. Turner, H. J. Geist, S. B. Agbola, A. Angelsen, J. W. Bruce, O. T. Coomes, R. Dirzo, G. Fischer, C. Folke, P. S. George, K. Homewood, J. Imbernon, R. Leemans, X. B. Li, E. F. Moran, M. Mortimore, P. S. Ramakrishnan, J. F. Richards, H. Skanes, W. Steffen, G. D. Stone, U. Svedin, T. A. Veldkamp, C. Vogel and J. C. Xu. 2001. The causes of land-use and land-cover change: moving beyond the myths. *Global Envir. Change Human Pol. Dim.* 11: 261–269.

Lamont, B. B. and N. J. Enright. 2000. Adaptive advantages of aerial seed banks. *Plant Spec. Biol.* 15: 157–166.

Lande, R. 1996. Statistics and partitioning of species diversity, and similarity among multiple communities. *Oikos* 76: 5–13.

Langdon, P. 1997. *A Better Place to Live: Reshaping the American Suburb*. University of Massachusetts Press, Amherst, MA.

Larcher, W. 1995. *Physiological Plant Ecology*, 3rd ed. Springer-Verlag, Berlin.

Larcher, W. and H. Bauer. 1981. Ecological significance of resistance to low temperatures. In *Encyclopedia of Plant Physiology*, Vol. 12A, O. L. Lange, P. S. Nobel, C. B. Osmond and H. Ziegler (eds.), pp. 403–437. Springer-Verlag, Berlin.

La Roi, G. H. and R. J. Hnatiuk. 1980. The *Pinus contorta* forests of Banff and Jasper National Parks: a study in comparative synecology and syntaxonomy. *Ecol. Monogr.* 50: 1–29.

Latham, R. E. and R. E. Ricklefs. 1993. Global patterns of tree species richness in moist forests: energy-diversity theory does not account for variation in species richness. *Oikos* 67: 325–333.

Laurance, W. F. 2000a. Do edge effects occur over large spatial scales? *Trends Ecol. Evol.* 15: 134–135.

Laurance, W. F. 2000b. Mega-development trends in the Amazon: implications for global change. *Envir. Monit. Assess.* 61: 113–122.

Laurance, W. F., L. V. Ferreira, J. M. Rankin-de Merona and S. G. Laurance. 1998. Rain forest fragmentation and the dynamics of Amazonian tree communities. *Ecology* 79: 2032–2040.

Laurance, W. F., M. A. Cochrane, S. Bergen, P. M. Fearnside, P. Delamonica, C. Barber, S. D'Angelo and T. Fernandes. 2001. The future of the Brazilian Amazon. *Science* 291: 438–439.

Lavery, P. B. and D. J. Mead. 1998. *Pinus radiata*: a narrow endemic from North America takes on the world. In *Ecology and Biogeography of Pinus*, D. M. Richardson (ed.), pp. 432–449. Cambridge University Press, Cambridge.

Lavoral, S. and S. Cramer. 1999. Plant functional types and disturbance dynamics. *Journal of Vegetation Science* 10: 603–730 (Special Feature).

Lavorel, S. and P. Chesson. 1995. How species with different regeneration niches coexist in patchy habitats with local disturbances. *Oikos* 74: 103–114.

Law, R. and A. R. Watkinson. 1987. Response-surface analysis of two-species competition: An experiment on *Phleum arenarium* and *Vulpia fasciculata*. *J. Ecol.* 75: 871–886.

Lawton, J. H. 1999. Are there general laws in ecology? *Oikos* 84: 177–192.

Lean, J. and D. A. Warrilow. 1989. Simulation of the regional climatic impact of Amazon deforestation. *Nature* 342: 411–413.

Lechowicz, M. J. 1995. Seasonality of flowering and fruiting in temperate forest trees. *Can. J. Bot.* 73: 175–182.

Lee, D. 1997. Iridescent blue plants. *Am. Sci.* 85: 56–63.

Lee, D. W. and R. Graham. 1986. Leaf optical-properties of rain-forest sun and extreme shade plants. *Am. J. Bot.* 73: 1100–1108.

Legendre, P. and Marie-J. Fortin. 1989. Spatial pattern and ecological analysis. *Vegetatio* 80: 107–138.

Legendre, P. and L. Legendre. 1998. *Numerical Ecology*, 2nd ed. Elsevier Science BV, Amsterdam.

Leibold, M. A. 1995. The niche concept revisited: mechanistic models and community context. *Ecology* 76: 1371–1382.

Leibold, M. A. 1999. Biodiversity and nutrient enrichment in pond plankton communities. *Evol. Ecol. Res.* 1: 73–95.

Leigh, R. A. and A. E. Johnston. 1994. *Long-Term Experiments in Agricultural and Ecological Sciences*. CAB International, Oxford.

Lejeune, K. D. and T. R. Seastedt. 2001. *Centaurea* species: the forb that won the west. *Cons. Biol.* 15: 1568–1574.

Lerdau, M. and J. Gershenzon. 1997. Allocation theory and chemical defense. In *Plant Resource Allocation*, F. A. Bazzaz and J. Grace (eds.), pp. 265–277. Academic Press, London.

Levin, D. A. 1990. The seedbank as a source of genetic novelty. *Am. Nat.* 135: 563–577.

Levin, S. A. 1989. Challenges in the development of a theory of community and ecosystem structure and function. In *Perspectives in Ecological Theory*, J. Roughgarden, R. M. May and S. A. Levin (eds.), pp. 242–255. Princeton University Press, Princeton, NJ.

Levin, S. A. 1992. The problem of pattern and scale in ecology. *Ecology* 73: 1943–1967.

Levine, J. M. 2000. Species diversity and biological invasions: relating local process to community pattern. *Science* 288: 852–854.

Levine, J. M. and C. M. D'Antonio. 1999. Elton revisited: a review of evidence linking diversity and invasibility. *Oikos* 87: 15–26.

Levins, R. 1969. Some demographic and genetic consequences of environmental heterogeneity for biological control. *Bull. Entomol. Soc. Amer.* 15: 237–240.

Lewis, H. and M. Lewis. 1955. The genus *Clarkia*. *Univ. Calif. Publ. Bot.* 20: 241–392.

Lewis, M. C. 1969. Genecological differentiation of leaf morphology in *Geranium sanguineum* L. *New Phytol.* 68: 481–503.

Lewis, M. C. 1970. Physiological significance of variation in leaf structure. *Sci. Prog.* 60: 25–51.

Lewis, M. C. 1972. The physiological significance of variation in leaf structure. *Sci. Prog.* 60: 25–51.

Lewontin, R. C. 1970. The units of selection. *Annu. Rev. Ecol. Syst.* 1: 1–18.

Lichtenthaler, H. K., C. Buschmann, M. Doll, H.-J. Fietz, T. Bach, U. Kozel, D. Meier and U. Rahmsdorf. 1981. Photosynthetic activity, chloroplast ultrastructure, and leaf characteristics of high-light and low-light plants and of sun and shade leaves. *Photosyn. Res.* 2: 115–141.

Liebhold, A. M. and J. Gurevitch. 2002. Integrating the statistical analysis of spatial data in ecology. *Ecography* (in press).

Liebhold, A. M., R. E. Rossi and W. P. Kemp. 1993. Geostatistics and geographic information systems in applied insect ecology. *Annu. Rev. Entomol.* 38: 303–327.

Likens, G. E., C. T. Driscoll, D. C. Buso, T. G. Siccama, C. E. Johnson, G. M. Lovett, T. J. Fahey, W. A. Reiners, D. F. Ryan, C. W. Martin and S. W. Bailey. 1998. The biogeochemistry of calcium at Hubbard Brook. *Biogeochemistry* 41: 89–173.

Liu, J., J. M. Chen, J. Cihlar and W. M. Park. 1997. A process-based boreal ecosystem productivity simulator using remote sensing inputs. *Remote Sens. Envir.* 62: 158–175.

Livingston, R. B. and M. L. Allessio. 1968. Buried viable seed in successional field and forest stands, Harvard Forest, Massachusetts. *Bull. Torrey Bot. Club* 95: 58–69.

Lodge, D. M. 1993. Biological invasions: lessons for ecology. *Trends Ecol. Evol.* 8: 133–137.

Lonsdale, W. M. 1990. The self-thinning rule: dead or alive? *Ecology* 71: 1373–1388.

Lonsdale, W. M. 1999. Global patterns of plant invasions and the concept of invasibility. *Ecology* 80: 1522–1536.

Loreau, M., S. Naeem, P. Inchausti, J. Bengtsson, J. P. Grime, A. Hector, D. U. Hooper, M. A. Huston, D. Raffaelli, B. Schmid, D. Tilman and D. A. Wardle. 2001. Biodiversity and ecosystem functioning: current knowledge and future challenges. *Science* 294: 804–808.

Louda, S. M. 1982. Distribution ecology: variation in plant recruitment over a gradient in relation to insect seed predation. *Ecology* 52: 25–41.

Louda, S. M. and M. A. Potvin. 1995. Effect of inflorescence-feeding insects on the demography and lifetime fitness of a native plant. *Ecology* 76: 229–245.

Louda, S. M. and J. E. Rodman. 1996. Insect herbivory as a major factor in the shade distribution of a native crucifer (*Cardamine cordifolia* A. Gray, bittercress). *J. Ecol.* 84: 229–237.

Louda, S. M., D. Kendall, J. Connor and D. Simberloff. 1997. Ecological effects of an insect introduced for the biological control of weeds. *Science* 277: 1088–1090.

Lovett Doust, L. 1981. Population dynamics and local specialization in a clonal perennial *Ranunculus repens* I. The dynamics of ramets in contrasting habitats. *J. Ecol.* 69: 743–755.

Ludwig, J. A. and J. F. Reynolds. 1988. *Statistical Ecology*. John Wiley & Sons, New York.

Lvovitch, M. I. 1973. The global water balance. *Eos* 54: 28–42.

Lynch, D. K. and W. Livingston. 1995. *Color and Light in Nature*. Cambridge University Press, Cambridge.

Lyons, E. E. and T. W. Mully. 1992. Density effects on flowering phenology and mating potential in *Nicotiana alata*. *Oecologia* 91: 93–100.

Lyons, S. K. and M. R. Willig. 1999. A hemispheric assessment of scale dependence in latitudinal gradients of species richness. *Ecology* 80: 2483–2491.

Ma, K. M., B. J. Fu, X. D. Guo and H. F. Zhou. 2000. Finding spatial regularity in mosaic landscapes: two methods integrated. *Plant Ecol.* 149: 195–205.

MacArthur, R. H. 1960. On the relative abundance of species. *Am. Nat.* 94: 25–36.

MacArthur, R. H. 1965. Patterns of species diversity. *Biol. Rev.* 40: 510–533.

MacArthur, R. H. 1972. *Geographical Ecology*. Princeton University Press, Princeton, NJ.

MacArthur, R. H. and E. O. Wilson. 1967. *The Theory of Island Biogeography*. Princeton University Press, Princeton, NJ.

Mack, M. C. and C. M. D'Antonio. 2001. Alteration of ecosystem nitrogen dynamics by exotic plants: a case study of C_4 grasses in Hawaii. *Ecol. Appl.* 11: 1323–1335.

Mack, R. N., D. Simberloff, W. M. Lonsdale, H. Evans, M. Clout and F. A. Bazzaz. 2000. Biotic invasions: causes, epidemiology, global consequences and control. *Ecol. Appl.* 10: 689–710.

Mackey, R. L. and D. J. Currie. 2001. The diversity-disturbance relationship: is it generally strong and peaked. *Ecology* 82: 3479–3492.

Magnuson, J. J., R. H. Wynne, B. J. Benson and D. M. Robertson. 2001. Lake and river ice as a powerful indicator of past and present climates. *Verh. Internat. Verein. Limnol.* 27: 2749–2756.

Magurran, A. E. 1988. *Ecological Diversity and Its Measurement.* Princeton University Press, Princeton, NJ.

Mahall, B. E. and F. H. Bormann. 1978. A quantitative description of the vegetative phenology of herbs in a northern hardwood forest. *Bot. Gaz.* 139: 467–481.

Mal, T. K., J. Lovett Doust and L. Lovett Doust. 1997. Time-dependent competitive displacement of *Typha angustifolia* by *Lythrum salicaria. Oikos* 79: 26–33.

Malcom, S. B. and M. P. Zalucki. 1996. Milkweed latex and cardenolide induction may resolve the lethal plant defence paradox. *Entomol. Exper. Appl.* 80: 193–196.

Malhi, Y. and J. Grace. 2000. Tropical forests and atmospheric carbon dioxide. *Trends Ecol. Evol.* 15: 332–337.

Mandujano, M. D., C. Montana, I. Mendez and J. Golubov. 1998. The relative contributions of sexual reproduction and clonal propagation in *Opuntia rastrera* from two habitats in the Chihuahuan Desert. *J. Ecol.* 86: 911–921.

Manly, B. F. J. 1992. *The Design and Analysis of Research Studies.* Cambridge University Press, Cambridge.

Marland, G., T. A. Boden and R. J. Andres. 2001. Global, regional, and national CO_2 emissions from fossil-fuel burning, cement production, and gas flaring: 1751–1998 (revised July 2001). Trends Online: A Compendium of Data on Global Change. Carbon Dioxide Information Analysis Center, Oak Ridge National Laboratory, U.S. Department of Energy [http://cdiac.esd.ornl.gov/ndps/ndp030.html].

Marschner, H. 1995. *Mineral Nutrition of Higher Plants*, 2nd ed. Academic Press, London.

Marshall, D. L. 1988. Postpollination effects on seed paternity: mechanisms in addition to microgametophyte competition operate in wild radish. *Evolution* 42: 1256–1266.

Marshall, D. L. and M. W. Folsom. 1991. Mate choice in plants: an anatomical to population perspective. *Annu. Rev. Ecol. Syst.* 22: 37–63.

Martin, P. S. and R. G. Klein. 1984. *Quaternary Extinctions: A Prehistoric Revolution.* University of Arizona Press, Tucson, AZ.

Martinez, M. L. and P. Moreno-Casasola. 1998. The biological flora of coastal dunes and wetlands: *Chamaecrista chamaecristoides* (Colladon) I. & B. *J. Coastal Res.* 14: 162–174.

Mather, A. S. and C. L. Needle. 2000. The relationships of population and forest trends. *Geogr. J.* 166: 2–13.

Mauseth, J. D. 1988. *Plant Anatomy.* Benjamin/Cummings, Menlo Park, CA.

Mauseth, J. D. 1991. *Botany: An Introduction to Plant Biology.* Saunders College Publishers, Philadelphia, PA.

May, R. M. 1973. *Stability and Complexity in Model Ecosystems.* Princeton University Press, Princeton, NJ.

Maybury, K. 1999. *Seeing the Forest and the Trees: Ecological Classification for Conservation.* The Nature Conservancy, Arlington, VA.

Mayewski, P. A., G. H. Denton and J. J. Hughes. 1981. The Late Wisconsin ice sheet in North America. In *The Last Great Ice Sheets*, G. H. Denton and J. J. Hughes (eds.), pp. 67–170. Wiley, New York.

Mayo, D. G. 1996. *Error and the Growth of Experimental Knowledge.* University of Chicago Press, Chicago.

Mazer, S. J. 1989. Ecological, taxonomic, and life history correlates of seed mass among Indiana Dune angiosperms. *Ecol. Monogr.* 59: 153–175.

Mazer, S. J. and L. M. Wolfe. 1992. Planting density influences the expression of genetic variation in seed mass in wild radish (*Raphanus sativus* L.: Brassicaceae). *Am. J. Bot.* 79: 1185–1193.

McCulloch, C. E. 1985. Variance tests for species associations. *Ecology* 66: 1676–1681.

McCune, B. and M. J. Mefford. 1995. *PC-ORD. Multivariate Analysis of Ecological Data, Version 2.0.* MjM Software Design, Gleneden Beach, OR.

McGinley, M. A., D. H. Temme and M. A. Geber. 1987. Parental investment in offspring in variable environments: theoretical and empirical considerations. *Am. Nat.* 130: 370–398.

McIlveen, J. F. R. 1992. *Fundamentals of Weather and Climate.* Chapman and Hall, London.

McIntosh, R. P. 1985. *The Background of Ecology.* Cambridge University Press, Cambridge.

McLaughlin, S. B. and R. Wimmer. 1999. Tansley Review No. 104: Calcium physiology and terrestrial ecosystem processes. *New Phytol.* 142: 373–417.

McNaughton, S. M. 1983. Serengeti grassland ecology: the role of composite environmental factors and contingency in community organization. *Ecol. Monogr.* 53: 291–320.

McNeilly, T. 1968. Evolution in closely adjacent plant populations. III. *Agrostis tenuis* on a small copper mine. *Heredity* 23: 99–108.

McNeilly, T. and J. Antonovics. 1968. Evolution in closely adjacent plant populations. IV. Barriers to gene flow. *Heredity* 23: 205–218.

Mead, R. 1988. *The Design of Experiments: Statistical Principles for Practical Application.* Cambridge University Press, Cambridge.

Menges, E. S. 1992. Stochastic modeling of extinction in plant populations. In *Conservation Biology: The Theory and Practice of Nature Conservation, Preservation, and Management*, P. L. Fiedler and S. K. Jain (eds.), pp. 253–275. Chapman and Hall, New York.

Milne, B. N. 1992. Spatial aggregation and neutral models in fractal landscapes. *Am. Nat.* 139: 32–57.

Mitchell-Olds, T. and J. Bergelson. 1990. Statistical genetics of an annual plant, *Impatiens capensis*. I. Genetic basis of quantitative variation. *Genetics* 124: 407–415.

Mitman, G. 1992. *The State of Nature.* University of Chicago Press, Chicago, IL.

Mittelbach, G. G., C. F. Steiner, S. M. Scheiner, K. L. Gross, H. L. Reynolds, R. B. Waide, M. R. Willig, S. I. Dodson and L. Gough. 2001. What is the observed relationship between species richness and productivity? *Ecology* 82: 2381–2396.

Mitton, J. B. and M. C. Grant. 1996. Genetic variation and the natural history of quaking aspen. *Bioscience* 46: 25–31.

Mohr, H. and P. Schopfer. 1995. *Plant Physiology.* Springer-Verlag, Berlin.

Mojonnier, L. 1998. Natural selection on two seed-size traits in the common morning glory *Ipomoea purpurea* (Convolvulaceae): patterns and evolutionary consequences. *Am. Nat.* 152: 188–203.

Morrison, J. A. 1996. Infection of *Juncus dichotomus* by the smut fungus *Cintractia junci*: An experimental field test of the effects of neighbouring plants, environment, and host plant genotype. *J. Ecol.* 84: 691–702.

Morrow, P. A. and V. C. LaMarche Jr. 1978. Tree-ring evidence for chronic insect suppression of productivity in subalpine *Eucalyptus. Science* 201: 1244–1246.

Mueller-Dombois, D. and H. Ellenberg. 1974. *Aims and Methods of Vegetation Ecology.* John Wiley & Sons, New York.

Muller, C. H. 1969. Allelopathy as a factor in ecological process. *Vegetatio* 18: 348–357.

Mulroy, T. W. and P. W. Rundel. 1977. Annual plants: adaptations to desert environments. *Bioscience* 27: 109–114.

Murdoch, W. W. 1966. Community structure, population control, and competition: a critique. *Am. Nat.* 100: 219–226.

Mutch, R. 1970. Wildland fires and ecosystems: a hypothesis. *Ecology* 51: 1046–1051.

Myers, R. L. 1990. Scrub and high pine. In *Ecosystems of Florida*, R. L. Myers and J. J. Ewel (eds.), pp. 150–193. University of Central Florida Press, Orlando, FL.

Naeem, S. 2000. Reply to Wardle et al. *Bull. Ecol. Soc. Amer.* 81: 241–246.

Naeem, S., K. Hakansson, J. H. Lawton, M. J. Crawley and L. J. Thompson. 1996. Biodiversity and plant productivity in a model assemblage of plant species. *Oikos* 76: 259–264.

Naeem, S., F. S. Chapin III, R. Costanza, P. R. Ehrlich, F. B. Golley, D. V. Hooper, J. H. Lawton, R. V. O'Neill, H. A. Mooney, O. E. Sala, A. J. Symstad and D. Tilman. 1999. Biodiversity and ecosystem functioning maintaining natural life support processes. Issues in Ecology No. 4, Ecological Society of America, Washington DC.

Naveh, Z. and A. S. Lieberman. 1994. *Landscape Ecology: Theory and Application*, 2nd ed. Springer-Verlag, New York.

Neilson, R. P. 1986. High-resolution climatic analysis and southwest biogeography. *Science* 232: 27–34.

Neilson, R. P. 1995. A model for predicting continental-scale vegetation distribution and water balance. *Ecol. Appl.* 5: 362–385.

Newman, E. I. 1973. Competition and diversity in herbaceous vegetation. *Nature* 244: 310–311.

Newman, E. I. 1978. Allelopathy: adaptation or accident? In *Biochemical Aspects of Plant and Animal Coevolution*, J. B. Harborne (ed.), pp. 327–342. Academic Press, London.

Newsham, K. K., A. H. Fitter and A. R. Watkinson. 1995. Multi-functionality and biodiversity in arbuscular mycorrhizas. *Trends Ecol. Evol.* 10: 407–411.

Nicholson, M. 1990. Henry Allan Gleason and the individualistic hypothesis: the structure of a botanist's career. *Bot. Rev.* 56: 91–161.

Niemelä, P. and W. J. Mattson. 1996. Invasion of North American forests by European phytophagous insects: legacy of the European crucible? *Bioscience* 46: 741–753.

Niklas, K. 1997. *The Evolutionary Biology of Plants*. University of Chicago Press, Chicago.

Nobel, P. S. 1983. *Biophysical Plant Physiology and Ecology*. W. H. Freeman, New York.

Nores, M. 1999. An alternative hypothesis for the origin of Amazonian bird diversity. *J. Biogeogr.* 26: 475.

Odening, W. R., B. R. Strain and W. C. Oechel. 1974. The effect of decreasing water potential on net CO_2 exchange of intact desert shrubs. *Ecology* 55: 1086–1094.

Odum, E. P. 1969. The strategy of ecosystem development. *Science* 164: 262–270.

Odum, E. P. 1971. *Fundamentals of Ecology*, 3rd ed. W. B. Saunders, Philadelphia, PA.

Odum, H. T. 1983. *Systems Ecology*. John Wiley, New York.

Odum, H. T. 1988. Self-organization, transformity, and information. *Science* 242: 1132–1139.

Oksanen, L., S. D. Fretwell, J. Arrunda and P. Niemela. 1981. Exploitation ecosystems in gradients of primary productivity. *Am. Nat.* 131: 424–444.

O'Neill, R. V., D. L. DeAngelis, J. B. Waide and T. F. H. Allen. 1986. *A Hierarchical Concept of Ecosystems*. Princeton University Press, Princeton, NJ.

Oosting, H. J. 1942. An ecological analysis of the plant communities of Piedmont, North Carolina. *Am. Midl. Nat.* 28: 1–126.

Oosting, H. J. and M. E. Humphreys. 1940. Buried viable seeds in a successional series of old fields and forest soils. *Bull. Torrey Bot. Club* 67: 253–273.

Orians, G. H. 1986. Site characteristics promoting invasions and system impact of invaders. In *Ecology of Biological Invasions of North America and Hawaii*, H. A. Mooney and J. A. Drake (eds.), pp. 133–148. Springer-Verlag, New York.

Orlean, S. 2000. *The Orchid Thief*. Ballentine Reader's Circle, New York.

Pacala, S. W. 1986a. Neighborhood models of plant population dynamics. 2. Multispecies models of annuals. *Theor. Pop. Biol.* 29: 262–292.

Pacala, S. W. 1986b. Neighborhood models of plant population dynamics. 4. Single-species and multispecies models of annuals with dormant seeds. *Am. Nat.* 128: 859–878.

Pacala, S. W. 1987. Neighborhood models of plant population dynamics. 3. Models with spatial heterogeneity in the physical environment. *Theor. Pop. Biol.* 31: 359–392.

Pacala, S. W. and J. A. Silander Jr. 1985. Neighborhood models of plant population dynamics. 1. Single-species models of annuals. *Am. Nat.* 125: 385–411.

Pärtel, M., M. Zobel, K. Zobel and E. van der Maarel. 1996. The species pool and its relation to species richness: evidence from Estonian plant communities. *Oikos* 75: 111–117.

Pagani, M., K. H. Freeman and M. A. Arthur. 1999. Late Miocene atmospheric CO_2 concentrations and the expansion of C_4 grasses. *Science* 285: 876–879.

Parayil, G. and F. Tong. 1998. Pasture-led to logging-led deforestation in the Brazilian Amazon: The dynamics of socio-environmental change. *Global Envir. Change Human Pol. Dim.* 8: 63–79.

Parton, W. J., J. W. B. Stewart and C. V. Cole. 1988. Dynamics of C, N, P and S in grassland soils: a model. *Biogeochemistry* 5: 109–131.

Parton, W. J., J. M. O. Scurlock, D. S. Ojima, T. G. Gilmanov, R. J. Scholes, D. S. Schimel, T. Kirchner, J. C. Menaut, T. Seastedt, E. G. Moya, A. Kamnalrut and J. I. Kinyamario. 1993. Observations and modeling of biomass and soil organic-matter dynamics for the grassland biome worldwide. *Global Biogeo. Cycles* 7: 785–809.

Paruelo, J. M. and W. K. Lauenroth. 1996. Relative abunances of plant functional types in grasslands and shrublands of North America. *Ecol. Appl.* 6: 1212–1224.

Patterson, B. D. and W. Atmar. 1986. Nested subsets and the structure of insular mammalian faunas and archipelagos. *Biol. J. Linnaean Soc.* 28: 65–82.

Pearson, L. C. 1995. *The Diversity and Evolution of Plants*. CRC Press, Boca Raton, FL.

Peet, R. K. and N. L. Christensen. 1987. Competition and tree death. *Bioscience* 37: 586–595.

Pellew, R. A. P. 1983. Impacts of elephant, giraffe, and fire upon the *Acacia tortilis* woodlands of the Serengeti. *Afr. J. Ecol.* 21: 41–74.

Pennycuick, C. J. and N. C. Kline. 1986. Units of measurement for fractal extent, applied to the coastal distribution of bald eagle nests in the Aleutian Islands, Alaska. *Oecologia* 68: 254–258.

Peralta, P. and P. Mather. 2000. An analysis of deforestation patterns in the extractive reserves of Acre, Amazonia from satellite imagery: a landscape ecological approach. *Inter. J. Remote Sens.* 21: 2555–2570.

Pfaff, A. S. P. 1999. What drives deforestation in the Brazilian Amazon? Evidence from satellite and socioeconomic data. *J. Envir. Econ. Manag.* 37: 26–43.

Pfunder, M. and B. A. Roy. 2000. Pollinator-mediated interactions between a pathogenic fungus, *Uromyces pisi* (Pucciniaceae), and its host plant, *Euphorbia cyparissias* (Euphorbiaceae). *Am. J. Bot.* 87: 48–55.

Phillips, D. L. and D. J. Shure. 1990. Patch-size effects on early succession in southern Appalachian Forests. *Ecology* 71: 204–212.

Pianka, E. R. 1966. Latitudinal gradients in species diversity: a review of concepts. *Am. Nat.* 100: 65–75.

Pianka, E. R. 1970. On *r*- and *K*-selection. *Am. Nat.* 104: 592–597.

Pickett, S. T. A. and M. L. Cadenasso. 1995. Landscape ecology: spatial heterogeneity in ecological systems. *Science* 269: 331–334.

Pickett, S. T. A. and J. N. Thompson. 1978. Patch dynamics and the design of nature reserves. *Biol. Conserv.* 13: 27–37.

Pickett, S. T. A. and P. S. E. White. 1985. *The Ecology of Natural Disturbance and Patch Dynamics.* Academic Press, Orlando, FL.

Pickett, S. T. A., S. L. Collins and J. J. Armesto. 1987. Models, mechanisms and pathways of succession. *Bot. Rev.* 53: 335–371.

Pickett, S. T. A., J. Kolasa and C. G. Jones. 1994. *Ecological Understanding.* Academic Press, San Diego, CA.

Pielou, E. C. 1975. *Ecological Diversity.* Wiley, New York.

Pielou, E. C. 1977. *Mathematical Ecology*, 2nd ed. Wiley, New York.

Pielou, E. C. 1991. *After the Ice Age.* University of Chicago Press, Chicago.

Pigliucci, M. 2001. *Phenotypic Plasticity: Beyond Nature and Nurture.* Johns Hopkins University Press, Baltimore, MD.

Pilon-Smits, E. A. H., H. 't Hart, J. W. Maas, J. A. N. Meesterburrie, R. Kreuler and J. van Brederode. 1991. The evolution of crassulacean acid metabolism in *Aeonium* inferred from carbon isotope composition and enzyme activities. *Oecologia* 91: 548–553.

Pimentel, D. 1993. *World Soil Erosion and Conservation.* Cambridge University Press, Cambridge.

Piñero, D., M. Martínez-Ramos and J. Sarukhán. 1984. A population model of *Astrocaryum mexicanum* and a sensitivity analysis of its finite rate of increase. *J. Ecol.* 72: 977–991.

Pirnat, J. 2000. Conservation and management of forest patches and corridors in suburban landscapes. *Land. Urb. Plan.* 52: 135–143.

Pitman, N. C. A., J. Terborgh, M. R. Silman and P. Nuñez V. 1999. Tree species distribution in an upper Amazonian forest. *Ecology* 80: 2651–2661.

Platenkamp, G. A. J. and R. G. Shaw. 1993. Environmental and genetic maternal effects on seed characters in *Nemophila menziesii*. *Evolution* 47: 540–555.

Platt, W. J. and S. L. Rathbun. 1993. Dynamics of an old-growth longleaf pine population. In *The Longleaf Pine Ecosystem: Ecology; Restoration and Management*, S. M. Herman (ed.), pp. 275–297. Proceedings of the Tall Timbers Fire Ecology Conference, No. 18.

Platt, W. J. and I. M. Weis. 1977. Resource partitioning and competition within a guild of fugitive prairie plants. *Am. Nat.* 111: 479–513.

Platt, W. J., G. W. Evans and S. J. Rathbun. 1988. The population dynamics of a long-lived conifer (*Pinus palustris*). *Am. Nat.* 131: 491–525.

Poole, R. W. and B. J. Rathcke. 1979. Regularity, randomness, and aggregation in flowering phonologies. *Science* 203: 470–471.

Pope, K. O., J. M. Rey Benayas and J. F. Paris. 1994. Radar remote sensing of forest and wetland ecosystems in the Central American Tropics. *Remote Sens. Envir.* 48: 205–219.

Popper, K. R. 1959. *The Logic of Scientific Discovery.* Hutchinson & Co., London.

Portnoy, S. and M. F. Willson. 1993. Seed dispersal curves: behavior of the tail of the distribution. *Evol. Biol.* 7: 25–44.

Powell, E. A. 1992. Life history, reproductive biology and conservation of the Mauna Kea silversword, *Argyroxiphium sandwicense* DC (Asteraceae): an endangered plant of Hawaii. Ph.D. Dissertation. University of Hawaii.

Powers, J. S., M. E. Sollins and J. A. Jones. 1999. Plant-pest interactions in time and space: a Douglas-fir bark beetle outbreak as a case study. *Landscape Ecol.* 14: 105–120.

Prentice, I. C., P. J. Bartlein and T. Webb III. 1991. Vegetation and climate change in eastern North America since the last glacial maximum. *Ecology* 72: 2038–2056.

Preston, F. W. 1962. The canonical distribution of commonness and rarity. *Ecology* 43: 185–215.

Priestley, D. A. 1986. *Seed Aging: Implications for Seed Storage and Persistence in the Soil.* Comstock Publishing, Ithaca, NY.

Proctor, M., P. Yeo and A. Lack. 1996. *The Natural History of Pollination.* Timber Press, Portland Oregon.

Putz, F. E. 1983. Treefall pits and mounds, buried seeds, and the importance of soil disturbance to pioneer trees on Barro Colorado Island, Panama. *Ecology* 64: 1069–1074.

Pyne, S. J., P. L. Andrews and R. D. Laven. 1996. *Introduction to Wildland Fire*, 2nd ed. Wiley, New York.

Qian, H. and R. E. Ricklefs. 1999. A comparison of the taxonomic richness of vascular plants in China and the United States. *Am. Nat.* 154: 160–181.

Qian, H., P. S. White, K. Klinka and C. Chourmouzis. 1999. Phytogeographic and community similarities of alpine tundras of Chang-baishan Summit, China, and Indian Peaks, USA. *J. Veg. Sci.* 10: 869–882.

Rabinowitz, D. 1981. Seven forms of rarity. In *The Biological Aspects of Rare Plant Conservation*, H. Synge (ed.), pp. 205–217. John Wiley, Chicester, UK.

Rabinowitz, D., S. Cairns and T. Dillon. 1986. Seven forms of rarity and their frequency in the flora of the British Isles. In *Conservation Biology: The Science of Scarcity and Diversity*, M. E. Soulé (ed.), pp. 182–204. Sinauer Associates, Sunderland, MA.

Raffa, K. 1991. Induced defensive reactions in conifer-bark beetle systems. In *Phytochemical Induction by Herbivores*, D. W. Tallamy and M. J. Raupp (eds.), pp. 245–276. John Wiley, New York.

Raffa, K. F. and A. A. Berryman. 1987. Interacting selective pressures in conifer-bark beetle systems: A basis for reciprocal adaptations. *Am. Nat.* 129: 234–262.

Raffaele, E. and T. T. Veblen. 1998. Facilitation by nurse shrubs of resprouting behavior in a post-fire shrubland in northern Patagonia, Argentina. *J. Veg. Sci.* 9: 693–698.

Rapoport, E. H. 1982. *Areography: Geographical Strategies of Species, Vol. 1.* Pergamon, New York.

Raunkiaer, C. 1934. *The Life Forms of Plants and Statistical Plant Geography.* Clarendon, Oxford.

Raven, J. A. 2002. Selection pressures on stomatal evolution. *New Phytol.* 153: 371–386.

Raven, P. H., R. F. Evert and S. E. Eichhorn. 1999. *Biology of Plants*, 6th ed. W. H. Freeman and Company, New York.

Reader, R. J., S. D. Wilson, J. W. Belcher, I. Wisheu, P. A. Keddy, D. Tilman, E. C. Morris, J. B. Grace, J. B. McGraw, H. Olff, R. Turkington, E. Klein, Y. Leung, B. Shipley, R. Vanhulst, M. E. Johansson, C. Nilsson, J. Gurevitch, K. Grigulis and B. E. Beisner. 1994. Plant competition in relation to neighbor biomass: an intercontinental study with *Poa pratensis*. *Ecology* 75: 1753–1760.

Redmann, R. E. 1975. Production ecology of grassland plant communities in western North Dakota. *Ecol. Monogr.* 45: 83–106.

Reekie, E. G. and F. A. Bazzaz. 1987. Reproductive effort in plants. 1. Carbon allocation to reproduction. *Am. Nat.* 129: 876–896.

Reich, P. B., M. B. Walters and D. S. Ellsworth. 1992. Leaf lifespan in relation to leaf, plant, and stand characteristics among diverse ecosystems. *Ecol. Monogr.* 62: 365–392.

Reiser, M. 1993. *Cadillac Desert: The American West and its Disappearing Water.* Penguin Books, New York.

Rejmánek, M. 1989. Invasibility of plant communities. In *Biological Invasions*, J. A. Drake, H. A. Mooney, F. diCastri, R. H. Groves, G. J. Kruger, M. Rejamnek and M. Williamson (eds.), pp. 369–388. John Wiley & Sons, New York.

Rejmánek, M. and D. M. Richardson. 1996. What attributes make some plant species more invasive? *Ecology* 77: 1655–1661.

Rendig, V. V. and H. M. Taylor. 1989. *Principles of Soil-Plant Interrelationships*. McGraw-Hill, New York.

Rey Benayas, J. M. and K. O. Pope. 1995. Landscape ecology and diversity patterns in the seasonal tropics from Landsat TM imagery. *Ecol. Appl.* 5: 386–394.

Rey Benayas, J. M. and S. M. Scheiner. 2002. Plant diversity, biogeography, and environment in Iberia: Patterns and possible causal factors. *J. Veg. Sci.* 13: (in press).

Rey Benayas, J. M., S. M. Scheiner, M. Garcia Sánchez-Colomer and C. Levassor. 1999. Commonness and rarity: theory and application of a new model to Mediterranean montane grasslands. *Cons. Ecol.* 3: 5. [http://www.consecol.org/vol3/iss1/art5]

Rice, E. L. 1974. *Allelopathy*. Academic Press, New York.

Richards, A. J. 1986. *Plant Breeding Systems*. Chapman and Hall, London.

Richards, M. B., R. M. Cowling and W. D. Stock. 1997. Soil factors and competition as determinants of the distribution of six fynbos Proteaceae species. *Oikos* 79: 394–406.

Ricklefs, R. E. 2000. Rarity and diversity in Amazonian forest trees. *Trends Ecol. Evol.* 15: 83–84.

Ricklefs, R. E. and D. Schluter. 1993. *Species Diversity in Ecological Communities*. University of Chicago Press, Chicago, IL.

Robinson, D., A. Hodge, B. S. Griffiths and A. H. Fitter. 1999. Plant root proliferation in nitrogen-rich patches confers competitive advantage. *Proc. Royal Soc. London B* 266: 431–435.

Rohde, K. 1992. Latitudinal gradients in species diversity: the search for the primary cause. *Oikos* 65: 514–527.

Rohde, K. 1997. The larger area of the tropics does not explain latitudinal gradients in species diversity. *Oikos* 79: 169–172.

Rohde, K. 1998. Latitudinal gradients in species diversity. Area matters, but how much? *Oikos* 82: 184–190.

Rosenzweig, M. L. 1968. Net primary productivity of terrestrial communities: Prediction from climatological data. *Am. Nat.* 102: 67–74.

Rosenzweig, M. L. 1971. Paradox of enrichment: destabilization of exploitation ecosystems in ecological time. *Science* 171: 385–387.

Rosenzweig, M. L. 1992. Species diversity gradients: we know more and less than we thought. *J. Mammol.* 73: 715–730.

Rosenzweig, M. L. 1995. *Species Diversity in Space and Time.* Cambridge University Press, Cambridge.

Rosenzweig, M. L. and Z. Abramsky. 1986. Centrifugal community organization. *Oikos* 46: 339–348.

Rosenzweig, M. L. and Z. Abramsky. 1993. How are diversity and productivity related? In *Species Diversity in Ecological Communities*, R. E. Ricklefs and D. Schluter (eds.), pp. 52–65. University of Chicago Press, Chicago, IL.

Rosenzweig, M. L. and M. V. Lomolino. 1997. Who gets the short bits of the broken stick? In *The Biology of Rarity: Causes and Consequences of Rare-Common Differences*, W. E. Kunin and K. J. Gaston (eds.), pp. 63–90. Chapman & Hall, London.

Rosenzweig, M. L. and E. A. Sandlin. 1997. Species diversity and latitudes: listening to area's signal. *Oikos* 80: 172–176.

Ross, M. A. and J. L. Harper. 1972. Occupation of biological space during seedling establishment. *J. Ecol.* 60: 77–88.

Rossotti, H. 1983. *Colour: Why the World isn't Grey*. Princeton University Press, Princeton, NJ.

Rothman, D. H. 2002. Atmospheric carbon dioxide levels for the last 500 million years. *Proc. Nat. Acad. Sci.* 99: 4167–4171.

Rothstein, D. E. and D. R. Zak. 2001. Photosynthetic adaptation and acclimation to exploit seasonal periods of direct irradiance in three temperate, deciduous-forest herbs. *Func. Ecol.* 15: 722–731.

Rovira, A. D., C. D. Bowen and R. C. Foster. 1983. The significance of rhizosphere microflora and mycorrhizas in plant nutrition. In *Inorganic Plant Nutrition* (*Encyclopedia of Plant Physiology*, new series, Vol. 15B), A. Läuchli and R. L. Bieleski (eds.), pp. 61–93. Springer-Verlag, Berlin.

Ruddiman, W. F. and J. E. Kutzbach. 1989. Forcing of late Cenozoic northern hemisphere climates by plateau uplift in southeast Asia and the American southwest. *J. Geophys. Res.* 94: 18409–18427.

Rübel, E. 1936. Plant communities of the world. In *Essays in Geobotany in Honor of William Albert Setchel*, T. H. Goodspeed (ed.), pp. 263–290. University of California Press, Berkeley, CA.

Rumney, G. R. 1968. *Climatology and the World's Climates*. Macmillan, New York.

Runkle, J. R. 1985. Disturbance regimes in temperate forests. In *The Ecology of Natural Disturbance and Patch Dynamics*, S. T. A. Pickett and P. S. White (eds.), pp. 17–33. Academic Press, Orlando, FL.

Sala, O. E., F. S. Chapin III, J. J. Armesto, E. Berlow, J. Bloomfield, R. Dirzo, E. Huber-Sanwald, L. F. Huenneke, R. B. Jackson, A. Kinzig, R. Leemans, D. M. Lodge, H. A. Mooney, M. Oesterheld, N. L. R. Poff, M. T. Sykes, B. H. Walker, M. Walker and D. H. Wall. 2000. Global biodiversity scenarios for the year 2100. *Science* 287: 1770–1774.

Salanti, E. and P. B. Vose. 1984. Amazon Basin: a system in equilibrium. *Science* 225: 129–138.

Salisbury, E. J. 1942. *The Reproductive Capacity of Plants*. G. Bell & Sons, Ltd., London.

Salzman, A. 1985. Habitat selection in a clonal plant. *Science* 228: 603–604.

Sanders, H. L. 1968. Marine benthic diversity: A comparative study. *Am. Nat.* 102: 243–282.

Sarukhán, J. and M. Gadgil. 1974. Studies on plant demography: *Ranunculus repens* L., *R. bulbosus* L., and *R. acris* L. III. A mathematical model incorporating multiple modes of reproduction. *J. Ecol.* 62: 921–936.

Schaal, B. A. 1980. Measurement of gene flow in *Lupinus texensis*. *Nature* 284: 450–451.

Schaffer, W. M. and M. L. Rosenzweig. 1977. Selection for optimal life histories. II: Multiple equilibria and the evolution of alternative reproductive strategies. *Ecology* 58: 60–72.

Schaffer, W. M. and M. V. Schaffer. 1979. The adaptive significance of variations in reproductive habit in the Agavaceae. II. Pollinator foraging behavior and selection for increased reproductive expenditure. *Ecology* 60: 1051–1069.

Schaffer, W. N. 1983. On the application of optimal control theory to the general life history problem. *Am. Nat.* 121: 418–431.

Scheiner, S. M. 1988. The seed bank and above-ground vegetation in an upland pine-hardwood succession. *Mich. Bot.* 27: 99–106.

Scheiner, S. M. 1989. Variable selection along a successional gradient. *Evolution* 43: 548–562.

Scheiner, S. M. 1992. Measuring pattern diversity. *Ecology* 73: 1860–1867.

Scheiner, S. M. 1993. Genetics and evolution of phenotypic plasticity. *Annu. Rev. Ecol. Syst.* 24: 35–68.

Scheiner, S. M. and J. Gurevitch. 2001. *Design and Analysis of Ecological Experiments*, 2nd ed. Oxford University Press, New York, NY.

Scheiner, S. M. and J. M. Rey-Benayas. 1994. Global patterns of plant diversity. *Evol. Ecol.* 8: 331–347.

Scheiner, S. M. and J. M. Rey-Benayas. 1997. Putting empirical limits on metapopulation models for terrestrial plants. *Evol. Ecol.* 11: 275–288.

Scheiner, S. M., S. B. Cox, M. R. Willig, G. G. Mittelbach, C. Osenberg and M. Kaspari. 2000. Species richness, species-area curves, and Simpson's paradox. *Evol. Ecol. Res.* 2: 791–802.

Schemske, D. W. and C. Horvitz. 1984. Variation among floral visitors in pollination ability: a precondition for mutualism specialization. *Science* 225: 519–521.

Schemske, D. W. and C. C. Horvitz. 1988. Plant-animal interactions and fruit production in a neotropical herb: A path analysis. *Ecology* 69: 1128–1137.

Schimper, A. F. W. 1903. *Plant-Geography upon a Physiological Basis.* Clarendon Press, Oxford.

Schlesinger, W. H. 1997. *Biogeochemistry: An Analysis of Global Change*, 2nd ed. Academic Press, New York.

Schlesinger, W. H., J. T. Cray and F. S. Gilliam. 1982. Atmospheric deposition processes and their importance as sources of nutrients in a chaparral ecosystem of southern California. *Water Resour. Res.* 18: 623–629.

Schlesinger, W. H., J. G. Reynolds, G. L. Cunningham, L. F. Huenneke, W. M. Jarrell, R. A. Virginia and W. G. Whitford. 1990. Biological feedbacks in global desertification. *Science* 247: 1043–1048.

Schlichting, C. D. 1986. The evolution of phenotypic plasticity in plants. *Annu. Rev. Ecol. Syst.* 17: 667–693.

Schluter, D. and R. E. Ricklefs. 1993. Convergence and the regional component of species diversity. In *Species Diversity in Ecological Communities*, R. E. Ricklefs and D. Schluter (eds.), pp. 230–240. University of Chicago Press, Chicago, IL.

Schmalzel, R. J., F. W. Reichenbacher and S. Rutman. 1995. Demographic study of the rare *Coryphantha robbinsorum* (Cactaceae) in southeastern Arizona. *Madroño* 42: 332–348.

Schmitt, J. 1993. Reaction norms of mophological and life-history traits to light availability in *Impatiens capensis. Evolution* 47: 1654–1668.

Schmitt, J. and R. D. Wulff. 1993. Light spectral quality, phytochrome and plant competition. *Trends Ecol. Evol.* 8: 47–51.

Schmitt, J., A. C. McCormac and H. Smith. 1995. A test of the adaptive plasticity hypothesis using transgenic and mutant plant disabled in phytochrome-mediated elongation

responses to neighbors. *Am. Nat.* 146: 937–953.

Schmitt, S. F. and R. J. Whittaker. 1996. Disturbance and succession on the Krakatau Islands, Indonesia. In *Dynamics of Tropical Communities*, D. M. Newbery, H. H. T. Prins and N. D. Brown (eds.), pp. 515–548. Blackwell Press, Oxford.

Schnitzer, S. A. and W. P. Carson. 2001. Treefall gaps and the maintenance of species diversity in a tropical forest. *Ecology* 82: 913–919.

Schulte, A. and D. Ruhiyat. 1998. *Soils of Tropical Forest Ecoytems: Characteristics, Ecology, and Management.* Springer-Verlag, Berlin.

Schwartz, M. W. and D. Simberloff. 2001. Taxon size predicts rates of rarity in vascular plants. *Ecol. Letters* 4: 464–469.

Schwinning, S. and J. R. Ehleringer. 2001. Water-use tradeoffs and optimal adaptations to pulse-driven arid ecosystems. *J. Ecol.* 89: 464–480.

Schwinning, S. and G. A. Fox. 1995. Competitive symmetry and its consequences for plant population dynamics. *Oikos* 72: 422–432.

Scott, L. 2002. Grassland development under glacial and interglacial conditions in southern Africa: review of pollen, phytolith and isotope evidence. *Palaeogeogr. Palaeoclim. Palaeoecol.* 177: 47–57.

Scowcroft, P. G. and J. Jeffrey. 1999. Potential significance of frost, topographic relief, and *Acacia koa* stands to restoration of mesic Hawaiian forests on abandoned rangeland. *For. Ecol. Manag.* 114: 447–458.

Sessions, L. and D. Kelly. 2002. Predator-mediated apparent competition between an introduced grass, *Agrostis capillaris*, and a native fern, *Botrychium australe* (Ophioglossaceae), in New Zealand. *Oikos* 96: 102–109.

Shaver, G. R. and F. S. Chapin III. 1991. Production-biomass relationships and element cycling in contrasting arctic vegetation types. *Ecol. Monogr.* 61: 1–31.

Shaver, G. R., W. D. Billings, F. S. Chapin III, A. E. Giblin, K. J. Nadelhoffer, W. C. Oechel and E. B. Rastetter. 1992. Global change and the carbon balance of arctic ecosystems. *Bioscience* 42: 433–441.

Shinozaki, K. and T. Kira. 1956. Intraspecific competition among higher plants. VII. Logistic theory of the C-D. effect. *J. Inst. Polytech. Osaka City Univ.* 7: 35–72.

Shipley, B. 2000. *Cause and Correlation in Biology.* Cambridge University Press, Cambridge.

Shipley, B. and P. A. Keddy. 1994. Evaluating the evidence for competitive hierarchies in plant communities. *Oikos* 69: 340–345.

Shmida, A. and S. Ellner. 1984. Coexistence of plant species with similar niches. *Vegetatio* 58: 29–55.

Shmida, A. and M. V. Wilson. 1985. Biological determinants of species diversity. *J. Biogeogr.* 12: 1–20.

Shukla, J., C. Nobre and P. Sellers. 1990. Amazon deforestation and climate change. *Science* 247: 1322–1325.

Shurin, J. B. and E. G. Allen. 2001. Effects of competition, predation, and dispersal on species richness at local and regional scales. *Am. Nat.* 158: 624–637.

Siegel, S. 1956. *Nonparametric Statistics for the Behavioral Sciences.* McGraw-Hill, New York.

Siegert, M. J., J. A. Dowdeswell, John-I. Svendsen and A. Elverhoi. 2002. The Eurasian Arctic during the last ice age. *Am. Sci.* 90: 32–39.

Silvertown, J. 1987. *Introduction to Plant Population Ecology*, 2nd ed. Longman, Harlow, UK.

Silvertown, J. W. and J. Lovett Doust. 1993. *Introduction to Plant Population Biology.* Blackwell Scientific Publications, Oxford.

Simberloff, D. and P. Stiling. 1996. How risky is biological control? *Ecology* 77: 1965–1974.

Simberloff, D. S. and E. O. Wilson. 1970. Experimental zoogeography of islands. A two-year record of colonization. *Ecology* 51: 934–937.

Simms, E. L. 1990. Examining selection on the multivariate phenotype: plant resistance to herbivores. *Evolution* 44: 1177–1188.

Simpson, D. A. 1984. A short history of the introduction and spread of *Elodea* Michx. in the British Isles. *Watsonia* 15: 1–9.

Sims, D. A. and S. Kelley. 1998. Somatic and genetic factors in sun and shade population differentiation in *Plantago lanceolata* and *Anthoxanthum odoratum. New Phytol.* 140: 75–84.

Skog, J. E. 2001. Biogeography of Mesozoic leptosporangiate ferns related to extant ferns. *Brittonia* 53: 236–269.

Skole, D. and C. Tucker. 1993. Tropical deforestation and habitat fragmentation in the Amazon-satellite data from 1978 to 1988. *Science* 260: 1905–1910.

Slobodkin, L. C., F. E. Smith and N. G. Hairston. 1967. Regulation in terrestrial ecosystems, and the implied balance of nature. *Am. Nat.* 101: 109–124.

Smallwood, P. D., M. A. Steele and S. H. Faeth. 2001. The ultimate basis of the caching preferences of rodents, and the oak-dispersal syndrome: tannins, insects, and seed germination. *Am. Zool.* 41: 840–851.

Smeck, N. E. 1985. Phosphorus dynamics in soils and landscapes. *Geoderma* 36: 185–199.

Smith, C. and S. D. Fretwell. 1974. The optimal balance between size and number of offspring. *Am. Nat.* 108: 499–506.

Smith, H. and G. C. Whitelam. 1990. Phytochrome, a family of photoreceptors with multiple physiological roles. *Plant Cell Environ.* 13: 695–707.

Smith, M. D. and A. K. Knapp. 1999. Exotic plant species in a C_4-dominated grassland: invasibility, disturbance and community structure. *Oecologia* 120: 605–612.

Smith, S., J. D. B. Weyers and W. G. Berry. 1989. Variation in stomatal characteristics over the lower surface of *Commelina communis* leaves. *Plant Cell Environ.* 12: 653–659.

Sneath, P. H. A. and R. R. Sokal. 1973. *Numerical Taxonomy*. W. H. Freeman & Co., San Francisco.

Snedecor, G. W. and W. G. Cochran. 1989. *Statistical Methods*. Iowa State University Press, Ames, IA.

Sokal, R. R. and F. J. Rohlf. 1995. *Biometry*. W. H. Freeman, New York.

Stanton, M. L. and C. Galen. 1997. Life on the edge: adaptation versus environmentally mediated gene flow in the snow buttercup, *Ranunculus adoneus*. *Am. Nat.* 150: 143–178.

Stearns, S. C. 1976. Life-history tactics: a review of the ideas. *Quart. Rev. Biol.* 51: 3–47.

Steele, M. A., P. D. Smallwood, A. Spunar and E. Nelsen. 2001. The proximate basis of the oak dispersal syndrome: Detection of seed dormancy by rodents. *Am. Zool.* 41: 852–864.

Stephenson, A. G. and J. A. Winsor. 1986. *Lotus corniculatus* regulates offspring quality through selective fruit abortion. *Evolution* 40: 453–458.

Stern, D. I. and R. K. Kaufmann. 1998. Annual Estimates of Global Anthropogenic Methane Emissions: 1860–1994. Trends Online: A Compendium of Data on Global Change. Carbon Dioxide Information Analysis Center, Oak Ridge National Laboratory, U.S. Department of Energy. [http://cdiac.esd.ornl.gov/trends/meth/ch4.htm].

Stevens, G. C. 1992. The elevational gradient in altitudinal range: an extension of Rapoport's latitudinal rule to altitude. *Am. Nat.* 140: 893–911.

Stevens, L., C. J. Goodnight and S. Kalisz. 1995. Multilevel selection in natural populations of *Impatiens capensis*. *Am. Nat.* 145: 513–526.

Stevens, M. H. H. and W. P. Carson. 1999. Plant density determines species richness along an experimental fertility gradient. *Ecology* 80: 455–465.

Stiling, P., A. M. Rossi, B. Hungate, P. Dijkstra, C. R. Hinkle, W. M. Knott III and B. Drake. 1999. Decreased leaf-miner abundance in elevated CO_2: reduced leaf quality and increased parasitoid attack. *Ecol. Appl.* 9: 240–244.

Stohlgren, T. J., D. Binkley, G. W. Chong, M. A. Kalkhan, L. D. Schell, K. A. Bull, Y. Otsuki, G. Newman, M. Bashkin and Y. Son. 1999. Exotic plant species invade hot spots of native plant diversity. *Ecol. Monogr.* 69: 25–46.

Stowe, L. C. 1979. Allelopathy and its influence on the distribution of plants in an Illinois old-field. *J. Ecol.* 67: 1065–1085.

Stowe, L. G. and M. J. Wade. 1979. The detection of small-scale patterns in vegetation. *J. Ecol.* 67: 1047–1064.

Strauss, S. Y. 1997. Floral characters link herbivores, pollinators, and plant fitness. *Ecology* 78: 1640–1645.

Strong, D. R., J. H. Lawton and R. Southwood. 1984. *Insects on Plants: Community Patterns and Mechanisms*. Harvard University Press, Cambridge, MA.

Strong, D. R., J. L. Maron, P. G. Connors, A. Whipple, S. Harrison and R. L. Jefferies. 1995. High mortality, fluctuation in numbers, and heavy subterranean insect herbivory in bush lupine, *Lupinus arboreus*. *Oecologia* 104: 85–92.

Strong, D. R., H. K. Kaya, A. V. Whipple, A. L. Child, S. Kraig, M. Bondonno, K. Dyer and J. L. Maron. 1996. Entomopathogenic nematodes: Natural enemies of root-feeding caterpillars on bush lupine. *Oecologia* 108: 167–173.

Sultan, S. E. 1987. Evolutionary implications of phenotypic plasticity in plants. *Evol. Biol.* 21: 127–178.

Swetnam, T. W. and J. L. Betancourt. 1990. Fire-Southern Oscillation relations in the southwestern United States. *Science* 249: 1017–1020.

Swetnam, T. W., C. D. Allen and J. L. Betancourt. 1999. Applied historical ecology: using the past to manage for the future. *Ecol. Appl.* 9: 1189–1206.

Symstad, A. 2000. A test of the effects of functional group richness and composition on grassland invasibility. *Ecology* 81: 99–109.

Taiz, L. and E. Zeiger. 1998. *Plant Physiology*, 2nd ed. Sinauer Associates, Sunderland, MA.

Takenaka, A., K. Takahashi and T. Kohyama. 2001. Optimal leaf display and biomass partitioning for efficient light capture in an understory palm, *Licuala arbuscula*. *Func. Ecol.* 15: 660–668.

Tansley, A. G. 1917. On competition between *Galium saxatile* L. (*G. hercynicum* Weig.) and *Galium sylvestre* Poll. (*G. asperum* Schreb.) on different types of soil. *J. Ecol.* 5: 173–179.

Tansley, A. G. 1935. The use and abuse of vegetational concepts and terms. *Ecology* 16: 284–307.

Tansley, A. G. 1939. *The British Islands and their Vegetation*. Cambridge University Press, Cambridge.

Tansley, A. G. and R. S. Adamson. 1925. Studies of the vegetation of the English chalk. *J. Ecol.* 13: 177–223.

Teeri, J. A. 1969. The phytogeography of subalpine black spruce in New England. *Rhodora* 71: 1–6.

Teeri, J. A. and L. G. Stowe. 1976. Climatic patterns and the distribution of C_4 grasses in North America. *Oecologia* 23: 1–12.

Terashima, I. 1992. Anatomy of nonuniform leaf photosynthesis. *Photosyn. Res.* 31: 195–212.

Terborgh, J. 1973. On the notion of favorableness in plant ecology. *Am. Nat.* 107: 481–501.

Terborgh, J., R. B. Foster and P. Nuñez V. 1996. Tropical tree communities: a test of the non-equilibrium hypothesis. *Ecology* 77: 561–567.

ter Braak, C. T. F. 1986. Canonical correspondence analysis: a new eigenvector technique for multivariate direct gradient analysis. *Ecology* 67: 1167–1179.

Theodose, T. A. and W. D. Bowman. 1997. The influence of interspecific competition on the distribution of an alpine graminoid: evidence for the importance of plant competition in an extreme environment. *Oikos* 79: 101–114.

Thomas, A. S. 1960. Changes in vegetation since the advent of myxomatosis. *J. Ecol.* 48: 287–306.

Thomas, A. S. 1963. Further changes in vegetation since the advent of myxomatosis. *J. Ecol.* 51: 177–183.

Thompson, J. N. 1978. Within-patch structure and dynamics in *Pastinaca sativa* and resource availability to a specialized herbivore. *Ecology* 59: 443–448.

Thrall, P. H. and J. Antonovics. 1995. Theoretical and empirical studies of metapopulations: population and genetic dynamics of the *Silene-Ustilago* system. *Can. J. Bot.* 73(Suppl.): 1249–1258.

Tilman, D. 1982. *Resource Competition and Community Structure*. Princeton University Press, Princeton, NJ.

Tilman, D. 1987. Secondary succession and the pattern of plant dominance along experimental nitrogen gradients. *Ecol. Monogr.* 57: 189–214.

Tilman, D. 1988. *Plant Strategies and the Dynamics and Structure of Plant Communities*. Princeton University Press, Princeton, NJ.

Tilman, D. 1996. Biodiversity: population versus ecosystem stability. *Ecology* 77: 350–363.

Tilman, D. 1997. Distinguishing between the effects of species diversity and species composition. *Oikos* 80: 185.

Tilman, D. 1999. Ecological consequences of biodiversity: a search for general principles. *Ecology* 80: 1455–1474.

Tilman, D. and J. A. Downing. 1994. Biodiversity and stability in grasslands. *Nature* 367: 363–365.

Tilman, D. and S. Pacala. 1993. The maintenance of species richness in plant communities. In *Species Diversity in Ecological Communities: Historical and Geographical Perspectives*, R. E. Ricklefs and D. Schluter (eds.), pp. 13–25. University of Chicago Press, Chicago.

Tilman, D., J. Knops, D. Wedin, P. Reich, M. Ritchie and E. Siemann. 1997. The influence of functional diversity and composition on ecosystem processes. *Science* 277: 1300–1302.

Tilman, D., C. L. Lehman and C. E. Bristow. 1998. Diversity-stability relationships: statistical inevitability or ecological consequence? *Am. Nat.* 151: 277–282.

Tischendorf, L. and L. Fahrig. 2000. On the usage and measurement of landscape connectivity. *Oikos* 90: 7–19.

Titus, J. G. and C. Richman. 2001. Maps of lands vulnerable to sea level rise: modeled elevations along the US Atlantic and Gulf coasts. *Climate Res.* 18: 205–228.

Tonsor, S. J. 1985. Intrapopulational variation in pollen-mediated gene flow in *Plantago lanceolata* L. *Evolution* 39: 775–782.

Tuljapurkar, S. 1990. *Population Dynamics in Variable Environments.* Springer-Verlag, New York.

Turesson, G. 1922. The species and the variety as ecological units. *Hereditas* 3: 100–113.

Turner, M. G. 1989. Landscape ecology: the effect of pattern on process. *Annu. Rev. Ecol. Syst.* 20: 171–197.

Turner, M. G. and R. H. Gardner. 1991. *Quantitative Methods in Landscape Ecology.* Springer-Verlag, New York.

Turner, R. M. 1990. Long-term vegetation changes at a fully protected Sonoran Desert site. *Ecology* 71: 464–477.

U.S. Department of Agriculture. 1941. *Climate and Man. Agricultural Yearbook.* U.S. Department of Agriculture, Washington, DC.

Vajda, V., J. I. Raine and C. Hollis J. 2001. Indication of global deforestation at the Cretaceous-Teriary boundary by New Zealand fern spike. *Science* 294: 1700–2351.

van der Maarel, E. and M. T. Sykes. 1993. Small-scale plant species turnover in a limestone grassland: the carousel model and some comments on the niche concept. *J. Veg. Sci.* 4: 179–188.

VanderMeulen, M. A., A. J. Hudson and S. M. Scheiner. 2001. Three evolutionary hypotheses for the hump-shaped productivity-diversity curve. *Evol. Ecol. Res.* 3: 379–392.

VanderPutten, W. H. and B. A. M. Peters. 1997. How soil-borne pathogens may affect plant competition. *Ecology* 78: 1785–1795.

van Noordwijk, A. J. and G. de Jong. 1986. Acquisition and allocation of resources: their influence on variation in life history tactics. *Am. Nat.* 128: 137–142.

Veblen, T. T. and D. H. Ashton. 1978. Catastrophic influence on the vegetation of the Valdivian Andes, Chile. *Vegetatio* 36: 149–167.

Venable, D. lawrence. 1985. The evolutionary ecology of seed heteromorphism. *Am. Nat.* 126: 577–595.

Venable, D. L. and C. Pake. 1999. Population ecology of Sonoran Desert annual plants. In *The Ecology of Sonoran Desert Plants and Plant Communities*, R. H. Robichaux (ed.), pp. 115–142. University of Arizona Press, Tucson.

Via, S., R. Gomulkiewicz, G. De Jong, S. M. Scheiner, C. D. Schlichting and P. Van Tienderen. 1995. Adaptive phenotypic plasticity: consensus and controversy. *Trends Ecol. Evol.* 10: 212–217.

Vitousek, P. M. and W. A. Reiners. 1975. Ecosystem succession and nutrient retention: a hypothesis. *Bioscience* 25: 376–381.

Vitousek, P. M., C. M. D'Antonio, L. L. Loope and R. Westbrooks. 1996. Biological invasions as global environmental change. *Am. Sci.* 84: 468–478.

Vitousek, P. M., J. D. Aber, R. W. Howarth, G. E. Likens, P. A. Matson, D. W. Schindler, W. H. Schlesinger and D. G. Tilman. 1997. Human alteration of the global nitrogen cycle: sources and consequences. *Ecol. Appl.* 7: 737–750.

Vogler, D. W. and S. Kalisz. 2001. Sex among the flowers: the distribution of plant mating systems. *Evolution* 55: 202–204.

Vogt, K. A., C. C. Grier, C. E. Meier and R. L. Edmonds. 1982. Mycorrhizal role in net primary production and nutrient cycling in *Abies amabilis* ecosystems in western Washington. *Ecology* 63: 370–380.

von Helversen, D. and O. von Helversen. 1999. Acoustic guide in bat-pollinated flower. *Nature* 398: 759–760.

Walker, L. R. 1999. Patterns and process in primary succession. In *Ecosystems of Disturbed Ground*, L. R. Walker (ed.), pp. 585–610. Elsevier, Amsterdam.

Waloff, N. and O. W. Richards. 1977. The effect of insect fauna on growth, mortality, and natality of broom, *Sarothamnus scoparius. J. Appl. Ecol.* 14: 787–798.

Wand, S. J. E., G. F. Midgley, M. H. Jones and P. S. Curtis. 1999. Responses of wild C_4 and C_3 grass (Poaceae) species to elevated atmospheric CO_2 concentration: a meta-analytic test of current theories and perceptions. *Global Change Biol.* 5: 723–741.

Wardle, D. A., O. Zackrisson, G. Hornberg and C. Gallet. 1997. Biodiversity and ecosystem properties. *Science* 278: 1867–1869.

Wardle, D. A., K. I. Bonner, G. M. Barker, G. W. Yeates, K. S. Nicholson, R. D. Bardgett, R. N. Watson and A. Ghani. 1999. Plant removals in perennial grassland: vegetation dynamics, decomposers, soil biodiversity, and ecosystem properties. *Ecol. Monogr.* 69: 535–568.

Wardle, D. A., M. A. Huston, J. P. Grime, F. Berendse, E. Garnier, W. K. Lauenroth, H. Setala and S. D. Wilson. 2000. Biodiversity and ecosystem function: an issue in ecology. *Bull. Ecol. Soc. Amer.* 81: 235–239.

Waring, R. H. and S. W. Running. 1998. *Forest Ecosystems: Analysis at Multiple Scales*, 2nd ed. Academic Press, New York.

Wartenberg, D., S. Ferson and F. J. Rohlf. 1987. Putting things in order: A critique of detrended correspondence analysis. *Am. Nat.* 129: 434–448.

Waser, N. M., L. Chittka, M. V. Price, N. M. Williams and J. Ollerton. 1996. Generalization in pollination systems, and why it matters. *Ecology* 77: 1043–1060.

Watt, A. S. 1947. Pattern and process in the plant community. *J. Ecol.* 35: 1–22.

Weaver, J. E. and F. E. Clements. 1929. *Plant Ecology.* McGraw-Hill, New York.

Webb, C. J., W. R. Sykes and P. J. Garnock-Jones. 1988. *Flora of New Zealand*, Vol. IV. Botany Divison, DSIR, Christchurch, New Zealand.

Weier, T. E., C. R. Stocking and M. G. Barbour. 1974. *Botany: An Introduction to Plant Biology*, 5th ed. Wiley, New York.

Weiner, J. 1990. Asymmetric competition in plant populations. *Trends Ecol. Evol.* 5: 360–364.

Weller, D. E. 1987. A reevaluation of the −3/2 power rule of plant self-thinning. *Ecol. Monogr.* 57: 23–43.

Weller, D. E. 1991. The self-thinning rule: Dead or unsupported? A reply to Lonsdale. *Ecology* 72: 747–750.

West-Eberhard, M. J. 1989. Phenotypic plasticity and the origins of diversity. *Annu. Rev. Ecol. Syst.* 20: 249–278.

Westman, W. E. and R. K. Peet. 1985. Robert H. Whittaker (1920–1980): The man and his work. In *Plant Community Ecology: Papers in Honor of Robert H. Whittaker*, R. K. Peet (ed.), pp. 5–30. Dr W. Junk, Dordrecht, The Netherlands.

Westoby, M. 1998. A leaf-height-seed (LHS) plant ecology strategy scheme. *Plant and Soil* 199: 213–227.

Westoby, M., D. Falster, A. Moles, P. Vesk and I. Wright. 2002. Plant ecological strategies: some leading dimensions of variation between species. *Annu. Rev. Ecol. Syst.* 33: (in press).

Wheeler, B. D. and S. C. Shaw. 1991. Aboveground crop mass and species richness of the principal types of herbaceous rich-fen vegetation of lowland England and Wales. *J. Ecol.* 79: 285–301.

White, J. 1985. The thinning rule and its application to mixtures of plant populations. In *Studies on Plant Demography*, J. White (ed.), pp. 291–309. Academic Press, London.

White, P. S. and S. T. A. Pickett. 1985. Natural disturbance and patch dynamics: an introduction. In *The Ecology of Natural Disturbance and Patch Dynamics*, S. T. A. Pickett and P. S. White (eds.), pp. 3–13. Academic Press, Orlando, FL.

Whitham, T. G. and S. Mopper. 1985. Chronic herbivory: impacts on architecture and sex expression of pinyon pine. *Science* 228: 1089–1091.

Whittaker, R. H. 1954. The ecology of serpentine soils. I. Introduction. *Ecology* 35: 258–259.

Whittaker, R. H. 1956. Vegetation of the Great Smoky Mountains. *Ecol. Monogr.* 26: 1–80.

Whittaker, R. H. 1960. Vegetation of the Siskiyou Mountains, Oregon and California. *Ecol. Monogr.* 30: 279–338.

Whittaker, R. H. 1966. Forest dimensions and production in the Great Smoky Mountains. *Ecology* 47: 103–121.

Whittaker, R. H. 1970. *Communities and Ecosystems*. Macmillan, New York.

Whittaker, R. H. 1972. Evolution and measurement of species diversity. *Taxon* 21: 213–251.

Whittaker, R. H. 1975. *Communities and Ecosystems*, 2nd ed. Macmillan, New York.

Whittaker, R. H. 1977. Evolution of species diversity in land communities. *Evol. Biol.* 10: 1–67.

Whittaker, R. H. and W. A. Niering. 1975. Vegetation of the Santa Catalina Mountains, Arizona. V. Biomass, production, and diversity

along the elevation gradient. *Ecology* 56: 771–790.

Wied, A. and C. Galen. 1998. Plant parental care: conspecific nurse effects in *Frasera speciosa* and *Cirsium scopulorum*. *Ecology* 79: 1657–1668.

Wielicki, B. A., T. Wong, R. P. Allan, A. Slingo, J. T. Kiehl, B. J. Soden, C. T. Gordon, A. J. Miller, S. Yang, D. A. Randall, F. Robertson, J. Susskind and H. Jacobowitz. 2002. Evidence for large decadal variability in the tropical mean radiative energy budget. *Science* 295: 841–844.

Williamson, M. and A. Fitter. 1996. The varying success of invaders. *Ecology* 77: 1661–1666.

Willig, M. R. and S. K. Lyons. 1998. An analytical model of latitudinal gradients of species richness with an empirical test for marsupials and bats in the New World. *Oikos* 81: 93–98.

Willson, M. F. 1983. *Plant Reproductive Ecology*. Wiley, Chichester, UK.

Wilson, C. and J. Gurevitch. 1995. Plant size and spatial pattern in a natural population of *Myosotis micrantha*. *J. Veg. Sci.* 6: 847–852.

Wilson, D. S. 1980. *The Natural Selection of Populations and Communities*. Benjamin/Cummings, Menlo Park, CA.

Wilson, S. D. and P. A. Keddy. 1986. Species competitive ability and position along a natural stress disturbance gradient. *Ecology* 67: 1236–1242.

Winn, A. A. 1991. Proximate and ultimate sources of within-individual variation in seed mass in *Prunella vulgaris* (Lamiaceae). *Am. J. Bot.* 78: 838–844.

Wiser, S. K., R. B. Allen, P. W. Clinton and K. H. Platt. 1998. Community structure and forest invasion by an exotic herb over 23 years. *Ecology* 79: 2071–2081.

Wolda, H. 1981. Similarity indices, sample size and diversity. *Oecologia* 50: 296–302.

Wollkind, D. J. 1976. Exploitation in three trophic levels: an extension allowing intraspecies carnivore interaction. *Am. Nat.* 110: 431–447.

Woods, K. D. and M. B. Davis. 1989. Paleoecology of range limits: beech in the Upper Peninsula of Michigan. *Ecology* 70: 681–696.

Woodward, F. I. 1987. *Climate and Plant Distribution*. Cambridge University Press, Cambridge.

Woodward, F. I. and W. Cramer. 1996. Plant functional types and climatic change. *J. Veg. Sci.* 7: 305–430 (Special Feature).

Woodwell, G. M. 1970. Effects of pollution on the structure and physiology of ecosystems. *Science* 168: 429–433.

Wright, D. H. 1983. Species-energy theory: An extension of species-area theory. *Oikos* 41: 496–506.

Wright, D. H., D. J. Currie and B. A. Maurer. 1993. Energy supply and patterns of species richness on local and regional scales. In *Species Diversity in Ecological Communities: Historical and Geographical Perspectives*, R. E. Ricklefs and D. Schluter (eds.), pp. 66–77. University of Chicago Press, Chicago.

Wright, D. H., B. D. Patterson, G. M. Mikkelson, A. Cutler and W. Atmar. 1998. A comparative analysis of nested subset patterns of species composition. *Oecologia* 113: 1–20.

Wright, S. 1931. Evolution in Mendelian populations. *Genetics* 16: 97–159.

Wright, S. 1968. *Evolution and the Genetics of Populations*. Vol. 1. *Genetic and Biometric Foundations*. University of Chicago Press, Chicago.

Wright, S. 1969. *Evolution and the Genetics of Populations*. Vol. 2. *The Theory of Gene Frequencies*. University of Chicago Press, Chicago.

Wu, J. and O. L. Loucks. 1995. From balance of nature to hierarchical patch dynamics: a paradigm shift in ecology. *Quart. Rev. Biol.* 70: 439–466.

Wuethrich, B. 2000. Conservation biology: combined insults spell trouble for rainforests. *Science* 289: 35–37.

Yanai, R. D., T. G. Siccama, M. A. Arthur, C. A. Federer and A. J. Friedland. 1999. Accumulation and depletion of base cations in forest floors in the northeastern United States. *Ecology* 80: 2774–2787.

Yoda, K., T. Kira, H. Ogawa and K. Hozumi. 1963. Self-thinning in overcrowded pure stands under cultivated and natural conditions. *J. Biol. Osaka City Univ.* 14: 107–129.

Young, T. P. and C. K. Augspurger. 1991. Ecology and evolution of long-lived semelparous plants. *Trends Ecol. Evol.* 6: 285–289.

Zar, J. H. 1999. *Biostatistical Analysis*, 4th ed. Prentice-Hall, Upper Saddle River, NJ.

Zobel, K. and J. Liira. 1997. A scale-independent approach to the richness vs. biomass relationship in ground-layer plant communities. *Oikos* 80: 325–332.

Index

Page numbers in *italics* indicate material in an illustration.

agamospermy, 148
C$_3$ and C$_4$ distribution of, 32–34, 427–428
as clonal organisms, 95–96
effects of climate on distribution, 427–428
elevated levels of carbon dioxide and, 442
heavy-metal tolerance, 102–103
physical defenses, 224
plant-herbivore interactions, 229
Grasslands
alpine, *383*, 401
chalk grasslands, 219–220
C$_4$ plants in, 30
fire and, 376–377
human impact on, 459
nutrient cycling and functional group diversity, 315–316
pampas, 365
physiognomy of, 376–377
species diversity and invasive species, 283–284
tropical savanna, 398–399
See also Temperate grasslands
Gravitational potential, 42, 44
Grazers, 213
Grazing
as disturbance, 264
impact on deserts, 400
impact on savannas, 398
impact on temperate grasslands, 396–397
impact on tropical rainforests, 386
Great Basin desert, 400, 434
Great Salt Lake, 434
Greenhouse effect, 354–355
biotic consequences, 447–451
causes and dynamics of, 442–444
evidence for, 444–445
future predictions, 445–446
Greenhouse gases, 443, 444, 445
Grime, Philip, 173–174, 197
Gross photosynthetic production (GPP), 303, 307
Gross primary production (GPP), 303, 307, 440
Ground ivy, 202
Ground pine, *423*
Ground-truthing, 335–336
Groundwater, 66
Group selection, 110–111
Growth. *See* Plant growth; Population growth
Growth forms, 145–146. *See also* Plant growth
Grubb, Peter, 197, 265
Guard cells, 21–22, 50, *51*
"Guerilla" growth, 146
Guilds, 236
Gulf Stream, 363
Gutiérrez, Javier, 221
Guzmania monostachya, 31
Gymnosperms

Mesozoic era, 425
Pleistocene species extinctions, *411*
sexual reproduction, 150
tracheids, 55
See also Conifers
Gynodioecy, 157
Gynomonoecy, 157
Gypsy moths, 263
Gyres, 363–364

Habitat fragmentation
effects on communities, 345–347
global climate change, 449
impact on biodiversity, 458–461
Habitat fragments
connectivity, 347–348
edge effects, 347
nestedness, 348
Habitats
defined, 9
effects of global climate change on, 449
plant competition in, 205–208
Hadley cells, 362, *364*
Haeckel, Ernst, 10
Hair grass, 207–208
Hairs, 223–224
Hakea, 164
H. sericea, 417
Halogenated hydrocarbons, 444
Halophytes, 46–47, 49
Hanski, Illka, 342
Haplopappus squarrosus, 216
Harper, John, 11, 96, 117, 118, 168
Harrison, Susan, 342
Hartwick Pines State Park, *392*
Hastings, James, 135
Hawaiian silverswords, 167, 172
Hawaii, precipitation patterns, 365, *366*
Hawthornes, 113
Heat, 354
Heath bedstraw, 202–203
Heather, 457
Heaths, 401, 416
Heavy-metal tolerance, 102–103
Hederagenenin, *227*
Hedychium gardnerianum, 148
Helianthemum chamaecistus, 220
Helianthus annuus, 169
Hemicryptophytes, 145, 248
Hemiparasites, 230
Hemiptera, 221
Hemlock-white pine-northern hardwood forest, 390, 412
"Hen-and-chicks," *31*
Hepialus californicus, 216–217
Herbivory, 263
apparent plant competition and, 191
in biological control, 217–219
chronic, 216
defined, 213
effects on fitness, 230
effects on flowering, 181
effects on individual plants, 213–214, *215*

effects on plant communities, 219–223
effects on populations, 214–219
HSS hypothesis of, 222–223
interactions with pathogens, 232
introduced and domesticated herbivores, 219–221
leaf nutrient levels and, 224
by native herbivores, 221–222
plant-herbivore interactions, 228–230
plant spatial distribution and, 216–217
types of herbivore behavior, 219
Heritability, 88, 90–94
Heritable traits, 87
Hermaphroditism, 157
Herrera, Carlos, 217
Hesperis, 114
Hesperolinon, 161
Hesperostipa neomexicana, 207, *208*
Heterogeneity
significance in ecology, 9
species diversity and, 287–288
Heterohabditis hepialus, 217
Heteromeles arbutifolia, 396
Heterostyly, 160
Hett, Joan, 134
Hexagonal fan experimental designs, 195–196
Hickories, 253, *254*, 390
Hieracium, 148
Hierarchical patch dynamics, 340
Highways, Amazonian forest destruction and, 463, *464*
Histisols, *71*
Hooper, David, 315
Hordeum vulgare, 45
Horizons, of soils, 67, *68*
Horse chestnut, 390
Horsetails, 224, 423
Horvitz, Carol, 154
Hot deserts, *383*, 399–400
Houle, David, 176
Howard, Timothy, 276
HSS hypothesis, 222–223
Hubbard Brook Experimental Forest, 251, 267, *298*
Human population growth, 462–463
Humboldt, Alexander von, 353, 406
Humidity, 361
Huntly, Nancy, 219
Hurricanes, 128, 133, 262
Hutchinson, G. Evelyn, 87, 275
Huxman, Travis, 30
Hybridization, 113, *114*
Hybrid swarms, 113
Hydraulic lift, 74
Hydrogen, in plant nutrition, *74*
Hygrophytes, 46
Hylobius transversovittatus, 218
Hylocomium splendens, 179
Hypotheses, 3
null model, 408
testing, 6–8
Hyraceum, 429